HANDBOOK OF NEUROCHEMISTRY

VOLUME VI

ALTERATIONS OF CHEMICAL EQUILIBRIUM IN THE NERVOUS SYSTEM

HANDBOOK OF NEUROCHEMISTRY

Edited by Abel Lajtha

Volume I	Chemical Architecture of the Nervous System
Volume II	Structural Neurochemistry
Volume III	Metabolic Reactions in the Nervous System
Volume IV	Control Mechanisms in the Nervous System
Volume V	Metabolic Turnover in the Nervous System
Volume VI	Alterations of Chemical Equilibrium in the Nervous System
Volume VII	Pathological Chemistry of the Nervous System

HANDBOOK OF NEUROCHEMISTRY

Edited by Abel Lajtha

New York State Research Institute
for Neurochemistry and Drug Addiction
Ward's Island
New York, New York

VOLUME VI

ALTERATIONS OF CHEMICAL EQUILIBRIUM IN THE NERVOUS SYSTEM

PLENUM PRESS • NEW YORK–LONDON • 1971

Library of Congress Catalog Card Number 68-28097

SBN 306-37706-3

ISBN 978-1-4615-7177-3 ISBN 978-1-4615-7175-9 (eBook)
DOI 10.1007/978-1-4615-7175-9

Softcover reprint of the hardcover 1st edition 1971

A Division of Plenum Publishing Corporation
227 West 17th Street, New York, New York 10011

United Kingdom edition published by Plenum Press, London
A Division of Plenum Publishing Company, Ltd.
Davis House (4th Floor), 8 Scrubs Lane, Harlesden, NW10 6SE, England

Contributors to this volume:

Bernard W. Agranoff	University of Michigan, Ann Arbor, Michigan (page 203)
V. A. Benignus	Department of Neurology and Psychiatry, The University of Texas Medical Branch, Galveston, Texas (page 1)
Edward L. Bennett	Laboratory of Chemical Biodynamics, Lawrence Radiation Laboratory, University of California, Berkeley, California (page 173)
R. G. Benton	Department of Neurology and Psychiatry, The University of Texas Medical Branch, Galveston, Texas (page 1)
Doris H. Clouet	New York State Narcotic Addiction Control and Commission Testing and Research Center, Brooklyn, New York (page 479)
Jill E. Cremer	Biochemical Mechanisms Section, Toxicology Research Unit, Medical Research Council, Woodmansterne Road, Surrey, England (page 311)
Wolf-Dietrich Dettbarn	Department of Pharmacology, Vanderbilt University School of Medicine, Nashville, Tennessee (page 423)
John Dobbing	Department of Child Health, Clinical Sciences Building, University of Manchester, Manchester, England (page 255)
Enrique Egaña	Laboratory of Neurochemistry–Institute of Experimental Medicine, School of Medicine, University of Chile, P.O.B. 3170, Santiago, Chile (page 525)
Sabit Gabay	Biochemical Research Laboratory, Veterans Administration Hospital, Brockton, Massachusetts (page 325)
Elliot S. Gershon	Neuropsychopharmacology Laboratory, Massachusetts Mental Health Center, Department of Psychiatry, Harvard Medical School, Boston, Massachusetts (page 357)

Holger Hydén	Institute of Neurobiology, Faculty of Medicine, University of Göteborg, Göteborg, Sweden (page 221)
Paul W. Lange	Institute of Neurobiology, Faculty of Medicine, University of Göteborg, Göteborg, Sweden (page 221)
G. M. Lehrer	Department of Neurology, The Division of Neurochemistry, The Mount Sinai School of Medicine of the City University of New York, New York, New York (page 267)
Richard A. Lovell	Department of Psychiatry, University of Chicago, Chicago, Illinois (page 63)
H. S. Maker	Department of Neurology, The Division of Neurochemistry, The Mount Sinai School of Medicine of the City University of New York, New York, New York (page 267)
L. J. Mullins	Department of Biophysics, University of Maryland School of Medicine, Baltimore, Maryland (page 395)
M. K. O'Heeron, Jr.	Department of Pediatrics, The University of Texas Medical Branch, Galveston, Texas (page 1)
Giuseppe Porcellati	Department of Biological Chemistry, University of Pavia, Pavia, Italy (page 457)
D. A. Rappoport	Department of Pediatrics, The University of Texas Medical Branch, Galveston, Texas (page 1)
Mark R. Rosenzweig	Department of Psychology, University of California, Berkeley, California (page 173)
Joseph J. Schildkraut	Neuropsychopharmacology Laboratory, Massachusetts Mental Health Center, Department of Psychiatry, Harvard Medical School, Boston, Massachusetts (page 357)

Mogens Schou	The Psychopharmacology Research Unit, Aarhus University Psychiatric Institute, Risskov, Denmark (page 387)
P. A. Shore	Department of Pharmacology, University of Texas Southwestern Medical School, Dallas, Texas (page 349)
U. B. Singh	Department of Biochemistry, All-India Institute of Medical Sciences, New Delhi, India (page 29)
Stephen I. Szara	Section on Psychopharmacology, Division of Special Mental Health Research, National Institute of Mental Health, St. Elizabeths Hospital, Washington, D.C. (page 441)
G. P. Talwar	Department of Biochemistry, All-India Institute of Medical Sciences, New Delhi, India (page 29)
N. M. Trieff	Department of Pharmacology and Toxicology, The University of Texas Medical Branch, Galveston, Texas (page 1)
Georges Ungar	Baylor College of Medicine, Houston, Texas (page 241)
Henrik Wallgren	Research Laboratories of the State Alcohol Monopoly (Alko), Helsinki, Finland (page 509)
Arthur Yuwiler	Neurobiochemistry Laboratory, Veterans Administration Center, and Department of Psychiatry, University of California Center for the Health Sciences, Los Angeles, California (page 103)

PREFACE

It has been recognized for more than a thousand years that the function of the brain, like the function of the other organs of the body, is determined by its physical, chemical, and biological properties. Evidence that even its highest functions could be explained by these properties was gathered only in recent years, however; these findings, which clearly have to be confirmed by a great deal of further experimental evidence, indicate that most, if not all, of the functions of the brain are based on its biochemical and biophysical mechanisms. This at first hearing may sound rather simple, but the ability to understand learning, emotion, perhaps even creativity, on biological terms may well be the most important scientific discovery of all time. Few pieces of knowledge can influence our future health and well-being to the degree that understanding of mental mechanisms will.

It has been clearly shown in many ways in the previous volumes of this Handbook that from the biochemical or neurochemical point of view the brain is one of the most active organs. The brain seems stable and in some respects permanent; this is evidence not of inactivity but of carefully controlled homeostasis, of dynamic rather than static equilibrium, with most components undergoing metabolic alterations. What this dynamic equilibrium entails is that brain possesses plasticity; its very active biological processes are likely to impart some stability, the possibility of repairing damage, certainly the possibility of adapting to varying situations and conditions. These dynamic processes, however, also give this organ a vulnerability not always fully realized. The mere existence of dynamic processes brings with it the possibility of influencing such processes. The present volume emphasizes the vulnerability of the nervous system, and makes one wish that in the present time, when so many try to experiment with substances or conditions that influence the brain or mental states, this vulnerability of the nervous system would be more widely appreciated and respected. This heuristic aim, however, is beyond the scope of the present volume, which is designed more for researchers engaged in prying the secrets of biological mechanisms from the nervous system. These chapters clearly demonstrate that cerebral chemical processes can be influenced in a number of ways, and can undergo changes that in turn will influence the functions of the nervous system. Surely this is one of the most exciting fields of neurochemistry, and one that needs many dedicated investigators. The number of problems waiting to be investigated is great; these chapters not only show which ones have been investigated, but also indicate a great many unmapped areas that need to be explored.

New York, New York
December 1970.

Abel Lajtha

CONTENTS

Chapter 5

Chapter 6

Chapter 7

Chapter 19

Chapter 20

Chapter 1

CHEMOSENSORY STIMULATION

D. A. Rappoport

Department of Pediatrics

N. M. Trieff

Department of Pharmacology and Toxicology

M. K. O'Heeron, Jr.

Department of Pediatrics

and

V. A. Benignus and R. G. Benton

Department of Neurology and Psychiatry
The University of Texas Medical Branch
Galveston, Texas

OLFACTION*

When quiet reigns and somnolence
O'ertakes the reach of eye and ear
One sense remains alert and clear
To cerebrating semiconsciousness.

On breath the subtle emanations rise
To cells unique in their direct ascent
To hippocampal cortex, there to scent
Aladdin dreams or playful Eros' drive.

Each odor its specific reference
Unclassed, its exteroceptive role
That points reaction to a distant goal,
Prime, in an evolutionary sense.

The streams of dream and love life both are bound
By odorous Nature through a law profound.

J. G. Sinclair (1959)

* Reprinted from *Texas Repts. Biol. Med.* **17**, 1968.

I. INTRODUCTION

Animals perceive chemical components in their environment, either by the common chemical sense, taste, or olfaction. The chemovector may be in the form of a gas, liquid, solid, or solute in aqueous media emanating from food, predators, potential mates, or from noxious compounds. Of the three chemoceptors, the common chemical sense exhibits the minimal discriminatory properties as judged by response of the organism since it is presumed that the sensor can detect the presence of an irritant but cannot sense its nature.(1) In contrast to this, the gustatory and olfactory systems can differentiate the qualitative characteristics of the chemovector and it has been presumed that olfaction is even more discriminatory than the gustatory sense; however, this impression may have to be reassessed.(1,2) The olfactory system has a higher sensitivity than the gustatory.(1,2)

In higher chordates, the gustatory and olfactory systems interact with their respective chemovectors in a similar manner such that, at times, it is difficult to distinguish whether the chemical is a sapid component or an odorivector. These sensory organs, however, differ anatomically in location and innervation and most of the chemical stimuli will elicit response from only one of the two systems. For example, in the marine environment, both olfactory and gustatory stimuli are dissolved in water and they are distinguished by animals, such as fish, either by the olfactory receptors in the nares or by the gustatory crypts in the mouth or those on the surface of the face and head, and in some fish, these crypts cover the entire body.(3,4) The octopus (*Octopus vulgaris*) has chemical receptors at the tip of the tentacles, but whether these are olfactory or gustatory is strictly semantic.(5) There is no reason to believe that the gustatory behavioral responses, as exhibited by higher chordates, are similar throughout the animal kingdom as a whole. Lenhoff reported that glutathione initiated feeding response in *Hydra littoralis* which he could not categorize as either olfactory, gustatory, or hormonal.(6) Also, the octopus has shown sensitivities to stimuli traditionally defined as gustatory which approached those of olfaction in chordates.(5) Perhaps the time has come to take an inverse view of gustation beginning with the less complex chemosensory mechanisms of lower animals(5) rather than the more complex mechanisms of higher animals. Lenhoff(6) found a coelenterate (*H. littoralis*) in which the glutathione response involves no coding or filtering mechanism, but instead consists of an all-or-none activation of a specific receptor site. In addition, gluthathione initiates a complex series of contractions and relaxations and, like hormones, it is effective at low concentrations (10^{-9} M–10^{-7} M). However, other hydroids are insensitive to glutathione but are sensitive to proline (10^{-6} M), thereby demonstrating a degree of selectivity in chemoreception.(6)

Although chemoreception has not been studied extensively in most phyla of marine metazoan invertebrates, the existing evidence from the selected phyla that have been studied indicate that the concept of taste is not distinct from olfaction in these organisms.[7] Chemoreception in some schizocoeles, such as marine annelids, has been shown to occur over the entire body surface.[8] Spawning in the common bay oyster (*Crassostrea virginica*) is initiated in the female by the presence, in the surrounding water, of oyster sperm from the same species.[9] That this response is not due to some ingested factor was shown in the failure to extinguish the response by plugging the mouth and esophagus.[9] The tips of *O. vulgaris* tentacles have been shown to respond to three sapid chemicals: hydrochloric acid, quinine, and sucrose.[5] However, this animal responded to changes in hydrochloric acid concentration below the sensitivity of a *p*H meter, thereby placing the level of response within the acceptable range of olfaction.[5]

In the enterocoelus animal group, in the subclass echinoderm, the holotheroids, asteroids, and echinoids have a general type of neurosensory cell which serves as both a tango- and chemoreceptor covering the entire body surface.[10]

Even in some chordates, gustatory chemoreception is not confined to the oral area.[5,11,12] The sea catfish, for example, has taste buds over the entire body surface.[11] Only the terrestrial tetrapod chordates have the gustatory chemoreception confined to the oral area.

In man, the gustatory epithelium at the tip of the tongue is mainly sensitive to sweet and salt, and to a lesser degree, to sour and bitter stimuli.[3,13,14] The lateral margins of the tongue are highly sensitive to sour but also respond to salt. The base of the tongue is responsive to bitter components. Various regions of the tongue are responsive to a variety of chemical stimuli with varying sensitivities. Taste buds in man and other mammals are found on the tongue, palate, interior facial pillars, larynx and the pharynx.[3,13–15] Even in frog tadpoles the presence of lingual premetamorphic papillae on the tongue has been shown.[16]

Taste buds appear as gustatory pores on the epithelial surface beneath which there are ovoid crypts containing closely packed columnar receptor neurons. These are bipolar neurons with apical microvilli protruding through the gustatory pore. The myelinated axons extend from the receptor bases toward the basement membrane where they become demyelinated. The fibers then converge to form plexi which separate by size into two groups. Small fibers of less than 0.5 μ in diameter are found in one group of plexi, while the other contains large fibers. The gustatory fibers from the anterior two-thirds of the tongue join to the chorda tympanic branch of the facial nerve, while fibers from the posterior one-third of the tongue run in the glossopharyngeal nerve.[2,3,12–14,17]

Beidler[18] found that the gustatory cells are continuously replaced from surrounding undifferentiated epithelial cells and that their life span is on the average 10 days. The new cells, originating at the rim of the taste buds, move towards the center of the gustatory pore, thus replacing the old gustatory cells. Although the manner by which the new taste cells are innervated is not known, Beidler[3,18] suggested that the newly forming taste cells at the rim of the gustatory pore are innervated by a branch from the nearest nerve fiber.

The olfactory membrane or mucosa consists of two types of cells: sustentacular and receptor. The former are columnar with microvilli projecting into the nasal cavity and are interspersed between receptor cells. The latter are bipolar neurons with 6–12 cilia which extend from the top into the nasal cavity while from the base of the cells extends an axon, passing through a basal plexis to join other axons to form neuronal fibers. The olfactory nerve fibers pass through the cribiform plate as they join the olfactory bulb.[2,14,17,19,20] The surface of the olfactory epithelium is continuously bathed by viscous material containing mucous or acidic mucopolysaccharides as in the toad, viper, and cat, whereas, in the triton, lizard, and salamander, this fluid consists mainly of mucoproteins and polysaccharides.[19,20]

It has been established that the nerve fibers from the olfactory receptor cells extend continuously into the olfactory bulb and that no synapse has been found between the receptor cells and the glomeruli of the bulb with which the fibers connect.[2,14,17,19,20] These fibers are myelinated. In many vertebrates, the olfactory mucosa is pigmented either yellow, dark brown, or black.[19,20] No such pigments have been found in the teleosts and in urodeles.[19,20] In cattle, the pigments consist mainly of carotinoids.[20,21] Some investigators have suggested that pigments are involved in the olfactory process,[20] and most recently Rosenberg *et al.*[22] demonstrated a very ingenious model using carotinoids compressed between electrodes in which volatile chemicals, known to be odorants, have significantly altered the conductivity through the carotinoid, stimulating the process of olfaction. However, the absence of pigments in the teleost argues against a direct function of pigment in olfaction.[20,21]

Ottoson[19] was the first to show that a saline-agar electrode placed on the olfactory mucosa of a frog initiated a slow monophasic negative potential, designated as the "electroolfactogram" (EOG), when a puff of air containing an odor was passed over this surface. It has been shown that the EOG has a fast rising phase and a slow exponential decline. It is not certain whether EOG represents a generator potential, although some have made this claim.[19,20]

In addition to the innervation of the olfactory mucosa through the olfactory receptors, the trigeminal nerves have access to the nasal mucosa through the ophthalmic and the maxillary nerve branches. Although these trigeminal nerves act as pain receptors, they are also responsive to odors.[19] It has been shown that odorants stimulate both the olfactory and trigeminal receptors, but there is a difference in the sensitivity of these receptors to odorants and in some situations, such as in the presence of ethylphenylalcohol, the nasal trigeminal receptors are 100 times more sensitive than the olfactory receptors.[19]

Some investigators consider the origin of all the sensory systems to be a modification of the motile cilia or flagella, irrespective of the great diversity in the structure and function of the receptor cells in the various sensory organs of vertebrates and invertebrates.[23,24] It is recognized that in the more primitive forms cilia serve in both locomotory and sensory functions.[23] Thus, Vinnikov considers that the rods and cones in photoreceptors are modified flagella which have been shown to constitute the visual cells of mollusks, annelids, echinoderms, and coelenterates.[23,24] Cilia-like processes are present in the olfactory cells in vertebrates and anthropods. Similarly, in the auditory system of the mammals, there are stiff hairs (stereocilia) and motile hairs (kinocilia). Vinnikov also points out that ciliary structures are present in the labyrinthine vestibule of vertebrates, in the receptor organ of gravity (utriculus), of vibration (sacculus), and of angular acceleration (ampulla cristae of semicircular canals), as well as in the lateral line organs of elasmobranches and teleosts.[23,24]

II. OLFACTION

The olfactory systems of mammals, insects, and birds require that the odorant be dispersed in the air which flows over the nasal receptors. It is reasonable to expect that only those substances which possess a certain degree of volatility will have significant odors. This has been discussed by a number of investigators such as Moncrieff,[1] and Roderick.[25] Roderick[25] has noted that in a homologous series of organic compounds, the odor passes through a maximum such that the higher members of the series are odorless. This is primarily due to lowered volatility (lower vapor pressure) with increasing molecular weight. This explains why foods are so much more aromatic when hot than cold; the increased temperature volatilizes all of the components to a larger extent and leads to a stronger odor.

Only seven elements have an odor and these all exist in diatomic or polyatomic molecules, being members of periodic groups V, VI, or VII. These are: P_4, As_4, O_3, F_2, Cl_2, Br_2, I_2, and they exhibit two types of odors: a halogen odor from F_2, Cl_2, Br_2, and I_2 (Group VII) and a garlic odor from P_4, As_4 (Group V). Some people detect an acrid odor in ozone (Group VI); others do not. Salts are odorless. Covalent compounds of nonmetallic elements often have odors which are usually unpleasant. Water and carbon dioxide are odorless, perhaps because they are always present, again leading to olfactory fatigue.[1,25]

In a homologous series of organic compounds, the odor intensity passes through a maximum such that higher members of the series are odorless. Two opposing factors are operative: (1) decreasing volatility with increasing chain length, leading to diminishing odor intensity; (2) greater lipid solubility (also greater solubility in epithelial mucus) with increasing chain length, leading to increasing odor intensity. In a series of compounds with lower chain lengths, the first factor predominates, while at higher chain lengths, the second factor prevails.[25]

Compounds with similar structures usually have similar odors, but they may have very different odors, as shown below[25]:

Positional aryl isomers:

m-R–C_6H_4–N=C=S, pungent odor; *p*-R–C_6H_4–N=C=S, sweet anise odor (R = –CH_3, –Cl, –Br, –OCH_3)

Macrocyclic compounds:

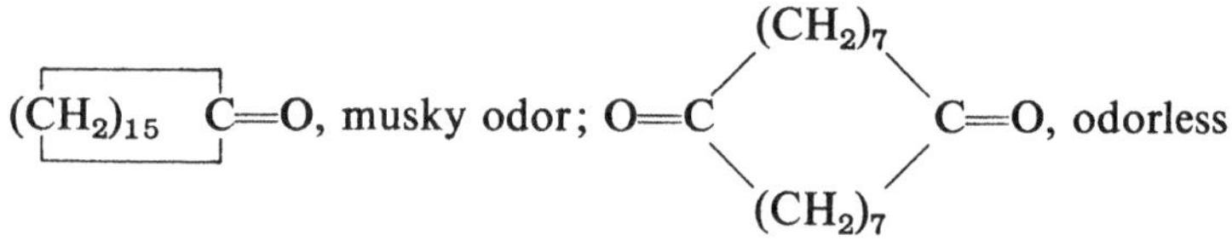

On the contrary, compounds having fairly different structures may have similar odors as illustrated below[1,25]:

Compounds with bitter almond odor:

H—C≡N, hydrogen cyanide; 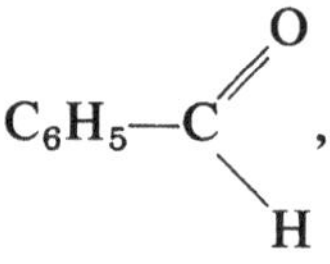, benzaldehyde; 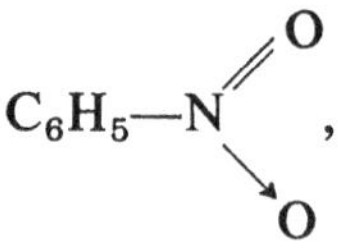, nitrobenzene; C_6H_5—C≡N, benzonitrile

Compounds with camphor odor:

Cl_3C–CCl_3, hexachloroethane; $C_{10}H_{26}O$, camphor; $(CH_3)_3C_6H_2CH_3$, 2,4,5-trimethyl toluene

Very reactive compounds usually have strong odors and unsaturation enhances this odor, producing an irritant odor. There is only a limited correlation of odor with various functional groups. Nevertheless, there are certain functional groups which are called osmophores or odoriphores.

Those which produce pleasant odors are called euosmophores:

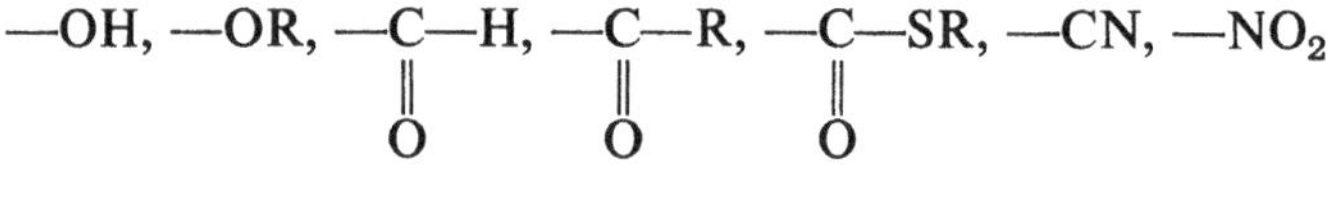

Those producing unpleasant odors are called cacosmophores:

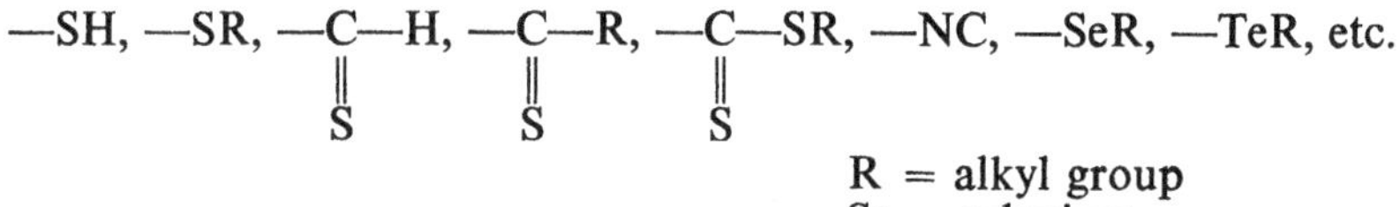

R = alkyl group
Se = selenium
Te = tellurium

For alicyclic compounds, odor is more dependent on size of the ring than on substituents. Odor of aromatic compounds depends more on positions than on nature of substituents. Most perfumes are oxygen-containing organic compounds, but with a conspicuous absence of nitrogen.

Structurally similar isomers usually have similar odors. Positional isomers of benzene have greatly different odors. Geometrical isomers differ in quality and intensity of odors, although the differences are usually not great. Generally, odors of optical isomers are either the same or only slightly different. Some general observations relating odors to functional groups have been reported.[1,25] A tertiary carbon frequently induces a camphoraceous odor; the introduction of a hydroxyl group frequently depresses or abolishes odor; etherification of the hydroxyl group usually restores the odor; esters have fragrant or fruity odors; nitrogen compounds frequently have an "animal" odor; compounds in which an element functions at a valency lower than its maximum usually have offensive odors, e.g., hydrogen sulfide and the isonitriles; branching of a chain usually enhances the odor; the position of a side chain as well as position isomerism in the benzene ring affect the odor.[1,26]

Various attempts have been made to explain how the olfactory system assesses qualitative differences between the various odors. That is, why does one odor smell fragrant, another putrid, and still another camphoraceous? Most of the theories promulgated suggest that the differences in "signals" to the brain are introduced at the receptor level, although amplification or modulation of the initial signal can occur at any location between receptor and the central nervous system.

In a series of papers, Amoore[27-36] has clearly detailed his steric theory of olfaction. Basically, his theory in its original form stated that there are seven different types of molecular olfactory receptors. Each kind of receptor is complementary in size, shape, and electrical properties to the odorous molecule which fits into it and excites it. Seven primary odors were thus defined as those which each fit well into only one receptor. These primary odors were suggested to be: ethereal, camphoraceous, musky, floral, minty, pungent, and putrid, the first five of which are categorized on the basis of size and molecular shape (steric configuration), while the last two, pungent and putrid odors, depend on molecular charge.

The notion that certain molecules are smelled when they fit into certain complementary receptor sites on the olfactory nervous system is derived from Lucretius and was reformulated by Moncrieff.[1,37] It is quite like the enzyme-substrate interaction and antibody-antigen combination. The importance of the steric configurations of odorants, as opposed to the nature of the functional groups, was also forecast by Pauling[38] and Timmermans.[39,40]

In some recent papers, Amoore[27,29] determined "similarity of odor quality" values compared with odor standards using organoleptic testing with human panels [41] A recent procedure, employed by Amoore[27] compares by graphical method the molecular silhouette of the odorant with the silhouettes of standard compounds representative of each odor class.

The parameter, measuring "the similarity of shape" is calculated and plotted. When "similarity of odor quality" values were plotted versus "similarity of shape" values (with each in turn as the independent variable), a "highly significant ($P < 0.001$) linear correlation" was obtained.[27] Recently, this technique has been computerized.[32]

All "normal" people can smell unless they have suffered damage to the brain or olfactory process. Since the olfactory mammalian process includes the olfactory epithelium, the "olfactory bulb, and those areas in the central nervous system which receive secondary olfactory fibers from the olfactory bulb,"[42] it is clear the damage to any of these or accessory areas could lead to a partial or complete anosmia. There are cases of selective anosmia in which there is an inability of the individual to detect certain odors. Thus, selective anosmia to hydrogen cyanide is thought to be a genetic effect.[43] Amoore[29,33,34] has recently studied the question of "specific anosmia." Following a suggestion of Guillot,[44–46] it was hypothesized that specific anosmia would correspond to the inability or insensitivity to an odor. On the basis of odor threshold measurements of 400 people with an apparently normal sense of smell, ten persons were found to have a statistically significant higher odor threshold (higher perception concentration) for the "sweaty" odor of isobutyric acid. This specific anosmia was further studied by testing the thresholds of these individuals to 18 related compounds. Isovaleric acid was found to be the compound for which the difference in olfactory threshold between the anosmic and the normal group was the largest. According to Amoore, the isovaleric molecule represents the nearest approximation to what is a specific odor stimulant for an aliphatic fatty acid receptor. The three parameters: molecular size, shape, and nature of functional groups, were shown to be necessary for this specific anosmia to the "sweaty" odor to prevail. Whatever olfactory detection mechanism is responsible for signaling the presence of the "sweaty" odor, it has the following specificities: a fairly wide tolerance, interacting with straight-chain molecules from C_4 to C_7 selectively but somewhat with C_1 to C_{10}; a distinct preference for certain branched-chain acids, primarily with isovaleric acid; the presence of a carboxylic acid group. A highly significant correlation was found by Amoore[31,33,34] between differences in olfactory thresholds and the molecular similarity of the fatty acid compared to the isovaleric acid structure.

While specific anosmia represents the decreased sensitivity in detecting an odor, general anosmia has been postulated to result from first nerve loss, and complete anosmia appears to be implicated to loss of first and fifth nerve function.[36] On the other hand, a general hyperosmia has been discovered in individuals suffering from cystic fibrosis.[36] Such individuals have an extremely high olfactory threshold (approximately 10,000 times the normal). The relation between the different conditions and the approximate range of sensitivities or thresholds have been tabulated by Amoore.[36]

The main contributions of Amoore can be briefly summarized: (1) He has demonstrated by organoleptic testing that the steric configuration, size, and functional group are essential elements for an odorivector. (2) He has demonstrated the existence of one specific anosmia, fatty acid partial anosmia, and indicated other related odorivectors to which anosmics have a higher threshold. His results implicate receptors which have a specificity for a class of compounds (fatty acids) and closely related compounds with differential affinity for each member of the group.

The olfactory nerve cell possesses a surface which is divided into delicate hair-like filaments[47,48] thought by some to be ultimate sensory process of the receptor cells. The large surface area is suggestive of adsorption as at least a first step in the olfactory process.

Early work performed by Moncrieff[1,49] in which odorants were passed through a sheep's head from a freshly decapitated sheep demonstrated the existence of reversible adsorption. His observations appeared to demonstrate some specificity for different odorivectors. He also concluded that the odorant was not altered during the olfactory process.

Davies and Taylor[50] proposed an adsorption theory of odors based on the use of a novel model for the olfactory nerve cell membrane—namely the erythrocyte membrane. Demonstrating that a variety of odorous substances accelerated the hemolysis of erythrocytes by saponin, they related this to the olfactory thresholds of the odors. Thus, they found that a plot of the logarithm of the hemolytic accelerating power versus the logarithm of the olfactory threshold yields essentially a straight line. Some of the compounds acting as hemolytic accelerators have also been shown to produce leakage of potassium ions across the erythrocyte membrane.[51]

Davies and Taylor[50] then postulated that the odorant molecules distort the olfactory cell membrane (creating a rupture in the red cell membrane), thus permitting an exchange of sodium and potassium ions. This exchange of ions initiates the neural impulse according to the general theory of Hodgkin and Katz.[52] Weaknesses in this proposed model and in the interpretation are the assumption that the olfactory cell membrane and the erythrocyte membrane are similar in structure, and the use of the log-log regression plot which tends to force a curve into a linear form and to reduce scatter.

Rideal and Taylor[53] found that hemolytic acceleration is a surface phenomenon, the effect being simply proportional to the number of accelerator molecules adsorbed on the membrane, and their observations suggest that size, shapes, and flexibilities of the penetrating molecules are also important. Davies and Taylor[54] and Davies[55] assessed the relative importance of adsorption and of molecular morphology in olfaction and derived the following equation from Langmuir's adsorption isotherm:

$$\log OT + \log K_{L/A} = \frac{-4.64}{p} + \frac{\log p!}{p} + 21.19$$

This allows calculation of the olfactory threshold for any odorivector; OT is the olfactory threshold, $K_{L/A}$ is the adsorption constant for molecules passing from air to the lipid–water interface, and p is the number of molecules that must be concentrated on one site of the cell surface to cause a response. The greater the number of molecules, p, required to cause a stimulus, the lower the "puncturing" ability, of which $1/p$ is a measure. Values of the adsorption constant, $K_{L/A}$, are obtained from measurements of interfacial tension lowering at a petroleum ether–water interface.[56]

Davies and Taylor[54] maintain that the quantitative success of their theory in the calculation of olfactory thresholds for man supports the notion that adsorbed odorant molecules initiate the nervous impulse by causing localized permeability changes in the cell membrane. To be a strong odorant, a substance must possess a large value of $K_{L/A}$ and a low value of p. Few molecules possess both characteristics since decreasing p through the introduction of branched chains causes $K_{L/A}$ to decrease significantly. However, artificial musk (trinitrotertiarybutylxylene) is a strong odorant since the molecule is strongly adsorbed and $1/p$ is relatively high. The theory of Davies and Taylor explains olfactory fatigue in that it is represented by a simple depolarization phenomenon in which the cell fails to build up the "equilibrium ion concentrations in the periods between successive inspirations" of odorants well above the threshold.

In an extension of his adsorption theory, Davies[57] proposes a modification designed to explain the qualities of odors. The revised theory describes the stimulation process as follows: the odorant molecule, strongly adsorbed, penetrates the lipid cell membrane, causes membrane rupture, and after a certain time, the molecule desorbs again either into the interior of the cell or into mucus outside. Before the membrane has "healed," Na^+ and K^+ leak through the hole, the area of which is initially the cross section of the adsorbed "oriented" molecules, approximately 50–64 Å^2. The hole is repaired after a short time (approximately 10^{-4} sec) depending on the surface pressure, rigidity, and viscosity of the lipid film. Then, the cell is ready for the initiation of another impulse providing that an odorant molecule penetrates the membrane and desorbs. The crucial point is that the time during the desorption must be long enough to allow penetration of ions through the hole.

Using the theory of surface viscosity and postulating values for the surface pressure gradient, the surface viscosity of the membrane, and the geometry of the hole, Davies and Rideal[58] approximated the time for desorption. For different receptor sites the varying surface viscosities and surface pressure gradients lead to differing times of membrane repair. Thus, for one site, certain large slow-moving molecules will desorb so slowly that the repair will occur directly after the molecule desorbs, not allowing time for ion exchange to occur; conversely, rapidly moving small molecules may desorb fast enough to permit ion exchange and olfactory neuronal response. The time of desorption is related to the free energy of desorption and the absolute temperature.

Davies[57] suggested that position on a plot of molecular cross-sectional area versus the true energy of desorption is related to the odor quality. He found that using this graphic procedure "ethereal" components adjoined the camphorous odorants and the latter were contiguous to both "aromatic" and "pepperminty" odorants. This interpretation implicates the molecular cross-sectional area and free energy of desorption as important but not the sole parameters of odor quality.

Dravnieks[59] has noted that the upper limit of interaction energies is imposed by desorption rates. If one assumes that there is no energy of activation in adsorption, then the activation energy for the desorption process is numerically equal to the energy of adsorption. Consequently, if the energy of adsorption is too large, the desorption process will be very slow as a result of the high activation energy. Dravnieks[59] estimates from a consideration of the transition-state theory of Eyring *et al.*[60,61] that desorption times exceed a few seconds when the adsorption energy approaches 18–20 kcal/mole. When only partial charges are present on the odorants, partial ionic bonds can be formed on the receptor as in the case of strong nucleophilic (putrid) or electrophilic (pungent) odorants. Dravnieks[59] notes that van der Waals–London dispersion forces and hydrogen bonds are operative in odorant-receptor interactions. He suggests also[59,62] that a charge-transfer complex might be involved in odor sensing; the actual process, however, is not known.

Dravnieks[59] proposed a mechanism for initiation of the neuron signal due to combination of odorant and olfactory receptor, without any direct experimental evidence. He suggests a layered structure of proton or electron-transfer complexes forming concave adsorption sites at the junction of the donor and acceptor molecules in the receptor site where the odorant facilitates this transfer. These sensors act as nonlinear capacitors, their electrical nonlinearity changing when an odorant molecule is adsorbed. Serving as passive transducers, the sensors report the presence of the odorant in the site through the generation of new or the disappearance of the original intermodulated frequencies.

III. GUSTATION

Gustation is commonly defined in terms of taste, which itself is usually described as a peculiar sensation produced by the tongue, and is capable of distinguishing between sweet, sour, salt and bitter.[14] However, taste receptor cells are not necessarily confined to the tongue in all animals[3,5,11,12] or even in humans.[63–65] The classical idea of the localization of the four traditional taste sensations[64,66] has also been questioned.[67] Therefore, a more general definition of gustation is the ability to perceive a change in the chemical environment as arbitrarily distinguished from olfaction by the anatomical location of the receptors.

Linnaeus in 1754 classified taste into eleven groups which were later condensed to six groups by Wundt in 1865.[65] Later Wundt reduced his six categories to the current four: sweet, bitter, salt, and sour. As a result, much of the modern research has been concerned with attempting to describe the mechanism by which a particular stimulus is perceived in one of the four taste areas. However, the taste regions described on the human tongue, for instance, when reinvestigated using more sensitive methods, show no distinct boundaries but a gradual phasing of one taste area into another.[63]

One of the traditional characteristics of taste is the apparent lack of discrimination within classes of stimuli eliciting a similar behavioral response; i.e., sugar, bitter, salt, or sour.[1] However, recent work has shown that enantiomers of sugars[68–70] and amino acids[72] can cause quite different responses. Verkade[73] reported that in a homologous series of the 4,*N*-acetoxyphenylureas, the hydroxy compound has a bitter taste, the methoxy compound is intensely sweet, the ethoxy compound is still sweeter, while the *N*-propoxy compound is tasteless. The 1,*N*-acetyloxy-2-amino-4-nitrobenzenes also exhibit a similar phenomenon.[73] Ferguson and Lawrence[66] have also shown that minor structural changes in the chemovector may induce marked changes in taste.

Since most of the detailed biochemical observations have been made on higher chordates, particularly eutherian mammals, one can only postulate a similarity of receptor mechanisms throughout the animal subkingdom. However, behavioral work on diverse selected groups such as hydrazoans,[6] cephalopods,[5] and insects[73] seem to favor such a hypothesis.

Dastoli and his co-workers[74–76] have succeeded in isolating and partially characterizing two receptor proteins of similar size; one from bovine tongue that responds to sweet and another from porcine tongue which combines with bitter components. Each of these proteins was found to be specific for either sugar or bitter but not sensitive to both.[74] One criticism of this study is that the authors used a nonspecific and insensitive technique, refractometry, for measurement of interaction of receptor protein and chemovector. In the future, more receptor proteins may be found in addition to the two reported by Dastoli,[75] such as for salt and sour taste areas, since at least one organic compound, L-glucose, is known to elicit a salt response and several organic acids elicit sour responses far greater than is expected from the hydrogen ion concentration alone.[18] As indirect support of the work of Dastoli, Hidaka and Yokota[77] found that the application of mercuric chloride to the tongue abolished the taste response to sugars while the sweet taste response of glycine was unaffected.

The gustatory perception of some chemicals such as quinine or 6,*N*-propylthiouracil (PTU) appears to be the result of an all-or-none response. Rubin *et al.*[78] have shown a discontinuous taste pattern of binary solutions of quinine and PTU. As a result, Gander *et al.*[79] have proposed a "multiple site chemoreceptor model of taste perception." According to this model, the drug is adsorbed sequentially, first to the binding site(s) and then to the translating site. It has been postulated that when sequential adsorption occurs on proteins, it is accompanied by configurational changes in the protein as each site is occupied with a chemical stimulant.(79) In fact, the two proteins isolated by Dastoli and his co-workers(74) may be some of the primary receptors suggested by Gander *et al.*(79)

Further evidence in favor of the enzyme model of Gander *et al.*(79) comes from the histochemical finding of alkaline phosphatases in human taste pores(80,81) and ATPase activity in the taste buds of rats.(82) Although this in itself might not be significant, interference with oxidative phosphorylation by 2,4-dinitrophenol or thyroxine was shown to abolish taste in a frog's tongue. Administration of ATP restored the taste, but further administration of ATP in turn reduced the taste response.(83) In addition, reports of work done on frog(84) and albino rats(85,86) indicate that inhibition of cholinesterase may play a part in bitter taste. In fact, Tsuchiza and Aoki(86) found that known bitter substances such as quinine hydrochloride, methantheline bromide, L-methylephedrine hydrochloride, sodium phenobarbital, and sodium dehydrochlorohydrate did inhibit cholinesterase activity in the rat taste buds. More evidence comes from the finding that copper is a necessary activator of taste receptors. Henkin *et al.*(87) and Keiser *et al.*(88) found that patients with normal taste response treated with penicillamine, known to chelate copper, lost their sense of taste but not that of smell. However, administration of copper apparently restored the gustatory senses.(87) On the other hand, patients with Wilson's disease typically lose the taste sense which is restored upon the advent of penicillamine therapy.(87) It has been postulated that the penicillamine treatment liberates some of the large tissue deposits of copper available for the gustatory senses,(87) whereas in the individual only small concentrations of copper ion are present in the plasma and most of this is chelated by the penicillamine.

In spite of the general assumption that taste is an all-or-none phenomenon involving the combination of the stimulant with taste receptors, recent work with sugars and amino acids has shown a fairly high degree of discrimination.(68,69,71) For instance, Pangborn and Chrisp(68) have shown different degrees of sweet taste sensitivities between enantiomers of the same sugar, while Steinhardt *et al.*(89) showed that α-D-mannose elicited a sweet response and β-D-mannose a bitter response. Boyd and Matsudara(69) found that L-glucose produced a salt response from their subjects. The response to L-mannose was inconclusive, possibly due to the enantiomeric mixture used.(89) Vantag and Egli(71) discovered that the L forms of leucine, phenylalanine, tryptophan, and tyrosine tasted bitter, while their enantiomers tasted sweet.

In general, both the specific chemical groups and stereoconfiguration are responsible for certain tastes judging from homologous chemical series in which a taste spectrum from bitter to sweet to no taste at all was elicited.(66) Although the above considerations are concerned mainly with the sweet taste, the principles apply also to bitter taste.(72)

Much of the gustatory research to date has had implicit in it the assumption that the four traditional taste responses of salt, sweet, bitter and sour defined over 100 years ago involve separate and discrete, though similar, receptor mechanisms. This assumption has never been verified. Additional evidence of the arbitrariness of the assumption comes from the work of von Bekesy[(63)] who showed a large degree of overlapping of the traditional taste zones on the human tongue. Therefore, it seems reasonable to expect at least in terrestrial tetrapods, a continuous taste spectrum which has been divided into four apparent zones by the preconditioned behavior responses of civilized man. In light of recent findings, it appears that what is experienced as taste responses or taste types represents arbitrary discrete portions, determined by acquired behavior patterns, from an apparently continuous gustatory spectrum. The stimuli are received at the membrane surface by protein receptor molecules[(74–76)] sensitive to various charge distribution patterns within the stimulant molecule. Of course, the charge distribution patterns of the sapid stimuli are determined by the tertiary steric conformation of the stimulant. The apparently continuous spectrum is produced by overlapping sensitivities of the protein receptors. However, some taste responses are impossible to categorize, such as beryllium and lead which elicit a sugar response,[(1)] while L-glucose tastes like salt. Some stimulants elicit in man responses from both gustatory and olfactory senses. Thus, chloroform, an airborne chemical, is tasted rather than smelled as sweet and 3,4-dichloroaniline and 2,5-dichloroaniline have similar sensory thresholds for both olfaction and gustation.[(90)]

Beidler[(18)] developed a theory of taste stimulation in which he assumes adsorption of the chemovector on the receptor sites under condition of equilibrium rather than steady state. The influence of the stimulant on the taste receptor was measured by using single nerve fibers from the gustatory cells for recording electrical responses. Although the electrical response from a single nerve fiber varied with time, the summation of these responses from a number of nerve fibers yielded a steady level of saturation of taste response, indicating that the degree of response depended on the combination of the stimulus with the receptor. He deduced that the magnitude of the neuronal response was directly proportional to the number of receptor sites occupied by the stimulant (R) and that the maximum neuronal response which can be elicited at a high concentration of the stimulant occurs when all the sites are filled (R_s). Thus, he formulated the following equation which is basically an analogue of the Langmuir adsorption isotherm[(18)] and the Michaelis–Menten equation for enzyme kinetics[(91)]: $C/R = C/R_s + 1/KR_s$ where R is the magnitude of the electrical response to the chemical stimulant at concentration C, R_s is the maximum response at the high concentration of the stimulant and K is the equilibrium constant for the combination of stimulants to the receptor site. He called this the fundamental taste equation which relates the magnitude of the response to the concentration of the stimulus. The K can be determined by measuring the magnitude of the electrical response to two different concentrations of the stimulus. The facts which underly this equation were summarized by Beidler based on his experimental studies. (1) Taste receptors may respond within 30 msec when stimulated by a chemovector, (2) The taste receptors respond to a large range of the stimulant concentration. (3) Use of nonphysiological chemovectors such as sodium cyanide, strychnine, and strong acids elicit gustatory response from taste receptors without damaging the receptor cells.(92) (4) The chemovectors elicit a rapid and steady rate of response from taste receptors where the magnitude of the response is a function of the concentration of the applied stimulus.(93) (5) The gustatory response to many chemovectors remains constant over a long period of stimulation. (6) Gustatory response rapidly declines following a water rinse. (7) The taste response is almost independent of temperature between 20–30°C and is also independent of *p*H between 3–11. (8) Different species reveal different responses to cationic and anionic stimulants.(18,94) Beidler deduced that the speed of the olfactory response to stimulants and the fact that toxic chemicals stimulate the receptors without damaging them indicates that the chemovector does not enter the interior of the receptor cell when it initiates excitation. He also deduced from insensitivity of taste to temperature and from the observation that toxic stimulants can be removed by water rinse that enzymes are probably not involved in the interaction of receptor with stimuli. He concluded that the stimuli are bound to the receptors by adsorptive forces. His data also show that since a single gustatory cell is sensitive to a number of chemical stimulants and to a variety of taste qualities that a number of different adsorptive sites exist on the surface of each taste cell and this suggests that a variety of given taste stimulants may be adsorbed to a given number of different sites without interaction between them. In his more recent work,(18) he concludes that many of the anionic sites in proteins and phospholipids of the receptor cell membrane bind H^+ and others bind alkali cations. Not all of the ions bound to the anionic sites on the gustatory membrane elicit taste response, since some cations in combination with anionic sites only function to change the net charge density of the receptor membrane. He also finds that anions influence the magnitude of the gustatory response elicited by cations by decreasing the net positive surface charge. Hence, the relative number of both anionic and cationic sites of the gustatory membrane and the nature of the specific ion involved in the stimulating solution establishes the magnitude of the taste response and determines whether the response is inhibitory or excitatory.(18)

IV. IMPULSE PROPAGATION AND MEMBRANE CHANGES

In eutherian animals, the gustatory and olfactory receptors are modified neurons which are uniquely responsive to chemovectors. It is presumed that in both chemosensory systems the propagated impulse is initiated by the conformation changes resulting from the interaction of chemovector with the receptor molecule on the surface of the excitatory membrane.[3,12,18–20,57,95–97] A recent concept (the induced fit theory of Koshland and Kirtley[98]) of protein conformation changes when enzyme proteins or receptor proteins interact with substrates or modifiers, offers a mechanistic view on the changes induced in such membranes as are encountered in chemosensory receptor cells. It thus accounts for the entrance of extracellular sodium into the receptor cell, initiating depolarization and the subsequent events. The protein is visualized as having greater flexibility and capability for assuming different tertiary structures than formerly recognized, when it is complexed with a small molecule. Here the chemical entity induces a conformational change in the protein when it interacts with it in a position other than the active site, by altering the position of the groups in the active site. This, in turn, induces conformational changes along the protein backbone, eventually changing the structure of the membrane of which it is a constituent. Under this concept of an induced fit theory, Koshland and Kirtley attempt to show how the binding of a component to an external site on the membrane induces a conformational change in a component protein which activates an enzyme site on the inner surface of the membrane. Thus, in the absence of this component or modifier, the enzyme can interact with the substrate inside the cell, but be inactive. In the presence of the modifier, however, the enzyme is activated by a conformational change and thus becomes active. The reverse condition may also be true where the substrate is acting as a feedback control. This illustrates how a chemical or modifier on the outside of the cell can induce conformation changes of membrane proteins and activate an enzyme within the inner surface of the membrane without having the chemical or modifier consumed or altered in any reaction, a situation simulating the effect of a chemovector, whether odorant or gustatant, on the receptor cell membrane when it initiates a sensory response. A similar model has been proposed by Gander *et al.*[79] for gustation. This conformational change distorts the membrane configuration, permitting extracellular Na^+ to enter the cell interior, initiating depolarization of the membrane. Subsequently, with the influx of Na^+, the charge on the inner surface of the membrane becomes more positive than the outer surface and Na^+ conductance exceeds that of K^+; ultimately the membrane becomes depolarized. At this stage, it is generally considered that Na^+ conductance is greatly reduced and K^+ begins to leak out of the cell into the extracellular fluid in sufficient amounts to repolarize the membrane. At this moment, the receptor cell has, in the region of the membrane, an abnormally high concentration of Na^+ in the cytoplasm and conversely a high K^+ content extracellularly. This condition activates the "sodium pump," which, at the expense of energy derived from ATP, actively transfers Na^+ from the interior of the cell to the extracellular fluid and simultaneously facilitates the transfer of extracellular K^+ into the cell.

The mechanism of the ion transfer facilitated by the sodium pump is generally envisioned as follows[95,99,100]: The Na^+–K^+ dependent ATPase system, contained within the receptor membrane, is believed to consist of two enzymes, one of which is a protein kinase which requires Na^+ and phosphorylates protein in the reaction

$$\text{Protein} + \text{Mg ATP} \xrightarrow[Na^+]{\text{kinase}} \text{P-protein} + \text{ADP} + Mg^{2+}$$

The resulting phosphorylated protein (P-protein) is presumed to be the carrier for Na^+ and K^+. Thus, the phosphoprotein carriers combine with the respective ions, carrying the Na^+ from the interior of the cell to the extracellular compartment and concomitantly K^+ from the extracellular fluid into the cell. The respective ions are liberated in these compartments through the action of a second enzyme, phosphoprotein phosphatase, which catalyzes

$$Na^+ - \text{P-protein} + H_2O \xrightarrow[K^+]{\text{phosphatase}} \text{Protein} + Na^+ + P_i,$$

where P_i is inorganic phosphate ion. A similar reaction occurs in K^+ transfer. Thus, the Na^+–K^+ ATPase system carries out the following overall reaction:

$$\text{MgATP} + \text{Protein} \xrightarrow{\text{Activated by } Na^+} \text{Protein-phosphate} + \text{ADP} + Mg^{2+}$$

$$\text{Protein-phosphate} \xrightarrow[\substack{\text{inhibited by} \\ \text{ouabain and oligomycin}}]{\text{Activated by } K^+} \text{Protein} + P_i$$

$$\text{ATP} \longrightarrow \text{ADP} + P_i$$

The relocation of the Na^+ extracellularly and of the K^+ within the cell, re-establishes the resting state of the excitatory receptor cell.[95,99,100]

The above interpretation of impulse propagation, associated with biochemical changes within the membrane of the receptor cell, has been derived from studies mainly of the kidney, brain, and red cell and none of these observations have been reported using chemosensory epithelia. There is general agreement among investigators concerned with ion transport and membrane depolarization that the mechanism summarized above represents a universal mechanism within excitatory neurons during stimulation.[95,99,100]

A more recent interpretation of the structure and behavior of neuronal membranes has implicated the interface between the glia and the neurons as being a structural extension of the neuronal membrane, participating in the neuronal function of impulse propagation; thus the cell coat is part of the supermolecular organization of the plasma membrane.[101] Also, the intracellular space between neurons and glia is thus a structural matrix of the neuronal membrane and not a free dilute solution of solutes.[101]

Although the plasma membrane was considered as a bileaflet structure with two outer layers of protein and an inner bilayer of lipids as originally proposed by Danielli and Davson[51,102] and further defined by Robertson,[103] it is now suggested that the bileaflet membrane model represents one possible structure.[104–106] The cell membrane may have other forms such as globular bilayers in different cells,[101,104–106] the underlying assumption being that, under various conditions, part of the linear bileaflet forms of the membrane may become globular bilayers and that this transition from one form to another is an innate property of the plasma membrane, dependent on the changes and conditions impinged upon it, such as action potentials, from the outside (Fig. 1A).[101,107] The conformational change in these membranes can be detected by various techniques such as circular dichroism or optical rotatory dispersion.[108] Nuclear magnetic resonance has shown that the proteins within the membrane are relatively fixed but the lipids have a considerable degree of movement.[109]

The outer coat of the neuronal membrane contains glycolipids and glycoproteins. The glycolipids are electrically neutral, and the cerebrosides represent a class of glycolipids which are found only in glial cells and myelin membranes. The neuronal membranes contain glycoproteins, represented by gangliosides. The gangliosides have two hydrocarbon tails and a long branched chain polar head composed of oligosaccharides containing one or more sialic acid (*N*-acetyl or *N*-glycoyl derivatives of neuraminic acid), and these are found only in neurons, not in glial elements.[110,111] The importance of these gangliosides in neuronal function is illustrated by the observation that brain tissue *in vitro* under anoxia loses its property of electrical excitability, mainly due to the loss of sialic residues from its ganglioside. Addition of gangliosides to such a preparation restores its excitability.[101,110] These gangliosides have negative charges and can bind Na^+, K^+, and Ca^{2+} which bind at the carboxyl groups of the sialic acid.[110] The glycoproteins represent the second class of components in addition to the gangliosides that constitute the components of the cell coat on the outside surface of the plasma membrane. The glycoproteins extend into the intracellular space between the neuron and glia. The neurons

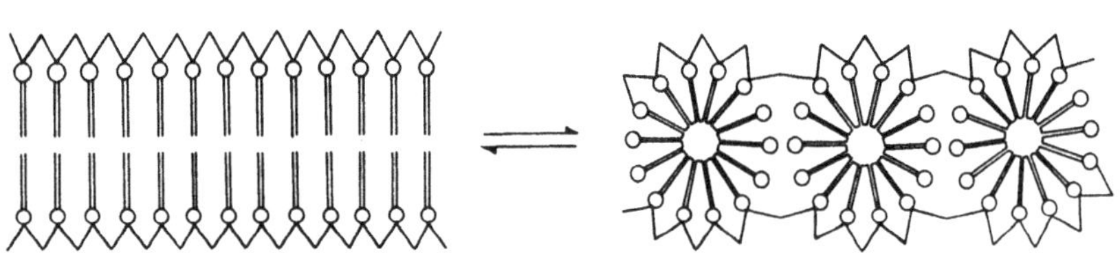

Fig. 1(A). Schematic representation of the membrane transition from the bilayer to the globular form (From Lehninger.[101])

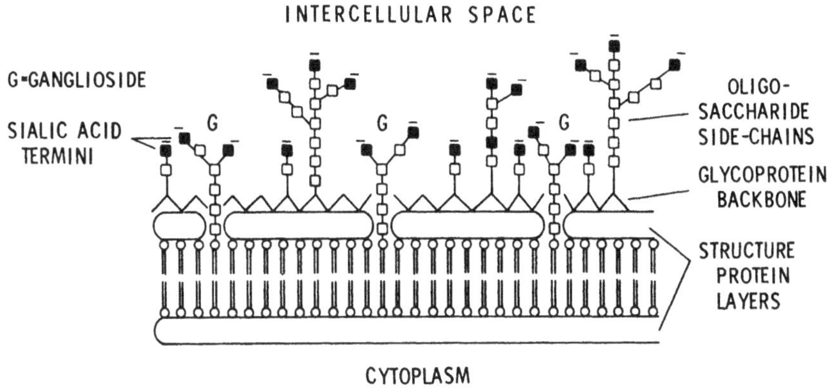

Fig. 1(B). A schematic drawing representing a section of the membrane depicting the protruding molecular components in the intracellular space from the outer surface of the membrane. (From Lehninger.[101])

contain sialic acid residues which contribute to the negative charge of the neuronal surface. Figure 1B represents a model of the neuronal plasma membrane with the attached glycoproteins and gangliosides as depicted by Lehninger.[101] It is suggested that the glycoproteins are superimposed on a monolayer of protein, thus forming the neuronal and plasma membrane. From this model, it is possible to visualize the vectorial transport of ionic solutes.

Although the concept of Na^+–K^+ exchange as the present concept of the basis for neuronal excitation and the propagation of action potentials was first formulated by Hodgkin and Huxley,[111] Tasaki has shown[112–114] that neither Na^+ nor K^+ is required on the inside or outside of a squid giant axon to support excitability. These investigators found the infusion of squid axon with Li^+, Rb^+, Cs^+, or $NH_4{}^+$ as phosphate can maintain excitability and elicit an action potential if the axon is externally bathed with Ca^{2+}, Ba^{2+}, or Sr^{2+}. He also showed that if the internal K^+ is replaced by Na^+ the membrane is still excitable, contrary to the Hodgkin and Huxley[111] concept. These findings have led Tasaki and co-workers to propose that in the resting state the outer region of the axon membrane presumably contains Ca^{2+}, and on stimulation (electrical or chemical), the Ca^{2+} is replaced by K^+ derived from the interior of the cell.[112–114] This ion exchange presumably causes a change in the conformation of the membrane subunits which increases membrane conductance and allows Na^+ to enter the cell and K^+ to diffuse out. At the end of the action potential, the membrane-bound K^+ is displaced by Ca^{2+} again and the membrane is returned to its original conformation in the resting state. The binding sites for Ca^{2+} are very likely the membrane gangliosides and actually only a small fraction of the membrane area need undergo loss of Ca^{2+} to cause its all-or-none excitatory response.

It has also been discovered that tetrodotoxin (toxin of the ovaries of the Japanese puffer-fish and a similar one from California salamander) in very small amounts, blocks the exchange of univalent cations for divalent cations.[114] Its effect is only on the outer surface membranes. Moore and Narahashi[115,116] have shown that only a few sites are tetrodotoxin-sensitive, confirming Tasaki's view that cation exchange of only a few molecular units in the membrane are sufficient to convert a very much greater number of units into conformational changes in the membrane to permit passage of Na^+ and K^+.

From this brief discussion of the more recent findings, it appears that in the chemosensory systems the binding of the chemical stimulant to the receptor site in the neuronal membrane may elicit the following changes: (1) binding of the chemical to the receptor causes displacement of Ca^{2+} and induces a conformational change in the receptor protein; (2) the Ca^{2+} site is replaced by Na^+ and the conformational change in the receptor protein distorts the membrane, possibly causing transition of membrane from linear to globular bileaflet form and thus allowing Na^+ to penetrate and enter the cell. The subsequent events are those which have been described above in the conventional Na^+–K^+ exchange and the concomitant depolarization and repolarization of the membrane.

V. SENSORY INFORMATION ACCRUAL IN THE BRAIN

Consideration of the chemosensory systems would be incomplete without some mention of the recent developments and implications of the chemical changes induced in the CNS by the coded afferent impulses originating from these and other peripheral sensory systems.[117–124]

Review of the diverse interpretations of the biochemical events in the brain which represent information accrual and storage has culminated in the development of a working hypothesis in which the accrual of conditioned information is considered to be a two-step molecular biosynthetic process. The first step, the transduction of the information from coded afferent impulse into nuclear RNA, must be a rapid phenomenon occurring within the speed at which the impulses reach the neurons in the brain. The resulting RNA probably contains the information of the afferent impulse in the form of base sequences (identical to the genetic code). It appears that this RNA most likely represents the immediate information engram which can serve as a template for the synthesis of protein. The second step in the two-step process occurs at much lower speed, requiring reinforcement of afferent information in order to obtain sufficient protein synthesis. Conceptually, it is suggested that protein represents the long-term storage of information. The information encoded in the protein is in the form of sequential linkage of specific amino acids, each amino acid arising as a product of the triplet code in RNA presumably identical to the genetic code.

In order to maintain a clear concept which would permit experimental testing of the mechanisms involved in the accrual of information transmitted by afferent impulses from the sensors, the following concepts may be deemed as a working hypothesis for information accrual by the brain. First, it is conceivable that sensory information is stored essentially in the nucleus of the brain cells since it is a stable environment in which genetic information is retained for the life span of the organism. Secondly, irrespective of the accepted mechanism responsible for the transfer of coded impulses along the axons and dendrites of the neurons, it is believed here that the impulse is also transmitted through the cytoplasm in the brain cells, possibly via the endoplasmic reticulum, thus reaching the nucleus. In the nucleus the impulse initiates *de novo* synthesis of RNA, thus transcribing the information from the impulse into the sequence of bases within this polymer. The latter subsequently acts as a template to transcribe this information into a specific sequence of amino acids in a newly formed protein. This mechanism specifies that an RNA synthesizing enzyme is present which does not depend upon DNA as a template, hence differs from RNA polymerase. The only known enzyme which can synthesize RNA, and has been found in animal tissue,[125–127] is polynucleotide phosphorylase[128] which acts without requiring either DNA or RNA as templates and which uses nucleoside diphosphates as substrates. It is possible that some other enzyme yet to be discovered may perform this function. However, designation of polynucleotide phosphorylase for this role avoids presumptive assignments to an imaginary entity and allows experimental approach to the concept.

The schematic diagram in Fig. 2 illustrates the concept of information processing in the brain cell nuclei. It is assumed that information originating from the sensory organs is transmitted to the neuronal nuclei in the form of an afferent impulse which contains the information in a frequency modulated signal spelled out in the diagram as DEF. When this coded afferent impulse reaches the neuronal nuclei, a particulate component of the nuclei receives this signal and this activates an enzyme within this particle capable of synthesizing an RNA in the absence of a template or primer. Such an enzyme is polynucleotide phosphorylase complexed in a transducer particle which, in response to the frequency modulation of the afferent impulse, can select one specific nucleotide diphosphate from the surrounding pool and, hence synthesize an RNA whose base sequence is determined by frequency components in the afferent impulse. In Fig. 2, this particle is designated as the transducer since it converts the coded afferent impulse into a coded RNA. This transduction occurs rapidly, at speeds equal to or greater than the speed with which afferent impulses enter the neuronal nuclei.

The resulting coded RNA than acts as a template for the biosynthesis of a protein,[129] through the action of nuclear ribosomes and adjuvant

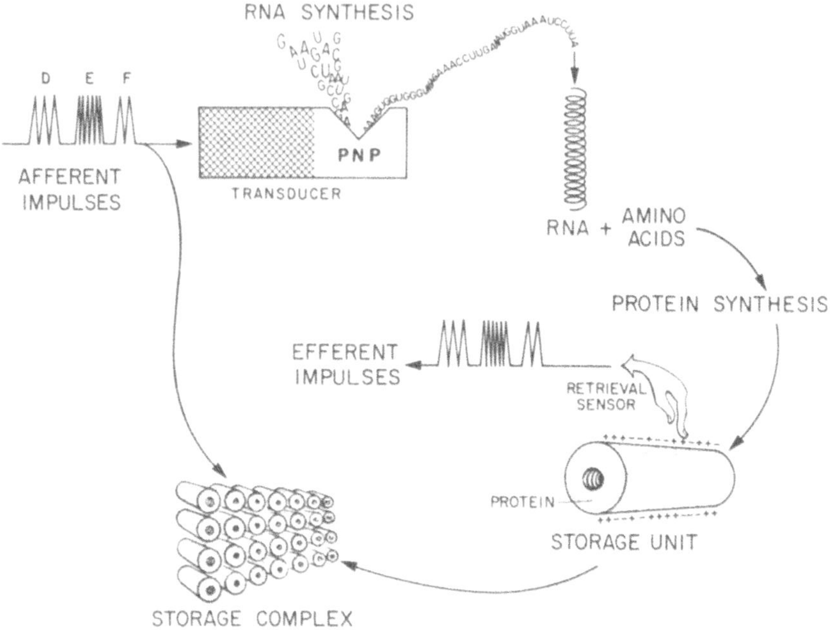

Fig. 2. Scheme of the concept of information processing in the neuronal nuclei.

enzymes, which will contain the information in the encoded RNA. However, information represented by the sequence of amino acids in this newly synthesized protein is important from the standpoint of charge distribution on the surface of the protein due to the amino acid side chains. Thus, the nature of the charged and the uncharged side groups on the surface of the protein may represent the code originally presented to the nuclei as frequency components in the afferent impulse. It is assumed that the newly synthesized protein forms a tight helix constituting the storage unit, as indicated in Fig. 2. It is visualized that the surface charges and their distribution on this protein particle will be a permanent coded memory unit from which the information can be retrieved by an electrical sensor (indicated in the diagram) whose identity, at present, cannot be depicted, and thus an efferent impulse can be generated at any time when retrieval is necessary. A group of these storage units representing particulate proteins, fixed in a relative position to each other and in a temporal sequence dependent on the time of information accrual, represent stored memory. These units can be imagined as permanent entities which are destroyed only when the neurons are disrupted by physical injury.

It is conceived in this hypothesis that the afferent impulse entering the neuronal nuclei is initially bifurcated, with one branch of the impulse immediately scanning the storage complex and rapidly noting whether the new information in this impulse has previously been experienced by the presence or absence of a corresponding storage unit. If such information has been accrued and duly stored, then that component of the impulse, which would normally pass through the transducer, will not pass through the transducer and no new RNA will be formed. This is essentially a conservation mechanism. Further examination of this concept indicates that both the transducer and the storage unit are postulated to be insoluble particles. This concept also presupposes that the transducer will show enhanced polynucleotide phosphorylase activity only under the influence of an electrical impulse. Thus presumably some fraction of the insoluble particles isolated from neuronal nuclei should exhibit pronounced polynucleotide phosphorylase activity under direct impulse stimulation. There is no evidence for this concept, except for some published data by Rappoport and Daginawala[130] which can be interpreted by this hypothetical scheme.

ACKNOWLEDGMENTS

This work was supported in part by a Robert A. Welch Foundation Grant, No. H-180 and a Public Health Service Grant, DHEW 1 R01 NB 07707.

VI. REFERENCES

1. R. W. Moncrieff, *The Chemical Senses*, 3rd ed., Chemical Rubber Co. Press, Cleveland, Ohio (1967).
2. L. M. Beidler, Comparison of gustatory receptors, olfactory receptors and free nerve endings, *Cold Spring Harbor Symp. Quant. Biol.* **30**:191–200 (1965).
3. Y. Zotterman, The neuronal mechanism of taste, *Progr. Brain Res.* **23**:139–154 (1967).
4. J. Bardach, M. Fujiya, and A. Holl, *in Olfaction and Taste* 2 (T. Hayashi, ed.), pp. 647–665, Pergamon Press, Oxford (1967).
5. M. J. Wells, Taste by touch: some experiments with octopus, *J. Exptl. Biol.* **40**:187–193 (1963).
6. H. M. Lenhoff, Behavior, hormones and hydra, *Science* **161**:434–442 (1968).
7. J. A. Colin Nicol, *The Biology of Marine Animals*, John Wiley and Sons, New York (1967).
8. R. H. Whittaker, New concepts of kingdoms of organisms, *Science* **163**:150–160 (1969).
9. P. S. Galtsoff, The American oyster *Crassostrea virginica, Fishery Bulletin of the Fish and Wildlife Service* (1964) p. 64.
10. L. H. Hyman, *The Invertebrates: Echinodermata the Coelomate Bilatera*, Vol. 4, McGraw-Hill Book Co., New York (1955).
11. C. K. Weichert, *Anatomy of the Chordates*, McGraw-Hill Book Co., New York (1958).

12. B. Oakley and R. M. Benjamin, Neural mechanisms of taste, *Physiol. Rev.* **46**:173–211 (1966).
13. R. I. Henkin and R. L. Christiansen, Taste localization on the tongue, palate and pharynx of normal man, *J. Appl. Physiol.* **22**:316–230 (1967).
14. H. D. Patton, *in Physiology and Biophysics* (T. C. Ruch and H. D. Patton, eds.), Vol. 19, pp. 364–378, W. B. Saunders, Philadelphia (1965).
15. G. Borg, H. Diamant, B. Oakley, L. Strom, and Y. Zotterman, *in Olfaction and Taste* 2 (T. Hayashi, ed.), pp. 253–264, Pergamon Press, Oxford (1967).
16. D. L. Hammerman, Lingual premetamorphic papillae as larval taste structures in frogs, *Nature* **215**:98–99 (1967).
17. C. J. Herrick, *in An Introduction to Neurology*, 4th ed., pp. 215–223, W. B. Saunders, Philadelphia (1916).
18. L. M. Beidler, *in Olfaction and Taste* 2 (T. Hayashi, ed.), pp. 509–534, Pergamon Press, Oxford (1967).
19. D. Ottoson, Some aspects of the function of the olfactory system, *Pharm. Rev.* **15**:1–42 (1963).
20. D. G. Moulton and L. M. Beidler, Structure and function in the peripheral olfactory system, *Physiol. Rev.* **47**:1–52 (1967).
21. D. G. Moulton, Pigment and the olfactory mechanism, *Nature* **195**:1312–1314 (1962).
22. B. Rosenberg, T. N. Misra, and R. Switzer, Mechanism of olfactory transduction, *Nature* **217**:423–427 (1968).
23. Y. A. Vinnikov, Structural and cytochemical organization of receptor cells of the sense organs in the light of their functional evolution, *J. Evolutionary Biochem. Biophysics* **1**:67 (1965).
Y. A. Vinnikov, Structural and cytochemical organization of receptor cells of the sense organs in the light of their functional evolution, *Fed. Proc.* **25**:T34–T42 (1966).
24. Y. A. Vinnikov, Principles of structural, chemical, and functional organization of sensory receptors, *Cold Spring Harbor Symp. Quant. Biol.* **30**:293–299 (1965).
25. W. R. Roderick, Current ideas on the chemical basis of olfaction, *J. Chem. Ed.* **43**:510 (1966).
26. R. W. Moncrieff, *The Chemical Senses*, 1st ed., John Wiley and Sons, New York (1946).
27. J. E. Amoore, Psychophysics of odor, *Cold Spring Harbor Symp. Quant. Biol.* **30**:623–637 (1965).
28. J. E. Amoore and D. Venstrom, Sensory analysis of odor qualities in terms of the stereochemical theory, *J. Food Sci.* **31**:118–128 (1966).
29. J. E. Amoore and D. Venstrom, *in Olfaction and Taste* 2 (T. Hayashi, ed.), pp. 3–17, Pergamon Press, Oxford (1967).
30. J. E. Amoore, *in Symposium on Foods: The Chemistry and Physiology of Flavors* (H. W. Schultz, E. A. Day, and L. M. Libbey, eds.), pp. 119–147, Avi Publ. Co., Westport, Connecticut (1967).
31. J. E. Amoore, Specific anosmia: A clue to the olfactory code, *Nature* **214**:1095–1098 (1967).
32. J. E. Amoore, G. Palmieri, and E. Wanke, Molecular Shape and Odor: Pattern Analysis by PAPA, *Nature* **216**:1084–1087 (1967).
33. J. E. Amoore, *in Theories of Odor and Odor Measurement* (N. Tanyloac, ed.), pp. 71–81, Robert College, Istanbul, Turkey (1968).
34. J. E. Amoore, D. Venstrom, and A. R. Davis, Measurement of specific anosmia, *Percept. and Motor Skills* **26**: 143–164 (1968).

35. D. Venstrom and J. E. Amoore, Olfactory threshold in relation to age, sex or smoking, *J. Food Sci.* **33**: 264–265 (1968).
36. J. E. Amoore, Molecular basis of odor, Lecture Notes, University of California Extension Short Course (April 25–27, 1968), Berkeley, Copyright, The Regents of the University of California.
37. R. W. Moncrieff, What is odor? A new theory, *Am. Perfumer* **54**:453 (1949).
38. L. Pauling, Analogies between antibodies and simpler chemical substances, *Chem. Eng. News* **24**:1064 (1946).
39. J. Timmermans, Un nouvel état mesomorphe. Les cristaux organiques plastiques, *J. Chim. Phys.* **35**:331 (1938).
40. J. Timmermans, Odour and chemical constitution, *Nature* **174**:235 (1954).
41. H. G. Schutz, A Matching-Standards Method for Characterizing Odor Qualities, *Ann. N.Y. Acad. Sci.* **116**:517–526 (1964).
42. A. H. Lohman and H. J. Lammers, On the structure and fibre connections of the olfactory centres in mammals, *Progr. Brain. Res.* **23**:65–82 (1967).
43. R. L. Kirk and N. S. Stenhouse, Ability to smell solutions of potassium cyanide, *Nature* **171**:698–699 (1953).
44. M. Guillot, Anosmies partielles et odeurs fondamentales, *Compt. Rend. Soc. Biol.* **226**:1307–1309 (1948).
45. M. Guillot, Sur quelques caractères des phénomènes d'anosmie partielle, *Compt. Rend. Soc. Biol.* **142**:161–162 (1948).
46. M. Guillot, Aspect pharmaco dynamique de quelques problems liés a l'olfaction, *Actualites Pharmacd.* **9**:21–34 (1956).
47. H. Engstrom and G. Bloom, The structure of the olfactory region in man, *Acta Otolaryngol.* **43**:11–21 (1953).
48. G. Bloom and H. Engstrom, The structure of the epithelial surface in the olfactory region, *Exptl. Cell Res.* **3**:699–701 (1952).
49. R. W. Moncrieff, The sorptive properties of the olfactory membrane, *J. Physiol.* **130**:543–558 (1955).
50. J. T. Davies and F. H. Taylor, A model system for the olfactory membrane, *Nature* **174**:693–694 (1954).
51. H. Davson and J. F. Danielli, Studies on the permeability of erythrocytes, V., *Biochem. J.* **32**:991–1001 (1938).
52. A. L. Hodgkin and B. Katz, The effect of sodium ions on the electrical activity of the giant axon of the squid, *J. Physiol.* **108**:37–77 (1949).
53. E. K. Rideal and F. H. Taylor, On Haemolysis and Haemolytic Acceleration, *Proc. Roy. Ser. B,* **148**:450–464 (1958).
54. J. T. Davies and F. H. Taylor, The role of adsorption and molecular morphology in olfaction: The calculation of olfactory thresholds, *Biol. Bull.* **117**:222–238 (1959).
55. J. T. Davies, The mechanism of olfaction, *Symp. Soc. Exptl. Biol.* **16**:170–179 (1962).
56. D. A. Haydon and J. N. Phillips, Gibbs equation and the surface equation of state for soluble ionized monolayers in absence of added electrolyte at the oil-water interface, *Trans. Faraday Soc.* **54**:698–704 (1958).
57. J. T. Davies, A theory of the quality of odours, *J. Theoret. Biol.* **8**:1–7 (1965).
58. J. T. Davies and E. K. Rideal, *in Interfacial Phenomena*, 2nd ed., pp. 6, 253, Academic Press, New York (1963).
59. A. Dravnieks, Physico-chemical basis of olfaction, *Ann. N.Y. Acad. Sci.* **116**:429–439 (1964).
60. H. Eyring, H. Gershinowitz, and C. E. Sun, Absolute rate of homogeneous atomic reactions, *J. Chem. Phys.* **3**:786 (1935).

61. S. Glasstone, K. Laidler, and H. Eyring, *in The Theory of Rate Processes*, p. 100, McGraw-Hill Book Co., New York (1941).
62. A. Dravnieks, *in Olfaction and Taste* 2 (T. Hayashi, ed.), pp. 89–107, Pergamon Press, Oxford (1967).
63. G. von Bekesy, Taste theories in the chemical stimulation of a single papillae, *J. Appl. Physiol.* **21**:1–9 (1966).
64. R. I. Henkin, The role of taste and disease in nutrition, *Rev. Nutrition Res.* **28**:71–87 (1967).
65. I. Hornstein and R. Teranishi, The chemistry of flavor, *Chem. Eng. News* **45**:92–108 (1967).
66. L. N. Ferguson and A. R. Lawrence, The physical chemical aspects of the sense of taste, *J. Chem. Educ.* **35**:436–443 (1958).
67. R. Fischer, F. Griffin, and M. A. Rockey, Gustatory chemoreception in man: Multidisciplinary aspects and perspectives, *Perspect. Biol. Med.* **9**:549–577 (1966).
68. R. N. Pangborn and R. B. Chrisp, Gustatory responses to anomeric sugars, *Experientia* **22**:612–615 (1966).
69. W. C. Boyd and S. Matsubara, Different tastes of enantiomorphic hexoses, *Science* **137**:669 (1962).
70. R. S. Shallenberger, T. E. Acree, and W. E. Guild, Configuration confirmation in sweetness of hexose animers, *J. Food Sci.* **30**:560–563 (1965).
71. J. Solms, L. Vuataz, and R. H. Egli, The taste of the L and D amino acids, *Experientia* **21**:692–694 (1965).
72. P. E. Verkade, On organic compounds with a sweet and/or bitter taste, *Illpharmico Ed. Sc.* **23**:248–291 (1967).
73. J. Boeck, K. E. Kaissling, and D. Schneider, Insect olfactory receptors, *Cold Spring Harbor Symp. Quant. Biol.* **30**:263–280 (1965).
74. F. Dastoli, The chemistry of taste, *New Scientist* **29**: 465–467 (1968).
75. F. R. Dastoli, D. V. Lopiekes, and S. Price, A sweet sensitive protein from bovine taste buds: Purification and partial characterization, *Biochemistry* **8**:1160–1164 (1968).
76. F. R. Dastoli and S. Price, Sweet sensitive protein from bovine taste buds: Isolation and assay, *Science* **154**:905–907 (1966).
77. I. Hidaka and S. Yokada, Taste receptor stimulation by sweet tasting substances in the carp, *Jap. J. Physiol.* **17**:652–666 (1967).
78. T. R. Rubin, F. Griffin, and R. Fischer, A physical chemical treatment of taste thresholds, *Nature* **195**:362–364 (1962).
79. J. E. Gander, F. Griffin, and R. Fischer, A multiple site chemoreceptor model, *Arch. Int. Pharmacodyn.* **51**:540–551 (1967).
80. M. T. Rakhawy, Alkaline phosphatases in the epithelium of the human tongue and a possible mechanism of taste, *Acta Anatomy* **55**, 323–342 (1963).
81. G. H. Bourne, Alkaline phosphatase in taste buds and nasal mucosa, *Nature* **161**:445–446 (1948).
82. T. Iwayama and O. Nada, Histochemically demonstrable ATPase activity in the taste buds of the rat, *Exptl. Cell Res.* **46**:607–608 (1967).
83. G. Y. Yur'eva, Role of macroenergetic compounds in the function of taste receptors: effect of the substances altering the energy metabolism on the functional activity of taste receptors, *Vestn. Mosk. Univ. Ser.* **6**, *Biol. Pochvoved.* **22**:21–26 (*CA* **49**: 126801) (1967).
84. K. Sakai, Chemical transmission in taste fiber endings. I. The action of acetylcholinesterase on bitter taste, *Chem. Pharm. Bull.* **12**:1159–1163 (1964).

85. K. Sakai, Chemical transmission in taste fiber endings. II. The effect of chominesterase inhibitor on taste, *Chem. Pharm. Bull.* **13**:304–307 (1965).
86. S. Tsuchiya and T. Aoki, Cholinesterase activities in the gustatory region of the rat tongue and their inhibition by bitter tasting substances, *Tohoku, J. Exptl. Med.* 91:41–52 (1967).
87. R. I. Henkin, R. H. Keiser, I. A. Jaffee, I. Sternlieb, and I. H. Schienberg, Decreased taste sensitivity after D-Penicillamine reversed by copper administration, *Lancet* **ii**:1268–1270 (1967).
88. H. R. Keiser, I. L. Henkin, F. C. Bartter, and A. Sjoerdsma, Loss of taste during therapy with penicillamine, *J. Am. Med. Assoc.* **203**:381–383 (1968).
89. R. G. Steinhardt, A. D. Calvin, and E. A. Dodd, Taste structure correlation with Alpha-D-Mannose and Beta-D-Mannose, *Science* **135**:367–368 (1962).
90. N. V. Rusakov, Dichloranaline features of biological action and hygienically safe levels in water reservoirs, *Gig. Sanit.* **33**:8–12 (1968).
91. L. Michaelis and M. L. Menten, Die Kinetik der Invertinwirkung, *Biochem. Z.* **49**:333–369 (1913).
92. L. M. Beidler, A theory of taste stimulation, *J. Gen. Physiol.* **38**:133–139 (1954).
93. L. M. Beidler, Properties of chemoreceptors of tongue of rat, *J. Neurophysiol.* **16**:595–607 (1953).
94. L. M. Beidler, I. Y. Fishman, and C. W. Hardiman, Species differences in taste responses, *Am. J. Physiol.* **181**:235–239 (1955).
95. R. Whittam, *in The Neurosciences* (C. C. Quarton, T. Melnechuk, and F. O. Schmitt, eds.), pp. 313–325, Rockefeller Press, New York (1967).
96. L. M. Beidler, *in Taste Receptor Stimulation* (J. A. V. Butler, H. E. Huxley, and R. E. Zirkle, eds.), Vol. 12, pp. 107–151, Pergamon Press, New York (1962).
97. C. J. Duncan, *The Molecular Properties and Evolution of Excitable Cells*, Pergamon Press, New York (1967).
98. D. E. Koshland, Jr. and M. E. Kirtley, *in Major Problems in Developmental Biology* (M. Locke, ed.), pp. 217–249, Academic Press, New York (1966).
99. R. W. Albers, Biochemical Aspects of Active Transport, *Ann. Rev. Biochem.* **36**:727–756 (1967).
100. A. Schwartz and H. Matsui, *in Secretory Mechanisms of Salivary Glands* (L. H. Schneyer and C. A. Schneyer, eds.), pp. 75–98, Academic Press, New York (1967).
101. A. L. Lehninger, The neuronal membrane, *Neurosci. Res. Prog. Bull.*, 6 *Suppl.*, 15–26 (1968).
102. J. F. Danielli and H. Davson, A contribution to the theory of permeability of thin films, *J. Cell. Comp. Physiol.* **35**:495–508 (1935).
103. J. D. Robertson, The Molecular Structure and Contact Relationships of Cell Membranes, *Progr. Biophys. Chem.* **10**:343–418 (1960).
104. H. Fernandez-Moran, *in The Neurosciences* (G. C. Quarton, T. Melnechek, and F. O. Schmitt, eds.), pp. 281–304, Rockefeller Univ. Press, New York (1967).
105. J. D. Robertson, Granulo-fibrillar and globular structure in unit membrane, *Ann. N.Y. Acad. Sci.* **137**:421–440 (1966).
106. A. A. Benson, On the orientation of lipids in chloroplasts and cell membranes, *J. Am. Oil Chem. Soc.* **43**:265–270 (1966).
107. P. Mueller and D. O. Rudin, Action Potentials Induced in Bimolecular Lipid Membranes, *Nature* **217**:713–719 (1968).
108. D. F. Wallach and A. Gordon, Lipid Protein interactions in cellular membranes, *Fed. Proc.* **27**:1263–1268 (1968).
109. D. Chapman, V. B. Kamat, J. deGier, and S. A. Penkett, Nuclear magnetic resonance spectroscopic studies of erythrocyte membranes, *Nature* **213**:74–75 (1967).

110. D. M. Derry and L. S. Wolfe, Gangliosides in isolated neurons and glial cells, *Science* **158**:1450–1452 (1967).
111. A. L. Hodgkin and A. F. Huxley, Properties of nerve axons. I. Movement of sodium and potassium ions during nervous activity, *Cold Spring Harbor Symp. Quant. Biol.* **17**:43–52 (1952).
112. I. Tasaki, *Nerve Excitation: A Macromolecular Approach*, C. C. Thomas, Springfield, Ill. (1968).
113. A. Watanabe, I. Tasaki, and L. Lerman, Bi-ionic action potentials in squid giant axons internally perfused with sodium salts, *Proc. Natl. Acad. Sci.* **58**:2246–2260 (1967).
114. A. Watanabe, I. Tasaki, I. Singer, and L. Lerman, Effect of tetrodoxin on excitability of squid giant axons in sodium-free media, *Science* **155**:95–97 (1967).
115. J. W. Moore and T. Narahashi, Tetrodotoxin's highly selective blockage of an ionic channel, *Fed. Proc.* **26**:1655–1663 (1967).
116. J. W. Moore, T. Narahashi, and T. I. Shaw, An upper limit to the number of sodium channels in nerve membrane?, *J. Physiol.* **188**:99–105 (1967).
117. H. Hyden, *in The Neurosciences* (G. C. Quarton, T. Melnechuk, and F. O. Schmitt, eds.), pp. 765–771, Rockefeller Univ. Press (1967).
118. G. P. Talwar, B. K. Goel, S. P. Chopra, and B. D'Monte, *in Macromolecules and Behavior*, pp. 71–88, Appleton-Century-Crofts, New York (1966).
119. T. H. Bullock, Representation of information in neurons and sites for molecular participation, *Neurosci. Res. Prog. Bull.* **6** Suppl.:4–14 (1968).
120. L. B. Flexner, Loss of memory in mice as related to regional inhibition of cerebral protein synthesis, *Texas Repts. Biol. Med.* **24**:3–19 (1966).
121. B. W. Agranoff and R. E. Davis, *in Physiological and Biochemical Aspects of Nervous Integration* (F. D. Carlson, ed.), pp. 309–327, Prentice-Hall, Englewood Cliffs, N.J. (1968).
122. L. B. Adair, J. E. Wilson, J. W. Zemp, and E. Glassman, Brain function and macromolecules. III. Uridine incorporation into polysomes of mouse brain during short-term avoidance conditioning, *Proc. Natl. Acad. Sci.* **61**:606–613 (1968).
123. J. Z. Young, Is there an addressed memory in the nervous system?, *Recent Adv. Biol. Psychiat.* **10**:179–198 (1968).
124. S. H. Barondes and H. D. Cohen, Comparative effects of cycloheximide and puromycin on cerebral protein synthesis and consolidation of memory in mice, *Brain Res.* **4**:44–51 (1967).
125. R. J. Hilmore and L. A. Heppel, Polynucleotide phosphorylase in liver nuclei, *J. Am. Chem. Soc.* **79**, 4810–4811 (1957).
126. A. A. Hakim, Polynucleotide phosphorylase from human sperm and urine, *Enzymologia* **21**:81–95 (1960).
127. K. Yagi, T. Ozawa, and H. Konogi, Occurrence of polynucleotide phosphorylase in atypical epithelioma of rat, *Nature* **184**:1939 (1959).
128. M. Grunberg-Manago and S. Ochoa, Enzymatic synthesis and breakdown of polynucleotides; polynucleotide phosphorylase, *J. Am. Chem. Soc.* **77**:3165–3166 (1955).
129. J. H. Frenster, V. G. Allfrey, and A. E. Mirskey, *In Vitro* incorporation of amino acids into the proteins of isolated nuclear ribosomes, *Biochim. Biophys. Acta* **47**:130–137 (1961).
130. D. A. Rappoport and H. F. Daginawala, Changes in nuclear RNA of brain induced by olfaction in catfish, *J. Neurochem.* **15**: 991–1006 (1968).

Chapter 2

EXCITATION

G. P. Talwar and U. B. Singh

Department of Biochemistry
All-India Institute of Medical Sciences
New Delhi, India

I. INTRODUCTION

Excitability is a trait characteristic of most (if not all) cells, though the form that it may assume varies. There are differences in the sensitivity, type, mode, and magnitude of response, but it is basically an inherent property of matter and is manifest at different levels depending on the degree of organization. The forms of excitation familiar to biologists stem mostly from the organization of macromolecules in aggregate structures. Macromolecules individually have also the ability to respond to environmental factors, a property that is dependent on shifts in conformation brought about by their combination with chemicals (ions, metals, metabolites, substrates, effector agents, drugs, etc.) and/or induced by physical agents (such as temperature, pressure, radiations, and other forms of energy inputs).

In the central nervous system, the term excitation usually refers to events leading to the firing of the neuron or the generation of action potentials. The electrophysiological aspects of the phenomenon have been extensively studied and are adequately reviewed in books and journals of neurophysiology. This chapter will be mostly confined to a discussion of the chemical and molecular correlates of processes leading to excitation and their effects on cellular metabolism. The information available on these topics is diffused and far from complete. The present review will not therefore claim to be comprehensive in its bibliographic coverage.

II. DYNAMICS OF IONS AND EXCITABILITY IN THE NERVE TISSUE

A. Resting Potentials

The nerve membrane in resting condition has a positive charge on the outside surface and a negative charge on the inside, leading to the creation of a potential difference across the membrane. Pioneering studies of Hodgkin,[1] Huxley,[2] and others (see Refs. 1, 2 for cross references) have lent substance to the belief that the resting membrane potentials are primarily caused by the selective permeability of the membrane to the flow of K ions. The concentration of potassium inside the cell is severalfold higher than its concentration in the extracellular medium. There is thus the tendency of potassium ions to move from the more concentrated solution to the less concentrated spaces outward. The equilibrium potential of K^+ (defined in the sense of internal potential minus external potential) is related to the concentrations (activities) of K^+ inside $[K]_i$ and outside $[K]_o$ and is given by

$$V_k \simeq \frac{RT}{F} \log_e \frac{[K]_o}{[K]_i}$$

This equation expresses very nearly the resting membrane potentials. It is computed that when an axoplasm having a concentration of potassium equal to 400 mM is separated by a membrane permeable to K^+ and the concentration of K^+ in the other compartment is 20 mM, there would be a development of a potential of 75 mV negative to the external solution.[1] This is of the same order as the actual potentials encountered in some nerve membranes.

1. *Permeability of the Resting Membrane to Cations*

Several mechanisms may be implicit in the selective permeability properties of the resting membranes. Amongst these the pore size and the charge on the membrane are important determining factors.[3,4]

a. The Pore Size. The nerve membranes in resting condition carry a charge and have pores with an effective diameter of about 3 Å, i.e., between that of the hydrated diameters of Na^+ and K^+. The hydrated radii of K^+ are reported to be 2.2 Å and that of Na^+ 3.4 Å.[3] The larger size of hydrated Na^+ prevents its flow inwards to any substantial extent, which otherwise would be directed by the ionic concentration gradients in the resting conditions. It is only when the pore size is altered by physical or chemical agents that the membrane becomes appreciably permeable to Na^+. The smaller size of the K^+, on the other hand, permits its flow across the membrane. There is no precise information so far on the structural organization of the membrane constituents that could explain the selective permeability properties of the resting membrane to K^+. The organization of proteins and lipids in lattices resembling liquid crystals possibly creates discrete spaces for passage and interaction of the ions and reactants on both sides of the membrane.

Recent studies on ion movements induced by some cyclic antibiotics in mitochondrial membranes have yielded interesting data which appear to be pertinent to the problem of selective cation permeability in the nerve membrane. Valinomycin, a circular peptide of 12 amino acid residues, permits K^+ but not Na^+ to penetrate the membrane.[173] Another macrocyclic tetrolide, nonactin, forms a complex with K^+, K^+ being bound by the anionic groups of the molecule within its "doughnut hole" configuration.[174] The effect of these cyclic molecules on cation dynamics in the mitochondria is believed to be exercised by their location in the membrane which creates ion-specific tunnels or channels, permitting the selective passage of K^+ or Na^+.

Another highly interesting recent report[175] demonstrates the ability of a charged circular molecule macrocyclic antibiotic, alamethicin, to impart electrical excitability to a synthetic phospholipid bilayer separating two aqueous phases having different concentrations of K^+. Such synthetic membranes were found to simulate the electrical behavior of a wide variety of natural membranes, e.g., they show an action potential on electrical stimulation and an increased cation conductance. It has been postulated that a group of five or six alemethicin molecules, each binding a K^+, may aggregate into a stack which may extend through the membrane. Stimulation results in directional opening and vectorial discharge of K^+. This type of synthetic artificial membranes suggests a model. It is not unlikely that stacks of polypeptides (cyclic or otherwise) deeply embedded in phospholipid bilayers of membrane and having selective properties of binding Na^+ or K^+ may serve as channels for flow of these ions. Some of these channels may be closed in resting conditions, and by bringing about configurational alterations, stimuli may open such gates.

b. Charge. The nature of the fixed charge on the walls of the pores would be another important factor controlling permeability.[4,5] Fixed negative charges would repel negatively charged particles and so would cause the pores to become impermeable to anions. Conversely, fixed positive charges would counteract the permeability of the membrane to cations. The creation of a negative charge on the inside of the membrane as a result of the extrusion of K^+ in the resting membrane and leaving behind of the nonpermeable anions becomes in turn a restrictive force for further expulsion of K^+, owing to the attraction exercised by the negatively charged anions for the positively charged K^+.

According to the Davson–Danielli model, the membranes are composed of a double leaflet of lipids and proteins. Tobias[6] believes that the charge on the membrane is contributed to a great extent by phospholipids. Phosphatidyl choline has a pK of about 1 for phosphoric acid residue and a pK of about 14 for the strongly basic group of choline. Phosphatidyl serine has three ionizable groups—the phosphoric acid residue, the weakly acidic carboxyl group of serine (pK about 4.5) and the weakly amino residue of serine (pK about 10). Sialic acid residues, gangliosides, sulfolipids, galactolipids, and mucoproteins would also be contributing to the charge on the nervous tissue membrane in an important manner.

2. *Selective Concentration of Potassium in the Cells*

It is a well-established fact that the concentration of potassium is 10–20-fold higher in most cells than in extracellular medium. The concentration of potassium in the nerve tissue is particularly high (~ 30–40-fold). What could be the reason for the high concentration of potassium in the cells? No definite answer is available to this question for the time being. The uptake of several metabolites by the cells is accompanied by an intake of potassium into the cells. Glutamate is particularly reported to be associated with the transfer of potassium in the nerve tissue.[7–9] In addition to a continuous intake of this ion along with metabolites, it is likely that cells have constituents which can bind or store potassium. This ability should be particularly developed in the brain tissue which has a high content of intracellular potassium. Along with Z. Iqbal and Shail Sharma, we have assayed the content of sodium and potassium in the developing chick embryo brain at different stages.[10] There is no electrical activity detectable in the chick embryo brain earlier than the fourth day. Both the spontaneous and evoked electrical activities in the embryo brain mature after the twelfth day of age. There is a sharp rise in the content of potassium in the embryo brain between the sixth and eighth day of age. This increase is obviously related to the development of mechanisms capable of concentrating this cation in the tissue. Further work on this system should help to put in evidence the "potassium-philic" proteins, acidic lipids, and other anionic structures which are involved in the concentration of potassium in the nervous tissue.

There is evidence to suggest that acidic lipids have the property of binding cations.[11,12] The avidity of the lipids is particularly high for binding of divalent cations and only at high concentrations are the univalent cations bound.[13] Woolley and Campbell[14] have reported that 50-fold higher concentrations of Na^+ or K^+ were required to displace half of Ca^{2+} bound to lipids of stomach or spinal cord. In case of cerebrosides which were good K^+ binders, 16 μmoles of K^+ were needed to displace half of 4.5 μmoles of Ca^{2+}. Potassium bound to the anionic groups of lipids can be reversibly displaced by other ions such as sodium, magnesium, and calcium.

The ensemble of these observations indicate that the anionic groups in lipids are involved in the binding and transfer of cations in general. It is also possible that these compounds mediate the transport of cations from one aqueous phase to the other through the lipidic environment. Woolley and Campbell[15] have shown the transport of Ca^{2+} from an aqueous phase through a lipid solvent, which is promoted by serotonin in collaboration with lipids extracted from nerves or smooth muscles. However, there are reasons to believe that acidic lipids may not be the unique class of compounds involved in the retention of potassium in the nervous tissue. Tracer studies show that about 20% of the brain potassium does not exchange with the environmental potassium,[16] which would imply that either this fraction of the cation is held in nonionic bonds and is not freely exchangeable, or that this fraction of the total potassium is sequestered behind a potassium-impermeable barrier. The cation bound to lipids is essentially exchangeable; this is an indirect argument for a "nonlipid" form of this fraction of potassium. Another observation pointing to the need for further search of alternate potassium-binding molecules in the nerve cells is the report by Katzman and Wilson,[17] who found that the phospholipid–cation complex extracted at low temperatures (−45° to −55°C) had a Na:K ratio of 1.8, whereas the Na:K ratios in the whole brain tissue were 0.6. The extraction procedure used by these workers avoids the alteration of cations bound to lipids *in situ* and prevents the exchange of inorganic cations between the lipid-bound and other compartments during extraction. If this extract is representative of the cation phospholipid complexes in the brain, it would imply the presence in cells of lipids which bind Na^+ better than K^+. As the total concentration of K^+ is higher in cells, there should be other substances present in the cells which have the ability to sequester potassium preferentially.

B. Excitation: Flux of Sodium

The resting nerve membrane is very little permeable to Na^+. Excitatory events lead to a change in the permeability properties of the membrane, causing a momentary flux of Na^+, a local depolarization followed by a reversal of polarity, which generates an action potential. At 20°C, the net entry of Na^+ in one impulse amounts to $3\text{–}4 \times 10^{-12}$ mole/per cm^2. A similar quantity of potassium ions leave the fiber during an impulse. It is estimated that an entry of about 20,000 sodium ions occurs per square micron of the surface in one impulse.[1,2]

The presence of a concentration gradient of sodium from outside to inside would be an essential requirement for the generation of action potentials. If the sodium in an extracellular compartment is replaced by choline chloride or by glucose, the action potentials, but not the resting potentials are reduced.[1] If all the external Na^+ is removed, the axon becomes reversibly inexcitable. The effect of varying external concentrations of sodium has been studied in a number of excitable systems such as frog muscle,[18] myelinated nerve,[19] Purkinje fibers of the heart,[20] and crustacean nerve.[21] In each case similar results have been obtained. There may be some exceptions to this general statement, where the mechanisms are different. In the crab muscles, the entry of Ca^{2+} or divalent cations provides the inward current instead of Na^+.[22] In the plant cell, *Chara*, an exit of chloride ions from the vacuolar sap is the primary process.[23] Except for these two exceptional cases, the inward flux of sodium ions from the extracellular spaces provides the trigger for generation of the impulse, and for that to occur there should be (1) a higher concentration of sodium ions in the extracellular environment and (2) sodium should normally be excluded by the resting membrane from free entry. The latter property may depend on charge pore size attributes of the resting membrane.

It will not be out of place to mention some recent observations on the squid giant axon by Tasaki and colleagues[176] which question the generally believed hypothesis that the excitation process requires as the first event a specific increase in Na^+ permeability and influx of Na^+. In their experiments, axons perfused internally with the phosphate salts of the univalent cations Li^+, Rb^+, or Cs^+ or substituted ammonium ions can maintain excitability and show an action potential if the external medium contains a divalent cation such as Ca^{2+}, Ba^{2+} or Sr^{2+}. There is no unique requirement for Na^+ or K^+ for excitability. On the other hand, the presence of a divalent cation (even unphysiological) in the external medium is essential. They propose that in the resting state, a divalent cation (probably Ca^{2+}) would be bound to a few sites (probably sialic acid residues of gangliosides) on the outer side of the membrane. On stimulation, some of the bound Ca^{2+} is replaced by univalent cations (K^+) derived from the internal membrane. This ion-exchange process is supposed to induce a change in the conformation of membrane subunits, which permits Na^+ to diffuse in and K^+ to diffuse out. In the recovery phase Ca^{2+} again regains its previous binding sites displacing the univalent cation. Some support for this hypothesis accrues from the potentiality of tetrodotoxin to block the action potentials. This toxin prevents the exchange of univalent cations for divalent cations at anionic binding sites, preceding the influx of Na^+ and efflux of K^+ once excitation has taken place.[177] The site of action of tetrodotoxin is on the outer side of the axonal membrane, as its perfusion inside has little effect.[178]

C. Recovery

Excitation involves changes in the resting equilibrium concentrations of cations. The signal change is in the permeability of the membrane to sodium. The ascending phase of action potential is accompanied by an intracellular flux of sodium. A local depolarization leads to a momentary reversal of polarity with the interior of the membrane assuming a positive charge. There is a simultaneous efflux of potassium ions.

The recovery of the cell to resting equilibrium conditions involves the expulsion of sodium ions, a process which is accompanied by an intake of potassium ions. A number of mechanisms have been proposed (see Ref. 24) but the most probable mechanism involved in this vital task appears to be a membrane-linked complex enzyme system which utilizes a mole of ATP for expulsion of an equivalent of Na^+ and uptake of K^+. The enzyme was initially discovered by Skou in the erythrocytes[25] but its presence in the nerve membrane[26,27] and membrane fractions from mammalian brain has also been demonstrated. Most of the enzyme activity is recovered in the conventional "microsomal" fraction which consists of the synaptosomes, nerve endings, boundary membranes of the cells, dendrites of the tissue and endoplasmic reticulum.[28–30] The system requires Mg^{2+}, Na^+, and K^+ for full activation. The activity of the enzyme was found to be highest in the gray matter of the cerebral cortex.[31] The enzyme activity recovered in the "microsomal" fractions is sufficient to handle the order of Na^+ and K^+ movements in the tissue. The enzyme seems to possess two sites of binding for alkali metals: One at which K^+ increases activity, and another at which Na^+ has preferential binding and increases activity but at which excess K^+ could inhibit by competing with Na^+. Activity within 70% of maximal was given by quite a wide range of combinations of Na^+ and K^+ concentrations (between 2–110 mM K^+ and 40–145 mM Na^+).[24] The low K^+ concentrations to which the system was sensitive were closer to those of extracellular than of intracellular fluids. If Mg^{2+} and high enough amounts of K^+ were present, the sodium concentration to which it was sensitive were those of intracellular rather than extracellular fluids. The system is, in short, most sensitive to Na^+ concentrations encountered within the cell and to K^+ concentrations present in the extracellular medium. Na^+ shifts the K_m of the enzyme for its substrate ATP. With 3 mM Mg^{2+} and 30 mM K^+, 20 mM ATP was required for half maximal rate, but with the addition of 100 mM Na^+ to the system, half maximal rate was reached with 8.7 mM ATP.[29] The mechanism of action of the enzyme was considered by Chignell and Titus[32] to be a two-step reaction: a Na^+-dependent transfer of the terminal phosphate to the protein of microsomal ATPase and a subsequent loss of inorganic phosphate on addition of K^+ to the medium.

There is the possibility of the presence of additional mechanisms for the uptake of K^+. The uptake of several substrates (such as glutamic acid) is accompanied by K^+, but the rate of uptake in presence of glutamic acid is of the order of 15–20 μeq K/g of the tissue/hr. This would be a too slow reaction to handle K^+ of the order of 500–600 μeq/g/hr liberated during excitation.[33] The Mg–Na–K-activated ATPase in the brain microsomal fractions hydrolyzes 900–1200 μmoles of phosphate per gram of the tissue per hour, which would be sufficient to transport the 500–600 μeq K released during activity.

Hertz[34,35] has postulated a special role of neuroglia in the transport of cations. It has been shown that the fall in tissue respiration quotient in a sodium-free medium or the increase in respiratory rate of brain slices in a medium enriched with K^+ is primarily owing to the effect of these ions on the respiratory activities of isolated glial cells.[36] Potassium also appears to induce a stimulation of the active transport of other ions in brain slices.[8]

III. EXCITABLE MEMBRANES

A. Changing Notions on Structure

Cell membranes form a major part of the cell mass. In some cells they may account for up to 80% of the dry weight.[37] They have an active metabolism with turnover rates as high as 100% per hour.[38] Membranes have a key role in the excitation events.

The membranes are composed of lipids and proteins in defined structural patterns, for which various models have been proposed.[39,40] One of the earliest (1935) was the familiar Danielli–Davson concept,[41] which postulated that a lipid bilayer would be stabilized by the adsorption of two monolayers of hydrophobic proteins, one on each side of the lipid leaflet. The "unit membrane" hypothesis of Robertson[42,43] has merits but appears to be an oversimplification to account for the widely varying properties of different biological membranes. Membranes are not only structural components of cells, they are also active functional elements. The structure of membranes is, therefore, probably more varied and complicated than that contemplated in simple lipid double leaflet models. Lucy[44] proposed a micellar organization in membranes. According to him, the lipids would be partially arranged in a manner resembling the organization of colloidal molecules in a spherical micelle. Within the membrane, globular micelles would be in dynamic equilibrium with the bimolecular leaflet disposition of these molecules, i.e., transitions in lipid structure (micellar–double leaflet) would occur within membranes.

The lipids (in micellar or leaflet form) attached to proteins (or glycoproteins) would be present in at least two broad conformational states: either in form of a globular constellation or as a bimolecular leaflet of lipids and proteins. Other globular proteins (hormones, etc.) could fit in at certain points in the plane of the lattice in place of a globular constellation of lipids and proteins. This model offers a degree of plasticity. The proportion of lipid molecules in micellar configuration may vary from membrane to membrane and even within one membrane, depending on the environmental factors. Reversible transitions between a micellar arrangement and the bimolecular leaflet could influence in an important manner the properties of the membrane. Green and Purdue[45] have proposed also a structure of membrane which conceives of a repeating sequence of globular lipoprotein units. The electron micrographs of very thin sections of membranes often reveal a globular structure.[46]

Plasma membranes of animal cells generally have attached carbohydrate moieties. Some membrane-bound gangliosides have been identified with "receptor" function for serotonin in the gastrointestinal tract.

The loading of carbohydrate moieties on proteins takes place after the synthesis of the polypeptide chain is completed. The enzymes involved in this function are present in the endoplasmic reticulum.

Duncan[3] has put forward the hypothesis that the membranes of excitable tissues have a mechanoenzyme system localized on the walls of the membrane pores. The mechanoenzyme is an ATPase system which is capable of undergoing changes in the physical conformation and/or charge distribution as a function of the balance between the ATPase and its substrate ATP (and other nucleotides).

B. Conformational Alterations

A large number of biological lipoprotein membranes respond *in vivo* as well as *in vitro* to the binding of specific ligands by some modification of their properties, which reflects perhaps a rearrangement of the membrane organization and presumably of the conformation of repeating units.[47-51] Changeux *et al.*[52] have discussed the cooperative properties of biological membranes in terms of their lattice structure. Mathematical formulations have been proposed for the relationship between the conformational transitions of its repeating units and the binding of ligands.

Attempts have been made to monitor changes in the molecular conformation of membrane constituents during an excitatory episode. Tasaki *et al.*[53] have very recently reported changes in fluorescence, light scattering, and birefringence in invertebrate nerves during excitation. They employed a dye (8-anilinonaphthalene-l-sulfonate) which fluoresces under ultraviolet illumination when bound to hydrophobic sites of the macromolecules. The fluorescence in nerves from crabs and lobsters increased during excitation of the nerve. Light scattering and birefringence methods have also been used by Cohen *et al.*[54] to detect changes at molecular level in the nerve during activity. The changes in birefringence were observed to be directly dependent on the potential difference across the axon membrane and arose in radially oriented molecules associated with the membrane.

C. Some Cell Products Causing Excitation

1. *Acetylcholine*

It has been long recognized that some neurohumors such as acetylcholine mediate the transmission of the nerve impulse across synaptic junctions. Acetylcholine is present at the nerve endings in a bound form. As and when it is liberated in a free state, it manifests its action on the postsynaptic membrane presumably by altering the permeability of the membrane to cations. It is believed that a proteinic "receptor" for acetylcholine is present on the membrane.[55] The receptor has not been isolated so far, though various arguments for its presence have been given (see Ref. 56 for review). Nachmansohn[56] has advanced the view that acetylcholine is released within the membrane by excitation. It is then recognized by the acetylcholine receptor present within the membrane. The binding induces a conformational change, thereby releasing possibly Ca^{2+} ions bound to carboxyl groups of the protein(s). Ca^{2+} ions are known to be involved in the excitability of nerve and muscle fibers. Ca^{2+} ions may further cause conformational changes in phospholipids and other components of the membrane. The consequence of these chains of reaction is a change in the permeability of the membrane to ions, permitting the movement of 20,000–40,000 ions across the membrane per molecule of acetylcholine released. The fact that the signal is amplified prompts the hypothesis that a chain of reactions is probably involved.

Free acetylcholine is rapidly hydrolyzed by a powerful membrane-linked enzyme acetylcholinesterase which by removal of acetylcholine restores the conformation of the membrane to the preexcitation state. The enzyme has been purified from the electric organs of the electric fish.[57] It has a molecular weight of 260,000 and has four subunits of about equal molecular weight. It has been speculated[3] that cholinesterase may also be the "receptor" for acetylcholine. Enzymes have a high degree of specificity for combination with their substrates, hence it would be natural for cholinesterase to bind acetylcholine. The enzyme–substrate complex would quickly lead to the hydrolysis of the substrate (acetylcholine) and return of the enzyme to its native conformation. However, evidence to the contrary has been reported[57,58] on grounds of differential sensitivity of the acetylcholine "receptor" and cholinesterase to sulfhydryl blocking agents.

2. *Amino Acids*

During the past few years, some dicarboxylic amino acids and their derivatives have been reported to have excitatory action on spinal neurons.[59,60] The excitatory action of the amino acids is not restricted to the spinal neurons, but has also been observed in the ventral group of nuclei of the feline thalamus.[61] In these experiments, L-glutamic, DL-3-aminopropane sulfonic (homocysteic), and DL-*N*-methyl aspartic acid were applied iontophoretically at different depths of the nervous tissue. The neurons at a depth up to 6 mm below the dorsal surface of the fornix were powerfully excited by DL-homocysteic and *N*-methyl aspartic acids but to a much less extent by glutamate. They were also unaffected by acetylcholine. The neurons below the 6-mm level became much more sensitive to glutamate and acetylcholine and rather less to the other two amino acids. These observations indicate the possible involvement of several compounds other than acetylcholine in causation of excitatory events. The sensitivity of thalamic neurons to glutamate has also been found to vary widely between one region and another.

The mechanism of action of amino acids is not known. It is believed that they cause depolarization of the membrane by reaction with specific receptor sites.[62,63]

Excitatory agents such as acetylcholine and amino acids are interesting examples of an intercellular communication system, where the signal is transferred from one cell to the other through a substance or product secreted by the cell.

IV. ONTOGENESIS OF ELECTRICAL ACTIVITY

A. Developing Brains of Different Species

The cellular elements (neurons and glia) in the brain are not only specialized cells in terms of their metabolism but acquire also additional properties in the course of their organization and interaction. The manifestation of spontaneous and evoked electrical activity is a characteristic trait of a functionally developed nervous system. The stage at which these attributes appear in a developing nervous system varies from animal to animal. In chick embryo[64] it has been observed that the neurons of the cortical surface start spontaneous firing early in the incubation period of the egg at 37°C. Occasional slow electroencephalographic waves with intermittent silent intervals are first recorded on the fourth day of incubation. These occasional slow waves gradually become regular and increase in amplitude with age. By the thirteenth day, fairly well-marked low-voltage fast activity appears which becomes prominent in periods after the thirteenth day of age. It has been also reported that application of strychnine produces spikes in chick embryo EEG only after the age of 12 days.[65] The developing chick embryo up to day 12 and more specially between day 4 and 12 offers a good experimental system for investigation of biochemical and biophysical parameters which develop during this period and enable the tissue to manifest both the spontaneous and evoked electrical activity.

In rats the EEG activity was detectable on the twenty-first day of gestation. This becomes more rhythmic and regular during the first week of postnatal life.[66] Similarly, the electrical activity in pigeons matures also in the postnatal period.[67] A suggestion has been made that the onset of electrical activity takes place in the prenatal period in those animals which are mobile and have to feed themselves soon after birth, whereas in others who stay in the nest for a while and are fed, the electrical activity matures only in the postnatal period.[67] The delayed maturation of electrical activity may in fact be a reason for their lack of functional mobility at the time of birth.

B. Possible Factors Related to Ontogenesis of Electrical Activity

1. *Organization and Establishment of Contacts*

Schade[68] has observed the ramification of axons and the growth of dendritic plexuses in the cortical cells at the time when spontaneous electrical activity appears in the developing rabbits. Tuge *et al.*[67], however, observe that there is no coincidence between the onset of EEG and primary sensory responses ontogenically. The EEG appears in guinea pigs prior to the time of the initiation of primary sensory response. The question is therefore open whether the onset of electrical activity depends mainly on the establishment of circuitry or on the developments within cells. Possibly both factors are involved, and are interrelated.

2. *Cation Permeability*

A number of morphological and biochemical changes are observed to take place at a critical stage of development when the permeability of the cortex to sodium changes.[69] According to Jasper *et al.*[70] this period coincides with the appearance of electrical activity in these animals. The close temporal relationship between the appearance of electrical activity and the increase in permeability of the membrane to sodium (on encounter with an excitatory situation) indicates the importance of this property in the development of functional maturity. It is possible that the cell membrane acquires new protein(s) or other constituents at this stage which make possible the altered response to sodium. It may also be pointed out that the capacity to concentrate potassium in the brain rises sharply between 6–8 days in chick embryo.[10] The induction of potassium-retaining mechanisms would be a prerequisite to the establishment of the resting membrane potentials.

3. *Nissl Granules and Galactolipids*

Observations have been made by some workers that the spontaneous electrical activity in developing chick embryo brain appears to correspond with the appearance of Nissl granules in the tissue.[67] There are also reports of a sudden increase in galactolipids in the chick embryo brain between the tenth and the thirteenth day.[71] It remains to be investigated whether these metabolic changes in the developing nervous system exercise any determinant role in the ontogenesis of electrical activity.

4. *Na-K-Activated ATPase*

ATPase is inherently linked with the transport of cations across the membrane. The activity of this enzyme is low in the chick embryo brain before the eighth day. There is a rapid rise between the eighth and twelfth day, after which date the specific activity of the enzyme in the 105,000*g* pellet stays constant till the hatching of the egg.[72] The fact that the activity of this enzyme assumes an optimal status before the spontaneous and evoked electrical activity mature may well indicate a possible role in the manifestation of EEG. The increase in activity between the eighth and twelfth days is not apparently due to the removal of an inhibitor, but represents a net synthesis of the enzyme during this period. The enzyme is subject to inhibition by ouabain and Ca^{2+} and has other properties similar to the enzyme in the post-13-day brain. It is optimally active at an Na^+ concentration of 100 mM and K^+ concentration of 20 mM, but differs from the neonatal rat brain enzyme in having an optimal activity at 37°C[72] instead of 45–50°C reported for the rat brain enzyme.[73]

This enzyme is induced in the chick embryo brain at an earlier period than in the developing rat brain. The electrical activity also matures in chick brain earlier than in rat brain. A recent report[66] has described the appearance of the Na^+–K^+ stimulated, ouabain-sensitive ATPase on the twenty-first day of gestation in the rats. The activity reaches the adult levels by the twelfth day postpartum. Between the sixteenth day of gestation and the first day after birth, a marked trend toward structural maturation of the synaptic components is also observed. The EEG activity is detectable on the twenty-first day of gestation, and becomes more rhythmic and regular during the first week of postnatal life in these animals.

5. *Other Membrane-Linked constituents*

While the appearance and synthesis of Na-K-ATPase bear a temporal relationship to the onset of electrical activity, it is quite likely that additional components may be required for the full expression of the properties of spontaneous and evoked electrical activity. Proteinic constituents extracted from the synaptosomes and nerve ending fractions of the 6-day and 13-day-old chick brains show differences when analyzed on polyacrylamide gels (Z. Iqbal, N. Jaffery, and G. P. Talwar, unpublished data). It is likely that other membrane-bound proteins such as cholinesterase may also be required in nerve cells for the functional maturation of the electrical activity patterns.

V. GENERAL EFFECTS OF EXCITATORY ACTIVITY ON METABOLISM OF THE NERVOUS TISSUE

A. Energy Metabolism

Brain has an active oxidative metabolism. It accounts for approximately one-fifth of the total consumption of oxygen by the body in a basal state.[74] The respiratory quotient of the cells in gray matter is the highest of all other types of cells in mammals.[75]

The principal nutrient taken up by the brain is glucose from the arterial blood. At an average it is computed that 20 μmoles of glucose are taken up by each gram of nervous tissue per hour.[76] Using radioactive glucose, it has been found that only a part of the glucose-^{14}C radioactivity is converted into CO_2 in the first instance; the rest of it shuttles through pools of keto acids, dicarboxylic amino acids, and perhaps other compounds.[77] Though glucose is the main nutrient utilized by the brain normally, Abood and Geiger[78] were able to maintain cat brain preparations on glucose-free perfusion fluid, provided some nucleosides like uridine and cytidine were added to the medium. The role of these nucleosides in such experiments is not yet interpretable. They may have a part in utilization of structural compounds such as galactolipids in an emergency situation of glucose deprivation.

Kety *et al.*[79] measured the blood flow and oxygen consumption of the brain in resting subjects and in those engaged in hard thinking on mathematical problems. They could not detect any notable difference in the brain metabolism in these two states. The inference may, however, be cautiously drawn. Even in basal state of rest, a fairly high number of neurons are active in different areas of the brain, with the result that an increment in a small area(s) of the brain may not become perceptible when viewed as a part of the whole brain metabolism. The nerve tissue slices show an increased uptake of oxygen when stimulated by electric currents.[80] There is also a breakdown of ATP and phosphocreatine in these conditions of induced electrical activity. A continouus supply of ATP is also required for the operation of the "sodium pump" which will be activated during the excitation of the tissue.

B. Phospholipid Metabolism

Brain is particularly rich in phospholipids. These compounds are constituents of all cell membranes. Phospholipids like phosphatidic acid, phosphoinositidyl, etc., have a rapid turnover, but their role in the transport of substances is a subject of controversy.[81] Other phospholipids are relatively stable. In myelin sheath the phospholipids may have a long life and may not be renewed throughout the life span of the animal.[82]

1. *Effect of Acetylcholine on Phospholipid Metabolism*

Although acetylcholine is usually considered as an excitatory factor, under certain circumstances it can also act as a depressant.[83] Hokin and Hokin[84,85] showed that low concentrations of acetylcholine (10^{-3}–10^{-4}M) stimulate the uptake of ^{32}P into the phospholipids of secreting sites of pancreas. Similar results were also obtained on brain slices and in isolated sympathetic ganglia. The most significantly affected lipids of the brain cortex were phosphatidic acid, phosphatidyl inositol, and phosphatidyl choline.[86,87] An increase in the uptake of orthophosphate-^{32}P in these fractions of phospholipids was observed in isolated sympathetic ganglia on addition of acetylcholine.[88] At levels approximating that of the brain *in vivo*, acetylcholine was observed to stimulate the incorporation of ^{32}P in the phosphatidic acid of the "microsomal" fraction of the brain.[89]

Larrabee and Leicht[90] have shown that the labeling of phosphatidyl inositol by orthophosphate-^{32}P is increased in presynaptic terminals by the nerve impulse entering the isolated sympathetic ganglia of animals. Hokin is, however, of the view that the stimulation of labeling of these phospholipids takes place in the postsynaptic structures.[91] It is implied that synaptic transmitter acting on the postsynaptic cell causes an increase in the incorporation of ^{32}P into phospholipids.

2. *Phospholipid Metabolism in Convulsions and Other Excitatory States*

The metabolism of phospholipids has been studied in various other types of situations. Convulsions induced by electric shock or chemicals (pentylenetetrazole, nikethamides, etc.) cause a decrease of radioactivity in brain phospholipids.[92–94] The results suggest the predominance of catabolic over anabolic reactions in phospholipid metabolism during hyperactivity of the brain.

During stimulation of moderate intensity the turnover rate of phospholipids increases. Vladimirov[95] observed an increase in the turnover of brain phospholipids on stimulation of the rat paws electrically for 3 hr.

The metabolism of phospholipids has also been studied in cases of conditioned reflex developed in response to an auditory stimulus. The stimulation by electric bell, buzzer, or whistle accompanied by alimentary reinforcement, caused a considerable increase in the turnover of phospholipids in the auditory area. No change in the phosphorus metabolism was seen in motor or visual areas of the cortex or in medulla oblongata during these experiments.[96]

In the resting state of animals (e.g., during natural sleep) a decrease in the turnover of phospholipids is observed in the visual and motor areas with only a slight decrease in the auditory area.[95,97]

C. RNA Metabolism

There are a large number of reports showing changes in the content and/or composition of RNA of the neural tissue under a wide variety of conditions. At times the observations are conflicting, which emphasizes the complexity of the problem. It is difficult to state at this stage whether there is any *direct* relation between RNA and the functional activity of the brain, even though changes in RNA metabolism have been noticed in different situations.

1. *Moderate Stimulation*

Pioneer studies of Hydén and collaborators (see for review Ref. 98) have shown an increase in the content of RNA of the neurons as a result of stimulation. There are also reports suggesting a reciprocal relationship between the RNA in the neurons and the glia. The RNA content of the nerve cells increased in the lateral vestibular nuclei, while that of the glial cells decreased.[99] Electrical stimulation of the superior sympathetic ganglion in the cat for 3 hr induced a marked increase in the nucleic acids and proteins of the cytoplasm of the neurons. The RNA content of the glial satellite cells was at the same time shown to decrease.[100] These observations suggest a close interrelation between the neurons and glia in their metabolic activities. There may be a transfer of RNA between the two cell types. Danneholt and Bråttgard[101] have shown an active synthesis of RNA in the glial cells, which was in fact found to be twice as rapid as in the nerve cells of hypoglossal nucleus during a 4-hr pulse experiment.

Vestibular stimulation activated the synthesis of RNA in the lateral vestibular nuclei,[99] i.e., the effect is on cells in specific areas of the brain responding to the particular type of stimulus. It has further been reported that learning of a new task, i.e., balancing on a wire,[102] or change of handedness,[103] causes not only an increase in the nuclear RNA of the corresponding cells, but also that this RNA (synthesized during a new learning task) has a different base ratio as compared to the RNA synthesized in response to simple vestibular stimulation.

Similar observations have been made by Rappoport *et al.* (see for review Ref. 104) in olfactory lobes of the fishes. Morpholine, camphor, and other odorants induced an increase in the nuclear RNA of the olfactory area of the fish cortex. Furthermore, the base ratios of RNA were different for different types of odorants used in the perfusion water.[105]

2. *Intensive Excitation*

Numerous investigators have reported a decrease of RNA content of the brain during experimentally induced convulsions or as a result of intensive stimulation of various types. A fall in RNA of rat brain was observed in severe convulsions induced by metrazol[106-108] or by electroshocks.[109] Hydén[110] reported an intracellular loss of pentose nucleic acids of the brain after convulsions induced by insulin. Electrical stimulation of spinal ganglion cells or intense muscular work resulted also in a considerable decrease of pentose nucleic acids and protein content of the corresponding cells.[111] Acoustic stimulation was observed to cause a decrease in the nucleoproteins of the nerve cells.[112] Stimulation in cats through afferent nerves led to an accumulation of acid-soluble nitrogenous compounds.[113,114] The nucleic acid nitrogen decreased, whereas acid-soluble nonprotein nitrogen increased. These chemical changes were proportional to the number of effective stimuli and were reversible at rest. It has also been observed in rats that strenuous physical exercise (compulsive swimming) induced a decrease of the nucleic acids, proteins, and orcinol-positive substances.[115] Motor exhaustion produced a marked decrease of the total dry mass, including the RNA-containing fraction per unit area, in the motor root cells.[116] Attardi [117] reported a significant decrease of the RNA content of the Purkinje cells after muscular exhaustion. Large pyramidal cells of the cortex were found to be chromophobic to basophilic dyes in a patient who died of a series of grand-mal seizures.[118] There exists therefore, fairly extensive experimental data to indicate that the RNA content of the brain falls during convulsions and also in conditions of sustained intensive neuronal stimulation.

The decrease in RNA content of the brain is not a generalized phenomenon; other organs such as liver do not show any significant change in RNA during convulsions.[107] It is also observed that most of the decrease in brain RNA during convulsions is confined to the particulate fractions of the tissue sedimented at 800*g* in 10 min in 0.32 M sucrose.[107]

The depletion of RNA in the brain following intensive stimulation may be due to any one or a combination of the following reasons:

A slowing down or arrest of the biosynthesis of RNA
Increased breakdown of RNA
Transfer of RNA to other compartments

The fact that the ATP content of the nervous tissue is lowered during excitation would favor the first viewpoint, namely, a slowing down of the biosynthesis of RNA. A continuous synthesis of RNA at the normal rates would require the availability of adequate amounts of the four types of nucleoside triphosphates. A more detailed discussion of this subject has been published earlier.[119]

3. *Significance of Excitatory Events on RNA Metabolism*

Though changes in content or composition of RNA have been observed in many cases, there is not as yet adequate evidence to state that these changes represent either unique or specific events in transduction of electrical impulses into molecular processes or molecular engrams. A few instances have also come up where sustained electrical activity did not cause any alteration in the content of RNA. In stretch receptor preparation of the lobster, spike potentials over prolonged periods did not cause any change in the RNA content of the tissue.[120] There is also a report[121] in which massive doses of actinomycin D inhibiting up to 96% of the synthesis of RNA in the brain of mice over a period of several hours did not impair the functional capacity of the mice to learn a new maze. A comprehensive review giving all the facts and fallacies on brain RNA has recently been published.[122]

D. Protein Metabolism

Brain has a fairly active turnover of proteins, a fact which was perceived only on injection of radioactive precursors intracisternally. The systemic injections of radioactive amino acids had earlier given the impression of a slow incorporation of these amino acids into proteins in the brain, owing perhaps to their low rate of passage across the blood–brain barrier.

An average human brain has approximately 10 billion neurons and about 8–10 times as many glial cells. A high percentage of these cells are functionally active in all situations. Furthermore, the rapidity of propagation and conduction of electrical impulses also contributes to a rapid wear and tear of the cellular components, and enzymes involved in the overall activity. There would therefore be a substantial and quick turnover of at least some of the proteinic constituents.

1. *The Secretory Function of Nerve Cells*

Neurons in hypothalamus (and perhaps other areas of the brain) have also a secretory function. A wide variety of short-length polypeptide "release-factors" have been identified which are produced in specific nuclei of the hypothalamus and which activate the synthesis and release of hormones from the anterior pituitary. These "release factors" are in a way instrumental in the integration of nervous and endocrine activities in the body. Their release occurs in response to appropriate "messages" received by the neurons. The presence of "receptors" for various steroid hormones in groups of neuronal nuclei has been indicated.[123] The posterior pituitary hormones, i.e., the octapeptides oxytocin and vasopressin (ADH) are produced by the neurons. The release of these hormones occurs on receipt of specific stimuli, i.e., an increase in the osmotic pressure of blood in states of dehydration is perceived by the "osmoreceptors" in the hypothalamus which trigger the secretion of vasopressin or ADH. ADH acts on the kidney tubules, increasing the absorption of water. This familiar example illustrates two points: (1) the synthesis of proteins and polypeptides by neurons (at least of the hypothalamus), a function analogous to endocrine glands, and (2) the existence of a direct relationship between the nervous activity and protein metabolism in the nerve cells.

2. *Involvement of Proteins in the Consolidation of Experiences or Long-Term Memory*

Rapid advances made in molecular genetics during the past two decades have led to theories on the nucleic acids and proteins playing a possible role in storage of information in the brain. An analogy has apparently been drawn from the known ability of these macromolecules for encoding the genetic, enzymatic, and immunological memory. A number of good reviews are available on the subject besides some chapters in the present series.[122,124–131]

Psychologists have distinguished two types of memory, viz., short-term and long-term memory. While short-term memory may reside in reverberating circuits, it is highly likely that long-term memory has a structural basis.

The initial processes connected with the processing of an electrical impulse in the nerve cell (and glia) are poorly understood. One may postulate a molecular structure of labile nature, which is perturbed by a change in electric fields or by electrochemical factors such as concentration of electrolytes and other charged molecules, *p*H, etc. This conformation change may activate a single or chain of metabolic events, thus rendering the transduction of an electric message into chemical components.

The impulse signal has a time constant in the range of milliseconds. The "consolidation" of the experience usually requires minutes and sometimes hours. If electric shock is applied after a few seconds of learning, it does not interfere in the consolidation process, which suggests that the information content of the impulse is fixed and processed beyond the point sensitive to electrical shocks after about 10 sec. There is a transfer of information from an electrical or electrochemical pattern to a spatial pattern within this short time. The synthesis of macromolecules may well be involved. In bacteria, enzymes are induced within minutes. A polynucleotide chain of about 3×10^6 mol. wt. can grow in 1 sec in *Escherichia coli* and a protein of about 10^4 mol. wt. can be synthesized in 1 sec. These orders of magnitude fulfill the theoretical temporal requirements for fixation of electrical information into structural macromolecular components.

It is further observed that if inhibitors of protein synthesis are given immediately after learning trials, the experience is not fixed as memory.[124] On the other hand, if the inhibitors are given 1 hr after the learning trials, no interference is observed in consolidation of the learned experience. The inhibitors of protein synthesis do not suppress the learning act. These experiments help to distinguish two distinct phases, the initial phase of learning independent of the synthesis of macromolecules and the consolidation phase in which synthesis of macromolecules such as proteins is required.

3. *The Effect of Light on Protein Metabolism of the Visual System*

Reports in literature show that the development of the visual system in several animals is influenced by afferent stimuli.

a. Retina. Light activates markedly the RNA and protein metabolism of the retina in rabbits during early postnatal development. When normal light stimulation was available to the rabbit pups, the mass or the proteins per unit volume of retina increased 100%. On depriving the animals of light during the early postnatal period, the changes in RNA and protein contents were arrested, and the chemical differentiation of the retina stopped.[132] It has also been reported[133] that frogs when exposed to light after a long period of darkness displayed signs of active protein synthesis in the nuclei of retinal ganglion cells within a few seconds. The volume of mitochondria and the number of ribonucleoprotein particles also increased. Succinic dehydrogenase activity of frog retina increased on illumination with bright intermittent light.[134] A significant increase in the activity of succinate oxidase in rods and cones of the isolated retinal layer of rats has been observed on transition from darkness to light.[135] An increase in acetylcholinesterase activity in the eye exposed to light as compared to the control kept in darkness has also been reported.[136]

b. Occipital Cortex. The overall weight and protein content of the cerebral cortex increased in rats exposed to sensory stimulation as compared to the control animals.[137] The importance of adequate afferent stimulation for normal maturation of visual cortex has also been demonstrated. It has been reported that the diameters of the nuclei in cortical layers II, III, and IV are smaller in mice with inherited retinal dystrophy than in the normals.[138] The protein content of the visual cortex (areas 17, 18, and 19) increases markedly in 2 days following the opening of the eyes in rabbit pups.[139,140] Prior blinding of the animals by severing of the optic nerve prevents this expected rise in the occipital cortex proteins to a great extent, while other areas of the brain such as cerebellar cortex and hippocampus are not influenced in a parallel manner. Other areas of the cerebral cortex are affected, but the change is of a relatively lower order as compared to areas 17, 18, and 19.[140]

The visual cortex of mammals and primates presents a high degree of complexity. Five different types of neurons have been described on the basis of their response to light and darkness.[141] Type A neurons do not react to retinal afferent impulses. They constitute about half of the total visual cortex neurons in cats and give a continuous background discharge independent of retinal stimuli. Types B and D are antagonistic neurons, B neurons being activated by light-on and D by light-off. C neurons are inhibited after light-on and -off, and E neurons show a short preexcitatory inhibition by light-on followed by activation and stronger activation by light-off. In the context of this intricate organization pattern, it is difficult to design an experimental system in which all neurons of the visual cortex can be fixed in "activated" or "inhibited" state. It is also clear that what one is measuring in a given situation is the resultant of several components.

The exposure of rabbits or monkeys to a rhythmically flickering light activated the incorporation of intracisternally administered radioactive precursors into proteins of the occipital cortex.[140,142] The controls in these experiments were the animals maintained in total darkness during the same period. The flicker frequency of the light is an important factor. Normal continuous light or a light source of a flicker frequency of 2 per second does not give values higher than in the control animals kept in darkness. A higher incorporation is observed when the light is rhythmically flickering and has a flicker frequency of 7 per second. Other parameters, such as the intensity of light, its color, and the effect of the duration of exposure of the animal to light source, have also been studied.[142]

The flicker frequency of the light is a critical factor and may well be responsible for the failure to obtain a higher incorporation of radioactive leucine in chicks.[143]

It may, however, be mentioned that experiments on pigeons have also been carried out in our laboratory and we have failed to obtain a consistent higher incorporation of systemically administered radioactive amino acids into proteins of the optic lobe corresponding to the eye exposed to light. There are obviously two assumptions made in these experiments.

(1) It is presumed that the optic lobe is the area in the pigeon's brain processing visual information in a manner analogous to the occipital cortex in mammals.

(2) There exists a *complete* decussation of the visual tract in birds, so that it is certain that the eye exposed to light is communicating impulses to *only one* of the optic lobes and that the contralateral lobe is not receiving anything directly or through the association areas of the cortex.

Metzger *et al.*[144,145] studied the effect of unilateral visual stimulation on synthesis of cortical proteins in each hemisphere of the split-brain monkeys. No change in relative rates of protein synthesis were observed by them in the two halves of the monkey brain. The reasons for discrepancy in results of these workers and our laboratory[140,142] are not quite clear.

There are differences in the experimental protocol and procedures which may be pertinent.

The radioactive precursor used by these workers was tritiated water, in contrast to radioactive amino acids employed by us. The former would introduce labeling of all constituents by exchange reactions and may not measure uniquely the rate of *de novo* protein synthesis. If tritiated water has been presumed to label the amino acids, its incorporation into proteins would be dependent on not only the rate of incorporation of amino acids into proteins, but also the rate at which the amino acids are synthesized using tritiated water as a precursor.

The experimental conditions of various individual experiments are not explicit in the legends to the tables.[145] It is apparent that the investigators have employed two sorts of approaches. In one set they have exposed the animal sequentially for 10 min each to light of a frequency of 2, 4, 6, and 8 flashes/sec. Light of an appropriate flicker frequency has not been used continuously for a long enough period. In our experiments, a 45-min exposure to a flicker frequency of 7/sec was considered essential to obtain stimulation.[142]

In another set of experiments the animals were exposed to light of a flicker frequency of 2/sec for 4 min, which was followed by a 1-min presentation of flashes at 8/sec. Twice during this minute, a 10 mA shock of 0.1 min duration was given to the animal. The cycle was repeated for 3 hr. Our experience shows that light of a flicker frequency of 2/sec is ineffective in stimulating beyond the high basal levels.[142] The possibly noxious effect of shocks should not be ruled out.

The duration of 3 hr appears to be too long. In our experiments the increase is obtained only up to 2 hr of continuous exposure of the animal to light. Beyond this period the effect is decreased.[142]

4. *Occipital Cortex Proteinic Fractions Stimulated by Exposure to Light*

An increased radioactivity is found in both the soluble and particulate proteinic fractions of the occipital cortex of the monkey on exposure of the animal to a light of 1614 lumens/m^2 and a flicker frequency of 7/sec for 45 min. There appears to be a generalized activation of the turnover of proteins.

The soluble proteins (105,000*g* supernatant) have been fractionated on DEAE-cellulose followed by a further resolution of each fraction on polyacrylamide gels. A couple of low molecular weight fast-moving acidic proteins have been identified whose turnover is rapid and which show a higher rate of labeling in animals exposed to light.[146] These proteins give an immunological cross reaction with anti-S100 serum.

VI. EFFECTS OF IONS ON METABOLIC REACTIONS

Earlier discussion in this chapter has amply demonstrated that excitation of nerve tissue is inherently linked with a dynamic transaction of cations. A number of effects on metabolism of the tissue during or as a result of excitation have also been described. It appears therefore opportune to examine whether cations such as Na^+ and K^+ can influence in any selective manner various metabolic reactions in the tissues, as these ions are in a way the vehicle of the excitation events.

A. Enzymes Strongly Activated by Potassium

In the literature on the properties of enzymes,[147,148] a large number of enzymes have been shown to be sensitive to a smaller or greater extent to various types of ions. There are at least 23 enzymes known in animal, bacterial, or plant cells which are activated 10–20-fold by potassium.[149] Among these are six enzymes of the glycolytic pathway connected ultimately with the energy metabolism. These are fructokinase, 6-phosphofructokinase, pyruvate phosphokinase, phosphotransacetylase, malic enzyme, acetyl Co-A synthetase, etc. There are several enzymes connected with the synthesis of various metabolites such as carbamyl phosphate, S-adenosyl methionine synthetase, inosine 5′-phosphodehydrogenase, AICAR transformylase, etc., which are strongly activated by potassium ions. Na^+, K^+, and Mg^{2+} in concentrations up to 100 mM stimulate choline acetyl transferase activity.[150] At high concentration Na^+ and K^+ continue to stimulate acetylcholine synthesis whereas Ca^{2+} and Mg^{2+} produce marked inhibition. Tyrosine-activating enzymes[151] are strongly stimulated by K^+. Similarly, an enzyme supposed to be involved in the turnover of messenger RNA, i.e., phosphodiesterase, is activated strongly by potassium.[152] The phosphodiesterase has also a role in regulation of the concentrations of cyclic nucleotides which are the chemical "messengers" involved in the action of several hormones on animal cells.[153]

The stimulatory effect of K^+ on various enzymes is presumably brought about by the alterations induced by this ion in the conformation of proteins, as has been demonstrated for an enzyme in pyrrole synthesis.[154] This protein can differentiate between K^+ and Na^+.

The selective activation of various enzymes by changes in the concentrations of K^+ and Na^+ in defined cellular compartments can bring about changes in the metabolic activities of the tissue.

B. Ion-Induced Stimulation of Respiration in Neural Tissue

It is well established that brain slices react to an increase in the concentration of K^+ (or Cs^+, Rb^+, Li^+, NH_4^+, or choline$^+$) with an augmentation of 50–100% in their rate of oxygen uptake.[155] This response is specific for brain and muscle and is not given by all tissues.[8,156] K^+ stimulates the oxygen uptake and phosphorylation of ADP into ATP in the rat brain mitochondria. K^+ has no effect on the coupled respiration of mitochondria from rat heart and brain.[157] The response is optimal in gray matter. At least 20 mM potassium is required to evoke the increase in respiration; 50 mM K^+ gives maximum stimulation. A certain minimum concentration of sodium is required. There is of course the requirement for the substrate. The ionic stimulation is prevented by pharmacologically active concentrations of several drugs, e.g., barbiturates, phenothiazines, and ethanol, which have little or no effect on the unstimulated oxygen uptake.[158]

Potassium does not give an increase in the uptake of oxygen in homogenates of brain tissue.[158,159] Recent experimental evidence indicates that isolated nerve cells do not respond to high potassium concentrations. It is the glial cells which show an increase in their rate of oxygen uptake.[36,160]

Several analogies are found in the metabolic changes induced by K^+ and those induced by electrical stimulation. It is conceivable that the latter may in fact be due to a release of potassium from the cells.

C. Effect of Ions on Metabolism of Phospholipids

The addition of 0.1 M KCl to the incubation medium or the exclusion of $CaCl_2$ from the medium stimulates the incorporation of (^{32}P) into phospholipids in rat brain cortex slices.[161] The turnover of high energy phosphate compounds also increases under these conditions, even though their steady-state concentrations may show some decline. The presence of sodium ions in the incubating medium is essential for the manifestation of potassium effect on increased labeling of phospholipids. The osmotic strength of the medium has also apparently an important role; 100 mM K^+ increased the specific radioactivity of the (^{32}P) phosphoproteins and phospholipids if K^+ was added to hypertonic medium. No effect of addition of 100 mM K^+ was obtained under isotonic conditions.[162]

D. Effect of Ions on Transport of Amino Acids

The uptake and release of amino acids is influenced by ions. Lajtha *et al.*[163] have shown an absolute requirement for sodium in the medium for uptake of a large number of amino acids (L- and D-lysine, L- and D-leucine, amino isobutyrate, etc.) by the slices of whole mouse brain *in vitro*. Sodium could not be replaced by Li, choline, or potassium. The ionic composition of the medium was important for optimal uptake. The omission of potassium, Ca^{2+}, Mg^{2+}, and phosphate inhibited uptake, while excess of K^+ was also inhibitory. A close relationship between the transport of Na^+ and amino acids has also been reported in other systems.[163] An essential requirement for K^+ in the medium has also been shown by Tsukada *et al.*[164] When K^+ was removed from the medium, amino acid accumulation was almost completely inhibited. Ca^{2+} appears to have a depressing effect; its omission from the medium increased the transport of amino acids.

The uptake of some amino acids such as D- or L-glutamic acid is accompanied by an uptake of potassium.[164] On the other hand, GABA or β-alanine as neutral amino acids were accumulated actively without considerable changes in the distribution of electrolytes.

Increases in the local concentration of cations may also be a signal for release of some neurohumors. An increase of K^+ concentration produced a rapid release of GABA from brain slices *in vitro*.[165]

E. Effect of Ions on RNA and Protein Metabolism

All cells need potassium for growth; all growing and actively metabolizing cells accumulate potassium. The relationship between the rate of growth and the potassium concentration has been most clearly shown in a mutant of *E. coli*—B207. The K^+ concentration in this strain can be altered and is dependent on the concentration of this ion in the medium. It is observed that there occurs an increase of about three-fold in the rate of growth of these cells when the potassium concentration of the medium is raised from 0.5 mM to 2.5 mM.[166] When these cells were incubated in a sodium salt medium, which causes a gradual depletion of potassium from the cells, the ability of the cells to synthesize RNA from radioactive guanine was increased.[167]

A stimulation of protein synthesis by ions in a cell-free system derived from liver was first reported by Sachs.[168] Potassium at an optimum concentration of 50 mM has been used in the cell-free systems employing synthetic messengers (such as poly U) by Nirenberg and Matthaei[169] and Lengyel *et al.*[170]

A stimulation of the synthesis of proteins is caused by a decrease in concentration of sodium and or an increase in the concentration of potassium. The positive effect of potassium is not localized on the formation of aminoacyl tRNA but at subsequent steps in the transfer of activated amino acids to the nascent polypeptide chain.[149] It has been suggested that potassium ions may influence in an important manner the binding of transfer-RNA and messenger-RNA to ribosomes.

F. Effect of Electrolytes on Memory and Learning

A few isolated reports have appeared in the literature implicating electrolytes in processes influencing learning and retention of memory. Sachs[171] investigated the effects of intraventricular injections of Ca^{2+} (22.5 μ equivalent from a 0.25% solution) and potassium (25 μ equivalent from 0.375% solution) on avoidance learning in cats. The rate of learning of the animals after treatment with KCl was superior to that of the control groups, whereas the rate of learning of the animals treated with $CaCl_2$ was depressed. It has been suggested that the facilitatory or depressing effects of these electrolytes may be due to the action of these ions on the activity of neural structures adjacent to the ventricles and in particular the hippocampus.

Flexner[172] has recently reported that intracerebral injections of saline are effective in removing the puromycin-induced block in the expression of memory in mice which were given a maze learning task 1 or more days prior to the administration of puromycin.

ACKNOWLEDGMENTS

The work in the department has been supported by research grants from the Indian Council of Medical Research, World Health Organization, Geneva, and Population Council, Inc., New York.

VII. REFERENCES

1. A. L. Hodgkin, The ionic basis of nervous conduction, *Science* **145**:1148–1154 (1964).
2. A. F. Huxley, Excitation and conduction in nerve: Quantitative analysis, *Science* **145**:1154–1159 (1964).
3. C. J. Duncan, *The Molecular Properties and Evolution of Excitable Cells*, Pergamon Press, New York (1967).
4. J. C. Eccles, Ionic mechanisms of post synaptic inhibition, *Science* **145**:1140–1147 (1964).
5. J. Boistel and P. Fatt, Membrane permeability change during inhibitory transmitter action in crustacean muscle, *J. Physiol.* **144**:176–191 (1958).
6. J. M. Tobias, A chemically specified molecular mechanism underlying excitation in nerve: a hypothesis, *Nature* **203**:13–17 (1964).
7. C. Terner, L. V. Eggelston, and H. A. Krebs, Role of glutamic acid in the transport of potassium in brain and retina, *Biochem. J.* **47**:139–149 (1950).
8. L. Hertz, Potassium effects on ion transport in brain cells, *J. Neurochem.* **15**:1–16 (1968).
9. G. Takagaki, S. Hirano, and Y. Nagata, Some observations on the effect of D-glutamate on the glucose metabolism and the accumulation of potassium ions in brain cortex slices, *J. Neurochem.* **4**:124–134 (1959).
10. Z. Iqbal, S. K. Sharma, and G. P. Talwar (to be published).
11. J. Folch, M. Lees, and G. H. Sloane-Stanley, *in Metabolism of the Nervous System* (D. Richter, ed.), pp. 174–181, Pergamon Press, New York (1957).
12. R. O. Hayden, B. Garoutte, J. Wagner, and R. B. Aird, Binding of radioactive sodium, potassium and bromide in guinea pig brain homogenates, *Proc. Soc. Exp. Bio. Med.* **107**:754–760 (1961).
13. U. Breyer, Competitive binding of Na, K, Mg and Ca ions to cerebroside sulphuric acid, *J. Neurochem.* **12**:131–133 (1965).
14. D. W. Woolley and N. K. Campbell, Tissue lipids as ion exchangers for cations and the relationship to physiological processes, *Biochim. Biophys. Acta* **57**:384–385 (1962).
15. D. W. Woolley and N. K. Campbell, Serotonin receptors II. Calcium transport by crude and purified receptor, *Biochim. Biophys. Acta* **40**:543–544 (1960).

16. F. J. Brinley, *in International Review of Neurobiology* (C. C. Pfeiffer and J. R. Smythies, eds.), Vol. 5, pp. 183–242, Academic Press, New York (1963).
17. R. Katzman and C. E. Wilson, Extraction of lipid and lipid cation from frozen brain tissue, *J. Neurochem.* 7:113–127 (1961).
18. W. L. Nastuk and A. L. Hodgkin, The electrical activity of single muscle fibres, *J. Cell. Comp. Physiol.* **35**:39–73 (1950).
19. A. F. Huxley and R. Stampfli, Effect of potassium and sodium on resting and action potentials of single myelinated nerve fibres, *J. Physiol. (London)* **112**:496–508 (1951).
20. M. H. Draper and S. Weidmann, Cardiac resting and action potentials recorded with an intracellular electrode, *J. Physiol. (London)* **115**:74–94 (1951).
21. J. C. Dalton, Effect of external ions on membrane potentials of a lobster giant axon, *J. Gen. Physiol.* **41**:529–542 (1958).
22. P. Fatt and B. L. Ginsborg, The ionic requirements for the production of action potentials in crustacean muscle fibres, *J. Physiol. (London)* **142**:516–543 (1958).
23. C. T. Gaffey and L. J. Mullins, Ion fluxes during the action potential in *Chara*, *J. Physiol. (London)* **144**:505–524 (1958).
24. H. McIlwain, *in Chemical Exploration of the Brain* pp. 174–189, Elsevier, Amsterdam (1963).
25. J. C. Skou, Enzymatic basis for active transport of Na^+ and K^+ across cell membrane, *Physiol. Rev.* **45**:596–617 (1965).
26. J. C. Skou, Influence of some cations on an adenosine triphosphatase from peripheral nerves, *Biochim. Biophys. Acta* **23**:394–401 (1957).
27. J. C. Skou, Further investigations on a Mg^{2+} and Na^+-activated adenosine triphosphatase, possibly related to the active, linked transport of Na^+ and K^+ across the nerve membrane, *Biochim. Biophys. Acta* **42**:6–23 (1960).
28. J. Jarnefelt, Sodium-stimulated adenosine triphosphatase in microsomes from rat brain, *Biochim. Biophys. Acta* **48**:104–110 (1961).
29. A. Schwartz, H. S. Bachelard, and H. McIlwain, The sodium activated adenosine triphosphatase activity and other properties of cerebral microsomal fractions and subfractions, *Biochem. J.* **84**:626–637 (1962).
30. J. C. Skou, Preparation from mammalian brain and kidney of the enzyme system involved in active transport of Na^+ and K^+, *Biochim. Biophys. Acta* **58**:314–325 (1962).
31. S. L. Bonting, K. A. Simon, and N. M. Hawkins, Studies on sodium, potassium-activated adenosine triphosphatase 1. Quantitative distribution in several tissues of cat, *Arch. Biochem.* **95**:416–423 (1961).
32. C. F. Chignell and E. Titus, The effect of hydroxylamine on a Na^+ and K^+ requiring adenosine triphosphatase from beef brain, *Proc. Nat. Acad. Sci. (Wash.)* **56**:1620–1624 (1966).
33. J. T. Cummins and H. McIlwain, Electrical pulses and the potassium and other ions of isolated cerebral tissues, *Biochem. J.* **79**:330–341 (1961).
34. L. Hertz, *in Biological Basis of Medicine* (E. E. Bitter, ed.), Vol. 5, pp. 3–37, Academic Press, London (1969).
35. L. Hertz, Possible role of neuroglia: A potassium mediated neuronal-neuroglial-neuronal impulse transmission system, *Nature* **206**:1091–1094 (1965).
36. L. Hertz, Neuroglial localization of potassium and sodium effects on respiration in brain, *J. Neurochem.* **13**:1373–1387 (1966).
37. J. S. O'Brien, Cell membranes—composition: structure: function, *J. Theor. Biol.* **15**:307–324 (1967).
38. L. Gross, Active membranes for active transport, *J.Theor. Biol.* **15**: 298–306 (1967).

39. V. P. Whittaker, Structure and function of animal cell membranes, *Br. Med. Bull.* **24**:101–106 (1968).
40. A. D. Bangham and D. A. Haydon, Ultrastructure of membranes: Biomolecular organization, *Br. Med. Bull.* **24**:124–126 (1968).
41. J. F. Danielli and H. A. Davson, A contribution to the theory of permeability of thin films, *J. Cell. Comp. Physiol.* **5**:495–508 (1935).
42. J. D. Robertson, Some features of the ultrastructure of reptilian skeletal muscle, *J. Biophys. Biochem. Cytol.* **2**:369–379 (1956).
43. J. D. Robertson, *in Regional Neurochemistry* (S. S. Kety and J. Elkes, eds.), p. 497, Pergamon Press, Oxford (1961).
44. J. A. Lucy, Ultrastructure of membranes: Micellar organization, *Br. Med. Bull.* **24**:127–129 (1968).
45. D. E. Green and J. F. Purdue, Membranes as expressions of repeating units, *Proc. Nat. Acad. Sci. U.S.* **55**:1295–1302 (1966).
46. F. S. Sjöstrand, A new ultrastructural element of the membranes in mitochondria and of some cytoplasmic membranes, *J. Ultrastruct. Res.* **9**:340–361 (1963).
47. J. Kavanau, *in Structure and Function in Biological Membranes*, Holden-Day, Inc. New York (1965).
48. J. Kavanau, *in Recent Progress in Surface Science* (J. F. Danielli, K. G. Pankhurst, and A. C. Riddiford, eds.), Vol. I, Academic Press, New York (1964).
49. J. Del Castillo, A. Rodrigrez, C. A. Romero, and V. Sanchez, Lipid films as transducers for detection of antigen-antibody and enzyme-substrate reactions, *Science* **153**:185–188 (1966).
50. M. Delbruck and W. Reichardt, *in Cellular Mechanisms in Differentiation and Growth* (D. Rudwick, ed.) Princeton University Press (1956).
51. A. L. Lehninger, Water uptake and extrusion by mitochondria in relation to oxidative phosphorylation, *Physiol. Rev.* **42**:467–517 (1962).
52. J. P. Changeux, J. Thiery, Y. Tung, and C. Kittel, On the co-operativity of biological membranes, *Proc. Nat. Acad. Sci. U.S.* **57**:335–341 (1967).
53. I. Tasaki, A. Watanabe, R. Sandlin, and L. C. Carnay, Changes in fluorescence, turbidity and birefringence associated with nerve excitation, *Proc. Nat. Acad. Sci. U.S.* **61**:883–888 (1968).
54. L. B. Cohen, R. D. Keynes, and B. Hille, Light scattering and birefringence changes during nerve activity, *Nature* **218**:438–441 (1968).
55. E. Schoffeniels and D. Nachmansohn, Isolated single electroplax preparation. 1. Effect of acetylcholine and related compounds. *Biochim. Biophys. Acta* **26**:1–15 (1957).
56. D. Nachmansohn, Proteins in bioelectricity: The control of ion movements across excitable membranes, *Proc. Nat. Acad. Sci. U.S.* **61**:1034–1041 (1968).
57. G. D. Webb, Affinity of benzoquinonium and ambenonium derivatives for the acetylcholine receptor, tested on the electroplax, and for acetylcholinesterase in solution, *Biochim. Biophys. Acta.* **102**:172–184 (1965).
58. A. Karlin, Chemical distinctions between acetylcholinesterase and the acetylcholine receptor, *Biochim. Biophys. Acta.* **139**:358–362 (1967).
59. D. R. Curtis, J. W. Phillis, and J. C. Watkins, Chemical excitation of spinal neurons, *Nature* **183**:611–612 (1959).
60. D. R. Curtis, J. W. Phillis, and J. C. Watkins, The chemical excitation of spinal neurons by certain acidic amino acids, *J. Physiol.* **150**:656–682 (1960).
61. H. McLennan, R. D. Huffman, and K. C. Marshall, Patterns of excitation of thalamic neurons by amino acids and by acetylcholine, *Nature* **219**:387–388 (1968).

62. D. R. Curtis and J. C. Wilkins, The excitations and depression of spinal neurons by structurally related amino acids, *J. Neurochem.* **6**:117–141 (1960).
63. D. R. Curtis and J. C. Watkins, Acidic amino acids with strong excitatory actions on mammalian neurons, *J. Physiol.* **166**:1–14 (1963).
64. K. N. Sharma, S. Dua, B. Singh, and B. K. Anand, Electroontogenesis of cerebral and cardiac activities in the chick embryo, *Electroenceph. Clin. Neurophysiol.* **16**:503–509 (1964).
65. E. Jr. Garcia-Austt, Development of electrical activity in cerebral hemispheres of the chick embryo, *Proc. Soc. Exp. Biol. Med.* **86**:348–352 (1954).
66. A. A. Abdel-Latif, J. Brody, and H. Ramahi, Studies on sodium-potassium adenosine triphosphatase of the nerve endings and appearance of electrical activity in developing rat brain, *J. Neurochem.* **14**:1133–1141 (1967).
67. H. Tuge, Y. Kanayama, and C. H. Yueh, Comparative studies on the development of EEG, *Jap. J. Physiol.* **10**:211–220 (1960).
68. J. P. Schade, Maturational aspects of EEG and of spreading depression in rabbit, *J. Neurophysiol.* **22**:245–257 (1959).
69. L. B. Flexner and J. B. Flexner, Biochemical and physiological differentiation during morphogenesis. IX. The extracellular and intracellular phases of the liver and cerebral cortex of the fetal guinea pig as estimated from distribution of chloride and radio sodium, *J. Cell. Comp. Physiol.* **34**:115–127 (1949).
70. H. H. Jasper, C. S. Bridgman, and L. Carmichael, An ontogenetic study of cerebral electrical potentials in the guinea pig, *J. Exp. Psychol.* **21**:63–71 (1937).
71. P. Mandel, R. Bieth, and R. Stoll, Distribution of various lipid fractions in the brain of the chick embryo during the second half of the incubation period, *Compt. Rend. Soc. Biol. (Paris).* **143**:1224–1226 (1949).
72. N. Zaheer, Z. Iqbal, and G. P. Talwar, Metabolic parameters of ontogenesis of electrical activity in the brain, *J. Neurochem.* **15**:1217–1224 (1968).
73. F. E. Samson and D. J. Quinn, Na^+-K^+-activated ATPase in rat brain development, *J. Neurochem.* **14**:421–427 (1967).
74. N. A. Lassen, Cerebral blood flow and oxygen consumption in man, *Physiol. Rev.* **39**:183–238 (1959).
75. H. McIlwain, *in Biochemistry of the Central Nervous System*, pp. 53–55, J. & A. Churchill, Ltd., London (1966).
76. H. McIlwain, *in Biochemistry of the Central Nervous System*, p. 62, J. & A. Churchill, Ltd., London (1966).
77. M. K. Gaitonde, S. A. Marchi, and D. Richter, The utilization of glucose in the brain and other organs of the cat, *Proc. Roy. Soc. B.* **160**:124–136 (1964).
78. L. G. Abood and A. Geiger, Breakdown of proteins and lipids during glucose-free perfusion of the cat's brain, *Amer. J. Physiol.* **182**:557–560 (1955).
79. S. S. Kety, Biochemistry and mental function, *Nature* **208**:1252–1257 (1965).
80. H. McIlwain and P. Joanny, Characteristics required in electrical pulses of rectangular time-voltage relationships for metabolic change and ion movements in mammalion cerebral tissues, *J. Neurochem.* **10**:313–323 (1963).
81. G. B. Ansell, *in Advances in Lipid Research* (R. Paoletti and D. Kritchevsky, eds.), Vol. 3, pp. 139–170, Academic Press, New York (1965).
82. A. N. Davison and J. Dobbing, Phospholipid metabolism in nervous tissue. 3. The anatomical distribution of metabolically inert phospholipid in the central nervous system, *Biochem. J.* **75**:571–574 (1960).
83. P. B. Bradley and J. H. Wolstencroft, Excitation and inhibition of brain stem neurons by noradrenaline and acetylcholine, *Nature* **196**:840 (1962).

84. M. R. Hokin and L. E. Hokin, Enzyme secretion and the incorporation of ^{32}P into phospholipids of pancreas slices, *J. Biol. Chem.* **203**:967–977 (1953).
85. M. R. Hokin and L. E. Hokin, Effect of acetylcholine on phospholipids in the pancreas, *J. Biol. Chem.* **209**:549–558 (1954).
86. L. E. Hokin and M. R. Hokin, Acetylcholine and the exchange of inositol and phosphate in brain phosphoinositide, *J. Biol. Chem.* **233**:818–821 (1958).
87. H. Yoshida and J. H. Quastel, Effects of metabolic inhibitors on potassium and acetylcholine stimulated incorporation of phosphate into phospholipids of rat brain cortex slices, *Biochim. Biophys. Acta.* **57**:67–72 (1962).
88. M. R. Hokin, L. E. Hokin, and W. D. Shelp, The effects of acetylcholine on the turnover of phosphatidic acid and phosphoinositide in sympathetic ganglia, and in various parts of the central nervous system *in vitro*, *J. Gen. Physiol.* **44**:217–226 (1960).
89. L. E. Hokin and M. R. Hokin, Acetylcholine and the exchange of phosphate in phosphatidic acid in brain microsomes, *J. Biol. Chem.* **233**:822–826 (1958).
90. M. G. Larrabee and W. S. Leicht, Metabolism of phosphatidyl-inositol and other lipids in active neurons of sympathetic ganglia and other peripheral nervous tissues, *J. Neurochem.* **12**:1–13 (1965).
91. L. E. Hokin, Effects of acetylcholine on the incorporation of ^{32}P into various phospholipids in slices of normal and denervated superior cervical ganglia of the cat, *J. Neurochem.* **13**:179–184 (1966).
92. R. M. C. Dawson and D. Richter, Effect of stimulation on the phosphate esters of the brain, *Amer. J. Physiol.* **160**:203–211 (1950).
93. C. Torda, Effects of corticotropin and various convulsion inducing agents on the ^{32}P content of brain phospholipids, nucleoproteins and total acid soluble phosphorus compounds, *Amer. J. Physiol.* **177**:179–182 (1954).
94. N. V. Zakharov and R. L. Orlyanskaya, The effect of excitation and convulsions induced by nekethamide on the metabolism of protein and phosphates in the brain of rats, *Vaprosy, Med. Khim.* **6**:249–253 (1960).
95. G. Ye. Vladimirov, T. N. Ivanova, and N. I. Pravdina, The effect of the functional state upon the metabolism of phosphorus compounds in the brain tissue, *Biokhimia* **19**:578–585 (1954).
96. A. V. Palladin, *in Problems of the Biochemistry of the Nervous System* (A. V. Palladin, ed.), pp. 303–319, Pergamon Press, London (1964).
97. A. A. Smirnov, Phosphorus metabolism in the cortex of the cerebrum in dogs during sleep and wakefulness, *Dokl. Akad. Nauk S.S.S.R.* **101**:913–916 (1955).
98. H. Hydén, *in Handbook of Neurochemistry* (A. Lajtha, ed.), Vol. III, Plenum Press, New York.
99. H. Hydén and P. Lange, *in Regional Neurochemistry* (S. S. Kety and J. Elkes, eds.), pp. 190–199, Pergamon Press, New York (1961).
100. L. Z. Pevzner, Topochemical aspects of nucleic acids and protein metabolism within the neuron-neuroglia unit of the superior cervical ganglion, *J. Neurochem.* **12**:993–1002 (1965).
101. B. Daneholt and S. O. Bråttgard, A. comparison between RNA metabolism of nerve cells and glia in the hypoglossal nucleus of the rabbit, *J. Neurochem.* **13**:913–921 (1966).
102. H. Hydén and E. Egyhazi, Nuclear RNA changes of nerve cells during a learning experiment in rats, *Proc. Nat. Acad. Sci. U.S.* **48**:1366–1373 (1962).
103. H. Hydén and E. Egyhazi, Changes in RNA content and base composition in cortical neurons of rats in a learning experiment involving transfer of handedness, *Proc. Nat. Acad. Sci. U.S.* **52**:1030–1035 (1964).

104. D. A. Rappoport, *in Handbook of Neurochemistry* (A. Lajtha, ed.), Vol. III, Plenum Press, New York.
105. D. A. Rappoport and H. F. Daginawala, Changes in nuclear RNA of brain induced by olfaction in catfish, *J. Neurochem.* **15**:991–1006 (1968).
106. B. Sadasivudu and G. P. Talwar, Changes in phosphorylated compounds and proteins of rat brain during metrazole convulsions, *Bull Nat. Inst. Sci. (India)* **19**:237–248 (1960).
107. V. S. Chitre, S. P. Chopra, and G. P. Talwar, Changes in the ribonucleic acid content of the brain during experimentally induced convulsions, *J. Neurochem.* **11**:439–449 (1964).
108. V. S. Chitre and G. P. Talwar, Correlation of electrical activity of brain with metabolic parameters. II. Pentose nucleic acid content of isolated cerebral cortex during various phases of electrical activity following topical application of metrazol, *Indian J. Med. Res.* **51**:80–91 (1963).
109. E. L. Noach, J. J. Bunk, and A. Wijling, Influence of electro-shock and phenobarbital on nucleic acid content of rat brain cortex, *Acta Physiol. Pharmacol. Neerl.* **11**:54–69 (1962).
110. H. Hydén, Protein and nucleotide metabolism in the nerve cell under different functional conditions, *Symp. Soc. Expt. Biol.* **1**:150–162 (1947).
111. H. Hydén, Protein metabolism in the nerve cell during growth and function, *Acta Physiol. Scand.* **6** Suppl. 17:1 (1943).
112. C. A. Hamberger, H. Hydén, and G. Nilsson, The correlation between cytochemical changes in the cochlear ganglion and functional tests after acoustic stimulation and trauma, *Acta Otolaryng. Suppl.* **75**:124–133 (1949).
113. A. Geiger and S. Yamasaki, Cytidine and uridine requirement of the brain, *J. Neurochem.* **1**:93–100 (1956).
114. A. Geiger, S. Yamasaki, and R. Lyons, Changes in nitrogenous components of brain produced by stimulation of short duration, *Amer. J. Physiol.* **184**:239–243 (1956).
115. R. Vrba and J. Folbergrova, Observations on endogenous metabolism in brain *in vitro* and *in vivo*, *J. Neurochem.* **4**:338–349 (1959).
116. G. Gomirato, Quantitative evaluation of the metabolic variations in the spinal motor root cells, studied by biophysical method and following adequate stimulation (muscular fatigue). Action on metabolism of vitamin B_{12}, *J. Neuropath. Exp. Neurol.* **13**:359–368 (1954).
117. G. Attardi, An ultramicrospectrophotometric study of the Purkinje cells of the albino rat, *Experientia* **9**:422–424 (1953).
118. L. Einarson and E. Krogh, Variations in the basophilia of nerve cells associated with increased cell activity and functional stress, *J. Neurol. Neurosurg. Psychiat.* **18**:1–12 (1955).
119. G. P. Talwar, B. K. Goel, S. P. Chopra, and B. D'Monte, *in Macromolecules and Behavior* (J. Gaito, ed.), pp. 71–88, Appleton-Century-Crofts, New York (1966).
120. W. Grampp and J. E. Edström, The effect of nervous activity on ribonucleic acid of the crustacean receptor neuron, *J. Neurochem.* **10**:725–732 (1963).
121. H. D. Cohen and S. H. Barondes, Further studies of learning and memory after intracerebral actinomycin-D, *J. Neurochem.* **13**:207–211 (1966).
122. G. P. Talwar, *in Biological Basis of Medicine* (E. E. Bitter, ed.), Vol. 5, 77–109 Academic Press, London (1969).
123. R. D. Lisk and M. Newlon, Estradiol: Evidence for its direct effect on hypothalamic neurons, *Science* **139**:223–224 (1963).
124. B. W. Agranoff, Memory and protein synthesis, *Sci. Am.* **216**:115–122 (1967).

124a. R. B. Roberts and L. B. Flexner, The biochemical basis of long-term memory, *Quart. Rev. Biophysics* **2**:135–173 (1969).
125. S. H. Appel, *in Enzymes in Mental Health* (G. J. Martin and B. Kisch, eds.), pp. 186–193, J. B. Lippincott Co., Philadelphia (1966).
126. W. Dingman and M. B. Sporn, The incorporation of 8-azaguanine into rat brain RNA and its effect on maze learning by the rat: an enquiry into the biochemical basis of memory, *J. Psychiat. Res.* **1**:1–11 (1961).
127. W. Dingman and M. B. Sporn, Molecular theories of memory, *Science* **144**:26–29 (1964).
128. M. Eigen, *in Neuroscience Research Symposium Summaries* (F. O. Schmitt and T. Melnechuk, eds.), pp. 267–277, M.I.T. Press, Cambridge, Mass. (1966).
129. J. Gaito, *in Macromolecules and Behavior* (J. Gaito, ed.), pp. 89–102, Appleton-Century-Crofts, New York (1966).
130. E. Roberts, Models for correlative thinking about brain behaviour and biochemistry, *Brain Res.* **2**:109–144 (1966).
131. R. G. Smith, Magnesium pemoline: lack of facilitation in human learning, memory and performance tests, *Science* **155**:603–605 (1967).
132. S. O. Bråttgard, The importance of adequate stimulation for the chemical composition of ganglion cells, *Exp. Cell Res.* **2**:693–695 (1951).
133. Iu. S. Chentsov, V. L. Boroviagin, and V. Ia. Brodskii, Submicroscopic morphology of ganglion neurons of the retina as a reflection of some characteristics of their metabolism, *Biofizika* **6**:590–595 (1961).
134. T. P. Lukashevich, Lokalizatsiia i aktivnost' degidraz v mitokhondriiakh ellipsoidov fotoretseptorov pozvonochnykh, *Dokl. Akad. Nauk. SSSR* **156**:1436–1439 (1964).
135. M. H. Epstein and J. S. O'connor, Enzyme changes in isolated retinal layers in light and darkness, *J. Neurochem.* **13**:907–911 (1966).
136. P. H. Glow and S. Rose, Effects of light and dark on the acetylcholinesterase activity of the retina, *Nature* **202**:422–423 (1964).
137. M. R. Rosenzweig, E. L. Bennett, M. C. Diamond, S. Wu, R. W. Slagh, and E. Saffran, Influences of environmental complexity and visual stimulation in development of occipital cortex in rats, *Brain Res.* **14**:427–445 (1969).
138. L. Gyllensten and J. Lindberg, Development of the visual cortex of mice with inherited retinal dystrophy, *J. Comp. Neurol.* **122**:79–82 (1964).
139. G. P. Talwar, S. P. Chopra, and B. K. Goel, Correlation of functional activity of brain with metabolic parameters: RNA and protein metabolism of occipital cortex in relation to its activation by light stimulus, *VI Int. Congr. Biochem., New York* **5**:419 (1964).
140. G. P. Talwar, S. P. Chopra, B. K. Goel, and B. D'Monte, Correlation of the functional activity of the brain with metabolic parameters. III. Protein metabolism of the occipital cortex in relation to light stimulus, *J. Neurochem.* **13**:109–116 (1966).
141. R. Jung, Excitation, inhibition and coordination of cortical neurons, *Exp. Cell Res. Suppl.* **5**:262–271 (1958).
142. U. B. Singh and G. P. Talwar, Effect of flicker frequency of light and other factors on the synthesis of proteins in the occipital cortex of monkey, *J. Neurochem.* **14**:675–680 (1967).
143. J. Altman, *in Macromolecules and Behavior* (J. Gaito, ed.), pp. 103–126, Appleton-Century-Crofts, New York (1966).
144. H. P. Metzger, M. Cuenod, A. Grynbaum, and H. Waelsch, Visual stimulation: Lack of effect on protein synthesis in the monkey brain, *Life Sci.* **5**:1115–1120 (1966).

145. H. P. Metzger, M. Cuenod, A. Grynbaum, and H. Waelsch, The use of tritium oxide as a biosynthetic precursor of macromolecules in brain and liver, *J. Neurochem.* **14**:99–104 (1967).
146. U. B. Singh and G. P. Talwar, Identification of a protein fraction in the occipital cortex of the monkey rapidly labelled during the exposure of the animal to rhythmically flickering light, *J. Neurochem.* **16**: 951–959 (1969).
147. M. Dixon and E. C. Webb, *in Enzymes*, Academic Press, New York (1958).
148. G. L. Cantoni, *in Comparative Biochemistry*, Vol. 1, Academic Press, New York (1960).
149. M. Lubin, *in The Cellular Functions of Membrane Transport* (J. F. Hoffman, ed.), pp. 193–211, Prentice-Hall, Englewood Cliffs, N.J. (1964).
150. D. Morris and S. Tucek, The influence of inorganic cations on the activities of choline acetyltransferase, phosphate acetyltransferase and acetylphosphatase, *J. Neurochem.* **13**:333–345 (1966).
151. R. W. Holley, E. F. Brunngraber, F. Saad, and H. H. Williams, Partial purification of the threonine- and tyrosine-activating enzymes from rat liver and the effect of potassium ions on the activity of the tyrosine enzyme, *J. Biol. Chem.* **236**:197–199 (1961).
152. P. F. Spahr and D. Schlessinger, Breakdown of messenger ribonucleic acid by a potassium activated phosphodiesterase from *E. coli*, *J. Biol. Chem.* **238**:PC 2251 (1963).
153. E. W. Sutherland, I. Oye, and R. W. Butcher, *in Recent Progress in Hormone Research* (G. Pincus, ed.), Vol. 21, pp. 623–646, Academic Press, New York (1965).
154. D. Shemin and D. L. Nandi, The effect of K^+ ions on the conformation of an enzyme: Mechanism of pyrrol synthesis, *Proc. Nat. Acad. Sci. US* **61**:1154 (1968).
155. L. Hertz and M. Schou, Univalent cations and the respiration of brain-cortex slices, *Biochem. J.* **85**:93–104 (1962).
156. L. Hertz and T. Clausen, Effect of potassium and sodium on respiration: Their specificity to slices from certain brain regions, *Biochem. J.* **89**:526–533 (1963).
157. A. R. Krall, M. C. Wagner, and D. M. Gozansky, Potassium ion stimulation of oxidative phosphorylation by brain mitochondria, *Biochem. Biophys. Res. Commun.* **16**:77–81 (1964).
158. J. J. Ghosh and J. H. Quastel, Narcotics and brain respiration, *Nature* **174**:28–31 (1954).
159. M. M. de Piras and J. A. Zadunaisky, Effect of potassium and ouabain on glucose metabolism by frog brain, *J. Neurochem.* **12**:657–661 (1965).
160. S. P. R. Rose, Ionic accumulation and membrane properties of enriched preparation of neurons and glia from mammalian cerebral cortex, *J. Neurochem.* **14**:373–375 (1967).
161. M. Brossard and J. H. Quastel, Studies of the cationic and acetylcholine stimulation of phosphate incorporation into phospholipids in rat brain cortex, *in vitro*, *Canad. J. Biochem. Physiol.* **41**:1243–1256 (1963).
162. J. Durell and P. J. Heald, The effect of potassium ion concentration on phosphate metabolism in cerebral slices, *J. Neurochem.* **9**:71–79 (1962).
163. S. Lahiri and A. Lajtha, Cerebral amino acid transport *in vitro*—1. Some requirements and properties of uptake, *J. Neurochem.* **11**:77–86 (1964).
164. Y. Tsukada, Y. Nagata, S. Hirano, and T. Matsutani, Active transport of amino acids into cerebral cortex slices, *J. Neurochem.* **10**:241–256 (1963).
165. Y. Machiyama, R. Balazs, and D. Richter, Effect of K^+-stimulation on GABA metabolism in brain slices *in vitro*, *J. Neurochem.* **14**:591–594 (1967).

166. M. Lubin and H. L. Ennis, On the role of intracellular potassium in protein synthesis, *Biochim. Biophys. Acta* **80**:614–631 (1964).
167. H. L. Ennis and M. Lubin, Dissociation of ribonucleic acid and protein synthesis in bacteria deprived of potassium, *Biochim. Biophys. Acta* **50**:399–402 (1961).
168. H. Sachs, A stabilized enzyme system for amino acid incorporation, *J. Biol. Chem.* **228**:23–39 (1957).
169. M. W. Nirenberg and J. H. Matthaei, The dependence of cell-free protein synthesis in *E. coli* upon naturally occurring or synthetic polyribonucleotides, *Proc. Nat. Acad. Sci. U.S.* **47**:1588–1602 (1961).
170. P. Lengyel, J. F. Speyer, and S. Ochoa, Synthetic polynucleotides and the amino acid code, *Proc. Nat. Acad. Sci. U.S.* **47**:1936–1942 (1961).
171. E. Sachs, "The role of brain electrolytes in learning and retention", unpublished doctoral dissertation, Univ. Rochester, Rochester, New York (1962).
172. L. B. Flexner and J. B. Flexner, Intracerebral saline: Effect on memory of trained mice treated with puromycin, *Science* **159**:330–331 (1968).
173. B. C. Pressman, Induced active transport of ions in mitochondria, *Proc. Nat. Acad. Sci. U.S.* **53**:1076–1083 (1965).
174. B. T. Kilbourn, J. D. Dunitz, L. A. R. Pioda, and W. Simon, Structure of the K^+ complex with nonactin, a macrotetrolide antibiotic possessing highly specific K^+ transport properties, *J. Mol. Biol.* **30**:559–563 (1967).
175. P. Mueller and D. O. Rudin, Action potentials induced in biomolecular lipid membranes, *Nature* **217**:713–719 (1968).
176. A. Watanabe, I. Tasaki, and L. Lerman, Bi-ionic action potentials in squid giant axons internally perfused with sodium salts, *Proc. Nat. Acad. Sci. U.S.* **58**:2246–2252 (1967).
177. A. Watanabe, I. Tasaki, I. Singer, and L. Lerman, Effects of tetradotoxin on excitability of squid giant axons in sodium free media, *Science*, **155**:95–97 (1967).
178. J. W. Moore, T. Narahashi, and T. I. Shaw, An upper limit to the number of sodium channels in nerve membranes, *J. Physiol.* **188**:99–105 (1967).

Chapter 3

SOME NEUROCHEMICAL ASPECTS OF CONVULSIONS

Richard A. Lovell*

Department of Psychiatry, The University of Chicago
Chicago, Illinois

I. INTRODUCTION

A. Convulsions from the Point of View of Neurochemistry

This chapter deals with recent work relevant to the neurochemical basis of convulsive activity of the central nervous system. It has been written chiefly from the viewpoint of neurochemistry, a viewpoint which considers such activity as representative of or associated with certain critical chemical changes in neurons. While it is recognized that the question of cause and effect with regard to these neurochemical changes and the hyperactivity of neurons in convulsion is important and perhaps even crucial, this brief review is directed rather toward the question of the role of certain endogenous amines and amino acids in the neurochemical mechanisms which underlie convulsive neuronal activity. Certain neurophysiological studies have been included because they are relevant to an adequate understanding of these neurochemical mechanisms.

B. Scope and Purpose of this Chapter

Thus the scope of this review is limited to recent neurochemical, and some neurophysiological, research on the role of brain amino acids and amines in convulsions. Some general biochemical investigations of metabolic changes in the brain responsible for or associated with convulsions have also been discussed. The pharmacology of convulsions—the large literature on convulsant and anticonvulsant drugs—has been largely omitted here. Research on the mechanism of action of convulsant or

* Supported in part by The Schweppe Foundation and in part by State of Illinois Public Welfare Grant 17-302.

anticonvulsant drugs is discussed in this chapter only if it has thrown light on the role of brain amino acids and amines in seizure mechanisms.

The purpose of this chapter is twofold: to illustrate the kinds of studies and information which characterize the neurochemical approach to the problem of convulsions, and to provide a literature basis for further study and investigation of the neurochemical basis of convulsions. More specifically, the author's intention has been to illustrate how the study of convulsions can illuminate the role of brain amines and amino acids in the regulation or moderation of central neuronal activity. Thus, in this point of view, convulsive activity is regarded more as hyperactivity of the nervous system rather than as a pathological condition. Yet neurochemical studies of the kind surveyed here are obviously relevant to pathological conditions accompanied by convulsions.

C. Review of Reviews

The literature on the neurochemistry of convulsions has been rather thoroughly reviewed in previous publications. Hence, in this chapter, only work subsequent to the review of this subject by Wolfe and Elliott(1) is considered (with some exceptions). Other reviews which are directly relevant to convulsions and which should be noted are those by Tower,(2) by Kreindler,(3) and by Tews and Stone.(4) There are also several reviews of the literature on vitamin B_6 and on GABA which ought to be consulted in any study of convulsions. Two excellent reviews on vitamin B_6 and its role in the central nervous system exist.(5,6) For background on GABA, see the reviews by Elliott and Jasper(7) and by Elliott(8); also very helpful and more current is that by Roberts and Kuriyama.(9) The pharmacology of GABA and related amino acids has been carefully reviewed by Curtis and Watkins.(10) Other relevant reviews are those on the pharmacology of inhibition(11) and on iontophoretic studies of central neurons.(12) For background information on the amines, a useful starting point is the recent review by Bloom and Giarman.(13)

II. BIOCHEMICAL STUDIES OF CONVULSIONS

Even a cursory look at some of the review articles just surveyed will reveal the extent to which the biochemical approach has contributed to our knowledge of convulsive phenomena. There has been a rather vast number of biochemical studies of convulsions: biochemical studies of epilepsy, investigations of experimental "epilepsy" in animals induced by drugs or lesions or sound, research on the biochemical changes in the brain which precede, accompany, or result from convulsions, etc. These studies are aimed at the discovery of the biochemical "characteristics" of convulsions or of the neurochemical defects in the brain, which may or may not be causally related to convulsions. While the limitations of space prevent an adequate review of this work here, it may be useful to present examples of some of the more significant experiments in order to illustrate the biochemical approach to the study of convulsive phenomena and the kinds of information which it yields.

A good example is seen in the experiments which showed that if seizures were induced in unanesthetized rats or mice by electroshock, there was a massive increase in the metabolic rate during the first few seconds after the shock and a marked acceleration of glucose into the brain. These changes were accompanied by a marked fall in ATP and phosphocreatine,(14–16) a result which confirmed earlier work of Dawson and Richter. Anesthesia prevented most of the clinical symptoms of electroshock but did not necessarily prevent the metabolic evidence of massive neuronal discharge; hence, it appeared that there was a tendency for a dissociation of ATP and phosphocreatine levels from the presence or absence of clinical seizures. This notion was supported also by hypoglycemic seizures in mice—after a large dose of insulin, the animals passed through all the stages of hypoglycemia but showed no decrease in either ATP or phosphocreatine in the brain. Whatever the cause of seizures, the lack of high energy phosphate was apparently not a factor.(16) Studies of cerebral metabolites in the brain of mice after seizures induced by flurothyl showed that the brain has a remarkable ability to adjust to a new steady state following the marked increase in glycolytic rate in the convulsive state, as well as the ability to maintain adequate supplies of energy reserves and oxygen.(17) Similarly, in an attempt to determine whether cerebral blood flow becomes insufficient to meet the demands of oxidative metabolism during convulsive activity, impressive evidence of the brain's ability to adjust systemic blood pressure and the tonus of its own blood vessels to meet its metabolic demands was obtained.(18) Convulsions were induced by pentylenetetrazole and electrical stimulation in dogs and monkeys. The cerebral blood flow was found to increase greatly, but the results indicated that the normal autoregulation of cerebral circulation was suspended and that the increased blood flow possibly followed the increased blood pressure. It was suggested that this was possibly mediated by a receptor in the cerebral extravascular tissue or by a direct action of metabolites on the cerebral vascular bed.(18) Increased metabolic demand after strychnine or ischemia caused by temporary interruption of the cerebral circulation in rabbits caused a dilation of the pial arteries and a contraction of the lumina of cortical arteries. Thus, the pial arteries are at least in part responsible for the increased blood flow to the cortex.(19)

Stone and his co-workers achieved a classification of the different types of seizures on the basis of specific neurochemical patterns of changes in the amino acids and other substances in the brain of anesthetized dogs.(4,20–23) In a further study, status epilepticus was induced in dogs for 30–40 min by repeated doses of pentylenetetrazole; the following pattern of changes resulted. There were significant increases in brain alanine, arginine, GABA, glycine, histidine, leucine, lysine, phenylalanine, serine, tyrosine, and valine, and also ammonia. There were decreases in glutamate and aspartate, as well as large decreases in the ganglioside fraction of brain lipids and decreases in the fraction which contains lecithins and sphingomyelins.[24]

Certain neurochemical factors of importance in seizure mechanisms have been brought to light by recent research. One such factor is the apparent increase in permeability of the blood–brain barrier which is associated with convulsions. There is an apparently enhanced vascular-extravascular exchange of radioactive sulfur (as sulfate) at sites of electrical activity or discharge in the cat brain in seizure states induced by pentylenetetrazole or strychnine.[25–27] Also in the cat, increased penetration of labeled sulfate into the lateral geniculate was observed after photic stimulation and into the interior colliculi after auditory stimulation.[28] Another factor of possible importance in seizure mechanisms is the effect of changes in serum osmolarity. Strong solutions of certain substances such as GABA or succinate or sucrose, injected intraperitoneally, will postpone or prevent convulsions caused by picrotoxin or pentylenetetrazole or oxygen at high pressure, and it has been suggested that this protective effect is due to increased osmolarity of the serum and a concomitant dehydration of the brain which may be partly responsible for the elevation of seizure threshold.[29] On the other hand, serum hyperosmolarity and brain dehydration may induce seizures in conditions where there is a potentially epileptogenic process in some cortical areas.[30] Intraventricular administration of hypertonic sodium chloride to unanesthetized cats may induce seizure activity,[31] and when the lateral ventricle of cats was perfused with CSF which was high in potassium, seizure activity in the hippocampus resulted.[32]

III. THE ROLE OF CERTAIN BRAIN AMINO ACIDS AND AMINES IN CONVULSIVE MECHANISMS

The main concern of this chapter is the relationship of certain important amino acids and amines in the brain to neuronal excitability. The amino acids of chief interest in this regard are glutamate and GABA, and the amines are norepinephrine (NE) and 5-hydroxytryptamine (5-HT). There is much evidence—a considerable portion is outlined below—to support the view that these endogenous substances are involved in the maintenance of appropriate levels of excitability in neurons and hence of normal brain function. Thus, in considering the neurochemical processes which underlie convulsive phenomena, the assumption is that these substances participate in the control or moderation of the level of neuronal firing and that the mechanisms of this participation are somehow upset in the chain of events which culminates in seizure activity. The question to be taken up at this point is precisely how GABA, glutamate, and the amines are involved in these mechanisms. What light on their role in convulsions can be derived from recent studies?

A. Relationship of the Glutamic Acid-GABA System to Convulsions

1. *Excitatory Effects of Glutamic Acid*

The technique of microelectrophoresis* has proved to be extremely useful, particularly in studies analyzing chemical transmission in the CNS and work aimed at elucidating the mechanism of action of drugs which have central effects. While the large number of these essentially neurophysiological studies cannot be reviewed here, some of the more recent research may be considered for its contribution to the understanding of the neurochemical events underlying convulsive phenomena. The powerful excitatory effects of L-glutamate have been demonstrated in the cortex, hippocampus, thalamus, and subthalamic areas of the rat,[36,37] in the hippocampus of rabbits,[38] and in the hippocampus,[39] pyriform cortex,[40] pericruciate cortex,[41] cuneate nucleus,[42] and spinal cord[43] of the cat. While it is true that L-glutamate is not the most potent of the excitatory amino acids—homocysteic acid is more potent and *N*-methyl-D-aspartic acid is the most potent[43,41]—glutamate is the most interesting since, as Krnjević points out, it has a very strong and quickly reversible action, it is found in the brain in large quantities, and its excitatory effect on cortical neurons occurs with as little as 10^{-15} mole.[34] In a thorough study of receptors on goldfish Mauthner neurons, Diamond has shown that L-glutamate depolarizes these neurons and hence excites them but only if applied to the outside of the neuron; intracellular injection of L-glutamate was ineffective.[44] Our understanding of the mechanism of the excitatory effect of acidic amino acids like glutamate has been aided by the study of Bradford and McIlwain, who investigated the ionic basis for the depolarization of the central neurons. On the basis of their data they suggested that glutamate increases the permeability to sodium by five times that to potassium.[45] In this connection, it is interesting to note that Watkins had proposed, on the basis of the similarity in structure and charge distribution of acetylcholine, GABA and glutamic acid to the polar head groups of lecithin, phosphatidyl ethanolamines and phosphatidyl serine, that the neuronal membrane may contain complexes between these lipids and protein and that the action of GABA and glutamate may cause dissociation of these complexes, a dissociation which results in permeability changes in the membrane.[46] An earlier exhaustive study of structure-activity relationships relevant to the effects of excitatory and depressant amino acids on the neuronal membrane led to the conclusion that GABA and glutamic acid probably

* For evaluation of the technique of microelectrophoresis, or iontophoresis, see the short reviews by Curtis,[33] Krnjević,[34] and Bradley and Wolstencroft[35]; for a more recent and more extensive critique of iontophoretic studies of central neurons, see the review article by Salmoiraghi and Stefanis.[12] The latter review is particularly valuable for its detailed presentation of precautions for interpretation of the results of microelectrophoresis and for its evaluation of the evidence provided by this technique for chemical transmitters in the CNS.

act at the same or very similar sites which are considered to be two-point (two-charge) receptors.[47] In this view the difference in the resultant permeability changes would be associated with the additional carboxyl group of the excitatory amino acid, glutamate. This free anionic, α-carboxyl group of glutamate would favor additional cationic transport across the membrane and could be responsible for sodium ions entering the cell. The study of Bradford and McIlwain provides some experimental support for this view.

2. *Inhibitory Effects of GABA*

The early history of factor I, the inhibitory factor discovered byFlorey in mammalian brain extracts, is reviewed by Elliott and Jasper.[7]* The identity of factor I and GABA were subsequently definitely established.[8,48,49] The inhibitory effects of GABA on central neurons have been intensively studied, and have been effectively demonstrated in numerous laboratories. For example, the rapid and reversible depressant effects of GABA on electrical activity in the cerebral cortex were demonstrated by Iwama and Jasper[50] and by Purpura *et al.*[51] A depressant action of GABA on cortical neurons was also demonstrated in the rat[52] and in the surgically isolated (main circulation intact) cerebral cortex in the dog.[53] By the use of isotopically labeled potassium, Brinley *et al.* provided evidence for some depolarizing effect of GABA on the cortex by showing a small but significant increase in the rate of potassium outflux from the cortical surface in the rabbit.[54] Bindman *et al.* showed that if measures were taken to ensure that the whole thickness of the gray matter was exposed to GABA, then almost all electrical activity in the cortex was abolished.[55] This confirmed that the effect of GABA is a nonspecific blocking of all neuronal activity, and that the positive waves at the surface after GABA are not necessarily due to hyperpolarization.[55]

More recently, GABA has been shown to be a potent depressant of the cortical neuronal firing in the pyriform cortex[40] and in the pericruciate cortex[41,56,57] of the cat. Krnjević and his co-workers found that GABA mimicked the synaptic inhibitory effect in hyperpolarizing and in lowering the resistance of the membrane of pericruciate cortical neurons in the cat.[56,57] Both the inhibitory synaptic potential and the action of GABA were found to be reversible by chloride ion applied iontophoretically. Also, the effects of GABA were observed only when it was applied extraneuronally; intracellular injections were ineffective. Krnjević thinks, on the basis of this evidence, that GABA could be the main inhibitory trans-

* In their very thorough review already cited,[10] Curtis and Watkins make note (p. 348) of the historical significance of factor I as follows: "Due recognition should be made of the historical importance of 'factor I,' from the first descriptions of which arose much of the increased interest in the pharmacology of amino acids responsible for the work summarized in this review."

mitter in the cortex.* In this connection, recent demonstrations of the presence of interneurons in the cerebral cortex are relevant to the analysis of the effects of GABA and to the problem of convulsions. From a study of the effects of strychnine and homocysteic acid, administered by microelectrophoresis in the postcruciate gyrus of the cat, it was concluded that the excitation of cells other than pyramidal tract neurons was evidently required for producing depression; it was suggested that these cells were inhibitory and excitatory interneurons.[(70)] Krnjević *et al.* found that sodium L-glutamate, administered iontophoretically to cat, rabbit, or monkey neurons, caused an excitation which could be inhibited by single cortical shocks. Thus, they considered that there existed a widespread system of intracortical interneurons which can be activated by direct or indirect stimulation of the cortex to cause a powerful inhibition, different from spinal inhibition. Various drugs, cholinergic and adrenergic antagonists, anesthetics, and convulsants, administered iontophoretically or intravenously, had no effects on this cortical inhibition, but GABA depressed cortical neurons. None of the pharmacological agents tested interfered with this depressant action of GABA.[(71–74)] In the cat, GABA depressed both spontaneous firing and the neuronal firing induced by L-glutamate in the hippocampus,[(39)] and blocked all forms of spontaneous and evoked activity in the cuneate nucleus.[(42)]

Recent studies of inhibition in the cerebellum have provided a strong case for GABA being a transmitter there. Ito and his colleagues found that stimulation of the anterior lobe of the cerebellum produced inhibitory postsynaptic potentials, recorded intracellularly from ipsilateral Deiters' neurons in the cat. There was a close temporal correlation between activation of the Purkinje cells by the climbing fiber responses and the inhibition

* The question of whether GABA is an inhibitory transmitter in the mammalian CNS is still the subject of considerable controversy. Curtis,[(41,10)] and also Eccles,[(58)] have argued against concluding that GABA is the inhibitory transmitter on the grounds that it has no hyperpolarizing effects on spinal motoneurones or cortical neurons and that its depressant action is not suppressed by strychnine. The opposite view is held by Krnjević[(56,57)] and by Salmoiraghi and Stefanis.[(12)] There is also a considerable body of evidence for the view that GABA is the inhibitory transmitter in crustaceans. There is not space to review this, but see for example the studies of Dudel[(59,60)] and the Takeuchis[(61–64)] in which GABA has been shown to mimic the action of the inhibitory transmitter at the crayfish neuromuscular junction; the studies of the Harvard group, Kravitz and co-workers, should also be mentioned in this connection. They have shown that GABA occurs only in inhibitory axons in the lobster and that its presence there is due to the enzyme, glutamic decarboxylase, the activity of which is eleven times higher in inhibitory axons than in excitatory axons.[(65)] The high level of GABA is stabilized by high glutamic decarboxylase activity and its inhibition by GABA.[(66)] Further studies of the physiological organization of neurons in a lobster ganglion have confirmed the axon work; in this ganglion, there are two populations of cell bodies, inhibitory cells with high GABA and excitatory cells with low GABA (the glutamate content is approximately the same in the two types).[(67)] Another important finding was that of the release of GABA in response to inhibitory nerve stimulation demonstrated in the lobster neuromuscular junction.[(68)] GABA may also be a neural transmitter in the cockroach.[(69)]

of the Deiters' neurons. Thus Deiters' neurons receive abundant inhibitory synapses directly from the cerebellar Purkinje cells.[75,76] GABA depressed both inhibitory and excitatory postsynaptic potentials, often blocked the spike potentials while it increased membrane conductance, and caused membrane hyperpolarization. Such evidence is consistent with the conclusion that GABA is a transmitter in this system.[77]

Finally, in a study of the effects of GABA and L-glutamate, applied by iontophoresis, on the sensitivity of the soma and lateral dendrite membrane of the Mauthner neuron in goldfish, Diamond found that GABA inhibited by increasing the membrane conductance but had little effect on the resting potential, while glutamate depolarized and its effects could result in impulse firing.[44] Neither GABA nor glutamate was effective when released intracellularly. Diamond had previously reported a difference in sensitivity to GABA between the Mauthner cell body and distant (lateral) dendritic regions.[78] However, with improved and more refined techniques, this conclusion has been shown to be erroneous; even the distal region of the lateral dendrite, which receives excitatory club endings, is highly sensitive to GABA. The more recent report[44] provides experimental clarification of this problem, as well as an experimental basis for the concept of dendritic remote inhibition.

3. *GABA and Convulsions*

a. Correlative Studies: Effects of GABA on Convulsive Phenomena. The inhibitory effect of GABA on central neurons as revealed by iontophoresis has to some extent been corroborated by systemic injections or by direct placement in the brain by one of several techniques. To verify the depressant effects of GABA on the CNS, researchers have endeavored to demonstrate either a depressant effect on the animal or a protective effect against pathological convulsions or experimentally induced convulsions. Intravenous or intraperitoneal GABA protected against drug-induced convulsions in monkeys.[79] Intracisternal GABA suppressed seizures induced by intracisternal L-glutamate and pyridoxal 5-phosphate.[80] Intravenous GABA produced sleep in chicks before the development of the blood–brain barrier and protected against pentylenetetrazole convulsions in 1-day-old chicks.[81] Earlier studies of Berl and Purpura and co-workers on the cat cerebral cortex, which had shown that intravenous GABA suppressed the epileptiform activity resulting from freezing lesions in the cortex,[82,83] have been confirmed. By the use of ^{14}C-labeled GABA, it was verified that penetration of free GABA into the brain in the lesion area is responsible for the suppression of the epileptiform activity.[84] GABA reverses the polarity of strychnine potentials recorded from the cortical surface of the anesthetized cat.[85] In unanesthetized cats, the effects of GABA and strychnine on responses elicited by means of direct cortical stimulation or on responses evoked either by peripheral stimulation or by single shocks in the lateral posterior nucleus of the thalamus were studied; topical GABA depressed the superficial neocortical structures which resulted in hyperexcitability of deeper structures.[86] In cats with epileptogenic lesions (aluminum oxide) in the hippocampus, it was found that epileptiform activity developed in the homologous region of the opposite hemisphere. Topical GABA inhibited the spontaneous epileptiform activity and diminished the amplitude of the surface-negative wave of responses evoked by means of stimulation to the olfactory bulb.[87] GABA also has been observed to protect against audiogenic seizures in mice, in one case if applied directly to the exposed cortex,[88] in another if administered systemically.[89]

Wood and Watson and their group in Canada have found that GABA, injected intraperitoneally, protects against convulsions induced by oxygen at high pressure (OHP).[90–92] It has been confirmed that previous intraperitoneal injections of strong solutions of GABA (molar solutions in 0.9% sodium chloride) protect against seizures induced in rats by OHP, and also by pentylenetetrazole and picrotoxin; such injections do not protect against strychnine convulsions, however.[29] It was noted above that this "protective" effect of solutions administered intraperitoneally may in fact be due to an increased seizure threshold resulting from increased serum osmolarity and brain dehydration. Since there is a minimum convulsion time below which any increase in oxygen pressure has no further effect, Paton argued that OHP convulsions are due to the poisoning of some enzyme system, which would result in either an accumulation of a convulsant substrate or a depletion of an anticonvulsant product.[93] He also concluded that the site of initiation of OHP convulsions is subcortical. The subcortical involvement was confirmed by Rucci *et al.*, who observed OHP convulsions in rats in which the cerebral cortex had been removed.[94]

Although there was a report that glutamine synthetase activity is inhibited by OHP,[95] Wood and Watson believe that OHP convulsions may involve a derangement in GABA metabolism. This view is based on the following experimental findings: (1) rats exposed to 100% oxygen at a pressure of 6 atm for 33 min either convulsed severely or mildly or had no convulsions, but all three groups showed a decrease in brain GABA relative to unexposed control animals, the decrease being most pronounced (35%) in those rats which convulsed severely[96]; (2) there was no decrease in brain GABA after exposure to air at high pressure or to 100% oxygen at ambient pressure, and brain concentrations of glutamate, aspartate, and α-amino acids were not altered[96]; (3) GABA transaminase (GABA-T) activity in rat brain homogenates was similar in air or oxygen, whereas glutamic decarboxylase (GAD) activity was significantly inhibited in the presence of pure oxygen[97]; (4) the similarities observed between OHP convulsions and thiosemicarbazide (TSC) convulsions (long latency, the protective effect of GABA injected intraperitoneally, the lack of effect on

GABA-T, the inhibition of GAD, the decrease in brain GABA*)[101]; (5) the susceptibility to OHP convulsions of various mammalian species (mouse, rat, hamster, guinea pig, rabbit) was observed to vary inversely with the stability of the brain GABA concentration under the influence of OHP.[102] Although inhibition of GAD and decreased brain GABA may play a role in OHP seizures, the mechanism is more complex, as indicated by the finding that rats with elevated brain GABA, induced by the administration of aminooxyacetic acid (AOAA), showed a susceptibility to OHP convulsions equal to that of control rats.[103] Thus, a derangement in GABA metabolism probably involves a complex acceleration of the GABA shunt pathway, compounded by a decreased rate of synthesis of GABA from glutamic acid.[102] The effect of OHP on GABA enzymes may involve the oxidation of active sulfhydryl groups[79]; lipid peroxidation may also be a factor in the effect of OHP.[104] DeFeudis and Elliott determined the GABA content of rat brain taken before and after convulsions from animals subjected to OHP with and without protection by prior injection of hyperosmotic solutions and from animals treated with convulsant drugs; they confirmed the observation of Wood and Watson that the GABA content of the brain decreases in animals that suffer convulsions during treatment with OHP. Prior intraperitoneal injections of hyperosmotic solutions prevented this decrease; hence, when it occurs it appears to be a concomitant of the convulsions induced by OHP.[105]

There is other evidence which reveals the high sensitivity of the GABA system in brain to oxygen. Hypoxia also affects GABA metabolism. Earlier, Elliott and Van Gelder had shown that hypoxia increased the rat brain GABA concentration[106]; this was confirmed in the brains of rats which had been quick-frozen in liquid nitrogen—exposure to a mixture of 5% oxygen and 95% nitrogen for 30 min caused an elevation in rat brain GABA of 35%.[49] Hypoxia induced by ligation of the carotid arteries caused significant increases in GABA in the prosencephalon (cerebral cortex, basal ganglia, and diencephalon) of young rats, particularly at 25 days of age, when the animals also showed pronounced seizures.[107] However, hypoxia induced by carotid ligation is complicated by other factors, such as restriction of the supply of glucose to the brain.

* It should be noted here that, although it seems generally accepted that thiosemicarbazide (TSC) lowers brain GABA concentration as a result of GAD inhibition, the possibility still remains that the effect of TSC on brain GABA concentration is a methodological artifact caused by the rapid postmortem increase in GABA in the control group, an increase which is partially or completely blocked by TSC.[49,98] A consideration and discussion of the importance and implications of this possibility, in relation to the anticonvulsant effects of vitamin B_6, was presented by Baxter at a recent conference on "Vitamin B_6 in Metabolism of the Nervous System," New York, November 7–9, 1968.[99] It is interesting that the paper of Killam and Bain,[100] which is often referred to as the source of the finding that TSC lowers brain GABA, does not in fact include an experimental demonstration of this finding, although it does report inhibition of GAD by TSC and a decrease in rat brain GABA induced by semicarbazide.

Recently, Wood and Watson showed that exposure to 8% oxygen for 10 min is sufficient to elevate brain GABA significantly and that hypoxia raises brain GABA in the mouse, hamster, rat, guinea pig, and rabbit, with maximum elevation occurring at 60 min and the GABA falling off gradually thereafter.[108,109] Wood has suggested that GABA possibly plays a homeostatic role in brain metabolism in the hypoxic condition.[108]

Another complicating factor in the interpretation of the effects of OHP is the pulmonary damage which, however, is not unique to OHP convulsions. It is also found to accompany convulsions induced by pentylenetetrazole, picrotoxin, etc. Bean *et al.* concluded that the pulmonary damage is due to neuroendocrinogenic factors in the autonomic component of the seizures. This view is based on evidence from a study of OHP convulsions in rats in which it was observed that anesthesia prevented both convulsions and pulmonary damage, and that sympatholytic and antiepinephrine drugs prevented the pulmonary damage but not the somatic component of the seizures (skeletal muscular reaction), whereas curarization prevented the somatic component of the seizures but not the pulmonary damage.[110,111]

b. Involvement of the Glutamate-GABA System in Convulsive Mechanisms: Metabolic Considerations. Although there may be some sort of "derangement" or alteration of GABA metabolism associated with convulsive phenomena, the question arises whether such an alteration is integrally related to the basic mechanism of the convulsive process. Experimental studies relevant to this problem may be arbitrarily organized on the basis of the following central areas: (1) GABA synthesis (GAD); (2) GABA-T activity; (3) pyridoxal phosphate; (4) compartmentation of glutamate-GABA metabolism.

(*i*) *Glutamic acid decarboxylase* (*GAD*). According to Roberts and Eidelberg, steady-state concentrations of GABA in the brain may be regulated by the activity of GAD.[112] It was seen above that inhibition of GAD is associated with OHP convulsions. The activity of mouse brain GAD in the presence of pure oxygen was shown to be reduced to 50% of its normal activity in air, probably because of the oxidation of the sulfhydryl groups, the presence of which in the enzyme is well established.[113] Utley observed that anthranilic hydroxamic acid (AHA), given daily to rats, produced convulsions on the third day, when the brain GABA concentration was reduced to 56% of normal.[114] Inhibition of GAD, both *in vivo* and *in vitro*, was observed; both seizures and the reduction in brain GABA were reversed by intravenous pyridoxal. Such examples of the use of drugs which affect GAD in attempts to demonstrate a correlation between GABA production, or GABA concentration, and convulsions could be multiplied. There have also been attempts to demonstrate changes in GABA concentration in different parts of the brain before, during, or after convulsions—for example, picrotoxin seizures in mice and rabbits[115] and in monkeys[116]—with some success, but the complexities of such an approach, of interpretation of the resulting data, make any definitive conclusions hazardous. These complexities are well illustrated by the studies of glutamic acid hydrazide (GAH) by Tapia and his co-workers.[117-121] GAH was found to inhibit both GAD and GABA-T *in vitro* (mouse brain homogenates), and brain homogenates from GAH-treated mice had reduced GABA-T and GAD activity.[118] GAH also caused convulsions in mice, as well as a two- to fourfold *elevation* of brain GABA concentration. A single convulsant dose was shown to produce maximum inhibition of GAD at the time when the convulsion began; at this point, brain GABA was elevated and GABA-T was also inhibited.[119] Similar effects resulted from the simultaneous administration of GAH and pyridoxal phosphate or from the administration of the hydrazone of GAH and pyridoxal phosphate (PPGH), namely convulsions, a *reduction* of brain GABA concentration, and inhibition of GAD.[117-120] Although the inhibition of GAD *in vitro* is reversed by pyridoxal phosphate, PPGH stimulates the GAD activity of brain homogenates.[120] Tapia *et al.* concluded that GAD inhibition, independently of the total GABA concentration in brain, and the rate and site of GABA synthesis may be important factors in the production of some types of convulsions.[120,121] Also, the effects of GAH and pyridoxal phosphate, administered simultaneously, are due to PPGH, formed *in vivo*, and these effects may be mediated by the inhibition of pyridoxal kinase in brain.[120] The interaction of hydrazides with the pyridoxal phosphate system *in vivo* is discussed in fuller detail below. Thus, there is an experimental basis for the view that GAD does indeed fulfill a regulatory role with regard to neuronal activity. In addition to the above, there is also the experimental fact that mouse brain GAD is inhibited by chloride ions in the range of concentration comparable to its extracellular and intracellular concentrations *in vivo*. It has been suggested that the variation in chloride ion content at nerve endings may regulate GAD activity and thus determine the proportions of glutamate and GABA which might be released on neuronal stimulation.[122] It is interesting that the mouse brain enzyme is different from bacterial GAD[123] in this regard; the bacterial enzyme is activated by chloride. GAD may play a regulatory role in the lobster nervous system also: while glutamate, α-ketoglutarate, and GABA-T are present in similar amounts in both inhibitory and excitatory axons in the lobster, the GAD activity in inhibitory axons is about eleven times that in excitatory axons.[124] The lobster enzyme is different from the mammalian in that it is inhibited by its product, in its absolute requirement for potassium ions, and in being a soluble enzyme.[125] The combination of high GAD activity and product inhibition may provide a means of stabilizing the high GABA concentration found in inhibitory axons in the lobster.[124]

(*ii*) *GABA transaminase* (*GABA-T*). A central point which arose early in the history of the relationship of brain GABA to convulsions was the question whether inhibitors of GABA-T (agents which would elevate brain GABA concentrations) have anticonvulsant effects. It has been known for some years that convulsions occur in animals with normal brain concentrations of GABA,[106] and even with elevated brain GABA.[126] Yet certain inhibitors of GABA-T do have anticonvulsant properties. For example, pyrrolidone derivatives are effective anticonvulsants in rats and mice, and also inhibit the GABA-T activity of brain homogenates.[127] The bulk of the work on this question has centered around the interesting substance, aminooxyacetic acid (AOAA). This drug is a competitive inhibitor of GABA-T,[128] and hence produces a marked elevation of brain GABA.[128,129] It also has anticonvulsant action against thiosemicarbazide convulsions in rats, mice, and cats,[130] and in dogs.[23] The protective activity of a series of aminooxy acids against thiosemicarbazide (TSC) convulsions was found to parallel their potency in inhibiting GABA-T *in vivo*, and aminooxyacetic acid proved to be the most potent in this regard.[131] However, the anticonvulsant properties of AOAA did not appear to be due (at least not simply) to the increased brain GABA concentration, since optimal protection against TSC was observed in the first 3 hr after administration of the drug, whereas the maximum elevation in brain GABA came at 6–8 hr after administration.[130] Furthermore, in a reevaluation of brain GABA elevation after AOAA in relation to seizure susceptibility in the mouse, Kuriyama *et al.* observed a decreased electroshock seizure incidence during the first 90 min after AOAA, but when the GABA increase was maximal (about 6 hr), the seizure incidence was back to normal.[132] The same was true for the incidence of convulsions induced by strychnine, pentylenetetrazole, or picrotoxin. There was no increase in brain glutamate after AOAA. As Kuriyama *et al.* suggested, the GABA system may not be the only inhibitory system in the CNS, and it may be the balance between inhibitory and excitatory systems which is important in the control of neuronal activity.

GABA-T activity has been quite extensively studied in mammalian brain,[133–137] and in the lobster nervous system.[138] Related to these studies are those of succinic semialdehyde dehydrogenase in mammalian brain,[139–142] and in the lobster.[143] These enzymes complete the "GABA shunt" pathway from glutamate to GABA to succinate which may serve as auxiliary to the normal cycle of oxidative metabolism. The question of the contribution of the metabolic flux through this pathway to the total oxidative metabolism in brain has been controversial.* Recent studies, however, have established that the contribution of the GABA shunt to the overall oxidation of glucose or the turnover through the tricarboxylic acid cycle is relatively minor, not more than about 10%, under normal conditions.[148–152] Whether this contribution is increased in convulsive

* Although the original report of this contribution as being as much as 40%[144] has not been verified, the idea is still current in the literature—see, for example, references (135), (145), and (147). More correct statements of the situation can also be found, as for example in reference (146), namely that about 40% of the oxidation of glutamate may proceed through the GABA shunt pathway.

conditions is a question still unanswered; the experimental study of this question would doubtless provide some important clues to the metabolic aspects of the moderating function of GABA in the CNS.

(*iii*) *Pyridoxal phosphate.* The history of vitamin B_6 deficiency and the involvement of vitamin B_6 in GABA metabolism and in convulsive mechanisms is complex, and not only because of the fact that pyridoxal phosphate is the coenzyme for so many enzymes in the brain and indeed throughout the body. The pharmacological aspects of vitamin B_6 have been thoroughly reviewed by Holtz and Palm,[5] and Roberts *et al.* have provided a general overview of the relationships between vitamin B_6 and GABA, as well as general cautions with regard to any attempted correlations of the physiological parameters of convulsions with brain GABA concentrations.[153] There are a few salient facts which are central to the matter of the involvement of the various metabolic forms of vitamin B_6 in convulsions, and whether this involvement implicates the GABA system in convulsive mechanisms. First, there is the fact that vitamin B_6 deficiency in the mouse causes a significant decrease in brain GABA (along with certain other changes in brain amino acids and urea), a change which can be reversed by the administration of the vitamin.[154] Second, convulsions in the rat (whether induced by hydrazides, pentylenetetrazole, or electroshock) are associated with a decrease in brain pyridoxal phosphate of 50–60%.[155] Third, hydrazones of pyridoxal phosphate have coenzyme activity for GAD comparable to that of pyridoxal phosphate itself, at least *in vitro*[120,156–158]; when added to brain homogenates, they stimulate formation of GABA. Fourth, certain hydrazones of pyridoxal phosphate are convulsants *in vivo*. Thus, the hydrazone of isoniazid and pyridoxal 5-phosphate causes convulsions in the cat and inhibits both GAD and GABA-T *in vivo*[159]; other examples of convulsant hydrazones are the γ-glutamyl hydrazones,[120] and the hydrazones of 1,1-dimethylhydrazine and pyridoxal or pyridoxal phosphate.[160] Finally, pyridoxal hydrazones are potent inhibitors of brain pyridoxal kinase[161,162]; the inhibitory effects of certain hydrazides on glutamic decarboxylase *in vivo* may be due to the effective removal of coenzyme from GAD apoenzyme through inhibition of pyridoxal kinase (see below).

Consideration of the above experimental facts reveals the complexity of the interrelationship between GABA metabolism and convulsions. In no series of studies has this complexity been better illustrated than in the work on the hydrazines—hydrazine itself, and its methyl derivatives, monomethyl-(MMH), 1,1-dimethyl-(UDMH) and 1,2-dimethyl-(SDMH) hydrazine. All except SDMH are convulsants. Hydrazine elevates brain GABA,[49,163–165] while all the methyl derivatives lower it.[164,165] In the rat, both hydrazine and its methyl derivatives raise brain 5-HT and norepinephrine (NE),[165] and lower brain lactate.[166] Hydrazine elevates blood lactate and pyruvate, inhibits gluconeogenesis, and produces marked hypoglycemia.[167,168] Minard and his group have studied UDMH convulsions in rats. Although the periodic convulsions induced by UDMH were accompanied by a gradual diminution of brain glycogen (increased phosphorylase activity), Minard *et al.* concluded that the glycogen concentration was not closely associated with the occurrence of convulsions.[169] Appropriate doses (intracerebral or intraperitoneal) of pyridoxine, pyridoxal, or pyridoxal phosphate are effective antidotes of UDMH convulsions.[170] Minard and Mushahwar found that the total radioactivity accumulated in the brain after intraperitoneal administration of 2-^{14}C-glucose was decreased both before and after convulsions induced by UDMH—glucose utilization in the brain and incorporation of label into brain amino acids and other metabolites were diminished.[171] Since neither the amounts nor the labeling of metabolites fluctuated in the convulsive and interictal recovery periods, it was concluded that neither the synthesis nor the amounts of brain amino acids in whole brain reflect the onset of and recovery from convulsions. Thus, at least in one important kind of drug-induced convulsive activity, this was direct experimental evidence (beyond the determination of whole brain amino acid concentrations) for what has often been referred to in such terms as "the lack of correlation between brain GABA levels and convulsions." Minard carried this line of investigation a step further and compared UDMH-treated rats with vitamin B_6-deficient rats.[172] The induction period for the first UDMH convulsion proved to be a sensitive index for B_6 deficiency in rats—it is shortened more in the vitamin deficiency state than are the periods between subsequent convulsions, and furthermore, brain pyridoxal phosphate decreases to the same minimal level in B_6-deficient and UDMH-treated rats. On the basis of these findings, Minard formulated the following hypothesis to explain the mechanism of hydrazine convulsions. Such convulsions do not occur until pyridoxal phosphate falls to a critical level within some compartment of the brain; once the first convulsion has occurred, a series of convulsions ensues. The periodicity of the subsequent convulsions is controlled by a mechanism which does not involve detectable changes in the minimum amount of pyridoxal phosphate.[172] Thus, the convulsions are not caused directly by UDMH, but rather result from a compartmental deficiency of pyridoxal phosphate and the consequent effect on this control mechanism. Available experimental evidence concerning the mechanism of convulsions induced by hydrazides, such as thiosemicarbazide, is not inconsistent with this hypothesis. It is clear that hydrazones which are known to inhibit pyridoxal kinase would certainly affect such a control mechanism in a way which could also result in convulsions. This hypothesis also indicates the importance of compartmentation in the regulation of neuronal excitability, an importance which has been suggested by other research.

(*iv*) *Compartmentation of glutamate–GABA metabolism.* In any consideration of possible alterations in the metabolism of the glutamate–GABA system in convulsive activity, it is well to remember that it may be the balance between GABA and glutamate at the appropriate site (or in the appropriate compartment) which is one of the critical factors in convulsive mechanisms.[173,174] The real question, though, is how such a balance would be maintained functionally *in vivo*. This is part of the larger question of how physiologically active amines and amino acids are "stored," or "bound" or "occluded," i.e., kept in a physiologically inactive state. The so-called "binding" of endogenous substances is related to compartmentation; indeed, they may be different manifestations of the same phenomenon, a phenomenon which may be an important factor in the neurochemical mechanisms by which glutamate and GABA moderate neuronal activity. From this point of view, compartmentation is relevant to the problem of convulsions. Here the author's intention is only to indicate some of the important lines of research which have provided evidence for compartmentation as a basis for conceptualization of convulsive mechanisms.

Experimental foundation for the concept of glutamate compartmentation in brain came from the laboratory of Professor Heinrich Waelsch. The concept was based on the fact that within a short time after intravenous administration of ^{14}C-glutamate to rats or mice, the specific activity of plasma glutamine was greater than that of either glutamate or glutamine in the brain or liver,[175] and that shortly after intracisternal administration of ^{14}C-glutamate, brain glutamine had a considerably higher specific activity than did its precursor, glutamic acid.[176] It was subsequently observed that this compartmentation apparently follows a developmental pattern in the cat,[177] and that the compartments develop at a different rate in different brain regions.[178] Compartmentation of glutamate metabolism in brain has recently been demonstrated *in vitro*, with guinea pig brain cortex slices.[179] Further evidence solidifying the case for compartmentation of the glutamate–GABA system and the citric acid cycle intermediates has come from studies of the incorporation of radioactivity into these substances from labeled glucose, both *in vivo* and *in vitro*. A thorough investigation and discussion of the factors affecting the incorporation of label from ^{14}C-glucose into brain amino acids and steady-state concentrations of brain amino acids can be found in the work of Balazs, Haslam, and Krebs.[148,180–182] On the basis of his study of the metabolism of labeled glucose in human brain, which provided evidence for the existence of a small, very active pool of glutamate in equilibrium with the large total brain glutamate pool, Sacks postulated that this small pool is *within* the citric acid cycle and that it also includes the remainder of the GABA shunt (GABA and succinic semialdehyde).[145] Thus, in this hypothetical scheme, the pathway glutamate → GABA → succinic semialdehyde was placed in the direct oxidative pathway as part of the citric acid cycle. However, the rapid labeling of amino acid pools, especially the marked incorporation of label into brain glutamate, reflects certain special features of brain metabolism which are not related to glutamate metabolism specifically. The pathway of glucose oxidation cannot be inferred from the evidence of rapid glutamate labeling along, since this can be explained by the phenomenon of exchange transamination.[148,180]

Additional evidence for the existence of metabolic compartments for the metabolism of the glutamate-GABA system in rat brain has come from the *in vivo* work of Gaitonde *et al.*[183,184] and of Cremer who studied the labeling of amino acids in brain from uniformly labeled glucose both *in vivo* and *in vitro* (cortex slices) and obtained good agreement between the two systems.[185]* This work was confirmed and extended in a study of the precursors of glutamate and aspartate in rat brain *in vivo* by O'Neal and Koeppe[187]; evidence for compartmentation of the free amino acids in sheep brain was also obtained.[188] Studies of the interconversion of glutamate and aspartate in guinea pig brain cortex slices have provided further evidence consistent with the theory of the compartmentation of brain amino acid metabolism.[189,190]

B. Amines and Convulsions

Despite rapid developments in neuropharmacology and the elaboration of its neurochemical basis in recent years, the problem of how the brain amines are involved in the control of behavior is still largely unsolved. Thus it is hardly surprising that little is understood about the role of amines in convulsive phenomena. Yet, while this has indeed been a noticeable lacuna in our knowledge, we have begun to progress toward the goal of understanding the involvement of amines in neurochemical mechanisms of convulsions. Microelectrophoretic studies have contributed significantly to the elucidation of the mechanisms of monoamine action in the brain. The effects of monoamines on central neurons have been extensively studied.[13] For example, monoamine action in the cat thalamus has been investigated,[191] and the excitatory action of 5-HT on cortical neurons, and its antagonism by LSD and 2′-(3-dimethylaminopropylthio)cinnamanilide (cinanserin), were systematically studied in both unanesthetized and anesthetized cats.[192]

By means of topical application of cholinomimetic and sympathomimetic substances to the rat somatosensory cortex, Malcom *et al.* obtained evidence for interrelationships between cholinergic and adrenergic inhibition, and offered the hypothesis of a cortical cholinergically activated noradrenergic inhibitory system.[193] While such research is helping us to understand how the amines are involved in the control of normal neuronal

* In contrast to the findings of Gaitonde *et al.* and Cremer, Minard and Mushahwar found that the specific activity of brain GABA was less than that of glutamate after administration of labeled glucose to rats. They suggested that the postulated highly active glutamate pool may be detectable only as a result of autolytic changes in the brain following decapitation, because the specific activity of GABA relative to that of glutamate was significantly higher in brains obtained from warm decapitated animals than in those from intact animals which were directly frozen in liquid nitrogen.[171,186] Since there are additional reasons for underscoring the importance of such direct quick-freezing in avoiding methodological artifacts, it is clear that data which result from experiments concerned with convulsive phenomena need to be judged in the light of what is known about the effects of direct quick-freezing.

function, direct studies of their involvement in convulsive mechanisms are now also yielding valuable information.

1. *Changes in Brain Amines Associated with Convulsions*

Exposure to OHP has been reported to lower NE in both mouse[194] and rat[195,196] brain; brain 5-HT was decreased in the mouse[194] and increased in the rat.[196] Rat brain 5-HT, histamine, and NE have been reported to be increased by two- or threefold after electroshock or pentylenetetrazole convulsions.[197] 5-HT is increased in the rat brain (especially in cortex) at a time just prior to convulsions, following treatment by hydrazine, methylhydrazine, or UDMH; SDMH also raised brain 5-HT, although it did not produce convulsions.[165] Single or repeated flurothyl convulsions did not result in any alteration of whole brain 5-HT, NE, or DA in the rat.[198] However, Kety *et al.* observed that repeated electroconvulsive shocks induced a sustained increase in the synthesis and utilization of NE in the brain of the rat.[199] Share and Melville obtained evidence in anesthetized, vagotomized cats for the conclusion that sympathetic cardiovascular changes by intraventricular picrotoxin or tyramine are mediated by brainstem norepinephrine.[200] The connection with convulsions is not clear since convulsions were observed after picrotoxin but not after tyramine. The convulsions which result from the administration of fluorinated precursors of the brain amines (*m*-fluorophenylalanine, *m*-fluorotyrosine, 5-fluorotryptophan) are apparently unrelated to amine metabolism as such but rather are due to the formation of the potent toxic compound, fluoroacetate.[201,202]

2. *Effect of Pharmacological Alteration of Amines on Seizure Susceptibility*

The idea that certain amines may be important factors in the determination of seizure susceptibility was advanced some time ago, on the basis of the experimental fact that reserpine, which releases brain amines, enhances convulsions, while MAO inhibitors, which increase the amines, protect against convulsions.[203] Reserpine shortened the time of onset of OHP convulsions in the mouse, while nialamide increased it.[196] Reserpine also is known to antagonize the anticonvulsant effect of carbonic anhydrase inhibitors and of carbon dioxide in both rats and mice.[204,205] In mice, norepinephrine appears to be required for the anticonvulsant effect of carbonic anhydrase inhibitors, since NE depletion is the first significant change associated with a reduced anticonvulsant effect, and the repletion of NE after reserpine treatment is associated with a restoration of the anticonvulsant effect; iproniazid prevents the antagonistic action of reserpine and depletion of NE but not of dopamine.[206] A series of papers on the effect of reserpine and other amine depletors on seizure susceptibility and brain amines in rats and mice led Pfeifer and Galambos to the view that NE is more important than 5-HT in the regulation of seizure susceptibility.[207-209] However, whether the effect of reserpine and MAO inhibitors on neuronal excitability or seizure susceptibility is due to alterations in the amines is still an open question.

Reserpine has been reported to decrease brain GABA in mice,[210] rats,[211,212] and monkeys,[213] although there have also been reports of no effect[214] or of an increase in the rat[215] and the mouse.[216] Since both reserpine and the convulsive hydrazines (e.g., UDMH) lower homocarnosine, a dipeptide of GABA and histidine, the question arises whether homocarnosine is a storage form of GABA and whether the effects of reserpine on seizure susceptibility somehow involve GABA storage or binding. MAO inhibitors also influence brain GABA concentration—increases in brain GABA have been observed in the rat after iproniazid[49,106] and tranylcypromine.[49] Popov, Matthies, and their group found that unsubstituted arylalkylhydrazines, particularly phenelzine (phenylethylhydrazine) and phenylpropylhydrazine, increase brain GABA and aspartate in the mouse and the rat without any change in brain glutamate.[211,217-219] The increase in GABA after phenelzine was observed in all of 11 regions of the rat brain[218]; it is probably due to inhibition of GABA-T.[220] Popov and Matthies observed no such effect with tranylcypromine, however[211]; in fact, they found that the increase in brain GABA produced by phenelzine or phenylpropylhydrazine could be prevented by pretreatment of the rats with other MAO inhibitors such as tranylcypromine.[218,219] The increase of brain GABA after AOAA was unaffected by such pretreatment. At higher doses, phenelzine inhibited 5-hydroxytryptophan decarboxylase and produced convulsions in the rat, but the occurrence of convulsions as the dose was increased did not correlate with the enzyme inhibition, at least as measured in whole brain homogenates.[221]

Other indications of a possible role of the brain amines in the control of neuronal excitability have come from the study of audiogenic seizures. For example, Schlesinger *et al.* found significant differences in brain 5-HT and NE in mice of certain seizure-susceptible strains (e.g., DBA/2J), relative to seizure-resistant mice (C57B1/6J), at the age when the seizure incidence was maximal (21 days).[222] Furthermore, they observed that agents which lower the brain amines (e.g., reserpine, α-methyltryosine, *p*-chlorophenylalanine) increased susceptibility to audiogenic seizures, while agents which increase the brain amines or GABA (e.g., 5-hydroxytryptophan, MAO inhibitors, AOAA) had a protective effect against such seizures.[223] MAO inhibitors have been reported to inhibit various components of audiogenic seizures.[224] Lehmann also observed that MAO inhibitors tend to protect against audiogenic seizures in mice, while reserpine increases their severity.[225] The involvement of brain 5-HT and NE in seizure mechanisms is being studied[226] in mutant strains of mice which are susceptible to audiogenic seizures as a result of gene substitutions on a constant genetic background. From similar studies of the neurochemical basis of seizures in these mice, data have been obtained which suggest that changes in hippocampal ATPase activity and developmental alterations of whole brain GAD activity may be involved in the seizure mechanism.[227,228] Furthermore, pharmacological manipulation of the endogenous brain GABA concentration during the early postnatal period can alter the subsequent development of seizure behavior, and such manipulation has also revealed that the biosynthesis of GAD in the mouse brain is repressible by GABA.[229]

The fact that electroshock seizure latencies are apparently well correlated with brain amine concentration in widely different strains and genera of mice, shorter latencies occurring in mice with low brain amine concentrations and longer latencies occurring in mice with higher concentrations,[230] further supports the case for amine involvement in neuronal control. The inverse relationship between pyridoxal kinase activity and the concentrations of brain NE, dopamine, and 5-HT, recently reported to exist in the rabbit brain,[231] opens another avenue which needs to be explored. Are the amines involved in controlling the synthesis of pyridoxal phosphate? Any mechanism which would control production of pyridoxal phosphate could be very important in the regulation of the complex of neurochemical changes which govern neuronal excitability.

In the light of what we do know about the compartmentation of amine and amino acid storage and metabolism in the brain, the critical factors operative in the production of neuronal hyperactivity may indeed be concerned with mechanisms of "binding," processes by which potentially active amines and amino acids are maintained at those sites or in those states in which they are either physiologically active or functionally inactive. Thus, whatever processes govern the binding or storage of these endogenous substances may be the critical element(s) in the stimulation and/or inhibition of neuronal activity at the convulsive level.

C. Binding Studies

Considered in view of the potent physiological effects on neuronal activity of certain amino acids and amines when they are in the proper state or at the proper site, these binding studies provide experimental evidence, both structural and neurochemical, for the existence of mechanisms in brain tissue for controlling the action of these endogenous substances. Such studies have helped us to understand how GABA and the amines may exert their physiological action, as well as how the brain "handles" them, i.e., prevents them from action by uptake, storage and/or metabolism. Previous studies have shown that binding may be involved in mechanisms associated with convulsions or mechanisms involved in the control of the level of activity in the nervous system. For example, Takahashi *et al.* obtained evidence which suggested that the rate of liberation of free acetylcholine from the bound form is a factor in the intensity of seizures.[232] Data obtained from a study of *ep* mice indicated that changes in the osmotically labile fraction of bound acetylcholine are responsible for the decrease or increase of acetylcholine which accompanies convulsions or anesthesia, respectively.(233) In rats killed at the time of convulsions induced by methionine sulfoximine, DeRobertis *et al.* found that there was no change in the whole brain dopamine concentration, but that the free-bound relationship was inverted.(234) On the basis of his study of convulsions produced by anthranilic hydroxamic acid in rats, Utley suggested a correlation between the level of the bound form of brain GABA and the convulsions.(114) The factors which affect the binding of amino acids by brain tissue *in vitro* have been investigated; important factors are osmolarity of the medium, temperature, *p*H, salt concentration, etc.(235) There are apparently at least two forms of bound GABA in brain—a storage form, which is preserved in salt-free sucrose media, and a sodium-dependent binding, which is observed in Ringer phosphate or saline solutions. There may be very little freely diffusible GABA in the brain *in vivo*.(236) These different states of GABA have been discussed by Elliott in relation to its physiological role.(237)

The sodium-dependent binding of GABA is the most intriguing, in the light of what has been discovered with regard to its possible physiological significance. Strasberg and Elliott suggested that the sites of this binding could be the receptor sites at which GABA is bound when it is acting physiologically.(236) Roberts and his group have studied this sodium-dependent binding extensively (see the recent review article(9) for a summary and discussion of this research and for documentation). In a further extension of this work,(238) GABA, glutamate, and GAD were found to be associated with a mouse brain synaptic vesicle fraction, and the sodium-dependent binding of GABA to the membrane of these particles was observed. It was suggested that such binding mediates the transfer of GABA across the membrane and that this applies to most of the usual brain subcellular particles, including synaptosomes and synaptic vesicles. The particles may also be able to synthesize GABA as well as to "sequester" it.(238) The characteristics of the binding of GABA by brain subcellular particles have been further explored in relation to the binding of acetylcholine, NE, and 5-HT. In a study of the electrophoretic mobility of the particles,(239) it was learned that each of these substances shows a different pattern of binding. For example, the binding of GABA to a mouse brain synaptic vesicle fraction has an absolute requirement for sodium, whereas the binding of acetylcholine is inhibited by sodium. The binding of GABA is not affected by neuraminidase, whereas the binding of acetylcholine is susceptible to the action of neuraminidase.(240)

The achievement of the successful subcellular fractionation of brain tissue by means of differential and density-gradient centrifugation techniques has provided a sound experimental basis for the concept of compartmentation and has proved the existence of subneuronal organelles for the storage and metabolism of such active endogenous substances as GABA, glutamate, and the amines. It has also given us a functional basis for our ideas of the way in which these substances moderate neuronal activity. Thus, for example, the enzymes of GABA metabolism have been localized subcellularly—glutamic decarboxylase occurs in the synaptosomes or nerve ending particles of rat and guinea pig brain, although its localization has been the subject of some controversy. Fonnum has analyzed the apparently conflicting results of several laboratories.[241] GABA transaminase is predominantly a mitochondrial enzyme[136,242,243]; however, there is a recent report of multiple forms of this enzyme.[244] Brain tryptophan hydroxylase has been reported to be localized in synaptosomes,[245] but both soluble and particulate preparations have been obtained and the properties of the two enzyme preparations vary markedly.[246] Thus, it is still unknown whether tryptophan hydroxylase is primarily a cytoplasmic enzyme, or is simply released from nerve ending particles during homogenization.

There is now morphological (electron microscopic) evidence for amine storage in synaptic vesicles of nerve endings in the brain *in vivo*.[247–249] Centrifugation studies had localized the catecholamines[250–252] and 5-HT[253–254] in brain subcellular particulates. Synaptosomal preparations have proven especially useful for binding studies and the investigation of storage and release mechanisms *in vitro*. Thus, the binding of acetylcholine has been studied in synaptosomes,[255] a high-affinity binding component which binds 5-HT to rat brain nerve ending particles has been characterized,[256] and the influence of physiochemical factors and drugs on the binding of 5-HT further investigated.[257] The use of synaptosomal preparations to study the effects of drugs (particularly pharmacological antagonists such as LSD) on 5-HT binding has been fruitful in research on the central function of amines such as 5-HT.[258,259]

Uptake mechanisms can be looked upon as ways in which the brain handles active substances prior either to their storage or to their metabolic degradation. Specific uptake mechanisms have been investigated for GABA and glutamate[173,260–264] and for the amines, in slices[265–268] and in brain synaptosomes.[269–272] The transport mechanisms involved in this uptake in brain may also be a control mechanism for cerebral metabolite levels.[273] The suggestion has been made that glial cells in the brain may function in the removal of transmitters from the extracellular space in the cortex.[274]

While binding studies help to understand interactions at receptor sites, and perhaps also at enzyme sites, uptake studies *in vitro* should contribute to the elucidation of amine storage and release mechanisms. In future research on the mechanisms which control neuronal firing, and hence those which are involved in convulsive phenomena, it is of paramount importance to distinguish between uptake, binding, and storage, and especially between uptake and binding. Binding may be looked upon merely as the end result of the transport mechanisms involved in uptake, but whether this is accurate or not is a question of strategic importance. Clarification of interactions at receptor sites and whether they are the same as interactions at enzyme sites, and also of interactions at storage sites, is badly needed.* Synaptosomal preparations should continue to be of considerable use in this regard. Synaptosomes retain both the fine structure and the transmitter content of their original terminals, so that they may serve as model neurons or miniature cells for the study of neuronal membrane function.[278,279] They are apparently able to synthesize protein[280-282]—although there is also evidence that at least some of the protein of particulate components of the nerve ending is synthesized in the nerve cell body and transported to the nerve ending by axoplasmic flow[283,284]—to incorporate glucosamine into macromolecules,[285] to transport sodium ions,[286] and possess osmotically sensitive compartments.[287] It has been suggested that the convulsant effect of picrotoxin is due to competition of the drug with 5-HT receptors.[258] Neurophysiological studies of convulsive activity have indicated the importance of feedback mechanisms[288] and of a generalized increase in excitability of a hyperexcitable neuronal pool which is normally offset by inhibitory mechanisms.[289] Such studies highlight the need for further research (1) on the nature of receptor sites and of interactions at receptor sites and their role in neuronal control mechanisms, and (2) on the real meaning of the "free" and "bound" states of endogenous amines and amino acids.

IV. CONCLUDING REMARKS

This chapter has reviewed and illustrated recent neurochemical research bearing on the question of how certain endogenous brain amino acids and amines are implicated in the neurochemical mechanisms which underlie convulsive activity. These endogenous substances are prime "candidates" for chemical transmitter (whether excitatory or inhibitory) in the brain; the evidence for some of them is strong. Transmitters would obviously be "involved" in any type of neuronal activity, including convulsions; yet there remain questions concerning the mechanisms of storage and release and of effective removal. Beyond these, and since a

* An example of the kind of problem in which the distinction and clarification of these concepts (uptake, storage, binding, receptor sites, etc.) would be immensely helpful is that raised by some recent findings of Alivisatos and his group, namely, what is the physiological significance of the aldehyde intermediates produced *in vivo* by the action of MAO upon 5-HT, NE, or dopamine or by transamination, as in the case of succinic semialdehyde? Subcellular fractions from brain incorporate radioactivity from ^{14}C-5-HT or ^{14}C-tryptamine *in vitro*; the incorporated radioactivity is in the acid-insoluble fraction, and the immediate substrates for the incorporation are indoleacetaldehydes.[275] A similar incorporation occurs also *in vivo* after intraperitoneal administration of ^{14}C-5-hydroxytryptophan to mice[275] or after intraventricular injection of ^{14}C-5-HT into the mouse brain.[276] These findings may also be considered as an additional indication of the importance of compartmental mechanisms in the control of neuronal activity. They are also intriguing in view of the recent purification of a brain aldehyde dehydrogenase, separable from succinic semialdehyde dehydrogenase, the substrates of which are the aldehydes originating from 5-HT, NE, and dopamine.[277]

direct transmitter role for any of these substances has not yet been definitely demonstrated, there are also questions of whether and by what mechanisms these amino acids and amines control or moderate neuronal activity, a function which clearly cannot be strictly limited to neurochemical mechanisms. Not only the catecholamines, 5-HT, glutamate, and GABA, but also substances which may be derived from these, as well as other possible inhibitory or excitatory endogenous substances, as yet undiscovered or unrealized, must be considered in relation to neuronal control. Furthermore, there are different types of convulsions, whether induced by drugs or by some endogenous condition or defect, each with its own characteristic mechanisms. The neurochemical classification of convulsions provided by Stone and his co-workers has been valuable in helping to delineate these mechanisms. There are of course different approaches to the experimental study of convulsions in animals. One approach involves the analysis before, during, and after convulsions of the biochemical events within the neuron and its various compartments. This approach frequently involves the determination of changes in endogenous neuronal components and provides useful information, but there is always the problem of cause and effect in regard to the relationship of these biochemical events to the convulsive mechanism. Another approach is the analysis of convulsant drug action, as for example in the case of methionine sulfoximine. This approach may result in the dissociation of the convulsant effect from the neurochemical mechanism of drug action, although such an outcome is not insignificant in itself, and new knowledge of drug effects is almost always forthcoming. A third approach, the analysis of relationships between neurochemical changes and neurophysiological events in the course of the genesis, duration, and termination of the seizure, is directed toward the elucidation of the neurochemical control mechanisms.

A basic problem with regard to the neurochemical control of neuronal activity was raised by Elliott,(8) namely, the interrelationships of brain function and the metabolism and condition of brain GABA or glutamate (or some other amino acid). The question is whether the role of GABA in the control of neuronal activity is primarily neurophysiological (release from storage or binding sites and interaction with receptor sites), or neurochemical (involving the moderation of metabolic mechanisms, perhaps in some way via the GABA shunt pathway), or both. Such a question could be raised in an analogous way for 5-HT or NE (or some other amine). We have seen that GABA and the amines have definite and potent neurophysiological effects when applied extracellularly to neuronal membranes, that GABA may bring about a suppression or inhibition of convulsive activity when administered systematically or intracerebrally, and that there are mechanisms in brain for the storage or binding, for the release,(290,291) and for the reuptake or metabolism after release of both GABA and the amines. These are primarily neurophysiological phenomena. There are also neurochemical phenomena which are more suggestive of the kind of controlling or moderating influence which would be exercised via some metabolic pathway. Certain types of convulsions are apparently due to some kind of derangement or alteration in GABA metabolism. The metabolism of both GABA and the amines seems to be very finely attuned to changes in available oxygen and carbon dioxide. There is evidence that GAD fulfills a regulatory function in relation to synaptic events, and tryptophan hydroxylase may be analogously involved. GAD is apparently of strategic importance in audiogenic seizure mechanisms in certain mouse strains; studies of tryptophan decarboxylase or aromatic amino acid decarboxylase from this point of view ought to follow. The increased turnover of NE which results from repeated electroconvulsive shocks may not be surprising. Perhaps more important is the fact that agents which alter brain amine concentration affect seizure susceptibility, but whether this effect involves the amines directly or is the indirect result of some other mechanism is also an important, still-unanswered question. Whether the observed effects on the GABA system of these agents which affect the brain amines have anything to do with the changes in seizure susceptibility must also be determined.

Certain present "needs" of research in this area stand out as important avenues to be followed in the near future. The enzymes which produce and degrade GABA and the amines in the brain may be subject to regulatory influences by the concentrations of these substances in critical neuronal compartments. Hence further studies of compartmentation, of pool size, of the mechanisms which maintain critical compartments *in vivo* are indicated. The experimental distinction of uptake and binding in relation to amine and amino acid compartments is very important. Recent turnover and pool studies which utilize radioactive tracer techniques have begun to delineate the complex mechanisms by which the brain amines are involved in drug effects, in alterations in behavior, and in the effects of stress. Studies which utilize these techniques in the analysis of convulsive phenomena should be fruitful. Another important need is the elucidation of the real role of the GABA shunt pathway in brain. Is a reduction in turnover through this pathway responsible for certain types of convulsions, as has been suggested?[98] Is there an increase in flux through this pathway before, or after, or especially during convulsions? Perhaps we might take a clue for the investigation of some intermediary metabolic function for endogenous substances like GABA or 5-HT from the fact that they occur in plants and must fulfill some strictly metabolic role, however significant, in the plant cell. We have seen something of the difficulties and complexities of research into pyridoxal phosphate and its role in convulsions. Despite much work on the relationship of vitamin B_6 to CNS function, many genuine questions remain unanswered. The role of B_6 enzymes (both those which are involved in the metabolism of vitamin B_6 itself and those for which it serves as coenzyme) in convulsive mechanisms should be a subject of continuing research. Convulsive studies point up the need for the evaluation of certain weaknesses or defects in methodology and the need for caution in the interpretation of data in relation to the situation *in vivo*. Though always with us, this problem is perhaps more acute in certain phases or areas of research on convulsive phenomena. Finally, although this review has concentrated on the glutamate-GABA system and on 5-HT and NE, it must be clear that other equally important inhibitory or excitatory substances may well exist and ought to be looked for. Studies on the relationships of the GABA-glutamate, 5-HT, or NE systems to the role of acetylcholine could also prove unusually fruitful.

V. REFERENCES

1. L. S. Wolfe and K. A. C. Elliott, *in Neurochemistry* (K. A. C. Elliott, I. H. Page, and J. H. Quastel, eds.) 2nd ed., pp. 694–727, C. C. Thomas, Springfield, Ill. (1962).
2. D. B. Tower, *Neurochemistry of Epilepsy*, C. C. Thomas, Springfield, Ill. (1960).
3. A. Kreindler, *in Progress in Brain Research*, Vol. 19, pp. 168–181, Elsevier, Amsterdam (1965).
4. J. K. Tews and W. E. Stone, *in Progress in Brain Research* (W. A. Himwich and J. P. Schade, eds.) Vol. 16, pp. 133–163, Elsevier, Amsterdam (1965).
5. P. Holtz and D. Palm, Pharmacological aspects of vitamin B_6, *Pharmacol. Rev.* **16**: 113–178 (1964).
6. E. Roberts, J. Wein, and D. G. Simonsen, *in Vitamins and Hormones, Advances in Research and Applications* (R. S. Harris, I. G. Wool, and J. A. Loraine, eds.), Vol. 22, pp. 503–559, Academic Press, New York (1964).
7. K. A. C. Elliott and H. H. Jasper, Gamma-aminobutyric acid, *Physiol. Rev.* **39**:383–406 (1959).
8. K. A. C. Elliott, γ-Aminobutyric acid and other inhibitory substances, *Brit. Med. Bull.* **21**:70–75 (1965).
9. E. Roberts and K. Kuriyama, Biochemical-physiological correlations in studies of the γ-aminobutyric acid system, *Brain Res.* **8**:1–35 (1968).
10. D. R. Curtis and J. C. Watkins, The pharmacology of amino acids related to gamma-aminobutyric acid, *Pharmacol. Rev.* **17**:347–391 (1965).
11. D. R. Curtis, The pharmacology of central and peripheral inhibition, *Pharmacol. Rev.* **15**: 333–364 (1963).
12. G. C. Salmoiraghi and C. N. Stefanis, *in International Review of Neurobiology* (C. C. Pfeiffer and J. R. Smythies, eds.), Vol. 10, pp. 1–30, Academic Press, New York (1967).
13. F. E. Bloom and N. J. Giarman, *in Annual Review of Pharmacology* (H. W. Elliott, ed.), Vol. 8, pp. 229–258, Annual Reviews, Inc., Palo Alto, California (1968).
14. F. N. Minard and R. V. Davis, The effects of electroshock on the acid-soluble phosphates of rat brain, *J. Biol. Chem.* **237**:1283–1289 (1962).
15. L. J. King, G. M. Schoepfle, O. H. Lowry, J. V. Passonneau, and S. Wilson, Effects of electrical stimulation on metabolites in brain of decapitated mice, *J. Neurochem.* **14**:613–618 (1967).
16. L. J. King, O. H. Lowry, J. V. Passonneau, and V. Venson, Effect of convulsants on energy reserves in the cerebral cortex, *J. Neurochem.* **14**:599–611 (1967).
17. B. Sacktor, J. E. Wilson, and C. G. Tiekert, Regulation of glycolysis in brain, in situ, during convulsions, *J. Biol. Chem.* **241**:5071–5075 (1966).
18. F. Plum, J. B. Posner, and B. Troy, Cerebral metabolic and circulatory responses to induced convulsions in animals, *Arch. Neurol.* **18**:1–13 (1968).

19. G. I. Mchedlishvili, D. G. Baramidze, and L. S. Nikolaishvili, Functional behaviour of pial and cortical arteries in conditions of increased metabolic demand from the cerebral cortex, *Nature* **213**:506–507 (1967).
20. W. E. Stone, J. K. Tews, and E. N. Mitchell, Chemical concomitants of convulsive activity in the cerebrum, *Neurology* **10**:241–248 (1960).
21. S. H. Carter and W. E. Stone, Effect of convulsants on brain glycogen in the mouse, *J. Neurochem.* **7**:16–19 (1961).
22. J. K. Tews, S. H. Carter, P. D. Roa, and W. E. Stone, Free amino acids and related compounds in dog brain: Post-mortem and anoxic changes, effects of ammonium chloride infusion, and levels during seizures induced by picrotoxin and by pentylenetetrazol, *J. Neurochem.* **10**:641–653 (1963).
23. P. D. Roa, J. K. Tews, and W. E. Stone, A neurochemical study of thiosemicarbazide seizures and their inhibition by aminooxyacetic acid, *Biochem. Pharmacol* **13**:477–487 (1964).
24. K. E. Whisler, J. K. Tews, and W. E. Stone, Cerebral amino acids and lipids in drug-induced status epilepticus, *J. Neurochem.* **15**:215–220 (1968).
25. A. V. Lorenzo, C. F. Barlow, and L. J. Roth, Effect of Metrazole convulsions on ^{35}S entry into cat central nervous system, *Am. J. Physiol.* **212**:1277–1287 (1967).
26. A. V. Lorenzo and C. F. Barlow, Effect of strychnine convulsions upon the entry of S^{35} sulfate into the cat central nervous system, *J. Pharmacol. Exp. Therap.* **157**: 555–564 (1967).
27. R. W. P. Cutler, A. V. Lorenzo, and C. F. Barlow, *in Progress in Brain Research* (A. Lajtha and D. H. Ford, eds.), Vol. 29, pp. 367–380, Elsevier, Amsterdam (1968).
28. A. V. Lorenzo, C. Fernandez, and L. J. Roth, Physiologically induced alteration of sulfate penetration into brain, *Arch. Neurol.* **12**:128–132 (1965).
29. F. V. DeFeudis and K. A. C. Elliott, Delay or inhibition of convulsions by intraperitoneal injections of diverse substances, *Can. J. Physiol. Pharmacol.* **45**:857–865 (1967).
30. E. F. Vastola, M. Maccario, and R. Homan, Activation of epileptogenic foci by hyperosmolarity, *Neurology* **17**:520–526 (1967).
31. G. H. Glaser, Sodium and seizures, *Epilepsia* **5**:97–111 (1964).
32. E. C. Zuckermann and G. H. Glaser, Hippocampal epileptic activity induced by localized ventricular perfusion with high-potassium cerebrospinal fluid, *Exp. Neurol.* **20**:87–110 (1968).
33. D. R. Curtis, Actions of drugs on single neurones in the spinal cord and thalamus, *Brit. Med. Bull.* **21**:5–9 (1965).
34. K. Krnjević, Actions of drugs on single neurones in the cerebral cortex, *Brit. Med. Bull.* **21**:10–14 (1965).
35. P. B. Bradley and J. H. Wolstencroft, Actions of drugs on single neurones in the brain stem, *Brit. Med. Bull.* **21**:15–18 (1965).
36. F. A. Steiner and K. Ruf, Excitatory effects of L-glutamic acid upon single unit activity in rat brain and their modification by thiosemicarbazide and pyridoxal 5′-phosphate, *Helv. Physiol. Pharmacol. Acta* **24**:181–192 (1966).
37. F. A. Steiner and K. Ruf, Interactions of L-glutamic acid, thiosemicarbazide, and pyridoxal 5-phosphate at single unit level in rat brain, *Brain Res.* **3**:214–216 (1966).
38. A. Herz and G. Gogolak, Mikroelektrophoretische Untersuchungen am Septum des Kaninchens, *Arch. Ges. Physiol.* **285**:317–330 (1965).
39. T. J. Biscoe and D. W. Straughan, Microelectrophoretic studies of neurones in the cat hippocampus, *J. Physiol.* (*London*) **183**:341–359 (1966).
40. K. F. Legge, M. Randic, and D. W. Straughan, The pharmacology of neurones in the pyriform cortex, *Brit. J. Pharmacol.* **26**:87–107 (1966).

41. J. M. Crawford and D. R. Curtis, The excitation and depression of mammalian cortical neurones by amino acids, *Brit. J. Pharmacol.* **23**:313–329 (1964).
42. A. Galindo, K. Krnjević, and S. Schwartz, Micro-iontophoretic studies on neurones in the cuneate nucleus, *J. Physiol.* (*London*) **192**:359–377 (1967).
43. D. R. Curtis and J. C. Watkins, Acidic amino acids with strong excitatory actions on mammalian neurones, *J. Physiol.* (*London*) **166**:1–14 (1963).
44. J. Diamond (with an appendix by A. F. Huxley), The activation and distribution of GABA and L-glutamate receptors on goldfish Mauthner neurones: an analysis of dendritic remote inhibition, *J. Physiol.* (*London*) **194**:669–723 (1968).
45. H. F. Bradford and H. McIlwain, Ionic basis for the depolarization of cerebral tissues by excitatory acidic amino acids, *J. Neurochem.* **13**:1163–1177 (1966).
46. J. C. Watkins, Pharmacological receptors and general permeability phenomena of cell membranes, *J. Theoret. Biol.* **9**:37–50 (1965).
47. D. R. Curtis and J. C. Watkins, The excitation and depression of spinal neurones by structurally related amino acids, *J. Neurochem.* **6**:117–141 (1960).
48. E. Levin, R. A. Lovell, and K. A. C. Elliott, The relation of gamma-aminobutyric acid to Factor I in brain extracts, *J. Neurochem.* **7**:147–154 (1961).
49. R. A. Lovell and K. A. C. Elliott, The γ-aminobutyric acid and Factor I content of brain, *J. Neurochem.* **10**:479–488 (1963).
50. K. Iwama and H. H. Jasper, The action of gamma-aminobutyric acid upon cortical electrical activity in the cat, *J. Physiol.* (*London*) **138**:365–380 (1957).
51. D. P. Purpura, M. Giardo, and H. Grundfest, Selective blockade of excitatory synapses in the cat brain by γ-aminobutyric acid, *Science* **125**:1200–1202 (1957).
52. J. H. Mahnke and A. A. Ward, The effects of γ-aminobutyric acid on evoked potentials, *Exp. Neurol.* **2**:311–323 (1960).
53. R. H. Rech and E. F. Domino, Effects of gamma-aminobutyric acid on chemically and electrically evoked activity in the isolated cerebral cortex of the dog, *J. Pharmacol. Exp. Therap.* **130**:59–67 (1960).
54. F. J. Brinley, E. R. Kandel, and W. H. Marshall, Effect of gamma-aminobutyric acid (GABA) on K^{42} outflux from rabbit cortex, *J. Neurophysiol.* **23**:237–245 (1960).
55. L. J. Bindman, O. C. J. Lippold, and J. W. T. Redfearn, The nonselective blocking action of γ-aminobutyric acid on the sensory cerebral cortex of the rat, *J. Physiol.* (*London*) **162**: 105–120 (1962).
56. K. Krnjević and S. Schwartz, Is γ-aminobutyric acid an inhibitory transmitter? *Nature* **211**:1372–1374 (1966).
57. K. Krnjević and S. Schwartz, The action of γ-aminobutyric acid on cortical neurons, *Exp. Brain Res.* **3**:320–336 (1967).
58. J. C. Eccles, *The Physiology of Synapses*, Academic Press, New York (1964).
59. J. Dudel, Presynaptic and postsynaptic effects of inhibitory drugs on the crayfish neuromuscular junction, *Arch. Ges. Physiol.* **283**:104–118 (1965).
60. J. Dudel, The action of inhibitory drugs on nerve terminals in crayfish muscle, *Arch. Ges. Physiol.* **284**:81–94 (1965).
61. A. Takeuchi and N. Takeuchi, Localized action of gamma-aminobutyric acid on crayfish muscle, *J. Physiol.* (*London*) **177**:225–238 (1965).
62. A. Takeuchi and N. Takeuchi, The inhibitory action of γ-aminobutyric acid on neuromuscular transmission in the crayfish, *J. Physiol.* (*London*) **183**:418–432 (1966).
63. A. Takeuchi and N. Takeuchi, The permeability of the presynaptic terminal of the crayfish neuromuscular junction during synaptic inhibition and the action of γ-aminobutyric acid, *J. Physiol.* (*London*) **183**:433–449 (1966).

64. A. Takeuchi and N. Takeuchi, Electrophysiological studies of the action of GABA on the synaptic membrane, *Fed. Proc.* **26**:1633–1638 (1967).
65. E. A. Kravitz, P. B. Molinoff, and Z. W. Hall, A comparison of the enzymes and substrates of γ-aminobutyric acid metabolism in lobster excitatory and inhibitory axons, *Proc. Natl. Acad. Sci. U.S.* **54**:778–782 (1965).
66. E. A. Kravitz and D. D. Potter, A further study of the distribution of γ-aminobutyric acid between excitatory and inhibitory axons of the lobster, *J. Neurochem.* **12**:323–328 (1965).
67. M. Otsuka, E. A. Kravitz, and D. D. Potter, Physiological and chemical architecture of a lobster ganglion with particular reference to gamma-aminobutyrate and glutamate, *J. Neurophysiol.* **30**:725–752 (1967).
68. M. Otsuka, L. L. Iversen, Z. W. Hall, and E. A. Kravitz, Release of gamma-aminobutyric acid from inhibitory nerves of lobster, *Proc. Natl. Acad. Sci.* **56**:1110–1115 (1966).
69. G. A. Kerkut and R. J. Walter, The effect of L-glutamate, acetylcholine and γ-aminobutyric acid on the miniature end-plate potentials and contractures of the coxal muscles of the cockroach, *Periplaneta Americana*, *Comp. Biochem. Physiol.* **17**:435–454 (1966).
70. T. J. Biscoe and D. R. Curtis, Strychnine and cortical inhibition, *Nature* **214**:914–915 (1967).
71. K. Krnjević, M. Randic, and D. W. Straughan, Cortical inhibition, *Nature* **201**:1294–1296 (1964).
72. K. Krnjević, M. Randic, and D. W. Straughan, An inhibitory process in the cerebral cortex, *J. Physiol.* (*London*) **184**:16–48 (1966).
73. K. Krnjević, M. Randic, and D. W. Straughan, Nature of a cortical inhibitory process, *J. Physiol.* (*London*) **184**:49–77 (1966).
74. K. Krnjević, M. Randic, and D. W. Straughan, Pharmacology of cortical inhibition, *J. Physiol.* (*London*) **184**:78–105 (1966).
75. M. Ito and M. Yoshida, The origin of cerebellar-induced inhibition of Deiters neurones. I. Monosynaptic initiation of the inhibitory postsynaptic potentials, *Exp. Brain Res.* **2**:330–349 (1966).
76. M. Ito, K. Obata, and R. Ochi, The origin of cerebellar-induced inhibition of Deiters neurones. II. Temporal correlation between the trans-synaptic activation of Purkinje cells and the inhibition of Deiters neurones, *Exp. Brain Res.* **2**:350–364 (1966).
77. K. Obata, M. Ito, R. Ochi, and N. Sato, Pharmacological properties of the postsynaptic inhibition by Purkinje cell axons and the action of γ-aminobutyric acid on Deiters neurones, *Exp. Brain Res.* **4**:43–57 (1967).
78. J. Diamond, Variation in the sensitivity to gamma-aminobutyric acid of different regions of the Mauthner neurone, *Nature* **199**:773–775 (1963).
79. L. M. Kopeloff and J. G. Chusid, Pyridoxine and GABA as antagonists to drug-induced convulsions in monkeys, *J. Appl. Physiol.* **20**:1337–1340 (1965).
80. P. Wiechert and A. Herbst, Provocation of cerebral seizures by derangement of the natural balance between glutamic acid and γ-aminobutyric acid, *J. Neurochem.* **13**:59–64 (1966).
81. S. Kobrin and J. Seifter, ω-Amino acids and various biogenic amines as antagonists to pentylenetetrazole, *J. Pharmacol. Exp. Therap.* **154**:646–651 (1966),
82. S. Berl, D. P. Purpura, M. Girado, and H. Waelsch, Amino acid metabolism in epileptogenic and non-epileptogenic lesions of the neocortex (cat), *J. Neurochem.* **4**:311–317 (1959).

83. S. Berl, G. Takagaki, and D. P. Purpura, Metabolic and pharmacological effects of injected amino acids and ammonia on cortical epileptogenic lesions, *J. Neurochem.* **7**:198–209 (1961).
84. P. Strasberg, K. Krnjević, S. Schwartz, and K. A. C. Elliott, Penetration of blood-brain barrier by γ-aminobutyric acid at sites of freezing, *J. Neurochem.* **14**:755–760 (1967).
85. O. Feher, P. Halasz, and F. Mechler, The influence of γ-aminobutyric acid (GABA) on seizure potentials, *Epilepsia* **6**:47–53 (1965).
86. E. Crighel, The effects of GABA and strychnine on neocortical structures involved in the onset mechanism of epileptic activity, *Epilepsia* **7**:283–290 (1966).
87. R. Guerrero-Figueroa, F. DeBalbian Verster, A. Barros, and R. G. Heath, Cholinergic mechanism in subcortical mirror focus and effects of topical application of γ-aminobutyric acid and acetylcholine, *Epilepsia* **5**:140–155 (1964).
88. E. Ballantine, Effect of γ-aminobutyric acid on audiogenic seizure, *Colloq. Intern. Centre. Natl. Rech. Sci.* (*Paris*) No. **112**:447–451 (1963).
89. J. M. Pasquini, J. R. Salomone, and C. J. Gomez, Amino acid changes in the mouse brain during audiogenic seizures and recovery, *Exp. Neurol.* **21**:245–256 (1968).
90. J. D. Wood, W. J. Watson, and F. M. Clydesdale, γ-Aminobutyric acid and oxygen poisoning, *J. Neurochem.* **10**:625–633 (1963).
91. J. D. Wood and W. J. Watson, Molecular structure-activity relations of compounds protecting rats against oxygen poisoning, *Can. J. Physiol. Pharmacol.* **42**:641–646 (1964).
92. J. D. Wood, N. E. Stacy, and W. J. Watson, Pulmonary and central nervous system damage in rats exposed to hyperbaric oxygen and protection therefrom by gamma-aminobutyric acid, *Can. J. Physiol. Pharmacol.* **43**:405–410 (1965).
93. W. D. M. Paton, Experiments on the convulsant and anesthetic effects of oxygen, *Brit. J. Pharmacol.* **29**:350–366 (1967).
94. F. S. Rucci, M. L. Giretti, and M. LaRocca, Changes in electrical activity of the cerebral cortex and of some subcortical centers in hyperbaric oxygen, *Electroencephalog. Clin. Neurophysiol.* **22**:231–238 (1967).
95. Z. S. Gershenovich, A. A. Krichevskaya, and A. Kolousek, The effect of raised oxygen pressure and of methionine sulphoximine on the glutamine synthetase activity of rat brain, *J. Neurochem.* **10**:79–82 (1963).
96. J. D. Wood and W. J. Watson, γ-Aminobutyric acid levels in the brain of rats exposed to oxygen at high pressures, *Can. J. Biochem. Physiol.* **41**:1907–1913 (1963).
97. J. D. Wood and W. J. Watson, The effect of oxygen on glutamic acid decarboxylase and γ-aminobutyric-α-ketoglutaric acid transaminase activities in rat brain homogenates, *Can. J. Physiol. Pharmacol.* **42**:277–279 (1964).
98. H. Balzer, P. Holtz, and D. Palm, Untersuchungen uber die biochemischen Grundlagen der konvulsiven Wirkung von Hydraziden, *Arch. Exp. Pathol. Pharmakol.* **239**:520–552 (1960).
99. C. F. Baxter, Changes in gamma-aminobutyric-acid-shunt enzymes and substrates after administration of carbonyl reagents and Vitamin B_6 *in vivo*: an apparent discrepancy in assay techniques, *Ann. N.Y. Acad. Sci.* **166**:267–280 (1969).
100. K. F. Killam and J. A. Bain, Convulsant hydrazides I: *In vitro* and *in vivo* inhibition of vitamin B_6 enzymes by convulsant hydrazides, *J. Pharmacol. Exp. Therap.* **119**:255–262 (1957).
101. J. D. Wood, W. J. Watson, and N. E. Stacey, A comparative study of hyperbaric oxygen-induced and drug-induced convulsions with particular reference to γ-aminobutyric acid metabolism, *J. Neurochem.* **13**:361–370 (1966).

102. J. D. Wood, W. J. Watson, and A. J. Ducker, Oxygen poisoning in various mammalian species and the possible role of gamma-aminobutyric acid metabolism, *J. Neurochem.* **14**:1067–1074 (1967).
103. J. D. Wood and W. J. Watson, The effect of aminooxyacetic acid and hydroxylamine on rats breathing oxygen at high pressures, *J. Neurochem.* **12**:663–669 (1965).
104. J. D. Wood and W. J. Watson, Lipid peroxidation (thiobarbituric acid reacting material) and enzyme inhibition in rat brain homogenates exposed to oxygen at high pressure, *Can. J. Physiol. Pharmacol.* **45**:752–755 (1967).
105. F. V. DeFeudis and K. A. C. Elliott, Convulsions and the γ-aminobutyric acid content of rat brain, *Can. J. Physiol. Pharmacol.* **46**:803–804 (1968).
106. K. A. C. Elliott and N. M. Van Gelder, The state of Factor I in rat brain: the effects of metabolic conditions and drugs, *J. Physiol.* (*London*) **153**:423–432 (1960).
107. A. R. Dravid and L. Jilek, Influence of stagnant hypoxia (oligaemia) on some free amino acids in rat brain during ontogeny, *J. Neurochem.* **12**:837–843 (1965).
108. J. D. Wood, A possible role for γ-aminobutyric acid in the homeostatic control of brain metabolism under conditions of hypoxia, *Exp. Brain Res.* **4**:81–84 (1967).
109. J. D. Wood, W. J. Watson, and A. J. Ducker, The effect of hypoxia on brain γ-aminobutyric acid levels, *J. Neurochem.* **15**:603–608 (1968).
110. J. W. Bean and D. Zee, Influence of anesthesia and CO_2 on CNS and pulmonary effects of O_2 at high pressure, *J. Appl. Physiol.* **21**:521–526 (1966).
111. J. W. Bean, D. Zee, and B. Thom, Pulmonary changes with convulsions induced by drugs and oxygen at high pressure, *J. Appl. Physiol.* **21**:865–872 (1966).
112. E. Roberts and E. Eidelberg, *in International Review of Neurobiology* (C. C. Pfeiffer and J. R. Smythies, eds.), Vol. 2, pp. 279–332, Academic Press, New York (1960).
113. E. Roberts and D. G. Simonsen, Some properties of glutamic decarboxylase in mouse brain, *Biochem. Pharmacol.* **12**:113 134 (1963).
114. J. D. Utley, The effects of anthranilic hydroxamic acid on rat behaviour and rat brain γ-aminobutyric acid, norepinephrine, and 5-hydroxytryptamine concentrations, *J. Neurochem.* **10**:423–432 (1963).
115. S. Saito and Y. Tokunaga, Some correlations between picrotoxin-induced seizures and γ-aminobutyric acid in animal brain. *J. Pharmacol. Exp. Therap.* **157**:546–554 (1967).
116. I. A. Sytinskii and Nguyen-Thi-Thinh, The distribution of γ-aminobutyric acid in the monkey brain during picrotoxin-induced seizures, *J. Neurochem.* **11**:551–556 (1964).
117. G. H. Massieu, R. Tapia, H. Pasantes, and B. G. Ortega, Convulsant effect of L-glutamic acid-γ-hydrazide by simultaneous treatment with pyridoxal phosphate, *Biochem. Pharmacol.* **13**:118–120 (1964).
118. R. Tapia, H. Pasantes, B. G. Ortega, and G. H. Massieu, Effects *in vivo* and *in vitro* of L-glutamic-γ-hydrazide on metabolism of some free amino acids in brain and liver, *Biochem. Pharmacol.* **15**:1831–1845 (1966).
119. R. Tapia, H. Pasantes, M. Perez de la Mora, B. G. Ortega, and G. H. Massieu, Free amino acids and glutamate decarboxylase activity in brain of mice during drug-induced convulsions, *Biochem. Pharmacol.* **16**:483–496 (1967).
120. R. Tapia, M. Perez de la Mora, and G. H. Massieu, Modifications of brain glutamate decarboxylase activity by pyridoxal phosphate-γ-glutamyl hydrazone, *Biochem. Pharmacol.* **16**:1211–1218 (1967).
121. R. Tapia and J. Awapara, Formation of γ-aminobutyric acid in brain of mice treated with L-glutamic acid-γ-hydrazide and pyridoxal phosphate-γ-glutamyl hydrazone, *Proc. Soc. Exp. Biol. Med.* **126**:218–221 (1967).

122. J. P. Susz, B. Haber, and E. Roberts, Purification and some properties of mouse brain L-glutamic decarboxylase, *Biochemistry* **5**:2870–2877 (1966).
123. R. Shukuya and G. W. Schwert, Glutamic acid decarboxylase. I. Isolation procedures and properties of the enzyme, *J. Biol. Chem.* **235**:1649–1652 (1960).
124. E. A. Kravitz, P. B. Molinoff, and Z. W. Hall, A comparison of the enzymes and substrates of γ-aminobutyric acid metabolism in lobster excitatory and inhibitory axons, *Proc. Natl. Acad. Sci. U.S.* **54**:778–782 (1965).
125. P. B. Molinoff and E. A. Kravitz, The metabolism of γ-aminobutyric acid (GABA) in the lobster nervous system—glutamic decarboxylase, *J. Neurochem.* **15**:391–409 (1968).
126. C. F. Baxter and E. Roberts, Demonstration of thiosemicarbazide-induced convulsions in rats with elevated brain levels of γ-aminobutyric acid, *Proc. Soc. Exp. Biol. Med.* **104**:426–427 (1960).
127. G. Carvajal, M. Russek, R. Tapia, and G. Massieu, Anticonvulsive action of substances designed as inhibitors of γ-aminobutyric acid-α-ketoglutaric acid transaminase, *Biochem. Pharmacol.* **13**:1059–1069 (1964).
128. D. P. Wallach, Studies on the GABA pathway—I. The inhibition of γ-aminobutyric-α-ketoglutaric acid transminase *in vitro* and *in vivo* by U-7524 (amino-oxyacetic acid, *Biochem. Pharmacol.* **5**:323–331 (1961).
129. C. F. Baxter and E. Roberts, Elevation of γ-aminobutyric acid in brain by selective inhibition of γ-aminobutyric-α-ketoglutaric acid transaminase, *J. Biol. Chem.* **236**:3287–3294 (1961).
130. J. P. DaVanzo, M. E. Grieg, and M. A. Cronin, Anticonvulsant properties of amino-oxyacetic acid, *Am. J. Physiol.* **201**:833–837 (1961).
131. E. L. Schumann, L. A. Paquette, R. V. Heinzelman, D. P. Wallach, J. P. DeVanzo, and M. E. Grieg, The synthesis and γ-aminobutyric acid transaminase inhibition of aminooxy acids and related compounds, *J. Med. Pharm. Chem.* **5**:464–477 (1962).
132. K. Kuriyama, E. Roberts, and M. Rubinstein, Elevation of γ-aminobutyric acid in brain with aminooxyacetic acid and susceptibility to convulsive seizures in mice: a quantitative reevaluation, *Biochem. Pharmacol.* **15**:221–236 (1966).
133. A. Waksman and E. Roberts, Purification and some properties of mouse brain γ-aminobutyric-α-ketoglutaric acid transaminase, *Biochemistry* **4**, 2132–2139 (1965).
134. F. N. Pitts, Jr., C. Quick, and E. Robins, The enzymic measurement of γ-aminobutyric-α-oxoglutaric transaminase, *J. Neurochem.* **12**:93–101 (1965).
135. J. J. Sheridan, K. L. Sims, and F. N. Pitts, Jr., Brain γ-aminobutyrate-α-oxoglutaric transaminase. II. Activities in 24 regions of human brain, *J. Neurochem.* **14**:571–578 (1967).
136. A. Waksman, M. K. Rubinstein, K. Kuriyama, and E. Roberts, Localization of γ-aminobutyric-α-oxoglutaric acid transaminase in mouse brain, *J. Neurochem.* **15**:351–357 (1968).
137. K. L. Sims, J. Witztum, C. Quick, and F. N. Pitts, Jr., Brain 4-aminobutyrate:2-oxoglutarate aminotransferase. Changes in the developing rat brain, *J. Neurochem.* **15**:667–672 (1968).
138. Z. W. Hall and E. A. Kravitz, The metabolism of γ-aminobutyric acid in the lobster nervous system. I. GABA-glutamate transaminase, *J. Neurochem.* **14**:45–54 (1967).
139. F. N. Pitts, Jr. and C. Quick, Brain succinate semialdehyde dehydrogenase—I. Assay and distribution, *J. Neurochem.* **12**:893–900 (1965).
140. F. N. Pitts, Jr. and C. Quick, Brain succinate semialdehyde dehydrogenase—II. Changes in the developing rat brain, *J. Neurochem.* **14**:561–570 (1967).

141. A. L. Miller and F. N. Pitts, Jr., Brain succinate semialdehyde dehydrogenase—III. Activities in twenty-four regions of human brain, *J. Neurochem.* **14**:579–584 (1967).
142. C. Kammeraat and H. Veldstra, Characterization of succinate semialdehyde dehydrogenase from rat brain, *Biochim. Biophys. Acta* **151**:1–10 (1968).
143. Z. W. Hall and E. A. Kravitz, The metabolism of γ-aminobutyric acid (GABA) in the lobster nervous system—II. Succinic semialdehyde dehydrogenase, *J. Neurochem.* **14**:55–61 (1967).
144. G. M. McKhann, R. W. Albers, L. Sokoloff, O. Mickelsen, and D. B. Tower, *in Inhibition in the Nervous System and γ-Aminobutyric Acid* (E. Roberts, ed.), pp. 169–181, Pergamon Press, New York (1960).
145. W. Sacks, Cerebral metabolism of doubly labeled glucose in humans *in vivo*, *J. Appl. Physiol.* **20**:117–130 (1965).
146. M. K. Campbell, H. R. Mahler, W. J. Moore, and S. Tewari, Protein synthesis systems from rat brain, *Biochemistry* **5**:1174–1184 (1966).
147. R. P. Schmidt and B. J. Wilder, *Epilepsy*, Contemporary Neurology Series, Vol. 2, p. 110, F. A. Davis Co., Philadelphia (1968).
148. R. J. Haslam and H. A. Krebs, The metabolism of glutamate in homogenates and slices of brain cortex, *Biochem. J.* **88**:566–578 (1963).
149. R. Balazs, D. Biesold, and K. Magyar, Properties of rat brain mitrochondria preparations. Respiratory control, *J. Neurochem.* **10**:685–708 (1963).
150. M. Bacila, A. P. Campello, C. H. M. Vianna, and D. O. Voss, The respiratory chain of rat cerebrum and cerebellum mitochondria. Respiration and oxidative phosphorylation *J. Neurochem.* **11**:231–242 (1964).
151. B. Haber, The effect of hydroxylamine and aminooxyacetic acid on the cerebral in vitro utilization of glucose, fructose, glutamic acid, and γ-aminobutyric acid, *Can. J. Biochem.* **43**:865–876 (1965).
152. Y. Machiyama, R. Balazs, and T. Julian, Oxidation of glucose through the γ-aminobutyrate pathway in brain, *Biochem. J.* **96**:68P–69P (1965).
153. E. Roberts, J. Wein, and D. G. Simonsen, *in Vitamins and Hormones, Advances in Research and Applications* (R. S. Harris, I. G. Wool, and J. A. Loraine, eds.), Vol. 22, pp. 503–559, Academic Press, New York (1964).
154. J. K. Tews and R. A. Lovell, The effect of a nutritional pyridoxine deficiency on free amino acids and related substances in mouse brain, *J. Neurochem.* **14**:1–7 (1967).
155. F. Bilodeau, Effects of stimulants of the central nervous system on the pyridoxal phosphate content of the rat brain, *J. Neurochem.* **12**:671–678 (1965).
156. P. Gonnard and S. Fenard, Cerebral glutamic acid decarboxylase and pyridoxal-5-phosphate hydrazones, *J. Neurochem.* **9**:135–142 (1962).
157. P. Gonnard, J. Duhault, and S. Fenard, Action of α-methylhydrazinodopa and of its phospho-5′-pyridoxal hydrazone on amino acid decarboxylases, *J. Neurochem.* **11**:819–824 (1964).
158. P. Gonnard and J. Duhault, Action of cyanacetyl-hydrazone of phospho 5′-pyridoxal on the cerebral glutamate decarboxylase. Modification of electrocorticographic record in the rat, *J. Neurochem.* **13**:407–412 (1966).
159. V. Bonavita, R. Guarneri, and P. Monaco, Neurophysiological and neurochemical studies with the isonicotinoylhydrazone of pyridoxal 5-phosphate, *J. Neurochem.* **11**:787–792 (1964).
160. A. Furst and W. R. Gustavson, A comparison of alkylhydrazines and their B_6-hydrazones as convulsant agents, *Proc. Soc. Exp. Biol. Med.* **124**:172–175 (1967).
161. D. B. McCormick, M. E. Gregory, and E. E. Snell, Pyridoxal phosphokinases. I. Assay, distribution, purification and properties, *J. Biol. Chem.* **236**:2076–2084 (1961).

162. D. B. McCormick and E. E. Snell, Pyridoxal phosphokinase. II. Effects of inhibitors, *J. Biol. Chem.* **236**:2076–2084 (1961).
163. E. W. Maynert and H. K. Kaji, On the relationship of brain γ-aminobutyric acid to convulsions, *J. Pharmacol. Exp. Therap.* **137**:114–121 (1962).
164. M. A. Medina, The in vivo effects of hydrazines and vitamin B_6 on metabolism of γ-aminobutyric acid, *J. Pharmacol. Exp. Therap.* **140**:133–137 (1963).
165. T. Uchida and R. D. O'Brien, The effects of hydrazines on rat brain 5-hydroxytryptamine, norepinephrine and gamma-aminobutyric acid, *Biochem. Pharmacol.* **13**:725–730 (1964).
166. P. S. Miller and R. D. O'Brien, Effects of hydrazine and alkylhydrazines on carbohydrate metabolism of rat brain, *Biochem. Pharmacol.* **13**:1096–1098 (1964).
167. S. R. Fortney, Effect of hydrazine on liver glycogen, arterial glucose, lactate, pyruvate and acid-base balance in the anesthetized dog, *J. Pharmacol. Exp. Therap.* **153**:562–568 (1966).
168. S. R. Fortney, D. A. Clark, and E. Stein, Inhibition of gluconeogenesis by hydrazine administration in rats, *J. Pharmacol. Exp. Therap.* **156**:277–284 (1967).
169. F. N. Minard, C. H. Kang, and I. K. Mushahwar, The effect of periodic convulsions induced by 1,1-dimethylhydrazine on the glycogen of rat brain, *J. Neurochem.* **12**:279–286 (1965).
170. C. L. Geake, M. L. Barth, and H. H. Cornish, Vitamin B_6 and the toxicity of 1,1-dimethylhydrazine, *Biochem. Pharmacol.* **15**:1614–1618 (1966).
171. F. N. Minard and I. K. Mushahwar, The effect of periodic convulsions induced by 1,1-dimethylhydrazine on the synthesis of rat brain metabolites from [2-^{14}C]glucose, *J. Neurochem.* **13**:1–11 (1966).
172. F. N. Minard, Relationships among pyridoxal phosphate, Vitamin B_6-deficiency, and convulsions induced by 1,1-dimethylhydrazine, *J. Neurochem.* **14**:681–692 (1967).
173. K. A. C. Elliott and N. M. Van Gelder, Occlusion and metabolism of γ-aminobutyric acid by brain tissue, *J. Neurochem.* **3**, 28–40 (1958).
174. P. Wiechert and A. Herbst, Provocation of cerebral seizures by derangement of the natural balance between glutamic acid and γ-aminobutyric acid, *J. Neurochem.* **13**:59–64 (1966).
175. A. Lajtha, S. Berl, and H. Waelsch, Amino acid and protein metabolism of the brain—IV. The metabolism of glutamic acid, *J. Neurochem.* **3**:322–332 (1959).
176. S. Berl, A. Lajtha, and H. Waelsch, Amino acid and protein metabolism—VI. Cerebral compartments of glutamic acid metabolism, *J. Neurochem.* **7**:186–197 (1961).
177. S. Berl, Compartmentation of glutamic acid metabolism in developing cerebral cortex, *J. Biol. Chem.* **240**:2047–2054 (1965).
178. S. Berl and D. P. Purpura, Regional development of glutamic acid compartmentation in immature brain, *J. Neurochem.* **13**:293–304 (1966).
179. S. Berl, W. J. Nicklas, and D. D. Clarke, Compartmentation of glutamic acid metabolism in brain slices, *J. Neurochem.* **15**:131–140 (1968).
180. R. Balazs and R. J. Haslam, Exchange transamination and the metabolism of glutamate in brain, *Biochem. J.* **94**:131–142 (1965).
181. R. Balazs, Control of glutamate oxidation in brain and liver mitochondrial systems, *Biochem. J.* **95**:497–508 (1965).
182. R. Balazs, Control of glutamate metabolism. The effect of pyruvate, *J. Neurochem.* **12**:63–76 (1965).
183. M. K. Gaitonde, D. R. Dahl, and K. A. C. Elliott, Entry of glucose carbon into amino acids of rat brain and liver *in vivo* after injection of uniformly ^{14}C-labelled glucose, *Biochem. J.* **94**:345–352 (1965).

184. M. K. Gaitonde, Rate of utilization of glucose and 'compartmentation' of α-oxoglutarate and glutamate in rat brain, *Biochem. J.* **95**:803–810 (1965).
185. J. E. Cremer, Amino acid metabolism in rat brain studied with ^{14}C-labelled glucose, *J. Neurochem.* **11**:165–185 (1964).
186. F. N. Minard and I. K. Mushahwar, Synthesis of γ-aminobutyric acid from a pool of glutamic acid in brain after decapitation, *Life Sci.* **5**, 1409–1413 (1966).
187. R. M. O'Neal and R. E. Koeppe, Precursors in vivo of glutamate, aspartate and their derivatives in rat brain, *J. Neurochem.* **13**:835–847 (1966).
188. R. M. O'Neal, R. E. Koeppe, and E. I. Williams, Utilization *in vivo* of glucose and volatile fatty acids by sheep brain for the synthesis of acidic amino acids, *Biochem. J.* **101**:591–597 (1966).
189. G. Simon, J. B. Drori, and M. M. Cohen, Mechanism of conversion of aspartate into glutamate in cerebral-cortex slices, *Biochem. J.* **102**:153–162 (1967).
190. G. Simon, M. M. Cohen, and J. F. Berry, Conversion of glutamate into aspartate in guinea-pig cerebral-cortex slices, *Biochem. J.* **107**:109–111 (1968).
191. J. W. Phillis and A. K. Tebecis, The responses of thalamic neurones to iontophoretically applied monoamines, *J. Physiol. (London)* **192**:715–745 (1967).
192. M. H. T. Roberts and D. W. Straughan, Excitation and depression of cortical neurones by 5-hydroxytryptamine, *J. Physiol. (London)* **193**:269–294 (1967).
193. J. L. Malcolm, P. Saraiva, and P. J. Spear, Cholinergic and adrenergic inhibition in the rat cerebral cortex, *Intern. J. Neuropharmacol.* **6**:509–527 (1967).
194. M. D. Faiman and A. R. Heble, The effect of hyperbaric oxygenation on cerebral amines, *Life Sci.* **5**:2225–2234 (1966).
195. J. Haggendal, The effect of high pressure air or oxygen with and without carbon dioxide on the catechol amine levels of rat brain, *Acta Physiol. Scand.* **69**:147–152 (1967).
196. J. Haggendal, Effect of hyperbaric oxygen on monoamine metabolism in central and peripheral tissues of the rat, *Eur. J. Pharmacol.* **2**:323–325 (1968).
197. L. Kato, B. Gozsy, P. B. Roy, and V. Groh, Histamine, serotonin, epinephrine and norepinephrine in the rat brain, following convulsions, *Intern. J. Neuropsychiat.* **3**:46–51 (1967).
198. M. W. Adler, S. Sagel, S. Kitagawa, T. Segawa, and E. W. Maynert, The effects of repeated fluorothyl-induced seizures on convulsive thresholds and brain monoamines in rats, *Arch. Intern. Pharmacodyn.* **170**:12–21 (1967).
199. S. S. Kety, F. Javoy, A. M. Thierry, L. Julou, and J. Glowinski, A sustained effect of electroconvulsive shock on the turnover of norepinephrine in the central nervous system of the rat, *Proc. Natl. Acad. Sci. U.S.* **58**:1249–1254 (1967).
200. N. N. Share and K. I. Melville, Involvement of brain stem norepinephrine in picrotoxin and tyramine-induced central sympathetic stimulation, *Arch. Intern. Pharmacodyn.* **153**:267–282 (1965).
201. B. K. Koe and A. Weissman, Convulsions and elevation of tissue citric acid levels induced by 5-fluorotrytophan, *Biochem. Pharmacol.* **15**:2134–2136 (1966).
202. A. Weissman and B. K. Koe, *m*-Fluorotyrosine convulsions and mortality: relationship to catecholamine and citrate metabolism, *J. Pharmacol. Exp. Therap.* **155**:135–144 (1967).
203. D. J. Prockop, P. A. Shore, and B. B. Brodie, Anticonvulsant properties of monoamine oxidase inhibitors, *Ann. N.Y. Acad. Sci.* **80**:643–651 (1959).
204. W. D. Gray, C. E. Rauh, and R. W. Shanahan, The mechanism of the antagonistic action of reserpine on the anticonvulsant effect of inhibitors of carbonic anhydrase, *J. Pharmacol. Exp. Therap.* **139**:350–360 (1963).

205. W. D. Gray and C. E. Rauh, The anticonvulsant action of carbon dioxide: interaction with reserpine and inhibitors of carbonic anhydrase, *J. Pharmacol. Exp. Therap.* **163**:431–438 (1968).
206. W. D. Gray and C. E. Rauh, The anticonvulsant action of inhibitors of carbonic anhydrase: relation to endogenous amines in brain, *J. Pharmacol. Exp. Therap.* **155**:127–134 (1967).
207. A. K. Pfeifer and E. Galambos, Action of α-methyldopa on the pharmacological and biochemical effect of reserpine in rats and mice, *Biochem. Pharmacol.* **14**:37–40 (1965).
208. A. K. Pfeifer and E. Galambos, The effect of reserpine, α-methyl-*m*-tyrosine, prenylamine, and guanethidine on Metrazole convulsions and the brain monoamine level in mice, *Arch. Intern. Pharmacodyn.* **165**:201–211 (1967).
209. A. K. Pfeifer and E. Galambos, The effect of (±)-*p*-chloroamphetamine on the susceptibility to seizures and on the monoamine level in brain and heart of mice and rats, *J. Pharm. Pharmacol.* **19**:400–402 (1967).
210. D. Palm, H. Balzer, and P. Holtz, Amino acid and carbohydrate metabolism of brain after reserpine, *Intern. J. Neuropharmacol.* **1**:173–177 (1962).
211. N. Popov and H. J. Matthies, Die Wirkung von Monoaminoxydase-Hemmstoffen und Reserpin auf den γ-Aminobuttersaure-Gehalt des Rattenhirns, *Acta Biol. Med. Ger.* **18**:91–98 (1967).
212. F. D. Marshall, Jr. and W. C. Yockey, The effect of various agents on the levels of homocarnosine in rat brain, *Biochem. Pharmacol.* **17**:640–642 (1968).
213. S. I. Singh and C. L. Malhotra, Amino acid content of monkey brain. III. Effects of reserpine on some amino acids of certain regions of monkey brain, *J. Neurochem.* **11**:865–872 (1964).
214. H. H. Tallan, *in Amino Acid Pools* (J. T. Holden, ed.), pp. 465–470, Elsevier, Amsterdam (1962).
215. B. C. Bose, R. Vijayvargiya, A. Q. Saifi, and S. K. Sharma, Effect of reserpine on amino acids in brain, *Arch. Intern. Pharmacodyn.* **146**:114–118 (1963).
216. S. S. Bhattacharya, K. Kishor, P. N. Saxena, and K. P. Bhargava, A neuropharmacological study of γ-aminobutyric acid, *Arch. Intern. Pharmacodyn.* **150**:295–305 (1964).
217. H. J. Matthies and N. Popov, Die Bedeutung der chemischen Struktur von Monoaminoxydase-Hemmstoffen für ihre Wirkung auf den γ-Aminobuttersäure-Gehalt des Rattenhirns, *Acta Biol. Med. Ger.* **18**:617–624 (1967).
218. N. Popov, W. Pohle, V. Roesler, and H. Matthies, Wirkung von Phenelzin auf den Gehalt an γ-Aminobuttersäure im 11 Regionen des Rattenhirns, *Acta Biol. Med. Ger.* **20**:365–370 (1968).
219. H. Matthies and N. Popov, Die Beeinflussung der Wirkung von Phenelzin, Phenylpropylhydrazin und Aminooxyessigsäure auf den Gehalt an γ-Aminobuttersäure des Rattenhirns durch Monoaminoxydase-Hemmstoffe, *Acta Biol. Med. Ger.* **20**:371–378 (1968).
220. N. Popov, W. Pohle, and H. Matthies, Einfluss von Phenelzin und Aminooxyessigsäure auf die γ-Aminobuttersäure, Glutaminsäure-Dekarboxylase und γ-Aminobuttersäure-α-Ketoglutarsäure-Transaminase in verschiedenen Regionen des Rattenhirns, *Acta Biol. Med. Ger.* **20**:509–516 (1968).
221. A. Feldstein and C. M. Sidel, Phenelzine-induced convulsions and alterations in the conversion of 5-HTP-^{14}C to serotonin-^{14}C in vivo, *Biochem. Pharmacol.* **15**:111–117 (1966).
222. K. Schlesinger, W. Boggan, and D. X. Freedman, Genetics of audiogenic seizures. I. Relation to brain serotonin and norepinephrine in mice, *Life Sci.* **4**:2345–2351 (1965).

223. K. Schlesinger, W. Boggan, and D. X. Freedman, Genetics of audiogenic seizures II. Effects of pharmacological manipulation of brain serotonin, norepinephrine and *gamma*-aminobutyric acid, *Life Sci.* **7**:437–447 (1968).
224. N. Plotinkoff, J. Huang, and P. Havens, Effect of monoamine oxidase inhibitors. on audiogenic seizures, *J. Pharm. Sci.* **52**:172–173 (1963).
225. A. Lehmann, Audiogenic seizures data in mice supporting new theories of biogenic amines mechanisms in the central nervous system, *Life Sci.* **6**:1423–1431 (1967).
226. B. E. Ginsburg, R. A. Lovell, and D. X. Freedman, unpublished data.
227. B. E. Ginsburg, *in Behavior-Genetic Analysis* (J. Hirsch, ed.), pp. 135–153, McGraw-Hill, New York (1967).
228. B. E. Ginsburg, J. S. Cowen, S. C. Maxson, and P. Y. Sze, Neurochemical effects of gene mutations associated with audiogenic seizures, *in Progress in Neuro-Genetics* (A. Barbeau and J. R. Brunette, eds.), pp. 695–701, Excerpta Medica, Amsterdam (1969).
229. P. Y. Sze, Neurochemical factors in the development of genetically determined susceptibility to audiogenic seizures in the mouse, Ph.D. dissertation, Department of Biochemistry, The University of Chicago (1969).
230. C. L. Scudder, A. G. Karczmar, G. M. Everett, J. E. Gibson, and M. Rifkin, Brain catecholamines and serotonin levels in various strains and genera of mice and a possible interpretation for the correlations of amine levels with electroshock latency and behavior, *Intern. J. Neuropharmacol.* **5**:343–351 (1966).
231. M. S. Ebadi, R. L. Russell, and E. E. McCoy, The inverse relationship between the activity of pyridoxal kinase and the level of biogenic amines in rabbit brain, *J. Neurochem.* **15**:659–665 (1968).
232. R. Takahashi, T. Nasu, T. Tamura, and T. Kariya, Relationship of ammonia and acetylcholine levels to brain excitability, *J. Neurochem.* **7**:103–112 (1961).
233. M. Kurokawa, Y. Machiyama, and M. Kato, Distribution of acetylcholine in the brain during various states of activity, *J. Neurochem.* **10**:341–348 (1963).
234. E. DeRobertis, O. Z. Sellinger, G. Rodriguez de Lores Arnaiz, M. Alberici, and L. M. Zieher, Nerve endings in methionine sulphoximine convulsant rats, a neurochemical and ultrastructural study, *J. Neurochem.* **14**:81–89 (1967).
235. K. A. C. Elliott, R. T. Kahn, F. Bilodeau, and R. A. Lovell, Bound γ-aminobutyric and other amino acids in brain, *Can. J. Biochem.* **43**:407–416 (1965).
236. P. Strasberg and K. A. C. Elliott, Further studies on the binding of γ-aminobutyric acid by brain, *Can. J. Biochem.* **45**:1795–1807 (1967).
237. K. A. C. Elliott, γ-Aminobutyric acid and other inhibitory substances, *Brit. Med. Bull.* **21**:70–75 (1965).
238. K. Kuriyama, E. Roberts, and T. Kakefuda, Association of the γ-aminobutyric acid system with a synaptic vesicle fraction from mouse brain, *Brain Res.* **8**:132–152 (1968).
239. J. Vos, K. Kuriyama, and E. Roberts, Electrophoretic mobilities of brain subcellular particles and binding of γ-aminobutyric acid, acetylcholine, norepinephrine, and 5-hydroxytryptamine, *Brain Res.* **9**:224–230 (1968).
240. K. Kuriyama, E. Roberts, and J. Vos, Some characteristics of binding of γ-aminobutyric acid and acetylcholine to a synaptic vesicle fraction from mouse brain, *Brain Res.* **9**:231–252 (1968).
241. F. Fonnum, The distribution of glutamate decarboxylase and aspartate transaminase in subcellular fractions of rat and guinea-pig brain, *Biochem. J.* **106**:401–412 (1968).
242. L. Salganicoff and E. DeRobertis, Subcellular distribution of the enzymes of the glutamic acid, glutamine, and γ-aminobutyric acid cycles in rat brain, *J. Neurochem.* **12**:287–309 (1965).

243. R. Balazs, D. Dahl, and J. R. Harwood, Subcellular distribution of enzymes of glutamate metabolism in rat brain, *J. Neurochem.* **13**, 897–905 (1966).
244. A. Waksman and M. Bloch, Identification of multiple forms of aminobutyrate transaminase in mouse and rat brain: subcellular localization, *J. Neurochem.* **15**:99–105 (1968).
245. D. G. Grahame-Smith, The biosynthesis of 5-hydroxytryptamine in brain, *Biochem. J.* **105**:351–360 (1967).
246. D. Robinson, W. Lovenberg, and A. Sjoerdsma, Subcellular distribution and properties of rat brain stem tryptophan hydroxylase, *Arch. Biochem. Biophys.* **123**:419–421 (1968).
247. G. K. Aghajanian and F. E. Bloom, Localization of tritiated serotonin in rat brain by electron microscopic autoradiography, *J. Pharmacol. Exp. Therap.* **156**:23–30 (1967).
248. G. K. Aghajanian and F. E. Bloom, Electron microscopic localization of tritiated norepinephrine in rat brain: Effect of drugs, *J. Pharmacol. Exp. Therap.* **156**:407–416 (1967).
249. F. E. Bloom and G. K. Aghajanian, An electron microscopic analysis of large granular synaptic vesicles of the brain in relation to monoamine content, *J. Pharmacol. Exp. Therap.* **159**:261–273 (1968).
250. J. Glowinski and L. Iversen, Regional studies of catecholamines in the rat brain. III. Subcellular distribution of endogenous and exogenous catecholamines in various brain regions, *Biochem. Pharmacol.* **15**:977–987 (1966).
251. J. Glowinski, S. Snyder, and J. Axelrod, Subcellular distribution of H^3-norepinephrine in the rat brain: effect of reserpine and amphetamine, *J. Pharmacol. Exp. Therap.* **152**:282–292 (1966).
252. Y. Gutman and H. Weil-Malherbe, The intracellular distribution of brain catecholamines, *J. Neurochem.* **14**:619–625 (1967).
253. A. Pellegrino De Iraldi, L. M. Zieher, and G. J. Etcheverry, *in Advances in Pharmacology* (S. Garattini and P. A. Shore, eds.), Vol. 6, Part A, pp. 257–270, Academic Press, New York (1968).
254. I. A. Michaelson, *in Advances in Pharmacology* (S. Garattini and P. A. Shore, eds.), Vol. 6, Part A, pp. 271–274, Academic Press, New York (1968).
255. R. M. Marchbanks, Exchangeability of radioactive acetylcholine with the bound acetylcholine of synaptosomes and synaptic vesicles, *Biochem. J.* **106**:87–95 (1968).
256. R. M. Marchbanks, Serotonin binding to nerve ending particles and other preparations from rat brain, *J. Neurochem.* **13**:1481–1493 (1966).
257. C. D. Wise and H. W. Ruelius, The binding of serotonin in brain. A study in vitro of the influence of physicochemical factors and drugs, *Biochem. Pharmacol.* **17**:617–631 (1968).
258. R. M. Marchbanks, Inhibitory effects of lysergic acid derivatives and reserpine on 5-HT binding to nerve ending particles, *Biochem. Pharmacol.* **16**:1971–1979 (1967).
259. J. A. Rosecrans, R. A. Lovell, and D. X. Freedman, Effects of lysergic acid diethylamide on the metabolism of brain 5-hydroxytryptamine, *Biochem. Pharmacol.* **16**:2011–2021 (1967).
260. Y. Tsukada, Y. Nagata, S. Hirano, and T. Matsutani, Active transport of amino acid into cerebral cortex slices, *J. Neurochem.* **10**:241–256 (1963).
261. R. Blasberg and A. Lajtha, Substrate specificity of steady-state amino acid transport in mouse brain slices, *Arch. Biochem. Biophys.* **112**:361–377 (1965).
262. R. Blasberg and A. Lajtha, Heterogeneity of the mediated transport systems of amino acid uptake in brain, *Brain Res.* **1**:86–104 (1966).

263. R. Nakamura and M. Nagayama, Amino acid transport by slices from various regions of the brain, *J. Neurochem.* **13**:305–313 (1966).
264. L. L. Iversen and E. A. Kravitz, The metabolism of γ-aminobutyric acid (GABA) in the lobster nervous system—uptake of GABA in nerve-muscle preparations, *J. Neurochem.* **15**:609–620 (1968).
265. S. M. Schanberg, A study of the transport of 5-hydroxytryptophan and 5-hydroxytryptamine (serotonin) into brain, *J. Pharmacol. Exp. Therap.* **139**:191–200 (1963).
266. K. J. Blackburn, P. C. French, and R. J. Merrills, 5-hydroxytryptamine uptake by rat brain in vitro, *Life Sci.* **6**:1653–1663 (1967).
267. S. B. Ross and A. L. Renyi, Uptake of some tritiated sympathomimetic amines by mouse brain cortex slices in vitro, *Acta Pharmacol. Toxicol.* **24**:297–309 (1966).
268. S. B. Ross and A. L. Renyi, Accumulation of triatiated 5-hydroxytryptamine in brain slices, *Life Sci.* **6**:1407–1415 (1967).
269. E. W. Maynert and K. Kuriyama, Some observations on nerve-ending particles and synaptic vesicles, *Life Sci.* **3**:1067–1087 (1964).
270. J. D. Robinson, J. H. Anderson, and J. P. Green, The uptake of 5-hydroxytryptamine and histamine by particulate fractions of brain, *J. Pharmacol. Exp. Therap.* **147**:236–243 (1965).
271. T. Segawa and I. Kuruma, The influence of drugs on the uptake of 5-hydroxytryptamine by nerve-ending particles of rabbit brain stem, *J. Pharm. Pharmacol.* **20**:320–322 (1968).
272. T. Segawa, I. Kuruma, K. Takatsuka, and H. Takagi, The influence of drugs on the uptake of 5-hydroxytryptamine by synaptic vesicles of rabbit brain stem, *J. Pharm. Pharmacol.* **20**:800–801 (1968).
273. A. Lajtha, *in Progress in Brain Research* (A. Lajtha and D. N. Ford, eds.), Vol. 29, pp. 201–218, Elsevier, Amsterdam (1968).
274. K. Krnjević and S. Schwartz, Some properties of unresponsive cells in the cerebral cortex, *Exp. Brain Res.* **3**:306–319 (1967).
275. S. G. A. Alivisatos and F. Ungar, Incorporation of radioactivity from labeled serotonin and tryptamine into acid-soluble material from subcellular fractions of brain. I. The nature of the substrate, *Biochemistry* **7**:285–292 (1968).
276. S. G. A. Alivisatos, F. Ungar, S. S. Parmar, and P. K. Seth, Monoamine oxidase dependent labeling *in vivo* of mouse brain by ^{14}C-serotonin, *Biochem. Pharmacol.* **17**:1993–1995 (1968).
277. V. G. Erwin and R. A. Deitrich, Brain aldehyde dehydrogenase. Localization, purification, and properties, *J. Biol. Chem.* **241**:3533–3539 (1966).
278. V. P. Whittaker, The storage of transmitters in the central nervous system, *Biochem. J.* **109**:20P–21P (1968).
279. V. P. Whittaker, Synaptic transmission, *Proc. Natl. Acad. Sci. U.S.* **60**:1081–1091 (1968).
280. L. Austin and I. G. Morgan, Incorporation of ^{14}C-labeled leucine into synaptosomes from rat cerebral cortex in vitro, *J. Neurochem.* **14**:377–387 (1967).
281. I. G. Morgan and L. Austin, Synaptosomal protein synthesis in a cell-free system, *J. Neurochem.* **15**:41–51 (1968).
282. M. K. Gordon, K. G. Bench, G. G. Deanin, and M. W. Gordon, Histochemical and biochemical study of synaptic lysosomes, *Nature* **217**:523–527 (1968).
283. S. H. Barondes, On the site of synthesis of mitochondrial protein of nerve endings, *J. Neurochem.* **13**:721–727 (1966).
284. S. H. Barondes, Further studies of the transport of protein to nerve endings, *J. Neurochem.* **15**:343–350 (1968).

285. S. H. Barondes, Incorporaton of radioactive glucosamine into macromolecules at nerve endings, *J. Neurochem.* **15**:699–706 (1968).
286. C. M. Ling and A. A. Abdel-Latif, Studies on sodium transport in rat brain nerve-ending particles, *J. Neurochem.* **15**:721–729 (1968).
287. R. M. Marchbanks, The osmotically sensitive potassium and sodium compartments of synaptosomes, *Biochem. J.* **104**:148–157 (1967).
288. P. Gloor, L. Sperti, and C. L. Vera, A consideration of feedback mechanisms in the genesis and maintenance of hippocampal seizure activity, *Epilepsia* **5**:213–238 (1964).
289. R. N. Straw and C. L. Mitchell, The effect of pentylenetetrazole on bioelectrical activity recorded from the cat brain, *Arch. Intern. Pharmacodyn.* **168**:456–466 (1967).
290. R. I. Katz, T. N. Chase, and I. J. Kopin, Evoked release of norepinephrine and serotonin from brain slices: inhibition by lithium, *Science* **162**:466–467 (1968).
291. J. F. Mitchell, M. J. Neal, and V. Srinivasan, The release of ^{3}H-gamma-aminobutyric acid (GABA) from rat cerebral cortex, *Brit. J. Pharmacol.* **34**:661P–662P (1968).

Chapter 4

STRESS

Arthur Yuwiler

Neurobiochemistry Laboratory, Veterans Administration Center and Department of Psychiatry, University of California Center for the Health Sciences Los Angeles, California

I. INTRODUCTION

Stress research is confounded by a host of factors, not the least of which is the term stress itself. Stress, like the word love, is broadly understood but poorly defined. In biology, it generally refers to a physiological displacement from homeostasis after a stimulus, termed a stressor, which is followed by a series of presumably restorative processes called the stress response. Even this vague generality of undefined terms is not universally agreed upon and, in the literature, not only are stress and stressor often used interchangeably,[1,2] but also stress is sometimes defined by exclusion[3]

TABLE I

Adrenocortical Indices

1. Primary	Pituitary ACTH	Decreases
	Blood ACTH	Increases
	Adrenal corticoids	Increases
	Adrenal ascorbic acid	Decreases
	Adrenal cholesterol	Decreases
	Adrenal weight	Increases
	Blood corticoids	Increases
2. Secondary	Blood eosinophiles	Decreases
	Blood lymphocytes	Decreases
	Thymus weight	Decreases
3. Tertiary	Blood cholesterol	Increases
	Blood catecholamines	Increases
	Urinary catecholamines	Increases
	Blood free fatty acids	Increases
	Tissue enzymatic activities	Increases and decreases

as all the nonspecific changes that occur following a stimulus, and sometimes by inclusion, explicitly referring to stimuli-specific changes. In an attempt to establish a more specific referent, it has been customary to identify stress by some index of adrenocortical activation because of Selye's brilliant demonstration that a wide variety of noxious stimuli actuate the adrenal cortex and that adrenocortical hormones are essential, in higher organisms, for successful adaptation to the noxious stimuli. But, as noted by Selye, this identification, too, is inadequate, for obviously organisms without adrenals respond to noxious stimuli. Further, the adrenals are essentially a way-station, albeit of considerable importance, between receipt of a stimulus and metabolic adaptation; adrenal activity is controlled in a complex manner from above, and adrenal hormones are involved in even more complex metabolic processes below. Whatever the limitation of adrenocortical stimulation as a stress index, however, some knowledge of adrenal physiology is necessary for any study of the stress literature.

II. PHYSIOLOGY

A. Hypothalamic–Pituitary–Adrenal Interrelations

The activity of the adrenal gland is under the trophic control of the pituitary (Fig. 1). In response to appropriate stimuli, the anterior pituitary secretes adrenocorticotrophic hormone (ACTH) which in turn stimulates the adrenal cortex to synthesize and release adrenal corticoids. Although the transplanted and denervated pituitary gland can secrete low levels of adrenocorticotrophic hormone,[4] it is generally agreed that the central nervous system is a necessary mediator of the pituitary response to stress and that the hypothalamus is intimately involved.[5] Contributions from

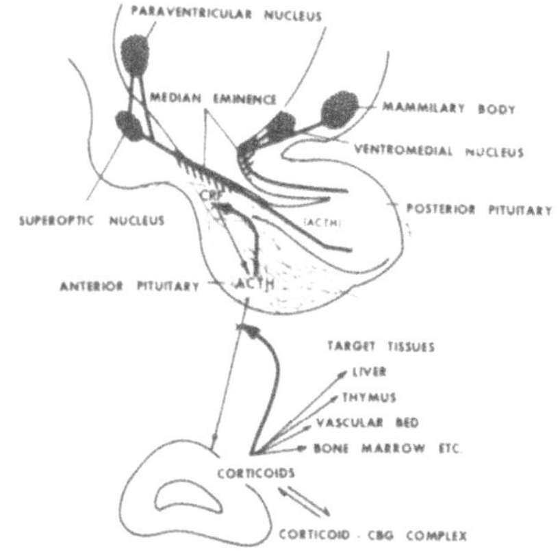

Fig. 1. Relationship between brain, pituitary, adrenals, and target tissues.

other brain areas are less clear. Hippocampus,[6,7] reticular formation,[8,9] rhinencephalon, temporal and occipital cortex,[10] and limbic cortex,[11] all modify the pituitary response to stress[12-14]; indeed, some areas, such as the reticular formation, may exert a tonic restraint upon the pituitary gland and its production of ACTH.[15]

The median eminence region of the hypothalamus is especially involved in control of ACTH production by the pituitary, and stimulation of this region increases[16,17] while lesions decrease[18,19] its formation. This influence of the median eminence is via humoral rather than neural control. The release of ACTH is, in fact, controlled by another trophic hormone, or set of hormones, called corticotropin-releasing factor, or CRF,[20-22] thought to be secreted by neurosecretory cells in the periventricular and superoptic regions. These discharge in the median eminence and are thought to be carried to the anterior pituitary by a venous network called the pituitary portal system. At least three active peptides with CRF activity have been isolated from the median eminence region.[22] Although these peptides are quite active, their concentrations are low and their source, the median eminence, is an extremely small structure, so that insufficient material has as yet been isolated to permit a full exposition of peptide structure. The amino acid sequence of one of the active CRF peptides α_2-CRF, however, appears closely related to another mammalian hormone, α-MSH (α-melanocyte-stimulating hormone), while another, β-CRF, resembles both vasopressin and MSH.[23-25]

While CRF is necessary for mediation of pituitary ACTH release following some stressors, it may not be required for all. Indeed, different hypothalamic sites[26-28] may be involved in mediating different "ACTH-induced" responses of the adrenal such as the increase in adrenocortical hormone and the depletion of adrenal ascorbic acid under different stress conditions.

Although the major portion of released ACTH comes from the anterior pituitary, the presence of an "ACTH" in the posterior pituitary has also been shown.[29] It is not known whether these two ACTH's are chemically identical, but there does seem to be some selectivity in their release. Some types of stressors, such as loud sounds, seem to selectively release ACTH from the posterior pituitary while most stressors release ACTH from the anterior pituitary.[30] That there is stress specificity in regional release is not too surprising. As mentioned previously, the anterior pituitary is richly perfused by a venous network. In contrast, the posterior pituitary is richly innervated and the vascular bed is more modest. It is at least conceivable that an interplay between neural and humoral control on ACTH release is involved in the posterior pituitary, whereas humoral control alone is operative in the anterior pituitary. Indeed, Fortier[31,32] posited just such a separation of "neurotropic" and "humoral" systems for ACTH release some years ago and it has since received some experimental support.[33,34] Whether this pituitary ACTH distribution is related to function, however, has yet to be demonstrated.

In addition to ACTH in both the anterior and posterior pituitary, it now appears that ACTH may also be present in the hypothalamus itself.[35] The significance of this is unclear. It is possible, however, that ACTH exerts a feedback control on CRF release and that the ACTH found in the hypothalamus is that portion of the circulating ACTH which is actively inhibiting further CRF release. It is of interest that a small vascular flow from the anterior pituitary to the hypothalamus has recently been observed,[36] so that the possibility exists of direct flow of ACTH from the pituitary to control CRF release.

CRF is not the only humoral substance which can release ACTH. Vasopressin[37-40] and serotonin,[41] for example, elicit adrenal corticoid secretion by both direct action on the adrenal and by an indirect action on ACTH production.

The mechanism whereby CRF promotes synthesis and release of ACTH is unclear. ACTH itself is a polypeptide with 39 amino acids and molecular weight of 4500. The first 24 amino acids from the amino terminal serine are essential for activity and are found in purified ACTH from all species thus far examined. Species differences have been found in the next nine amino acids of the chain involving interchanges between alanine, glycine, serine, leucine, glutamic acid, aspartic acid, and glutamine. The terminal end of the molecule from the 25 amino acid to the C-terminal phenylalanine is not necessary for activity but is common to all natural ACTH's studied[42-46] (Cf. Fig. 2).

ACTH exerts its primary effect on the adrenal cortex. However, it also appears to stimulate release of unesterified fatty acids from epididymal fat pads,[47] to stimulate melanin formation and induce pigment dispersion of melanophores.[48] The latter actions are not too surprising since only small substitutions in the amino acid sequence of the first 20 amino acids of the essential portion of the ACTH molecule are required to make the

```
Human
  ACTH     S-Ty-S-M-GA-H-Ph-Ar-Tr-Gl-Ly Pr-V-Gl-Ly-Ly-Ar-Ar-Pr-V—
Ox         S ,,  ,, ,,  ,,  ,, ,,  ,,  ,,  ,,  ,,  ,, ,, ,,  ,,  ,,  ,,  ,,  ,,—
Pig        S ,,  ,, ,,  ,,  ,, ,,  ,,  ,,  ,,  ,,  ,, ,, ,,  ,,  ,,  ,,  ,,  ,,—
Sheep      S ,,  ,, ,,  ,,  ,, ,,  ,,  ,,  ,,  ,,  ,, ,, ,,  ,,  ,,  ,,  ,,  ,,—
        O
        ‖
αMSH CH3-C—S ,,  ,, ,,  ,,  ,, ,,  ,,  ,,  ,,  ,,  ,, ,, ,,  ,,  ,,  ,,  ,,  ,, NH2

Human
  ACTH     —Ly-V-Ty-Pr-AA-Al-Gl-GA-AA-GTM-S-Al-GA-Al-Ph-Pr-Le-GA-Ph
Ox         —,,  ,, ,,  ,, ,,  Gl-GA-Al-GA-AA  -,, ,,-GTM- ,, ,,  ,, ,,  ,,  ,,
Pig        —,,  ,, ,,  ,, ,,  Gl-Al- ,,   ,, ,,     -Le-  ,,   ,, ,,  ,, ,,  ,,  ,,
Sheep      —,,  ,, ,,  ,, ,,  Al-,,   ,,   ,,  GA-Al-S-GTM- ,, ,,  ,, ,,  ,,  ,,
```

Fig. 2. Relationship between vertebrate ACTH's and α-MSH.

sequence identical to large portions of the α-MSH and β-MSH molecules. The action of ACTH on epididymal fat pads seems to be mediated by activation of a specific[49] adenyl cyclase on the plasma membrane of the fat cell.[50]

ACTH appears to elicit a number of behavioral effects in addition to its metabolic effects. ACTH is reported to delay extinction of the conditioned avoidance response[51,52] and, when injected directly into cerebrospinal fluid or brain ventricles, to produce sexual excitement[53] and a peculiar behavioral syndrome of stretching and yawning.[54] These effects are also produced by MSH and their mechanism is unknown although they are not mediated through the adrenal corticoids since they can be demonstrated in adrenalectomized animals.

Upon reaching the adrenal cortex, ACTH promotes the synthesis and release of adrenocortical hormones by a process that may involve cyclic AMP. Adenyl cyclase, the enzyme forming cyclic AMP, is found in the adrenal membrane, its activity is stimulated by ACTH,[55] and cyclic-3′5-AMP is rapidly produced.[56] This action is not blocked by cycloheximide,[57] suggesting that protein synthesis is not involved, a view consistent with the rapidity of the reaction. However, stimulated but not basal steroid synthesis is prevented by puromycin,[58,59] chloramphenicol, and cycloheximide,[60,61] suggesting that protein synthesis may be involved in this step. This is somewhat surprising since adrenal corticoid responses to ACTH are very rapid and detectable within minutes. RNA synthesis does not appear to play a major role. Actinomycin does not inhibit steroidogenesis,[58,60,62,63] although ACTH does stimulate RNA synthesis.[64,65]

B. Adrenocortical Hormones

1. *Synthesis and Physiology*

Adrenocortical hormones fall into three groups: (1) Those with an oxygen on position 11 of the steroid nucleus which affect carbohydrate and protein metabolism but have little effect on salt and water regulation. (2) Those without an oxygen at position 11 which affect electrolyte and water regulation but have little influence on protein or carbohydrate metabolism. (3) Aldosterone, which paradoxically does have a hydroxyl group at position 11, but is the most potent of the electrolyte regulating hormones. These compounds are shown in Fig. 3.

The complex biochemical transformations in the biosynthesis and metabolism of the adrenal corticoids have been summarized elsewhere[66,67] and cannot be repeated here. Essentially, however, the stimulated adrenal cortex secretes adrenocortical hormones into the bloodstream. About 30% circulates as free steroid.[68] Of the remainder, 50% is bound to a low affinity, high capacity carrier associated with albumin and 20% to a high affinity, low capacity globulin of 52,000 molecular weight[69] called "transcortin"[70] or CBG (corticoid binding globulin).[71–73] The unbound

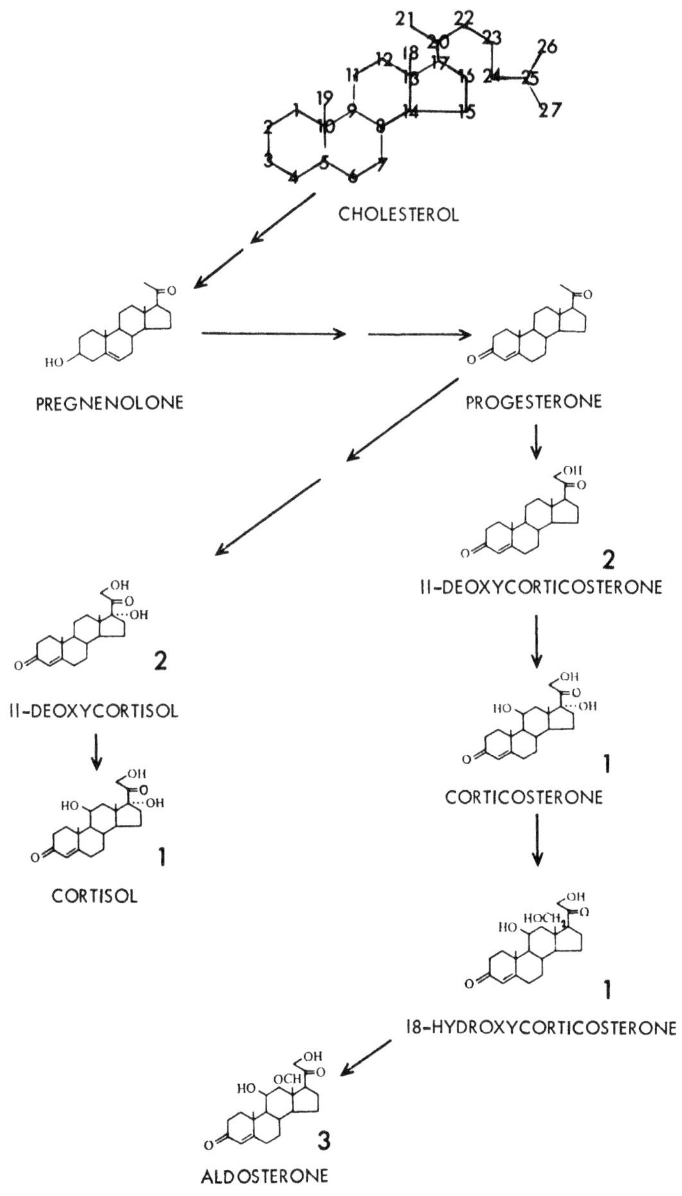

Fig. 3. Adrenal steroid interrelationships. Glucocorticoids labeled 1, mineralocorticoids labeled 2, and aldosterone labeled 3.

steroid is thought to be the biologically active component of the system. The steroid–protein complex, in turn, may serve a buffer capacity, supplying additional unbound steroid as either free steroid is removed or as conditions change the binding constants between steroid and protein. The steroid–protein binding is markedly sensitive to temperature. Steroid–protein binding in serum, for example, is decreased by increasing temperature in and about the normal range.[68] This could be of physiological significance where homeothermic processes are disturbed, such as in the local hyperthermia attendant on tissue damage and during wound healing.

Species differ markedly in the kind of steroids secreted by the adrenals. Some 30 steroids have been isolated from adrenal tissue and the particular composition varies between species. The major glucocorticoid components however, are cortisol and corticosterone and possibly 18-hydroxycorticosterone, while aldosterone is the major mineralocorticoid. Following a stressor, glucocorticoids are released rapidly into the bloodstream, reaching peak values in 15 min. Corticoid levels just as promptly return to baseline if the stressor was acute in its effect. Such would be the case with an injection of small volumes of saline, for example. On the other hand, stressors such as fasting, cold, or laparotomy lead to more sustained elevations in serum corticoids.

Release of ACTH from the pituitary and response of the adrenal to ACTH is influenced by the level of circulating corticoids.[74–76] The

TABLE II

Compounds in Adrenal Venous Blood

Cortisol	Dog, human, guinea pig
Corticosterone	Cat, dog, human, guinea pig, rat
11-deoxycortisol	Dog, human
Δ^4-Androstenedione	Human, dog
Dehydroepiandrosterone	Human
Dehydroepiandrosterone sulfate	Human
17-Hydroxyprogesterone	Human, dog
17-Hydroxypregnenolone	Dog, human
11β-Hydroxy-Δ^4-androstenedione	Human, dog
18-Hydroxy-11-deoxycorticosterone	Rat
11-Ketoprogesterone	Dog
11β-Hydroxyprogesterone	Dog
Cortisone	Cat, guinea pig
11-Dehydrocorticosterone	Cat, guinea pig
Progesterone	Dog
Aldosterone	Dog, monkey
20α-Hydroxypregnen-3-one	Bovine
Pregnenolone	Human

extent of this influence is subject to some controversy,[78–81] but systemically administered natural or synthetic corticoids or direct implantation of corticoids into the hypothalamus or reticular formation[74,82–85] decrease ACTH secretion and the response of the adrenal to administered ACTH. Presumably, this action involves inhibition of CRF release[86] either directly or by a secondary ACTH feedback, as mentioned earlier. While these effects of pharmacological doses of corticoids are acknowledged, their relevance to adrenal responses in stress is less clear. However, it does appear that adrenal sensitivity may be modified by prior stress history which alters the physiological "set" of the hypothalamic–pituitary control system.[87,88] Since animals may differ in the set point of their "corticostats,"[87] considerable variation might be expected between laboratories, depending upon the prior stress histories of the experimental animals.

Peripherally administered corticoids enter both the brain[89] and the pituitary[90] and changes in local permeability to corticoids may be involved in establishing this "set" point. Further, hypothalamic tissue can convert cortisol to cortisone and other metabolites[91] and this conversion too may be influenced by earlier stimuli. More importantly, corticoid feedback inhibition appears to block ACTH release by some stressors but not others.[14,92] Yates[92] has divided stressors in Class I and Class II stimuli on the basis of their ability to elicit ACTH release despite prior administration of the synthetic long-acting steroid, dexamethazone. Class I stimuli consist of stressors such as ether, burns, electric shock, and laparotomy, in which ACTH release is blocked by prior corticoid treatment, and Class II by stressors such as hemorrhage, intestinal traction, cervical dislocation and anoxia, in which blockade does not occur. On this basis he has elaborated a very elegant model of glucocorticoid control.[92]

This raises the important point that all stressors cannot be regarded in the same light but rather are related to each other as games are related; by a set of overlapping phenomena in which all elements are not necessarily present in all members of the class. This will become more evident in subsequent sections.

The picture developed to this point, then, is that stressors are perceived centrally and responded to by release of tropic hormones from the median eminence region of the hypothalamus which stimulate release of ACTH by the pituitary. ACTH then stimulates the adrenal cortex to release adrenocortical hormones which are partly bound by serum proteins and which circulate throughout the body. Both CRF release and ACTH release are under some degree of feedback control.

2. *Peripheral Effects*

Adrenocortical hormones have a wide variety of actions on target tissues, some of which are summarized in Table III. Generally, these actions are thought to be of restorative importance, permitting return of the organism to homeostatic equilibrium. The time for these effects varies

TABLE III
Effects of Glococorticoids

A. General effects
- Brain
  - Increased glucose
  - Increased glycogen
- Liver
  - Increased glycogen
  - Increased glucose production
  - Increased ketone body production
  - Increased urea production
  - Increased amino acid uptake
  - Increased RNA synthesis
  - Increased protein synthesis
- Adipose tissue
  - Increased FFA release
  - Decreased glucose utilization
- Muscle
  - Decreased glucose utilization
- Lymphatic tissue
  - Decreased glucose utilization
  - Decreased nucleic acid synthesis

B. Enzymatic effects
- Brain
  - Tryptophan-5-hydroxylase
  - Glycerolphosphate dehydrogenase
  - 5-Hydroxytryptophan decarboxylase (?)
- Adrenal
  - Phenylethanolamine-*N*-methyltransferase
- Liver
  - In carbohydrate metabolism
    - Aldolase
    - Fructose-1,6-diphosphatase
    - Glucose-6-phosphatase
    - Glycogen synthetase
    - Phosphoenolpyruvate carboxykinase
    - Phosphoglucomutase
    - Phosphoglyceraldehyde dehydrogenase
    - Pyrivate carboxylase
  - In amino acid metabolism
    - Alanine aminotransferase
    - Glutamic acid dehydrogenase
    - 5-Hydroxytryptophan decarboxylase (?)
    - Serine dehydratase
    - Threonine dehydratase
    - Tryptophan-5-hydroxylase
    - Tryptophan aminotransferase
    - Tryptophan oxygenase
    - Tyrosine aminotransferase

from hours to weeks and there is as yet no uniform picture encompassing the biological significance of all of these changes and no explanation of their importance in meeting the biological challenges of specific stressors. In addition, information on the effects of adrenocorticoids is derived largely from responses to pharmacological levels of corticoids, to changes attendant on adrenalectomy together with reversal of these changes by administered corticoids, and to pathological changes accompanying adrenal insufficiency, as in Addison's disease, or adrenal hyperplasia, as in Cushing's syndrome. Whether the same complex of reactions occur during stress or are even common to all stresses is unknown.

The actions of adrenocortical hormones are remarkably tissue specific and a set of changes produced in one tissue may be quite opposite to those produced in another tissue. Thus, in liver, adrenocortical hormones increase hepatic urea formation(93) as a result of rapid amino acid catabolism, increase uptake of amino acids and other nutrients,(94,95) increase liver protein nitrogen(96,97) at expense of protein depletion of other tissues, and increase liver RNA without changing DNA.(98) The increase in hepatic protein following corticoid administration is due, in part, to hormonal induction of a variety of enzymes important in gluconeogenesis, glycolysis, urea formation, and amino acid catabolism. Induction of these enzymes is preceded by an increase in liver RNA nucleotidyl transferase,(99) increased RNA polymerase,(100) increased RNA turnover, and an increase in the absolute amount of RNA.(101) Glucocorticoids strongly affect liver glucose metabolism, and, indeed, such effects are perhaps primary to their adaptive function.

Glucocorticoids, *in vitro*, inhibit glycogen formation by liver.(102,103) However, *in vivo* this is overcome by gluconeogenesis both from amino acids and compounds such as lactate, pyruvate, and succinate.(104,105) This gluconeogenesis is not dependent upon new enzyme synthesis,(106) even though synthesis of a variety of key gluconeogenic enzymes such as glucose-6-phosphatase,(107) fructose-1,6-diphosphatase,(108,109) phosphoenolpyruvate carboxykinase,(110) and pyruvate carboxylase(111) is enhanced by glucocorticoid treatment. It has been suggested that glucocorticoids act somewhere between pyruvate and triose phosphate in gluconeogenesis and enhance glucose production(112,113) sufficiently so that they elevate blood glucose(114,115) and increase liver glycogen deposition(115,116) despite the inhibitory action of the glucocorticoids themselves on glycogen formation. This increase in glycogen represents gluconeogenesis and not simply reduced glucose utilization by this tissue.(117)

What holds true for liver, however, is not necessarily true for other tissues. While liver RNA and protein concentration are increased by corticoids, RNA in both thymus and spleen is decreased while protein is decreased in thymus but not in spleen.(118) The marked involution of thymus and other lymphatic tissue(119) upon glucocorticoid treatment dramatically demonstrates the tissue selectivity of corticoid actions. Involution of these tissues appears due to a suppression of DNA synthesis and an inhibition of mitosis,[120] resulting in a marked decrease in lymphocytes in the peripheral circulation.[121] The number of blood eosinophiles also declines,[121] apparently owing to destruction of these cells in bone marrow.[122] On the other hand, polymorphonuclear neutrophiles increase[123] despite depressed mitotic activity,[124] probably as a result of increased cellular half-life.

Because these hemocytological changes can be conveniently detected both in experimental animals and man, they have had wide use in stress studies, particularly in the older literature. Advantage has also been taken of the ionic effects of adrenocorticoids. Although aldosterone is the most important of these in regulating water and ion metabolism, the glucocorticoids themselves are also active. Thus, glucocorticoids lower blood calcium by inhibiting calcium uptake from the gastrointestinal tract.[125] Bone formation is also inhibited by glucocorticoids, although this effect is probably due less to changes in calcium uptake than to direct inhibition of the synthesis of the mucopolysaccarides which comprise the ground substance for bone.[126] Urinary excretion of phosphate is also enhanced by the glucocorticoids,[125] probably by decreasing resorption in the renal tubules. Like aldosterone, glucocorticoids in small doses reduce sodium excretion and increase potassium excretion.[127] However, whereas aldosterone affects all mechanisms of sodium uptake, the active glucocorticoids appear to primarily stimulate sodium-potassium exchange.[128] These effects of glucocorticoids on ion metabolism are rather general among tissues and entrance of sodium into, and potassium out of, intracellular compartments is facilitated by these hormones. Cortisol and cortisone both also enhance water diuresis.

It should be pointed out that while glucocorticoids are under the control of ACTH, aldosterone secretion is regulated by a more complex mechanism. Primary control involves the proteolytic kidney enzyme, renin, which is released into the bloodstream where it reacts with the blood protein angiotensinogen to produce angiotensin. Angiotensin,[129] together with plasma sodium and potassium concentration, then exerts a major regulatory role on aldosterone release. ACTH also appears to be involved[130] despite earlier reports[131] that it was without influence.

This brief tabulation of adrenocortical actions on liver metabolism, blood cytocomposition, and ion metabolism encompasses most of the peripheral changes which have been used experimentally to detect adrenocortical activation. Numerous other actions exist, however, which are important to stress reactions but which cannot be detailed here although they deserve mention. Thus, adrenocorticoids increase gastric secretion,[132] uropepsin excretion,[133] and produce, or reactivate, gastric ulcers.[134] Autonomic reactivity is potentiated, cardiac output is increased,[135] circulatory responses to norepinephrine and epinephrine are potentiated, and capillary resistance is increased.[136] It has recently been suggested that corticoids act in microregulation of capillary beds by increasing local formation of histamine from induced histidine decarboxylase while simultaneously potentiating catecholamine vasoconstriction.[137] In addition, adrenocorticoids inhibit inflammatory reactions,[138] a property which has led to widespread therapeutic application, although the detailed mechanisms involved are not fully understood. They also inhibit antibody production,[139–141] impair migration of phagocytic cells,[142,143] and impair phagocytosis itself.[144] As might be expected from this enumeration, resistance to infection is decreased.[145]

3. *Central Effects*

Curiously, little is known about the direct effects of adrenocortical hormones upon the brain itself, although a considerable clinical and experimental literature exists on behavioral changes in adrenal malfunction. Adrenal insufficiency in Addison's disease is accompanied by behavioral changes such as depression, apathy, instability, and apprehension[146] together with changes in EEG.[147,148] The reverse case of adrenal hyperactivity in Cushing's disease is accompanied by euphoria, depression, and sometimes overt psychosis.[149,150] Administered corticoids also mimic these behavioral changes in some individuals and, in addition, cause abnormal EEG's,[151] alter evoked potentials in the diencephalon,[152] and block spontaneous activity in some neurons, particularly those of the periventricular gray of the third ventricle.[153] This latter effect may reflect participation of these neurons in the glucocorticoid feedback inhibition of ACTH release discussed earlier. ACTH administered to man produces similar behavioral responses as corticoids[154] and psychotic responses have been elicited in some small groups of individuals.[155]

Those behavioral effects in man are primarily personality changes and are more difficult to discern in animals in whom personality can only be inferred from performance. As a result, the literature on animal responses to glucocorticoid treatment is unclear. However, ACTH itself seems to inhibit extinction of avoidance behavior[156] and the first 10 amino acids of ACTH seem to be an essential grouping for eliciting this behavioral response. Both synthetic peptides with this amino acid sequence and MSH produce behavioral effects similar to those of ACTH.[157] There is not universal accord on these findings, however.[158]

The mechanism for any of the behavioral effects either of insufficient or excess corticoids is not known. Those related to adrenal insufficiency or after adrenalectomy may be the result of ionic changes and are complicated by difficulties in regulating salt balance in the absence of mineralocorticoids. The contribution of the glucocorticoids involved in stress to these ionic shifts is likely of minor significance. However, in salt-maintained adrenalectomized animals, brain sodium appears to be increased[159–161] and sodium distribution between extra- and intracellular spaces may be altered.[162] Potassium content seems to be unchanged although some conflict exists over whether[163,164] or not[165] potassium turnover is increased. Glucocorticoids do play some role in these changes. Chronic or acute administration of glucocorticoids or ACTH to intact animals does lower seizure threshold, the effect being minimal with corticosterone and maximal with cortisone or 17-hydroxycorticosterone,[160] and this alteration in seizure threshold may reflect altered sodium compartmentalization produced by these compounds.[164] Although ionic shifts occur in adrenalectomy, they do not alter activities even of such ion-sensitive enzymes as brain sodium-potassium activated ATPase.[165]

The behavioral changes produced by ACTH and glucocorticoids could be secondary to peripheral actions of these agents or could be direct effects on brain. Glucocorticoids do occur in brain, and tracer experiments show rather specific regional distributions. Normal levels in whole brain are quite low (1.6 μg of cortisol and an equivalent amount of corticosterone/100 g of rabbit brain),[166] but may nonetheless be significant. Further, the brain quite actively metabolizes corticoids, but whether this is a detoxification phenomenon or is a functional conversion to a compound more active in brain is uncertain. For example, cortisol is acetylated by brain in the presence of acetyl coenzyme A,[167] and it can be oxidized to cortisone in an NAD requiring reaction.[168] Similarly, corticosterone is converted to 11-dehydrocortisone and 11-β-hydroxyandrostenedione is converted to androstene.[169] These latter conversions are carried out quite rapidly by gliomas and glia may be the responsible cells for these reactions in whole brain tissue.[169]

Direct metabolic effects of glucocorticoids on brain are limited. Neither ACTH nor cortisone appear to alter oxygen consumption in man[170-172] although hydrocortisone may inhibit oxygen uptake by rat brain mitochondria *in vitro*.[173] In adrenalectomized animals cortisone increases whole brain carbohydrate and/or glycogen content.[174,175] In intact animals these effects are less clear and appear to depend somewhat upon the route of administration and the steroid dose. Subcutaneously administered hydrocortisone at high dose levels (100 mg/kg) elevates brain glucose twofold, slightly increases creatine phosphate, glucose-6-phosphate, and ATP, but glutamate and glycogen are unaltered.[176] On the other hand, infusion of cortisol, particularly if given with glutamate, reportedly increases brain glycogen[177] as does either continuous or acute injection of modest levels of hydrocortisone, (i.e., 3.0 mg/kg).[175,178] Continuous hydrocortisone injection for 12 days decreases brain ATP and creatine phosphate while increasing glycogen. The rise in glycogen precedes the fall in ATP.[178]

Although glucocorticoids are effective inducers of a variety of enzymatic activities in peripheral tissues, particularly liver, brain is relatively inert to these inductive effects. The best established enzymatic change is that of α-glycerolphosphate dehydrogenase which is elevated by ACTH or cortisol and decreased by adrenalectomy or hypophysectomy.[179] This is a relatively specific change, and other dehydrogenases such as isocitric dehydrogenase, malate dehydrogenase, and lactate dehydrogenase are unaffected by administered steroids. The consequences of this change in α-glycerolphosphate dehydrogenase are not fully established but are likely to involve alterations in lipid composition since α-glycerolphosphate dehydrogenase is intimately involved in phospholipid biosynthesis.[180-181]

Continuous hydrocortisone treatment has been reported to increase brain 5-hydroxytryptophan decarboxylase activity in young animals[182] and elevations of this enzyme in peripheral tissues following continuous steroid administration also have been reported.[183] Alterations in brain 5-hydroxytryptophan decarboxylase by glucocorticoids has not been found by all investigators, however.[189] Although this is an important enzyme, catalyzing the decarboxylation of both 5-HTP and DOPA,[185] the normal level of enzymatic activity is well in excess of need. Conceivably, however, corticoid alterations in the activity of this enzyme could have significance under very special conditions if, indeed, the enzyme is elevated by corticoid treatment.

Tryptophan hydroxylase is normally the rate-limiting enzyme in serotonin synthesis.[186] This enzyme has recently been reported to be decreased in brain following adrenalectomy and the decrease can be partly reversed by administration of glucocorticoids.[187] Further, chronic administration of glucocorticoids has also been reported to elevate liver tryptophan hydroxylase activity.[188] Although serotonin synthesis is primarily dependent upon centrally hydroxylated tryptophan, peripheral supplies of 5-hydroxytryptophan are also required to maintain normal levels. As a result, glucocorticoid action on both the central and peripheral tryptophan hydroxylases might be expected to influence brain serotonin. Data supporting this expectation conflict. Early reports, using a bioassay technique for serotonin and complicated by operative stress effects, suggested that adrenalectomy and hypophysectomy decreased cerebral serotonin.[189,190] However, other studies reported an increase in serotonin[191,192] and still others failed to find any effect of adrenalectomy or hypophysectomy on serotonin levels.[184,193-196]

The influence of administered adrenocorticoids on serotonin levels is also unclear. Chronic administration of hydrocortisone has been reported both to increase brain serotonin levels[197,198] and to be without effect.[199,184] Chronic corticoid treatment has also been reported to block the increase in serotonin normally following tryptophan administration.[184] Acute hydrocortisone administration is reported to cause a fall in brain serotonin.[200]

Presumably these conflicting reports result from the variety of tryptophan metabolizing enzymes induced by hydrocortisone and the subtle alterations in the degrees of induction that may occur if other hormones are elicited by the stress of the treatment. Thus, tryptophan oxygenase activity is also elevated by hydrocortisone as are liver tyrosine and tryptophan aminotransferase. Even brain tyrosine aminotransferase shows some marginal increase.[201] Hormonal induction of these enzymes, in turn, can be influenced by diet and by other hormones such as glucagon and growth hormone released in specific stresses. The circadian rhythms exhibited by some of the enzymes add a further complication so that balances between different catabolic routes of tryptophan metabolism may be in constant flux, and this flux is reflected in the circadian rhythms of serotonin in some tissues. Indeed, these studies on serotonin formation, even in the uncomplicated experimental condition of administering pharmacologic levels of hormone, illustrate the complexity of dissecting the shifting patterns of metabolic currents which are inherent in the biological changes following the more complex signal of a stressor.

Brain levels of dicarboxylic acids and γ-aminobutyric acid (GABA) also appear to be influenced by glucocorticoids and these changes may be relevant to the altered seizure thresholds of adrenalectomized and corticoid-treated animals. Cortisone[(202)] or cortisol[(164)] treatment of intact rats produces a fall in GABA[(202)] and glutamic acid.[(164)] Glutamine remains constant[(202)] or increases.[(164)] Although glutamic acid decreases upon a single injection of glucocorticoids, levels of this amino acid are increased upon chronic glucocorticoid treatment. This may reflect its role as an energy source in the acute conditions, and as an intermediary in gluconeogenesis in the chronic one.

Paradoxically, alterations in these amino acids following adrenalectomy are not simple reversals of the changes produced by glucocorticoids. Rather, glutamic acid, glutamine, and GABA all decrease after adrenalectomy[(202–204)] as does aspartic acid.[(204)] Indeed, adrenalectomy is followed by a general decrease in free amino acid pools in brain.[(202)] The activities of glutamic acid decarboxylase and γ-aminobutyric acid transaminase[(203)] also decrease. These changes in amino acid levels and enzymatic activity probably reflect more general alterations in cerebral metabolism and protein synthesis accompanying adrenalectomy. Administration of adrenocorticoids, particularly deoxycorticosterone, or cortisol, partly reverses some of these changes in adrenalectomy. For example, both hormones prevent the slight fall in glutamine following adrenalectomy but only deoxycorticosterone reverses the fall in GABA and partly restores glutamic acid levels.[(202)]

Perhaps the most dramatic effects of glucocorticoids on brain are the marked changes in brain size and composition following treatment of the neonate with corticoids. A single injection of corticoids on the first day of life markedly retards whole body growth[(205,206)] and even more markedly affects brain size, particularly cerebellar size.[(205)] These weight changes are accompanied by decreases in brain DNA indicating a reduction in cellular content.[(205,207)] Brain RNA, protein, and cholesterol are also reduced.[(206,207)] Functional changes accompany these changes in cell number and cellular composition. Thus, corticoid-treated animals have premature eye opening but visual maturation is delayed, as evidenced by a delay in development of the evoked response to light.[(208)] The decrease in body growth appears due to decreased cellular sensitivity to growth hormone and is of some clinical importance since chronic corticoid treatment of children with status asmaticus leads to retardation in growth.

Corticoids and ACTH can pass the placental barrier[209,210] and conceivably, severe maternal stress could have effects on the developing fetus, thus lending some credence to the old wives' tale of a relationship between maternal stress and aberrant birth. This fascinating subject of neonatal corticoid influences has been carefully reviewed by Schapiro[211] to whom the reader is referred for further information.

C. Adrenal Catecholamines

If stressors are disturbers of homeostatic peace, then in higher animals at least, adrenal corticoids act both as restorers of equilibrium and metabolic planners for future onslaughts. However, their major biological actions are slow in comparison to the immediacy of events and are less geared to the crisis of the moment than to those of the immediate future. Biological responses to the immediate threat are thus governed by other systems. Further, different stressors obviously present different challenges and require different metabolic answers. What is good to meet cold may be deadly in meeting heat. Were corticoids to act alone and were they to inexorably elicit the responses tabulated above, they could hardly serve to generate the metabolic flexibility required by the organism for survival. Metabolic flexibility does exist, however, and is achieved by interactions between adrenal corticoids and other systems and, perhaps, even by interactions between different adrenal corticoids themselves. Growth hormone is released in fasting[212] and thyroid hormone by cold,[213] and these interact with adrenocortical hormones at target tissues to elicit different metabolic melodies from the otherwise constant tone of adrenocortical action. Each of these hormones may have actions that are antagonistic, synergistic, or independent of adrenal cortical actions at specific

TABLE IV

Hormones, Other Than Corticoids, Involved in Stress Responses

	Hunger	Thirst	Cold	Surgical	Fear	Anxiety	Shock
Growth hormone	+[a]	0	+(?)	+	+	0	+
Thyroxine	0	0	+	0	0	0	−+[b]
Catecholamines	0	0	+	0	+	+	+
Glucagon	+	0	0	0	0	0	0
Aldosterone	0	+	0	0	0	0	0

[a] + = increase; − decrease; 0 no change.
[b] Decrease in acute; increase in chronic.

sites, and their interplay is enormously complex. Thus, growth hormone inhibits induction of tyrosine aminotransferase[214] while adrenal corticoids stimulate it. Growth hormone markedly stimulates ornithine decarboxylase,[215] an action which is augmented by thyroxine but unaffected by adrenal steroids. Thyroid hormone inhibits liver monoamine oxidase and catechol-*O*-methyltransferase[216] and, perhaps, thereby augments the effects of epinephrine[217] which is also released by a variety of stressors. Malic enzyme is decreased in fasting[218] but stimulated by thyroxine[219] so that a cold and hungry animal may be under different controls than one that is cold or hungry alone.

The variety of hormones which may be summoned by particular stressors, the metabolic complexity of hormonal actions, and the modifications in such actions as a result of the metabolic state of the animal when stressed, on the one hand permit the metabolic flexibility essential for life and on the other present enormously complex phenomena for investigation. Further, stressors may indirectly alter rates of hormonal degradation or synthesis, thereby altering the balance of hormonal interactions. For example, the adrenal steroids are metabolically interrelated so that small shifts in controlling enzymatic activities may change ratios of precursor and product, both of which are hormonally active compounds. While studies have not yet been carried out on possible qualitative and quantitative differences in the admixture of adrenal steroids produced by different stressors, quantitative and qualitative differences in the effects of different adrenal steroids are known, and these could have some significance in stress responses. For example, deoxycorticosterone decreases liver alanine aminotransferase activity[220] perhaps by blockade of ACTH release[221] without altering tyrosine aminotransferase[222] or tryptophan oxygenase[223] while hydrocortisone induces all three enzymes. A stress-induced shift in the rate of conversion of 11-deoxycorticosterone to corticosterone could specifically alter such enzymatic processes and more generally shift metabolism toward either deoxycorticosterone-mediated actions on electrolyte and water metabolism or corticosterone effects on carbohydrate and protein metabolism.

It is beyond the scope of this chapter to detail in depth the myriad endocrinological, physiological, and biochemical properties of all the hormones involved in stress reactions and to explore their interrelationships in detail. Specific aspects, therefore, will be considered in discussion on specific stressors. The adrenosympathetic system requires some special comment, however, both because many acute stress responses are mediated by this system and because this system plays such a large role in the stress literature, especially that portion devoted to man.

The framework for viewing emergency "fight or flight" reactions was first fully formulated by Cannon[224] following earlier demonstrations that stimulation of sympathetic nerve endings releases norepinephrine. Certain tissues contain epinephrine as well and in some the epinephrine is contained within histologically distinct chromaffin cells. Of greatest interest to stress studies are the catecholamines in the adrenal medulla. In this tissue catecholamines are stored in dense granules(229,230) in a complex involving 4 ATP's per catecholamine together with unidentified protein and lipid.(231) Whether more than one type of granule is involved in the storage of the amines is uncertain although histochemical and quantitative differences in the relative quantities of amines under certain conditions(232–234) suggest that this may be the case. The existence of a relatively stable and a labile pool of adrenal catecholamines has also been posited.(235) Norepinephrine is stored in granules in sympathetic nerve endings(236–238) but these granules differ from those in the adrenal medulla(239) although, in both, ATP and the catecholamines seem to exist in the same stoichiometric relationship.

The ratio of epinephrine to norepinephrine and the absolute level of catecholamines in the adrenal medulla appears to differ between species.(240) Epinephrine predominates in all cases. In the rabbit the ratio of epinephrine to norepinephrine is 50:1, in the rat 10:1, in man 5:1, and in the cat 5:2. Total catecholamine content of the adrenal is about 1.2 mg/gm in the rat and about 0.6 mg/gm in man.

The synthesis of catecholamines is reviewed elsewhere in this series and will not be discussed in detail here. All of the enzymes required for synthesis are present in the sympathetic nerve endings. Tyrosine hydroxylase, the rate-limiting step in norepinephrine synthesis(241–242) converts tyrosine to 3,4-dihydroxyphenylalanine (DOPA) which is, in turn, decarboxylated to 3,4-dihydroxyphenylethylamine (dopamine) by 5-HTP-DOPA decarboxylase. Dopamine is transported to cytoplasmic particles(243) containing dopamine-β-oxidase(244) which converts it to norepinephrine. In addition, some vesicles in the adrenal appear to contain the enzyme phenethanolamine-*N*-methyltransferase which methylates the free amine on norepinephrine to form epinephrine.(245) At least two sets of controls exist on this system. Tyrosine hydroxylase is inhibited by its end product norepinephrine(242) to provide a negative-feedback loop. *In vivo*, this control is probably exercised by free rather than particle-bound norepinephrine. Second, glucocorticoids,(246) insulin and glucagon,(247) stimulate *N*-methyltransferase, thereby permitting some regulation of the epinephrine to norepinephrine ratio. This is of interest both because it relates catecholamine synthesis and hormonal status and also because glucagon and insulin, which are normally considered to act in opposition, both exhibit the same action on this enzyme, as they do also in inducing liver tyrosine aminotransferase.

Upon neural stimulation, both epinephrine and norepinephrine are released into the circulation by a process involving influx of calcium into the storage vesicle.(248) This is accompanied by release of ATP(249) and dopamine-β-oxidase.(250) Once released, the catecholamines are rapidly bound by peripheral tissue,(251) degraded, or excreted as either free or conjugated catecholamine. The major urinary product is 3-methoxy-4-hydroxymandelic acid (VMA) which is normally excreted at a level of about

5 mg/day.[252] The remaining urinary products are present in much lesser amounts: 3,4-dihydroxymandelic acid, 100 μg/day;[253] normetanephrine, 100 μg/day; metanephrine, 50 μg/day;[254] free norepinephrine, 40 μg/day; free epinephrine, 5 μg/day; conjugated norepinephrine, 100 μg/day and conjugated epinephrine, 15 μg/day.[255] Degradation after release takes place primarily in liver by reaction with liver catechol-*O*-methyltransferase to produce metanephrine and normetanephrine which are then oxidized to VMA.[256]

Catecholamine taken up into tissue appears to go to two pools, one of which turns over rapidly and produces *O*-methylated derivatives, and a slower one which produces deaminated derivatives.[251,257] The faster pool is believed to represent metabolically active amine which is degraded by catechol-*O*-methyltransferase in the synaptic space while the slower pool is thought to be inactivated within the tissue by mitochondrial monoamine oxidase.

Blood norepinephrine levels are tonically maintained at a very low level (about 0.3 μg/liter)[258] by continual secretion and loss from noradrenergic neurons. Epinephrine levels wax and wane as a function of periodic bursts from the adrenal medulla. Although many of the pharmacological actions of epinephrine resemble the physiological responses in acute stress, it is only in very extreme stress that epinephrine really contributes to these responses, but, instead, the increased heart rate, heart stroke, the rise in systolic pressure and the diversion of blood flow to the muscles are manifestations of activation of the sympathetic nervous system.[259] Rather, epinephrine exerts its primary influence on stress adaptation by metabolic means.

One of the most important of these metabolic actions is to stimulate the formation of glucose from glycogen. In common with various other hormones, epinephrine stimulates adenylcyclase activity[200] and production of cyclic 3′5′-AMP.[261] This occurs in many tissues including brain.[262] In liver, adenylcyclase activates diphosphorylase kinase[263] which phosphorylates phosphorylase a to form phosphorylase b. Phosphorylase b then acts to convert glycogen to glucose.[264] In muscle a similar set of reactions occurs but this tissue lacks glucose-6-phosphatase so that lactic acid accumulates.[265] The end result of these reactions is an increase in blood glucose and blood lactic acid. The circulating lactic acid is picked up by the liver and converted to glucose by an epinephrine-stimulated gluconeogenesis.[266] Further, epinephrine stimulates release of free fatty acids from adipose tissue,[267,268] heart,[269] and diaphragm.[270] In adipose tissue this release of fatty acids is also thought to be mediated by cyclic AMP activation of lipase.[271] Epinephrine's effectiveness in stimulating release of free fatty acids from adipose tissue is markedly reduced by adrenalectomy[272] and restored by administration of adrenocorticoids, but only if given prior to administration of epinephrine.[273]

The catecholamines, then, serve in acute stresses to prepare the organism for action, by activation of the noradrenergic sympathetic system, and to provide the energetic resources to carry out such actions, by the metabolic effects of epinephrine.

III. STRESSORS

Conceptually, stressors can be divided into those which compromise the organism and require restorative processes and those which threaten but which are innocuous. The real physiological stressors of the natural environment, hunger, thirst, infection, exhaustion, cold, heat, tissue damage, etc., are encompassed in the first category, while the laboratory stressors of shock, shaking, immobilization, handling, etc., comprise the second. Experimentally, this distinction is much harder to make, particularly in the case of the animal. Humans in a stress study on hunger, for example, are generally secure in the belief that their hunger is temporary and that death by starvation is not part of the experiment. The rat has no such assurance. To the extent that stressors elicit different responses, animal studies are necessarily more contaminated by cross stressor responses. These, of course, occur in man as well since anxiety before the experiment may commingle sleep deprivation, for example, with the scheduled stressor, and anger at the discomfort or anxiety over physical responses may also contaminate the scheduled experiment.

Although the subsequent sections are grouped according to the stressor, it should be clear that the literature reported may be influenced by stressor interactions. An attempt will be made throughout to differentiate acute from chronic conditions since responses change with adaptation. Indeed, the stress response is an attempt to restore homeostatic balance and if it is successful in restoring a new level of homeostasis, the organism is no longer compromised and the stress response is unnecessary even though the stressors appear uncomfortable to the investigator. Age, too, is an important variable in stress adaptation but it cannot be fully specified here, beyond identifying young and old, because of the complexity of the literature. Very early postnatal life is of particular importance, however, and will be discussed in specific contexts.

A. Psychological Stressors

Some stressors directly and physically compromise the organism but the majority are either threats of potential danger or symbols of past dangers. These loosely constitute the psychological stressors and they represent a diverse group of stimuli united only in their ability to generate emotional disequilibrium. As in all stimulus response systems, discernment of the stimuli is essential to the response. Stimulus recognition is both automatic and relatively uniform in response to physiological stressors. This is not true for the psychological stimuli. An overdue traffic citation, a family quarrel, an angry boss, or an overdue chapter elicit anxiety, anger, fear, or guilt only to the extent that the stimulus is seen as an emotional threat. A psychopath may take delight in stimuli that for others generate horror, and a schizophrenic may be more or less stressed than a control subjected to the same experimental manipulation. Indeed, studies on stress responses of the mentally ill are often difficult to interpret because of the problem in assigning emotional meaning to a presumed stressor.

The difference between stimulus and stressor also prevents easy comparisons between animal and human studies. Man may respond to purely psychological stimuli and verbally recount his psychological responses for comparison with his physiological ones. An animal's responses must be inferred from his performance and the elicitation of such performance often invokes physiological stressors in conjunction with psychological ones. A rat made anxious by a forced response is only interested in making the response if he is to be rewarded (with food, for example) or punished (shock avoidance, for example). In the first instance, he must be made hungry or thirsty to perform, and in the second, he must be trained to anticipate an unpleasant stimulus. In both, physiological parameters of hunger or thirst or pain are interwoven into the psychological parameters of anxiety or fear or confusion.

In this section, then, human and animal studies have been separated and separate stressors discussed somewhat apart from each other.

1. *Man*

Stress is assumed to play a vital role in both mental and physical illness and most stress research has been directed to evaluating and elucidating this role. Paradoxically, although man is the relevant animal, he is the least suitable for biological studies, while yet the most suitable for psychological ones. While the psychologist can attempt to reach the mind by the relatively direct route of observation and conversation, the biologist must take a more devious path to the brain through rivers of biological fluid. The trip is not without its conceptual and practical hazards.

The prime biological fluids for examination are blood and urine. The process of drawing blood constitutes a stressor to many individuals, so that the stress of observational techniques is superimposed on that of the experimental stimulus. Collection of a urine sample is less stressful to the subject, but the sample itself is a solution collected, at least internally, over time, and the temporal relationships between internal production and external secretion are not always simple, and even less often known.

Just as the variables examined are limited to those which do not compromise the subject, the stimuli employed are limited to those with transitory and, hopefully, harmless effects. In effect, these stimuli are usually situational and are difficult to precisely define, to dissect into component stressors, or to quantify. When field studies are carried out on individuals subject to physical stressors, such as subjects on army maneuvers, stimuli are generally composites of one or more of the stressors of hunger, cold, exhaustion, injury and fear, so that responses are complexes of different adaptations. Finally, because man not only experiences various stressors but anticipates them as well, relating data to actuality rather than eventuality is difficult, and this is only compounded by man's ability to shift his mental presence from the reality of his actual environment to the fantasy of an imagined or anticipated one. Anticipatory responses are not limited to man, of course, Rats warned of an impending shock by a tone lose more weight than their equally shocked counterparts,[(274)] but man is more rapidly conditioned and the form and extent of anticipation more subtle. As a result of these and other factors, biological studies on man are seldom clean with regard to stimuli, uniform with regard to individual bits of data, or clear with regard to interpretation. Studies do not fall into easy categories and, consequently, the following discussion must be more general than specific.

The wide variety of situations which have been used in studying stress responses in man can be roughly divided into five categories: (1) Those demanding responsible decisions in the absence of overt physical activity. (2) Those in which responsibility is coupled with physical action. (3) Those in which the subject is an impotent observer in a dangerous situation. (4) Those in which the subject is a vicarious participant. (5) Those involving a personalized test of ability. An example of the first category is that of an airline pilot called upon to make a difficult landing; of the second, a member of an athletic team; of the third, an airline passenger in a storm; of the fourth, a spectator at a movie or a sports event; and of the fifth, a scholastic examination or a pyschological test. These situations differ in degree of necessary participation, degree of control, and importance of the outcome. The airline pilot must not make an error but he has some control over the situation. The passenger cannot prevent the error but the consequences are personally serious. The ball player shares control and, though important, the consequences of failure are not dangerous. The spectator is involved but not concerned. The student is threatened but not endangered.

Individual variation in such situations is great, probably dependent on each individual's estimation of the degree of personal involvement, extent of the threat, anticipated consequences, and degrees of personal control. General trends, however, exist.

Serum and urinary corticoids are released in a wide variety of psychological situations[(275–282)] involving fear and anxiety. Generally, these changes can be related to the specific experimental situation but at times the relationship can be deceptive. For example, one study,[(283)] using subjects about to leave the hospital, involved responses to traumatic and neutral movies. Surprisingly, corticoid elevations were observed in both cases. This was explained as a true stress response to the traumatic movie but a displaced stress response to the neutral one reflecting the subject's anxiety at leaving the hospital situation; an anxiety which was allowed to emerge during the innocuous neutral situation.

Although quantitative alterations in corticoid levels are taken as the stress index, qualitative changes in corticoid composition as a function of different stressor components in the stress situation have not been thoroughly investigated. One report of altered corticoid composition has appeared,[284] however, and more work in this area needs to be done.

A number of other stress-related variables are altered by emotionally charged situations. Plasma cholesterol increases in subjects exposed to such varied stimuli as occupational overload or examinations in medical school.[285-287] Serum lipoproteins[288] and plasma free fatty acids[289-292] rise in anxiety and fear probably because of concurrent elevations in catecholamines. These various indices do not always move in concert, however. Indeed, levels of plasma free fatty acids appear to correlate with anxiety but not hostility, triglyceride levels with both anxiety and hostility, and cholesterol levels with hostility alone.[293] However, most real stress situations are sufficiently potent to call forth whole sets of emotional responses and the dissection of what response belongs to what emotional component is exceedingly difficult.

Blood lymphocyte levels are lowered by situations such as race driving[294] and college examinations,[295] ion excretion is altered by psychological tasks like rotary pursuit or stressful interviews,[282,296,297] and even water excretion is changed by stressful situations.[298] This last is of considerable practical importance in complicating comparisons of urinary constituents under different conditions, since the rate of excretion or the urinary concentration of a substance may be differently modified depending upon whether the compound studied is excreted in a volume-dependent or volume-independent manner.[299]

Catecholamine excretion changes in many stressful situations[300-303] and some psychophysiological[304] and biochemical studies suggest that the ratio of excreted norepinephrine to epinephrine is related to the direction of aggression or to relative differences in aggression and anxiety provoked by the stimulus. For example, it has been reported that individuals with high drive excrete more catecholamine and have higher epinephrine to norepinephrine ratios than their less ambitious counterparts.[305] Studies on athletes awaiting action and those involved in action suggest that active aggressive displays are associated with increased urinary norepinephrine excretion while anxiety is associated with increased epinephrine excretion.[306,307] This view has received some support[308] from a variety of studies. Circulating norepinephrine levels appear correlated more with the degree than with the kind of emotional response,[309] but while urinary excretion of epinephrine and norepinephrine intercorrelate well, only epinephrine correlates with 17-hydroxycorticoid excretion.[310] This correlation between urinary epinephrine and urinary corticoid excretion levels would agree with the posited relationship between anxiety and both corticoid and epinephrine excretion.

Not all stimuli eliciting elevations in catecholamine excretion are necessarily unpleasant, nor is the catecholamine ratio a clear discriminator of emotional state. Subjects viewing emotionally neutral movies have less than normal catecholamine excretion, but both funny and tragic movies elevate excretion of epinephrine and norepinephrine.[311] Further, the excretion of epinephrine was greatest when anger was a strong component in the emotional response.

Only a small fraction of the catecholamine produced in the organism is excreted as free amine. A much larger proportion is excreted as metanephrine and normetanephrine, and these compounds appear to move in unison, being low during quiescence and high during agitation. Their ratio remains relatively constant in both conditions.[312]

Under some conditions and with some individuals, stressful situations may also alter thyroid function. Reports of elevated levels of protein-bound iodine following examinations for medical school,[313] and viewing of stressful movies[314] have appeared. This response, however, may be biphasic since one study[315] reports an initial decrease in protein-bound iodine immediately following a traumatic interview which was followed by a slow increase in levels. The generality of these changes to other stressors has yet to be established, however.

2. *Shock*

As an experimental stressor, electric shock has the considerable advantage that duration and intensity of administered stimuli may be regulated within fairly narrow limits and that the requisite equipment is readily available. It has the disadvantage that the degree of control of stressor is more apparent than real and that the stressor itself is not a simple one. Although delivered voltage or amperage may be controlled, resistance is variable depending upon animal size and positioning and the appearance of unintended conductors like feces and urine. The animal who freezes and takes the full period of shock is stressed differently than the animal who jumps, rolls, and bridges to escape. The experience itself is a compounding of fear, pain, muscular exertion, sometimes transitory anoxia, and sometimes transitory anorexia.

Despite these difficulties, shock has been used as a stressor per se, a tool for memory disruption, a model for electroconvulsive therapy, and a side product of neurophysiological studies, particularly those on reward and punishment systems.

Like several other stimuli, electric shock is an ineffective stressor during the early postnatal stress-nonresponsive period.[211] Thereafter, electric shock results in the usual adrenocortical activation characterizing stress. Thus, serum corticoids[316] and adrenal corticoids[211] are elevated, adrenal size is increased and blood eosinophiles drop.[317] Urinary sodium and potassium are not altered,[319] suggesting that mineralocorticoids are not involved in the response. As might be expected from the fear component of the stressor, adrenal catecholamines are released[224] and blood glucose is increased[318] probably by a direct action of epinephrine on conversion of glycogen stores to glucose since adrenalectomy abolished the elevation in blood glucose.[318] Further, blood glucose is markedly elevated by other stressors having a strong fear component, such as restraint[319,320] or fighting[321,322] to the extent that renal threshold is exceeded and glucosuria results.

In contrast to stressors such as cold or handling, little qualitative adaptation occurs upon repeated shock treatment, at least as measured by these indices.[323]

Shock has recently been used to evaluate the dual questions of whether stressors modify central nervous system metabolism of monoamines, and whether such modifications represent an involvement of monoamine-containing neurons in the mediation of central and peripheral responses to stress. Mediation would imply activation of such neurons and this should be reflected in increased transmission and, thereby, either alterations in the steady-state levels of the monoamines or an increase in their turnover. Such changes alone, of course, do not demonstrate primary mediation, since monoamine metabolizing enzymes may themselves be altered by hormones or changes elicited during stress which would in turn modify amine levels or turnover.

In contrast to results with other stressors, results obtained using shock have been relatively consistent. Levels of norepinephrine appear unchanged by low level foot shock (1.6 mA)[324] although norepinephrine levels are decreased by more strenuous foot shock (2–5 mA).[325–327] At both intensities of shock, norepinephrine turnover is enhanced[324,326,327] as a result of increases in both synthesis and degradation, although with the higher intensity shock, degradation appears to exceed synthesis. Remarkably enough, however, the drop in norepinephrine can be prevented by pretreatment with inhibitors of monoamine oxidase, but not inhibitors of catechol-*O*-methyltransferase.[326,327] Since transynaptic metabolism is thought to involve *O*-methylation, these results present something of a paradox. They are consistent with the finding that deaminated products of normetanephrine are the predominant metabolites during shock with a decrease in normetanephrine at low shock intensities[327] and a small increase in normetanephrine at higher levels.[324] This could indicate that the primary action of shock is to release norepinephrine from vesicles intraneuronally where it is subject to intracellular deamination by mitochondrial monoamine oxidase and that the increase in turnover does not represent increased noradrenergic innervation but rather release of tyrosine hydroxylase from feedback inhibition. This, together with increased degradation of unbound norepinephrine, would lead to both increased turnover and a rise in deaminated metabolites. Some degree of adaptation has been noted in that norepinephrine turnover in shock is accelerated by previous shocks or shock of higher intensity.[324]

Alterations in dopamine levels are less clear. Decreased dopamine levels following shock have been reported[325] although in most studies levels and turnover appear unchanged.[324,326,327] Foot shock administered to animals pretreated with α-methyl-*p*-tyrosine, which blocks tyrosine hydroxylase activity, leads to a greater fall in dopamine than that accounted for by the action of the inhibitor alone, suggesting that this monoamine, too, is under dynamic control and that its catabolism is also accelerated by shock.(327)

Like dopamine, serotonin levels are essentially unaltered by foot shock,(327) but turnover appears enhanced, as determined both by the increased rate of conversion of labeled tryptophan to serotonin(328) and by an increased formation of brain 5-hydroxyindole acetic acid.(327) In contrast to norepinephrine, prior shock experience does not alter the extent of serotonin turnover.(328) Acetylcholine levels are not altered by foot shock.(326)

Although electroconvulsive shock could be considered a more severe form of foot shock, it does differ in the degree of transitory anoxia and muscular exertion as well. Electroshock may increase the permeability of the blood–brain barrier to norepinephrine and other agents.(329) Chronic electroconvulsive treatment in rats increases brain weight, brain protein content and total acetylcholinesterase.(330) Subsequent behavior is markedly impaired.(330,331)

Handling parameters also differ significantly between shock studies and electroconvulsive treatment, and, as has been pointed out,(333) the usual control groups may not be appropriate to studies on electroconvulsive shock. Handling in electric shock studies is usually limited to placing animals into, and removing them from, the shock box. In studies on electroconvulsive shock, electrodes must be directly placed on the head, and the extent and immediacy of handling adds another stress dimension. Finally, rapid metabolic shifts occur during the convulsive episode which are not paralleled by events during foot shock. As an example, glucose uptake by brain is retarded during convulsions and increased thereafter.(334) For the period of the seizure itself, acetylcholine rapidly falls;(335) ammonia production increases;(336) brain glycogen, ATP, and glucose decrease;(337) and oxygen consumption and blood flow increase.(338) Restorative processes occur during the postictal phase.

Because of these differences it is not too surprising that the extent and direction of monoamine changes in electroconvulsive shock differ somewhat from the changes with foot shock, or that findings are more discrepant in studies on electroconvulsive shock, particularly because of differences between experimentors in what is selected as a suitable control group. Thus, brain norepinephrine levels after electroconvulsive shock have been reported to be unchanged initially and then to fall slightly,(332) to rise initially and then to slowly fall,(339) and, after chronic treatment, to rise initially in brainstem and, after some hours, to be increased in telencephalon and diencephalon.(333,340) Turnover is reported to be increased and persistent(340) and brainstem and cortical tyrosine hydroxylase activity is increased.(340a) Amine changes are regionally specific and within regions the temporal pattern appears to vary for each amine(333) so that some of these conflicts may be more apparent than real. Dopamine levels are elevated by even a single electroconvulsive treatment, as is homovanilic acid, the primary dopamine catabolite.[341] Serotonin levels are modestly increased in most studies,[333,339,340,342] but not in all,[343,345] but the increase appears rather slow after a single electroconvulsive treatment and more rapid if the animal has received a series of prior shock treatments.[339] Many of these changes, however, may be related to the handling stress rather than the electroconvulsive shock itself.[333] Histamine levels also reportedly slowly increase following electroconvulsive treatment and the rate of increase is unaffected by prior exposure to electroconvulsive sessions.[344]

Despite the intense stimulation of the central nervous system resulting from electric shock, initial turnover of proteins and nucleic acids only marginally increases,[345] but glutaminase and ammonia concentrations increase while glutamine synthetase activity is unchanged.[346]

Electrical self-stimulation, although apparently pleasurable since it is self-induced and sufficiently rewarding to make animals work for it, is also a stressor. In the monkey self-stimulation is accompanied by increased plasma and urinary corticoids, increased urinary epinephrine, and, less consistently, increased urinary norepinephrine as well.[347] In the rat, adrenal weights are elevated[348] and brain norepinephrine is decreased whether animals are self-stimulating in reward centers or are instrumentally stimulated in punishment areas.[349] Such catecholamine changes reportedly occur only if emotionality is elicited. However, direct electrical stimulation in some brain areas alters adrenocortical activity,[355] so that it is difficult to determine whether adrenocortical changes are associated with the change in emotional status or are the consequence of a spread of stimulation to areas involved in CRF release.

3. *Restraint*

Restraint is another commonly used laboratory stressor which has the apparent advantage of being relatively simple. Again, this simplicity is more apparent than real. At best, this stressor is compounded of fear, isometric muscular activity, and muscle cramps due to specific positioning. In acute experiments plasma[351] and adrenal[352] corticoids are elevated, and, in contrast to stressors such as hunger or shock, some degree of adaptation to experience occurs. Experimentally, this poses some problem since the magnitude of the stressor varies with experience and, more subtly, with differences in positioning and handling between experiments and experimentors. Chronic restraint (continuous restraint for 24–48 hours) leads to involution of the thymus and adrenal hypertrophy[353] and to the production of gastric ulcers.[354] The degree of ulceration and ulcer location varies with species.[355] Mice and rats show a high ulcer incidence, guinea pigs have a lower incidence and hamsters a very low incidence of ulceration. Rabbits and monkeys ordinarily have little or no ulceration upon restraint.

Chronic restraint, of course, is also complicated by concomitant voluntary or involuntary fasting, which varies both with accessibility of food and with the emotional state of the animal.

Metabolically, peripheral tissues, such as liver and heart, show increased oxidation of glucose, pyruvate, α-ketoglutarate and succinate[356] during acute restraint, and heart cardiolipin is decreased.[357] Both of these effects disappear as the restraint episodes are repeated.

Tyrosine aminotransferase activity in liver is increased by chronic immobilization.[358] This increase is dependent upon the adrenals and only occurs in animals showing development of ulcers, although the rise in aminotransferase activity precedes ulcer formation. Tryptophan oxygenase activity is also increased twofold by 6 hr of restraint and increased somewhat more after 24 hr. Adrenalectomy did not abolish this effect of restraint although the enzymatic response was somewhat lessened.[359] Since tryptophan oxygenase is known to be inducible by both hormone and substrate, these results suggest that substrate induction may be involved in this enzymatic response to restraint.

Several reports of brain monoamine changes in restraint have appeared. Short-term restraint (7 min) increased brain norepinephrine levels in mice,[360] particularly in mice housed singly in isolation cages. On the other hand, brain norepinephrine levels of mice, guinea pigs, or rats restrained for 2 or more hours decreased by 10–20%.[356,361,363] Changes were most marked in cerebellum and hypothalamic areas[361] and central noradrenergic neurons appeared depleted histochemically.[363] On the whole, inhibitor studies indicate a more rapid depletion of norepinephrine from noradrenergic neurons.[363] However, prior housing conditions do appear important in this phenomenon, and mice raised in isolation reportedly show a faster fall in norepinephrine upon restraint than those raised in grouped environments.[360]

Acute restraint also appears to result in an increase in dopamine levels in brain,[360] while chronic restraint leads to a fall.[362] Serotonin changes appear more complex. Both long[360] and short term[362] exposure to restraint appear to elevate serotonin. However, it has also been reported that serotonin is decreased after 3 hr of restraint but returns to normal after 6 hr of restraint.[363] Strangely enough, combining shock and restraint eliminates any restraint-dependent alterations in serotonin levels.[361] Further, forced positioning of animals without direct physical restraint other than continually replacing the animal in position, is reported to lead to a decrease in serotonin in cortex and hippocampus.[364] Levels of GABA may be slightly increased by restraint.[361]

4. *Isolation and Aggression*

The relationship between early environment and adult behavior will be discussed in detail elsewhere in this series. However, mention will be made here of environment as a stressor and of stress parameters in aggressive behavior. These two are conjoined because early environmental manipulations are often employed to modify aggression experimentally.

Animals reared in isolation differ considerably from their grouped littermates and this is often an unspecified experimental variable between laboratories and between studies in the same laboratory. The degree and kind of behavioral and biochemical changes vary with the age at which environmental housing is initiated and the species employed. After 4 days in isolation weanling mice (17–21 days old) have a higher activity level[365] than group-housed littermate controls and are socially dominant.[366,367] When isolated as young adults they are more susceptible to ulcers,[368] have higher blood glucose and liver glycogen levels,[369] and are less sensitive to amphetamine,[370] but not cocaine.[371] Indeed, drug sensitivity progessively declines with the number of animals housed together[372] and is modified by age.[373] Adrenal weights and heart weights are also smaller in isolates[372] and, upon rehousing, social dominance correlates inversely with adrenocortical weight and plasma corticosterone.[374] Brain norepinephrine levels in isolates are normally found to be higher,[375] and dopamine turnover is increased although brain levels of dopamine are the same as those in grouped animals.[376]

Brain serotonin levels are unchanged in isolated animals but serotonin turnover is decreased[377] and brain 5-hydroxyindoleacetic acid is low.[378] In contrast, simply aggregating group-housed mice with strangers results in a decrease in brain norepinephrine and an increase in 5-hydroxyindoleacetic acid, while dopamine and serotonin levels remained constant.[379]

In general, then, grouping appears to be a stressor for mice, and isolated male animals are less stressed and more aggressive. The degree of aggression in mice is dependent upon strain[378] and sex.[380] Female mice, for example, do not become aggressive upon isolation and do not show some of the amine changes, such as decreased serotonin, found in aggressive males.[378]

When placed with group-housed animals, isolated mice are aggressors. Exposure of the grouped animal to such aggression results in increased 5-hydroxytryptophan decarboxylase activity in frontal cortex and a decreased activity in amygdala, and these changes become more pronounced as the experience is repeated.[381] Monoamine oxidase activity initially increases and then falls in the hypothalamus, and falls in the amygdala and the frontal cortex, only to rise again after 14 days of fighting.[382] Acute fighting causes an increase in serotonin in brains of the aggressive mice with little change in norepinephrine.[383] Daily fighting for 10 days, however, elevates whole brain dopamine and whole brain norepinephrine without a change in brain serotonin, and adrenal catecholamines and adrenal weight increase.[384] Witnessing fighting is reported to lower norepinephrine levels in mice.[385]

On the other hand, rats raised in isolation have slightly smaller brain weights,[386–389] particularly cortical weights[387,389] than littermates raised in "enriched" group environments despite generally larger body weights among isolates. This change in weight is reportedly due to increased thickening of cortical folds(390) and an increased enrichment of oligodendroglia(391) in environmentally enriched animals rather than less rapid development in isolates. Further, neuronal size is reportedly increased(392) in the group-housed animals. Adrenal weights of isolates are higher,(386) as is the activity of liver tyrosine aminotransferase, but not of tryptophan oxygenase, tryptophan hydroxylase, or phenylalanine hydroxylase.(386) Increased levels of serum corticoids are sometimes(393) but not always(386) found. Whole brain levels of acetylcholinesterase remain constant in the isolated animals while brain monoamine oxidase levels decrease.(394) However, whole brain acetylcholinesterase is somewhat greater in environmentally enriched animals(386,389) due primarily to an increase in cortical acetylcholinesterase.(389,393) Brain lactic dehydrogenase activity is unchanged by housing.(395) Brain norepinephrine levels are higher in isolates than in environmentally enriched animals,(386) and the major elevations appear to be in the region of the caudate nucleus.(396)

In contrast to mice, isolated rats are less aggressive than grouped animals, and, indeed, in encounters between isolated and grouped animals, the grouped animal generally becomes dominant. Behaviorally, isolated rats exhibit considerable freezing behavior and perform less well on a variety of relatively complex behavioral tasks than grouped animals.(389)

In the case of the rat, then, although norepinephrine levels are raised, this rise is not associated with increased aggression. The mechanism for the increased norepinephrine levels is obscure. Estimation of activities of enzymes involved in norepinephrine biosynthesis (tyrosine hydroxylase and 5-HTP-DOPA decarboxylase) or degradation (catechol-*O*-methyl transferase and monoamine oxidase) in various brain regions failed to reveal any clear differences accounting for the regional changes in norepinephrine.(396)

The association between stress and the norepinephrine change is also unclear. In unpublished experiments, norepinephrine levels of isolated rats are initially lower than that of group-housed rats but rise steadily with time to a higher level. These changes are concomitant with an increase in adrenal weight and corticoid content in isolated rats and an increased activity of tyrosine aminotransferase in liver. This latter measure returns toward baseline as brain norepinephrine levels in isolates increase beyond the level found in grouped animals. This would be in accord with an initial stress response upon isolation in rats with a fall in norepinephrine accompanying signs of adrenal activation followed by a slower compensatory overshoot in norepinephrine content.

Serotonin levels(378,386) and turnover(378) are the same in isolated and grouped animals as are brain dopamine levels.(386)

Isolated dogs also show abnormal behavior in adulthood as well as regional alterations of glutamic acid, glutamine, GABA and aspartic acid concentrations(397) particularly in amygdala, hippocampus, and caudate.

The literature on effects of early isolation and group environment is not without contradictions. These arise in part from strain and species differences and in part from the experimental condition itself. Because animals can be differentially housed early in life and maintained in specific housing conditions indefinitely, long-term effects of stress can be examined. However, the age of onset of treatment is of considerable importance. Animals weaned early may be more severely stressed than those weaned later. In part, this may result from the nutritional problem in switching from liquid to solid food in very young animals. Further, temperature regulation is not firmly established at early ages and isolated animals may suffer from cold even in constant temperature rooms. Both of these factors, in turn, may modify eating patterns and there is now ample evidence that early nutritional deficits may markedly alter subsequent development. In strains of animals not free of chronic respiratory disease, death rates among isolated rats are invariably higher than among group-housed rats, possibly as a result of synergism between infection and stress. The survivors represent a selected subpopulation which may not be representative of the collective biology of the initial group.

Finally, isolation itself is generally a relative rather than absolute condition. In most instances isolation cages are not soundproofed so that noises of other rats, sounds of running water, laboratory experiments, etc., all add stimuli and decrease the degree of isolation. At best, cages must be occasionally cleaned and the very manner of cleaning may modify the biology and behavior of the isolates. All of these factors must therefore be recognized in evaluating this aspect of the stress literature.

5. *Miscellaneous Stressors*

A number of other situations have been employed in psychological stress studies. Sound, swimming to exhaustion, shaking, and centrifugation, to mention a few, have been employed singly or in combination with other stressors. The vast, but spotty, literature in this area cannot be covered in detail, but some selected portions are illustrative of general problems of stress and therefore deserve mention.

Intense sound would not overtly appear to call forth metabolic displacements. However, animals exposed to high intensity auditory stimuli show an increase in adrenal weight and a decrease in adrenal cholesterol and blood glutathione.[398] Perhaps of more importance, gravid female rats or rabbits exposed to 10 min of a medley of high intensity sound per hour produced a much higher incidence of subnormal offspring[399] than normally housed females. This finding is of some relevance to studies on long-term consequences of stress and may be even more important to man faced with increasing noise as a by-product of civilization.

Swimming to exhaustion has been widely employed to study brain monoamine changes in stress as a result of the initial finding by Barchas and Freedman[400] that brain serotonin increased while norepinephrine decreased as a result of such treatment. These reports have been confirmed[401] and it also appears that norepinephrine uptake by brain is retarded in this stress. These changes in brain norepinephrine are analogous to those occurring when shock or restraint are employed as stressors. Contributions to these changes from the physiological effects of exhaustion are difficult to assess since exhaustion alone produces central effects. For example, brain histidine levels are lowered.[402]

Results obtained using a shaker as a stressor illustrate two important points. Rats exposed to this continuous earthquake show marked adrenal activation, large and persistent increase in plasma and adrenal corticoids, and a decrease in adrenal ascorbic acid.[403] Despite these changes, neither tyrosine aminotransferase nor tryptophan oxygenase levels increase, although both are elevated by injection of hydrocortisone.[403] Indeed, this stressor blocks the effects of administered hydrocortisone on tyrosine aminotransferase induction.[404] Shaking adrenalectomized animals leads to an actual lowering of tyrosine aminotransferase, and this inhibitory action is partly abolished by adrenalectomy and completely abolished by hypophysectomy together with adrenalectomy.[405] These results suggest that this stress activates a mechanism, probably pituitary in origin, which represses activation of tyrosine aminotransferase and prevents the inductive effects of hydrocortisone on this enzyme. More generally, they demonstrate that stress does not lead to invariant enzymatic responses.

A second point emerging from such studies is that infant and adult enzymatic responses to the same stressor are not necessarily identical. Infant rats shaken by this procedure do not show an elevation in corticoids until about 8 days of age. Tyrosine aminotransferase activity, however, is elevated in infant animals 4 and 8 days of age even though this same stressor blocks elevations in the adult when corticoids are raised.[406] The crossover ages between tyrosine aminotransferase changes and corticoid changes have not been determined with this stressor. Similar inhibition of this enzyme has been obtained in adult animals using laparotomy[407] or electric shock as a stressor.

B. Physiological Stressors

Psychological stressors are often contaminated, purposefully or unknowingly, by physiological stressors. Animals subjected to stressful situations are often anorexic, may have exacerbations of otherwise subsymptomatic infections, may suffer from tissue damage and undergo transitory temperature changes. For this reason, some information on physiological stressors is of importance even to those primarily interested in psychological stress.

Physiological stressors are no more complex than psychological ones, but they are more clearly defined, and, as a result, considerably more information is available. In the brief reviews to follow, no attempt has been made to fully explore all of the available information on the biological effects of these stressors, but rather the reviews are limited to those parameters of some relevance to the general subject of stress or illustrative of particular experimental problems.

1. *Cold*

Because rates of reaction are temperature-dependent, the ability of homeothermic animals to maintain a relatively narrow internal temperature range independent of external temperature has given them a considerable selective evolutionary advantage. The price paid for this, however, is the requirement for physiologic shifts to meet the challenge of changing temperature. Generally, these shifts are designed to alter heat loss and change metabolism to alter heat production. In the case of exposure to cold, this means a transitory increase in nonuseful work, such as shivering, changes in vascular supply, and alterations in metabolism. These, in turn, require a shift in homeostasis from one level to another, and, in principle, cold should be a stressor. Once appropriate adjustments are made, however, the altered metabolism is no longer a stress response, but rather an expression of a new level of homeostasis, and the animal is temperature adapted until such time as the temperature again changes and a new metabolic shift is required. Distinctions, then, must be made between acute and chronic effects of this stress which are different from those, say, in hunger where no final satisfactory adjustment to the absence of food can be made.

The changes accompanying this stressor will be discussed only with regard to adult homeothermic animals. Neonates of warm-blooded animals are born poikilothermic and are able to survive temperatures intolerable to the adult.[408] Indeed, vasomotor mechanisms for heat regulation develop in concordance with the increasing need to maintain constancy of temperature.[409,410] This period of cold-bloodedness varies from species to species being only hours in the case of eider ducks to weeks for the rat. During such early poikilothermic periods, cold is not an equal stressor for the neonate and the adult. The newborn rat, for example, does not respond to cold with adrenocortical activation for the first 13–16 days of life,[411,412] and this refractory period is rather general for many stressors and has been termed the "stress-nonresponsive period" (SNR) by Schapiro *et al.*[413]

Acute exposure of adult rodents to cold raises peripheral corticoids, adrenal corticoids,[414] urinary excretion of corticoids,[415] pituitary ACTH content,[415] decreases adrenal ascorbic acid,[414] adrenal cholesterol,[416] and circulating eosinophiles,[417] and increases adrenal weight.[418] By the criterion of adrenal activation, then, acute cold is a stressor.

This pattern is not uniform for all species, however. Exposure of the human to abrupt cold results in only a slight rise in plasma cortisol after 1 hour's exposure,[419] and urinary 17-hydroxycorticoids are unchanged.[420] There is no change in corticoids in man if temperature is lowered gradually.[421] Adrenal weight of the spider monkey increases somewhat by exposure to cold (but not heat),[422] the rate of 17-hydroxycorticosteroid secretion by the adrenal of cold exposed dogs is only elevated if the dog is exposed to extremely low temperatures around $-50°C$,[423] and the quail responds to cold with elevated secretion of adrenocorticoids but no change in adrenal ascorbic acid.[424]

The effects of cold, even in rodents, are markedly altered by the time of exposure, the rapidity of onset, and the age of the animal. Adrenal ascorbic acid initially drops, returns to normal in 48 hours after cold exposure and continues to increase to a steady peak level 160% above normal after 10 days exposure.[425] This increase in ascorbic acid is not limited to the adrenal alone but occurs in other tissues as well,[426] although only adrenal ascorbic acid is depleted by the stress of cold.[427] In contrast, blood corticoid levels of the rat acutely rise upon cold exposure and only slowly return to baseline. Blood levels of corticoids are normal by day 18, but the relative rates of formation of corticosterone and 18-hydroxycorticosterone are altered, so that their ratio differs from that of controls.[428] Finally, adrenocortical activation in response to cold is much more marked in young rats than in older ones[429,430] due, in part, to the difference in their metabolic reserves.

The autonomic nervous system is rapidly activated by cold,[224] and there is an immediate release of catecholamines[431] from both the adrenal medulla[432] and extramedullary sources.[433] This response is rather general between species and occurs in the dog[434] and man,[435] as well as the rat.[433] Epinephrine excretion increases more markedly than norepinephrine excretion during the period of initial challenge, suggesting marked discharge from the adrenal medulla. With time, however, epinephrine excretion begins to decline while tonic stimulation of the automatic nervous system continues and norepinephrine levels remain high.[436] Normetanephrine excretion is markedly elevated as well,[437] attesting to the continual release and metabolism of norepinephrine during cold. High catecholamine levels in the adrenal parallel the increase in total urinary catecholamine,[438] and, indeed, the nonshivering thermogenesis of the cold-adapted rat appears to be mediated by the catecholamines,[439] and most probably by norepinephrine itself.

These changes in catecholamines are initially accompanied by an acute fall in pituitary growth hormone, which, however, returns to normal within 3 hr.[440,441] This could result from a transient localized hypoglycemia in sensitive hypothalamic areas since growth hormone is released in hypoglycemia. However, blood glucose levels were not altered at the time of growth hormone release. Depression of blood glucose during chronic cold exposure can occur, and with it there is an accompanying fall in insulin levels.[442] Thyroid hormone is also released during the initial phase of cold exposure[443,444] and secretion of pituitary thyrotropic hormone is increased.[445] The persistence of these changes in cold is somewhat controversial.[446,447]

These endocrinological shifts are accompanied by a series of physiologic shifts, although the exact relationship between the two is unclear. The onset of cold triggers rapid sympathetic vasoconstriction[448] so as to control heat loss. Heat production is rapidly accelerated both in rats[449] and in man,[450,451] and metabolic rate increases[452] at the expense of metabolic reserves.[453,454] Norepinephrine mediation of lipase activity,[455] which requires the presence of glucocorticoids,[456] accelerates release of free fatty acids from adipose tissue,[457] plasma free fatty acid levels markedly increase,[456,458–460] triglycerides[461] and other lipids accumulate in liver[462] awaiting degradation, and, as might be expected, total body fat rapidly declines.[454] The release of epinephrine promotes conversion of glycogen to glucose and glycogen sites quickly dwindle.[463] Throughout this period there is an increased excretion of urinary nitrogen[464] attesting to the contribution of protein to the catabolic fire.

This outpouring of metabolic activity often increases acetyl CoA levels faster than it can be utilized by the Krebs cycle, fatty acid biosynthesis, or other metabolic pathways. As a result, acetone, acetoacetic acid, and β-hydroxybutyrate accumulate, and ketosis develops.[465,466] This increase of coenzyme A levels[467] in tissues other than the adrenal inhibits synthesis of fatty acids while promoting their oxidation,[468] and this process, together with the decline in reduced NADPH following the decrease in intracellular glycogen,[469] decreases formation of lipid from acetate.[470]

Not surprisingly, levels of metabolic enzymes are altered during cold stress and adaptation. Within 6 hr histidine decarboxylase activity increases[471] and in 3 to 4 weeks there are increased hepatic activities of arginase, glutamic-oxalacetic aminotransferase, tyrosine aminotransferase, and tryptophan oxygenase.[472] All but the latter two are attributable to the increased food consumption accompanying cold adaptation. The rise of tryptophan oxygenase occurs after the first 4 hr of cold exposure and is not abolished by prior adrenalectomy,[359] so that this rise is not due to adrenocortical induction of this enzyme.

During the process of adaptation, short chain amino acids, such as alanine, glycine, and serine, as well as urea and creatinine, are excreted in increased amounts.[473] By the time of adaptation, glucose-6-phosphatase activity in liver and both glucose-6-phosphatase and glucokinase activity in muscle are elevated, while activities of glucose-6-phosphate dehydrogenase and 6-phosphogluconate dehydrogenase decrease.[474] Thus, formation of glucose from glucose-6-phosphate is facilitated and reflected in the increased utilization of glucose by the cold-adapted rat.[475] Hexose-monophosphate shunt activity is depressed at this time, but the depression disappears with longer cold exposure[476] and adaptation. During the acute phase, however, NADPH accumulates, and as we have said, fatty acid synthesis is decreased.

Cold, per se, seems to have only a limited influence on the central nervous system. Acute cold does not change brain serotonin levels[477,478] although urinary 5-hydroxyindole acetic acid excretion becomes elevated within a few days and this elevation persists thereafter.[479] Histamine excretion increases to peak values in 2 months of exposure and then slowly declines.[479,480] Similarly, levels of brain dopamine and norepinephrine stay essentially constant, although small rises after a month of cold exposure may occur.[477-478] Although levels of brain catecholamines appear unchanged, catecholamine turnover is accelerated.[481] Brain monoamine oxidase activity has been reported to be lowered by cold, although cholinesterase activity is unchanged.[482] Finally, brain glutamic acid is decreased somewhat during acute cold exposure[483] but returns to normal after longer periods of time.[484]

2. *Heat*

Thermal stress, of course, poses quite different problems to the organism than cold stress. Heat dissipation, and not heat conservation, becomes critical, and subcutaneous vasodilation substitutes for vasoconstriction. Heat production must be decreased, and the need for fuel input decreases so that food consumption drops.[485] Heat does elicit a rise in adrenal corticoids, however,[486-488] although the rise is of modest proportions and is age dependent.[488] Adrenal size increases with chronic exposure.[438] Adrenal norepinephrine content is unchanged by heat, but adrenal epinephrine levels may increase slightly.[489] Aldosterone secretion increases,[490] but although aldosterone has some effects on enzymes such as succinoxidase, which increases in activity during heat exposure, these enzymatic changes do not depend upon the aldosterone secretion.[491]

In general, heat exposure produces more of a behavioral change than cold exposure. The apathy, lethargy, and decreased reaction time[492] during the heat of summer, for example, are familiar to all, and, while these effects are obviously of value in reducing heat production, little is known about their central control. Neither whole brain[493] nor hypothalamic norepinephrine levels are affected by heat as a stressor. However, increased fluorescence of noradrenergic neurons in the central nervous system has been found[494] suggesting that local conservation mechanisms may be activated. Marginal decreases in brain serotonin levels have been reported,[493] and brain glutamic acid declines in this stress as it does in acute cold.[483]

Generally, then, exposure to temperature changes elicits transitory stress responses during the adaptation from one temperature range to another. Physiologic and metabolic activities of animals adapted to different temperature ranges differ, but once adaptation has occurred, these differences can no longer be truly regarded as stress responses even though they may inadvertently be included in stress studies. It is only when adaptation cannot occur, because, for example, the temperature of exposure is not compatible with life, food is not available to supply the increase in metabolic needs, or hormonal function is inadequate, that temperature continues to be a stressor until death.

3. *Hunger*

The progression from the sensation of hunger through the metabolic adaptations in fasting to the final terminus of starvation is altogether too common in the world of nature and the world of man. The metabolic machine must be fueled and the disappearance of fuel signals the death of the machine. The adult organism stores a considerable reserve of metabolic fuel and, with few exceptions, short bursts of fasting neither threaten the organism nor generate impossible discomfort. What is true for the adult is not true for the infant, however. Inadequate nutrition in early extrauterine[(495)] or intrauterine life[(496)] may lead to irreversible alterations in brain. Permanent decreases in total brain weight, brain DNA, and brain protein[(497)] have been observed in animals reared under nutritional deprivation while, in humans, body and head growth is decreased. The sociological implications of these studies are obvious!

Since large metabolic shifts occur during fasting, and homeostasis is disturbed, fasting should be a potent physiological stressor. Indeed, blood and adrenal corticoid levels rise 18–20 hr after food is removed from rats normally maintained on an *ad lib* diet.[(499,500)] On the other hand, excretion of 17-ketosteroids and 17-hydroxycorticoids of normal[(501,502)] or obese[(503)] men are unchanged during the first days of fasting and tend to decrease[(504,505)] upon longer periods of fasting. Plasma cortisol levels are unchanged throughout.[(504,505)]

Both biological and psychological factors may enter into this difference in adrenocortical responses of man and the rat.

Biologically, small animals, such as the rat, have a larger surface-to-volume ratio than larger animals and consequently lose heat more easily and must expend proportionately more energy to maintain body temperature. Similarly, physical work is a function of distance, time and mass, and only the latter changes with animal size. As a consequence, small animals expend relatively more energy in carrying out normal exploratory behavior than do larger animals. Even the relative proportion of the life span encompassed in a fixed period of fasting is greater for small than large animals. For example, 1 week of fasting for a rat represents 1/100 of its life span. An equivalent period for a man would be nearly $8\frac{1}{2}$ months.

Psychologically, men are certain they will not be starved to death in the experiment. The rat has no such assurance. Indeed, the physiological consequences of anxiety may play a greater role in the early adrenocortical response to fasting than the alteration in metabolic demand.

Other hormonal systems, in addition to the adrenal cortex, are involved in hunger. It is fairly clear that growth hormone is released in fasting both in man[(506)] and the rat.[(507)] Blood glucagon levels in man were found to be elevated in some,[(508,509)] but not all[(510)] studies in fasting, which may reflect individual variations in the ability to maintain blood sugar levels.[511] Insulin levels, of course, are normally low since blood glucose either remains constant or falls. On the other hand, excretion of epinephrine, norepinephrine, and 4-hydroxy-3-methoxymandelic acid (VMA) increases in both normal and obese humans upon fasting.[512] Catecholamine changes are largest in fasting obese subjects, but decreases in blood glucose are more pronounced in subjects of normal weight.

The mechanism for metabolic shifts in fasting differ between phyla, but the general pattern is much the same. Glycogen and glucose stores are consumed, metabolism shifts from glycolysis to gluconeogenesis, lipid catabolism increases, synthesis of long chain fatty acids decreases, and general protein synthesis decreases.

After 24 hr of fasting, 30% of brain glycogen of the rat is gone, as is 85% of liver glycogen.[513] With this dwindling of carbohydrate reserves, lipids and proteins are mobilized and, after the first period of adaptation, lipid becomes the predominant energy source. Thus, the metabolic machine is normally maintained by a balanced metabolism of fatty acid and carbohydrate in a "caloric homeostasis"[514] which is reflected in reciprocal blood levels of these materials. When glucose is high, blood fatty acids are low, and conversely a decrease in blood glucose is accompanied by an increase in fatty acids. Fatty acids are oxidized at rates roughly proportional to their blood levels.[515,516] As carbohydrate stores, in the form of glycogen in liver and muscle, are consumed, the organism shifts to fat metabolism to provide $\frac{4}{5}$ of the required fuel, and protein metabolism to supply the remaining $\frac{1}{5}$.[517,518] The brain, in particular, is the great consumer of glucose,[518] and it is to meet this glucose requirement that gluconeogenesis is primarily carried out. However, with continued fasting, the brain can adapt to utilize γ-hydroxybutyrate and acetoacetic acid in sufficient quantities to reduce the extent of protein catabolism.[519] The precise mechanism for this adaptation is unclear, but it is of some importance in at least delaying burning down the cellular walls to maintain the cellular fire.

These metabolic shifts require enzymatic participation. However, estimations of enzymatic activities can be deceptive in the fasted animal because of large concomitant weight changes in organ size and composition. For example, liver decreases by 50% in the rat after 3 days of fasting, while soluble protein/gram liver increases slightly. As a result, a wholly inert protein, being neither synthesized nor degraded, will show an apparent twofold increase in activity when calculated per gram of liver, and something a little less than that if activity is expressed per gram of protein.

Nonetheless there are marked changes in a number of glycolytic and gluconeogenic enzymes during fasting. In the rat, the gluconeogenic enzymes, glucose-6-phosphatase,[520] phosphoenolpyruvate carboxykinase,[521] fructose-1,6,diphosphatase and pyruvate carboxylase[522] are either preferentially spared or actually synthesized in the midst of general protein catabolism. On the other hand, the important glycolytic enzymes glucokinase,[523,524] phosphofructokinase, and pyruvate kinase[525] fall in activity. Further, pyruvate kinase is inhibited by free fatty acids[526] NADH and ATP which, altogether, provide a multilevel control for shutting off glycolysis. Metabolism of glucose via the hexosemonophosphate shunt in liver is also diminished in fasting as a result of a fall in activity of glucose-6-phosphate dehydrogenase.[527]

Concomitantly with these changes, there is a fall in acetyl CoA carboxylase[528] which, together with fatty acid synthetase, forms long chain fatty acids from malonyl CoA. Malic enzyme and citrate-condensing enzyme also decrease in both adipose tissue and liver,[529,530] further inhibiting lipogenesis.

Upon prolonged fasting, heavier polysomes decrease with the formation of smaller polyribosomal chains[531] and a decrease in the number of ribosomes[532] accompanying accelerated RNA breakdown, particularly breakdown of ribosomal RNA.[533] The degree of polysomal aggregation seems dependent upon tryptophan levels[534] and the deaggregation in fasting may be due to an eventual reduction in cellular levels of this amino acid as a result of increased tryptophan oxygenase activity.[499,535] Tryptophan also appears to play a key role in promoting formation of an inactive form of phosphoenolpyruvate carboxykinase,[536] thereby inhibiting gluconeogenesis when typtophan levels are high and relieving the inhibition when tryptophan levels are low. As might be expected, the decrease in heavier polysomes is reflected in a general fall in protein synthesis.

As mentioned previously, control mechanisms for these changes vary with different species. For example, in the rat, fasting elevates the levels of an inactive form of phosphoenolpyruvate carboxykinase.[536] This does not occur in ruminants, like the sheep, however, who normally depend upon a high level of gluconeogenesis from propionate and bacterial protein. Instead, in these animals fasting increases pyruvate carboxylase.[537]

Even more marked than the interspecies variations in adaptations to fasting is the variation in organ responses within a species. Most of the changes discussed to this point have referred specifically to liver, adipose tissue, or muscle. In a sense, the rather extensive changes occurring in these tissues are designed to spare the brain.

The behavioral and biochemical consequences of avitaminosis are well known. Deficiencies in thiamine, carotene, pyridoxine, nicotinamide, biotin, and pantothenic acid, especially, lead to a variety of neurological as well as peripheral, problems ranging from the dementia of pellagra, though the neuritis of beriberi, to the impaired vision in night-blindness. However, avitaminosis is generally a resultant of continuous undernutrition rather than of the abrupt fasting in stress studies. If the effects of avitaminosis are set aside, the absence of caloric intake has remarkably little effect on the brain or upon behavior. Man has been particularly suitable for behavioral studies not only because of his rich behavioral repertoire but also because avitaminosis can be avoided by use of vitamin supplementation. Further, starvation can be carried out therapeutically for startling periods of time. For example, obese subjects have been maintained literally without food beyond vitamin supplements for 4 months,[511] with increased serum uric acid, decreased body potassium and protein, and, in some instances, slight drops in blood glucose as the only gross physiological changes, and some increase in immature pranks as the only psychological ones.[538]

Transient fasting in rats also seems to have relatively little effect on central nervous system physiology or behavior. Fasted animals initially show an increase in exploratory behavior and transient hunger is often used as a motivating factor in behavioral studies. Some central nervous system changes have been reported but it is difficult to know whether these are related directly to decreased caloric intake or are secondary to some degree of avitaminosis, since vitamin supplementation is seldom employed in such studies and little is known about the rate of vitamin depletion in starved animals.

Despite decreases in muscle and blood glucose levels, in fasting, brain glucose is relatively stable.[539] Uptake of inorganic phosphate, total brain phosphate and phospholipid levels decrease,[540] however, over 9 days of starvation. Activities of glutamic acid dehydrogenase and, to a lesser extent, lactic dehydrogenase in brain also decrease with prolonged starvation.[541] Free amino acid levels remain relatively stable, with small fluctuations, even upon prolonged fasting.[542] Although one brief report suggested that brain serotonin, but not norepinephrine or dopamine, increased in fasted mice maintained at 23°C, but not mice maintained at 34°C,[543] monoamine levels in the rat appear unaffected by fasting.[400] Permeability of the "blood–brain barrier system" to agents such as cocaine appears to be increased by fasting.[544]

The importance of fasting to the stress literature, then, lies less with the direct effects of this stressor on brain per se, than with the possibility that peripheral effects of starvation may confuse interpretations of findings in stress situations in which fasting is inadvertently included. Thus, animals exposed to continuous immobilization may be unable to eat and those exposed to continuous shock may voluntarily decrease consumption. Further, results obtained with fasting again emphasize the nonequivalence of stressors and the difficulty in generalizing from one stress situation to another.

4. *Other Stressors*

Hunger and cold are the two most common physiological stressors which are often unintentionally encountered in stress studies. Obviously, a variety of other stressors, such as thirst, exhaustion, tissue damage, bone fractures, and even spontaneous tumors and pregnancy, have been studied both as stressors in their own right and as contributors to other stress studies. No attempt can be made here to detail the variety of responses elicited by these stressors. Some mention must be made, however, of infection both because it is a frequent contaminant of stress studies in animals and because of its heuristic importance in again underlining the significant differences between stressors.

Unless special precautions are taken, as for example by Charles Rivers, laboratory rodents are frequently endogenously infected by *Mycoplasma mycoides* and viral murine pneumonia. Under the usual laboratory conditions, these infections are latent and the animals asymptotic. However, under chronic stress situations, the infections may become manifest. Most investigators are all too familiar with the symptoms of reddish nose, red-rimmed eyelids, and the cheesy, pustulant lungs of animals with "lung crud." The exacerbations of this disease by stress is due, in part, to the direct antiinflammatory and antiphagocytic properties of the adrenal corticoids elicited by the stressor. Not only may such infections contribute to confounding interpretation of results but, if overlooked, may lead to wholly misleading conclusions.

Infection, as such, also poses a peculiar dilemma for the body. On the one hand, it represents a direct threat to life and, by altering metabolic balance, should be a stressor. Further, corticoids stabilize cellular lysosomes[545] and modify local blood flow,[546] which afford the organism some protection from bacterial endotoxins. On the other hand, the normal immunosuppressive and antiinflammatory effects of corticoids facilitate pathological consequences of microbial invasions. As a result, the response to infection appears to consist of an uneasy compromise which is dependent upon the rate and extent of the metabolic displacements.

Thus, pneumococcal septicemia is treated as a stressor. Adrenal corticoids are elevated and the corticoid-sensitive enzymes tyrosine aminotransferase[547] and tryptophan oxygenase[548] increase in activity. All of these effects decline with time. On the other hand, bacterial pyelonephritis induced by *Streptococcus faecalis* is a slowly developing disease with formation of kidney lesions and a slow decline in renal function.[549] Neither adrenal steroids nor liver tyrosine aminotransferase, tryptophan oxygenase, nor other stress indices are altered in this disease beyond a slight elevation in serum corticoids during the initial septicemia.[550]

IV. CONCLUSION

Life consists of a series of micro and macro adaptations to a constantly changing internal and external environment. Life is, indeed, "one damn thing after another" and as a consequence stress is an ever-present condition. Schopenhauer remarked that "it is not what happens to a man that is important, it is what he thinks happened to him." In the same manner, it is not simply stress, but the kind and magnitude of the stressor, that alters life. This brief and relatively cursory survey of some of the effects of some stressors is intended to outline some of the complexities of the phenomena subsumed under the word stress, to suggest that generalizations from the effects of one stressor to all stressors may be misleading, and to point to some experimental and conceptual problems in stress studies. Commonalities, indeed, do exist between effects of different stressors, and it is on this basis that the operational definition is made. Yet, clearly, the operational definition of stress is not congruent with the common usage of that term. Not all things eliciting the operational responses are either noxious or deleterious. Further, many phenomena, both noxious and dangerous, do not fit the operational definition. The differences between effects of stressors, then, are as important as the similarities between them, just as individual differences between men are important even though the basic similarities between men are truly remarkable.

Each stress, then, represents a relatively unique pattern of responses designed to facilitate adaptation to a particular stressor. Further, many effects of stressors are age dependent. Indeed, aging itself is a slow continuous adaptation from one biological stage to another, and it is not surprising that there is an interaction between this slow adaptation and those required to meet specific challenges of stressors. What is true for the adult may not be true for the neonate, and the aged animal is different than the adolescent one. The persistence of the stress is a further factor to be recognized. Some stressors permit adaptation, while others preclude it. Metabolic processes of value in meeting an emergency situation may be deleterious or metabolically impossible on a chronic basis. Today's metabolic solution may be tomorrow's metabolic problem in the organism, just as the solution to today's needs may complicate fulfillment of tomorrow's desires in our daily lives. Just as men must compromise in attempting to reconcile conflicting needs, so the body must also choose from its finite repertoire of responses those most suitable to the many conflicting needs of the body, and the ideal solution for one set of problems may not be ideal for all.

The effects of stressors are also species dependent. Evolution has shaped each species to its ecological niche and adapted it to the particular stressors in that microenvironment. Heat is a lesser stressor to the gerbil than to the polar bear. The marmot is inured to the long fast of a winter hibernation, which would be impossible for the hummingbird. Information on the effects of stressors is generally limited to laboratory rodents, dogs, and primates. It is to be expected that alternate adaptations will be found in stress responses of other organisms.

Finally, the stressors discussed here have been considered as single entities. In the real world stressors are combined and the physiological shifts required for adaptation to such combinations may be quite different than the algebraic sum of the responses to the individual stressors. At present, the basic components of responses to individual stressors are only poorly defined and understood. An evaluation of the effects of all the combinations and permutations of the numerous natural stressors is clearly a work for the future.

V. REFERENCES

1. J. C. Coleman, *Abnormal Psychology and Modern Life*, Scott, Foresman and Co., Chicago, (1950).
2. H. G. Wolff, *Stress and Disease*, Charles C Thomas, Springfield, Ill. (1953).
3. H. Selye, *The Stress of Life*, McGraw-Hill Co., New York (1956).
4. W. F. Ganong, in *Comparative Endocrinology* (A. Gorbman, ed.) pp. 187–201, John Wiley and Sons, New York (1959).
5. W. F. Ganong and D. M. Hume, Absence of stress-induced and compensatory adrenal hypertrophy in dogs with hypothalamic lesions, *Endocrinology* **55**;474–483 (1954).
6. C. Kim and C. U. Kim, Effect of partial hippocampal resection on stress mechanism in rat, *Amer. J. Physiol.* **201**;337–340 (1961).
7. J. W. Mason, *in International Symposium on Reticular Formation of the Brain* (H. Jaspar, L. D. Proctor, R. S. Knighton, W. E. Noshoy, and R. T. Costello, eds.) pp. 645–662, Little Brown and Co., Boston (1958).
8. E. Anderson, R. W. Bates, E. Hawthorne, W. Haymaker, K. Knowlton, D. McK. Rioch, W. T. Spence, and H. Wilson, The effects of midbrain and spinal cord transection on endocrine and metabolic functions with postulation on a midbrain hypothalamico-pituitary activating system, *Rec. Prog. Horm. Res.* **13**;21–66 (1957).
9. E. Endröczi and K. Lissak, The role of the mesencephalon, diencephalon and archicortex in the activation and inhibition of the pituitary–adrenocortical system, *Acta Physiol. Acad. Sci. Hung.* **17**;39–55 (1960).
10. J. Setekleiv, O. E. Skaug, and B. R. Kaada, Increase of plasma 17-hydroxycorticosteroids by cerebral cortical stimulation and amygdaloid stimulation in the cat, *J. Endocr.* **22**;119–127 (1961).
11. K. M. Knigge, Adrenocortical response to stress in rats with lesions in hippocampus and amygdala, *Proc. Soc. Exp. Biol. Med.* **108**;18–21 (1961).
12. P. R. Bouman, J. H. Gaarenstroom, P. G. Smelik, and D. De Wied, Hypothalamic lesions and ACTH secretion in rats, *Acta Physiol, Pharmacol. Neerl.* **6**; 368–378 (1957).
13. M. A. Slusher and J. E. Hyde, Inhibition of adrenal corticosteroid release by brain stem stimulation in cats, *Endrocrinology* **68**;773–782 (1961).
14. R. Egdahl, Cerebral cortical inhibition of pituitary-adrenal secretion, *Endocrinology* **68**;574–581 (1961).
15. R. Egdahl, Adrenal cortical and medullary responses to trauma in dogs with isolated pituitaries, *Endocrinology* **66**;200–216 (1960).
16. B. K. Anand and S. Dua, Hypothalamic involvement in the pituitary adrenocortical response, *J. Physiol. London* **127**;153–156 (1955).
17. A. Goldfien and W. F. Ganong, Adrenal medullary and adrenal cortical response to stimulation of diencephalon, *Am. J. Physiol.* **202**;205–211 (1962).
18. W. E. Dean and R. Guillemin, Adrenal sensitivity to ACTH as a function of time after hypothalamic lesions and after hypophysectomy, *Proc. Soc. Exp. Biol. Med.* **103**;356–358 (1960).
19. S. M. McCann, Effect of hypothalamic lesions on the adrenal cortical response to stress in the rat, *Am. J. Physiol.* **175**;13–20 (1953).
20. M. Saffran, A. V. Schally, and B. G. Benfrey, Stimulation of the release of corticotropin from the adenohypophysis by a neurohypophysial factor, *Endocrinology* **57**;439–444 (1955).
21. R. Guillemen and B. Rosenberg, Humoral hypothalamic control of anterior pituitary: A study with combined tissue cultures, *Endocrinology* **57**;599–607 (1955).

22. A. V. Schally, R. N. Andersen, H. S. Lipscomb, J. M. Long, and R. Guillemen, Evidence for the existence of two corticotropin-releasing factors, α and β, *Nature* **188**;1192–1193 (1960).
23. A. V. Schally, C. Y. Bowers, and W. Locke, Neurohumoral functions of the hypothalamus (Review), *Amer. J. Med. Sci.* **248**;79–101 (1964).
24. A. V. Schally and R. Guillemin, Isolation and chemical characterization of a β-CRF from pig posterior pituitary glands, *Proc. Soc. Exp. Biol. Med.* **112**;1014–1017 (1963).
25. A. V. Schally, H. S. Lipscomb, and R. Guillemin, Isolation and amino acid sequence of (α_2-CRF) from hog pituitary glands, *Endocrinology* **71**:164–173 (1962).
26. M. A. Slusher, Dissociation of the adrenal ascorbic acid and corticosterone response to stress in rats with hypothalamic lesions, *Endocrinology* **63**:412–419 (1958).
27. N. W. Nowell, Studies on the activation and inhibition of adrenocorticotropin secretion, *Endocrinology* **64**:191–201 (1959).
28. P. G. Smelik, *Autonomic Nervous Involvement in Stress-Induced ACTH Secretion*, Born, Assen, Netherlands (1959).
29. C. Mialhe-Voloss, Posthyophyse et activité corticotrope, *Acta Endocr.*, Supp. **35**:9–96 (1958).
30. G. J. Rochefort, J. Rosenberger, and M. Saffran, Depletion of pituitary corticotropin by various stresses and by neurohypophyseal preparations, *J. Physiol.* **146**:105–116 (1959).
31. C. Fortier, Dual control of adrenocorticotropin release, *Endocrinology* **49**:782–788 (1951).
32. C. Fortier, *in Ciba Foundation Colloquia on Endocrinology* (G. E. W. Wolstenholme and M. P. Cameron, eds.) Vol. 4, pp. 124–135, Little, Brown Co., Boston (1952).
33. D. De Wied, The significance of the antidiuretic hormone in the release mechanism of corticotropin, *Endocrinology* **68**:956–970 (1961).
34. N. W. Nowell, Studies on the activation and inhibition of adrenocorticotropin secretion, *Endocrinology* **64**:191–201 (1959).
35. A. V. Schally, H. S. Lipscomb, J. M. Long, W. E. Dear, and R. Guillemin, Chromatography and hormonal activities of dog hypothalamus, *Endocrinology* **70**:478–480 (1962).
36. J. Szentagothai, B. Flerko, B. Mess, and B. Holasz, Hypothalamic control of the anterior pituitary, *Hungarian Academy of Sciences*, (*Budapest*) (1962) pp. 86–95.
37. J. G. Hilton, L. F. Scian, C. D. Weistermann, and D. R. Kruesi. Effect of synthetic lysine vasopressin on adrenocortical secretion, *Science* **129**:970–971 (1959).
38. H. C. Kwaan and H. J. Bartelstone, Corticotropin release following injections of minute doses of arginine vasopressin into the third ventricle of the dog, *Endocrinology* **65**:982–985 (1959).
39. B. L. Nichols, Jr., The role of antidiuretic hormone in corticotropin release, *Yale J.* Biol. Med. **33**:366–377 (1961).
40. K. Rezabek, Influence of pituitrin on corticotropin secretion, *Physiol. Bohemoslov* **10**:385–389 (1961).
41. A. S. Verdesca, C. D. Westermann, R. S. Crampton, W. C. Black, R. I. Nedeljkovic, and J. G. Hilson, Direct adrenocortical stimulatory effect of serotonin, *Am. J. Physiol.* **201**:1065–1067 (1961).
42. C. H. Li, Synthesis and biological properties of ACTH peptides, *Rec. Prog. Hormone Research* **18**:1–40 (1962).
43. C. H. Li, Perspectives in the biochemical endocrinology of adenohypophyseal hormones, *Bull. N.Y. Acad. Med.* **39**:141–155 (1963).

44. K. Hofmann, Chemistry and function of polypeptide hormones, *Ann. Rev. Biochem.* **31**:213–246 (1962).
45. K. Hofmann and H. Yajima, Synthetic pituitary hormones, *Rec. Prog. Hormone Research* **18**:41–88 (1962).
46. K. Hofmann, H. Yajima, N. Yanaihara, T. Liu, and S. Lande, The synthesis of a tricosapeptide possessing essentially the full biological activity of natural ACTH, *J. Am. Chem. Soc.* **83**:487–489 (1961).
47. J. E. White and F. L. Engel, Fat mobilization by purified corticotropin in the mouse, *Proc. Soc. Exptl. Biol. Med.* **102**:272–274 (1959).
48. F. Hu and W. Chavin, Induction of Melanogenesis *in vitro*, *Anat. Rec.* (Abstract) **125**:600 (1956).
49. H. P. Bär and O. Hecter, Adenylcyclase and hormone action. I. Effects of adrenocorticotropic hormone, glucagon, and epinephrine on the plasma membrane of rat fat cells, *Proc. Natl. Acad. Sci. U.S.* **63**:350–356 (1969).
50. M. Rodbell, Metabolism of fat cells. V. Preparation of "Ghosts" and their properties; adenyl cyclase and other enzymes, *J. Biol. Chem.* **242**:5744–5756 (1967).
51. D. De Wied, Inhibitory effect of ACTH and related peptides on extinction of conditioned avoidance behavior in rats, *Proc. Soc. Exptl. Biol. Med.* **122**:28–32 (1966).
52. B. Bohus and D. De Wied, Inhibitory and facilitory effect of two related peptides on extinction of avoidance behavior, *Science* **153**:318–320 (1966).
53. A. Bertolini, W. Vergoni, G. L. Gessa, and W. Ferrari, Induction of sexual excitement by the action of adrenocorticotropic hormone in brain, *Nature* **221**:667 (1969).
54. W. Ferrari, Behavioral changes in animals after intracisternal injection of adrenocorticotropic hormone and melanocyte-stimulating hormone, *Nature* **181**:925–926 (1958).
55. W. J. Reddy and J. M. Streeto, *in Functions of the Adrenal Cortex* (K. McKerns, ed.) Vol. I, pp. 601–622, Appleton-Century-Crofts, New York (1967).
56. J. M. Streeto and W. J. Reddy, An assay for adenyl cyclase, *Anal. Biochem.* **21**:416–425 (1967).
57. D. G. Grahame-Smith, R. W. Butcher, R. L. Ney, and E. W. Sutherland, Adenosine-3′5′-monophosphate (cyclic AMP) as the mediator of the action of ACTH on the adrenal cortex, *Clin. Res.* **15**:259 (1967) (Abstract).
58. J. J. Ferguson, Jr., Protein synthesis and adrenocorticotropin responsiveness, *J. Biol. Chem.* **238**:2754–2759 (1963).
59. J. J. Ferguson, Jr., Puromycin and adrenal responsiveness to adrenocorticotropic hormone, *Biochim. Biophys. Acta* **57**:616–617 (1962).
60. J. J. Ferguson, Jr., *in Functions of the Adrenal Cortex* (K. McKerns, ed.) Vol. I, pp. 463–478, Appleton-Century-Crofts, New York (1967).
61. R. V. Farese, Inhibition of the steroidogenic effect of ACTH and incorporation of amino acid into rat adrenal protein *in vitro* by chloramphenicol, *Biochim. Biophys. Acta* **87**:699–701 (1964).
62. J. J. Ferguson, Jr., and Y. Morita, RNA synthesis and adrenocorticotropin responsiveness, *Biochim. Biophys. Acta* **87**:348–350 (1964).
63. R. L. Ney, W. W. Davis, and L. D. Garren, Heterogeneity of template RNA in adrenal glands, *Science* **153**:896–897 (1966).
64. E. D. Bransome, Jr., and E. Chargaff, Synthesis of ribonucleic acids in the adrenal cortex: early effects of adrenocorticotropic hormone, *Biochim. Biophys. Acta* **91**: 180–182 (1964).
65. E. D. Bransome, Jr., *in Progress in Biomedical Engineering* (L. J. Fogel and F. W. George, eds.) pp. 7–28, Spartan Books, Washington, D.C. (1967).

66. L. T. Samuels and T. Uchikawa, *in The Adrenal Cortex* (A. B. Eisenstein, ed.) pp. 61–102, Little Brown and Co., Boston (1967).
67. R. S. Rosenfield, D. K. Fukushima, and T. F. Gallagher, *in The Adrenal Cortex* (A. B. Eisenstein, ed.) pp. 103–132, Little Brown and Co., Boston (1967).
68. W. H. Daughaday and I. K. Mariz, Corticosteroid-binding globulin: Its properties and quantitation, *Metabolism* **10**:936–950 (1961).
69. U. S. Seal and R. P. Doe, Purification, some properties and composition of the corticosteroid- and thyroxin-binding globulins from human serum, Proc. 2nd Internat. Cong. Endocrinol., pp. 325–328, London (1964).
70. W. R. Slaunwhite, Jr., and A. A. Sandberg, Transcortin, a corticosteroid binding protein of plasma, *J. Clin. Invest.* **38**:384–391 (1959).
71. W. H. Daughaday, Evidence for two corticosteroid binding systems in human plasma, (Abstract) *J. Lab. Clin. Med.* **48**:799–800 (1956).
72. W. H. Daughaday, Binding of corticosteroids by plasma proteins: III. The binding of corticosteroid and related hormones by human plasma and plasma protein fractions as measured by equilibrium dialysis, *J. Clin. Invest.* **37**:511–518 (1958).
73. U. S. Seal and R. P. Doe, Corticosteroid-binding globulin: I. Isolation from plasma of diethylstilbestrol-treated men, *J. Biol. Chem.* **237**:3136–3140 (1962).
74. G. Sayers and M. A. Sayers, Regulation of pituitary adrenocorticotropic activity during the response of the rat due to acute stress, *Endocrinology* **40**:265–273 (1947).
75. K. L. Snydor, Blood ACTH in the stressed adrenalectomized rat after intravenous injection of hydrocortisone, *Endocrinology* **56**:204–208 (1955).
76. C. Fortier, Effect of hydrocortisone on pituitary ACTH and adrenal weight in the rat, *Proc. Soc. Exp. Biol. Med.* **100**:16–19 (1959).
77. F. E. Yates, S. E. Leeman, D. W. Glenister, and M. F. Dallman, Interaction between plasma corticosterone concentration and adrenocorticotropin-releasing stimuli in the rat: Evidence of the reset of an endocrine feedback control, *Endocrinology* **69**:67–80 (1961).
78. P. G. Smelik, Failure to inhibit corticotropin secretion by experimentally-induced increases in corticoid levels, *Acta Endocrinol.* (Copenhagen) **44**:36–46 (1963).
79. J. R. Hodges and M. T. Jones, The effect of injected corticosterone on the release of adrenocorticotropic hormone in rats exposed to acute stress, *J. Physiol.* (*London*) **167**:30–37 (1963).
80. R. H. Egdahl. The acute effects of steroid administration on pituitary adrenal secretion in the dog, *J. Clin. Invest.* **43**:2178–2184 (1964).
81. F. G. Peron, F. Moncloa, and R. I. Dorfman, Studies on the possible inhibitory effect of corticosterone on corticosteroidogenesis at the adrenal level in the rat, *Endocrinology* **67**:379–388 (1960).
82. J. B. Richards and R. L. Pruitt, Hydrocortisone suppression of stress-induced adrenal 17-hydroxycorticosteroid secretion in dogs, *Endocrinology* **60**:99–104 (1957).
83. E. Endröczi, K. Lissak, and M. Tekeres, Hormonal feedback regulation of pituitary–adrenocortical activity, *Acta Physiol. Acad. Sci. Hung.* **18**:291–299 (1961).
84. P. G. Smelik and C. H. Sawyer, Effect of implantation of cortisol into the brain stem of pituitary gland on the adrenal response to stress in the rabbit, *Acta Endocrinol.* **41**:561–570 (1962).
85. I. Chowers, S. Feldman, and J. M. Davidson, Effects of intrahypothalamic crystalline steroids on acute ACTH secretion, *Am. J. Physiol.* **205**:671–673 (1963).
86. S. Schapiro, J. Marmorston, and H. Sobel, The steroid feedback mechanism, *Am. J. Physiol.* **192**:58–62 (1958).

87. S. Schapiro, E. Geller, and S. Eiduson, Corticoid response to stress in the steroid-inhibited rat, *Proc. Soc. Expl. Biol. Med.* **109**:935–937 (1962).
88. F. E. Yates and J. Urquhart, Control of plasma concentrations of adrenocortical hormones, *Physiol. Rev.* **42**:359–433 (1962).
89. B. S. McEwen, J. M. Weiss, and L. S. Schwartz, Selective retention of corticosterone by limbic structures in rat brain, *Nature* **220**:911–912 (1968).
90. S. Schapiro and J. Katz, Preferential concentration of C^{14} in posterior pituitary following hydrocortisone C^{14} injection, *Proc. Soc. Exptl. Biol. Med.* **102**:609–612 (1960).
91. K. B. Eik-Nes and K. R. Brizzee, Concentration of tritium in brain tissue of dogs given (1,2-$^3H_2$) cortisol intravenously, *Biochim. Biophys. Acta* **97**:320–333 (1965).
92. E. F. Yates, *in The Adrenal Cortex* (A. B. Eisenstein, ed.) pp. 133–184, Little, Brown and Co., Boston (1967).
93. F. L. Engel, A consideration of the roles of the adrenal cortex and stress in the regulation of protein metabolism, *Recent Progr. Hormone Res.* **6**:277–313 (1950).
94. A. D. Bass, J. W. Chambers, and A. A. Richtarik, The effect of hydrocortisone on AIB uptake by the isolated perfused rat liver, *Life Sci.* **4**:266–269 (1963).
95. H. N. Christensen, *in Metabolic Effects of Adrenal Hormones* (Ciba Foundation Study Group 6), p. 56, Little, Brown and Co., Boston (1960).
96. I. Clark, The effect of cortisone upon protein synthesis, *J. Biol. Chem.* **200**:69–76 (1953).
97. C. D. Kochakain and E. Robertson, Adrenal steroids and body composition, *J. Biol. Chem.* **190**:495–503 (1951).
98. R. H. Silber and C. C. Porter, Nitrogen balance, liver protein repletion and body composition of cortisone-treated rats, *Endocrinology* **52**:518–525 (1953).
99. N. Lang and C. E. Sekeris, Zum Wirkungsmechanismus der Hormone: III. Einfluss von Cortisol auf den Ribonucleinsäure und Proteinstoffwechsel in Rattenleber, *Z. Physiol Chem.* **339**:238–248 (1964).
100. O. Barnabei, B. Romano, G. di Bitonto, and V. Tomasi, Factors influencing the glucocorticoid-induced increase of ribonucleic acid polymerase activity of rat liver nuclei, *Arch. Biochem. Biophys.* **113**:478–486 (1966).
101. M. Feigelson, P. R. Gross, and P. Feigelson, Early effects of cortisone on nucleic acid and protein metabolism of rat liver, *Biochim. Biophys. Acta* **55**:495–504 (1962).
102. H. P. G. Seckel, The influence of various physiological substances on the glycogenolysis of surviving rat liver, *Endocrinology* **26**:97–101 (1940).
103. C. T. Teng, F. M. Sinex, H. W. Deane, and A. B. Hastings, Factors influencing glycogen synthesis by rat liver slices *in vitro*, *J. Cellular Comp. Physiol.* **39**:73–88 (1952).
104. R. A. Lewis, D. Kuhlman, C. Delbue, G. V. Koepf, and G. W. Thorn, The effect of adrenal cortex on carbohydrate metabolism, *Endrocrinology* **27**:971–982 (1940).
105. G. F. Koepf, H. W. Horn, C. L. Gemmill, and G. W. Thorn, The effect of adrenal cortical hormone on the synthesis of carbohydrate in liver slices, *Am. J. Physiol.* **135**:175–186 (1941).
106. P. D. Ray, D. O. Foster, and H. A. Lardy, Gluconeogenesis stimulated by hydrocortisone independently of "de novo" synthesis of enzymes, *Fed. Proc.* **23**:482 (1964).
107. G. Weber, R. L. Singhal, N. B. Stamm, E. A. Fisher, and M. A. Mentendiek, Regulation of enzymes involved in gluconeogenesis, *Advances Enzym. Regulat.* **2**;1–38 (1963).
108. G. Weber, S. K. Srivastava, and R. Singhal, Role of enzymes in homeostasis. VII. Early effects of corticosteroid hormones on gluconeogenic enzymes, ribonucleic acid metabolism and amino acid level, *J. Biol. Chem.* **240**:750–756 (1965).

109. L. C. Mokrasch, W. D. Davidson, and R. W. McGilvery, The response to glucogenic stress of fructose-1,6-diphosphatase in rabbit liver, *J. Biol. Chem.* **222**:179–184 (1956).
110. E. Shrago, H. A. Lardy, R. C. Nordlie, and D. O. Foster, Metabolic and hormonal control of phosphoenolpyruvate carboxykinase and malic enzyme in rat liver, *J. Biol. Chem.* **238**:3188–3192 (1963).
111. H. V. Henning, I. Seiffert, and W. Seubert, Cortisol induzierter Ansteig der Pyruvatcarboxylaseaktivität in der Rattenleber, *Biochim. Biophys. Acta* **77**:345–348 (1963).
112. J. Ashmore, A. B. Hastings, F. B. Nesbett, and A. E. Renold, Studies on carbohydrate metabolism in rat liver slices: VI. Hormonal factors influencing glucose-6-phosphatase, *J. Biol. Chem.* **218**:77–88 (1956).
113. I. D. Welt, D. Stetten, D. J. Ingle, and E. H. Morley, Effect of cortisone upon rates of glucose production and oxidation in the rat, *J. Biol. Chem.* **197**:56–66 (1952).
114. J. Ashmore, F. Stricker, W. C. Love, and G. Kilsheimer, Cortisol stimulation of glycogen synthesis in fasted rats, *Endocrinology* **68**:599–606 (1961).
115. C. N. H. Long, O. K. Smith, and E. G. Fry, *in Metabolic Effects of Adrenal Hormones* (Ciba Foundation Study Group No. 6) pp. 4–19, Little, Brown and Co., Boston (1960).
116. P. M. Hyde, Liver glycogen deposition after intravenous and intragastric administration of cortisol-4-C^{14} to rats, *Endocrinology* **61**:774–779 (1957).
117. J. Ashmore, G. F. Cahill, Jr., R. Hillman, and A. E. Renold, Adrenal cortical regulation of hepatic glucose metabolism, *Endocrinology* **62**:621–626 (1958).
118. P. Feigelson and M. Feigelson, *in Action of Hormones on Molecular Processes* (G. Litwack and D. Kritchevsky, eds.) pp. 218–233, John Wiley and Sons, New York (1964).
119. A. White, *in The Harvey Lectures*, (1947–1948) pp. 43–70, Charles C Thomas, Springfield, Ill. (1950).
120. W. Stevens, C. Colessides, and T. F. Dougherty, Effects of cortisol on the incorporation of thymidine-2-^{14}C into nucleic acids of lymphatic tissue from adrenalectomized CBA-mice, *Endocrinology* **76**:1100–1108 (1965).
121. D. H. Nelson, *in Bioassay of Anterior Pituitary and Adrenocortical Hormones* (Ciba Foundation Colloquia on Endocrinology) Vol. 5 (G. E. W. Westenholme, ed.) pp. 162–169, Little, Brown and Co., Boston (1953).
122. A. S. Gordon, Some aspects of hormonal influences upon the leucocytes, *Ann. N.Y. Acad. Sci.* **59**:907–927 (1955).
123. D. H. Nelson, A. A. Sandberg, J. G. Palmer, and F. H. Tyler, Blood levels of 17-hydroxycorticoids following the administration of adrenal steroids and their relation to circulating leukocytes, *J. Clin. Invest.* **31**:843–849 (1952).
124. G. Cardinali, G. Cardinali, B. M. DeCaro, A. H. Handler, and M. Aboul-Enein, Effect of high doses of cortisone on bone marrow cell proliferation in the Syrian hamster, *Cancer Res.* **24**:969–972 (1964).
125. H. C. Stoerk and R. N. Arison, *in Inflammation and Diseases of Connective Tissue* (L. C. Mills and J. H. Moyer, eds.) pp. 399–403, W. B. Saunders, Philadelphia (1961).
126. I. Clark, R. F. Geoffroy, and W. Bowers, Effects of adrenal cortical steroids on calcium metabolism, *Endocrinology* **64**:849–856 (1959).
127. D. J. Ingle, C. H. Li, and H. M. Evans, The effect of adrenocorticotropic hormone on the urinary excretion of sodium, chloride, potassium, nitrogen, and glucose in normal rats, *Endocrinology* **39**:32–42 (1946).

128. J. N. Mills, S. Thomas, and K. S. Williamson, The effects of intravenous aldosterone and hydrocortisone on the urinary electrolytes of the recumbent human subject, *J. Physiol. (London)* **156**:415–423 (1961).
129. W. F. Ganong, P. J. Mulrow, and G. Cera, Effect of angiotensin and ACTH on adrenocortical secretion in hypophysectomized nephrectomized dogs, Abst. Endocrinol. Soc. 43rd Meeting, New York (1961) p. 14.
130. J. O. Davis, *in The Adrenal Cortex* (A. B. Eisenstein, ed.) pp. 203–247, Little, Brown and Co., Boston (1967).
131. J. D. H. Slater, B. H. Barbour, H. H. Henderson, A. G. T. Casper, and F. C. Bartter, Influence of the pituitary and the renin-angiotensin system on the secretion of aldosterone, cortisol and corticosterone, *J. Clin. Invest.* **42**:1504–1520 (1963).
132. S. J. Gray, J. A. Benson, H. M. Spiro, and R. W. Reifenstein, Effects of ACTH and cortisone upon the stomach: Its significance in the normal and in peptic ulcer, *Gastroenterology* **19**:658–673 (1951).
133. H. M. Spiro and S. S. Milles, Clinical and physiologic implications of steroid-induced peptic ulcer, *New Engl. J. Med.* **263**:286–294 (1960).
134. D. S. Kahn, M. J. Phillips, and S. C. Skoryna, Healed experimental gastric ulcer in the rat; reulceration resulting from cortisone administration, *Am. J. Pathol.* **38**:177–188 (1961).
135. M. P. Sambhi, M. H. Weil, and V. N. Udhoji, Acute pharmacodynamic effects of glucocorticoids: Cardiac output and related hemodynamic changes in normal subjects and patients in shock, *Circulation* **31**:523–530 (1965).
136. B. W. Zweifach, E. Shorr, and M. M. Black, The influence of the adrenal cortex on behavior of terminal vascular bed, *Ann. N.Y. Acad. Sci.* **56**:626–633 (1953).
137. R. W. Schayer, Induced synthesis of histamine, microcirculatory regulation and the mechanism of action of the adrenal glucocorticoid hormone, *Progr. Allerg.* **7**:187–212 (1963).
138. D. M. Spain, *in Inflammation and Diseases of Connective Tissue* (L. C. Mills and J. W. Moyer, eds.) pp. 514–516, W. B. Saunders Co., Philadelphia (1961).
139. S. Raffel, *Immediate Hypersensitivities. Immunity*, Appleton-Century-Crofts, New York (1961).
140. M. Bjorneboe, E. M. Fischel, and H. C. Stoerk, The effect of cortisone and adrenocorticotrophic hormone on the concentration of circulating antibody, *J. Exp. Med* **93**:39–48 (1951).
141. S. Malkiel and B. J. Hargis, The effect of ACTH and cortisone on the quantitative precipitin reaction, *J. Immun.* **69**:217–221 (1952).
142. F. G. Germuth, Jr., B. Ottinger, and J. Oyama, The influence of cortisone on the evolution of acute infection and the development of immunity, *Bull. Johns Hopkins Hosp.* **91**:22–48 (1952).
143. R. J. Glaser, J. W. Berry, L. H. Loeb, W. B. Wood, and A. Hamlin, The effect of cortisone in streptococcal lymphadenitis and pneumonia, *J. Lab. Clin. Med.* **38**:363–373 (1951).
144. R. M. Dreendyke, E. M. Bradley, and S. N. Swisher, Studies on the effects of administration of ACTH and adrenal corticosteroids on erythrophagocytosis, *J. Clin. Invest.* **44**:746–753 (1965).
145. L. Thomas, Cortisone and infection, *Ann. N.Y. Acad. Sci.* **56**:799–814 (1953).
146. R. A. Cleghorn, Adrenal cortical insufficiency: Psychological and neurological observations, *Can. Med. Assoc. J.* **65**:449–454 (1951).
147. G. L. Engel and S. G. Margolin, Neuropsychiatric disturbances in Addison's disease and role of impaired carbohydrate metabolism in production of abnormal cerebral function, *Arch. Neurol. Psychiat.* **45**:881–883 (1941).

148. W. C. Hoffman, R. A. Lewis, and G. W. Thorn, The electroencephalogram in Addison's disease, *Bull. Johns Hopkins Hosp.* **70**:335–361 (1942).
149. R. A. Cleghorn, *in Hormones, Brain Function, and Behavior* (H. Hoagland, ed.) pp. 3–18, Academic Press, New York (1957).
150. J. D. Spillane, Nervous and mental disorders in Cushing's syndrome, *Brain* **74**:72–94 (1951).
151. G. H. Glaser, D. S. Kornfeld, and R. P. Knight, Intravenous hydrocortisone, corticotrophin and electroencephalogram, *A.M.A. Arch. Neurol. Psychiat.* **73**:338–344 (1955).
152. S. Feldman, J. C. Todt, and R. W. Porter, Effect of adrenocortical hormones on evoked potentials in the brain stem, *Neurology (Minneap.)* **11**:109–115 (1961).
153. K. Ruf and F. A. Steiner, Steroid sensitive single neurons in rat hypothalamus and midbrain; identification by microelectrophoresis, *Science* **156**:667–669 (1967).
154. S. Cobb, Some clinical changes in behavior accompanying endocrine disorders, *J. Nerv. Ment. Dis.* **130**:97–107 (1960).
155. H. P. Rome and F. J. Braceland, The psychological response to ACTH, cortisone, hydrocortisone and related substances, *Am. J. Psychiat.* **108**:641–651 (1952).
156. J. V. Murphy and R. E. Miller, The effect of adrenocorticotropic hormone (ACTH) on avoidance conditioning in the rat, *J. Comp. Physiol. Psychol.* **48**:47–49 (1955).
157. B. Bohus and D. DeWied, Inhibitory and facilitory effect of two related peptides on extinction of avoidance behavior, *Science* **153**:318–320 (1966).
158. J. L. Fuller, R. M. Chambers, and R. P. Fuller, Effect of cortisone and of adrenalectomy on activity and emotional behavior of mice, *Psychosomat. Med.* **18**:234–242 (1956).
159. P. S. Timiras, D. M. Woodbury, and L. S. Goodman, Effect of adrenalectomy, hydrocortisone acetate and desoxycorticosterone acetate on brain excitability and electrolyte distribution in mice, *J. Pharmacol. Exptl. Therap.* **112**:80–93 (1954).
160. D. M. Woodbury, Effect of hormones on brain excitability and electrolytes, *Rec. Progr. Hormone Res.* **10**:65–104 (1954).
161. V. D. Davenport, Relation between brain and plasma electrolytes and electroshock seizure thresholds in adrenalectomized rats, *Am. J. Physiol.* **156**:322–327 (1949).
162. R. H. Leiderman and R. Katzman, Effect of adrenalectomy, desoxycorticosterone and cortisone on brain potassium exchange, *Am. J. Physiol.* **175**:271–275 (1953).
163. J. R. Bergen, D. Stone, and H. Hoagland, Studies of brain potassium in relation to the adrenal cortex, *in Proc. Internat. Conf. on Peaceful Uses of Atomic Energy*, Vol. 12, 301–305 (1956).
164. D. M. Woodbury, P. S. Timiras, and A. Verdakis, *in Hormones, Brain Function and Behavior* (H. Hoagland, ed.) pp. 27–49, Academic Press, New York (1957).
165. B. B. Gallagher and G. H. Glaser, Seizure threshold, adrenalectomy and sodium potassium stimulated ATP*ase* in rat brain, *J. Neurochem.* **15**:525–528 (1968).
166. I. G. Fazekas and A. T. Fazekas, Paper chromatographic demonstration of corticosteroid fractions in rabbit organs and tissues, *Endocrinologie* **50**:130–138 (1966).
167. P. J. Thomas, Cortisol acetyltransferase from baboon brain, *Biochem. J.* **109**:695–696 (1968).
168. N. A. Peterson, I. L. Chaikoff, and C. Jones, The *in vitro* conversion of cortisol to cortisone by subcellular brain fractions of young and adult rats, *J. Neurochem.* **12**:273–278 (1965).
169. B. I. Grosser, 11-β-hydroxysteroid metabolism by mouse brain and glioma 261, *J. Neurochem.* **13**:475–478 (1966).

170. J. F. Schieve, P. Scheinberg, and W. P. Wilson, The effect of adrenocorticotropic hormone (ACTH) on cerebral blood flow and metabolism, *J. Clin. Invest.* **30**:1527–1529 (1951).
171. W. Sensenbach, L. Madison, and L. Ochs, The effect of ACTH and cortisone on cerebral blood flow and metabolism, *J. Clin. Invest.* **32**:372–380 (1953).
172. R. W. Alman and J. F. Fazekas, Effects of ACTH on cerebral blood flow and oxygen consumption, *Arch. Neurol. and Psychiat.* **65**:680–682 (1951).
173. A. N. Panov and A. Fonyo, Effect of corticoid hormones on rat-brain mitochondrial NADH-oxidase, *Acta Physiol. Acad. Sci. Hung.* **30**:7–13 (1966).
174. F. Vaccari and M. Rossanda, Sulle modificazioni indotte dal cortisone e dal desossicorticosterone sul glicogeno e sui carbuidrati totali del cervello nell' animale normale e surrenalectomizzato, *Boll. Soc. Ital. Biol. Sper.* **27**:734–738 (1951).
175. P. S. Timiras, D. M. Woodbury, and D. H. Baker, Effect of hydrocortisone acetate, desoxycorticosterone acetate, insulin, glucagon and dextrose, alone or in combination, on experimental convulsions and carbohydrate metabolism, *Arch. Int. Pharmacodyn Therap.* **105**:450–467 (1956).
176. J. H. Thurston and R. W. Pierce, Increase of glucose and high energy phosphate reserve in the brain after hydrocortisone, *J. Neurochem.* **16**:107–111 (1969).
177. K. G. Prasannan, R. Rajan, and Subrahmanyam, Effect of glutamic acid on the brain and liver glycogen of normal and cortisone treated rats, *Indian J. Med. Res.* **52**:208–212 (1964).
178. M. M. Kosmina, Effect of hydrocortisone on certain indexes of energy metabolism in the brain cortex of guinea pigs, *Fiziol. Zh.* **13**:252–255 (1967).
179. J. DeVellis and D. Inglish, Hormonal control of glycerolphosphate dehydrogenase in the rat brain, *J. Neurochem.* **15**:1061–1070 (1968).
180. E. P. Kennedy, Synthesis of phosphatides in isolated mitochondria, *J. Biol. Chem.* **201**:399–412 (1953).
181. A. Kornberg and W. E. Pricer, Jr., Enzymatic esterification of α-glycerophosphate by long chain fatty acids, *J. Biol. Chem.* **204**:345–357 (1953).
182. H. Oshika, The 5 hydroxytryptophan decarboxylase (5HTPD) activity in young animals, *Nippon Yakurigaku Zasshi* **62**:143–151 (1966) (Chem. Abstracts 67: 80259n).
183. V. E. Davis, Effect of cortisone and thyroxine on aromatic amino acid decarboxylation, *Endocrinol.* **72**:33–38 (1963).
184. N. S. Shah, S. Stevens, and H. E. Himwich, Effect of chronic administration of cortisone on the tryptophan induced changes in amine levels in the rat brain, *Arch. Int. Pharmacodyn Therap.* **171**:285–295 (1968).
185. A. Yuwiler, E. Geller, and S. Eiduson, Studies on 5-hydroxytryptophan decarboxylase. I. *In vitro* inhibition and substrate interaction, *Arch. Biochem. Biophys.* **80**:162–173 (1959).
186. E. M. Gal, M. Morgan, S. K. Chatterjee, and F. D. Marshall, Jr., Hydroxylation of tryptophan by brain tissue *in vivo* and related aspects of 5-hydroxytryptamine metabolism, *Biochem. Pharmacol.* **13**:1639–1653 (1964).
187. E. C. Azmitia and B. S. McEwen, Corticosterone regulation of tryptophan hydroxylase in midbrain of the rat, *Science* **166**:1274–1276 (1969).
188. R. A. Freedland, Factors affecting the activity *in vitro* of the tryptophan- and phenylalanine-hydroxylating systems, *Biochim. Biophys. Acta* **73**:71–75 (1963).
189. D. DeMaio, Influence of adrenalectomy and hypophysectomy on cerebral serotonin, *Science* **129**:1678–1679 (1959).
190. S. Sofer and C. J. Gubler, Studies on the effect of various procedures on the 5-OH tryptamine (5HT) levels in the brain of rats, *Fed. Proc.* **21**:340 (1962) (Abstract).

191. T. R. Put and J. W. Meduski, The effect of adrenalectomy on the 5-hydroxytryptamine metabolism in the rat, *Acta Physiol Pharmacol. Neerlandia* **11**:240–256 (1962).
192. A. K. Pfeifer, E. S. Vizi, E. Satory, and E. Galambos, The effect of adrenalectomy on the norepinephrine and serotonin content of the brain and on reserpine action in rats, *Experentia* **19**:482–483 (1963).
193. J. C. Towne and J. O. Sherman, Failure of acute bilateral adrenalectomy to influence serotonin levels in the rat, *Proc. Soc. Exptl. Biol. Med.* **103**:721–722 (1960).
194. R. H. Resnick, G. T. Smith, and S. J. Gray, Endocrine influences on tissue serotonin content of the rat, *Am. J. Physiol.* **201**:571–573 (1961).
195. S. Garattini, L. Lamesta, A. Mortari, V. Palma, and L. Valzelli, Pharmacological and biochemical effects of 5-hydroxytryptamine in adrenalectomized rats, *J. Pharm. Pharmacol.* **13**:385–388 (1961).
196. C. T. McKennee, P. S. Timiras, and W. B. Quay, Concentrations of 5-hydroxytryptamine in rat brain and pineal after adrenalectomy and cortisol administration, *Neuroendocrinol.* **1**:251–256 (1965/1966).
197. R. Kato and L. Valzelli, Cortisone e 5-idrossitriptamina cerebrale, *Boll. Soc. Ital. Sper.* **34**:1042–1404 (1958).
198. D. DeMaio and C. Marbbrio, The influence of certain hormones on the serotonin contents of the brain of the rat (normal and adrenalectomized), *Arch. Sci. Med.* **111**:369–373 (1961).
199. G. Curzon and A. R. Green, Effect of hydrocortisone on rat brain 5-hydroxytryptamine, *Life Sci.* **7**:657–663 (1968).
200. A. R. Green and G. Curzon, Decrease of 5-hydroxytryptamine in the brain provoked by hydrocortisone and its prevention by allopurinol, *Nature* **220**:1095–1097 (1968).
201. T. C. Butler, Cerebral and hepatic tyrosine alpha-oxoglutarate transaminase in the immature and adult rat, *Develop. Med. Child Neurol.* **7**:392–394 (1965).
202. A. Vernadakis and D. M. Woodbury, *in Inhibition in the Nervous System and Gamma-Aminobutyric Acid* (E. Roberts, C. F. Baxter, A. Van Harreveld, C. A. G. Wiersma, W. R. Adey, and K. F. Killam, eds.) pp. 242–247, Pergamon Press, New York (1960).
203. L. Pandolfo and S. Macaione, Influence of adrenalectomy on the activity of δ-aminobutyric acid and glutamic acid decarboxylase in the rat cerebral cortex, *Giorn. Biochem.* **13**:256–261 (1964).
204. V. C. Sutherland and M. Rikimaru, The regional effects of adrenalectomy and ethanol on cerebral amino acids in the rat, *Int. J. Neuropharmacol.* **3**:135–139 (1964).
205. E. Howard, Effects of corticosterone and food restriction on growth and on DNA, RNA and cholesterol content of brain and liver of infant mice, *J. Nutr.* **95**:111–121 (1968).
206. S. Schapiro, Some physiological, biochemical, and behavioral consequences of neonatal hormone administration—cortisol and thyroxine, *Gen. Comp. Endocrinol.* **10**:214–228 (1968).
207. E. Howard, Reductions in size and total DNA of cerebrum and cerebellum in adult mice after corticosterone treatment in infancy, *Exptl. Neurol.* **22**:191–208 (1968).
208. M. Salas and S. Schapiro, Hormonal influences upon the maturation of the rat brain's responsiveness to sensory stimuli, *Physiol. and Behavior* **5**:7–11 (1970).
209. C. J. Migeon, H. Prystowsky, M. M. Grumbach, and M. C. Byron, Placental passage of 17–hydroxycorticosteroids: Comparison of the levels in maternal and fetal plasma and effect of ACTH and hydrocortisone administration, *J. Clin. Invest.* **35**:488–493 (1956).

210. Y. Eguchi, Experimental studies on the adrenal cortex of the mouse fetus. II. Effects of various treatment of the mother on the adrenal of the fetus, *Bull. Osaka Med. Sch.*, Ser. B., **11**:29-36 (1961).
211. S. Schapiro, *in Early Experience and Behavior*, (G. Newton and S. Levine, eds.) pp. 198–257, Charles C Thomas, Springfield, Ill. (1968).
212. J. Roth, S. M. Glick, R. S. Yalow, and S. A. Berson, Hypoglycemia: A potent stimulus to secretion of growth hormone, *Science* **140**:987–988 (1963).
213. K. Brown-Grant, Changes in the thyroid activity of rats exposed to cold, *J. Physiol.* **131**:52–57 (1956).
214. S. Schapiro, Interaction between growth hormone and cortisol on the regulation of tyrosine transaminase activity, *Endocrinology* **83**:475–478 (1968).
215. D. H. Russell and S. H. Snyder, Amine synthesis in regenerating rat liver: Effect of hypophysectomy and growth hormone on ornithine decarboxylase, *Endocrinology* **84**:223–228 (1969).
216. A. D'lorio and C. Mavrides, *in Advances in Enzyme Regulation* (G. Weber, ed.) Vol. I, pp. 97–101, Macmillan Co., New York (1963).
217. J. Comsa, Adrenalin-thyroxine interaction in guinea pigs, *Am. J. Physiol.* **161**:550–553 (1950).
218. E. M. Wise, Jr. and E. G. Ball, Malic enzyme and lipogenesis, *Proc. Natl. Acad. Sci. U.S.* **52**:1255–1263 (1964).
219. H. M. Tepperman and J. Tepperman, Patterns of dietary and hormonal induction of certain NADP-linked liver enzymes, *Am. J. Physiol.* **206**:357–361 (1964).
220. F. Rosen, N. R. Roberts, and C. A. Nichol, Glucocorticoids and transaminase activity. I. Increased activity of glutamic-pyruvic transaminase in four conditions associated with gluconeogenesis, *J. Biol. Chem.* **234**:476–480 (1959).
221. H. R. Harding, F. Rosen, and C. A. Nichol, Inhibition of hepatic alanine transaminase activity in response to treatment with 11-deoxycorticosterone, *Proc. Soc. Exptl. Biol. Med.* **108**:96–99 (1961).
222. F. Rosen, H. R. Harding, R. Milholland, and C. A. Nichol, Glucocorticoids and transaminase activity. VI. Comparison of the adaptive increases of alanine- and tyrosine-α-ketoglutarate transaminases, *J. Biol. Chem.* **238**:3725–3729 (1963).
223. W. E. Knox and V. H. Auerbach, The hormonal control of tryptophan peroxidase in the rat, *J. Biol. Chem.* **214**:307–313 (1955).
224. W. B. Cannon, *Bodily Changes in Pain, Hunger, Fear and Rage*, Branford, Boston (1929).
225. O. Loewi, Über humorale Übertragbarkeit der Herznervenwirkung, *Pflügers Arch. Ges. Physiol.* **189**:239–242 (1921).
226. W. B. Cannon and J. E. Uridil, Studies on the conditions of activity in endocrine glands. VIII. Some effects on the denervated heart of stimulating the nerves of the liver, *Am. J. Physiol.* **58**:353–354 (1921).
227. U. S. von Euler, A specific sympathomimetic ergone in adrenergic nerve fibres (sympathin) and its relation to adrenaline and noradrenaline, *Acta Physiol. Scand.* **12**:73–97 (1946).
228. U. S. von Euler, Identification of the sympathomimetic ergone in adrenergic nerves of cattle (sympathin-N) with laevo-noradrenaline, *Acta Physiol. Scand.* **16**:63–74 (1948).
229. H. Blaschko and A. D. Welch, Localization of adrenaline in cytoplasmic particles of the bovine adrenal medulla, *Arch. Exp. Path. Pharmak.* **219**:17–22 (1953).
230. N.-A. Hillarp, S. Lagerstedt, and B. Nilson, The isolation of a granular fraction from the suprarenal medulla, containing the sympathomimetic catecholamines, *Acta Physiol. Scand.* **29**:251–263 (1953).

231. N.-A. Hillarp, *in Ciba Foundation Symposium on Adrenergic Mechanisms, 1960* (G. E. W. Wolstenholme and M. O'Connor, eds.) pp. 481–486, Churchill, London (1961).
232. N.-A. Hillarp, Different pools of catecholamines stored in the adrenal medulla, *Acta Physiol. Scand.* **50**:8–22 (1960).
233. M. H. Burgos, Histochemistry and electron microscopy of the three cell types in the adrenal gland of the frog, *Anat. Rec.* **133**:163–174 (1959).
234. N. Kirshner, Uptake of catecholamines by a particulate fraction of the adrenal medulla, *Science* **135**:107–108 (1962).
235. J. Axelrod, *in Symposium on the Clinical Chemistry of Monoamines* (H. Varley and A. H. Gowenlock, eds.) Vol. 2, pp. 5–18, Elsevier Publ. Co., Amsterdam (1963).
236. L. T. Potter and J. Axelrod, Properties of norepinephrine storage particles of the rat heart, *J. Pharmac. Exp. Ther.* **142**:299–305 (1963).
237. H. A. Campos and F. E. Shideman, Subcellular distribution of catecholamines in the dog heart. Effects of reserpine and norepinephrine administration, *Int. J. Neuropharmac.* **1**:13–22 (1962).
238. U. S. von Euler and N.-A. Hillarp, Evidence for the presence of noradrenaline in submicroscopic structures of adrenergic axones, *Nature, Lond.* **177**:43–45 (1956).
239. L. Stjärne, Studies of catecholamine uptake, storage and release mechanisms, *Acta Physiol. Scand.* **62**:Suppl. 228 (1964).
240. U. S. von Euler, *Noradrenaline*, Charles C Thomas, Springfield, Ill. (1956).
241. T. Nagatsu, M. Levitt, and S. Udenfriend, The initial step in norepinephrine biosynthesis, *J. Biol. Chem.* **239**:2910–2917 (1964).
242. S. Udenfriend, Tyrosine Hydroxylase (Second International Catecholamine Symposium, Milan), *Pharmacol. Rev.* **18**:43–52 (1966).
243. J. R. Crout, Pheochromocytoma, *Pharmacol. Rev.* **18**:651–657 (1966).
244. E. Y. Levin, B. Levenberg, and S. Kaufman, The enzymatic conversion of 3,4-dihydroxyphenylethylamine to norepinephrine, *J. Biol. Chem.* **235**:2080–2086 (1960).
245. J. Axelrod, Purification and properties of phenylethanolamine-*N*-methyl transferase, *J. Biol. Chem.* **237**:1657–1660 (1962).
246. R. J. Wurtman and J. Axelrod, Control of enzymatic synthesis of adrenaline in the adrenal medulla by adrenal cortical steroids, *J. Biol. Chem.* **241**:2301–2305 (1966).
247. A. E. Kitabchi, S. S. Solomon, and R. H. Williams, Stimulatory effects of insulin and glucagon on phenylethanolamine-N-methyl transferase (PNMT) of rat adrenal, *Proc. Soc. Exptl. Biol. Med.* **127**:296–300 (1968).
248. W. W. Douglas and R. P. Rubin, The role of calcium in the secretory response of the adrenal medulla to acetylcholine, *J. Physiol. (London)* **159**:40–57 (1961).
249. W. W. Douglas, The mechanism of the release of catecholamines from the adrenal medulla, *Pharmac. Rev.* **18**:471–480 (1966).
250. I. Kopin, Private communication.
251. J. Axelrod, H. Weil-Malherbe, and R. Tomchick, The physiological disposition of H^3-epinephrine and its metabolite metanephrine, *J. Pharm. Exp. Therap.* **127**:251–256 (1959).
252. J. J. Pisano, J. R. Crout, and D. Abraham, Determination of 3-methoxy-4-hydroxymandelic acid in urine, *Clin. Chem. Acta* **7**:285–291 (1962).
253. H. Weil-Malherbe, The estimation of 3,4-dihydroxymandelic acid in urine and its excretion by man, *J. Lab. Clin. Med.* **69**:1025–1035 (1967).
254. S. Brunjes, D. Wybenga, and V. J. Johns, Fluorometric determination of urinary metanephrine and normetanephrine, *Clin. Chem.* **10**:1–12 (1964).

255. Z. Kahane, A. H. Esser, N. S. Kline, and P. Vestergaard, Estimation of conjugated epinephrine and norepinephrine in urine, *J. Lab. Clin. Med.* **69**:1042–1050 (1967).
256. J. Axelrod, Metabolism of norepinephrine and other sympathomimetic amines *Physiol. Rev.* **39**:751–776 (1959).
257. L. G. Whitby, J. Axelrod, and H. Weil-Malherbe, The fate of H^3-norepinephrine in animals, *J. Pharmac. Exp. Ther.* **132**:193–201 (1961).
258. H. L. Price and M. L. Price, The chemical estimation of epinephrine and norepinephrine in human and canine plasma. II. A critique of the trihydroxyindole method, *J. Lab. Clin. Med.* **50**:769–777 (1957).
259. O. Celander, The range of control exercised by the sympathico-adrenal system, *Acta Physiol. Scand.* **32**:Suppl. 116, 1–132 (1954).
260. E. W. Sutherland, T. W. Rall, and T. Menon, Adenylcyclase. I. Distribution, preparation, and properties, *J. Biol. Chem.* **237**:1220–1227 (1962).
261. T. W. Rall and E. W. Sutherland, Adenyl cyclase. II. The enzymatically catalyzed formation of adenosine 3′,5′-phosphate and inorganic pyrophosphate from adenosine triphosphate, *J. Biol. Chem.* **237**:1228–1232 (1962).
262. L. M. Klainer, Y. M. Chi, S. L. Friedberg, T. W. Rall, and E. W. Sutherland, Adenylcyclase. IV. The effects of neurohormones on the formation of adenosine 3′5′-phosphate by preparations from brain and other tissues, *J. Biol. Chem.* **237**: 1239–1243 (1962).
263. G. A. Riley, Action of adenosine 3′,5′-phosphate in dog liver preparations, *Fed. Proc.* **22**:258 (1963) (Abstract).
264. E. W. Sutherland and C. F. Cori, Effect of hyperglycemic–glycogenolytic factor and epinephrine on liver phosphorylase, *J. Biol. Chem.* **188**:531–543 (1951).
265. C. F. Cori and A. D. Welch, The adrenal medulla, *J. Am. Med. Assoc.* **116**:2590–2596 (1941).
266. J. H. Exton and C. R. Park, The stimulation of gluconeogenesis from lactate by epinephrine glucagon and cyclic 3′,5′-adenylate in the perfused rat liver, *Pharmacol. Rev.* **18**:181–188 (1966).
267. R. S. Gordon, Jr. and A. Cherkes, Production of unesterified fatty acids from isolated rat adipose tissue incubated *in vitro*, *Proc. Soc. Exptl. Biol. Med.* **97**:150–151 (1958).
268. J. E. White and F. L. Engel, A lipolytic action of epinephrine and norepinephrine on rat adipose tissue *in vitro*, *Proc. Soc. Exptl. Biol. Med.* **99**:375–378 (1958).
269. D. R. Challoner and D. Steinberg, Metabolic effect of epinephrine on the QO_2 of the arrested isolated perfused rat heart, *Nature* **205**:602–603 (1965).
270. P. J. Randle, P. B. Garland, C. N. Hales, and E. A. Newsholme, The glucose fatty-acid cycle. Its role in insulin sensitivity and the metabolic disturbances of diabetes mellitus, *Lancet* **1**:785–789 (1963).
271. R. W. Butcher, R. J. Ho, H. C. Meng, and E. W. Sutherland, Adenosine 3′,5′-monophosphate in biological material. II. The measurement of cyclic 3′,5′-AMP in tissues and the role of cyclic nucleotide in the lipolytic response of fat to epinephrine, *J. Biol. Chem.* **240**:4515–4523 (1965).
272. E. Shafrir, K. E. Sussman, and D. Steinberg, Role of the pituitary and the adrenal in mobilizing free fatty acids and lipoproteins, *J. Lip. Res.* **1**:459–465 (1960).
273. R. P. Maickel and B. B. Brodie, Interaction of drugs with the pituitary adrenocortical system in the production of the fatty liver, *Ann. N.Y.Acad. Sci.* **104**:1059–1064 (1963).
274. J. P. Brady, D. R. Thornton, and D. DeFisher, Deleterious effects of anxiety elicited by conditioned preaversive stimuli in the rat, *Psychosomat. Med.* **24**:590–595 (1962).

275. E. L. Bliss, C. J. Migeon, C. M. H. Branch, and L. T. Samuels, Reaction of the adrenal cortex to emotional stress, *Psychosomat. Med.* **18**:56–76 (1956).
276. S. R. Hill, F. C. Goetz, H. M. Fox, B. J. Murawski, L. J. Krakauer, R. W. Reifanstein, S. J. Gray, W. J. Reddy, S. E. Hedberg, J. R. St. Marc, and G. W. Thorn, Studies on adrenocortical and psychological response to stress in man, *A.M.A. Arch. Internal Med.* **97**:269–298 (1956).
277. J. W. Mason, Psychological influences on the pituitary-adrenal cortical system, *Recent Progr. Hormone Res.* **15**:345–389 (1959).
278. H. Persky, S. J. Korchin, H. Basowitz, F. A. Board, M. Sabshin, D. A. Hamburg, and R. R. Grinker, Effect of two psychological stresses on adrenocortical function: Studies on anxious and normal subjects, *A.M.A. Arch. Neurol. Psychiat.* **81**:219–226 (1959).
279. J. M. Davis, R. Morrill, J. A. Fawcett, V. Upton, P. K. Bondy, and H. M. Spiro, Apprehension and elevated serum cortisol levels, *J. Psychosomat. Res.* **6**:83–86 (1962).
280. A. M. Connell, J. Cooper, and J. W. Redfearn, The contrasting effects of emotional tension and physical exercise on the excretion of 17-ketogenic steroids and 17-ketosteroids, *Acta Endocrinol.* **27**:179–194 (1958).
281. J. R. Fishman, D. A. Hamburg, J. H. Handlon, J. W. Mason, and E. Sachar, Emotional and adrenal cortical responses to a new experience, *Arch. Gen. Psychiat.* **6**:271–278 (1962).
282. G. Pincus and H. Hoagland, Adrenocortical responses to stress in normal men and those with personality disorders, *Am. J. Psychiat.* **106**:641–650 (1950).
283. R. W. Wadeson, J. W. Mason, D. A. Hamburg, and J. H. Handlon, Plasma and urinary 17-OHCS responses to motion pictures, *Arch. Gen. Psychiat.* **9**:146–156 (1963).
284. C. C. Jensen and J. I. Ek, The excretion of certain adrenal steroids during mental stress in healthy persons, *Acta Psychiat. Scand.* **38**:302–306 (1962).
285. M. Friedman, R. H. Rosenman, and V. Carroll, Changes in the serum cholesterol and blood clotting time in man subjected to cyclic variation of occupational stress, *Circulation* **17**:852–861 (1958).
286. C. B. Thomas and E. A. Murphy, Further studies on cholesterol levels in the Johns Hopkins medical students, the effect of stress at examination, *J. Chronic Dis.* **8**:661–668 (1958).
287. P. T. Westlake, A. A. Wolcox, M. Haley, and J. E. Peterson, Relationship of mental and emotional stress to serum cholesterol levels, *Proc. Soc. Exp. Biol. Med.* **97**:163–165 (1958).
288. S. M. Grundy and A. C. Griffin, Relationship of periodic mental stress to serum lipoprotein and cholesterol levels, *J. Am. Med. Assoc.* **171**:1794–1796 (1959).
289. M. D. Bogdonoff, E. H. Estes, Jr., and D. Trout, Acute effect of psychologic stimuli upon plasma non-esterified fatty acid levels, *Proc. Soc. Exp. Biol. Med.* **100**:503–504 (1959).
290. P. W. Cardon, Jr. and R. S. Gordon, Rapid increase of plasma unesterified fatty acids in man during fear, *J. Psychosom. Res.* **4**:5–9 (1959).
291. R. J. Havel and A. Goldfien, The role of the sympathetic nervous system in the metabolism of free fatty acids, *J. Lipid Res.* **1**:102–108 (1959).
292. M. D. Bogdonoff, E. H. Estes, Jr., W. R. Harlan, D. L. Trout, and N. Kirschner, Metabolic and cardiovascular changes during a state of acute central nervous system arousal, *J. Clin. Endocrin. and Metab.* **20**:1333–1340 (1960).
293. L. A. Gottschalk, J. M. Cleghorn, G. C. Gleser, and J. M. Iacono, Study of the relationship of emotion to plasma lipids, *Psychosomat. Med.* **27**:102–111 (1965).

294. J. W. Frost, R. L. Dryer, and K. G. Kohlstaedt, Stress studies on auto race drivers, *J. Lab. Clin. Med.* **38**:523–525 (1951).
295. A. Markkanen, A. Pekkarinen, K. Pulkkinen, and P. E. Simola, On the emotional eosenopenic reaction caused by examination, *Acta Physiol. Scand.* **35**:225–239 (1956).
296. G. Pincus, H. Hoagland, H. Freeman, and F. Elmadjian, Adrenal function in mental disease, *Rec. Prog. Hormone Res.* **4**:291–311 (1949).
297. A. Mittleman, L. P. Romanoff, G. Pincus, and H. Hoagland, Neutral steroid excretion by normal and schizophrenic men, *J. Clin. Endocrinol.* **12**:831–840 (1952).
298. W. W. Schottstaedt, W. J. Grace, and H. G. Wolff, Life situation, behavior, attitudes, emotions, and renal excretion of fluids and electrolytes, *J. Psychosomat. Res.* **1**:287–291 (1956).
299. A. Yuwiler, *in Schizophrenia: Current Concepts and Research* (D. V. Siva Sankar, ed.) pp. 600–613, PJD Publications, Ltd., Hicksville, N.Y. (1969).
300. L. Levi, *in An Introduction to Clinical Neuroendocrinology* (E. Bajusz, ed.) pp. 78–105, S. Karger, Basel, New York (1967).
301. L. Levi, Some experimental stress situations with simultaneous study of physiological and psychological variables, *Psychosom. Med.* **20**: 167 (1958).
302. L. Levi, A new stress tolerance test with simultaneous study of physiological and psychological variables, *Acta Endocrin.* **37**:38–44 (1961).
303. U. S. v. Euler, Quantitation of stress by catecholamine analysis, *Clin. Pharmacol. Therap.* **5**:398–404 (1964).
304. D. H. Funkenstin, S. H. King, and M. E. Drolette, The direction of anger during a laboratory stress-inducing situation, *Psychosom. Med.* **16**:404–413 (1954).
305. M. Friedman, S. St. George, S. O. Byers, and R. H. Rosenman, Excretion of catecholamines, 17-ketosteroids, 17-hydroxycorticoids and 5-hydroxyindole in men exhibiting a particular behavior pattern (a) associated with high incidence of clinical coronary artery disease, *J. Clin. Invest.* **39**:758–764 (1960).
306. F. Elmadjian, J. M. Hope, and E. T. Lamson, Excretion of epinephrine and norepinephrine under stress, *Recent Progr. Hormone Res.* **14**:513–553 (1958).
307. F. Elmadjian, J. M. Hope, and E. T. Lamson, Excretion of epinephrine and norepinephrine in various emotional states, *J. Clin. Endocr.* **17**:608–620 (1957).
308. A. J. Silverman, S. I. Cohen, B. M. Shmavonian, and N. Kirshner, *in Rec. Adv. Biol. Psychiat.* (J. Wortis, ed.) Vol. 3, pp. 104–117, Grune and Stratton, New York (1961).
309. P. F. Regan and J. Reilly, Circulating epinephrine and norepinephrine in changing emotional states, *J. Nerv. Ment. Dis.* **127**, 12–16 (1958).
310. G. C. Curtis, R. A. Cleghorn, and T. L. Sourkes, The relationship between affect and the excretion of adrenaline, noradrenaline and 17-hydroxycorticosteroids, *J. Psychosom. Res.* **4**:176–184 (1960).
311. L. Levi, The urinary output of adrenaline and noradrenaline during pleasant and unpleasant emotional states, *J. Psychosomat. Med.* **27**:80–85 (1965).
312. G. N. Nelson, M. Masuda, and T. H. Holmes, Correlation of behavior and catecholamine metabolite excretion, *Psychosom. Med.* **28**:216–226 (1966).
313. J. O. Tingley, A. W. Morris, and S. R. Hill, Studies on the diurnal variation and response to emotional stress of the thyroid gland, *Clin. Res.* **6**:134 (1958).
314. F. Alexander, G. W. Flagg, S. Foster, T. Clemens, and W. Blahd, Experimental studies of emotional stress: Hyperthyroidism, *Psychosomat. Med.* **23**:104–114 (1961).
315. B. S. Hetzel, D. S. de la Haba, and L. E. Hinkle, Jr., Life stress and thyroid function in human subjects, *J. Clin. Endocrin.* **12**:941 (1952) (Abstract).

316. S. B. Friedman, R. Ader, L. J. Grota, and T. Larson, Plasma corticosterone response to parameters of electric shock stimulation in the rat, *Psychosom. Med.* **29**:323–328 (1967).
317. W. P. Pare and T. E. McCarthy, Urinary sodium and potassium and prolonged environmental stress, *J. Genetic Psychol.* **108**:135–142 (1966).
318. O. F. Ehrentheil, L. J. Reyna, C. J. Adams, T. J. Giovanniello, and E. T. Chen, Blood sugar response to electric shock stress in normal adrenalectomized, and hypophysectomized white rats, *Diabetes* **16**:325–330 (1967).
319. R. Böhm and F. A. Hoffmann, Beiträge zur Kenntnis des Kohlehydratstoffwechsels, *Arch. Exp. Path. U. Pharm.* **8**:271–308 and 375–445, (1877/1878).
320. W. B. Cannon, A. T. Shohl, and W. S. Wright, Emotional glycosuria, *Am. J. Physiol.* **29**:280–287 (1911).
321. S. A. Barnett, J. C. Eaton, and H. Morag McCallum, Physiological effects of "social stress" in wild rats. II. Liver glycogen and blood glucose, *J. Psychosom. Res.* **4**:251–260 (1960).
322. O. F. Ehrentheil, L. J. Reyna, G. Yerganian, and E. T. Chen, Studies in stress glycosuria. I. Prolonged glycosuria in Chinese hamsters after repeated stress, *Diabetes* **13**:83–86 (1964).
323. R. Ader, S. B. Friedman, L. J. Grota, and A. Schaefer, Attenuation of the plasma corticosterone response to handling and electric shock stimulation in the infant rat, *Physiol. and Behav.* **3**:327–331 (1968).
324. A. M. Thierry, F. Javoy, J. Glowinski, and S. S. Kety, Effects of stress on the metabolism of norepinephrine, dopamine and serotonin in the central nervous system of the rat. I. Modifications of norepinephrine turnover, *J. Pharmacol. Exptl. Ther.* **163**:163–170 (1968).
325. H. C. Nielson and R. M. Fleming, Effects of electroconvulsive shock and prior stress on brain amine levels, *Exptl. Neurol.* **20**:21–30 (1968).
326. E. W. Maynert and R. Levi, Stress induced release of brain norepinephrine and its inhibition by drugs, *J. Pharmacol.* **143**:90–95 (1964).
327. E. L. Bliss, J. Ailion, and J. Zwanziger, Metabolism of norepinephrine, serotonin and dopamine in rat brain with stress, *J. Pharmacol. Exp. Therap.* **164**:122–134 (1968).
328. A.-M. Thierry, M. Fekete, and J. Glowinski, Effects of stress on the metabolism of noradrenaline, dopamine and serotonin (5HT) in the central nervous system of the rat. II. Modifications of serotonin metabolism, *European J. Pharmacol.* **4**:384–389 (1968).
329. S. Rosenblatt, J. D. Chanley, H. Sobotka, and M. R. Kaufman, Interrelationships between electroshock, the blood-brain barrier and catecholamines, *J. Neurochem.* **5**:172–176 (1960).
330. G. T. Pryor, L. S. Otis, M. K. Scott, and J. J. Colwell, Duration of chronic electroshock treatment in relation to brain weight, brain chemistry, and behavior, *J. Comp. Physiol. Psych.* **63**:236–239 (1967).
331. W. T. Greenough, J. K. Fulcher, A. Yuwiler, and E. Geller, Enriched rearing and chronic electroshock: Effects on brain and behavior in mice, *Physiol. Behav.* **5**: 371–373 (1970).
332. A. A. Schatalova and E. K. Antonov, Abst. Commun. Fifth Internat. Congress, Biochem., Moscow, 1961 (Oxford, Pergamon Press, 1961), p. 271.
333. R. K. Hinesley, J. A. Norton, and M. H. Aprison, Serotonin, norepinephrine and 3,4-dihydroxyphenylethylamine in rat brain parts following electroconvulsive shock, *J. Psychiat. Res.* **6**:143–152 (1968).

334. A. Geiger, S. Yamasaki, and R. Lyons, Changes in nitrogenous components of brain produced by stimulation of short duration, *Am. J. Physiol.* **184**:239–243 (1956).
335. D. Richter and J. Crossland, Variation in acetylcholine contents of the brain with physiological state, *Am. J. Physiol.* **159**:247–255 (1949).
336. D. Richter and R. M. C. Dawson, The ammonia and glutamine content of the brain, *J. Biol. Chem.* **176**:1199–1210 (1948).
337. J. R. Klein and N. S. Olson, Effect of convulsive activity upon the concentration of brain glucose, glycogen, lactate and phosphates, *J. Biol. Chem.* **167**:747–756 (1947).
338. C. F. Schmidt, S. S. Kety, and H. S. Pennes, The gaseous metabolism of the brain of the monkey, *Am. J. Physiol.* **143**:33–52 (1945).
339. L. Kato, B. Gozsy, P. B. Roy, and V. Groh, Histamine, serotonin, epinephrine and norepinephrine in the rat brain, following convulsions, *Internat. J. Neuropsychiat.* **3**:46–51 (1967).
340. S. S. Kety, F. Javoy, A. Thierry, L. Julou, and J. Glowinski, A sustained effect of electroconvulsive shock on the turnover of norepinephrine in the central nervous system of the rat, *Proc. Nat. Acad. Sci. U.S.* **58**:1249–1254 (1967).
340a J. M. Musacchio, L. Julou, S. S. Kety, and J. Glowinski, Increase in rat tyrosine hydroxylase activity produced by electroconvulsive shock *Proc. Nat. Acad. Sci. U.S.* **63**:1117–1119 (1969).
341. J. Engle, L. C. F. Hanson, B.-E. Roos, and L. E. Strömbergsson, Effect of electroshock on dopamine metabolism in rat brain, *Psychopharmacologia* **13**:140–143(1968).
342. S. Garattini and L. Valzelli *in Psychotropic Drugs* (S. Garattini and V. Giletti, eds.) pp. 428–436, Elsevier Publ. Co., Amsterdam (1957).
343. G. Bertaccini, Effect of convulsant treatment on the 5-HT content of brain and other tissues of the rat, *J. Neurochem.* **4**:217–222 (1959).
344. D. D. Bonnycastle, N. J. Giarman, and M. K. Paasonen, Anticonvulsant compounds and 5-HT in rat brain, *Brit. J. Pharmacol. Chemotherap.* **12**:228–231 (1957).
345. F. N. Minard and D. Richter, Electroshock-induced seizures and the turnover of brain protein in rat, *J. Neurochem.* **15**:1463–1468 (1968).
346. P. Wiechert and G. Göllnitz, Metabolic investigation of epileptic seizures, The activity of glutaminase and glutamine synthetase and ammonia metabolism before and during cerebral convulsions, *J. Neurochem.* **16**:317–322 (1969).
347. P. R. McHugh, W. C. Black, and J. W. Mason, Some hormonal responses to electrical self-stimulation in the *Macaca mulatta*, *Amer. J. Physiol.* **110**:109–113 (1966).
348. W. C. Black and P. R. McHugh, Adrenal and thymus weights in self-stimulating rats, *Proc. Eastern Psychol. Assoc.*, April 1964.
349. E. L. Bliss, B. B. Wilson, and J. Zwanziger, Changes in brain norepinephrine in self-stimulating and "aversive" animals, *J. Psychiat. Res.* **4**:59–63 (1966).
350. E. Endroczi and K. Lissak, Effect of hypothalamic and brain stem structure stimulation on pituitary-adrenocortical function, *Acta Physiol. Acad. Sci. Hung.* **24**:67–77 (1964).
351. A. De Schaepdryver, P. Preziosi, and U. Scapagnini, Brain monoamines and adrenocortical activation (Abstract), *Brit. J. Pharmacol.* **34**:663P–664P (1968).
352. E. Zimmerman and V. Critchlow, Effects of diurnal variation in plasma corticosterone levels on adrenocortical response to stress, *Proc. Soc. Exp. Biol. Med.* **125**:658–663 (1967).
353. H. Selye, Thymus and adrenals in the response of the organism to injuries and intoxications, *Brit. J. Exp. Pathol.* **17**:234–248 (1936).

354. G. Rossi, S. Bonfils, F. Lieffoogh, and A. Lambling, Technique nouvelle pour produire des ulceratations gastriques chez le rat blanc, *Compt. Rend. Soc. Biol.* **150**:2124–2126 (1956).
355. D. A. Brodie and H. M. Hanson, A study of the factors involved in the production of gastric ulcers by the restraint technique, *Gastroenterology* **38**:353–360 (1960).
356. L. Macho, M. Palkovic, L. Mikulaj, and R. Kvetnansky, Tissue metabolism in rats adapted to immobilization stress, *Physiol. Bohemoslov.* **17**:173–178 (1968).
357. P. Prioreschi and E. Peterman, Effect of forced restraint on heart phospholipids, *Can. J. Physiol. Pharmacol.* **45**:407–413 (1967).
358. O. Hanninen and K. Hartiala, The induction of liver tyrosine 2-oxoglutarate transaminase in rats by immobilization, *Acta Endocrinol.* **54**:85–90 (1967).
359. J. Nomura, Effect of stress and psychotropic drugs on rat liver tryptophan pyrrolase, *Endocrinol.* **76**:1190–1194 (1965).
360. B. L. Welch and A. S. Welch, Differential activation by restraint stress of a mechanism to conserve brain catecholamines and serotonin in mice differing in excitability due to prior isolation vs. group housing, *Nature* **218**:575–577 (1968).
361. E. L. Bliss and J. Zwanziger, Brain amines and emotional stress, *J. Psychiat. Res.* **4**:189–198 (1966).
362. A. DeSchaepdryver, P. Preziosi, and U. Scapagnini, Brain monoamines and adrenocortical activation, *Brit. J. Pharmacol.* **35**:460–467 (1969).
363. H. Corrodi, K. Fuxe, and T. Hokfelt, The effect of immobilization stress on the activity of central monoamine neurons, *Life Sci.* **7**:107–112 (1968).
364. W. T. Liberson, J. Bernsohn, A. Wilson, and V. Daly, Brain serotonin content and behavioral stress, *J. Neuropsychiat.* **5**:363–365 (1964).
365. W. B. Essman, Development of activity differences in isolated and aggregated mice, *Animal Behavior* **14**:406–409 (1966).
366. E. T. Uyeno and M. White, Social isolation and dominance behavior, *J. Comp. Physiol. Phychol.* **63**:157–159 (1967).
367. P. A. Janssen, A. H. Jageneau, and C. J. E. Niemegeers, Effects of various drugs on isolation-induced fighting behavior of male mice, *J. Pharmacol. Exptl. Therap.* **129**:471–475 (1960).
368. W. B. Essman, Gastric ulceration as a function of food deprivation in isolated vs. aggregated mice, *Psychonomic Sci.* **4**:251–252 (1966).
369. K. E. Moore, L. C. Sawdy, and S. R. Shaul, Effects of *d*-amphetamine on blood glucose and tissue glycogen levels of isolated and aggegated mice, *Biochem. Pharmacol.* **14**:197–204 (1965).
370. J. A. Gunn and M. R. Gurd, The action of some amines related to adrenaline. Cyclohexylalkylamines, *J. Physiol.* **97**:453–470 (1940).
371. F. Grabarits, H. Lal, and R. D. Chessick, Effects of cocaine on brain noradrenaline in relation to toxicity and convulsions in mice, *J. Pharm. Pharmacol.* **18**:131–132 (1966).
372. B. L. Welch and A. S. Welch, Graded effect of social stimulation upon *d*-amphetamine toxicity, aggressiveness and heart and adrenal weight, *J. Pharmacol. Exptl. Therap.* **151**:331–338 (1966).
373. H. I. Chernov, P. Furness, D. Partyka, and A. J. Plummer, Age, confinement and aggregation as factors in amphetamine toxicity in mice, *J. Pharm. Exptl. Therap.* **154**:346–349 (1966).
374. C. D. Louch and M. Higginbotham, The relationship between social rank and plasma corticosterone levels in mice, *Gen. Comp. Endocrinol.* **8**:441–444 (1967).
375. B. L. Welch and A. S. Welch, The effect of grouping on the level of brain norepinephrine in white Swiss mice, *Life Sci.* **4**:1011–1018 (1965).

376. B. L. Welch and A. M. Welch, *in Progress in Brain Research* (H. E. Himwich and W. A. Himwich, eds.) Vol. 8, pp. 201–206, Elsevier Publishing Co., New York (1964).
377. S. Garattini, E. Giacalone, and L. Valzelli, Isolation, aggressiveness and brain 5-hydroxytryptamine turnover, *J. Pharm. Pharmacol.* **19**:338–339 (1967).
378. E. Giacalone, M. Tansella, L. Valzelli, and S. Garattini, Brain serotonin metabolism in isolated aggressive mice, *Biochem. Pharmacol.* **17**:1315–1327 (1968).
379. E. L. Bliss and J. Ailion, Response of neurogenic amines to aggression and strangers, *J. Pharmacol. Exptl. Therap.* **168**:258–263 (1969).
380. L. Valzelli, *in Advances in Pharmacology* (S. Garattini and P. A. Shore, eds.) Vol. 5, pp. 79–108, Academic Press, London (1967).
381. B. E. Eleftheriou and R. L. Church, Brain 5-hydroxytryptophan decarboxylase in mice after exposure to aggression and defeat, *Physiol and Behav.* **3**:323–325 (1968).
382. B. E. Eleftheriou and K. W. Boehlke, Brain monoamine oxidase in mice after exposure to aggression and defeat, *Science* **155**:1693–1694 (1967).
383. A. S. Welch and B. L. Welch, Effect of stress and para-chlorophenylalanine upon brain serotonin, 5-hydroxyindoleacetic acid and catecholamines in grouped and isolated mice, *Biochem. Pharmacol.* **17**:699–708 (1968).
384. B. L. Welch and A. S. Welch, Sustained effects of brief daily stress (fighting) upon brain and adrenal catecholamines and adrenal, spleen and heart weights of mice, *Proc. Nat. Acad. Sci. U.S.* **64**:100–107 (1969).
385. A. S. Welch and B. L. Welch, Reduction of norepinephrine in the lower brainstem by psychological stimulus, *Proc. Natl. Acad. Sci. U.S.* **60**:478–481 (1968).
386. E. Geller, A. Yuwiler, and J. F. Zolman, Effects of environmental complexity on constituents of brain and liver, *J. Neurochem.* **12**:949–955 (1965).
387. D. Krech, M. R. Rosenzweig, and E. L. Bennett, Effects of environmental complexity and training on brain chemistry, *J. Comp. Physiol Psychol.* **53**:509–519 (1960).
388. M. R. Rosenzweig, E. L. Bennett, and D. Krech, Cerebral effects of environmental complexity and training among adult rats, *J. Comp. Physiol Psychol.* **57**:438–439 (1964).
389. M. R. Rosenzweig, Environmental complexity, cerebral change and behavior, *Am. Psychol.* **21**:321–332 (1966).
390. M. C. Diamond, D. Krech, and M. R. Rosenzweig, The effect of an enriched environment on the histology of the rat cerebral cortex, *J. Comp. Neurol.* **123**:111–119 (1964).
391. M. C. Diamond, F. Law, H. Rhodes, B. Lindner, M. R. Rosenzweig, D. Krech, and E. L. Bennett, Increases in cortical depth and glia numbers in rats subjected to enriched environment, *J. Comp. Neurol.* **128**:117–126 (1966).
392. M. C. Diamond, Extensive cortical depth measurements and neuronal size increases in the cortex of environmentally enriched rats, *J. Comp. Neurol.* **131**:357–364 (1967).
393. A. M. Hatch, G. S. Wiberg, Z. Zawidzka, M. Cann, J. M. Airth, and H. C. Grice, Isolation syndrome in the rat, *Toxicol. Appl. Pharmacol.* **7**:737–745 (1965).
394. M. Ruckebusch and C. Brunet-Tallon, Monoaminoxydases et cholinestérases du cerveau: influence de diverses aggressions, *Comp. Rend. Soc. Biol.* **160**:2131–2133 (1966).
395. E. Bennett, D. Krech, M. R. Rosenzweig, and H. Karisson, Cholinesterase and lactic dehydrogenase activity in the rat brain, *J. Neurochem.* **3**:153–160 (1958).
396. E. Geller and A. Yuwiler, *in Ontogenesis of the Brain* (L. Jílek and S. Trojan, eds.) pp. 277–284, Charles University Press, Prague (1968).

397. H. C. Agrawal, M. W. Fox, and W. A. Himwich, Neurochemical and behavioral effects of isolation-rearing in the dog, *Life Sci.* **6**:71–78 (1967).
398. P. Jurtshuk, Jr., A. S. Weltman, and A. M. Sackler, Biochemical responses of rats to auditory stress, *Science* **129**:1424–1425 (1959).
399. W. F. Geber and T. A. Anderson, Abnormal fetal growth in the albino rat and rabbit induced by maternal stress, *Biol. Neonat.* **11**:209–215 (1967).
400. J. D. Barachas and D. X. Freedman, Brain amines: Response to physiological stress, *Biochem. Pharmacol.* **12**:1232–1234 (1963).
401. E. Ruether, M. Ackenheil, and N. Matussek, Norepinephrine and serotonin in rat brain after stress, *Arzneim.-Forsch.* **16**:261–263 (1966).
402. J. N. Williams, Jr., P. E. Schurr, and C. A. Elvehjem, Influence of chilling and exercise on free amino acid concentration of rat brain, *J. Biol. Chem.* **182**:55–59 (1950).
403. E. Geller, A. Yuwiler, and S. Schapiro, Comparative effects of a stress and cortisol upon some enzymic activities, *Biochim. Biophys. Acta* **93**:311–315 (1964).
404. S. Schapiro, A. Yuwiler, and E. Geller, Stress-activated inhibition of the cortisol effect on hepatic transaminase, *Life Sci.* **3**:1221–1226 (1964).
405. S. Schapiro, A. Yuwiler, and E. Geller, Maturation of a stress-activated mechanism inhibiting induction of tyrosine transaminase, *Science* **152**:1642–1643 (1966).
406. S. Schapiro, E. Geller, and A. Yuwiler, Differential effects of a stress on liver enzymes in adult and infant rats, *Neuroendocrinology* **1**:138–143 (1965/1966).
407. E. Geller, A. Yuwiler, and S. Schapiro, Tyrosine Aminotransferase: Activation or repression by a stress, *Proc. Soc. Exptl. Biol. Med.* **130**:458–461 (1969).
408. R. A. McChance, The maintenance of stability in the newly born. I. Chemical exchange, *Arch. Dis. Child.* **34**:361–370 (1959).
409. P. A. Poczopko, Contribution to the studies on changes of energy metabolism during postnatal development, *J. Cell. Comp. Physiol.* **57**:175–184 (1961).
410. E. F. Adolph, The ontogeny of physiological regulation of heart rate in fetal infant and adult rats, *Am. J. Physiol.* **209**:1095–1105 (1965).
411. J. W. Jailer, The maturation of the pituitary adrenal axis in the newborn rat, *Endocrinology* **46**:420–425 (1950).
412. E. Irwin, A. R. Buchanan, B. B. Longwell, D. E. Holtkamp, and R. M. Hill, Ascorbic acid content of adrenal glands of young albino rats after cold and other stresses, *Endocrinol.* **46**:526–535 (1950).
413. S. Schapiro, E. Geller, and S. Eiduson, Neonatal adrenal cortical response to stress and vasopressin, *Proc. Soc. Exptl. Biol. Med.* **109**:937–941 (1962).
414. V. Jonec, The dynamic analysis of the mechanisms involved in acute activation of the pituitary-adrenocortical system by cold, *Can. J. Physiol. Pharmacol.* **42**:585–591 (1964).
415. K. M. Knigge, Neuroendocrine mechanisms influencing ACTH and TSH secretion and their role in cold acclimation, *Fed. Proc. 19, Suppl.* **5**:45–49 (1960).
416. M. W. Ralli and E. P. Dumm, Influence of ascorbic acid, pantothenic acid and protein on the resynthesis of adrenal cholesterol after stress, *Fed. Proc.* **13**:38 (1954).
417. O. Heroux and J. S. Hart, Cold acclimation and adrenocortical activity as measured by eosenophile levels, *Am. J. Physiol.* **178**: 453–456 (1954).
418. S. Katsh, G. F. Katsh, and P. Osher, Adrenal, pituitary and urinary ascorbic acid levels in rats subjected to hypothermic environment, *Am. J. Physiol.* **178**:457–460 (1954).
419. M. Suzuki, T. Tonoue, S. Matsuzaki, and K. Yamamoto, Initial response of human thyroid, adrenal cortex and adrenal medulla to acute cold exposure, *Can. J. Physiol. Pharmacol.* **45**:423–432 (1967).

420. G. W. Thorn and J. C. Laidlow, Studies on the adrenal cortical response to stress in man, *Trans. Am. Clin. Clymatol.* **65**:179–199 (1953).
421. W. F. Bernhard, The effects of hypothermia on the peripheral serum levels of free 17-hydroxycorticoids in the dog and in man, *Natl. Acad. Sci.—Natl. Res. Council Publ.* **451**:175–182 (1956).
422. R. R. J. Chaffee, J. R. Allen, M. Brewer, S. M. Horvath, C. Mason, and R. E. Smith, Cellular Physiology of cold and heat-exposed squirrel monkeys (*Saimiri Sciurea*) *J. Appl. Physiol.* **21**:151–157 (1966).
423. D. H. Nelson, R. M. Egdahl, and D. M. Hume, Corticosteroid secretion in the adrenal vein of the nonstressed dog exposed to cold, *Endocrinology* **58**:309–314 (1955).
424. M. X. Zarrow and J. T. Baldini, Failure of adrenocorticotropin and various stimuli to deplete the ascorbic acid content of the adrenal gland of the quail, *Endocrinology* **50**:555–561 (1952).
425. S. Katsh, G. F. Katsh, and P. Osher, Adrenal, pituitary and urinary ascorbic acid levels in rats subjected to hypothermic environment, *Am. J. Physiol.* **178**:457–461 (1954).
426. L. P. Dugal and M. Thérien, Ascorbic acid and acclimatization to cold environment, *Can. J. Res.* (Section 1) **25**:111–136 (1947).
427. G. Sayers, M. A. Sayers, T.-Y. Liang, and C. N. H. Long, The cholesterol and ascorbic acid content of the adrenal, liver, brain and plasma following hemorrhage, *Endocrinology* **37**:96–110 (1945).
428. R. Boulourd, Adrenocortical activity during adaptation to cold in the rat: role of Porter-Silber chromogens, *Fed. Proc. 25, Suppl. No. 4*, 1195–1199 (1966).
429. C. Duyck, Lack of correlation between body weight and corticosterone secretion in the rat, *Endocrinology* **75**:822–823 (1964).
430. Z. Hruza and O. Poupa, *in Handbook of Physiology Adaptation to the Environment*, Washington, D.C., Am. Physiol. Soc., (1964), Sect. 4. Chapt. 60, p. 939.
431. J. Leduc, Catecholamine production and release in exposure and acclimation to cold, *Acta Physiol. Scand., Supp. 183*, 1–101 (1961).
432. C. N. H. Long, The relation of cholesterol and ascorbic acid to the secretion of the adrenal cortex, *Prog. Hormone Res.* **1**:99–123 (1947).
433. M. Wada and E. Fuzii, Effect of severe cold upon the denervated heart of non-anaesthetized dogs and the epinephrine secretion, *Tohoku J. Exp. Med.* **37**:505–516 (1940).
434. M. Wada, M. Seo, and K. Abe, Further study of the influence of cold on the rate of epinephrine secretion from the suprarenal glands with simultaneous determination of blood sugar, *Tohoku J. Exp. Med.* **26**:381–411 (1935).
435. E. L. Arnett and D. T. Watts, Catecholamine excretion in men exposed to cold, *J. Appl. Physiol.* **15**:499–500 (1960).
436. R. Bertin and L. Chevillard, Urinary elimination of catecholamines in male Long Evans rats during adaptation to various temperatures, *Comp. Rend. Soc. Biol.* **161**:248–251 (1967).
437. G. E. Johnson, K. V. Flattery, and E. Schönbaum, The influence of methimazole on the catecholamine excretion of cold-stressed rats, *Can. J. Physiol. Pharmacol.* **45**:415–421 (1967).
438. L. Chevillard and R. Bertin, Adrenal catecholamine content of male Long Evans rats during adaptation to various temperatures, *Compt. Rend. Soc. Biol.* **161**:279–283 (1967).

439. E. Zeisberger and K. Brueck, Quantitative relation between NE effect and the extent of non shivering thermogenesis in guinea pigs, *Gesamte Physiol.* **296**:263–275 (1967).
440. L. Krulich and S. M. McCann, Influence of stress on the growth hormone (GH) content of the pituitary of the rat, *Proc. Soc. Exptl. Biol. Med.* **122**:612–616 (1966).
441. E. E. Müller, T. Saito, A. Arimura, and A. V. Schally, Hypoglycemia, stress and growth hormone release: Blockade of growth hormone release by drugs acting on the central nervous system, *Endocrinology* **80**:109–117 (1967).
442. L. V. Beck, D. S. Zaharko, S. C. Kalser, R. Miller, M. Van Germert, and R. Kunig, Variations in serum insulin and glucose in rats with chronic cold exposure, *Life Sci.* **6**:1501–1506 (1967).
443. M. Cottle and L. D. Carlson, Turnover of thyroid hormone in cold-exposed rats determined by radioactive iodine studies, *Endocrinology* **59**:1–11 (1956).
444. C. P. Leblond, J. Gross, W. Peacock, and R. D. Evans, Metabolism of radioiodine in the thyroids of rats exposed to high or low temperatures, *Am. J. Physiol.* **140**:671–676 (1944).
445. P. K. Bondy and M. A. Hagewood, Effect of stress and cortisone on plasma protein-bound iodine and thyroxine metabolism in rats, *Proc. Soc. Exper. Biol. Med.* **81**:328–331 (1953).
446. P. Starr and R. Roskelley, A comparison of the effects of cold and thyrotropic hormone on the thyroid gland, *Am. J. Physiol.* **130**:549–556 (1940).
447. R. Woods and L. D. Carlson, Thyroxine secretion in rats exposed to cold, *Endocrinology* **59**:323–330 (1956).
448. R. E. Smith, Cold acclimation—an altered steady state, *J. Am. Med. Assoc.* **179**:948–954 (1962).
449. F. Decopas, J. S. Hart, and O. Héroux, Energy metabolism of the white rat after acclimation to warm and cold environments, *J. Appl. Physiol.* **10**:393–397 (1957).
450. P. F. Iampietro, D. E. Bass, and E. R. Buskirk, Diurnal oxygen consumption and rectal temperature of man during continuous cold exposure, *J. Appl. Physiol.* **10**:398–400 (1957).
451. P. E. Scholander, H. T. Hammel, K. L. Andersen, and Y. Løyning, Metabolic acclimation to cold in man, *J. Appl. Physiol.* **12**:1–8 (1958).
452. F. Depocas and R. Masironi, Body glucose as fuel for thermogenesis in the white rat exposed to cold, *Am. J. Physiol.* **199**:1051–1055 (1960).
453. W. Cottle and L. D. Carlson, Adaptive changes in rats exposed to cold. Caloric exchange, *Am. J. Physiol.* **178**:305–308 (1954).
454. E. J. Masoro, C. L. Asuncion, R. K. Brown, and D. Rapport, Lipogenesis from carbohydrate in the negative caloric balance state induced by exposure to cold, *Am. J. Physiol.* **190**:177–179 (1957).
455. A. S. W. DeFreitas and F. Depocas, Fatty acid and glyceride glycerol synthesis from glucose during high rates of glucose uptake in the intact rat, *Can. J. Biochem.* **43**:437–450 (1965).
456. R. Paoletti, R. P. Maickel, R. L. Smith, and B. B. Brodie, *in Proc. First Intern. Pharmacol. Meeting*, (E. C. Howing and P. Lindgren, eds.) Vol. II, pp. 29–41, Macmillan, New York (1963).
457. E. Wertheimer, M. Hamosh, and E. Shafrir, Factors affecting fat mobilization from adipose tissue, *Am. J. Clin. Nutr.* **8**:705–710 (1960).
458. C. R. Treadwell, D. F. Flick, and G. V. Vahouny, Nutrition studies in the cold. II. Curative effect of cold on fat fatty livers, *Proc. Soc. Exptl. Biol. Med.* **97**:434–436 (1958).

459. P. G. Hanson and R. E. Johnson, Variation of plasma ketones and free fatty acids during acute cold exposure in man, *J. Appl. Physiol.* **20**:56–60 (1965).
460. S. Mallov, Cold effects in rat: Plasma and adipose tissue free fatty acids and adipose tissue lipase, *Am. J. Physiol.* **204**:157–164 (1963).
461. R. P. Maickel and K. Yamada, *in Advances in Enzyme Regulation* (G. Weber, ed.) Vol. I, pp. 235–245, Macmillan, New York (1963).
462. J. M. Felts and E. J. Masoro, Effect of cold acclimation on hepatic carbohydrate and lipid metabolism, *Am. J. Physiol.* **197**:34–36 (1959).
463. E. J. Masoro, A. I. Cohen, and S. S. Panagos, Effect of exposure to cold on some aspects of hepatic acetate utilization, *Am. J. Physiol.* **179**:451–456 (1954).
464. S. S. You, R. W. You, and E. A. Sellers, Effect of thyroidectomy, adrenalectomy and burning on the urinary nitrogen excretion of the rat maintained in a cold environment, *Endocrinology* **47**:156–161 (1950).
465. P. A. Mayes, A caloric deficiency hypothesis of ketogenesis, *Metabolism* **11**:781–799 (1962).
466. F. Sargent, II, and C. F. Consolazio, Stress and ketone body metabolism, *Science* **113**:631–633 (1951).
467. J. Campbell, G. R. Green, E. Schönbaum, and H. Socol, Effect of exposure to cold and of diet on coenzyme A levels in tissue, *Can. J. Biochem. Physiol.* **38**:175–185 (1960).
468. R. O. Brady, A. Mamoon, and E. R. Stadtman, The effects of citrate and coenzyme A on fatty acid metabolism, *J. Biol. Chem.* **222**:795–802 (1956).
469. E. J. Masoro, Depressed lipogenesis induced by cold stress, *Am. J. Physiol.* **199**:449–452 (1960).
470. E. J. Masoro, A. I. Cohen, and S. S. Panagos, Effect of exposure to cold on some aspects of hepatic acetate utilization, *Am. J. Physiol.* **179**:451 456 (1954).
471. R. W. Schayer, Relationship of induced histidine decarboxylase activity and histamine synthesis to shock from stress and from endotoxin, *Am. J. Physiol.* **198**:1187–1192 (1960).
472. G. J. Kalin and D. A. Vaughan, Alterations of protein metabolism during cold acclimation, *Fed. Proc.* **22**:862–866 (1963).
473. R. B. Mefferd, Jr., H. B. Hale, and H. H. Martens, Nitrogen and electrolyte excretion of rats chronically exposed to adverse environments, *Am. J. Physiol.* **192**:209–218 (1958).
474. J. P. Hannon, Intermediary glucose metabolism in the cold-acclimatized rat, *Fed. Proc. 19, Suppl.* **5**:100–105 (1960).
475. E. Page, L. M. Bavineau, and J. P. Lachange, Carbohydrate utilization in rats adapted to cold, *Rev. Can. Biol.* **14**:144–151 (1955).
476. R. E. Smith and A. S. Fairhurst, A mechanism of cellular thermogenesis in cold adaptation, *Proc. Natl. Acad. Sci. U.S.* **44**:705–711 (1958).
477. A. U. Ingenito, Effect of acute and prolonged cold exposure on brain amine depleting action of reserpine, *Arch. Int. Pharmacodyn. Therap.* **166**:324–332 (1967).
478. A. U. Ingenito and D. D. Bonnycastle, The effect of exposure to heat and cold upon rat brain catecholamine and 5-hydroxytryptamine levels, *Can. J. Physiol. Pharmacol.* **45**:733–743 (1967).
479. J. LeBlanc, Secretion and activity of histamine and serotonin during cold adaptation, *Am. J. Physiol.* **204**:520–522 (1963).
480. J. LeBlanc, Histamine and cold adaptation, *Proc. Soc. Exptl. Biol. Med.* **112**:25–26 (1963).

481. R. Gordon, S. Spector, A. Sjoerdsma, and S. Udenfriend, Increased synthesis of norepinephrine and epinephrine in the intact rat during exercise and exposure to cold, *J. Pharmacol. Exptl. Therap.* **153**:440–447 (1966).
482. M. Ruckebusch and C. Brunet-Tallon, Effect of various stresses on cerebral monoamine oxidase and cholinesterase, *Comp. Rend. Soc. Biol.* **160**:2131–2133 (1966).
483. V. V. Subba Rao and M. L. Gupta, Effect of heat and cold stress on brain glutamic acid, *Fed. Proc. 25, Suppl. 4,* 1185–1186 (1966).
484. G. Rindi and U. Ventura, Influence of adrenalectomy, of adrenal cortex hormones and of cold on the γ-aminobutyric acid and glutamic acid content of rat brain, *Ital. J. Biochem.* **10**:135–146 (1961).
485. H. B. Hale and R. B. Mefferd, Jr., Effect of adrenocorticotropin on temperature and pressure-dependent metabolic functions, *Am. J. Physiol.* **197**:1291–1296 (1959).
486. J. B. Richards and R. H. Egdahl, Effect of acute hyperthermia on adrenal 17-hydroxycorticosteroid secretion in dogs, *Am. J. Physiol.* **186**:435–439 (1956).
487. D. H. P. Streeten, J. W. Conn, L. H. Louis, S. S. Fajans, H. S. Seltzer, R. D. Johnson, R. D. Gittler, and A. H. Dube, Secondary aldosteronism: Metabolic and adrenocortical responses of normal men to high environmental temperatures, *Metabolism* **9**:1071–1092 (1960).
488. S. Kotby, H. D. Johnson, and H. H. Kibler, Plasma corticosterone response to elevated environmental temperature (34°C) and related physiological activities as influenced by age, *Life Sci.* **6**:709–719 (1967).
489. A. Cession-Fossion, J. G. Henrotte, P. Y. Duchesne, and M. L. Beaumariage, Adrenal and hypothalamic catecholamine levels in rats grown in warm environments, *Compt. Rend. Soc. Biol.* **160**:2218–2220 (1966).
490. K. I. Furman and G. Beer, Dynamic changes in sweat electrolytes composition induced by heat stress as an indicator of acclimatization and aldosterone activity, *Clin. Sci.* **24**:7–12 (1963).
491. E. Bedrak and V. Samoiloff, Comparative effects of aldosterone and heat acclimatization, *Can. J. Physiol. Pharmacol.* **45**:717–722 (1967).
492. D. C. Fraser and K. F. Jackson, Effect of heat stress on serial reaction time in man, *Nature* **176**:976–977 (1955).
493. I. Kuruma, Variations of intracerebral serotonin and catecholamine levels in febrile rabbits. II. Influence of heat congestion on intracerebral serotonin and noradrenaline contents, *Nippon Yakurigaku Zasshi* **62**:8–10 (1966).
494. A. Carlsson, A. Dahlstron, K. Fuxe, and M. Lindqvist, Histochemical and biochemical detection of monoamine release from brain neurons, *Life Sci.* **4**:809–816 (1965).
495. M. Winick and A. Noble, Cellular response with increased feeding in neonatal rats, *J. Nutrit.* **91**:179–182 (1967).
496. S. Zamenhof, E. VanMarthiens, and F. L. Margolis, DNA (cell number) and protein in neonatal brain: Alteration by maternal dietary protein restriction, *Science* **160**:322–323 (1968).
497. R. H. Barnes, Experimental animal approaches to the study of early malnutrition and mental development, *Fed. Proc.* **26**:No. 1, 144–147 (1967).
498. G. G. Graham, Effect of infantile malnutrition on growth, *Fed. Proc.* **26**:No. 1, 139–143 (1967).
499. A. Yuwiler, E. Geller, and S. Schapiro, Response differences in some steroid-sensitive hepatic enzymes in the fasting rat, *Can. J. Physiol. Pharmacol.* **47**:317–328 (1969).
500. G. G. Slater, Adrenal ascorbic acid, plasma and adrenal corticoid response to fasting in young rats, *Endocrinology* **70**:18–23 (1962).

501. A. Keys, Experimental introduction of psychoneuroses by starvation, *in Biology of Mental Health and Disease*, New York, Paul B. Hoeber, 1952, Chap. 30.
502. E. J. Kollar, G. G. Slater, J. O. Palmer, R. F. Docter, and A. J. Mandell, Measurement of stress in fasting men, *Arch. Gen. Psychiat.* **11**:113–125 (1964).
503. W. S. Bloom, Fasting as introduction to treatment of obesity, *Metabolism* **8**:214–220 (1959).
504. S. H. Schachner, R. G. Wieland, D. E. Maynard, F. A. Kruger, and G. J. Hamwi, Alterations in adrenal cortical functions in fasting obese subjects, *Metabolism* **14**:1051–1058 (1965).
505. A. L. Schultz, A. Kerlow, and R. A. Ulstrom, Effect of starvation on adrenal cortical function in obese subjects, *J. Clin. Endocrinol.* **24**:1253–1257 (1964).
506. J. Roth, S. M. Glick, R. S. Yalow, and S. A. Berson, Hypoglycemia: A potent stimulus to secretion of growth hormone, *Science* **140**:987–988 (1963).
507. R. C. Friedmann and S. Reichlin, Growth hormone content of the pituitary gland of starved rats, *Endocrinology* **76**:787–788 (1965).
508. R. H. Unger, A. M. Eisentraut, and L. L. Madison, The effects of total starvation upon the levels of circulating glucagon and insulin in man, *J. Clin. Invest.* **42**:1031–1039 (1963).
509. A. M. Lawrence, Radioimmunoassayable glucagon levels in man: Effects of starvation, hypoglycemia and glucose administration, *Proc. Nat. Acad. Sci.* **55**:316–320 (1966).
510. E. Samols, J. Tyler, C. Megyesi, and V. Marks, Immunochemical glucagon in human pancreas, gut, and plasma, *Lancet* **2**:727–729 (1966).
511. E. J. Drenick, M. E. Swendseid, W. H. Blahd, and S. G. Tuttle, Prolonged starvation as a treatment for severe obesity, *J. Am. Med. Assoc.* **187**:100–105 (1964).
512. W. Januszewicz, M. Sznajderman-Ciswicka, and B. Wocial, Urinary excretion of catecholamines in fasting obese subjects, *J. Clin. Endocrinol. Metab.* **27**:130–133 (1967).
513. R. Rajan, K. G. Prasannan, and Subrahmanyam, Brain glycogen in the fed and fasted state, *Ind. J. Med. Res.* **51**:703–707 (1963).
514. D. S. Fredrickson and R. S. Gordon, Jr., Transport of fatty acids, *Physiol. Rev.* **38**:585–630 (1958).
515. P. A. Mayes, A caloric deficiency hypothesis of ketogenesis, *Metabolism* **11**:781–799 (1962).
516. E. N. Bergman, K. Kon, and M. L. Katz, Quantitative measurements of acetoacetate metabolism and oxidation in sheep, *Am. J. Physiol.* **205**:658–662 (1963).
517. F. G. Benedict, A study of prolonged fasting, Carnegie Institute of Washington, publication 203 (1915).
518. G. F. Cahill, Jr., M. G. Herrera, A. P. Morgan, J. S. Soeldner, J. Steinke, P. L. Levy, G. A. Reichard, Jr., and D. M. Kipnis, Hormone–fuel interrelationships during fasting, *J. Clin. Invest.* **45**:1751–1769 (1966).
519. O. E. Owen, A. P. Morgan, H. G. Kemp, J. M. Sullivan, M. G. Herrera, and G. F. Cahill, Jr., Brain metabolism during fasting, *J. Clin. Invest.* **46**:1589–1595 (1967).
520. G. Weber and A. Cantero, Glucose-6-phosphatase studies in fasting, *Science* **120**:851–852 (1954).
521. E. Shrago, H. A. Lardy, R. C. Nordlie, and D. O. Foster, Metabolic and hormonal control of phosphoenolpyruvate carboxykinase and malic enzyme in rat liver, *J. Biol. Chem.* **238**:3188–3192 (1963).
522. G. Weber, R. L. Singhal, and S. K. Srivastava, *in Advances in Enzyme Regulation* (G. Weber, ed.) Vol. 3, pp. 43–75, Pergamon Press, New York (1965).

523. M. Salas, E. Viñuela, and A. Sols, Insulin-dependent synthesis of liver glucokinase in the rat, *J. Biol. Chem.* **239**:3535–3538 (1963).
524. C. Sharma, R. Manjeshwar, and S. Weinhouse, Effects of diet and insulin on glucose-adenosine triphosphate phosphotransferases of rat liver, *J. Biol. Chem.* **238**:3840–3845 (1963).
525. G. Weber, R. L. Singhal, N. B. Stamm, M. A. Lea, and E. A. Fisher, *in Advances in Enzyme Regulation* (G. Weber, ed.) Vol. 4, pp. 59–81. Pergamon Press, New York (1965).
526. G. Weber, M. A. Lea, H. J. Hird-Convery, and N. B. Stamm, *in Advances in Enzyme Regulation* (G. Weber, ed.) Vol. 5, pp. 261–302, Pergamon Press, New York (1967).
527. H. M. Tepperman and J. Tepperman, The hexosemonophosphate shunt and adaptive hyperlipogenesis, *Diabetes* **7**:478–485 (1958).
528. D. D. Hubbard, R. E. McCaman, M. R. Smith, and D. M. Gibson, The acetyl CoA-malonyl CoA condensation reaction in rat liver during starvation and fat-free refeeding, *Biochem. Biophys. Res. Commun.* **5**:339–343 (1961).
529. M. S. Kornacker and J. M. Lowenstein, Citrate and the conversion of carbohydrate into fat. The activities of citrate-cleavage enzyme and acetate thiokinase in livers of starved and refed rats, *Biochem. J.* **94**:209–215 (1965).
530. M. S. Kornacker and E. G. Ball, Citrate cleavage in adipose tissue, *Proc. Natl. Acad. Sci.* **54**:899–904 (1965).
531. T. E. Webb, G. Blobel, and V. R. Potter, Polyribosomes in rat tissues. III. The response of the polyribosome pattern of rat liver to physiologic stress, *Cancer Research* **26**:253–257 (1966).
532. R. M. Campbell and H. W. Kosterlitz, The effects of dietary protein, fat, and choline on the composition of the cell and the turnover of phospholipid and protein-bound phosphorus, *Biochim. Biophys. Acta* **8**:664–670 (1952).
533. T. W. Wikramanayake, F. C. Heagy, and H. N. Munro, Effect of level of energy intake on metabolism of RNA and phospholipin in different parts of the liver cell, *Biochim. Biophys. Acta* **11**:566–574 (1953).
534. H. Sidransky, D. S. R. Sarma, M. Bongiorno, and E. Verney, Effect of dietary tryptophan on hepatic polyribosomes and protein synthesis in fasted mice, *J. Biol. Chem.* **243**:1123–1132 (1968).
535. A. Yuwiler, L. Wetterberg, and E. Geller, Alterations in induction of tyrosine aminotransferase and tryptophan oxygenase by glucose pretreatment, *Biochim. Biophys. Acta* **208**, 428–433 (1970).
536. P. D. Ray, D. O. Foster, and H. A. Lardy, Paths of carbon in gluconeogenesis and lipogenesis. IV. Inhibition by L-tryptophan of hepatic gluconeogenesis at the level of phosphoenolpyruvate formation, *J. Biol. Chem.* **241**:3904–3908 (1966).
537. I. G. Jarrett, *in Progress in Endocrinology* (C. Gual and F. J. G. Ebling, eds.) pp. 176–183, Moutin Press, the Hague (1969).
538. E. Crumpton, D. B. Wine, and E. J. Drenick, Starvation: Stress or satisfaction, *J. Amer. Med. Assoc.* **196**:394–396 (1966).
539. J. Krawczynski, H. Bachelard, and R. Vrba, Concentration of glucose and other reducing substances in organs of fed and starved rats, *Acta Physiol. Polon.* **15**:367–372 (1964).
540. G. A. Gregory, J. W. Wagner, and B. Garoutte, Brain phosphate in fasting rats, *Proc. Soc. Exptl. Biol. Med.* **117**:929–930 (1964).
541. E. Hirschberg, D. Snider, and M. Osnos, *in Advances in Enzyme Regulation* (G. Weber, ed.) Vol. 2, pp. 301–310, Pergamon Press, New York (1964).

542. P. E. Schurr, H. T. Thompson, L. M. Henderson, J. N. Williams, Jr., and C. A. Elvehjem, Influence of fasting and nitrogen deprivation on the concentration of free amino acids in rat tissues, *J. Biol. Chem.* **182**:39–45 (1950).
543. R. M. Fleming, D. T. Masuoka, and W. G. Clark, Effect of fasting and refeeding on brain amines, *Fed. Proc.* **24**:265 (1965)) (Abstract).
544. C. Angel, Starvation, stress and the blood-brain barrier, *Dis. Nerv. System* **30**:94–97 (1969).
545. G. Weissman and L. Thomas, Studies on lysosomes. I. The effects of endotoxin, endotoxin tolerance, and cortisone on the release of acid hydrolases from a granular fraction of rabbit liver, *J. Exp. Med.* **116**:433–450 (1962).
546. D. Cavanagh and J. A. DeCenzo, Endotoxin shock. A promising new method of treatment based on animal studies, *Amer. J. Obstet. Gynec.* **85**:892–904 (1963).
547. G. E. Shambaugh, III, and W. R. Beisel, Endocrine influences on altered hepatic tyrosine transaminase activity during pneumococcal septicemia in the rat, *Endocrinology* **83**:965–974 (1968).
548. M. I. Rapoport, G. Lust, and W. R. Beisel, Host enzyme induction of bacterial infection, *Arch. Int. Med.* **121**:11–16 (1968).
549. L. B. Guze and G. M. Kalmanson, Pyelonephritis. III. Observations on the association between chronic pyelonephritis and hypertension in the rat, *Proc. Soc. Exp. Biol. Med.* **108**:496–498 (1961).
550. A. Yuwiler, E. Geller, S. Schapiro, and L. B. Guze, Adrenocortical responses in experimental pyelonephritis, *Proc. Soc. Exptl. Biol. Med.* **122**:465–468 (1966).

Chapter 5

CHEMICAL ALTERATIONS PRODUCED IN BRAIN BY ENVIRONMENT AND TRAINING*

Edward L. Bennett

Laboratory of Chemical Biodynamics
Lawrence Radiation Laboratory, University of California
Berkeley, California

and

Mark R. Rosenzweig

Department of Psychology, University of California
Berkeley, California

I. INTRODUCTION

Section II of this chapter treats neurochemical responses to environmental variables. By environment we mean relatively long-term conditions that cannot easily be reduced to specific sensory stimulation. Most of the environmental studies have dealt either with effects of isolation on biogenic amines or with effects of environmental enrichment or impoverishment on cholinergic enzymes. Rose has very recently discussed neurochemical correlates of learning and environmental change.[1a] Effects of sensory stimulation are discussed in the chapter by Talwar and Singh[1b] in this volume. Sensory stimulation has also been recently reviewed by Glassman.[2]

Section III focuses upon experiments designed to search for biochemical consequences of learning, usually in the hope that these may elucidate memory mechanisms. Most of this work involves measures of RNA and protein. One of the chief problems in this area is to distinguish effects of learning from those of other variables such as stress, sensory stimulation, and locomotor activity.

* The preparation of this manuscript was supported by the U.S. Atomic Energy Commission and by NSF Grant GB-5537.

Section IV presents a summary, discussion, and conclusions about the present state of investigations in this field.

II. ENVIRONMENTAL EFFECTS ON BRAIN CHEMISTRY

A. Biogenic Amines

Inasmuch as the biogenic amines—noradrenaline, dopamine, and serotonin—represent a closely related family, they will be discussed as a group rather than individually. The role of environment in changing both the behavior and neurochemistry of animals has been particularly exploited in studies of these biogenic amines. In part this interest derives from the hope that studies of causes and control of emotional behavior in laboratory animals may lead to human application. Specifically, we will consider studies indicating that environmental treatment which modifies neurochemical characteristics can produce in laboratory animals a model of aggressive behavior and that the effects of drugs on such animals are modulated by environmental conditions.

1. *Isolation and Aggression in Mice*

It has long been known that prolonged social isolation of mice can lead to aggressive behavior.[3,4] Thus, if young male albino Swiss or Albino CFI mice live in individual plastic cages for approximately 1 month and are subsequently regrouped two or more in a cage, vigorous fighting will ensue. Female Swiss mice do not exhibit this behavior upon regrouping after isolation. Male mice of various strains show different susceptibilities to this isolation-induced aggressiveness, and most research has been done with male albino Swiss mice. Black CBA and C57 mice do not become aggressive after prolonged isolation, nor does the Sprague–Dawley rat.[5] It has been reported that the rabbit[6] and the Wistar rat[7] will develop aggressiveness upon prolonged isolation.

Among other observations suggesting that the biogenic amines may be particularly responsive to the environmental conditions are the following: (1) Pentobarbitone is less effective in producing sleep in aggressive than in normal mice,[8] (2) dexamphetamine is more than twice as toxic in aggressive than in normal mice,[9] and (3) cerebral dopamine metabolism of isolated mice differs from that of grouped mice.[10] Welch and Welch, and Garattini and his co-workers have made the most extensive investigations of the effect of isolation or group housing on these neurotransmitters, and each has recently published excellent reviews of their work and the work of others.[11–13a]

2. *Relations Between Transmitter Metabolism and Aggressiveness*

Several investigators had reported in the early 1960s that strains of rats with low levels of serotonin either as measured in the whole brain or in the brainstem were more aggressive than strains with high levels.[14-16] These observations, coupled with pharmacological evidence relating catecholamines and serotonin to central nervous system activation and reactivity, led some investigators to infer that mice made aggressive by isolation would have lower levels of noradrenaline, dopamine, and serotonin than would control animals. Valzelli reported, however, that except for minor changes, the concentrations of neither serotonin nor the catecholamines were significantly modified by isolation.[17] Isolated mice did have slower rates of synthesis, release, and turnover of brain and adrenal catecholamines than did grouped mice, as shown by the greater decrease of catecholamines after administration of *dl*-α-methyltyrosine in the grouped mice (Table I).[19,26] This was supported by the finding that brain catecholamines (norepinephrine and dopamine), and also serotonin, were increased more in grouped mice than in isolates after administration of pargyline,[11] and brain serotonin was increased more after administration of tranylcypromine (Table II).[17,18] Giacalone *et al.* have made a detailed

TABLE I

Effect of Tyrosine Hydroxylase Inhibition on Brain and Adrenal Catecholamines in Male Mice Group-Housed and Individually Housed for 14 Weeks Subsequent to Weaning at Age of 4 Weeks[19]

	Isolated (± S.E.)	Group 10 (± S.E.)	$P <$
Number	30	28	
Body wt.	38.1 ± 0.8	33.0 ± 0.8	0.01
Left adrenal (mg)	2.08 ± 0.09	2.44 ± 0.07	0.01[a]
Brain noradrenaline			
Saline control (ng/g)	402 ± 14.8	456 ± 16.7	0.025[a]
α-Methyltyrosine decrease (% control)	58.4 ± 1.38	64.4 ± 1.96	0.01
Brain dopamine			
Saline control (ng/g)	788 ± 30.9	748 ± 29.1	n.s.
α-Methyltyrosine decrease (% control)	53.8 ± 0.99	62.0 ± 0.89	0.001
Adrenal adrenaline			
Saline control (μg/left adrenal)	3.66 ± 0.25	4.01 ± 0.33	n.s.[a]
α-Methyltyrosine decrease (% control)	20.9 ± 3.25	33.3 ± 4.16	0.01
Adrenal noradrenaline			
Saline control (μg/left adrenal)	0.44 ± 0.04	0.42 ± 0.10	n.s.
α-Methyltyrosine decrease (% control)	11.4 ± 10.28	14.3 ± 9.84	n.s.

[a] Differences are enhanced when values are adjusted to body weight by covariance analysis; on this basis, the adrenals of the grouped mice contain significantly more adrenaline ($P < 0.001$.)

TABLE II

Dynamic Aspects of 5-HT Metabolism in Brain of Isolated and Grouped Mice After Tranylcypromine Administration[17]

	Isolated mice	Grouped mice
$[5\text{-HT}]_0 \mu g/g$	0.84 ± 0.02	0.84 ± 0.04
$[5\text{-HIAA}]_0 \mu g/g$	0.29 ± 0.01[a]	0.38 ± 0.01
Rate constant of 5HIAA loss	1.42 ± 0.08	1.69 ± 0.12
5-HT turnover rate, μg/g/hour	0.38	0.59
5-HT turnover time, minutes	122	79

[a] $P = < 0.01$.

study of the concentration and turnover of brain serotonin, and the concentration of its metabolite, 5-hydroxyindoleacetic acid, as a function of time after isolation and the number of mice per group.[13] No change in concentration of serotonin was observed either in whole brain or in discrete areas of the brain as a result of variation in housing conditions. On the other hand, isolation did decrease the 5-hydroxyindoleacetic acid content within 1 day, and this decrease persisted for at least 1 month when the mice were maintained in their differential housing conditions. Although the brain serotonin did not differ, isolated male mice showed a decreased turnover rate and a concomitant increased turnover time of serotonin when compared to the normal grouped condition of 10 mice per cage. These differences were evident either when whole brain or specific regions were analyzed. The largest differences between isolated and grouped mice in turnover rates were observed in the corpora quadrigemina. In this area the rate was two-thirds as fast in the isolated mice as in the grouped mice. The diencephalon had the highest turnover rate and lowest turnover time for serotonin. Female mice did not show the differences in serotonin turnover rate or time as a result of isolation and also did not become aggressive. Similarly, isolation did not change the levels or turnover of serotonin in Sprague–Dawley and Wistar rats after 30 days, and these animals did not become aggressive. Although strains of male mice which did develop aggressive behavior also showed decreased brain serotonin metabolism and strains which did not become aggressive did not show a change in serotonin metabolism, the dissimilarity of the temporal development of change in serotonin turnover and aggressive behavior prompted the Italian investigators to doubt if the small change in serotonin turnover is responsible for the behavioral change. The development of aggressiveness is a time-dependent process, and its degree depends both upon time and the completeness of isolation. Periodic disturbance of the animals being tested will reduce the aggressiveness. However, attempts to correlate the change in either concentration or turnover of serotonin with the development of aggressiveness have not been successful.

On the other hand, Essman has reported that isolated male CF-1 and Swiss-Webster mice have elevated brain serotonin levels and an increased rate of brain serotonin turnover compared to grouped (5 per cage) mice.[20,21] Essman has suggested that these results, which differ from those generally reported by Welch and Garattini, may be due to the earlier age (21 days) at which he began isolation of the mice. However, it should be pointed out that Essman took the brain samples immediately after the mice had been tested for behavioral activity in a novel situation. Therefore, it seems likely that he did not determine the basal rate of metabolism but rather the result of the reaction to a novel situation. The results would then be a measure of the reactivity of the mice to this novel situation rather than a measure of the metabolism in the housing conditions in which the animals had been maintained.

Marcucci *et al.* investigated the effect of grouping and isolation on the concentrations of *N*-acetyl-L-aspartic acid (NAA) in male and female albino Swiss mice.[13a,22] A significant difference of NAA was found in the brains of isolated male mice where the content was about 80% of that of the grouped males. With the exception of the corpora quadrigemina, all other brain areas showed the decrease in NAA concentration. No change was found in the concentrations of asparatic and glutamic acids as a result of prolonged isolation.[13a,22a] Isolated female mice[22] or male rats[13a,22] which do not become aggressive after isolation did not show any change in NAA concentrations. This and other work suggests a relationship between NAA and serotonin metabolism.

3. *Effects of Immediate Environment on Biogenic Amines*

Fighting may modify the brain concentrations of serotonin, noradrenaline, and dopamine in brain, but the magnitude and directions of change appear to be dependent upon several factors including the duration and intensity of fighting. Thus, serotonin, noradrenaline, and dopamine increased as much as 30 to 50% when a pair of mice engaged in moderate fighting. Larger increases in biogenic amine levels were obtained by the administration of pargyline, a monoamine oxidase inhibitor. The effects of fighting plus pargyline were not additive—that is, the combined effect was less than the sum of the individual effects.[11] More vigorous fighting in groups of four appeared to lower norepinephrine and dopamine in the brainstem and to elevate serotonin in the midbrain.[23]

Eleftheriou and Boehlke[24] have presented evidence that the central monoamine oxidase levels of mice were modified by exposing individual C57 mice to trained fighters so that they experienced brief daily periods of fighting and defeat. The most marked changes were noted in the hypothalamus where the MAO activity initially increased by nearly 50% after

1 or 2 days of fighting and subsequently fell to a level approximately one-half of the original.

Even a psychological stimulus has been shown by Welch and Welch to alter brain amines.[11,25] Mice were first made aggressive or hypersensitive by isolation for 18 weeks. These mice were then allowed to observe for 1 hr other mice that were fighting nearby. Both serotonin and norepinephrine were elevated in whole brain of the witnesses. On the lower brainstem, however, the norepinephrine content of these hyperactive mice was lowered about 25%. Others have previously demonstrated that psychological events can modify peripheral physiological and endocrine function, and it seems reasonable to assume that these phenomena were mediated through the brain. However, this report by Welch and Welch is perhaps the first evidence for a specific neurochemical change following exposure of a naive animal to a mildly stimulating, purely psychological event—that is, watching other mice fight with the attendant sights, sounds, and odors.

In spite of the slower turnover of brain catecholamines in isolated mice as compared to grouped mice, restraint stress causes a greater elevation of catecholamines and serotonin in the isolated mice. If the biosynthesis of norepinephrine and dopamine is inhibited, stress facilitates the depletion in isolated mice and retards it in group-housed mice.[26]

Another study has revealed complex interrelations of stress and prior environment on effects of *p*-chlorophenylalanine, an inhibitor of serotonin biosynthesis, on serotonin levels.[27] The results led Welch and Welch to stress the unimportance of amine levels as such and the necessity of studying the functional pools and metabolic rates of brain biogenic amines. They believe that isolated mice are characterized by a relatively low basal level of general neuroendocrine function but that they are more reactive to a novel situation. Mechanisms exist in neural tissue to prevent large changes in the concentration of catecholamines after an increase in neural activity that would deplete the stores in the absence of such mechanisms. This conservative or homeostatic mechanism is better developed in mice accustomed to higher levels of stimulation, i.e., grouped mice, than in those less accustomed to stimulation, i.e., isolated mice. Another way of stating the difference is to say that grouped mice are emotionally less reactive to stress. Welch and Welch suggest "that during stress, mice which differ in behavioral reactivity because of differences in their previous environmental conditions, metabolize brain biogenic amines at different rates and also activate to different degrees temporal mechanisms which control the availability of amines during stress."[26] One mechanism to regulate the concentration operates by altering the activity of monoamine oxidase. As a result of these homeostatic mechanisms, determination of concentrations of catecholamines is a relatively insensitive method to observe changes brought about by environment; determinations of turnover rates are generally more meaningful.

Recently, it has been shown that various types of intermittent stress administered only once or twice daily can produce small sustained elevations in brain catecholamines. Thus, allowing male mice that have previously been made aggressive by long-term isolation to fight for only 5 min each day for either 10 or 14 consecutive days elevated brain norepinephrine and dopamine, enlarged the heart and adrenals, and caused dramatic increases in the adrenal content of catecholamines; the mice in these experiments were killed 24 hr after the last exposure to fighting.[27a]

4. *Isolation and Stress in Rats*

Geller, Yuwiler, and Zolman[28] have reported that isolated rats have markedly higher brain levels of norepinephrine than do group-enriched rats. This increase is localized in the caudate nucleus.[29] Dopamine concentration did not differ between the two groups. Although the activity of the synthetic enzymes (tyrosine hydroxylase and DOPA decarboxylase) and the degradative enzymes (monoamine oxidase and catechol-*O*-methyltransferase) often differed greatly in the two groups, most of the differences were not significant. The pattern of changes did not reveal a basis for the observed difference in norepinephrine concentration.

It should be pointed out that Geller and Yuwiler weaned and placed the rats into the isolated and grouped conditions at 18 days of age. This is an unusually early age for weaning of rats. Their isolated rats weighed less than the grouped-enriched rats, whereas in the studies of Rosenzweig *et al.* to be discussed below, isolated rats are typically about one-tenth heavier than the enriched rats. Křeček[30] has recently discussed the many important physiological, biochemical, and behavioral effects produced by weaning of rats prior to about 28 days of age; he terms this premature weaning. Premature weaning caused disturbances in higher nervous activity and behavior. Particularly striking were the alterations of function of the adrenals and sexual organs. Hypothalamic damage has also been suggested as a consequence of premature weaning. However, the generality of such effects among different strains or stocks of rats does not appear to have been tested; perhaps because of selective factors in colony breeding those stocks of rats more commonly used in the United States can be weaned without deleterious effect after about 21 days of age.

Riege and Morimoto have recently studied the effects of brief daily stress produced by tumbling on the concentration of biogenic amines (5-HT, NE, and DA) in rats which had been raised from weaning in either an isolated or a grouped, enriched environment.[31] The various conditions produced relatively small changes in concentrations of the biogenic amines. The most significant differences were found in comparison of the norepinephrine content of isolated-stressed rats with isolated nonstressed rats; daily stress produced an increase of 10 to 20% in norepinephrine concentration. Generally, biogenic amine concentrations did not show significant differences between the grouped-enriched and the isolated condition, and stress did not produce a significant additive effect. In contrast to the reports of Geller and Yuwiler, the total brain content of norepinephrine was not different when isolated rats were compared to grouped-enriched rats. In studies of this sort, turnover rates might reveal more differences between the groups than did concentrations.

B. Amino Acids

Numerous factors which influence the free amino acid composition of the brain have been summarized by Himwich and Agrawal.[32] It is surprising, then, that isolation as a factor appears to have been investigated by only two groups—in the mouse study using *N*-acetyl-L-aspartic, aspartic and glutamic acids, cited above,[13,22,22a] and in the dog.[33] Mongrel puppies were maintained for 1 week, beginning at 4 weeks of age, under conditions of partial sound-isolation and partial sensory deprivation, and were compared to littermates maintained in single cages in the animal house environment. The daily human contact of both groups during this 1 week period was minimal and amounted to less than 2 min/day. At the end of the study, behavioral observations were made for approximately 1 hr and then the brain samples were taken. The young puppies reared in partial isolation all showed behavioral abnormalities; they showed little emotional arousal and were hyperactive with a high level of random activity. One wonders to what extent the testing itself may have contributed to the observed changes in amino acid concentrations.

The four amino acids studied were glutamic acid, γ-aminobutyric acid, glutamine, and aspartic acid. Only small nonsignificant changes in concentrations of these amino acids were found in the cerebral cortex of isolated dogs; significant increases (20%) in glutamic acid and γ-aminobutyric acid were found in the thalamus–hypothalamus sample and significant decreases in the caudate nucleus. Glutamine content was significantly lower in both the superior colliculi and caudate nucleus (20 to 30%) for the dogs raised in isolation than for the controls. The content of these four amino acids increased about 20% in the hippocampus–amygdala region but none of the increases was statistically significant, with the possible exception of γ-aminobutyric acid.

In the case of early weaning, Himwich, Davis, and Agrawal[34] have shown that rats weaned at 15 days and sacrificed 7 days later had significantly increased cortical levels of glutamic acid, glutamine, γ-aminobutyric acid, aspartic acid and alanine, as compared to nonweaned control rats. Rats weaned at 19 days and sacrificed 3 days later had significant increases only in glutamine, aspartic acid, and alanine, and only small, nonsignificant changes were observed in cortical glutamic acid and γ-aminobutyric acid concentrations. In the midbrain a different pattern of change was observed. Rats weaned at 15 days showed significant decreases in glutamic acid, aspartic acid and threonine, and an increase in alanine concentration. Rats weaned at 19 days showed decreases of about 20% in glutamic acid, glutamine, γ-aminobutyric acid and alanine concentrations. It should be pointed out that the rats weaned at 15 days grew less than the controls and had smaller brains. These changes may be the result of the nonspecific stress accompanying early or premature weaning.

C. Nucleic Acids

Effects of differential environments on numerous biochemical and anatomical measures have been reported in a number of studies by Bennett, Diamond, Krech, and Rosenzweig beginning in 1960.[35] An extensive review of this work is now in press.[35a] These studies were undertaken to investigate brain mechanisms of learning and memory storage. In their "enriched condition" (EC), animals are housed in a group of 10 to 12 in a large cage with objects that are changed daily. A picture of rats in the enriched environment is seen in Fig. 1 of Bennett *et al.*[36] The pool of stimulus objects is shown in Fig. 1 of Rosenzweig and Bennett[37]; it includes such items as a ladder, rotating wheel, sponge, metal box, bottle, etc. In the "impoverished condition" (IC), animals are housed singly in cages without stimulus objects. Other aspects of the environments were similar for EC and IC, since all animals had food and water available at all times, and room temperature and light values were the same; indeed, in many of the more recent experiments, the EC and IC rats have been housed in the same room. In the EC cages, the presence of stimulus objects and of other animals both tend to increase the locomotor activity of each animal. Perhaps because of this increased activity, the EC animals gain weight less rapidly than the IC's.

When rats were assigned to the differential conditions at weaning (about 25 days of age) and kept there for 80 days, the EC animal showed significantly greater weight and thickness of cerebral cortex than the IC littermates. These differences were significantly greater in the occipital (visual) region of the cortex than in other cortical regions.[38] The hippocampus was also found to be thicker in EC than in IC littermates, but in six experiments the effect in hippocampal thickness was found to be less consistent and less reproducible than the effect in cortical thickness.[39] Diamond *et al.* showed that the EC rats had a greater glia-to-neuron ratio in the occipital cortex than did the rats raised in isolation for 80 days.[40] Subsequently, these authors showed that the number of glia was 14% greater and the glia-to-neuron ratio in the occipital cortex was 16% greater in EC rats than in IC rats.[41] These studies suggest an alteration in cell proliferation and in DNA formation produced by environment. Other studies of this group[35a] have shown that the total RNA content of the occipital cortex increases in EC, as does cortical weight, whereas the DNA concentration decreases significantly. These observations are consonant with the observation that enriched rats have larger neuronal perikarya.[42]

Similar research was done in the preweaning period by Malkasian.[43] He assigned either a single mother and three 6-day-old pups to a colony cage (unifamily environment—UFE) or placed three mothers, each with three pups, in an EC cage (multifamily environment in enriched condition—

MFE-EC). At 28 days of age, the MFE-EC rats showed, in comparison with UFE, significantly greater thickness of cerebral cortex and larger neuronal nuclei.

Altman and Das in a pilot study compared the incorporation of thymidine-^{3}H in 4-month-old rats that had been raised in enriched and isolated environments since weaning.[44,45] One week after the administration of the DNA precursor, the enriched group was reported to have 60% more labeled glial cells in the neocortex than did the isolated group; this difference was significant at beyond the 0.001 level. In other subcortical areas including amygdala and pyriform cortex, "dorsal thalamus," and inferior colliculus, the enriched group had 12 to 20% more labeled glial cells in a standard microscopic field, but these differences were not significant. When the increased area of the neocortex was taken into account, the sampled sections of neocortex were calculated to have an increase of 75% in the total number of labeled glial cells. It should be pointed out that each group consisted of only three animals, and it is not clearly stated if all were of the same sex. These rats were selected to be matched in body weights at the time of thymidine injection, but since rats raised in isolation typically become heavier than littermates in an enriched environment,[46] weight-matching groups at the end of an experiment may not be appropriate; in fact, body weight matching may have produced differences in brain measures.

Recently, newer studies by Altman have cast doubt on their original interpretation that the labeled cells represent an increase in the glial population.[47] Altman has proposed that the labeled cells are primarily undifferentiated elements migrating from the subependymal germinal zone of the lateral ventricle to the gray matter of the cortex. The increase in labeled cells seen in enriched rats 1 week after labeling may represent an increase in the proliferation and migration of undifferentiated cells whose fate is not presently known. If these cells should become glial cells in the cortex, these studies would be in harmony with those of Diamond *et al.*[41]

Recently, Walsh *et al.*[48] have also reported that the occipital cortex of rats raised in environmental complexity for 90 days is 5% thicker than rats raised in isolation. The hippocampus was also thicker by nearly 6%. Increased numbers of oligodendroglia and glial "intermediates" were found in the hippocampus, while the number of astroglia decreased. It should be noted that this paper reports the result of only one experiment with a total of 17 rats.

D. Proteins and Enzymes

When rats raised in an enriched environment were compared to rats raised in an isolated environment from weaning, there were differences not only in number of glial cells in the occipital cortex, size of the neuronal perikarya, weight and thickness of cortex, but also in the following enzymatic characteristics: (1) a slight increase in total acetylcholinesterase (AChE) activity in the cortex, but a clear decrease in AChE per unit of weight, (2) increases in both total cholinesterase (ChE) and ChE to weight, (3) an increase in the ChE to AChE ratio. Most of these differences were significantly greater in the occipital region of the cortex than in other cortical regions.[(38)] Since brain glial cells are relatively rich in ChE, while AChE is found especially in neurons, and since glial proliferation occurs in rats given complex experience,[(41)] changes in the ChE-to-AChE ratio may reflect mainly altered glial number and function. The magnitude and statistical significance of both chemical and anatomical changes in the occipital cortex are shown in Table III.

The influences of several environmental factors and of characteristics of the subjects on these effects were summarized recently[(38,39)]: Age of rats was not critical, since adults showed effects similar to those in weanlings. Thirty days of exposure gave results similar to 80 days. Even 2 hr of daily exposure for 30 days sufficed to produce typical EC-IC differences. Several strains of rats have been tested, and all yielded similar brain effects of environmental treatment. Generally, similar effects have also been found in other species—mice[(49)] and gerbils.[(37)] In spite of the fact that the largest effects occur in the occipital region, where the visual projection area is located, blinded or dark-raised rats nevertheless showed EC-IC effects and these remained larger in the occipital area than elsewhere in the cortex.[(38)]

Further characteristics of EC and IC rats have been studied in a limited number of experiments which have not tested the range of effective factors.

TABLE III

Effects of Differential Experience on Occipital Cortex of S_1 Rats Kept in Enriched or Impoverished Conditions from 25 to 105 Days of Age

Measure	percentage diff. (EC > IC)	*P*	Number (EC > IC)
Weight	6.4	<0.001	133/175
Total protein	7.8[a]	<0.001	25/32
Thickness	6.3	<0.001	45/52
Total AChE	2.2	<0.01	102/171
Total ChE	10.2	<0.001	118.5/132
Total hexokinase	6.9[b]	<0.01	17/21
DNA/mg	−6.1	<0.001	4/23
RNA/mg	−0.7	n.s.	10/23
RNA/DNA	5.9	<0.01	19/23
No. Neurons	−3.1	n.s.	7/17
No. Glia	14.0	<0.01	12/17
Perikaryon cross section	13.4	<0.001	11.5/13

[a] Weight difference 7.0% in these experiments.
[b] Weight difference 5.5% in these experiments.

In contrast to the significant changes in activity per unit of weight of AChE and ChE, choline acetyltransferase activity[(50)] and ACh concentration (unpublished) increased with enriched experience, but only in direct proportion to tissue weight. Total protein,[(36)] total activity of hexokinase[(51)] and dry weight[(52)] also increased in proportion to fresh tissue weight in the enriched condition. Control experiments have shown that the usual differences between EC and IC rats cannot be attributed to any of the following factors: handling, locomotion, accelerated maturation, extra-cage environment, hormonal mediation, and stress.[(31,39,53)]

Long-term exposure to a visual pattern (vertical stripes) or to a blank screen has been reported to bring about large changes in AChE-to-weight in the posterior cortex of female Holtzman rats. After 70 days in the experimental conditions, the pattern group of Singh *et al.*[(54)] showed *double* the AChE activity of the visually restricted group. This difference is both much larger and also opposite in direction to that reported just above with enriched experience. Unpublished experiments by W. Maki in our laboratory, using both S_1 strain male rats and female Holtzman rats, have failed to replicate these results.

Garattini *et al.*[(13a)] reported that the activity of choline acetyltransferase of grouped albino Swiss mice does not differ from that of isolated mice. Seven separate brain areas were analyzed.

In the same pilot study in which an increased formation of cells labeled with thymidine was reported in enriched rats by Altman and Das, no increase in uptake of leucine-^{3}H per unit area was found. In fact, evaluation of the results suggested an apparent decrease in labeling of brains from enriched rats.[(44,37)] Other studies of Altman have yielded similar results.[(45)] Increased leucine incorporation was found only in an experiment probably involving "stress" produced by enforced exercise of naive rats.[(45)] Altman concludes from this that his "studies to date do not provide support for the hypothesis that the presumed 'activation' of brain structures by behavioral manipulations . . . will affect the rate of incorporation of labeled amino acids into proteins in cell systems known to be associated with that function."[(45)] It should be pointed out that incorporation was measured only over hours, and after the animals had been in the differential environments for 10 weeks or more. It may well be that the period of differential incorporation had passed by this time. Also, if the EC animals are more accustomed than IC to variations in environment, this might account for the somewhat higher incorporation in IC rats after the injection treatment that was novel to them.

III. EFFECTS OF TRAINING ON BRAIN CHEMISTRY

A. Effects on Deoxynucleic Acid

If learning can be shown to involve changes in the quantity or quality of macromolecular synthesis, this, in accordance with modern ideas of molecular biology, may involve the derepression of genomal DNA. (Insofar as the studies described below have indicated modification in the amount or composition of RNA found in brain cells as a consequence of learning, modification of DNA transcription is suggested.) Implications of models involving DNA have been discussed by Bonner,[55] Hydén,[56,57] Gaito,[58] and Griffith and Mahler.[59] Consequences of changed function of DNA may be seen first in altered RNA or protein synthesis and perhaps subsequently in other brain constituents such as lipids and polysaccharides. No studies were noted of the effects of formal training on the quantity or composition of DNA. Measures of RNA, protein, and polysaccharides in relation to learning will be discussed in the following sections. (See also Hydén and Lange[60] in this volume.)

B. Effects on Ribonucleic Acid

1. *Changes in Amount of RNA and in Base Ratio*

The possible role of RNA in learning has been a major area of investigation since the pioneering research and speculations of Hydén first were reported in 1958. Many of the early studies involve the effects of stimulation rather than training, and have recently been reviewed by Hydén[61] and by Glassman.[2] Regarding learning, Hydén and associates have asked whether certain macromolecular changes occur in cells of the brain areas involved in specific learning tasks but do not occur in the same cells as a result of other types of physiological stimulation that cause increased neural activity.

Three types of learning situations have been used by Hydén and collaborators to study the effects on RNA content and composition of specific neurons.[56,60,62] In the first, rats were trained to use the nonpreferred paw to retrieve food far down a narrow glass tube. Control of each paw is located in the contralateral hemisphere of the brain. After 3 to 9 days of training, neurons from both sides of the cortex were analyzed. An increase in RNA per cell was found in the hemisphere controlling the newly trained paw. The magnitude of this change increased with duration of training. After 9 days there was twice as much RNA per neuron in the trained hemisphere as on the other side. The base composition of the RNA in the neurons was also determined. Early in training the newly formed RNA had a composition or base ratio that was claimed to be DNA-like, whereas late in the training the RNA formed had a composition said to be similar to the ribosomal RNA.

A second type of learning experiment involved training young rats to climb a 1-m wire ascending at a 45° angle from the floor of the cage to food on a small platform. The investigators analyzed Deiters' nerve cells in the vestibular nucleus which are stimulated in this training. These cells are characterized by a large cytoplasm in relation to the nucleus, so that analysis of the whole cell RNA primarily reflects the composition and amount of the cytoplasmic RNA. A 10% increase in total RNA was observed. When the nuclei were analyzed separately, an increase in ratio of adenine to uracil was observed in the nuclear RNA. Several types of control experiments involving vestibular stimulation and stress showed increases in the amount of RNA equal to or greater than that found in training, but the control treatments yielded no change in base ratio.

In this volume Hydén and Lange report preliminary data obtained by analysis of the nerve cells in the inferior temporal ventral cortex of *Macaca mulatta*.[60] These monkeys were trained by Mihailovíc *et al.*[60] The samples used by Hydén and Lange for RNA analyses were from the same material as used by Mihailovíc for protein analyses, to be mentioned later in this chapter. Three groups of monkeys were used: (1) an experimental group which performed a series of pattern discrimination tests over a 30-day period, (2) a control group in which the tasks and cues were randomized so that the subjects could not learn, and (3) a cage control group. The experimental group showed higher adenine than either of the other groups. In cytosine, the learning animals were lower than the active controls but virtually identical to the cage controls. It should be noted that no indication of the number of animals studied is given in this report.

Hydén concludes that learning activates the genome of both the glia and neurons engaged in the learning, resulting in increases in RNA content of both types of cells. He holds that mere sensory stimulation, in contrast, produces inverse or complementary changes in RNA of glia and neurons. Hydén suggests the RNA involved in learning to be of the chromosomal type, as it is rich in adenine and uracil, whereas the RNA synthesized during stimulation has ribosomal characteristics. Eventually the RNA synthesized in the neurons and glia during learning results in new protein synthesis and may give each neuron a unique protein pattern,[56,60] (see below). Although these experiments are very ingenious and the techniques employed are elegant, some problems of interpretation have been set forth by Glassman (p. 632).[2]

One of the problems frequently raised concerning Hydén's work is the thorny one of determining whether the effects are due to learning rather than to other aspects of the training procedure such as sensory input, attention, motor activity, stress, emotion and other psychological variables. We are sure that all who are working in this complex field will concede that it is very difficult to design any experiment or series of experiments that will unequivocally prove that the observed effects are due to learning.

Also of concern to many in Hydén's reports is the fact that although a large number of individual cells have been analyzed, the number of animals and thus the appropriate N for statistical comparisons is often small (3 to 7). The analytical techniques are difficult and laborious, and this hinders attempts to replicate in the Swedish laboratories, let alone in other laboratories.

Not only rats and monkeys but also planaria have been investigated by Hydén *et al.*[63] They reported that changes arose in base ratios of RNA from nerve cells of the head ganglion during conditioning of planaria.

Two of three stimulated control groups, however, showed similar base ratios although they did not learn. It was not possible, therefore, to associate the behavioral effect with specific changes in RNA base ratios.

Shashoua measured incorporation of a precursor into RNA after training goldfish in a novel and clever manner.(64) A small polystyrene float was attached to a suture at the ventral surface of the fish. The fish had to learn new swimming skills to overcome the buoyant force of the float, and learning curves were obtained. Typically, it required a naïve goldfish approximately 2½ hr on the first day to learn to swim again in a more or less normal fashion, at the usual angle, submerged beneath the water. On retest 1 day later only ⅓ hr was required. Orotic acid injected into the ventricle below the region of the optic tectum was used to assess the base composition of the RNA synthesized during learning. In control fish the uridine-^{14}C-to-cytosine-^{14}C ratio was 3.6, whereas in the trained fish the ratio was 6:9. Thus, training increased the uridine content by a factor of about 3 over the cytidine content. Shashoua recognized that the experimental group of goldfish was under constant stress and also had performed much work. To control for the possible effects of stress and work, he employed puromycin in an additional group. This did not affect learning during the training session, but no long-term memory or consolidation occurred; that is, on retest 1 day later a fish that had been given puromycin before original learning now required as long as on first trial to readapt to the float. No difference was noted in the experimental puromycin-treated fish as compared to the controls in the ratio of incorporation of orotic acid into the uridine and cytidine. The author concludes: "These data suggest that RNA base ratio changes accompany the acquisition stage of learning and imply that protein synthesis is required for the establishment of these changes."

2. *Increased Polysome to Ribosome Ratio*

Dellweg, Gerner, and Wacker(65) have also recently reported qualitative as well as quantitative changes in RNA of rat brain following training rather similar to one previously used by Hydén. Rats were required to climb from one compartment to another on a thin sloping rope twice daily for 8 to 10 days in order to get food. About 5 days were required to master this task. Except during the daily training period, control rats apparently lived in the same cage as the experimental animals, to stimulate their locomotor activity and to equate social experience. The content of ribosomal RNA in whole brain was slightly increased in the trained rats, whereas the amount of transfer RNA was not increased. The ratio of ribosomal RNA to transfer RNA was about 10% greater in trained than untrained animals. We feel that these results should be viewed with caution because only four values were reported for trained animals and because untrained and trained animals were apparently not analyzed at the same time. In another experiment trained rats were reported to have a 35% increase in ratio of polysomes to ribosomes (monosomes). This result was found in each of six paired comparisons. These authors also looked for alteration in the acceptor activity of the transfer RNA for four amino acids—lysine, leucine, phenylalanine, and valine. No differences were found between the naïve and trained rats. Dellweg, Gerner, and Wacker conclude that the increase in RNA depends upon training and that the greater total ribosomal and polysomal RNA leads to increased ability to synthesize protein.

3. *Increased Synthesis of RNA*

Wilson, Glassman, Zemp, Adair, and co-workers have published a series of papers in which brain RNA synthesis has been investigated in $C_{57}BL/6J$ mice doing short-term avoidance conditioning.[66–70] The trained mouse was conditioned to avoid an electric shock by jumping to a shelf when a light and a buzzer were presented for 3 sec prior to the onset of shock. Typically, 30 trials were given over a 15-min training period. Numerous types of controls have been used, but the most frequent control has been a yoked animal which received light, buzzer, and shock from which it could not escape. In order to measure RNA synthesis, radioactive uridine was injected into the frontal lobes 30 min prior to training. To correct for variation of injection and uptake of precursor into the soluble nucleotides, results were normalized to the specific activity of precursor uridylic acid.

Trained versus yoked mice showed an increased incorporation of uridine into the RNA of the nuclei (amounting to 40%) and in ribosomes (amounting to 65%), when the mice were sacrificed immediately after training (25 pairs). This difference was reduced when mice were sacrificed 15 min after the end of the training session and was not observed when mice were sacrificed 30 min after training had been completed. Adair *et al.* have shown by sucrose gradient techniques that an average of 46% more radioactivity is found in the polysome fraction of trained mice than in the yoked controls (43 pairs).[68] By gross dissection, the major site of increased incorporation was found to be the diencephalon and anatomically associated areas.

Recently, autoradiographic techniques have been used by these investigators to further study the localization of the increased RNA synthesis associated with the brief training experience.[69] Labeling was concentrated over the neuronal nuclei, and the only glial elements that were consistently labeled were the ependyma. In the mice which received brief training, nuclei in areas of the limbic system including olfactory and entorhinal cortices, hippocampus, amygdala, thalamus, hypothalamus, and mammillary body were labeled while the yoked control mice had little or no labeling in these areas. Conversely, in the outer layers of the neocortex, the trained mice had less radioactivity than did the yoked control mice.

The authors are cautious about the interpretation of their results, both behaviorally and biochemically. Although they have done several behavioral control experiments, they still cannot be sure that the increased incorporation of precursor into polysomal RNA relates directly to learning and memory, rather than to some other process such as necessity for recovery and maintenance of brain cells after response to stimulation. In a well-conceived series of experiments, mice were tested for uptake during subsequent sessions either in the same training or control condition, or under changed conditions. Only in cases where original learning occurred were differences found between trained and control animals.(70) This appears to rule out such factors as locomotor activity, emotional involvement, and attention to the stimuli. In another control experiment adrenalectomized animals showed as large an effect of conditioning as did normals, demonstrating that an adrenal stress response is not involved in the cerebral effect. On the biochemical side, Glassman and co-workers have pointed out that the increased radioactivity in the RNA may not be due to increased synthesis but rather may be due to alteration in the stability of the ribosomes. They suggested either change in synthesis of uridine or modification in permeability to RNA precursors as other factors that might have led to the observed difference.

Bowman and Strobel have recently shown that rats learning spatial reversals for a water reward in a Y-maze had increased pool sizes of intraveneously injected ^{3}H-cytidine in the nucleotide pools of pyriform cortex and hippocampus when compared to rats performing the previously learned task without reversal.(71) Increased incorporation of cytidine into RNA of hippocampus and pyriform cortex was also found, but after correction for the soluble cytidine-^{3}H pools in this area, only the hippocampus showed increased incorporation (22%). No differences were found after intracerebral injections of cytidine. Differences between the control and experimental rats were not found for the RNA measures in the cortex or subcortex after the hippocampus had been excluded. While admitting other possibilities of interpretation, these authors feel that the results are likely to be related to learning or memory processes because the control animals were engaging in a similar performance, except that reversal learning was not required of them.

4. *New Species of RNA*

The desirability of searching for new species of RNA formed as a result of learning has been discussed by Bonner(55) and by Gaito.(58) Recently, Machlus and Gaito have published brief reports indicating the formation of a unique species of RNA as a result of a behavioral task. The technique of successive competitive hybridization of RNA with DNA has been used.(72) In this method, it is assumed that if successive RNA-to-DNA hybridizations are carried out with identical RNA samples, the first hybridization will occupy all the homologous sites on the DNA; consequently, further trials will not allow further hybridization to occur. If, however, the second sample of RNA differs from the first, then additional hybridization will occur. This can readily be detected if the second RNA is radioactive. Typically in such experiments, both orders of presentation of appropriately labeled RNA'S are employed.

In the first of three papers, experimental rats were injected intracranially with orotic acid-^{3}H and control rats were injected with unlabeled orotic acid.[73] Seventy-five minutes later the experimental group was trained for 15 min on an active shock avoidance apparatus. Immediately afterward the rats were sacrificed, RNA was isolated by a phenol procedure, including hot phenol treatment. When hybridization was carried out first with RNA from untrained rats followed by hybridization from trained rats, significant hybridization of the RNA from trained animals was found. But when the hybridization was in the reverse order, no hybrid at all was detected from RNA isolated from the untrained animals. These results suggest that a unique RNA was formed during this training procedure, but other interpretations are possible. As the authors pointed out, the effect may have been the result of electric shock or of sensory stimulation or motor activity involved in the short-term avoidance task rather than being due to learning per se. In their second paper these authors have attempted to rule out motor activity as a major contributing factor in the formation of unique RNA in the shock avoidance task by using as a control, RNA isolated from rats forced to run on a treadmill. Again, evidence was found for unique species of RNA formed during the avoidance task.[74] In the third experiment an additional important control—shock—was included in the experimental design, and similar results were again obtained.[75] In this experiment uridine-5-^{3}H was used as the RNA precursor.

Demonstration of a unique species of RNA resulting from either stimulation or learning would be extremely important, but we have reservations about the validity of these experiments, for the following reasons. It is known that in mammalian systems, in contrast to bacterial systems, when the presaturation hybridization technique is used, a large excess of competitor RNA is required to inhibit the hybridization of short-term pulse-labeled RNA with DNA by more than 50–60%.[76,77] A large excess of total brain RNA would be required to approach saturation of the total DNA coding for rat brain messenger RNA.[78] This requirement for a large excess may be due to redundancy of DNA sequences. It seems to us unlikely that complete exclusion of the second hybridization would be observed with the RNA-to-DNA ratios (varying from 1 to 25) used for hybridization by Machlus and Gaito. Experiments by Dr. Kern von Hungen in our laboratory on hybridization of rat brain RNA reinforce our skepticism.[79]

C. Effects on Protein

It is to be expected that if formal training leads to the formation of more RNA, this will ultimately lead to more protein. The effects of inhibitors of protein synthesis implicate, but do not prove, the involvement of protein in memory. These inhibitor experiments are reviewed by Agranoff elsewhere in this series,[80] and also by Roberts and Flexner.[81] The latter authors conclude "that learning and memory can occur in spite of a high level of inhibition of protein synthesis and that some forms of memory do not involve protein synthesis." Current results would not appear to justify the statement that "protein synthesis is required for the establishment of long-term memory."[60]

There appears to be a paucity of experiments suggesting an increased production of protein as a result of learning. Ultimately, in view of the ability of proteins to control the rates of many reactions, as well as to serve as one of the structural components of cells, it would appear that modification of protein quantity or composition would be an important consequence of learning, although not necessarily the first consequence.

In this volume, Hydén and Lange describe a series of experiments in which protein synthesis has been measured in the pyramidal nerve cell of a region of the hippocampus of rats during the transfer of handedness.[60,82] This training procedure was very similar to that used by Hydén for the effect of training on RNA, described above, except that the biochemical measures were made on animals both during the acquisition phase of learning that is, after 4–5 days of training and after the rats had learned, 30 days after original training. Shortly before the last training session, leucine-^{3}H was injected into each lateral ventricle. The period between injection and the start of training is given both as half an hour[60] and as 1 hr[82] in what are presumably descriptions of the same experiment or otherwise identical experiments. Immediately after the one-half hour training session, the rats were sacrificed and the CA-3 region of the hippocampus was removed. The radioactivity in the protein of 300 pyramidal cells was compared to that of the leucine precursor in a larger sample of the surrounding tissue in order to correct for precursor specific activity. When the appropriate corrections were made for precursor specific activity, the specific activity of protein in the hippocampal nerve cells presumably engaged in learning was significantly higher (by a factor of approximately two) than that of control animals. This increase was found for both of the following groups: (1) rats trained for 5 consecutive days, and (2) rats given 5 days of training, then held without training for 14 days, and then given one more day of training. However, no increase in specific activity above the level of nontrained controls was found for rats given the schedule of the (2) group above, then held for an additional 14 days and finally given 3 more days of training. Hydén and Lange believe that rats in groups (1) and (2) were sacrificed while still in the learning phase but that the final group had completed acquisition before sacrifice. It is not apparent why the rats are said to be in different phases of acquisition since, on the only measure of learning given—average number of reaches per day—third training equals second training, and both are slightly less than the end of initial training.

Similar results were obtained on analysis of either the total protein of the nerve cells or of two acidic proteins isolated from acrylamide gel electrophoresis. It should be pointed out that the observed differences are based upon the corrected specific activities; if for any reason these corrections are not valid, quite different conclusions would be drawn, since the uncorrected activities fall in the opposite order. It is difficult from the description and data given to assess the validity of the correction procedures. The specific activity of the precursor leucine was three times as high in the control rat as it was in the contralateral sides of the experimental rats and approximately two times as high in the ipsilateral sides. It is curious that the specific activity of the precursor leucine was inversely related to the expected functional use of the nerve cells, and it is not clear if this may be related to the learning process.[82]

Bowman and Harding[83] have recently questioned several aspects of the original report including the validity of the correction for the free leucine-^{3}H concentration and the statistical tests used. They also questioned whether psychological factors other than learning may have caused the change. These criticisms have been responded to by Hydén and Lange,[84] in part by clarifying the methods used to correct for the ^{3}H-specific activities and in part by the description of the results of the additional group studied 30 days after initial training.

The problems of devising proper controls for biochemical effects attributed to learning are further exemplified by the two additional sets of experiments reported by Hydén and Lange.[60] The first of these deals with positive effects of sensory stimulation alone on brain measures, and the second raises the question of appropriate control treatment. On the former experiment there were three groups of rats: (a) those given presentations of light and tone stimuli paired with electric shock, so that the rats became conditioned to escape when the stimuli were given; (b) those given unpaired light and shock, so that they would be stimulated but could not become conditioned, and (c) those given presentation of light alone. All three groups—(a), (b) and (c)—differed in incorporation values from unstimulated controls, but the groups did not differ significantly from each other. Hydén and Lange conclude that in this experiment the brain effects could be attributed to attention of the experimental animals to the stimuli. They suggest that group (b), with unpaired stimuli, paid the most attention. Could effects of differential attention also apply to the handedness experiment? That is, could attention to a relatively novel situation explain the positive effects found early but not late in the course of reversing handedness?

The last of the experiments presented by Hydén and Lange utilized monkeys trained by Mihailovíc *et al.* It will be recalled that group (a) learned pattern discrimination, group (b) had tasks and cues randomized so that the monkeys could not learn, and group (c) were cage controls. Group (b), with unpaired stimuli, was found to show the lowest protein values and to differ most from the other two groups. The (a) and (c) groups yielded closely similar values. Here Hydén and Lange refer to group (b) as the appropriate controls for measuring the effects of learning in group (a). Here again, effects could be attributed to attention or to the emotional effects often produced in monkeys when they are confronted by an unsolvable task. Most importantly, no clear differences appeared between groups (a) and (c) where effects of learning would have been expected.

Autoradiographic evidence of increased protein synthesis during avoidance training in rats was found by Beach *et al.*[85] in a well-designed and carefully executed study. Littermate triplets were assigned to one of three groups—avoidance training, yoked control without opportunity to learn, and passive control. All rats were familiarized with the apparatus, light and buzzer, and mock injections for 10 days prior to the experimental session. The radioactive precursor was injected intraperitoneally 30 min prior to the start of the experimental session, and a trio was sacrificed only 3 min after the conditioned rat had made its first avoidance response. Leucine-^{3}H incorporation was significantly increased in the nuclei of neurons of hippocampal cells, differences approached statistical significance in entorhinal cortex and in the septal area but not in other brain areas or in liver. It is surprising that evidence of increased protein synthesis is found so quickly after the training session.

Gold *et al.*[86] have reported that brain protein is increased 15% within 2 hrs in young rats given 10 trials in a shock-escape task when compared to either cage control or exploration control groups. The control groups were substantially equal and were combined for statistical comparisons. No increase was found immediately after training. A 37% decrease in nonprotein nitrogen of the trained group was also reported. It should be noted that the "protein" fraction was the soluble material extracted by saline in 24 hr at 4°C which probably represented no more than 25% of the total protein of the brain. The measurement of nonprotein nitrogen represented the difference between total nitrogen and protein determinations and therefore it is not surprising that it decreased when protein increased, since total nitrogen of the groups did not differ. The RNA content of the groups was not statistically different. This group recently reported that no increases in RNA, protein, or total nitrogen were found when the experiment was repeated with old rats.[86a]

D. Effects on Brain Mucoids

Bogoch has recently suggested that brain mucoids—a complex and rather ill-defined group of substances with carbohydrate and protein, and/or lipid properties—may be involved in learning and memory.[87] Male pigeons were trained for a food reward to operate conditioning techniques. One group was sacrificed immediately after training; the second group was sacrificed from 3 to 11 months after training. Two acidic protein fractions were isolated by DEAE chromatography in increased amounts from the trained pigeons and appeared to be related to the phenomenon of learning.

Another group of proteins appeared to be related to the exposure to the training situation rather than to the phenomenon of learning. At least five distinct proteins uniquely increased by learning could be isolated by disc gel electrophoresis from the two fractions. These proteins were rich in bound carbohydrate of a heterogeneous nature. The increase in this fraction was considered by Bogoch to be supportive evidence for an experiential code hypothesis involving the brain mucoids. Evidence was presented that the carbohydrate composition of several of the mucoid fractions was dependent upon the behavioral state of the pigeon. Other studies showed that less incorporation of glucose-1-^{14}C into pigeon brain mucoids occurred when the pigeon was undergoing training than when it was resting, and that the peak incorporation was obtained at shorter times. It is not clear from the description of these studies to what extent the analyses of experimental and control groups were evenly matched in time. It is evident that this unique approach to the biochemistry of memory deserves critical examination.

IV. DISCUSSION AND CONCLUSIONS

The second section of this chapter described long-term studies aimed at elucidating brain mechanisms of aggressive or emotional behavior in biochemical terms. The main groups of investigators–Garattini(5,8,13,17,18) and collaborators in Italy and Welch and Welch(9–12,19,23,25–27) in the United States—have focused on biogenic amine systems. Although it has not been possible to relate aggressiveness to amine metabolism in a one-to-one manner, there seems to be much promise in this currently active field. It exemplifies the programmatic proposal of Haldane(88) to account for motivational states in terms of the accumulation or dissipation of substances in the central nervous system and in other organs—that is, in biochemical terms. Furthermore, the concepts arising from this research may help in understanding the action of drugs employed therapeutically with human subjects.

The third section dealt with studies aimed at elucidating biochemical mechanisms of learning and memory storage. These studies can be categorized into three main durations—those in which the period from start of training to removal of brain tissue lasted only from minutes to a few hours, those with a few training trials per day over a few days, and those with differential experience lasting for weeks. The longer periods have been employed since the late 1950s, whereas the short periods have been used only since the mid-1960s. Each time period was chosen for somewhat different possible cerebral effects of learning.

The long-term experiments of Bennett *et al.*(35–42,46,50–53) were undertaken on the premises that brief training would probably produce only small cerebral changes and that, since these would presumably be widely disseminated in the brain, they could be detected only if prolonged learning produced relatively large cumulative effects. Because of interest in synaptic processes, it was decided to measure enzymes of the cholinergic system. The results obtained to date may perhaps be interpreted more readily in terms of structural neural changes—which have also been found in these experiments—rather than as indications of altered synaptic function. Of course, these structural changes must themselves indicate changes in synthesis and biochemical function.

The intermediate-term experiments of Hydén[56,57,60-63] sought to enhance the level of changes by concentrating on presumably active sites of learning and, within such sites, on separate cell types. One region singled out for study was the vestibular nucleus of the brainstem, although most neurophysiologists would have expected only effects of stimulation but not of learning at this level of the nervous system. Initially these experiments focused on composition and amount of RNA and protein. Hydén's more recent experiments have employed radioactive precursors of nucleic acids and proteins to look for altered synthesis in nerve cells. His highly refined techniques for analyzing separately the chemistry of neural and glial cells has added to the power of this work, but the techniques are so demanding that few other investigators have sought to replicate these studies.

Short-term experiments have recently been undertaken by a number of groups including Wilson, Glassman and associates,[66-70] Bowman and Strobel,[71] Machlus and Gaito,[73-75] Gold *et al.*,[86,86a] and Beach *et al.*[85] Most of these studies have employed radioactive tracers in attempting to find a rapid increase of synthesis of proteins or nucleic acid during training to avoid or escape punishing electric shock. All of these reports indicate that rapid changes of synthesis do occur in this type of training situation. As a next step, some of the investigators are attempting to localize these changes either in relatively large parts of the brain, by analysis of dissected areas, or in particular brain structures or cell types, through the use of autoradiography.[69,85] Another direction is the attempt to characterize the structure and function of the molecules whose synthesis is altered during training. Experiments of this sort include hybridization of RNA and fractionation of RNA's and proteins.

In all of the research on learning and memory, a major problem is that of disentangling possible effects of learning from those of other variables that are inescapably present in the training situation. Most investigators are well aware of this problem. For example, Glassman[2] has discussed it, and he and his co-workers[70] have used a variety of control procedures to determine the importance of other variables. Hydén and Lange[60] take up this problem in their chapter in this volume, and Hydén has often considered it previously.[56,61,62,84] We have described a number of control procedures and tests of alternative hypotheses for effects of differential experience.[39] Nevertheless, the problem is difficult, as has been brought out at several points in this chapter, and it is not yet clear that any of the changes observed can be attributed directly to processes of learning. Certainly the first step is to show that differences in brain biochemistry can be brought about by behavioral manipulations. Then the factors responsible must be identified. There seems to be little point, however, in utilizing elaborate controls before a primary effect has been identified.

In spite of the recent appearance of a number of papers suggesting changes of neurochemistry during learning, the surface has barely been scratched. In fact, lest the reader be overly sanguine about the apparent concordance of results and conclusions from different laboraories, some caveats are in order: Some of the results are based on inadequate numbers of cases. In some studies the large N used for statistical analyses has been obtained by making numerous measures from each subject; the values obtained from the same subject cannot be considered to be independent for statistical purposes. Some of the biochemical methods are suspect since the values reported do not jibe with accepted standard values. Some studies lack obvious and necessary control procedures. Some studies apparently have not used coded samples to prevent analysts being biased by knowledge of group assignment of the subjects. In some studies analyses of the various groups were not run in parallel; in fact, in some cases they were run at quite different times, so that temporal variations from many causes could produce apparent group differences. It is surprising and worrisome that some of the most interesting studies have appeared as isolated reports with no follow-up, to date, from either the originating laboratory or by others. We would like to urge greater attention to replication of results, especially before publication, so that experimental results in the literature can be relied on as being factually correct. The interpretation of results will of course evolve with increasing information and new concepts, but for such progress to be achieved, a solid basis of experimental results is required.

The elucidation of brain mechanisms of motivation and learning in biochemical terms seems to us to be one of the most challenging areas of biological science. The techniques and results reviewed in this chapter represent a number of promising approaches to solution of these problems. We are confident that increasing numbers of investigators will take up these methods and devise more powerful techniques to bring about rapid progress leading to applicable knowledge and understanding in this socially important and intellectually stimulating field.

V. REFERENCES

1a. S. P. R. Rose, Neurochemical correlates of learning and environmental change, *FEBS Letters* **5**:305–312 (1969).

1b. G. P. Talwar and U. B. Singh, *Handbook of Neurochemistry* (A. Lajtha, ed.) Vol. 6, pp. 29–61, Plenum Press, New York (1971).

2. E. Glassman, The biochemistry of learning: An evaluation of the role of RNA and protein, *Ann. Rev. Biochem.* **38**:605–646 (1969).

3. W. C. Allee, Group organization among vertebrates, *Science* **95**:289–293 (1942).

4. C. Y. Yen, R. L. Stanger, and N. Millman, Ataractic suppression of isolation-induced aggressive behavior, *Arch. Intern. Pharmacodyn.* **123**:179–185 (1959).
5. L. Valzelli, *in Advances in Pharmacology* (S. Garattini and P. A. Shore, eds.) Vol. 5, pp. 79–108, Academic Press, New York (1967).
6. A. Wolf and E. Frhr. von Haxthausen, Zur Analyse der Wirkung einiger zentral-sedativer Substanzen, *Arzneimittel-Forsch.* **10**:50–52 (1960).
7. W. Bevan, Jr., W. L. Bloom, and G. T. Lewis, Levels of aggressiveness in normal and amino acid-deficient albino rats, *Physiol. Zool.* **24**:231–237 (1951).
8. S. Consolo, S. Garattini, and L. Valzelli, Sensitivity of aggressive mice to centrally acting drugs, *J. Pharm. Pharmacol.* **17**:594–595 (1965).
9. B. L. Welch and A. S. Welch, Graded effect of social stimulation upon *d*-amphetamine toxicity, aggressiveness, and heart and adrenal weight, *J. Pharmacol. Exptl. Therap.* **151**:331–338 (1966).
10. B. L. Welch and A. S. Welch, An effect of aggregation upon the metabolism of dopamine-1-^{3}H, *Prog. Brain Res.* **8**:201–206 (1964).
11. A. S. Welch and B. L. Welch, *in Physiology of Fighting and Defeat* (B. E. Eleftheriou and J. P. Scott, eds.) University of Chicago Press, Chicago (In press).
12. B. L. Welch and A. S. Welch, *in Proc. Int. Symp. on the Biology of Aggressive Behavior* (S. Garattini and E. B. Sigg, eds.) pp. 188–202, Excerpta Medical Foundation, Amsterdam (1969).
13. E. Giacalone, M. Tansella, L. Valzelli, and S. Garattini, Brain serotonin metabolism in isolated aggressive mice, *Biochem. Pharmacol.* **17**:1315–1327 (1968).
13a. S. Garattini, E. Giacalone, and L. Valzelli, *in Proc. Int. Symp. Aggressive Behavior* (S. Garattini and E. B. Sigg, eds.) pp. 179–187, Excerpta Medical Foundation, Amsterdam (1969).
14. P. C. Bourgault, A. G. Karczmar, and C. L. Scudder, Contrasting behavioral, pharmacological, neurophysiological, and biochemical profiles of C57B/6 and SC-I strains of mice, *Life Sci.* **8**:533–553 (1963).
15. A. G. Karczmar and C. L. Scudder, Behavioral responses to drugs and brain catecholamine levels in mice of different strains and genera, *Fed. Proc.* **26**:1186–1191 (1967).
16. J. W. Maas, Neurochemical differences between two strains of mice, *Science* **137**:621–622 (1962).
17. L. Valzelli, *in Proc. 5th Int. Cong. of Neuro-psycho-pharmacology, Washington, D.C., 1966*, pp. 781–788, Excerpta Medica International Congress Ser. No. 129 (1967).
18. S. Garattini, E. Giacalone, and L. Valzelli, Isolation, aggressiveness and brain 5-hydroxytryptamine turnover, *J. Pharm. Pharmacol.* **19**: 338–339 (1967).
19. B. L. Welch and A. S. Welch, Greater lowering of brain and adrenal catecholamines in group-housed than in individually-housed mice administered DL-α-methyltyrosine, *J. Pharm. Pharmacol.* **20**:244–246 (1968).
20. W. B. Essman and G. E. Smith, Behavioral and neurochemical differences between differentially housed mice, *Am. Zool.* **7**:793–794 (1967).
21. W. B. Essman, Differences in locomotor activity and brain serotonin metabolism in differentially housed mice, *J. Comp. Physiol. Psychol.* **66**:244–246 (1968).
22. F. Marcucci, E. Mussini, L. Valzelli, and S. Garattini, Decrease in *N*-acetyl-L-aspartic acid in brain of aggressive mice, *J. Neurochem.* **15**:53–54 (1968).
22a. F. Marcucci and E. Giacalone, *N*-Acetyl aspartic, aspartic, and glutamic acids brain levels in aggressive mice, *Biochem. Pharmacol.* **18**:691–692 (1969).
23. B. L. Welch and A. S. Welch, Fighting: Preferential lowering of norepinephrine and dopamine in the brainstem, concomitant with a depletion of epinephrine from the adrenal medulla, *Comm. Behav. Biol.* *A*3:125–130 (1969).

24. B. E. Eleftheriou and K. W. Boehlke, Brain monoamine oxidase in mice after exposure to aggression and defeat, *Science* **155**:1693–1694 (1967).
25. A. S. Welch and B. L. Welch, Reduction of norepinephrine in the lower brainstem by psychological stimulus, *Proc. Nat. Acad. Sci. U.S.* **60**:478–481 (1968).
26. B. L. Welch and A. S. Welch, Differential activation by restraint stress of a mechanism to conserve brain catecholamines and serotonin in mice differing in excitability, *Nature* **218**:575–577 (1968).
27. A. S. Welch and B. L. Welch, Effect of stress and *para*-chlorophenylalanine upon brain serotonin, 5-hydroxyindoleacetic acid and catecholamines in grouped and isolated mice, *Biochem. Pharmacol.* **17**: 699–708 (1968).
27a. B. L. Welch and A. S. Welch, Sustained effects of brief daily stress (fighting) upon brain and adrenal catecholamines and adrenal, spleen, and heart weights of mice, *Proc. Nat. Acad. Sci. U.S.* **64**:100–107 (1969).
28. E. Geller, A. Yuwiler, and J. Z. Zolman, Effects of environmental complexity on constituents of brain and liver, *J. Neurochem.* **12**:949–955 (1965).
29. E. Geller and A. Yuwiler, *in Ontogenesis of the Brain* (L. Jilek and S. Trojan, eds.) pp. 277–284, Charles University Press, Prague (1968).
30. J. Křeček, *in Biopsychology of Development* (E. Tobach, ed.) Academic Press, New York (1970). (In press.)
31. W. H. Riege and H. Morimoto, Effects of chronic stress and differential environments upon brain weights and biogenic amine levels in rats, *J. Comp. Physiol Psychol.* **71**:396–404.
32. W. A. Himwich and H. C. Agrawal, *in Handbook of Neurochemistry* (A. Lajtha, ed.) Vol. 1, pp. 33–52, Plenum Press, New York (1969).
33. H. C. Agrawal, M. W. Fox, and W. A. Himwich, Neurochemical and behavioral effects of isolation-rearing in the dog, *Life Sci.* **6**:71–78 (1967).
34. W. A. Himwich, J. M. Davis, and H. C. Agrawal, *in Recent Advances in Biological Psychiatry* (J. Wortis, ed.), Vol. 10, pp. 266–270, Plenum Press, New York (1968).
35. D. Krech, M. R. Rosenzweig, and E. L. Bennett, Effects of environmental complexity and training on brain chemistry, *J. Comp. Physiol. Psychol.* **53**:509–519 (1960).
35a. M. R. Rosenzweig, E. L. Bennett, and M. C. Diamond, *in Macromolecules and Behavior*, 2nd ed. (J. Gaits, ed.) Appleton-Century-Crofts. (In press.)
36. E. L. Bennett, M. C. Diamond, D. Krech, and M. R. Rosenzweig, Chemical and anatomical plasticity of brain, *Science* **146**:610–619 (1964).
37. M. R. Rosenzweig and E. L. Bennett, Effects of differential environments on brain weights and enzyme activities in gerbils, rats, and mice, *Develop. Psychobiol.* **2**:87–95 (1969).
38. M. R. Rosenzweig, E. L. Bennett, M. C. Diamond, S. Wu, R. W. Slagle, and E. Saffran, Influences of environmental complexity and visual stimulation on development of occipital cortex in rats, *Brain Res.* **14**:427–445 (1969).
39. M. R. Rosenzweig, *in Biopsychology of Development* (E. Tobach, ed.) Academic Press, New York (1970). (In press.)
40. M. C. Diamond, D. Krech, and M. R. Rosenzweig, The effects of an enriched environment on the histology of the rat cerebral cortex, *J. Comp. Neurol.* **123**:111–119 (1964).
41. M. C. Diamond, F. Law, H. Rhodes, B. Lindner, M. R. Rosenzweig, D. Krech, and E. L. Bennett, Increases in cortical depth and glia numbers in rats subjected to enriched environment, *J. Comp. Neurol.* **128**:117–126 (1966).
42. M. C. Diamond, Extensive cortical depth measurements and neuron size increases in the cortex of environmentally enriched rats, *J. Comp. Neurol.* **131**:357–364 (1967).

43. D. R. Malkasian, The morphological effects of environmental manipulation and litter size on the neonate rat brain, Ph.D. dissertation, University of California, Berkeley, 1969.
44. J. Altman and G. Das, Autoradiographic examination of the effects of enriched environment on the rate of glial multiplication in the adult rat brain, *Nature* **204**:1161–1163 (1964).
45. J. Altman, *in Macromolecules and Behavior* (J. Gaito, ed.) pp. 103–126, Appleton-Century-Crofts, New York (1966).
46. E. L. Bennett and M. R. Rosenzweig, *in Mind as a Tissue* (C. Rupp, ed.) pp. 63–86, Harper and Row, Hoeber Medical Division, New York (1968).
47. J. Altman, *in The Neurosciences, A Study Program* (G. C. Quarton, T. Melnechuk, and F. O. Schmitt, eds.) pp. 723–743, The Rockefeller Univ. Press, New York (1967).
48. R. N. Walsh, O. E. Budtz-Olsen, J. E. Penny, and R. A. Cummins, The effects of environmental complexity on the histology of the rat hippocampus, *J. Comp. Neurol.* **137**:361–366 (1969).
49. J. C. LaTorre, Effect of differential environmental enrichment on brain weight and on acetylcholinesterase and cholinesterase activities in mice, *Exptl. Neurol.* **22**:493–503 (1968).
50. E. L. Bennett, M. R. Rosenzweig, A. Orme, and S. Rosenbaum, Choline acetyltransferase activity differences in brain regions of ten rat strains, related to acetylcholinesterase activity, acetylcholine concentration, and environmental enrichment, *Fed. Proc.* **27**:836 (1968).
51. M. R. Rosenzweig, Effects of experience on brain chemistry and brain anatomy, *Atti. Accad. Naz. Lincei (Italy), Quaderno* **109**:43–63 (1968).
52. E. L. Bennett, M. R. Rosenzweig, and M. C. Diamond, Rat brain: Effects of environmental enrichment on wet and dry weights, *Science* **163**:825–826 (1969).
53. M. R. Rosenzweig, D. Krech, E. L. Bennett, and M. C. Diamond, *in Early Experience and Behavior* (G. Newton and S. Levine, eds.) pp. 258–298, Charles C Thomas, Springfield, Ill. (1968).
54. D. Singh, R. J. Johnston, and H. J. Klosterman, Effect of brain enzyme and behavior in the rat of visual pattern restriction in early life, *Nature* **216**:1337–1338 (1967).
55. J. Bonner, *in Macromolecules and Behavior* (J. Gaito, ed.), pp. 158–164, Appleton-Century-Crofts, New York (1966).
56. H. Hydén, *in The Neurosciences: A Study Program* (G. C. Quarton, T. Melnechuk, and F. O. Schmitt, eds.), pp. 765–771, The Rockefeller Univ. Press, New York (1967).
57. H. Hydén, Biochemical and molecular aspects of learning and memory, *Proc. Am. Phil. Soc.* **111**:326–342 (1967).
58. J. Gaito, *in Molecular Psychobiology*, Charles C Thomas, Springfield, Ill. (1966).
59. J. S. Griffith and H. R. Mahler, DNA ticketing theory of memory, *Nature* **223**:580–582 (1969).
60. H. Hydén and P. W. Lange, *in Handbook of Neurochemistry* (A. Lajtha, ed.) Vol. 6, pp. 221–239, Plenum Press, New York (1971).
61. H. Hydén *in The Neurosciences: A Study Program* (G. C. Quarton, T. Melnechuk, and F. O. Schmitt, eds.), pp. 248–270, The Rockefeller Univ. Press, New York (1967).
62. H. Hydén, *in Progress in Nucleic Acids Research* (J. N. Davidson and W. E. Cohn, eds.), Vol. 6, pp. 187–218, Academic Press, New York (1967).
63. H. Hydén, E. Egyházi, E. R. Johns, and F. Bartlett, RNA base ratio changes in planaria during conditioning, *J. Neurochem.* **16**:813–821 (1969).

64. V. E. Shashoua, RNA changes in goldfish brain during learning, *Nature* **217**:238–240 (1968).
65. H. Dellweg, R. Gerner, and A. Wacker, Quantitative and qualitative changes in ribonucleic acids of rat brain dependent on age and training experiments, *J. Neurochem.* **15**:1109–1119 (1968).
66. J. W. Zemp, J. W. Wilson, K. Schlesinger, W. O. Boggan, and E. Glassman, Brain function and macromolecules, I. Incorporation of uridine into RNA of mouse brain during short-term training experience, *Proc. Nat. Acad. Sci. U.S.* **55**:1423–1431 (1966).
67. J. W. Zemp, J. E. Wilson, and E. Glassman, Brain function and macromolecules, II. Site of increased labeling of RNA in brains of mice during a short-term training experience, *Proc. Nat. Acad. Sci. U.S.* **58**:1120–1125 (1967).
68. L. B. Adair, J. E. Wilson, J. W. Zemp, and E. Glassman, Brain function and macromolecules, III. Uridine incorporation into polysomes of mouse brain during short-term avoidance conditioning, *Proc. Nat. Acad. Sci. U.S.* **61**:606–613 (1968).
69. B. E. Kahan, M. R. Kriginan, J. E. Wilson, and E. Glassman, Brain function and macromolecules, VI. Autoradiographic analysis of the effect of a brief training experience on the incorporation of uridine into mouse brain, *Proc. Nat. Acad. Sci. U.S.* **65**:300–304 (1970).
70. L. B. Adair, J. E. Wilson, and E. Glassman, Brain function and macromolecules, IV. Uridine incorporation into polysomes of mouse brain during different behavioral experiences, *Proc. Nat. Acad. Sci. U.S.* **61**:917–922 (1968).
71. R. E. Bowman and D. A. Strobel, Brain RNA metabolism in the rat during learning, *J. Comp. Physiol. Psychol.* **67**:448–456 (1969).
72. D. Gillespie and S. Spiegelman, A quantitative assay for DNA-RNA hybrids with DNA immobilized on a membrane, *J. Mol. Biol.* **12**:829–842 (1965).
73. B. Machlus and J. Gaito, Detection of RNA species unique to a behavioral task, *Psychon. Sci.* **10**:253–254 (1968).
74. B. Machlus and J. Gaito, Unique RNA species developed during a shock avoidance task, *Psychon. Sci.* **12**:111–112 (1968).
75. B. Machlus and J. Gaito, Successive competition hybridization to detect RNA species in a shock avoidance task, *Nature* **222**:573–574 (1969).
76. V. P. Chiarugi, Changes of nuclear RNA in hepatomas as revealed by RNA/DNA hybridization, *Biochim. Biophys. Acta* **179**:129–135 (1969).
77. W. S. Riggsby and V. Merriam, RNA-DNA Hybridization: Demonstration in a mammalian system of competition by preincubation, *Science* **161**:570–571 (1968).
78. J. Stevenin, J. Samec, M. Jacob, and P. Mandel, Détermination de la fraction du génome codant pour les RNA ribosomiques et messagers dans le cerveau du rat adulte, *J. Mol.. Biol.* **33**:777–793 (1968).
79. K. von Hungen, Hybridization of rat brain RNA; Failure to confirm reports of new species of RNA induced with learning, *Am. Soc. Neurochem. Abstracts* (1970).
80. B. Agranoff, *in Handbook of Neurochemistry* (A. Lajtha, ed.) Vol. 6, pp. 203–219, Plenum Press, New York (1971).
81. R. B. Roberts and L. B. Flexner, The biochemical basis of long-term memory, *Quart. Rev. Biophys.* **2**:135–173 (1969).
82. H. Hydén and P. W. Lange, Protein synthesis in the hippocampal pyramidal cells of rats during a behavioral test, *Science* **159**:1370–1373 (1968).
83. R. E. Bowman and G. Harding, Protein synthesis during learning, *Science* **166**:199–200 (1969).
84. H. Hydén and P. W. Lange, Protein synthesis during learning, *Science* **164**:200–201 (1969).

85. G. Beach, M. Emmens, D. Kimble, and M. Lickey, Autoradiographic demonstration of biochemical changes in the limbic system during avoidance training, *Proc. Nat. Acad. Sci. U.S.* **62**:692–696 (1969).
86. A. M. Gold, H. Altschuler, H. H. Kleban, M. P. Lawton, and M. Miller, Chemical changes in rat brain following escape training, *Psychon. Sci.* **17**:37–38 (1969).
86a. H. Altschuler, M. Gold, M. H. Klebar, M. R. Lawton, and M. Miller, Neurochemical changes in the brain of aging albino rats resulting from avoidance learning, *Fed. Proc. Abstracts* **29**:658 (1970).
87. S. Bogoch, *The Biochemistry of Memory, with an Inquiry into the Function of Brain Mucoids*, Oxford University Press (1968).
88. J. B. S. Haldane, The sources of some ethological notions, *Brit. J. Anim. Behav.* **4**:162–164 (1956).

Chapter 6

MEMORY

Bernard W. Agranoff

University of Michigan
Ann Arbor, Michigan

I. INTRODUCTION

The concept of a molecular basis for memory is not new, The term engram, generally credited to Semon,[1] here refers to the physicochemical change in the brain which reflects a long-lasting acquired behavioral change. Its existence remains hypothetical, yet many inferences may be drawn from biochemical experiments designed to elucidate one or another aspect of its nature, and such experiments will be emphasized in this chapter. The field is clearly in its infancy. The day a biochemical, electrophysiological, or morphological correlation is indisputably established, we can be sure that the other disciplines will follow in hot pursuit. Such is the interdisciplinary approach to neurobiology!

A. Definitions

For present purposes, learning is defined operationally as the tendency of an organism to increase its probability of responding in some prescribed fashion. Memory is similarly defined operationally as the performance of the new behavior at some time following training. Long-term or permanent memory may last a lifetime and for convenience is usually tested rather arbitrarily a few days following training. While it becomes difficult to consider learning without memory and memory without learning, we find it convenient to equate learning with the formation of short-term memory, indicating responsive behavior from trial-to-trial over a few minutes or hours within a training session. We consider long-term or permanent memory the retention of this behavior from one training session to another generally some days later. There may well be a continuum of time constants of memories, but these extremes are at present useful.

B. The Brain and Evolution

Many textbook definitions of learning include the requisite that the new behavior be of biological value or significance to the experimental animal. I do not make that distinction here because I feel that the training tasks generally used in behavioral experiments are grossly artifactual. "Learning" results when an animal behaves in a way that pleases the investigator by responding in a way the investigator (hopefully) predicted would be logical. If we assume that the variety of possible responses that can be elicited from an experimental animal have arisen genetically as a result of natural selection and mutation, the survival of the species itself is the only means of expressing the biological significance of the way in which the animal has modified his behavior. As in the selection theory of antibody formation, the species interacts with its environment during phylogenesis and we can experimentally evoke only those behaviors that are in the repertoire of an individual of that species.[2] The wide variety of temporal patterning and hierarchical organization possible within the nervous system gives rise to a greater degree of flexibility than in any other biological system. The acquisition of a discrimination task in which a pigeon distinguishes vertical from horizontal bars probably reflects the same degree of adaptability as the formation of an antibody to Dacron, a substance never before encountered in the history of the species. In each instance, a large repertoire of responses together with feedback control systems provides a seemingly highly specific interaction with the environment.

C. Morphological Approaches

The modern era in the morphological approach to higher brain function began with Cajal who proposed a neuronal theory of development and learning.[3] More recent experiments on specific nerve regrowth[4] add credence to the concept of a high degree of specificity in neural connections. A simple conceptual model might invoke activation of a single cell to produce a behavioral change. Such simplistic models are improbable even though chemical changes have been reported in single cells as correlates of training.[5] The reasons we might not expect to find changes in single cells include first, the relatively small number of cells which ennervate the muscles leading to a behavioral response (common final pathway[6]). It is unlikely that multifunctional neurons would reflect grossly detectable changes upon acquisition by the organism of a single new behavior. Yet these cells are generally most accessible for electrophysiological and neurochemical studies. Second, there is electrophysiological evidence which indicates a probabilistic firing pattern for a given neuron in relation to a motor response.[7] Even in simple functions, networks of neurons may be operating in concert. It is nevertheless of interest that complex behavioral patterns have been elicited in a mollusk following stimulation of single cells.[8]

The extensive studies of Lashley[9] probably resulted in an "overkill" of neuronal (neurobiotaxic, connectionistic) thinking about learning and memory. In his experiments, rats were trained and retrained in various behavioral tasks in conjunction with surgical ablation of brain regions before or after their initial training. No specific associative tracts could be ablated. These experiments have since been criticized from a number of viewpoints, both conceptually and technically,[10] but there is no question that they served to direct physiologists away from reductionist approaches for many years. In the past decade, experiments with discrete lesions and stimulation[11] have again encouraged investigators to correlate complex behavior with brain anatomy. Ablation of various regions of the octopus brain has been reported to produce specific memory loss.[12] Studies on patients with epilepsy suggest that stimulation of discrete brain areas can evoke specific memories.[13] Bilateral hippocampal lesions in man have been reported to selectively block formation of long-term memory,[14] although the interpretation of such studies has been questioned.[15] Such a specific separation of short- and long-term memory by means of ablation studies in other species, including nonhuman primates, have not been successful.[16]

It has been argued that an increase in net environmental stimulation over a relatively long period of time should cause a change in brain structure or composition. Experiments of Rosenzweig *et al.*[17] indicate that handling, training, and general enrichment of the environment of rats for a few weeks results in brain growth, particularly in the thickness of the occipital cortex. In addition, the number of glia and dendritic spines as well as the amount of dendritic branching is said to be increased. While such increases are small (5–10%), they lend support to a connectionistic view of brain function. The Berkeley group has recently reported[18] that the changes seen in the occipital cortex do not persist—they regress after the enriched environment is withdrawn and may therefore reflect changes concomitant with the learning process rather than of stored memory itself. Altman[19] has recently reported increased length of the cerebrum as a result of handling. He has also observed neuronal replication following birth, evidenced by incorporation of labeled thymidine.[20] The development of synapses after birth in the mouse and the formation of new connections in tissue culture[21] have also been reported.

D. Electrophysiological Approaches

The evidence for electrophysiological correlates of learning have been recently reviewed elsewhere.[10] Correlations of electrical properties of the brain with learning and memory are of two general varieties; they involve simple (often invertebrate) neural nets which can be used as models of learning; or alternatively, electrophysiological correlates of training in the intact mammalian brain. Of the first variety, posttetanic potentiation[22] and more recently heterosynaptic facilitation have been studied.[23] Even in simple model systems, structural complexity does not yet permit an anatomical correlation of connections with the inferred electrical ones. These model systems do not require that gross behavior be mediated within a single cell but rather that the system itself show some of the formal properties of learning, namely, associative changes with appropriate temporal patterning of impulses.

An example of the other electrophysiological approach is the correlation of electroencephalographic patterns with stimulus presentation. For example, pairing of foot shock and auditory stimuli can result in changes in poststimulus histograms from selected regions of the cortex.[24] Changes in electroencephalographic rhythms have been reported with presentation of familiar visual forms.[25] Holographic theories of memory storage have been proposed to account for the apparent widespread storage of behavioral information in the brain.[26]

II. BIOCHEMICAL STUDIES

Molecular approaches to the study of memory are of two general classes: (a) a biochemical correlate of behavior is sought; or (b) agents having specific biological actions are administered and the effects on behavior are observed. The latter studies, while least direct, permit inferences about the nature of the biological concomitants in the intact animal—they permit us to search for the needle without worrying about the haystack.

A. Concomitant Changes

Hydén and Lange, in studies on the Deiter's cell nucleus of the rat, proposed that changes in total RNA as well as base ratios occur during training sessions involving balancing, as distinguished from vestibular exercise.[27] These studies can be criticized on the basis of the multifunctional nature of the Deiters' cell. It must accommodate many forms of learning. The changes observed would not seem to be specific for a given task. More recently, Hydén has studied reversal of handedness preference in rats. This task permits one side of the animal to be used as the control for the other. Initially, cortical neurons in a region believed to be involved in acquisition of handedness were studied, and changes in acrylamide gel protein patterns were reported.[27] More recently, this task has been studied in relation to biochemical changes in the hippocampus. Radioactive leucine incorporation into soluble protein separated on acrylamide as well as into total protein suggests that there is increased protein synthesis on both sides of the hippocampus with handedness training and that there is a trend for more incorporation on the side contralateral to the paw involved in the new training.[28]

Bogoch has reported changes in pigeon brain gel protein patterns following training in visual tasks.[29] The reported effects appear to reflect a gross alteration in the amount of a specific soluble band. Oddly, concomitant studies on incorporation of radioactivity into protein from ^{14}C-glucose indicate decreased labeling with training.

In learning experiments in the goldfish, Shashoua[30] studied incorporation of ^{14}C-orotic acid into uridine and cytidine moieties of RNA. While there was a variable amount of incorporation for each fish, the ratio of labeled UMP to CMP in RNA hydrolysates was relatively constant in resting control animals following a 4-hr pulse (3.6 ± 0.2). The experimental animals had a much higher ratio (6.9 ± 0.5). Injection of puromycin prior to training prevented the change in this ratio with training and also appeared to destroy memory of the task (adaptation of posture following attachment of a plastic float to the ventral surface). The ratio itself was extremely variable following the drug.

Zemp *et al.*[31] have studied uridine incorporation into brain RNA in the mouse during training. Mice were trained to jump to a shelf upon presentation of light and buzzer stimulus in order to avoid foot shock administered via grid on the bottom of the cage. "Quiet" control mice as well as "yoked" control animals were also used. The latter were in a neighboring training box and were subjected to the same light and shock as the experimental animal but differed in that they did not have a shelf to which they could escape. The foot shock current applied to both experimental and yoked control animal was terminated when the experimental animal jumped to the shelf. Both experimental and yoked controls thus received identical amounts of light, buzzer, and shock. ^{14}C-uridine or ^{3}H-uridine was injected into each animal intracerebrally under light anesthesia. Training began 30 min after injection and was completed by 45 min, when animals were killed. Pairs of ^{14}C and ^{3}H brains were combined and RNA was extracted. In these experiments the paired brains were either a quiet control and yoked control brain, a quiet control and an experimental brain, or an experimental and a yoked control brain. After normalizing for relative amounts of radioactivity in the UMP soluble pool, the $^{14}C/^{3}H$ ratio in RNA was calculated. In almost every instance, the experimental animal had more radioactivity in nuclear and ribosomal RNA than the yoked or resting control, the latter two being indistinguishable from one another. Dispersion of RNA on a phenol gradient did not dramatically distinguish a specific molecular weight class or a subcellular species which showed the greatest effect. The use of the double label was well suited to this phase of the experiment. Livers and kidneys from the paired animals did not show increased labeling in any group. In more recent studies, a more reproducible increase in the polysome fraction has been found, particularly in the monoribosome fraction.[32] Examination of brain regions indicated the greatest increases in the diencephalon. More extensive behavioral experiments[33] indicate that the increased labeling occurs only when the animal is acquiring a new task. Animals pulsed with isotope after having previously been trained do not show the increase. Animals subjected to the conditions of yoked control experiments or classical conditioning

(10 sec of foot shock in each trial with no possibility of jumping to a shelf) showed no increase in RNA labeling. Prior treatment with these latter conditions followed by a standard training session did not prevent the increase in RNA labeling during the training session. In other words, in each instance, it appears that only those sorts of training which result in formation of memory of the task resulted in increased RNA labeling. Prior adrenalectomy did not prevent acquisition nor did it affect the RNA stimulation.

1. *Molecular Considerations of Brain Plasticity*

A molecular counterpart of the plasticity experiments of Rosenzweig *et al.* might predict that with age, some slowly turning over component of brain would increase in amount. Cerebroside sulfate has been proposed as a candidate for such an accretion molecule.[34] Alternatively, the number of positive feedback loops for induction of a given enzyme, e.g., cholinesterase, might increase, as would the total amount of brain cholinesterase. Neither of these considerations are readily provable. Present-day methodology may not detect the very small molecular changes required to change behavior. We do not know how much chemical alteration should be expected for a given altered behavior. The hypothetical increment might be so insignificant that it could not be distinguished from endogenous levels of the relevant substance or enzyme. In addition, while we might expect one or another material to increase with age, we do not know that it should change during what we consider to be a training session. Animals may be constantly acquiring new behaviors for which we are not testing. The storage conditions in a laboratory may be as stimulatory for the hypothetical accretion process as is the training. Finally, we do not know for certain that covalent bonds are made when animals form new memories. The techniques we employ, such as isotopic labeling, macromolecular fractionation, enzyme assay, etc., deal with changes from one stable chemical state to another. A very important question, then, is: "What is the evidence that the changes in the brain relevant to memory formation involve chemical changes, particularly new covalent bonds?" Probably the best argument for the chemical nature of the engram is derived from the use of disruptive agents.

B. Disruption of Memory Formation

While the localization of memory in the brain has not been possible, there is considerable evidence about when it is formed, primarily from studies with agents which appear to disrupt the formation of memory or to interfere with it in some way.

1. *The Trace and the Engram*

These two hypothetical constructs have arisen for different reasons but are sometimes loosely interchanged. In associative pairing, as first demonstrated by Pavlov, a conditioned stimulus (CS) such as light, buzzer, bell, etc. precedes by some time interval (milliseconds or seconds) the unconditioned stimulus (US) such as food presentation, punishing shock, etc. For a given task, there is an optimal CS–US interval, and when the US precedes the CS, there is generally no demonstrable learning. From this it would seem that some central nervous system manifestation of the CS (the "*trace*") perseverates during the CS–US interval so that the association can form. It has been postulated that this manifestation is electrical in nature.[35,36]

Completely different sorts of observations have given rise to the concept of a fixation or consolidation of the total CS–US association (the engram) some time after the learning session. Consolidation is inferred from studies with various disruptive experiments.

Almost 100 years ago, Ribot[37] observed that when a man fell out of his carriage and was knocked unconscious, he did not upon recovery, recall events just preceding the fall—events of which he was undoubtedly aware just prior to the accident. He had a retrograde amnesia. Consolidation theory would interpret these events to mean that long-term memory ordinarily forms at some time following the acquisition period.

Studies with electroconvulsive shock in man and animals similarly show a period of disruptibility beginning immediately following learning. The actual consolidation time reported within a single species varies with the task used and is variously claimed to require seconds to hours for completion.[38] Once memory is fixed, convulsions or even cessation of all electrical activity does not destroy memory. It has thus been postulated that memory is first electrical in nature and later is converted to a stable chemical form. Our own experiments, with antimetabolites summarized below, indicate that disruption of macromolecular synthesis supports this notion. The agents do not block short-term memory formation. That is, the perseveration of the trace, reinforcement and selective retention of the desired response, all implicit in learning, are not blocked by various antimetabolites that block RNA or protein synthesis. The formation of long-term memory is, however, selectively disrupted, supporting the concept of a requirement of macromolecular synthesis for this process. While both trace and short-term memory formation are apparently insensitive to these drugs, they appear to serve different functions in memory formation. Qualitatively, it might seem that the duration of the trace is much briefer than short-term memory. While this is generally so, there are striking exceptions. Optimum latencies between stimulus and response of 250 msec are often quoted.[39] In experiments on food preference in the rat, however, over 12 hr can separate the CS (flavored water) from the US, an emetic agent.[40] Different types of learning may have different time constants. It may be advantageous for an animal to have a slow gustatory learning system. It has long been known that when a large rat population is not immediately destroyed by poison, the surviving population becomes inured. The long latency can have survival value.

2. *Consolidation vs. Interference*

On the heels of the consolidation hypothesis came an alternative explanation for the effect of disruptive agents. Was it not possible that the noxious aspect of the disruptive agent itself acted as a negative reinforcement? For example, an animal is trained to jump onto a shelf to avoid a mild punishing shock. Having learned the task, he is then given electroconvulsive shock. A few days later, he is tested and does not demonstrate evidence that he has learned the task. Like an untrained animal, he does not jump up to the shelf immediately. Consolidation theory explains this result by stating that the labile memory of the training was destroyed. Interference theory (also referred to as "competing responses" or "conflict of habits") would say the rat learned that if he did jump onto the shelf to avoid the mild punishing shock, he would receive an even greater punishment. He therefore has learned that jumping onto the shelf is not desirable. We cannot ask the animal why he performed the way he did, but there are indirect ways of testing him for the answer. In comparable human studies, e.g., posttraumatic retrograde amnesia, it appears that if the subject does have memory of incidents prior to trauma, he cannot retrieve it. Interpretation of the retrograde amnesia as disruption of consolidation or as interference rests on whether the information is under any circumstances retrievable. That is, even if the subject cannot recall the events in question, they may be repressed. Hypnosis, drug studies, and psychiatric interviews have not clarified this issue.

Jarvik has used a passive avoidance task to test the consolidation *vs.* interference hypothesis.[(41)] In such experiments, an animal is trained to remain in a restricted environment. For example, a mouse is kept in a small chamber. Given the opportunity, he ordinarily escapes into a larger, safer appearing one. In the experiment, the animal is given electroconvulsive shock as soon as he escapes. If electroconvulsive shock is punishing, on retraining we will expect the mouse not to show the escape behavior seen in the untrained mouse. Actually, shocked mice continue to escape, supporting the consolidation hypothesis.

3. *Studies with Antibiotics*

In 1961, Dingman and Sporn measured the rate of learning, in a swimming maze, of rats injected intracisternally with 8-azaguanine.[(42)] The same drug was used by Chamberlain, Rothschild, and Gerard[(43)] in a spinal asymmetry model of learning. The rate of onset of a nondisruptable asymmetry was reported to be delayed by this drug.

a. Mice. Flexner *et al.* injected puromycin subcutaneously into mice and reported no effect on learning or memory, although drowsiness was produced.[(44)] In subsequent experiments[(45)] the drug was injected under light anesthesia into three bilateral pairs of sites at various intervals following training of a shock avoidance discrimination task. Mice were trained to turn right or left in a maze in order to avoid a punishing shock applied via floor grids. Puromycin was injected 1–3 days after a single training session into bilaterial temporal sites and mice were retested a day later. They were found to have significant amnesia for the task. Frontal or intraventricular injections had no effect. Injection into all six sites several weeks after training however, caused loss of the trained response, while uninjected animals retained memory of the training. Since puromycin is reported to be a selective inhibitor of protein synthesis, it was tentatively concluded that protein synthesis is required for memory, and that memory is at first localized to the temporal and entorhinal cortex and subsequently becomes generalized. It was also concluded that by disrupting protein synthesis sufficiently over the entire brain, even old memories could be disrupted, indicating a maintenance stage of memory as well. In further studies[46] injections of acetoxycycloheximide, an even more potent protein synthesis inhibitor, did not result in amnesia, nor did the combination of acetoxycycloheximide and puromycin block memory. The amnesic effect of puromycin was apparently blocked by another protein inhibitor. A protein synthesis block did not appear to explain the amnesic effect of puromycin. These experiments gave rise to two different hypotheses[47,48] (1) *m*RNA is made during or after learning and is translated into a protein required for expression of the new behavior. The protein itself induces more of the behavior-specific *m*RNA. Puromycin blocks protein synthesis by splitting off the growing peptide chain, leaving *m*RNA bare and susceptible to nucleases. By the time the drug has worn off, neither protein nor *m*RNA specific to the new task remains and amnesia results. Acetoxycycloheximide, on the other hand, slows the rate of ribosomal travel along the *m*RNA and preserves the polysome and thereby *m*RNA so that when the drug ultimately wears off, the novel protein will be translated and the new behavior expressed. The hypothesis predicts that there will be a period during which memory will temporarily be unretrievable when acetoxycycloheximide is used, and this was reported experimentally.[49] The combination of acetoxycycloheximide and puromycin should likewise produce no permanent memory loss, since *m*RNA is preserved under these conditions. (2) The actual disruptive effector is peptidyl puromycin and not the block in protein synthesis. Acetoxycycloheximide, by blocking ribosomal travel also blocks the formation of peptidyl puromycin. Therefore, acetoxycycloheximide and the combination of puromycin and acetoxycycloheximide are not expected to block memory. Ultrastructural studies on mouse brain mitochondria indicate that puromycin alone is more destructive than the combination of puromycin and acetoxycycloheximide, in partial support of this thesis.[50]

A puzzling aspect of the Flexner studies is the report that intracranial injection of saline several months after puromycin has produced amnesia will reverse the amnesic effect. This claim throws into further doubt a possible relationship between consolidation and protein synthesis in the mouse. Peptidyl puromycin has been reported present in the mouse brain injection track[51,52] and to be liberated by the saline injection, although further biochemical verification seems in order.[53]

Similar studies have been performed by Cohen and Barondes. Their experiments differ in that puromycin is injected into the bitemporal sites of the mouse brain before training, rather than 1–3 days later.[54] These latter experiments are more consistent with the usual parameters of consolidation experiments. Initially puromycin, but not acetoxycycloheximide nor the combination puromycin and acetoxycycloheximide was reported to block memory of the Y-maze,[55] in agreement with the results from Flexner's posttrial injections. A further hypothesis was put forward: (3) Peptidyl puromycin is a convulsant and it is by virtue of its electrical effects that it blocks memory, not by the inhibition of protein synthesis by puromycin.[56] Subsequently, the same investigators found that acetoxycycloheximide *was* amnesic.[57] The previous failure was attributed to overtraining which overrode the protein blocking effects. Whatever the cause, the cycloheximides are now reported to be amnesic agents in the mouse and protein synthesis is again invoked in memory formation.[57]

b. The Goldfish Puromycin has been shown to block retention of the avoidance response in goldfish in a shuttlebox task in which light is coupled with punishing shock.[58] There was no effect on performance in previously trained fish. Block of memory by this drug in goldfish has since been demonstrated in other laboratories.[30,59] The drug was injected intracranially into unaesthetized fish by means of a microsyringe, leaving the underlying brain intact. An unusual aspect of this procedure was that the animals were completely alert during the disruptive process. Pretrial injection of the drug had no discernible effect on acquisition, but blocked memory tested 3 days later.[60] By training animals on day 1 of an experiment, storing them in home tanks for 3 days, and retesting them on day 4, both short-term and long-term memory could be measured. By varying the time after the last trial on day 1 in which puromycin is injected, we found that memory appears to be consolidated 1–3 hr following training with a number of agents including puromycin (170 μg), acetoxycycloheximide (0.2 μg),[61] actinomycin D (2.0 μg),[62] and electroconvulsive shock.[63] The antibiotics did not produce gross behavioral disturbance following their injection and animals demonstrated normal learning immediately thereafter. By retraining animals at various times after an antibiotic injected before training or immediately following the last trial on day 1, we have studied the decay of short-term memory.[64] This generally takes 2–3 days.

KCl is anomalous in both the consolidation time and its effect on short-term memory decay.[65] While protein synthesis is somewhat inhibited by its injection, KCl produces intense convulsions immediately following injection. Consolidation takes about 24 hr, while short-term decay appears complete in 6–8 hr.[65] If memory is first electrical and then chemical, it is strange that this convulsant agent can block memory at a time when agents purported to block the chemical stage of memory are no longer effective.

c. The Environmental Effect. In goldfish, the training task originally used required a 40 min period during which four blocks of five 1 min trials were administered. Each block of trials was separated by 5 min of rest in the apparatus. Puromycin given immediately following the twentieth trial appeared to obliterate all memory. Why did no memory become fixed during the training session? Thirty minutes after the last training trial, animals removed from their home storage tanks and injected with puromycin had significant amnesia on day 4. About one-half of the disruptible memory was already insusceptible to the drug. What then was different about 30 min post-trial in the home tank as opposed to about the same amount of time spent in the training apparatus? Davis *et al.*[64,65] showed that by maintaining fish in the shuttlebox for up to 3 hr following the twentieth trial, memory appeared to remain susceptible to puromycin, a time at which it would not ordinarily be at all susceptible. Removal of fish from the training environment appears to be prerequisite for the onset of fixation. We have postulated that a reasonable physiological mechanism for this trigger is the lowering of arousal.[66] This mechanism has apparent survival value and its genesis could be explained on an evolutionary basis. Similar environmental or "detention" effects have been observed.[67] Interference hypotheses[68] could also explain the environmental effect as follows. The training apparatus itself could have some aspects of the training situation and act as a reminder. The noxious elements of the puromycin injection when associated with the apparatus might produce learning in which the puromycin is a negative reinforcement. In experiments to examine this possibility, Davis and Klinger[64] subjected fish to the training environment for 25 min on day 2, a time at which no known disruptive agent can produce memory effects on day 4. Puromycin, acetoxycycloheximide, and actinomycin D were given immediately after the day 2 exposure to the training apparatus. No deficit was seen on day 4, supporting the concept that these agents are acting on consolidation of memory rather than as some nonspecific noxious stimulus. KCl however, given on day 2 in association with the training situation, produced a marked deficit on day 4. It might have some "interference" effects. It was further found that when animals were retested at some later time instead of day 4, the antimetabolites did produce some slight but significant losses in responding.[64] This apparent weakening of memory is under study.

d. Puromycin and Convulsions. Intracerebral puromycin injection has been shown to potentiate metrazol convulsions in the fish, while acetoxycycloheximide does not.[69] Puromycin aminonucleoside does not block protein synthesis nor does it block memory, but it does potentiate the metrazol convulsions. It therefore seems that this effect of puromycin and puromycin aminonucleoside are not related to the amnesic effects of protein blockers in goldfish. Acetoxycycloheximide does not protect against the puromycin potentiation, further indicating that peptidyl puromycin is not the convulsant in fish. In some ways, acetoxycycloheximide would seem to be a superior amnesic agent. On a per milligram basis it is about 1000 times more potent than puromycin in the fish and does not have puromycin's excitatory effect. While acetoxycycloheximide initially was shown to be amnesic in the fish and not in the mouse, it appears now that it is an effective agent in both animals. A major drawback is its current unavailability. Cycloheximide, a much weaker glutarimide derivative, is readily available and has been reported to be amnesic in the mouse.[70]

On the basis of agents thus far tested, we feel that formation of memory of shock avoidance in the fish requires normal protein synthesis. Actinomycin D, an RNA blocker, acts similarly. Arabinosyl cytosine (AraC), a DNA synthesis inhibitor, has no effect on acquisition or memory.[71] In fact, none of the agents tested block acquisition detectably.

e. The Rat. As yet, few studies on the effect of the antimetabolites have been reported in the rat. Effects of actinomycin D have been reported by Goldsmith[72] and by Meerson.[73] The use of various anticholinesterases in the rat have been examined by Deutsch.[74] Results of experiments with diisopropylfluorophosphate (DFP) are interpreted to mean there is an optimal level of acetylcholine in the brain to evoke memory. Thus, depending on the strength of the memory, DFP can strengthen a weak memory or depress a strong one. In general, it would appear that these agents are involved in the retrieval rather than storage of information in the brain.

f. Simple Systems. Horridge[75] showed that a leg position habit can be acquired in a headless cockroach preparation, mediated by the ventral ganglion. By attaching an electrode to the leg, the limb is given a mild electrical shock when it makes contact with a water solution. With increasing shocks, the limb tends to maintain itself so that the electrode does not touch the water. Training takes a few minutes and is retained for more than 1 hr. Brown and Noble have found that cycloheximide applied to the glanglion blocks protein synthesis[76] and also inhibits initial learning.[77] The drug does not appear to alter the motor aspects of the response. Whether this preparation has the equivalent of a short- and a long-term memory has yet to be established. The potential advantages of such a relatively simple system are great.

C. The Nature of Concomitant Biochemical Changes

All of these experiments taken together do not answer another sort of question. Are the biochemical changes seen in learning or the disruptive processes which affect memory reflecting actual changes in synaptic connections involved in relevant behavior, or are they simply epiphenomena of this process? If we accept that normal protein synthesis is required during the formation of memory, we still have not narrowed down the possibilities very much. Growth of new synaptic connections, induction of enzymes at the synapse, or a number of other protein-acquiring steps can be proposed. All of the above suggest that the changes in the brain during memory formation take place at the nerve endings. Ribosomes are reported to be absent from the axon hillock to the end of the axon, although some RNA and protein synthesis may occur in this region of the cell.[78] Ample evidence that protein is transported from the cell body to the axon is, however, also available.[79] On the postsynaptic side, in either axodendritic or axosomatic synapses, polyribosomes are nearby and could produce protein which would be immediately available at the synapse. The relevant connections in memory formation may take place in interneurons where the distance from the cell body to the synapse may be minute.

Changes in labeling of RNA and proteins and the action of blocking agents could both reflect an epiphenomenon of the informational process. There may be a memory fixing process which, like enzyme induction, requires *m*RNA and protein synthesis. This idea has been discussed elsewhere[2] and similar ones have been proposed.[33] The great appeal of the latter hypothesis for the biochemist is that it brings with it the hope that it will yield its secrets to present-day technical methodology.

D. Genetic Approaches

A great problem in studying biochemical correlates of memory is the individual variability. It may be genetic, or it might reflect prior experience of the individual experimental animal. Attempts to minimize genetic variability in mice have been explored by Bovet *et al.*[80] Strain differences in drug susceptibility appear to be extremely important. Our own laboratory has initiated experiments with *Poecilia formosa*, a naturally parthenogenetic fish.[81] A clone has been developed from a single individual with the idea of minimizing both biochemical and behavioral variability.[82]

Considerable interest has developed over the possible use of genetic mutants to systematically identify various component systems in behavioral processes. Behavioral mutants of *Drosophila* have been investigated.[83] The use of a rapidly reproducing nematode has also attracted considerable interest. Simple, reproducible assays of memory in nonchordates have not yet been used widely.

In summary, biochemical correlates of the formation of long-term memory are inferred from changes in brain metabolism as well as in the time relationships and nature of the effects of specific metabolic blocking of agents. Whether the relevant biochemical processes contain specific behavioral information or are epiphenomena of the information storing process is unknown.

III. REFERENCES

1. R. Semon, "Die Mneme als erhaltendes Prinzip in Wechsel des organischen Geschehens," Wilhelm Englemann, Leipzig (1904).
2. B. W. Agranoff, *in The Neurosciences: A Study Program* (G. C. Quarton, T. Melnechuk, and F. O. Schmitt, eds,) pp. 756–764, The Rockefeller University Press, New York (1967).

3. S. Ramon Y. Cajal, *Histologie du Systeme Nerveux de l'Homme et des Vertebres*, Vol. 2, Maloine, Paris (1909).
4. R. M. Gaze, Growth and differentiation, *Ann. Rev. Physiol.* **29**, 59–86 (1967).
5. H. Hydén, *The Neuron*, pp. 179–217, Elsevier, Amsterdam (1967).
6. C. S. Sherrington, *The Integrative Action of the Nervous System*, 2nd ed., Cambridge University Press, Cambridge (1947).
7. P. G. Nelson, *in The Neurosciences: A Study Program* (G. C. Quarton, T. Melnechuk and F. O. Schmitt, eds.), pp. 772–777, The Rockefeller University Press, New York (1967).
8. A. O. D. Willows, Behavioral acts elicited by stimulation of single, identifiable brain cues, *Science* **157**:570–574 (1967).
9. K. S. Lashley, *Brain Mechanisms and Intelligence: A Quantitative Study of Injuries to the Brain*, Chicago University Press, Chicago (1929).
10. E. R. Kandel and W. A. Spencer, Cellular neurophysiological approaches in the study of learning, *Physiol. Rev.* **48**:65–137 (1968).
11. J. Olds and P. Milner, Positive reinforcement produced by electrical stimulation of septal area and other regions of rat brain, *J. Comp. Physiol. Psychol.* **47**:419–427 (1954).
12. J. Z. Young, *A Model of the Brain*, Oxford University Press (Clarendon), London (1964).
13. W. Penfield, The interpretive cortex, *Science* **129**, 1719–1725 (1959).
14. B. Milner, *in Cognitive Process and the Brain* (P. Milner and S. Glickman, eds.), pp. 97, Van Nostrand, New York (1965).
15. M. Sidman, L. T. Stoddard, and J. P. Mohr, Some additional quantitative observations of immediate memory in a patient with bilateral hippocampal lesions, *Neuropsychologia* **6**:245–254 (1968).
16. M. P. Santacava and J. Delacour, Effets, chez le rat, des lesions de la zona incerta et des corps mamillaires, sur un conditionnement defensif, *Neuropsychologia* **6**:115–124 (1968).
17. M. R. Rosenzweig and A. L. Leiman, Brain functions, *Ann. Rev. Psychol.* **55**:98 (1968).
18. M. R. Rosenzweig, E. L. Bennett, and M. C. Diamond, Transitory components of cerebral changes induced by experience, *Proc. 75th Ann. Conv. Amer. Psychol. Assoc.* **2**:105–106 (1967).
19. J. Altman, R. B. Wallace, W. J. Anderson, and G. P. Das, Behaviorally induced changes in length of cerebrum in rats, *Develop. Psychobiol.* **1**:112–117 (1968).
20. J. Altman and G. D. Das, Autoradiographic and histological evidence of postnatal hippocampal neurogenesis in rats, *J. Comp. Neurol.* **124**:319–336 (1965).
21. M. B. Bunge, R. P. Bunge, and E. R. Peterson, The onset of synapse formation in spinal cord cultures as studied by electron microscopy, *Brain Res.* **6**:728–749, (1967).
22. J. C. Eccles, *in The Physiology of Synapses*, pp. 256, Academic Press, New York (1964).
23. E. R. Kandel and L. Tauc, Heterosynaptic facilitation in neurones of the abdominal ganglion of *Aplysia depilans*, *J. Physiol.* (*London*) **181**:1–27 (1965).
24. F. Morrell, *in The Neurosciences: A Study Program* (G. C. Quarton, T. Melnechuk, and F. O. Schmitt, eds.), pp. 452–469, The Rockefeller University Press, New York (1967).
25. E. R. John, *in The Neurosciences: A Study Program* (G. C. Quarton, T. Melnechuk, and F. O. Schmitt, eds.), pp. 690–704, The Rockefeller University Press, New York (1967).

26. P. T. Chopping, Holographic model of temporal recall, *Nature* **217**:781–782 (1968).
27. H. Hydén and P. W. Lange, A differentiation in RNA response in neurons early and late during learning, *Proc. Natl. Acad. Sci.* **53**:946–952 (1965).
28. H. Hydén and P. W. Lange, Protein synthesis in the hippocampal pyramidal cells of rats during a behavioral test, *Science* **159**:1370–1373 (1968).
29. S. Bogoch, *The Biochemistry of Memory*, Oxford University Press, New York (1968).
30. V. E. Shashoua, RNA changes in goldfish brain during learning, *Nature* **217**:238–240 (1968).
31. J. W. Zemp, J. E. Wilson, and E. Glassman, Brain function and macromolecules, II. Site of increased labeling of RNA in brains of mice during a short-term training experience, *Proc. Natl. Acad. Sci.* **58**:1120–1125 (1967).
32. L. B. Adair, J. E. Wilson, J. W. Zemp, and E. Glassman, Brain function and macromolecules, III. Uridine incorporation into polysomes of mouse brain during short-term avoidance conditioning. *Proc. Natl. Acad. Sci.* **61**:606–613 (1968).
33. L. B. Adair, J. E. Wilson, and E. Glassman, Brain function and macromolecules, IV. Uridine incorporation into polysomes of mouse brain during different behavioral experiences, *Proc. Natl. Acad. Sci.* **61**:917–922 (1968).
34. N. S. Radin, F. B. Martin, and J. R. Brown, Galactolipide metabolism, *J. Biol. Chem.* **224**:499–507 (1957).
35. R. W. Gerard, *in Brain Mechanisms and Learning* (J. F. Delafresnaye, A. Fessard, R. W. Gerard, and J. Konorski, eds.), pp. 21–35, Charles C Thomas, Springfield, Ill. (1961).
36. D. O. Hebb, *The Organization of Behavior*, Wiley, New York (1949).
37. T. Ribot, *Disorders of Memory*, Kegan, Paul, Trench and Co., London (1885).
38. S. E. Glickman, Perserverate neural processes and consolidation of the memory trace, *Psychol. Bull.* **58**:218–233 (1961).
39. A. W. Melton, Implications of short-term memory for a general theory of memory, *J. Verbal Learn. Verbal Behav.* **2**:1–21 (1963).
40. J. Garcia and R. A. Koelling, Relation of cue to consequence in avoidance learning, *Psychon. Sci.* **4**, 123 (1966).
41. C. A. Pearlman, S. K. Sharpless, and M. E. Jarvik, Retrograde amnesia produced by anesthetic and convulsant agents, *J. Comp. Physiol. Psychol.* **54**:109 (1961).
42. W. Dingman and M. B. Sporn, The incorporation of 8-azaguanine into rat brain RNA and its effect on maze-learning by the rat: An inquiry into the biochemical basis of memory, *J. Psychiat. Res.* **1**:1–11 (1961).
43. T. J. Chamberlain, G. H. Rothschild, and R. W. Gerard, Drugs affecting RNA and learning, *Proc. Natl. Acad. Sci.* **49**:918–924 (1963).
44. J. B. Flexner, L. B. Flexner, E. Stellar, G. de la Haba, and R. B. Roberts, Inhibition of protein synthesis in brain and learning and memory following puromycin, *J. Neurochem.* **9**:595–605 (1962).
45. J. B. Flexner, L. B. Flexner, and E. Stellar, Memory in mice as affected by intracerebral puromycin, *Science* **141**:57–59 (1963).
46. L. B. Flexner and J. B. Flexner, Effect of acetoxycycloheximide and of an acetoxycycloheximide-puromycin mixture on cerebral protein synthesis and memory in mice, *Proc. Natl. Acad. Sci.* **55**:369–374 (1966).
47. L. B. Flexner, J. B. Flexner, and R. B. Roberts, Memory in mice analyzed with antibiotics, *Science* **155**:1377–1384 (1967).
48. L. B. Flexner, J. B. Flexner, G. de la Haba, and R. B. Roberts, Loss of memory as related to inhibition of cerebral protein synthesis, *J. Neurochem.* **12**:535–541 (1965).

49. L. B. Flexner, J. B. Flexner, and R. B. Roberts, Stages of memory in mice treated with acetoxycycloheximide before or immediately after learning, *Proc. Natl. Acad. Sci.* **56**:730–735 (1966).
50. P. Gambetti, N. Gonatas, and L. B. Flexner, Puromycin: Action on neuronal mitochondria, *Science* **161**:900–902 (1968).
51. J. B. Flexner and L. B. Flexner, Restoration of expression of memory lost after treatment with puromycin, *Proc. Natl. Acad. Sci.* **57**:1651–1654 (1967).
52. L. B. Flexner and J. B. Flexner, Intracerebral saline: Effect on memory of trained mice treated with puromycin, *Science* **159**:330–331 (1968).
53. L. B. Flexner and J. B. Flexner, Studies on memory: The long survival of peptidyl-puromycin in mouse brain, *Proc. Natl. Acad. Sci.* **60**:923–927 (1968).
54. S. H. Barondes and H. D. Cohen, Puromycin effect on successive phases of memory storage, *Science* **151**:594–595 (1966).
55. S. H. Barondes and H. D. Cohen, Comparative effects of cycloheximide and puromycin on cerebral protein synthesis and consolidation of memory, *Brain Res.* **4**:44–51 (1967).
56. H. D. Cohen and S. H. Barondes, Puromycin effect on memory may be due to occult seizures, *Science* **157**:333–334 (1967).
57. S. H. Barondes and H. D. Cohen, Delayed and sustained effects of acetoxycycloheximide on memory in mice, *Proc. Natl. Acad. Sci.* **58**:157–164 (1967).
58. B. W. Agranoff and P. D. Klinger, Puromycin effect on memory fixation in the goldfish, *Science* **146**:952–953 (1964).
59. A. Potts and M. E. Bitterman, Puromycin and retention in the goldfish, *Science* **158**:1594–1596 (1967).
60. B. W. Agranoff, R. E. Davis and J. J. Brink, Memory fixation in the goldfish, *Proc. Natl. Acad. Sci.* **54**:788–793 (1965).
61. B. W. Agranoff, R. Davis, and J. J. Brink, Chemical studies on memory fixation in goldfish, *Brain Res.* **1**:303–309 (1966).
62. B. W. Agranoff, R. E. Davis, L. Casola, and R. Lim, Actinomycin D blocks formation of memory of shock-avoidance in goldfish, *Science* **158**:1600–1601 (1967).
63. R. E. Davis, P. J. Bright, and B. W. Agranoff, Effect of ECS and puromycin on memory in fish, *J. Comp. Physiol. Psychol.* **60**:162–166 (1965).
64. R. E. Davis and B. W. Agranoff, Stages of memory formation in goldfish: Evidence for an environmental trigger, *Proc. Natl. Acad. Sci.* **55**:555–559 (1966).
65. R. E. Davis and P. D. Klinger, Environmental control of amnesic effects of various agents in goldfish, *Physiol. Behav.* **4**:385–387 (1969).
66. B. W. Agranoff and R. E. Davis, *in Physiological and Biochemical Aspects of Nervous Integration* (F. D. Carlson, ed.), pp. 309–327, Prentice-Hall, Inc., Englewood Cliffs, N.J. (1968).
67. F. Robustelli, A. Geller, and M. E. Jarvik, Detention, electroconvulsive shock, and amnesia, *Proc. 76th Ann. Conv. Amer. Psychol. Assoc.* **3**:331–332 (1968).
68. D. J. Lewis and B. A. Maher, Neural consolidation and electroconvulsive shock, *Psychol. Rev.* **72**:225–239 (1965).
69. B. W. Agranoff, *in Protein Metabolism of the Nervous System* (A. Lajtha, ed.) (in press.)
70. S. H. Barondes and H. D. Cohen, Arousal and the conversion of "short-term" to "long-term" memory, *Proc. Natl. Acad. Sci.* **61**:923–929 (1968).
71. L. Casola, R. Lim, R. E. Davis, and B. W. Agranoff, Behavioral and biochemical effects of intracranial injection of cytosine arabinoside in goldfish, *Proc. Natl. Acad. Sci.* **60**:1389–1395 (1968).

72. L. J. Goldsmith, Effect of intracerebral actinomycin D and of electro-convulsive shock on passive avoidance, *J. Comp. Physiol. Psychol.* **63**:126–132 (1967).
73. F. Z. Meerson, R. I. Kruglikov, and I. A. Goryacheva, The effect of actinomycin 2703 on the formation of conditioned reflexes and the transition of short-term memory to long term memory, *USSR Acad. Sci. Doklady* **170**:741–744 (1966).
74. J. A. Deutsch, M. D. Hamburg, and H. Dahl, Anticholinesterase-induced amnesia and its temporal aspects, *Science* **151**:221–223 (1966).
75. G. A. Horridge, Learning leg position by the ventral nerve cord in headless insects, *Proc. Roy. Soc. Edinb. B* **157**:33–52 (1962).
76. B. M. Brown and E. P. Noble, Cycloheximide amino acid incorporation and learning in isolated cockroach ganglion, *Biochem. Pharmacol.* **17**:2371–2374 (1968).
77. B. M. Brown and E. P. Noble, Cycloheximide and learning in the isolated cockroach ganglion, *Brain Res.* **6**:363–366 (1967).
78. E. Koenig, Synthetic mechanisms in the axon.—IV. *In vitro* incorporation of [^{3}H] precursors into axonal protein and RNA, *J. Neurochem.* **14**:437–446 (1967).
79. J. J. Bray and L. Austin, Axoplasmic transport of ^{14}C proteins at two rates in chicken sciatic nerve, *Brain Res.* **12**:230–233 (1969).
80. D. Bovet, F. Bovet-Nitti, and A. Oliverio, Genetic aspects of learning and memory in mice, *Science* **163**:139–149 (1969).
81. B. W. Agranoff and R. E. Davis, *in The Central Nervous System and Fish Behavior* (D. Ingle, ed.), pp. 193–201, University of Chicago Press, Chicago (1968).
82. Available through R. Gossington, Rte. 2, Lake Worth, Florida 33460.
83. S. Benzer, Behavioral mutants of *Drosophila* isolated by countercurrent distribution *Proc. Natl. Acad. Sci.* **58**:1112–1119 (1967).

Chapter 7

DO SPECIFIC BIOCHEMICAL CORRELATES TO LEARNING PROCESSES EXIST IN BRAIN CELLS?*

Holger Hydén and Paul W. Lange

Institute of Neurobiology
Faculty of Medicine, University of Göteborg
Göteborg, Sweden

I. INTRODUCTION

The present paper will deal mainly with the biochemical changes observed in neurons during three different learning experiments. First, however, some experiments showing differences in the protein composition of nerve and glia cells will be described. Of the learning experiments, the first is a case of instrumental learning in rats, in which changes in the synthesis of three acidic neuronal proteins and in the RNA base composition of neurons occurred; the arguments that these changes are specifically related to the training and that they are an expression of increased gene activation will be presented. In the next study, the protein changes observed in brain cells during simple sensory conditioning in rats will be described, and it will be argued that these are due to an increased level of attention rather than to learning per se.

A. Neuronal and Glial Proteins

Five years ago, Moore and collaborators[1,2] described a brain-specific protein, called S100, because it is soluble in saturated ammonium sulfate. It is an acidic protein, has a molecular weight of 21,000, constitutes 0.1 % of the brain proteins and moves close to the anodal front in electrophoresis. It develops after 12 days postnatally in the rat and is present only in nervous tissue. Thirty per one-hundred moles of its amino

* The main content of this paper has also been published in *Proceed. Sympos. on Biology of Memory*, Sept. 1, 1969 (Tihany, G. Adám, ed.), Akadémiai Kiadó, the publishing house of the Hungarian Academy of Sciences, Budapest.

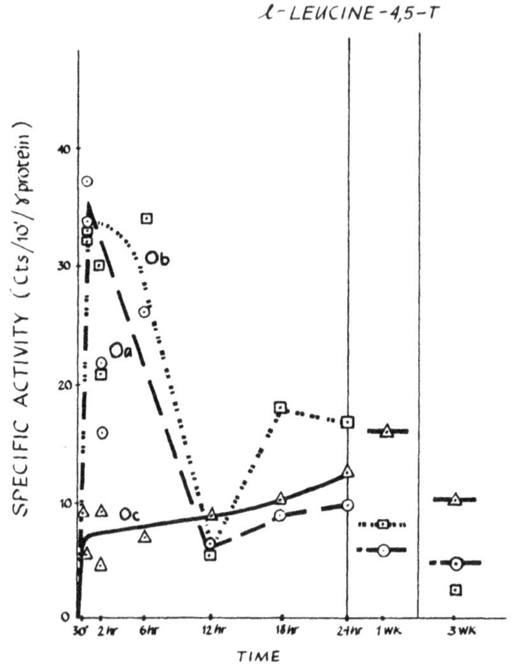

Fig. 1. Specific radioactivity of bands Oa, Ob, and Oc separated on 11.2% polyacrylamide gels as a function of time between isotope injection and sacrifice. Radioactivity was determined after combustion of slices of the polyacrylamide gels by liquid scintillation counting. Isotope: *l*-leucine-4,5 T.

acids are acidic. It contains 30% glutamic and aspartic acid. S100 can be further separated into at least three fractions, of which two have a high turnover and react immunologically with antiserum against S100.[3] The S100 protein is not linked to carbohydrates (Fig. 1).

Hydén and McEwen[4] have shown by antiserum precipitation reactions supported by the Coons technique[5] that S100 is mainly a glial protein which in nerve cells is found only in the nuclei. Recently, Benda and collaborators[6] confirmed its presence in glia and showed its ten-fold growth in a clonal strain of glial tumors. Perez and Moore[7] have also presented evidence that S100 is mainly a glial protein. Moore and Perez[8,9] have described another brain-specific protein which seems to be localized exclusively to the nerve cells and which has been named the 14-3-2 protein.

There is evidence for the existence of still other brain-specific soluble proteins. MacPherson[10] has described one in the β-globulin range, Kosinski[11] has described five soluble proteins, and Warecka and Bauer[12,13] recently described an α-glycoprotein rich in neuraminic acid, which develops 3 months after birth in man and is probably derived from glia. Bennett and Edelman[14] have purified and characterized still another acidic brain-specific protein.

B. An Immunological Study of Deiters' Nucleus

We have examined the properties of antibodies prepared against neurons and glia[15] obtained from Deiters' nucleus in the continuing attempt to identify brain-specific proteins in them. The antigens in brain cells presumably number in the order of hundreds; Huneeus-Cox[16,17] for instance, successfully prepared antisera against 11 antigens in preparations of squid axoplasm that did not include the external membranes. In our study antigen consisted of glial material dissected from the Deiters' nucleus of the rabbit by the freehand technique previously described.[18] The dissection was carried out at 4°C, with careful removal of capillaries and nerve cell bodies and processes; in this way, 3.2 mg of Deiters' nucleus glia was collected from 40 rabbits. The other antigen consisted of 1.3 g of whole Deiters' nucleus, containing both neurons and glia, dissected from 100 rabbits.

Each of these antigens was homogenized and mixed with both complete and incomplete Freund's adjuvant. A group of six rhesus monkeys weighing 3–3.5 kg was injected intramuscularly with 0.6 ml of one or the other emulsion once a week for 4 weeks. None ever showed neurological symptoms, or signs of tuberculosis. The animals were bled after 1 week. On day 44 each monkey received a booster injection of 0.2 ml of its antigen emulsion precipitated with $Al_2(SO_4)_3$, and was bled 1 week later. These sera were tested on Ouchterlony plates against extracts of glia and of Deiters' nucleus, and their precipitation activities against sucrose-Triton X-100 extracts of both glia and of Deiters' nucleus material were also evaluated. In addition, the micromethod for double diffusion in one dimension in glass capillaries previously described[4] was used as an assay system, the Coons[5] multiple layer indirect method for immunofluorescence applied to cryostat sections through the Deiters' nucleus, with evaluation of the specific fluorescence appearing in the nerve and glia cells was also used. Some samples of the antisera were absorbed in two or three steps with sucrose-Triton X-100 homogenates of glia and of rabbit spleen, while others were twice absorbed with rabbit spleen and then absorbed with glia.

Tables I to IV summarize some results of these studies. Both the anti-Deiters' nucleus and the antiglia sera formed well-defined precipitates with microgram per microliter amounts of their respective antigens (Table I). Table II shows that the antiglia serum formed precipitates with the glia but not with nerve cells obtained from Deiters' nucleus, and that no precipitates formed when normal rabbit serum was used against these antigens.

Table III shows the results of an antigen dilution study: homogenates of isolated nerve cells and of the same volumes of glial cells were tested against the antiglia antiserum in the dilution 1:512. Even when 300 isolated nerve cells were used, no precipitation was obtained, but glial homogenates gave well-defined precipitates.

Precipitates were obtained when the anti-Deiters' glia antiserum was tested against glia dissected from other parts of the brain, e.g., from the hypoglossal nucleus and from the spinal cord and cerebral cortex, but none appeared against homogenates of motor neurons, pyramidal nerve cells of the hippocampus and granular cells from the cerebellum, all containing from 3.5 to 0.01 μg of protein per microliter.

Antiserum against the whole Deiters' nucleus gave two precipitation lines with both glia and nerve cells as antigens. However, when this antiserum was absorbed with glia or with spleen, only the nerve cell homogenates gave precipitates (Table IV).

TABLE IA

Gel Precipitation Reactions (+) Between Antinucleus Deiters' Antiserum (1:512) and a Homogenate of Deiters' Nucleus

Antigen (μg/μl)	8.20	4.10	2.10	1.00	0.80	0.50	0.40	0.30	0.20	0.10	0.05	0.02	0.01
Reaction	–	–	–	–	+	+	+	+	+	+	–	–	–

TABLE IB

Gel Precipitation Reactions (+) Between Anti-Deiters' Glia Antiserum (1:512) and an Antigen Homogenate of Deiters' Glia

Antigen (μg/μl)	0.67	0.60	0.16	0.08	0.04	0.02
Reaction	+	+	+	–	–	–

TABLE II

Gel Precipitation Reactions (+) Between Anti-Deiters' Glia Antiserum and 0.9 μg of Protein Extracted from Nerve and Glia Cells Dissected from Deiters' Nucleus

	Protein from:	
Antiserum dilution	Nerve cell	Glia cell
1:64	–	+
1:128	–	+
1:256	–	+
1:512	–	+
1:1024	–	–
1:2048	–	–

[a] Normal serum controls negative in each case.

TABLE III

Precipitation Reaction Between Anti-Deiters' Glia Antiserum (1:512) and Homogenates of Deiters' Nerve Cells and Corresponding Volumes of Glia

	Deiters' neurons		Deiters' glia (same volume as nerve cells)	
Number of nerve cells	Calc. protein in 10^{-6} g	Ppt.	Calc. protein in 10^{-6} g	Ppt.
300	3.6	–	1.8	+2 ppt
150	1.8	–	0.9	+2 ppt
70	0.9	–	0.45	+2 ppt
60	0.72	–	0.36	+1 ppt
30	0.36	–	0.18	+1 ppt
15	0.18	–	0.09	–
6	0.09	–	0.045	–
3	0.045	–	0.022	–

[a] Neuronal protein estimates based on 12,000 $\mu\mu$g of protein per cell. Glial protein per unit volume estimated at 50% neuronal.

The results with the fluorescence technique matched those obtained with the immunodiffusion technique as summarized in these tables. Experiments were carried out according to the multiple layer method of Coons.[5,19] Five micron-thick cryostat sections through the lateral vestibular nucleus were first dried (sometimes left overnight in the refrigerator at +4°C) and subsequently fixed in cold acetone for 30 sec. After being washed for 5 min in the buffered saline, the sections were covered with the antiserum to be investigated for 30 min. After thorough washing (3 × 5 min) in a cold *p*H 7.1 phosphate-buffered saline, a goat–antimonkey globulin-γ-globulin conjugated with fluorescein isothiocyanate (Difco product) was applied to the sections for 30 min and the excess removed by repeated washing (again 3 × 5 min) in the buffer. Control sections heated with normal monkey serum and with conjugated γ-globulin only were regularly used with each experimental series. The sections were finally mounted in a small drop of buffered glycerol (9 parts of glycerol, 1 part of buffered saline) under a cover slip and immediately observed in a Zeiss fluorescence microscope. After the photographs were taken using high speed Ektachrome film with exposures varying from 1 to 5 sec, the sections were restained with Ehrlich's hematoxylin–eosin. The fields previously photographed were identified under the light microscope, and rephotographed in black and white, thus enabling comparison of conventional microscopical appearances of the structural details with fluorescent pictures.

TABLE IV

Number of Precipitation Lines After Absorption of Anti-Deiters' Nucleus Antiserum[a]

	Protein from	
	Nerve cells	Glia cells
Unabsorbed	2	2
Absorbed with glia	1	0
Absorbed with spleen	1	0

[a] Antigen: homogenates from 120 isolated nerve cells and corresponding amount of glia containing 1.6 μg used in each case. All dilutions tested (1:2, 1:4, 1:8, 1:16) gave the same result.

Antiserum to whole Deiters' nucleus when absorbed with glia, or with spleen, or with both, gave no fluorescence in glial cells, but did so in nerve cells; this fluorescence was localized to the outer rim of the cell body and to the dendritic processes, which could be traced through the section by their brilliant fluorescence, suggesting that the antigens were localized in the plasma membranes. The reaction was positive furthermore in the nerve cell nucleus, but not at the site of the nucleolus.

From these observations the following conclusions can be made. Neurons and glia differ with respect to antigen composition. This is an interesting finding from the point of view that both types of cells develop from the same type of ectodermal stem cell. The question is then whether the antigens are specific for the type of cell in which they occur. Judged by the absorption experiments, the neuronal antigens seem to be specific for that type of cell. It should be noted that the neuron-specific antigens were concentrated to the processes, to the outermost part of the cell and to the nucleus, and especially to the nuclear membrane.

It seemed clear, however, that the antigens in the glia were not glia-specific. They were localized all over in the cell body, but not in the nucleus. If the immunological organ specificity is considered, it seems to be due to the presence of antigens in the neurons.

On the other hand, glial cells possess protein which is confined only to the nerve tissue and which they share with neurons, namely the acidic S100 protein. The presence of this antigen cannot be demonstrated by the method used in this study to prepare immune sera.[20]

II. ALTERED PROTEIN SYNTHESIS DURING TRAINING—HANDEDNESS TRANSFER

A. Incorporation of ^{3}H-leucine into the Acidic Protein Fractions 4 and 5

If a narrow glass tube is arranged a few centimeters from the floor, filled in its lower third with protein pills (4 mm in diameter), and slightly tilted downward at its lower end, rats will reach down into the tube to retrieve the pills, one by one. They generally use either the left or right hand as they perform this task, and they can be induced to transfer this handedness.(21) When tested in free-choice reachings, all the rats in the present study showed clear preference for the left or right hand in 23 out of 25 reaches. A wall was then placed parallel to the glass tube so as to prevent use of the preferred paw; the rats began to retrieve the food pills with the nonpreferred paw. When given two training periods of 25 min per day in this situation, their performance, measured as the number of successful reaches per 25 min, increased linearly up to day 8. Performance curves were obtained on all rats used in our experiments and were similar to that shown in Fig. 2, which demonstrates the performance curve of a separate set of 12 rats during 16 days.

Once learned, this new behavior is retained for a long time. Since no stress (surgical, mechanical, or shock) is applied to induce the new behavior, this procedure has distinct advantages over other behavioral experiments used in rats.

To trace protein synthesis during this learning the rats, under fluothane anesthesia, received 60 μC if ^{3}H-leucine in 60 μl intraventricularly in both hemispheres half an hour before their final training period. Hippocampal nerve cell samples were then taken for analysis 15 min after the last training period. Nerve cells of the hippocampus were selected because (1) several

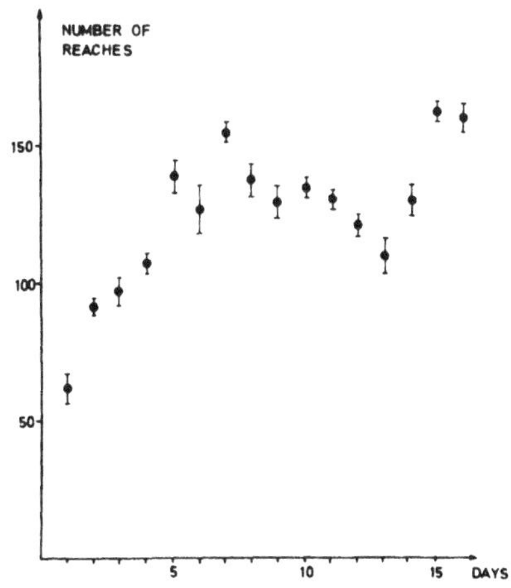

Fig. 2. Performance at reversal of handedness as the average number of successful reaches for 10 rats trained 2 × 25 min per day during 16 days.

clinical and behavioral studies have shown the importance of this structure for the formation of long-term memory (see, e.g., Refs. 22–26); (2) its bilateral destruction results in severe defects in learning and information of memory[22,23] (3) during attentative learning, impedance changes occur in the hippocampus;[26] and (4) no memory is formed if protein synthesis in the hippocampus is inhibited by 90%.[27,28]

The micromethod used for protein analysis was as follows. About 300 pyramidal cells from the CA3 region of the hippocampus, separately dissected out freehand on a cooling table, were analyzed for protein by a technique already described.[29]* An outline of this procedure is given in Fig. 3. The left side of the scheme gives the various steps leading to the value of the specific activities per amount of protein in each protein microfraction.

Since these specific activity values vary, because of variation in the local concentration of ^{3}H-leucine, the correction procedure shown on the right side of Fig. 3 was applied in order to allow a comparison of values from identical parts of both hemispheres or from different animals. This was accomplished in a separate experiment, where the relation between the uncorrected specific activities and the concentration of the free ^{3}H-leucine in the hippocampal nerve cells was determined and found to be linear. Dividing the specific activity values obtained by the values of the ^{3}H-leucine concentration determined locally allowed all specific activities to be compared at uniform free leucine-^{3}H concentration.

In an earlier study,[30] the incorporation of ^{3}H-leucine in the CA3 nerve cell protein fractions 4 and 5 (Fig. 4) was evaluated on the fifth day of training, i.e., on the linear, increasing part of the performance curve. The specific activities of these protein fractions were significantly greater in trained rats compared to control rats of the same age ($P < 0.005$), and there was some evidence for higher incorporation in the hippocampus contralateral to the training paw.

Protein fractions 4 and 5 presumably each contain several species of proteins, and there is no reason as yet to believe that the qualitative characteristics of the protein formed during training is specific for the process since no data as to the composition of these proteins exist. Nevertheless, it is pertinent to ask whether the increased synthesis of fractions 4 and 5 is specific for the training.

This we attempted to do in the present study by measuring fractions 4 and 5 in rats given 5, 7, and 10 training sessions according to the following schedule. A group of 24 rats was given 5 days of training. Five of these (Group I) received ^{3}H-leucine prior to the last training session and the

* One may ask why it seems necessary to struggle with such minute amounts of material and with the dissection of such small areas within the brain. As an answer, we would like to advocate the view that altered synthesis, if any, is more likely to be found in a uniform cell population from an area that is clearly involved functionally. In a mixed cell population from a whole brain such changes easily disappear in the background noise.

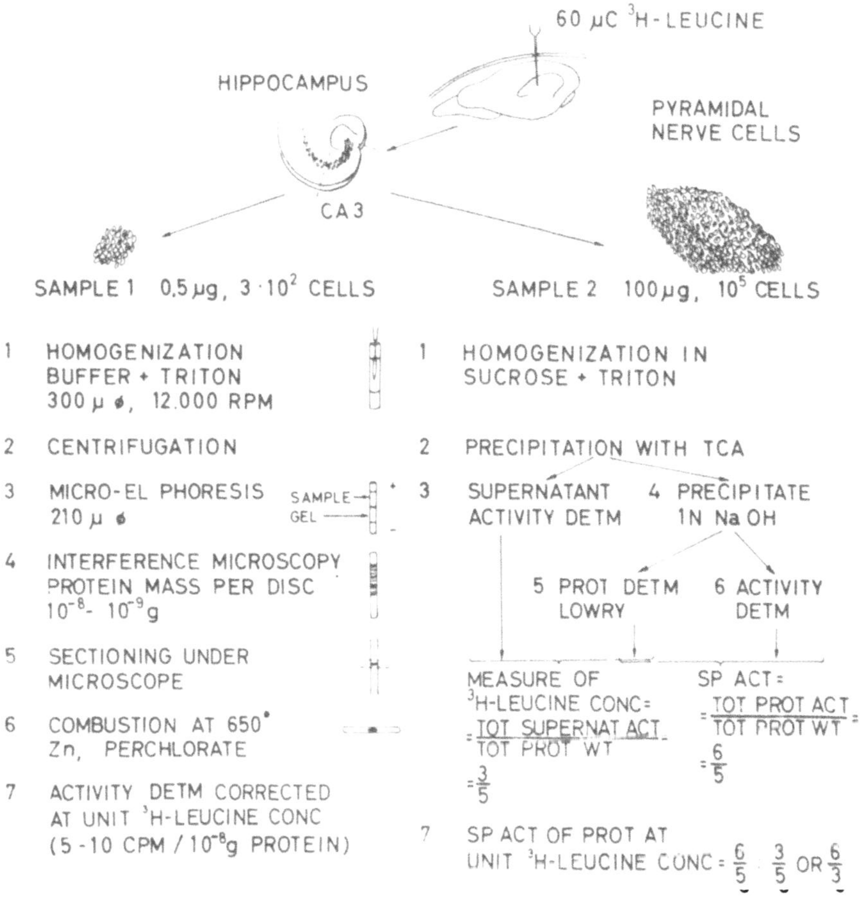

Fig. 3. Outline of the microdisc electrophoresis procedure for separation of 10^{-7} to 10^{-9} g of protein and evaluation of incorporation of radioactive amino acid into the individual fractions. Volume of sample is proportional to weight of total protein in sample.

CA3 hippocampal nerve cell material was taken for analysis as described above. The remaining animals were placed in cages, and given food and water ad lib. After 14 days, they were all subjected to two training periods of 25 min each; five of these animals (Group II) were given ^{3}H-leucine, and the CA3 nerve cell material was taken for analysis. The remaining rats (Group III) were returned to their cages for 14 additional days, then trained for 3 days with two training periods per day (each of 25 min), and, after ^{3}H-leucine injection, their hippocampal brain cells were taken for analysis. The controls were untrained rats of the same age of which 50% were littermates of the experimental animals.

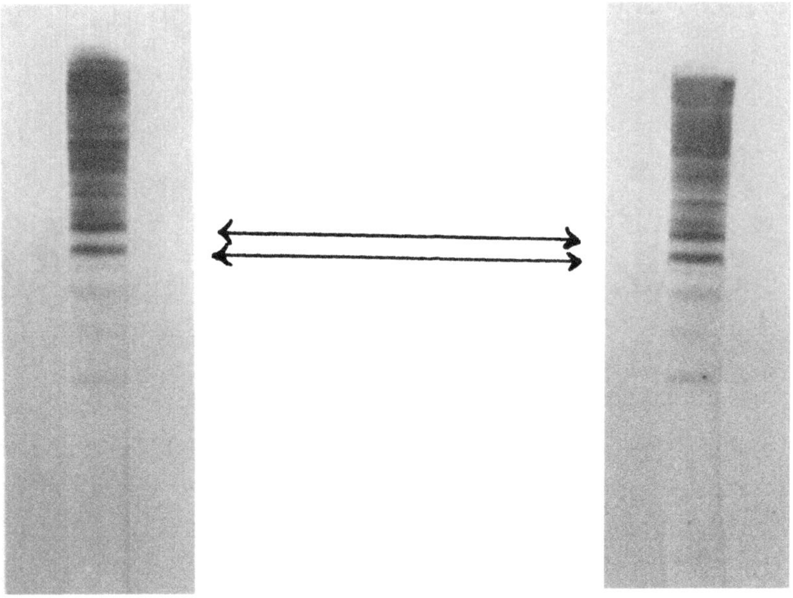

Fig. 4. Protein of pyramidal nerve cells of the hippocampus, CA3 region, separated on 400 μ diameter polyacrylamide gels, and stained with amido black. Fractions 4 and 5 from the anodal front are indicated by arrows.

The performance of the rats in the three groups is shown in Table V. Table VI demonstrates that the specific activities of protein fractions 4 and 5 were significantly increased after 5 and 7 training days but *not* after 10. The corrected specific activities (counts per minute per microgram) of protein fractions 4 and 5 differ from the corresponding values in a paper recently published[30]; this is due to a more refined separation technique which allowed a better separation of smaller amounts of the protein sample. The values found for the unseparated protein were, of course, not affected. The unseparated protein of the CA3 pyramidal nerve cells behaved like that

TABLE V

Average Number of Successful Reaches per Day for Rats Using the Nonpreferred Paw to Retrieve Food Pills from a Narrow Glass Tube

	Number of rats	Reaches
Group I (performance on day 5)	24	100
Group II (performance on day 14)	19	90
Group III (performance on day 30)	14	90

TABLE VI

Corrected Specific Activities of Hippocampal CA3 Nerve Cell Proteins, Both the Unseparated and Fractions 4 and 5[a]

	Fractions 4 and 5			Unseparated protein	
	Number of rats	Number of gels	Corrected specific activity	Number of samples	Corrected specific activity
Group I (training 5 days)	5	10	3.3 ± 0.40	10	14.20 ± 1.90
Group II (resumed training day 14)	5	10	3.9 ± 0.48	10	15.50 ± 1.90
Group IIA (half training time)	2	5	13.0 ± 0.60		
Group III (resumed training day 30)	14	35	1.8 ± 0.17	28	5.10 ± 0.58
Control	10	24	1.5 ± 0.16	20	6.00 ± 0.92

[a] Corrected specific activity refers to counts per minute per microgram ± standard error of the mean.

of fractions 4 and 5 in showing higher incorporation values in Groups I and II, but not in Group III (Table VI).

Following a chance observation, we made a study of the incorporation of ^{3}H-leucine by two rats which were subjected to half the initial training time allowed the other rats. Whereas Group I had 10 training sessions (2 × 25 min × 5 days), these 2 rats had 5 (2 × 25 min for 2 days, 1 × 25 minutes on day 3), and on day 14, they were given only a single 25-min training session before being killed for analysis. On training day 3 they had made 120 reaches, and in the final session 100 reaches. These animals, Group IIA in Table VI, gave a greater protein synthesis response than those receiving the longer training.

Since the incorporation of ^{3}H-leucine into the 4 and 5 protein fractions increased significantly during the training, it became important to know the relation of these fractions to the rest of the nerve cell protein in terms of ^{3}H-leucine incorporation. Protein of the CA3 nerve cells from Group I rats (trained for 5 days) was therefore separated on polyacrylamide gels, divided in four parts, and the radioactivity was determined in each part; as can be seen from Fig. 5, the radioactivity of protein fractions 4 and 5 is relatively high.

B. Increased Synthesis of S100 Protein

Both the electrophoretic pattern of the soluble CA3 nerve cell protein isolated in the experiment just described and micro-densitometer recordings (Fig. 7) made of 75 protein separations stained with brilliant blue showed

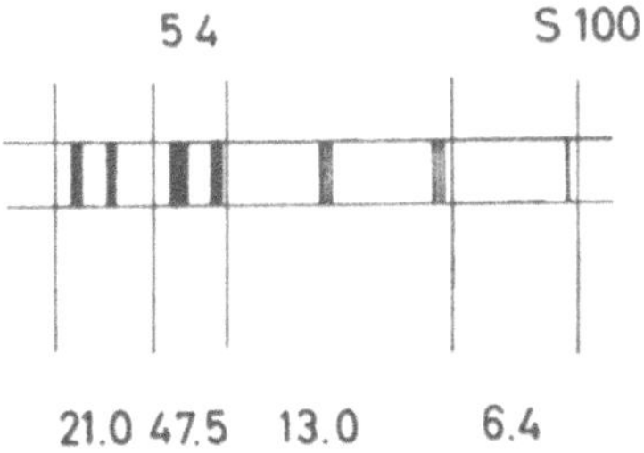

Fig. 5. Gels 400 μ in diameter, containing separated ^{3}H-labeled pyramidal nerve cell protein from the CA3 region, were cut in four pieces and the radioactivity was determined as counts per minute after combustion. Note that the radioactivity in protein fractions 4 and 5 is relatively great.

two protein bands at the front in the trained rats compared to only one in the controls (Figs. 6, 7, Table VII).

This protein fraction of the controls gave a positive immunological reaction when treated with antiserum against the S100 protein. Figure 7 shows that the amount of protein contained in the two anodal bands of trained rats was greater than the amount of protein contained in the one band of controls. Furthermore, when gel cylinders from experimental rats with two anodal front bands were immersed in saturated $(NH_4)_2SO_4$ solution for 20 min, the band closest to the anode disappeared, identifying it as S100. These facts—the electrophoretic localization of the new protein fraction, its disappearance in saturated $(NH_4)_2SO_4$, and the increased amount of front protein in the trained rats—suggest that brain-specific S100 protein increased in amount during training. This S100 protein was presumably localized in the nuclei of the hippocampal neurons.

At this point it seems appropriate to comment on the protein changes during the intermittent training over a period of one month.

The rats performed well both on the fifth, fourteenth, and thirtieth day, i.e., when they had received training for 5, 7, and 10 days. If the increased synthesis in the hippocampal nerve cells had been an expression

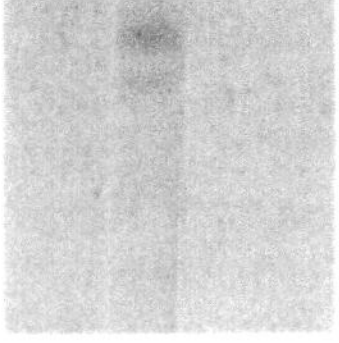

Fig. 6. Photographs of nerve cell protein of the hippocampus, CA3 region, separated on polyacrylamide gels 400 μ in diameter and stained by amido black. (left) From control rat, (right) from rat on day 5 of training with the nonpreferred paw. The acidic proteins migrate toward the bottom of the gel.

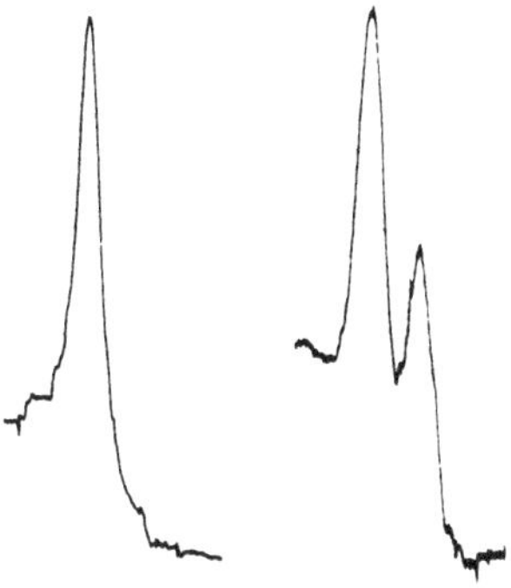

Fig. 7. Microdensitometric recording of the anodal front protein fractions shown in Fig. 6, control (left) and trained animal (right).

of increased and sustained neural function, then the increased incorporation values on the fifth day of initial training and on the fourteenth day of resumed training had presumably been found also when the rats were subjected to 3 days of resumed training 30 days after the initial training sessions. The fact that the incorporation values did not at that last stage differ from those of the controls is a strong indication that the observed increase in synthesis is correlated to learning processes occurring during the training. We would like to suggest the interpretation that when the novelty of the task has passed, the hippocampal nerve cells cease to respond with increased synthesis of this type of protein. A response may well occur in other parts of the brain.

TABLE VII

Frequency of Single and Double Front Anodal Protein Fractions in the Electrophoretic Pattern of 75 Polyacrylamide Gels from 23 Rats (7 Controls, 4 Resumed Training on 14th day, 12 Resumed Training on 14th day and on 30th day)

	Number of protein bands	
	One	Two
Control	20	0
Group II	5	10
Group III	20	20

It is even more striking that the S100 protein increased in amount during the learning to reverse hand since it is a brain-specific protein and thus can be expected to mediate specific brain functions.

Our interpretation of the result given above is that the increase of the S100 protein during reversal of handedness specifically relates the S100 protein to the learning processes. However, as we pointed out above, training involves several factors not related to learning per se. In the reversal of handedness experiments, the unspecific factors have been eliminated or reduced to a minimum. The motor and sensory activity, attention, motivation, and reward are equated between the experimental and control animals, and the stress involved in reversal of handedness is minimal. In view of these considerations, we used a technique which specifically related the S100 protein during reversal of handedness to learning per se.

A group of 8 rats were trained during 2 × 25 min per day for 3 days. Between the first and second training session of the fourth day, half of the rats were injected intraventricularly on both sides with 2 × 25 μg of antiserum against S100 in 2 × 25 μl. The other half of the group was similarly injected with the same amount and volume of antiserum against rat γ-globulin. The rats were slightly anesthetized with fluothane. The animals were trained a second session on the fourth day 45 min after the injections. After this treatment, the rats were trained for 3 further days with 2 × 25 min per day. The results are presented in Fig. 8. The number of reaches is plotted against the number of training days. Before injection of antisera, all rats followed an identical performance curve. After the injection of antiserum against rat γ-globulin, these rats followed a performance curve which was an extrapolation of the performance curve before the injection. The same was the case with a rat which was injected with the same volume of physiological NaCl.

By contrast, the rats injected with antiserum against S100 protein did not increase in performance, i.e., the number of reaches per day remained at the same values as those immediately before the injection. As is seen from the curves in Fig. 8, the difference in number of reaches between the two groups of injected rats is clearly significant.

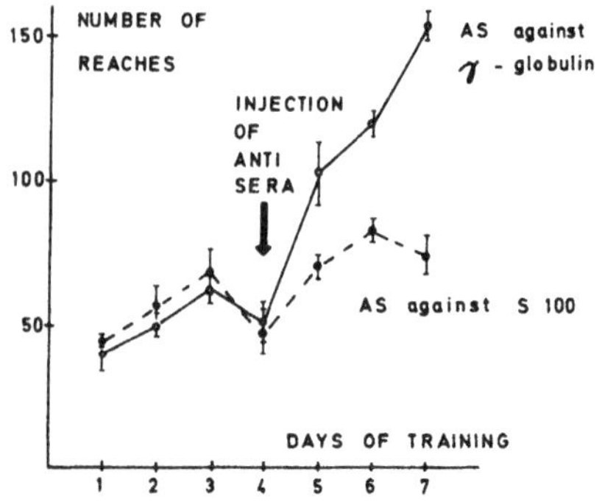

Fig. 8. Performance curves of two groups of rats, 6 experimental, 5 controls. One group injected with antiserum against S100, the other with antiserum against rat γ-globulin on the fourth day of training.

Another way to present the results is the following. For each rat, the sum of reaches for the first 3 training days is calculated as is also the sum of reaches for the last 3 training days. The number of reaches during the day of injection are thus not included in these sums. The ratio between the first and second sum is determined. The averages of this ratio are 0.73 ± 0.053 for the rats injected with S100 antiserum (6 rats), and 0.42 ± 0.033 for the control rats (5 rats). The difference between these ratios is highly significant ($P < 0.001$). It is obvious that the experimental rats show a decrease in learning capacity.

If instead the difference between the second and the first sum is calculated, you obtain for the rats injected with S100 antiserum 78 ± 18, and for the control rats 206 ± 19 reaches. The difference between these numbers (128) is highly significant ($P < 0.001$) leading to the same conclusion as above that the experimental rats show a decrease in learning capacity. Thus, the S100 protein is specifically correlated to learning processes.

III. SENSORY CONDITIONING TEST IN RATS

A. Incorporation of ^{3}H-leucine into Neuronal Protein

This experiment was designed to limit participation of such factors as motivation, motor activity, and reinforcement in an experiment involving change of behavior.[31] For that purpose we measured protein synthesis in rats subjected to paired and unpaired tone and light stimuli, and to light stimulus alone. A total of 80 rats were used in pilot tests, behavioral checks, and for final experiments. The rat was placed in a cage with a wired floor in a sound-absorbing dimly lit room for 20 min prior to the experiment. One group of rats received an acoustic signal of 1000 Hz (conditioning stimulus) followed by a visual (unconditioned) stimulus. The tone and light stimuli both lasted 0.2 sec and were presented automatically at a frequency of 6/min. Another group of rats was used as behavioral controls. In this group the tone and light stimuli were followed by an electric shock; in ten trials the rats learned to jump up to a shelf to escape the electric shock (criterion 8 out of 10) when the tone-light stimuli were presented. This test, a type of sensory conditioning described and discussed by Morrell,[32] demonstrated that a linkage had been formed between the two sensory areas. A third group of rats received tone and light stimuli distributed at random. A fourth group received only light stimulus at a frequency of 6/min.

The experimental rats were first bilaterally injected with ^{3}H-leucine intraventricularly during light fluothane anesthesia, then given sound–light or sound–light-at random stimulation for 15 min. The time lapse from the last injection to sampling of brain material was 40 min.

Both types of stimulation *increased* the incorporation of ^{3}H-leucine into the neuronal protein of the hippocampus but *decreased* it in the visual

TABLE VIII

Tone–Light Conditioning in Rats. Incorporation of ^{3}H-Leucine into Protein Isolated from Hippocampal Nerve Cell and Visual Cortex of Rats[a]

Stimuli	Number of samples	Incorporation of ^{3}H-Leucine	
		Visual cortex	Hippocampus
Paired	8	3.49 ± 0.23	3.43 ± 0.41
Unpaired	8	3.50 ± 0.23	4.45 ± 0.56
Light only	12	3.48 ± 0.18	
Control	8	4.46 ± 0.41	2.86 ± 0.41

[a] Data in cpm total protein/cpm total supernatant.

cortex (Table VIII). Control animals given light stimulus alone showed decreased incorporation in the visual cortex.

There are two findings in this sensory conditioning experiment which seem to exclude the possibility that the protein changes were correlated with learning processes during the conditioning. The first is the fact that light stimulus alone gave the same incorporation values of the visual cortex as did the paired and unpaired tone–light stimuli. The second circumstance is the fact that the tone–light stimuli distributed at random gave the highest incorporation of ^{3}H-leucine into the protein of the hippocampal nerve cells. Therefore, the conclusion is that the protein changes observed during the conditioning presumably are expression of increased attention or orientation reflexes. The finding that the incorporation values for the cells of the visual cortex in all three types of sensory experiments were lower than those of the controls, agrees with electrophysiological observations.

IV. DISCUSSION

The aim of the studies reported here has been to correlate protein changes in nerve cells (and glial cells) in particular parts of the brain with learning processes which occur during training. It seems evident that mapping the areas that respond with defined changes in protein fractions during behavioral experiments is prerequisite for a comprehensive theory on the mechanisms relating macromolecules in brain cells to storage and retrieval of information. The observations relating behavioral responses to synthesis and composition of RNA in nerve and glial cells[21,33–36] taken in conjunction with the observations on protein reported here, may be considered as a beginning which may eventually form the basis of such a theory.

It is interesting that the immunological study reported here brought out such clear differences in the antigen compositions of neurons and glia. This finding brings into question the matter of transfer of RNA from glia to neuron for which view there exists some evidence.[37] Such transfer could still take place even if the neuronal protein programmed by the glial RNA was not antigenic for rabbits challenged by our technique. S100 protein for instance, is not antigenic unless injected under special circumstances.

This S100 protein seems, however, to be definitely linked to learning, as is demonstrated especially well in the experiment where antiserum against S100 impaired learning while that against γ-globulin did not (Fig. 8). All factors, including that of stress, were identical for the control and experimental rats in this study. Before and after the antisera injections all were subjected to the same training program and the injections into the brain ventricles were carried out under identical conditions. Additional food was supplied to rats receiving S100 protein antiserum to compensate for the different amounts of reinforcement obtained. The result showing that only the S100 protein antiserum inhibited further learning seems clearly to link this brain-specific protein, S100, to learning processes occurring during training.

As for the nerve cell proteins 4 and 5, these are acidic even though their composition is still unknown. Their response during intermittent training spread over 1 month seems significant and is pertinent for the interpretation that the synthetic response was linked to learning processes within the training. The fact that the hippocampal nerve cells did not respond with increased synthesis of these proteins after the last training sessions (a month after the initial training) excludes the possibility that the increased protein synthesis during the two previous training sessions was merely an expression of increased motor activity, sensory activity, attention, or change in age.

The protein changes in the transfer of handedness experiment, a case of instrumental learning, can be compared with those in the sensory–sensory experiments, a case of classical conditioning. The instrumental case is a complicated type involving motor–sensory activities, motivation, and attention. The acidic–including S100–protein in hippocampal nerve cells rises during acquisition of behavior in this case, but during sensory conditioning no systematic change related to learning can be seen in either the hippocampal or the cortical cells. The protein synthetic response of hippocampal cells in this type of classical conditioning cannot therefore be equated with that taking place during instrumental learning. In sensory conditioning, the direction and magnitude of hippocampal protein synthesis changes seem only to follow the response of the cells to the sensory input, and to have no relation to the learning factors involved.

Thus both the light–tone and the at-random stimuli gave high incorporation values of ^{3}H-leucine into the hippocampal nerve cell protein. At visual cortex the incorporation values were equal and lower compared to the controls for all conditions of stimulation. It may therefore be tentatively concluded that in hippocampal cells the biochemical mechanisms taking place during instrumental learning differ from those during classical sensory conditioning.

ACKNOWLEDGMENT

We thank Dr. Robert Galambos for his generous help and advice with the preparation of the manuscript.

The studies reported in this paper were supported by Grant K68-11X-86 from the Swedish Medical Research Council and Riksbankens Jubileumsfond.

The antiserum prepared against the S100 protein was kindly provided by Dr. Lawrence Levine, Department of Bacteriology, Brandeis University, Waltham, Massachusetts.

V. REFERENCES

1. B. W. Moore, *Biochem. Biophys. Res. Comm.* **19**:739 (1965).
2. B. W. Moore and D. McGregor, *J. Biol. Chem.* **240**:1647 (1965).
3. B. S. McEwen and H. Hydén, *J. Neurochem.* **13**:823 (1966).
4. H. Hydén and B. S. McEwen, *Proc. Nat. Acad. Sci.* **55**:354 (1966).
5. A. H. Coons, *Ann. N.Y. Acad. Sci.* **69**:658 (1957).
6. P. Benda, J. Lightbody, G. Sato, L. Levine, and W. Sweet, *Science* **161**:370 (1968).
7. V. J. Perez and B. W. Moore, *J. Neurochem.* **15**:971 (1968).
8. B. W. Moore and V. J. Perez, *J. Immunol.* **96**:1000 (1966).
9. B. W. Moore and V. J. Perez, in *Physiological and Biochemical Aspects of Nervous Integration* (F. D. Carlson, ed.) p. 343, Prentice-Hall, Englewood Cliffs, N.J. (1968)
10. C. F. C. MacPherson and A. Liakopolou, *Fed. Proc.* **24**:272 (1965).
11. E. Kosinski and P. Grabar, *J. Neurochem.* **14**:273 (1967).
12. K. Warecka and H. Bauer, *J. Neurochem.* **14**:783 (1967).
13. K. Warecka and H. Bauer, *Deut. Z. Nervenheilk.* **194**:66 (1967).
14. G. S. Bennett and G. M. Edelman, *J. Biol. Chem.* **243**:6234 (1968).
15. L. Mihailovic and H. Hydén, *Brain Res.* **16**:243 (1969).
16. F. Huneeus-Cox, *Science* **143**, 1036 (1964).
17. F. Huneeus-Cox, H. L. Fernandez, and B. H. Smith, *Biophys. J.* **6**:675 (1966).
18. H. Hydén, *Nature* **184**:433 (1959).
19. A. H. Coons, *in General Cytochemical Methods* (J. F. Danielli, ed.), Vol. 1, p. 399, Academic Press, New York (1958).
20. L. Levine and B. W. Moore, *Neurosciences Res. Progr. Bull.* **3**, 18 (1965).
21. H. Hydén and E. Egyhazi, *Proc. Nat. Acad. Sci.* **52**:1030 (1964).
22. W. W. Meissner, *J. Psychiat. Res.* **4**:235 (1966).
23. R. G. Ojemann, *Neurosciences Res. Progr. Bull.* **4**:71 (1966).
24. W. Penfield, *Res. Publ. Assoc. Res. Nerv. Ment. Dis.* **30**:315 (1952).
25. P. Buresova, J. Bures, E. Fifkova, O. Vinogradova, and T. Weiss, *Exp. Neurol.* **6**:161 (1962).
26. W. R. Adey, R. T. Kado, J. Didio, and W. J. Schindler, *Exp. Neurol.* **7**:259 (1964).

27. S. H. Barondes and H. D. Cohen, *Proc. Nat. Acad. Sci.* **58**:157 (1967).
28. L. B. Flexner and J. B. Flexner, *Proc. Nat. Acad. Sci.* **60**:923 (1968).
29. H. Hydén and P. W. Lange, *J. Chromat.* **35**:336 (1968).
30. H. Hydén and P. W. Lange, *Science* **159**:1370 (1968).
31. H. Hydén, P. W. Lange, and D. Wood. (To be published.)
32. F. Morrell, *in The Neurosciences* (G. C. Quarton, T. Melnechuk, and F. O. Schmitt, eds.), p. 452, Rockefeller Univ. Press, New York (1967).
33. H. Hydén and E. Egyhazi, *Proc. Nat. Acad. Sci.* **48**:1366 (1962).
34. H. Hydén and E. Egyhazi, *Proc. Nat. Acad. Sci.* **49**:618 (1963).
35. H. Hydén and P. W. Lange, *Proc. Nat. Acad. Sci.* **53**:946 (1965).
36. V. E. Shashoua, *Nature* **217**:238 (1968).
37. H. Hydén and P. W. Lange, *Naturwiss.* **3**:64 (1966).

Chapter 8

CHEMICAL TRANSFER OF INFORMATION*

Georges Ungar

Baylor College of Medicine
Houston, Texas

I. INTRODUCTION

The molecular approach to the problem of learning, memory, and information processing in general has followed three main strategies: (1) detection of chemical changes in the brain that can be correlated with acquisition of information, (2) impairment of learning by inhibitors of RNA and protein metabolism and (3) transfer of information from one individual to another by purely chemical means.

The first two approaches are discussed in other chapters of this volume.[1,2] The present chapter summarizes the attempts made to date to transfer information from trained donors to naïve recipients by chemical means. Such attempts, if successful, provide incontrovertible evidence of a molecular mechanism in the processing of information by the nervous system.

These experiments are based on the same principle as many demonstrations of hormone and neurotransmitter action: production of a neural effect in the donor preparation and transfer of the effect to a receptor preparation. Application of this bioassay method to the chemical changes associated with learning must, however, take into account the greatly increased complexity of the situation. As is usual in areas that are just beginning to be explored, many of the factors on which the success of the experiments depends are poorly understood or completely unknown. This explains the somewhat irregular reproducibility of some of the results and contributes to the controversy over the issue.

The basic design of all the experiments includes the following steps: (1) training of donor animals, (2) preparation of extracts from the donors,

* The experimental work from the writer's laboratory referred to in this chapter was supported by USPHS grant no. MH-13361

(3) administration of the extract to suitably selected and prepared recipients, (4) testing of the recipients and (5) evaluation of the results. In the context of the present volume, the main emphasis will be on step 2, without, however, entirely omitting the nonchemical elements of the problem, without which it would lose its meaning.

II. TRANSFER OF LEARNED BEHAVIOR IN PLANARIANS

In 1962 McConnell[3] described experiments in which planarian worms (*Dugesia dorotocephala*) were trained in a Pavlovian conditioning paradigm. The conditioned stimulus (CS) was light and the unconditioned stimulus (US) was electric shock. As a result of the training, the subjects exhibited a contraction on presentation of the CS. The bodies of the trained donors were cut into pieces and fed to untrained recipients. A control group of recipients was fed fragments of untrained donors. The recipients of trained worms had a significantly higher rate of responses to the CS than did the controls. These results were confirmed by John,[4] Kabat,[5] Ragland and Ragland,[6] and extended to habituation by Westerman.[7]

The experiments of Corning and John[8] showed that ribonuclease added to the medium can, under certain conditions, wipe out memory in regenerating planarians. This prompted experiments of transfer in which, instead of cannibalism, the recipients were injected with an RNA preparation extracted from trained worms.[9,10]

These experiments stimulated a lively controversy in which doubt was cast on the trainability of planarians[11] and on the nature of the behavioral change observed in the recipients.[12] It seems that these theoretical objections have been adequately answered by the experiments of Jacobson *et al.*[10,13] and Block and McConnell.[14] The latter work showed that the transfer is stimulus-specific and eliminated the possibility of some vague nonspecific entity being transferred. For a detailed discussion see reviews by Corning and Riccio[15] and McConnell and Shelby.[16]

The only doubtful point remaining is the chemical nature of the active material responsible for the transfer. The evidence that it is RNA is based on observations of abolition of learning in planarians by addition of ribonuclease to the medium[8] and on the effect of "RNA-type" preparations injected into the worms.[9,10] The first evidence is indirect and the second is based on doubtful chemical arguments which will be examined later in connection with the mammalian work.

III. CHEMICAL TRANSFER IN VERTEBRATES

When the results of the planarian experiments had become known, it was inevitable that similar attempts would be made in higher organisms. In 1965 four publications appeared at intervals of a few weeks, in which experiments of transfer of learning were described in rats and mice.[17–20] Reinis[17] trained rats to open a gate leading to the food cup when a conditioned stimulus, light or sound, was presented. When the donors reached the criterion of 80% correct responses for three sessions, they were killed, their brains were extracted, and the extract was injected intraperitoneally to recipients. These were tested 48 hr after injection and for several days thereafter. Testing of the recipients was done identically with the training of the donors, with reinforcement. Under these conditions, the recipients of trained rat brain learned the task significantly faster than the controls, which had been treated with brain extract from untrained rats.

Fjerdingstad *et al.*[18] trained their donors for a dark-light discrimination task in a two-alley runway. The rats had been deprived of water and were rewarded with it when they selected the lighted runway. When the criterion of 10% or less errors was reached, the donors were killed and extracts made from their brains were injected intracisternally into the recipients, which were then subjected to the same training procedure as the donors. From the second day after injection, the recipients of trained brain performed significantly better than the controls.

Ungar and Oceguera–Navarro[19] produced habituation in the donor rats. This elementary form of learning consists of the suppression of an innate response, for example, startle response to a loud sound, by regular repetition of the stimuli. When the donors responded to 10% or less of the stimuli, their brains were removed and the extracts injected intraperitoneally into recipient mice. These, when first tested 24 hr later, responded to less than 50% of the stimuli, while the responses of the control mice, which had been given untrained brain, were over 90%. The differences between the two groups were always highly significant.

Babich *et al.*[20] trained donors to approach a food cup on hearing the click produced by the operation of the pellet dispenser. After a fixed training period of 5 days, the donors' brains were extracted and injected intraperitoneally into recipient rats. These were tested several times between 4 and 24 hr after injection and the authors found that recipients of trained brain approached the food cup on the click significantly more often than the controls.

The last-mentioned work, probably because it was the only one published in the U.S., became the most widely known and several attempts were made to replicate it.[21–26] Almost all these attempts failed and the reasons for the negative results will be examined below.

One of the replications of the experiments of Babich *et al.* by Rosenblatt *et al.*[26] was successful and pointed to some of the weaknesses of their procedure. These workers tried several other behavioral procedures[27,28] with varying success but their general conclusion was that a "wide variety of tasks . . . seem to be susceptible to transfer, and . . . the transfer is specific to the learned task."[27] This conclusion is now supported by the results of many experiments using instrumental conditioning,[29–34] spatial and brightness discrimination,[34-36] conditioned avoidance,[37-41] passive avoidance,[41,42] simple maze learning,[43] color and taste discrimination,[44] fixation of postural asymmetry[45] and other situations. The experimental data have been reviewed by Rosenblatt,[46] Ungar[47] and Ungar and Chapouthier.[48] Positive experiments have now been reported in close to 100 publications from 26 laboratories in Austria, Belgium, Canada, Czechoslovakia, Denmark, France, Germany, Holland, Hungary, the U.S.A. and the U.S.S.R.

Besides the transfer of habituation, mentioned above, several other types of experiments were performed in the author's laboratory. In one group of experiments the recipients were tested under conditions identical with the training of the donors, i.e., they were reinforced.[49] In these experiments the recipients of trained brain learned the task significantly faster than the controls injected with brain extracts taken from naive donors. This acceleration of learning, however, did not prove that actual transfer of information had taken place. In subsequent experiments, therefore, the recipients were tested without reinforcement so that their postinjection behavior could be determined only by the information carried by the brain material administered.

In the first of these experiments, brightness discrimination was used.[50,51] The donors were trained in a Y-maze to escape electric shock by running into the lighted arm of the maze. Recipients injected with brain extracts from these donors showed a significant tendency to run into the lighted arm, although both arms were shock-free. Similar experiments were done with spatial discrimination.[50,51] The donors were trained to escape into the left or right arm of the maze. In some experiments, the recipients showed a significantly higher proportion of runs into the arm to which the donors were trained. In other cases, however, the results were inconsistent or even reversed. Such reversals were observed by other workers using left-right discrimination[28,32] and it is probable that, in spite of its apparent simplicity, this paradigm is influenced by too many unknown factors to be a good bioassay.

The most recent experiments performed in this laboratory have been using two passive avoidance paradigms.[52,53] The first one, inspired by the success of the experiments of Gay and Raphelson,[41] is based on the innate preference of rodents for dark enclosures. The donors were trained to reverse this preference by receiving an electric shock when they entered the dark box. Recipients of their brain, although they were never shocked, exhibited a highly significant avoidance of the dark box. Animals injected with untrained or pseudotrained brain showed normal preference for the dark.[52] These experiments have been repeated on a large scale during the 2-year period of work leading to the isolation of the active substance. A total of 1540 recipient mice were used to test 149 brain extracts taken from close to 4000 donor rats. Before receiving the extract, the mice spent 72.6% ($\pm$8.3 S.D.) of the time in the dark. After injection, the proportion of time spent in the dark by the recipients of trained brain was 34.6 $\pm$ 9.6%

and by the controls that received untrained brain 66.7 ± 11.1%. The probability of this being due to chance is < 0.0001.

The second passive avoidance paradigm is based on the fact that when rodents are placed on a narrow platform above a wider area, they tend to step down. Donors were put on the platform and subjected to electric shock when they stepped down. Recipients of their brain showed a longer latency of step-down and spent more time on the platform than controls injected with untrained or pseudo-trained brain.[53]

The dark avoidance experiments have been replicated in several laboratories; two of these replications have been published.[54,55] A variant of this experiment using positive reinforcement has been successfully performed in this laboratory by Fjerdingstad.[56]

A comparison of the successful experiments with the negative results mentioned above[21–25] led to the recognition of a number of essential conditions described in detail in a critical review.[47] These were summarized as follows:

1. Donors should be trained for a period of between 6 and 12 days. They should be able to learn the task to a criterion of 90 percent or more correct responses; incompletely trained animals should not be used.

2. The initial experiments should be done with crude brain extracts and purification should be guided by assays of activity. All preparative procedures should be done at low temperature.

3. The recipients should belong to a species, strain and age group that can learn readily by training the task which is to be transferred to them. They should be screened individually so as to eliminate the subjects that do not normally exhibit the responses on which the transfer is based. If the response required deprivation in the donors, the recipients should be similarly deprived.

4. Usually, amounts equivalent to two or three recipient brains are necessary for significant transfer to take place but the optimal dose should be determined by trial and error.

5. Recipients should be tested at widely spaced intervals after injection. Tests should be done from a few hours after administration of the extract and continued daily for four to six days.

6. Recipients should be tested under blind conditions, in random order and preferably by automatic recording.

7. As a general and perhaps most important rule, all the conditions should be varied until the best procedure is found. Only after trying different modalities of training, a range of doses and varying the testing schedules of the recipients can one conclude that, in that particular case, transfer has not taken place.

Few of these conditions were fulfilled in the negative experiments. Consistently successful transfer requires extreme care and patience. This is not the type of work that can be done by inexperienced students with leftover animals and discarded pieces of equipment. It demands the organized and sustained effort of a team reasonably familiar with both the behavioral and biochemical aspects of the problem.

IV. CHEMICAL NATURE OF THE TRANSFER FACTORS

Several types of brain preparations were used in the experiments mentioned above. Good results were obtained by crude brain homogenate[17–29] and its supernatant fraction.[30,32] As a general rule, in the author's laboratory the preliminary experiments have been done with one of these crude preparations. In preparing the supernatant of the homogenate, one has to make a compromise between the thoroughness of the extraction of all the soluble material and the necessity to prevent a possible loss of active material by enzymatic destruction. In this laboratory, homogenates were stirred magnetically for 24 hr in an ice bath and centrifuged at 60,000*g* for 1 to 2 hr.

Most of the transfer experiments, however, were based on the assumption that the coded molecule and, consequently, the transfer factor, is an RNA.[18,20,37,41,42] The Danish group[18,36] extracted donor brains by the procedure of Laskov *et al.*[57] as modified by Haselkorn.[58] The frozen, powdered brain was treated with phenol in the presence of EDTA at *p*H 7.6. The water phase, containing RNA, was acidified to *p*H 5.5 and mixed with isopropanol. The resultant precipitate was centrifuged down, dissolved, dialyzed, and reprecipitated. This dry preparation was kept frozen until use. Shortly before injection, the preparation was dissolved in a buffered salt solution at 20 mg per ml and 50 μl of the solution was injected intracisternally into each rat.

The preparation used by Babich *et al.*[20] and, with minor modifications, by others,[37,41,42] was also based on phenol partition. RNA was precipitated out of the aqueous phase with $MgCl_2$ and ethanol. It was claimed that this procedure achieved the preparation of protein-free RNA in a high state of polymerization.[20] The Danish group, however, admitted the presence in their preparations of 5–50% protein.[36] Both methods yielded about 1 mg of "RNA" per gram of wet weight of brain. It would be unrealistic to claim a high degree of purity for these preparations; they probably contain many impurities in amounts sufficient to account for their activity.

Several groups of workers proceeding by successive stages of purification reached the conclusion that the transfer factors are peptides or proteins.[26,38,44,45,55] In the writer's laboratory the standard procedure has been to test the supernatant of the homogenate for dialyzability. In most situations, the activity was found in the dialyzate (habituation, spatial discrimination, both types of passive avoidance). In a few cases (conditioned avoidance, brightness discrimination) the transfer factors were not dialyzable; they were not submitted to further fractionation. The dialyzable substances were precipitated by acetone and ethanol and, after phenol partition, the activity was found in the phenol phase.

Treatment of these preparations by enzymes showed that they were resistant to RNase but could be destroyed by proteases. The transfer factor for sound habituation was inactivated by chymotrypsin but not by trypsin,[19] that for dark avoidance was attacked by trypsin and not by chymotrypsin,[52] while the material which transfers inhibition of step-down from platform was destroyed by both enzymes.[53] All three of these preparations were fractionated by gel filtration on Sephadex G–25 and their elution characteristics were compatible with the assumption that they consist of peptides with 8 to 16 amino acid residues.

The apparent contradiction between the results obtained with "RNA" preparations and those in which activity was destroyed by proteases prompted us to do the following experiment.[59] A sample taken from a pool of brain from rats trained for dark avoidance was extracted for peptides according to the standard procedure used in this laboratory.[52,53] Another sample from the same pool was extracted for RNA by the techniques mentioned above.[36] Both procedures yielded active preparations inactivated by incubation with trypsin. This suggested the possibility that the active material is a peptide forming a complex with RNA. This assumption was confirmed when we were able to separate the active peptide from the inactive RNA by dialysis at low *p*H.

It is possible that both peptides and RNA can serve as transfer factors, especially in the planarian experiments and by the intracisternal route in mammals. It is more difficult to assume that, after systemic administration, sequences of 20 to 30 nucleotides would remain intact and reach the brain across the blood–brain barrier.

There is no difficulty of the same order concerning peptides. Many of them are known to survive in the body fluids long enough to exert pharmacological or hormonal actions. They also produce fairly specific behavioral effects that can be explained only by a direct action on the brain,[60] after passage through the blood–brain barrier.

One has to postulate that each type of learned behavior is transferred by a specific factor or set of factors. This is based on the observations of specificity of transfer, as seen in habituation[49] and passive avoidance.[53] In the latter case, we have seen that there is no cross-transfer between the two passive avoidance situations and the two transfer factors have different chemical properties. Transfer of color discrimination represents further evidence of specificity.

The most urgent task seemed to be the isolation and identification of at least one of the transfer factors. We selected the factor capable of inducing dark avoidance because of the comparatively rapid training of the donors and the quantitative nature of the bioassay. After having collected about 5 kg of donor brain, we made an RNA extract following the procedure mentioned above. This extract was dialyzed at *p*H 3.7 and the dialyzate yielded about 1 mg of active material per gram of wet weight of brain. Further purification by gel filtration on Sephadex G–25 followed by thin layer chromatography, resulted in a unique spot, missing from identically processed preparations of untrained brain.

Elution of the spot yielded a material active at 100 ng per mouse. A microdansylation procedure[61] showed the presence of the following amino acids: Ala, Asp, Glu, Gly, Lys, Ser, and Tyr. The *N*-terminal group was found to be Ser. A quantitative amino acid analysis (for which we are indebted to W. C. Starbuck) indicated the following composition: 4 Glu, 3 Asp, 3 Gly, 2 Ser, Ala, Lys, Tyr. After incubation with trypsin, two fragments (T_1 and T_2) were obtained. T_1 contained Glu, Asp, Gly, Ser, and Lys, and T_2 gave Ala, Glu, Gly, Ser, and Tyr. The *N*-terminal group of both fragments was Ser.

The sequence obtained by a mass spectrometric method by D. M. Desiderio is as follows: Ser–Asp–Asn–Asn–Glu–Gln–Gly–Lys–Ser–Ala–Glu–Gln–Gly–Gly–TyrNH_2. We gave the name scotophobin to this pentadecapeptide (from the Greek skotos = dark and phobos = fear).[62] There are some uncertainties in the sequence given by mass spectrometry, particularly in the respective number and position of Glu, Asp, and their amides. Synthesis of scotophobin is now being attempted by B. Weinstein of the University of Washington in Seattle and it is hoped that the doubtful points will be clarified.

The potency of scotophobin is of the same order as that of other known peptides which are the most potent pharmacological agents on record. This suggests that peptides have extremely selective affinities for certain cell systems because their molecular structure contains highly specific instructions for their destination.

V. MOLECULAR MECHANISMS IN INFORMATION PROCESSING

There is at present quite an impressive array of evidence to support the intervention of molecular mechanisms in the processing of information by the nervous system. The problem, however, is still at the stage where the evidence is good enough to encourage the open-minded but not overwhelming enough to overcome the hard-core resistance.

The strongest reason for the opposition to the molecular hypothesis is the mistaken belief that it is incompatible with the highly organized structure of the nervous system. The idea of a chemical storage of information is somehow equated with the "mass," "molar," or "field" theories which tend to negate the existence of specific pathways.

It was emphasized in recent papers[63–65] that the extraordinary structural differentiation of the brain, which is the essential condition of its functions, is based on a mechanism of molecular recognition. The development of specific pathways is determined by chemical affinity between the constituent neurons.[66] This may represent the highest degree of differentiation of the coating substances by which cells of the same type recognize each other.[67]

Before any learning takes place, the nervous system is capable of responding to stimuli in a preordained manner. The program of these responses, inscribed in the innate organization of the brain, is undoubtedly under genetic control and operates, therefore, through the DNA–RNA–protein code. Elementary learning consists essentially in the creation of new connections between the existing pathways so that a given stimulus can evoke a new unprogrammed response.

It is economical to postulate that the acquisition of the new neural connections associated with learning utilizes the same principle as the differentiation of innate pathways. Basically the mechanism is similar to the one which allows a cholinoceptive neuron to recognize a cholinergic ending, but it is considerably more specific and requires molecular "markers" of a much higher information content than the transmitter amines. Peptide sequences are eminently suitable to play this role and comparatively small peptides could account for the specific connections linking together the many millions of pathways postulated in the human brain.[63–65]

The hypothesis assumes that new connections between existing pathways are created when neurons belonging to two or more of these pathways fire simultaneously or at very short intervals. This simultaneous firing, by altering the permeability at some points of contact, would allow an exchange of the marker substances and, consequently, open synapses to the flow of impulses. The whole process is accompanied by increased synthesis of the relevant pathway-specific molecules during the phase of acquisition. Some results of the transfer experiments suggest that only information acquired recently and under exceptionally intensive training can be transferred.[37,51,52] Attempts at transferring innate or long-established learned behavior have been unsuccessful, presumably because fixed synaptic connections require the presence of only minimal amounts of marking substances.

The chemical code of the marking and connecting molecules does not represent an "engram" in the sense of a unique molecule encoding a given experience. The coded substances could be compared to "signposts" which direct the traffic of impulses along the channels opened up by the learned experience. Such a system could be decoded only in terms of the existing organization of the brain just as traveling instructions have meaning only in terms of the existing highway network. The coded molecules, extracted from the brain of the donors, must, therefore, reach the equivalent pathways in the brain of the recipients and set up signposts along the trail created by the learning of the donors.

Chemical transfer represents a "playback" of the code. The phenomenon of transfer itself is of secondary importance and its possibilities are probably limited. The most significant point is that it can be used as a bioassay for the study and identification of the pathway-specific molecules and give us an insight into the code in which the brain handles information.

VI. REFERENCES

1. H. Hydén, *in Handbook of Neurochemistry* (A. Lajtha, ed.), Vol. 6, pp. 221–239, Plenum Press, New York (1970).
2. B. Agranoff, *in Handbook of Neurochemistry* (A. Lajtha, ed.), Vol. 6, pp. 203–219, Plenum Press, New York (1970).
3. J. V. McConnell, Memory transfer via cannibalism in planaria, *J. Neuropsychiat.* **3**:1–42 (1962).
4. E. R. John, *in Brain Function* (M. A. Brazier, ed.), Vol. II, pp. 161–182, University of California Press (1964).
5. L. Kabat, Transfer of training through ingestion of conditioned planarians by unconditioned planarians, *Worm Runner's Digest* **6**:23–37 (1964).
6. R. S. Ragland and J. B. Ragland, Planaria: interspecific transfer of a conditionability factor through cannibalism, *Psychon. Sci.* **3**:117–118 (1965).
7. R. A. Westerman, Somatic inheritance of habituation of responses to light in planarians, *Science* **140**:676–677 (1963).
8. W. C. Corning and E. R. John, Effect of ribonuclease on retention of conditioned response in regenerated planarians, *Science* **134**:1363 (1961).
9. A. Zelman, L. Kabat, R. Jacobson, and J. V. McConnell, Transfer of training through injection of "conditioned" RNA into untrained planarians, *Worm Runner's Digest* **5**:14–21 (1963).
10. A. L. Jacobson, C. Fried, and S. D. Horowitz, Planarians and memory: I. Transfer of learning by injection of ribonucleic acid, *Nature* **209**:599–601 (1966); II. The influence of prior extinction on the RNA transfer effect, *Nature* **209**:601 (1966).
11. E. L. Bennett and M. Calvin, Failure to train planarians reliably, *Neurosciences Research Program Bulletin* (July–August, 1964).
12. D. D. Jensen, Paramecia, planaria, and pseudo-learning, *Animal Behav., Suppl. 1* **13**:9 (1965).
13. A. L. Jacobson, S. D. Horowitz, and C. Fried, Classical conditioning, pseudoconditioning, or sensitization in the planarian, *J. Comp. Physiol. Psychol.* **64**:73–79 (1967).
14. R. A. Block and J. V. McConnell, Classically conditioned discrimination in the planarian, *Dugesia dorotocephala, Nature* **215**:1465–1466 (1967).
15. W. L. Corning and D. Riccio, *in Molecular Approaches to Learning and Memory* (W. L. Byrne, ed.), pp. 107–149, Academic Press, New York (1970).
16. J. V. McConnell and J. M. Shelby, *in Molecular Mechanisms in Memory and Learning* (G. Ungar, ed.), pp. 71–102, Plenum Press, New York (1970).
17. S. Reinis, The formation of conditioned reflexes in rats after the parenteral administration of brain homogenate, *Activ, Nerv. Sup.* **7**:167 (1965).
18. E. J. Fjerdingstad, T. Nissen, and N. H. Røigaard-Petersen, Effect of ribonucleic acid (RNA) extracted from the brain of trained animals on learning in rats, *Scand. J. Psychol.* **6**:1–5 (1965).

19. G. Ungar and C. Oceguera–Navarro, Transfer of habituation by material extracted from brain, *Nature* **207**:301–302 (1965).
20. F. R. Babich, A. L. Jacobson, S. Bubash, and A. Jacobson, Transfer of a response to naïve rats by injection of ribonucleic acid extracted from trained rats, *Science* **149**:656–657 (1965).
21. M. W. Gordon, G. G. Deanin, H. L. Leonhardt, and R. H. Gwynn, RNA and memory: a negative experiment, *Am. J. Psychiat.* **122**:1174–1178 (1966).
22. C. G. Gross and F. M. Carey, Transfer of learned response by RNA injection: failure of attempts to replicate, *Science* **150**:1749 (1965).
23. R. J. Kimble and D. P. Kimble, Failure to find "transfer of training" effects via RNA from trained rats injected into naïve rats, *Worm Runner's Digest* **8**:32–36 (1966).
24. M. Luttges, T. Johnson, C. Buck, J. Holland, and J. McGaugh, An examination of "transfer of learning" by nucleic acid, *Science* **151**:834–837 (1966).
25. W. L. Byrne, D. Samuel, E. L. Bennett, M. R. Rosenzweig, E. Wasserman, A. R. Wagner, R. Gardner, R. Galambos, B. D. Berger, D. L. Margules, R. L. Fenichel, L. Stein, J. A. Corson, H. E. Enesco, S. L. Chorover, C. E. Holt, III, P. H. Schiller, L. Chiappetta, M. E. Jarvik, R. C. Leaf, J. D. Dutcher, Z. P. Horovitz, and P. L. Carlson, Memory transfer, *Science* **153**:658–659 (1966).
26. F. Rosenblatt, J. T. Farrow, and W. F. Herblin, Transfer of conditioned responses from trained rats to untrained rats by means of a brain extract, *Nature* **209**:46–48 (1966).
27. F. Rosenblatt, J. T. Farrow, and S. Rhine, The transfer of learned behavior from trained to untrained rats by means of brain extracts I., *Proc. Nat. Acad. Sci.* **55**:548–555 (1966); The transfer of learned behavior from trained to untrained rats by means of brain extracts, II., *Proc. Nat. Acad. Sci.* **55**:787–792 (1966).
28. F. Rosenblatt, *in Molecular Approaches to Learning and Memory* (W. L. Byrne, ed.), pp. 195–242, Academic Press, New York (1970).
29. J. A. Dyal, A. M. Golub, and R. L. Marrone, Transfer effects of intraperitoneal injection of brain homogenates, *Nature* **214**:720–721 (1967).
30. A. M. Golub and J. V. McConnell, Transfer of response bias by injection of brain homogenates: a replication, *Psychon. Sci.* **11**:1–2 (1968).
31. W. L. Byrne and A. Hughes, Behavioral modification by injection of brain extract from trained donors, *Fed. Proc.* **26**:676 (1967).
32. Byrne, W. L. (ed.), *Molecular Approaches to Learning and Memory*, Academic Press, New York (1970).
33. O. A. Krylov, P. I. Kalyuzhnaya, and V. S. Tongur, Possible conveyance of conditioned connection by means of a biochemical substrate, *Zh. Vyssh. Nerv. Deiat. Pavlov.* **19**:286–291 (1969).
34. A. M. Golub, F. R. Masiarz, T. Villars, and J. V. McConnell, Incubation effects in behavior induction in rats, *Science* **168**:392–395 (1970).
35. O. L. Wolthuis, *in Molecular Approaches to Learning and Memory* (W. L. Byrne, ed.), pp. 285–294, Academic Press, New York (1970).
36. H. H. Røigaard-Petersen, T. Nissen, and E. J. Fjerdingstad, Effect of ribonucleic acid (RNA) extracted from the brain of trained animals on learning in rats. III. Results obtained with an improved procedure, *Scand. J. Psychol.* **9**:1–16 (1968).
37. G. Adam and J. Faiszt, Conditions for successful transfer effects *Nature* **216**:198–200 (1967).
38. G. Chapouthier and A. Ungerer, Effet de l'injection d'extraits de cerveau conditionné sur l'apprentissage, *C.R. Acad. Sci. Paris.* **267**:769–771 (1968).
39. E. J. Fjerdingstad, Memory transfer in goldfish, *J. Biol. Psychol.* **11**:20–25 (1969).

40. W. G. Braud, Extinction in goldfish; facilitation by intracranial injection of "RNA" from brains of extinguished donors, *Science* **168,** 1234–1236 (1970).
41. R. Gay and A. Raphelson, "Transfer of learning" by injection of brain RNA: a replication, *Psychon. Sci.* **8**:369–370 (1967).
42. Weiss, K., *in Molecular Approaches to Learning and Memory* (W. L. Byrne, ed.), pp. 325–334, Academic Press, New York (1970).
43. E. Rosenthal and S. B. Sparber, Transfer of a learned response by chick brain homogenate fed to naïve chicks, *The Pharmacologist* **10**, 168 (1968).
44. H. P. Zippel and G. F. Domagk, Versuche zur chemischen Gedächtnisübertragung von farbdressierten Goldfischen auf undressierte Tiere, *Experientia* **25**:938–940 (1969).
45. C. Giurgea, J. Daliers, and F. Mouravieff, Pharmacological studies on an elementary model of learning: the fixation of an experience at spinal level, *Abstracts, Fourth International Congress on Pharmacology* (Basel, Switzerland, July 14–18, 1969) Schwabe & Co., Basel, pp. 291–292 (1969).
46. F. Rosenblatt, *in Molecular Mechanisms in Memory and Learning* (G. Ungar, ed.), pp. 103–147, Plenum Press, New York (1970).
47. G. Ungar, *in Methods in Pharmacology* (A. Schwartz, ed.), Appleton-Century-Crofts, New York. (In press.)
48. G. Ungar and G. Chapouthier, Mécanismes moléculaires de l'utilisation de l'information par le cerveau, *L'Annee Psychologique.* (In press.)
49. G. Ungar, Chemical transfer of acquired information, *CINP Symposium* (Washington, 1966), pp. 169–175, Elsevier Publishing Co. Amsterdam (1970).
50. G. Ungar, Transfer of learned behavior by brain extracts, *J. Biol. Psychol.* **9**:12–27 (1967).
51. G. Ungar and L. N. Irwin, Transfer of acquired information by brain extracts *Nature* **214**:453–455 (1967).
52. G. Ungar, L. Galvan, and R. H. Clark, Chemical transfer of learned fear, *Nature* **217**:1259–1261 (1968).
53. G. Ungar, *in Protein Metabolism of the Nervous System* (A. Lajtha, ed.) pp. 571–585, Plenum Press, New York (1970).
54. O. L. Wolthuis, Inter-animal information transfer by brain extracts, *Arch. Int. Pharmacodyn.* **182**:439–442 (1969).
55. A. M. Golub, L. Epstein, and J. V. McConnell, The effect of peptides, RNA extracts, and whole brain homogenates on avoidance behavior in rats, *J. Biol. Psychol.* **11**:44–49 (1969).
56. E. J. Fjerdingstad, Chemical transfer of learned preference, *Nature* **222**:1079–1080 (1969).
57. R. Laskov, E. Margoliash, V. Z. Littauer, and H. Eisenberg, High molecular weight ribonucleic acid from rat liver, *Biochim. Biophys. Acta* **33**:275–284 (1959).
58. R. Haselkorn, Studies on infectious RNA from turnip yellow mosaic virus, *J. Molec. Biol.* **4**:357–367 (1962).
59. G. Ungar and E. J. Fjerdingstad, Chemical nature of the transfer factors; RNA or protein? *Proceedings of the Symposium on Biology of Memory* (Tihany, Hungary, September 1–4, 1969) Hungarian Academy of Science. (In press.)
60. D. DeWied and B. Bohus, Long term and short term effects on retention of a conditioned avoidance response in rats by treatment with long acting pitressin and α-MSH, *Nature* **212**:1484–1486 (1966).
61. V. Neuhoff, F. von der Haar, E. Schlimme, and M. Weise, Zweidimensionale Chromatographie von Dansyl-Aminosäuren im pico-Mol-Bereich, angewandt zur direkten Charakterisierung von Transfer-Ribonucleinsäuren, *Hoppe-Seylers Z. Physiol. Chem.* **350**:121–128 (1969).

62. G. Ungar, I. K. Ho, L. Galvan, and D. M. Desiderio, Isolation and identification of a specific behavior-inducing peptide extracted from brain, *Proceedings Western Pharmacology Society*. (In press.)
63. G. Ungar, Molecular mechanisms in learning, *Perspect. Biol. Med.* **11**:217–232(1968).
64. G. Ungar, *in Molecular Mechanisms in Memory and Learning* (G. Ungar, ed.), pp. 149–175, Plenum Press, New York (1970).
65. G. Ungar, Molecular mechanisms in information processing, *Int. Rev. Neurobiol.* **13,** 225–253.
66. R. W. Sperry, *in Biological and Biochemical Bases of Behavior* (H. H. Harlow and C. N. Wolsey, ed.), pp. 401–424, Univ. Wisconsin Press, Madison (1958).
67. A. A. Moscona, Studies on cell aggregation: demonstration of materials with selective cell-binding activity, *Proc. Nat. Acad. Sci. U.S.* **49**:742–747 (1963).

Chapter 9

UNDERNUTRITION AND THE DEVELOPING BRAIN

John Dobbing

Department of Child Health
University of Manchester
Manchester, England

I. INTRODUCTION

The importance of investigating the effects of undernutrition on the developing brain has become more widely recognized in the past few years. Unfortunately, a neurochemical consideration of the problem has scarcely begun. The aim of this chapter is to summarize the present state of knowledge, and an attempt will be made to interest neurochemical readers in the great need which exists for their participation in the debate.

As recently as the late 1960s it was still possible to collect the main body of opinion and knowledge from most of the many disciplines involved into one work under the general heading of "Malnutrition, Learning and Behaviour,"(1) and this remains the best general reference source for findings and attitudes up to the time it was written. It contains information about social and preventive medicine, ecological, and clinical aspects, as well as experimental findings from the laboratory approach to the problem. Other more recent reviews have dealt in detail specifically with the effects of undernutrition on the brain tissue, both at the time of the nutritional restriction and following attempted rehabilitation.(2,3)

The academic importance of the subject lies in its relation to the manner in which bodily, organ, and tissue growth is organized, and how the "program" of growth can be manipulated and distorted (sometimes permanently) by restrictive influences in certain early stages. However, this is clearly not a purely academic problem. There is evidence that the development of the brain can be interfered with in a manner which leaves permanent deficits in both its physical makeup and its functional activities in later life, and if this be true of human development, there can be few more important matters in the whole field of human nutrition and growth.

In the human species, whose distinction lies in qualities of higher intellectual function rather than in carcass weight or even athletic performance, these possibilities must be investigated very seriously.

Unfortunately, it is an inherent disadvantage for the neurochemical approach to the problem that brain tissue must be obtained for the test tube from experimentally treated subjects. This rules out direct experimental research in humans and drives us to the exploitation of the animal model. We must therefore, from the outset, seriously consider the validity of comparative studies as models for other species. Also, before we can test hypotheses derived from animals by observational research on human brains, we shall need to have much more quantitative data about human brain development than exists at the present time.

Finally, it must be stressed that at present we can only recognize *an association* between early undernutrition and later intellectual impairment. A direct *causative* relationship cannot yet be confidently implied, in view of the multiplicity of other factors in a growing child's environment which can modify its behavioral development.[4]

II. HISTORICAL BACKGROUND

The present awakening of interest in this subject is not the first. As always, it is salutory to read the conclusions reached by our ancestors, and to discover what was already well known in the early part of this century. This has recently[5] been summarized as follows:

> 1. The brain weight of an *adult* animal is scarcely affected by undernutrition, even when there is starvation to death and a reduction in body weight to one half its former value.
>
> 2. The brain weight is reduced by undernutrition in developing animals, but not so much as the weight of other tissues. In other words, the brain continues to grow in spite of quite severe undernutrition so that the weight of the underfed developing brain is less than that of the age controls but more than that of the body weight controls.
>
> 3. Since the brain weight during development is less affected than the body weight, the brain weight/body weight ratio, which normally declines as the animal matures, declines less rapidly in the underfed developing series. At any time during development, therefore, this ratio in an underfed animal will be higher than that of the age controls. For rather spurious mathematical reasons, however,[2] it may be almost exactly the same *in rats* as that of the weight controls.
>
> 4. It was also well known that certain developmental processes in the brain, especially myelination, proceeded almost normally in spite of undernutrition, whether measured histologically or by analytical chemistry.[5]

III. THE VULNERABLE PERIOD HYPOTHESIS

Most recent investigations have been concerned with the identification of periods of enhanced vulnerability at certain stages of brain growth, compared with the remarkable immunity of the adult brain to the effects of undernutrition. However severely the adult is starved, there are no detectable effects on either brain weight or composition. This has been verified for brain lipids and DNA,[6] and RNA, DNA, and protein nitrogen.[7,8] Some small alterations in brain amino acids have been reported.[9] However, all these findings refer to whole brain, and like all such measures, cannot exclude regional effects too small to influence whole brain values.

By contrast, even mild undernutrition at the time of the brain's fastest growth affects its composition, sometimes permanently. This has given rise to the "vulnerable period hypothesis" which is analogous to similar hypotheses in the developmental behavioral sciences.[10,11] In the general course of normal brain growth there is a transient period of rapid increase in rate, the "brain growth spurt." The weight curve is sigmoid, and so are the curves for increasing amounts of many constituents. In its simplest form, the hypothesis states that

1. The developing brain will be most vulnerable to restriction at the time of the brain "growth spurt."

2. At this time the restriction required to produce an effect may only be mild. It will have to be increasingly severe the further away from this period it is imposed; until ultimately the adult state is reached when the brain is virtually invulnerable.

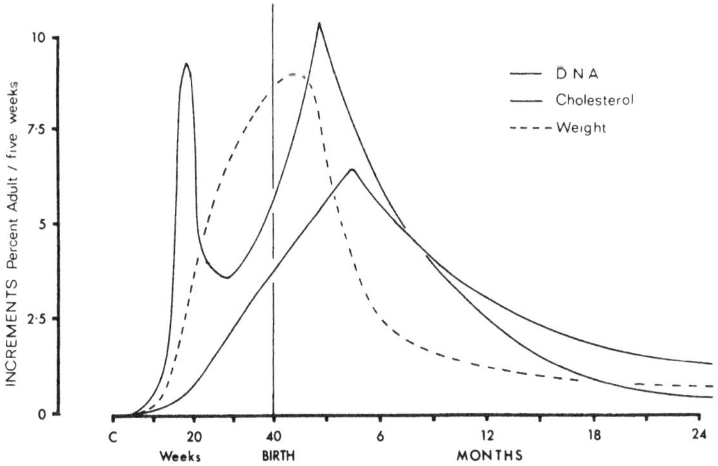

Fig. 1. Velocity curves showing incremental rates of DNA (two peaks), cholesterol (single peak), and fresh weight in whole human brain. Data from analysis of 200 normal human brains.[18] Note the bimodal curve for DNA, representing neuroblast followed by glial multiplication.

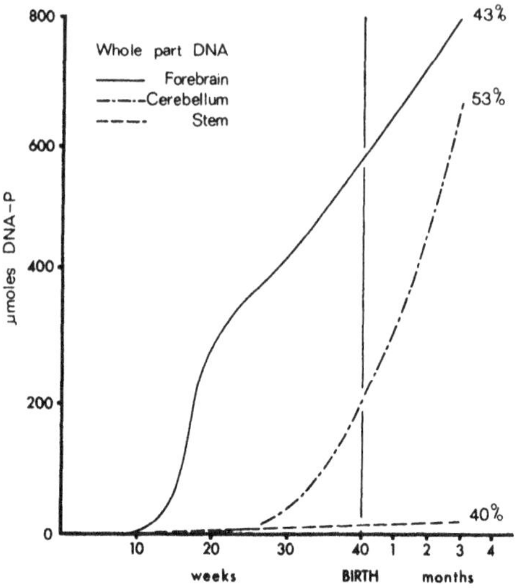

Fig. 2. Rising levels of DNA in human forebrain, cerebellum, and stem. Data from analysis of 200 normal human brains.[18] Percentage figures denote percent of adult values achieved at 3 postnatal months.

3. Restriction at the time of the "growth spurt" may result in deficits which may be irrecoverable.

This hypothesis is similar to that of Winick and Noble[12] who relate vulnerability with permanent sequelae to the period of most active *cell division*. It differs from Winick's and Noble's hypothesis in including other events within the "growth spurt," some of which do not occur at the same time as cell division (e.g., myelination).[13]

Thus stated, the hypothesis is deceptively oversimplified. The various major processes concerned with brain growth pass through their own "growth spurts" at different times, and each may be specifically vulnerable at these times (Fig. 1). There is evidence that two separate periods of rapid cell multiplication can be distinguished: one for neurons and a later one for glia[14] (Figs. 2 and 3). Furthermore, each of these individual growth spurts occurs at different times in different brain regions, and also at different rates according to the general rates obtaining in those regions. Such temporal and spatial heterogeneity[13] of the developing brain threatens to make testing of the hypothesis difficult, and it may therefore seem surprising that results obtained by examining the whole brain or whole, gross regions have actually produced substantial evidence in its favor.[2]

IV. NEUROCHEMICAL INDICES OF DEVELOPING STRUCTURE

The testing of the vulnerable period hypothesis demands a quantitative account of brain development in order both to identify the precise timing of the various "growth spurts," as well as to obtain a quantitative measure of deviation from the normally expected values. Except among those trained in quantitative developmental neuroanatomy, it is not sufficiently appreciated that the direct microscopic measuring of three-dimensional histological growth can be prohibitively laborious.[15] Counting all the cells in a single brain would take a large staff many months. It is scarcely surprising that histological methods have been temporarily abandoned in experiments involving hundreds of animals. At some considerable sacrifice of anatomical specificity, DNA estimation has replaced direct cell counting. Similarly, the chemical estimation of certain lipids has been a necessary substitute for measuring myelination.

In the former case, the total DNA in the whole organ or region can reasonably be regarded as an index of the total number of cells, but this does not distinguish between different cell types. The well-known tetraploidy of a tiny minority[16,17] of normal brain cells does not significantly upset this assumption. Fortunately, neuronal multiplication precedes glial, and there is recent evidence from developing human brain DNA that two DNA "growth spurts" can sometimes be distinguished representing the multiplication first of neuroblasts and then of glia.[14] However useful total DNA expressions can be, great caution must be exercised in the use of concentration expressions such as DNA per unit wet weight of tissue. This parameter normally *falls* during development in both the forebrain and brainstem, even though it may rise in the cerebellum. Such movements are not so much related to rates of cell division as to changes in factors contributing to the denominator of the concentration expression. The weight of the tissue increases mainly due to increases of cell size, growth and development of cell processes, and synthesis of "noncellular" myelin; and so the rising quantity of DNA can be diluted and the DNA/-unit wet weight fall even though cell numbers are still increasing rapidly. In the special case of the cerebellum, the multiplication is so extremely fast that this factor is overriding and even the concentration of DNA rises.[18]

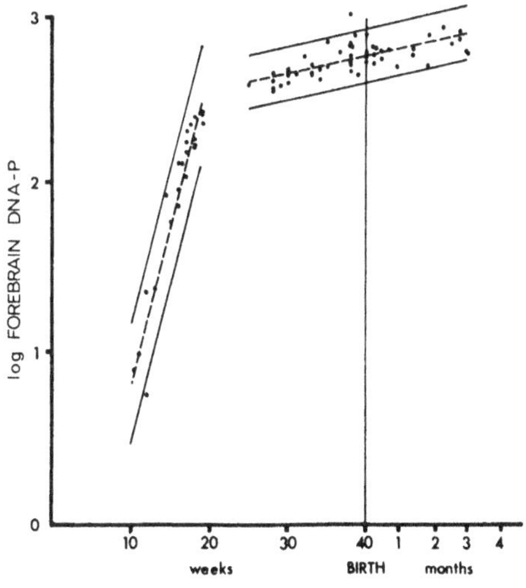

Fig. 3. Two phases of human forebrain DNA multiplication. The data from Fig. 2 have been analyzed in two groups, from 0–22 gestational weeks and from 25 gestational weeks to 3 postnatal months. Two exponentials are shown representing neuroblast followed by glial multiplication.[18] (Broken lines are calculated regression lines, solid lines are 95% confidence limits.)

Quantitative histological assessment of myelination is even more difficult than cell counting, and recourse must be had to the much simpler chemical estimation of myelin lipids. Unfortunately neurochemists have changed their ideas many times lately about the composition of myelin at different developmental stages, and it is difficult to choose a suitable myelin index substance. When dealing with so many samples it has proved useful to estimate cholesterol, even though it may not be a very characteristic myelin material. Indeed, in the early stages preceding myelination, it represents other membrane. Nevertheless it can be estimated in lipid extracts without prior fractionation or separation, and although its "growth spurt" precedes that of (for example) cerebroside, it bears a constant time-relation to the arrival of more typical myelin lipids. It is probably therefore a good, simple index of lipid accumulation.

In the case of a myelin lipid, the amount per unit fresh weight rather than the total organ quantity is likely to be the most useful measure of maturity. The total quantity in the tissue is too much dominated by the growth in tissue weight to be useful, and mere size of the brain is unlikely to be a useful parameter. As myelination proceeds, and the sheaths thicken, lipid is laid down largely at the expense of tissue water and the concentration (lipid/gram wet weight) increases. By contrast, as the brain grows in size and axons lengthen, the elongating portions become correspondingly myelinated. This second process makes much less contribution to the concentration value since both the amount of myelin and the weight increase at the same time. Thus maturation may be considered mainly to consist of the sheath-thickening processes and will hence be represented more meaningfully by increasing concentration.[13]

The use of these indices of brain growth is not, however, quite as time-saving as it may sound. Growth and development is a continuous process in time, and the interrelationships between weight, water content, cellularity, and myelination are constantly changing. The commonest source of error in describing the pattern of growth by these means is due to insufficient numbers of points being plotted on the time scale. For some normal control purposes it has been found necessary, for example, to obtain five or six brains every 2 days for the first 5 weeks in the rat before being able to draw velocity curves with any confidence that the normal range of biological variation has been appreciated. In addition, control animals must be selected as conforming to a predetermined normal pattern of body weight for age; otherwise the range of developmental ages obtainable for each chronological age becomes too wide, due to variations such as litter size and lactational performance.

V. INTERSPECIES COMPARISONS

In view of regional differences of timing, it is perhaps surprising that even when the whole brain is homogenized and its weight and content of

DNA and cholesterol expressed against age, it is still possible to detect a sequence of biologically meaningful events. In all species so far examined, the DNA "growth spurt," representing predominantly glial cell multiplication, precedes the cholesterol "growth spurt" of myelination. This is in complete accord with the histologically observed sequence of glial cell division preceding the manufacture by those cells of myelin.[15] The sequence has already been demonstrated in rats,[13] pigs,[19] guinea pigs,[20] and humans,[18] and from this it may be assumed that neither the general pattern of brain growth nor brain composition shows any species differences which are important for these purposes. The main difference to be allowed for in extrapolating from one species to another is that of the timing of birth in relation to the vulnerable period. There are great differences in brain maturity at birth, which occurs at the end of the brain "growth spurt" in guinea pigs, before it in rats, and during it in pigs and humans. One should never refer, for example, to a neonatal animal without specifying the species and without knowing its brain growth characteristics.

VI. UNDERNUTRITION AND DEVELOPING STRUCTURE

By using the above simple measures of structural development, the vulnerable period hypothesis has been crudely tested in developing brain. Most experiments have used the rat, since the brain "growth spurt" in this species is almost entirely encompassed by the suckling period of the first three postnatal weeks. Comparatively mild undernutrition can be imposed by random cross-fostering in artificially large litters. Thus the various permutations and combinations of different periods of undernutrition of varying severity can be arranged, and Table 1 shows those which have been tried.[2,12,21-23] It is possible to summarize the many results by stating that the only animals showing any lasting deficits as adults, in whole-part DNA or cholesterol concentration, are those whose undernutrition coincided with the brain growth spurt; i.e., those from groups II, III, and IV. Furthermore, those from groups II and IV show ultimate deficits as adults which appear to be little greater than those of group III. The deficits seem to be irrecoverable, except where rehabilitation is begun before the "growth spurt" is over.[21,24] Thus even though the hypothesis has so far been tested only very crudely, it has already been substantially supported.

Further powerful support comes from an analysis of the effects on those brain regions which grow faster than others. If vulnerability is related to rate of growth, then the cerebellum, which grows much faster than the rest of the brain,[19] should be differentially susceptible:[25] and so it has been shown to be.[21,26]

Attempts to demonstrate comparable results in man have suffered from the inherent difficulties in collecting autopsy material from underfed

TABLE I

Timing and Severity of Experimental Undernutrition in the Rat[a]

Group	Severe gestational	Moderate suckling	Severe weanling	Very severe adult
I	+	–	–	–
II	+————	+	–	–
III	–	+	–	–
IV	–	+————	+	–
V	–	–	+	–
VI	–	–	–	+

[a] + = Undernutrition, – = good nutrition.

children. The human underprivileged population is far less homogeneous than would be acceptable in an animal experiment. Observations have therefore been limited to relatively isolated cases. They have, of necessity, been compared with far too few controls, so that it has been difficult to take any range of normal values into account. Nevertheless, some gross deficits in DNA and lipid content have been reported[27,28] and these correspond well with values which were predictable from the animal models. It is perhaps churlish to complain of the inadequacy of the human data when it has been so difficult to collect. Perhaps the main difficulty, that of having a well-defined range of normal values, is now being overcome.[18,29,30]

VII. UNDERLYING MECHANISM OF THE VULNERABLE PERIOD

It would be easy to conclude from the above account that the "growth spurt" was vulnerable to nutritional restriction simply by virtue of the increased demand for structural precursors at that time. However, the proportion of the nutritional intake utilized in the development of structure is probably much too small to be so much influenced by restricted supplies, and the mechanism is likely to be otherwise.

There is evidence that there is a group of synthetic pathways during development which are transiently present, and served by transient enzyme systems. Thus cholesterol, which is mainly synthesized within the brain[31] during the myelination period, is not synthesized there to any appreciable extent at other times, presumably because the synthetic pathways are not available. The question arises whether the growth retardation produced by undernutrition (or by any other means) can delay and postpone the availability of such systems to a later time; or whether their appearance is strictly chronologically determined. In the former case, a poorly endowed suckling period succeeded by an *ad libitum* weanling diet would permit considerable "catch-up." In the latter, there would be a much more vulnerable once-and-for-all opportunity to operate a brain "growth spurt" strictly regulated by the calendar. Recent experimental investigations suggest the latter to be the case. If one constructs velocity curves of gain in brain weight, total DNA, and cholesterol in rats raised in growth restricting conditions, it is found in all three cases that the timing of the peak velocity is not altered compared with the controls. It is merely a lower peak and there is no resurgence or "catch-up" on restoring an *ad libitum* diet at weaning. It therefore seems as though the brain is incapable of varying the chronological timing of some of its major growth processes, and this is potentially of immense significance to our own species.[32]

This new concept of the nature of developmental vulnerability is capable of very much more meaningful investigation by neurochemists. It should, for example, be possible to identify enzymes whose role in growth is transient, and to investigate the effects on the timing of their peak activity in growth-restricted animals. Such an investigation has already been performed with respect to sulfokinase, and the idea of chronological immutability has once again been confirmed.[33]

Another consequence of this idea is that the vulnerable period hypothesis will apply as well to other developmental restrictions as to undernutrition. In other words, undernutrition does not produce its effects specifically by virtue of the restriction of supplies. The same periods of growth are likely to be vulnerable to such other diverse insults as inborn errors of metabolism,[34] hypothyroidism, and X-irradiation. Both clinical and experimental observations support this conjecture.

VIII. BEHAVIORAL EFFECTS

This chapter began by stressing the importance of the lasting effects of early undernutrition on brain performance. Our final and most difficult task is to consider this, both in the clinical human situation and in the experimental animal.

Clinically there is no doubt whatever that a poorer cerebral performance, amounting to a demonstrable reduction of human potential if not to frank mental subnormality, can be associated with a history of very early undernutrition.[35–38] It is quite another matter to imply direct causation, and even if causation could be demonstrated, it would still be necessary to show that the effect had been mediated through an alteration in the physical state of the brain. We are therefore a very long way from tracing the pathway of the association. However, in the present state of virtually complete ignorance of the link between tangible molecules and higher behavior, we need not be ashamed. It may already be significant that least some of the more easily measured constituent parameters have been shown to be permanently affected. We do not have to claim at this stage that these particular substances are of behavioral significance.

An obvious line of approach may seem to be the behavior of the experimental animal. Unfortunately this has so far yielded very conflicting results. One must be impressed with the exquisite sensitivity of developing behavior to permanent distortion. On the other hand, it has sometimes been reported that very gross manipulations of early experience in rats have produced no behavioral effects.[39] There is no doubt that growth restriction also produces a retardation of the course of functional development at the time, and several authors have reported long-term consequences.[40–44] It must probably be assumed that we have not yet discovered adequately sensitive behavioral tests appropriate to this problem, nor how to control the other environmental factors, some of which may tend to compensate for the effects of early undernutrition.

IX. REFERENCES

1. N. S. Scrimshaw and J. E. Gordon, eds. *Malnutrition, Learning and Behaviour*, M.I.T. Press, Boston (1968).
2. J. Dobbing, *in Applied Neurochemistry* (A. N. Davison and J. Dobbing, eds.) Blackwell, Oxford (1968).
3. M. Winick, Malnutrition and brain development, *J. Pediat.* **74**:667–679 (1969).
4. J. Cravioto, H. G. Birch, E. R. de Licardie, and L. Rosales, The ecology of infant weight gain in a preindustrial society, *Acta Paediat. Scand.* **56**:71–84 (1967).
5. J. Dobbing, *in Developmental Neurobiology* (W. A. Himwich, ed.) Charles C Thomas, Springfield, Ill. (1970).
6. J. Dobbing, *in Malnutrition, Learning and Behavior* (N. S. Scrimshaw and J. E. Gordon, eds.) M.I.T. Press, Boston (1968).
7. P. Mandel, M. Jacob, and L. Mandel, Étude sur le métabolisme des acides nucléiques. I. Action du jeûne protéique prolongé sur les deux acides nucléiques du foie, du rein et du cerveau. *Bull. Soc. Chim. Biol.* **32**:80–78 (1950).
8. P. Lehr and J. Gayet, Response of the cerebral cortex of the rat to prolonged protein depletion. I. Tissue weight, nitrogen, DNA and proteins, *J. Neurochem.* **10**:169–176 (1963).
9. P. Lehr and J. Gayet, Response of the cerebral cortex of the rat to prolonged protein depletion. II Free aspartic, glutamic and gamma-aminobutyric acids, *J. Neurochem.* **13**:805–810 (1966).
10. M. W. Fox, Neuro-behavioral ontogeny: a synthesis of ethological and neurophysiological concepts, *Brain Res.* **2**:3–20 (1966).
11. J. P. Scott, Comparative psychology and ethology, *Amer. Rev. Psychol.* **18**:1–40 (1967).
12. M. Winick and A. Noble, Cellular response in rats during malnutrition at various ages, *J. Nutr.* **89**:300–306 (1966).

13. A. N. Davison and J. Dobbing, *in Applied Neurochemistry* (A. N. Davison and J. Dobbing, eds.) Blackwell, Oxford (1968).
14. J. Dobbing and J. Sands, Timing of neuroblast multiplication in developing human brain, *Nature*, **226**:639–640.
15. J. P. M. Bensted, J. Dobbing, R. S. Morgan, R. T. W. Reid, and G. Payling Wright, Neurological development and myelination in the spinal cord of the chick embryo, *J. Embryol. Exp. Morph.* **5**:428–437 (1957).
16. L. W. Lapham, Tetraploid DNA content of Purkinje neurons of human cerebellar cortex, *Science* **159**:310–312 (1968).
17. E. Howard, Reductions in size and total DNA of cerebrum and cerebellum in adult mice after corticosterone treatment in infancy, *Exp. Neurol.* **22**:191–208 (1968).
18. J. Dobbing and J. Sands, Human brain development measured quantitatively, (*in preparation*).
19. J. W. T. Dickerson and J. Dobbing, Prenatal and postnatal growth and development of the central nervous system of the pig, *Proc. Roy. Soc. B.* **166**:384–395 (1967).
20. J. Dobbing and J. Sands, Growth and development of the brain and spinal cord of the guinea pig, *Brain Res.* **17**:115–123 (1970).
21. W. J. Culley and R. O. Lineberger, Effect of undernutrition on the size and composition of the rat brain, *J. Nutr.* **96**:375–381 (1968).
22. J. W. Benton, H. W. Moser, P. R. Dodge, and S. Carr, Modification of the schedule of myelination in the rat by early nutritional deprivation, *Pediatrics*, **38**:801–807 (1966).
23. J. Dobbing and E. M. Widdowson, The effect of undernutrition and subsequent rehabilitation on myelination of rat brain as measured by its composition, *Brain*, **88**:357–366 (1965).
24. M. Winick, I. Fish, and P. Rosso, Cellular recovery in rat tissues after a brief period of neonatal malnutrition, *J. Nutr.* **95**:623–626 (1968).
25. J. W. T. Dickerson and J. Dobbing, Some peculiarities of cerebellar growth, *Proc. Roy. Soc. Med.* **59**:1088 (1966).
26. H. P. Chase, W. F. B. Lindsley, and D. O'Brien, Undernutrition and cerebellar development, *Nature*, **221**:554–555 (1969).
27. M. Winick and P. Rosso, The effect of severe early malnutrition on cellular growth of human brain, *Pediat. Res.* **3**:181–184 (1969).
28. M. A. Fishman, A. L. Prensky, and P. R. Dodge, Low content of cerebral lipids in infants suffering from malnutrition, *Nature*, **221**:552–553 (1969).
29. E. Howard, D. M. Granoff, and P. Bujnovsky, DNA, RNA, and cholesterol increases in cerebrum and cerebellum during development of human fetus, *Brain Res.* **14**:697–706 (1969).
30. M. Winick, Changes in nucleic acid and protein content of the human brain during growth, *Pediat. Res.* **2**:352–355 (1968).
31. J. Dobbing, The entry of cholesterol into rat brain during development, *J. Neurochem.* **10**:739–742.
32. J. Dobbing and J. Sands, Chronological determination of brain growth (*in preparation.*)
33. H. P. Chase, J. Dorsey, and G. M. McKhann, The effect of malnutrition on the synthesis of a myelin lipid, *Pediatrics* **40**:551–560 (1967).
34. H. P. Chase and D. O'Brien, Effect of excess phenylalanine and of other amino acids on brain development in the infant rat, *Pediat. Res.* **4**:96–102 (1970).
35. V. Cabak and R. Najdanvic, Effect of undernutrition in early life on physical and mental development, *Arch. Dis. Child.* **40**:532–534 (1965).

36. J. Cravioto, E. R. de Licardie, and H. G. Birch, Nutrition, growth and neurointegrative development: an experimental and ecologic study, *Pediatrics* **38**:319–372 (1966).
37. M. B. Stoch and P. M. Smythe, Does undernutrition during infancy inhibit brain growth and subsequent intellectual development? *Arch. Dis. Child.* **38**:546–552 (1963).
38. M. B. Stoch and P. M. Smythe, The effect of undernutrition during infancy on subsequent brain growth and intellectual development, *South African Med. J.* **41**:1027–1030 (1967).
39. H. A. Guthrie, Severe undernutrition in early infancy and behavior in rehabilitated albino rats, *Physiol. Behav.* **3**:619–623 (1968).
40. R. H. Barnes, S. R. Cannold, R. R. Zimmerman, H. Simmons, R. B. MacLeod, and L. Krook, Influence of nutritional deprivations in early life on learning behaviour of rats as measured by performance in a water maze, *J. Nutr.* **89**:399–410 (1966).
41. D. F. Caldwell and J. A. Churchill, Learning ability in the progeny of rats administered a protein-deficient diet during the second half of gestation, *Neurology* **17**:95–99 (1967).
42. R. H. Barnes, Experimental animal approaches to the study of early malnutrition and mental development, *Fed. Proc.* **26**:144–147.
43. S. Frankova and R. H. Barnes, Influence of malnutrition in early life on exploratory behavior of rats, *J. Nutr.* **96**:477–484 (1968).
44. S. Frankova and R. H. Barnes, Effect of malnutrition in early life on avoidance conditioning and behavior of adult rats, *J. Nutr.* **96**:485–493 (1968).

Chapter 10

EFFECT OF ISCHEMIA

H. S. Maker and G. M. Lehrer

Department of Neurology
Division of Neurochemistry
The Mount Sinai School of Medicine of the City University of New York
New York, New York

I. INTRODUCTION

The metabolism of a tissue totally deprived of its blood supply changes immediately from a steady-state condition in which substrates are continually supplied and some end products and by-products continually removed to a "closed system."(1) Under such conditions: (1) oxidative phosphorylation and metabolism cease; (2) metabolism becomes dependent on endogenous substrate; (3) end products accumulate; (4) energy reserves decline; (5) there is a tendency of reactions toward equilibrium, and finally, (6) failure of function and eventually major structural disruption occur. The effect of ischemia on neural tissue has been examined mainly from two viewpoints: one, focusing on the clinical problems of cerebrovascular disease, with application of physiological or anatomical techniques, and the second, in order to elucidate biochemical mechanisms uses ischemia primarily as a method of modifying metabolism. In this review an attempt will be made to integrate the data obtained by both approaches in order to trace the pattern of events occurring during ischemia. Earlier reviews by Himwich,(2) Schneider,(3) and Stone(4) should be consulted for historical perspective. Because of space limitation, investigations conducted in the present decade and key publications which give access to the earlier literature will be emphasized.

II. METHODOLOGY

Before entering upon a discussion of available data, it is useful to subject the various methods which have been used in studying brain ischemia to critical evaluation. This is particularly important because workers in the field have not always considered crucial sources of error and methodological artifacts which must, of necessity, affect the validity and interpretation of the data obtained.

Measurements of metabolite levels in arterial and venous blood from the brain as well as cerebral blood flow allow calculations of the overall balance of these substances in normal and ischemic brain. These methods permit continuous and repeated measurements and are suitable for use in human subjects.[5,6] However, variations in techniques of blood sampling and of measuring cerebral blood flow, alterations in the conditions of measurement, e.g., sedation, etc., and assumptions of unmeasurable parameters complicate the interpretation of data. The reader is referred to recent reviews for complete discussion.[7]

Craniotomy makes possible the sampling of pure cerebral venous blood and the measurement of gas tensions, *p*H, sodium, and potassium at the cortical surface by means of ion-specific electrodes. These methods allow continuous recordings to be made from the same brain under both "normal" conditions and during ischemic change.[8] The effects of anesthesia, variations in properties of electrodes, and artifactual changes in the exposed cortex (temperature, spreading depression) should be considered in evaluating these results. Only net electrolyte changes are recorded in this technique and the values may differ considerably from those obtained from assays made by chemical methods.

The direct measurement of labile metabolites in brain is subject to errors resulting from changes occurring during the period between tissue sampling or loss of blood supply and the cessation of all metabolism produced by freezing or other means. When the brain is removed before freezing, early changes in labile metabolites are obscured.[9] The most rapid cessation of metabolism is obtained by freezing a thin layer of exposed brain tissue with a chilled heat conductor[10] or by pouring a refrigerant over the exposed surface and using only that portion of tissue which is frozen first.[11] These methods require anesthesia and craniotomy with concomitant difficulties in maintenance of respiration. A method for obtaining reproducible metabolite concentrations, approaching instantaneous normal values for whole brain, depends on freezing the entire animal. The brain is then removed and metabolites are extracted at low temperature.[1] Technical factors to be considered have been discussed.[12] Such a procedure is suitable for small rodents whose brain can be frozen within an acceptable period.

The freezing characteristics of tissue introduce an important variable into the investigation of rapid changes in brain metabolism. Several aspects of this problem have been reviewed in recent symposia.[13,14] Although bringing the temperature rapidly below zero greatly slows metabolism, some chemical reactions may continue at a detectable rate unless water and cations are immobilized in the solid state, that is, below the eutectic point of simple solutions. In complex salt solutions, as are found in tissue, there is a broad eutectic zone characterized by slow thermal and electrical resistance changes with cooling. Katzman and Wilson[15] found the resistance of a 5-mm brain section to increase progressively, to reach the value associated with the solid state at −46°C. This value might be lower than would be expected because of supercooling. The liquefaction of the aqueous phase begins at −33°C during warming. Katzman and Wilson[15] suggested that, in order to ensure absence of chemical activity, tissue temperature must be maintained below −33°, preferably below −46°C.

Temperatures at various levels from the surface of the brain have been recorded while animals were plunged into various refrigerants. When a whole mouse is plunged into liquid nitrogen, the temperature in the vicinity of the fourth ventricle reaches 0°C in 7 sec and −30°C within 35 sec.[16] Slightly longer freezing times were found when isopentane, chilled by Dry Ice-acetone was used as the refrigerant. In contrast, liquid helium at a lower temperature proved a much slower freezing agent. While the brain of a 35-g rat freezes (reaches 0°C) in 9–20 sec,[17] the surface of the brain of a 250-g rat plunged into liquid nitrogen does not drop below freezing until 24 to 40 sec have elapsed and deeper portions remain unfrozen for more than a minute.[18] A thin piece of cortex frozen between liquid nitrogen-chilled tongs made of brass blocks reaches 0° within 0.1 sec.[19] Apparently the temperature at a given depth tends to remain close to physiological before freezing and then declines sharply.[17,18] This phenomenon may indicate that some circulation is maintained to the depths while the outer parts are cooling, when the whole animal is frozen. The heart continues to beat for 30 sec after the upper half of a mouse is submerged in liquid oxygen.[20] A refrigerant of low volatility, good conductivity, and sufficiently low freezing point such as Freon 12 or 22, chilled to its freezing point, is preferable to a more volatile liquid at a lower temperature. The formation of gas bubbles about the object when it is frozen in a volatile refrigerant interferes with heat conduction and greatly slows cooling. In collaboration with Dr. Robert Katzman, the authors have measured the lactate concentrations at the surface and in deeper gray areas of mouse brain to determine the rate at which metabolism was slowed when different refrigerants were used. Lactate levels were 10–30% higher in the deeper regions when whole mice were frozen, indicating a gradient of freezing equivalent to 1–4 sec between freezing of the surface and the depth. The poorest results in terms of rapidity of stopping metabolism were obtained with liquid helium and liquid nitrogen. Metabolism (lactate formation) was more quickly arrested in isopentane chilled to the temperature of Dry Ice-acetone, despite its being at a higher temperature than the former agents. The best results were obtained using Freon 12 chilled to its freezing point (−154°) by liquid nitrogen.

Many studies have used the method of Levine[21] to produce ischemic lesions in rodent brain. This method involves occlusion of one carotid artery followed by a period of anoxia and is not suitable for measuring acute changes. A simple way to cause the complete cessation of blood flow to the brain is decapitation. However, with one exception,[9] most studies have found higher rates of change of labile brain metabolite levels during freezing following decapitation than when the animals are frozen whole.[1,12,22] This finding has been attributed to a massive neuronal discharge concurrent with decapitation which stimulates metabolism above resting levels; however, some of these differences could be explained by the possibility that some circulation to the brain is maintained during freezing of the whole animal without decapitation. Since a further increase in metabolic rate could be obtained by electrical stimulation of the isolated head,[23] it must be postulated that the brain is not maximally stimulated by decapitation. Nonetheless, some increase of metabolic flux with decapitation must always be considered in evaluating ischemic changes produced by this method.

A fall in brain temperature during an experimental procedure may slow the rate of metabolic change.[22] Temperature control therefore is important. The severed head of a small rodent cools rather rapidly. But, at room temperature, such fall is not sufficiently fast to cause a significant difference in the first one or two minutes.[1,24] For longer periods of observation, the severed head should be maintained at 37°C to make conditions more comparable to those *in vivo*. The body temperature of larger animals declines more slowly; the rabbit carcass, for instance, loses only 2° in 2 hr at room temperature.[25] Brain temperatures may follow a similar pattern.

Generally, anesthesia results in increases in brain glucose and glycogen and a slowing of metabolism before and during ischemia. It may also slow the changes induced by decapitation.[1,26,27] Barbiturates apparently both increase the amount of glucose extracted from the blood and decrease neuronal energy-requiring processes.[27] The levels of brain glucose and glycogen and, therefore, the total available metabolic energy reserve and eventual lactate production also vary with blood glucose levels.[26] Hence, blood glucose levels must be considered in evaluating ischemia studies.

Ischemia, as a stimulus of glycolysis under closed system conditions, has been utilized to locate points of metabolic control and the relationships among the various steps of carbohydrate metabolism.[28,29] Lowry *et al.*[1] have estimated steady-state metabolic fluxes in mouse brain from measurements of the early (within the first 30 sec) changes in metabolite levels during ischemia. They assume that the rate of high energy phosphate utilization ($\Delta{\sim}P$) is unchanged during the first few seconds after decapitation and reflects the rate of use before decapitation. The formula[1] for $\Delta{\sim}P$ is based on the expected yield of energy from each metabolite:

$$\Delta{\sim}P = 2\ \Delta\text{ATP} + \Delta\text{ADP} + \Delta\text{PC} + 2.9\ \Delta\text{glycogen}\ \text{(as glucosyl)} + 2\ \Delta\text{glucose}$$

The steady-state aerobic glucose flux can be calculated from the $\Delta{\sim}P$ value, assuming that glucose is the only source of energy and that under aerobic conditions 85% of glucose is oxidized to CO_2, yielding 38 ~P eq/mol and the remainder goes to lactate. The rate of glucose turnover required to supply the calculated $\Delta{\sim}P$ under these conditions is taken as the steady-state utilization rate (i.e., $\Delta{\sim}P/33$ = aerobic glucose).

Estimates of aerobic $\Delta{\sim}P$ by this method yield values somewhat higher than those expected from measurements of O_2 consumption[30] or glucose utilization[31] *in vivo*, possibly because the added stimulation of decapitation increases P utilization above basal rates. Furthermore, the lactate production during ischemia may lag behind carbohydrate disappearance[27] making it difficult to assess precisely the completeness of the carbohydrate energy yield. However, the method has distinct advantages in approximating aerobic steady-state fluxes in the brain by measurements under closed-system conditions without the introduction of artifacts necessary to measurements *in vitro*.

III. VASCULAR FACTORS

Circulatory factors modify the pattern and rate of development of ischemic changes. Under steady-state conditions brain tissue extracts from the blood primarily oxygen and glucose. Although other essential substances are also extracted, glucose is the major source of metabolic energy and for many structural requirements.[6] Despite this dependence on glucose, the brain at rest extracts only a fraction of the available glucose from the blood.[1,24] If the flow of blood through the brain is slowed, a larger fraction of blood glucose is extracted.[32] More glucose may also be extracted if oxygen is lacking.[33] Brain tissue, even at rest and well oxygenated, produces a small amount of lactate which is carried away by the venous blood.

When cerebral blood flow is impaired by a decline in cardiac output or vascular block, large brain vessels have the capacity to adjust their diameter so as to maintain a constant blood flow (autoregulation).[34] Only when this mechanism fails to maintain the blood supply do major tissue changes ensue. Although central neural regulation of vessel diameter may be a factor,[35] the most important mechanisms are probably local and related to the vessel's immediate environment. A mechanical myogenic response (relaxation) to a lessened perfusion pressure (Bayliss effect),[36] as well as metabolic changes caused by the decreased circulation are major modifiers of vessel diameter. A rise in tissue pCO_2 and fall in pO_2 both contribute to vasodilatation.[37-39] Although arterial pH has little effect on vessel diameter, the decline in pH of the extracellular fluid caused by the increase of tissue CO_2 and lactate may be the mechanism responsible for much of the metabolic vascular effect.[40,41] Another compensatory mechanism for decreased blood flow which probably precedes autoregulatory changes depends on the lower affinity of hemoglobin for oxygen at lower pH's (Bohr effect) and the added reduction of oxyhemoglobin by metabolic CO_2.[42]

If blood flow falls below a critical value (0.40 ml/g/min; mean arterial blood pressure 70 mm Hg), the autoregulatory mechanisms fail, resulting in ischemic changes in the tissue.[43] Siesjö and his collaborators[43a] have found an increasing lactate/pyruvate ratio in whole rat brain as systemic blood pressure is reduced below a critical level. Whole brain lactate rises before any change in high energy phosphate substances can be detected, suggesting an effective anaerobic metabolism during partial ischemia.[43a] When a local occlusion occurs, areas adjoining that which is totally ischemic have a paradoxically high blood supply in "excess" of need. An explanation may be found in the combination of decreased oxygen requirement and failure of circulatory autoregulation in the partly ischemic areas. This has been termed the "luxury perfusion syndrome."[44,45] Another explanation might be the opening of anastomotic thoroughfare channels.[46]

IV. METABOLIC CHANGES IN WHOLE BRAIN

Following a vascular occlusion, the cortical oxygen level declines to zero in from 30 to 150 sec while CO_2 increases. Soon after oxygen becomes undetectable the pCO_2 also declines.[37] Although the cessation of blood flow causes manifold changes both in the supply of some metabolites and the removal of others, the earliest tissue changes can be reproduced entirely by the lack of oxygen.[22,47] This implies that the early changes in metabolism are essentially due to failure of glycolysis to supply high energy phosphate needs. The normal distribution of oxygen within the brain is roughly described by the pericapillary tissue cylinder model of Krogh,[48] although some of its assumptions have been challenged.[49] This model relates oxygen tensions within the vascular system to those at various distances from a capillary. During oxygen lack, changes supposedly begin first in those areas which have the lowest oxygen tensions under normal circumstances. The postulated minimal pO_2 when metabolic alterations begin (11 mm Hg)[48] has been related to that required to saturate cytochrome oxidase.[50] Without oxygen for the respiratory chain, it might be assumed that the pyridine nucleotides linked to the chain remain reduced, causing a characteristic increase in the tissue fluorescence when excited by 340 nm light. At least 60% of the increased fluorescence of rat brain cortex during anoxia is sensitive to amytal and thus may be due to mitochrondrial NADH.[51] Mechanisms for the anaerobic oxidation of NADH are mainly localized in the cytosol and evidence from work with the grasshopper spermatid suggests that, at least in this cell, mitochondrial NAD can be completely reduced before any change in cytoplasmic NAD occurs.[51] Tissue fluorescence in anoxia increases much more rapidly in brain cortex than in kidney, as might be expected from the greater oxygen demand of the brain. By determining the fluorescence of reduced pyridine nucleotide in yeast cell suspensions as related to variations in oxygen concentration, Chance *et al.*[51] estimated that the initial change in pyridine nucleotide reduction occurs when the partial pressure of intracellular O_2 falls to 0.6 mm Hg and 80% of maximal reduction is reached at about 1/10 of that. Rat cerebral cortex undergoes a 1% increment in fluorescence when inspired gas contains 8% O_2 and a 30–60% increase in fluorescence occurs with 4% O_2.[51] If the fluorescence changes can be taken as indicators of intracellular O_2 concentrations, they indicate extremely efficient intracellular oxygen utilization and represent a much lower critical oxygen tension than suggested by the Krogh model. Since in neither case was oxygen tension measured directly, the minimal intracellular concentration of O_2 for function remains uncertain.

In the absence of respiration, oxidation of the NADH produced in glycolysis is mainly through lactate dehydrogenase and to a much lesser extent through α-glycerophosphate dehydrogenase. Direct measurements of the overall changes in pyridine nucleotide oxidation during ischemia were made by Lowry *et al.*[1] In the whole brains of 10-day-old mice, the steady-state level of NADH (29 μmoles/kg) is approximately 10% of that of NAD^+ (332 μmoles/kg). During an 8-min ischemic period the $NADH/NAD^+$ ratio increases, with a three to four-fold increase of NADH. The $NADPH/NADP^+$ ratio (0.6) remains constant, though total NADP(H) declines. Overall NAD^+ levels are still fairly high at a time when glycolysis is greatly stimulated.[1] However, overall pyridine nucleotide levels may fail to reflect the oxidation–reduction states of the various metabolic compartments. The redox states of such compartments may best be estimated from the ratios of oxidized and reduced metabolites in equilibrium with the pyridine nucleotides of these compartments (e.g., pyruvate/-lactate and dihydroxyacetone PO_4/α-glycerophosphate), provided that certain assumptions are met.[52–54] The redox state of the cytoplasmic compartment of whole brain may be best reflected in the lactate/pyruvate ratios to be discussed below (see p. 281).

A. Carbohydrate Metabolites

The rates of change of intermediate metabolites during early ischemia can yield information on the modes and regulation of metabolism occurring during ischemia.[28] For this purpose, the data are treated as if all metabolic elements were as free to react with one another as in a true solution. Any departure of metabolite relationships from those that would obtain under such conditions may indicate either systematic errors in assay or the failure of the assumption of homogeneous distribution (i.e., compartmentation).

At the time of cessation of blood flow, substrate levels would be sufficient to sustain aerobic brain metabolism for several minutes. However, in the absence of oxygen, nature provides for an increase in glycolysis to compensate for the relative inefficiency of glycolysis as a source of high energy intermediates. This rapidly depletes the brain of available energy-yielding substrates. Several groups have assayed intermediates of carbohydrate metabolism in whole mammalian brain under various conditions. Although a variety of methods were used even within the same laboratory, the overall patterns of change in metabolite levels during ischemia appear consistent. The relationships among metabolite levels to be described are derived primarily from data on whole brain.

1. *Citric Acid Cycle Intermediates*

During ischemia, rapid changes in citric acid cycle intermediates are expected to occur in relation to the cessation of steps linked to the reduction of NAD^+ (FAD), with a tendency toward metabolic equilibrium. However, partly because several of the intermediates are involved in reactions outside of the cycle, the changes during ischemia do not conform simply to the oxidative changes. Several intermediates of the pathway were examined by Goldberg *et al.*[55] in unanesthetized adult mouse brain frozen in the intact animal and 5, 10, and 30 sec after decapitation. As already discussed, decapitation causes an increase in metabolic flux above that of simple ischemia. With the onset of ischemia, pyruvate increases (91 to 135 μmoles/kg in 10 sec) and citrate declines about 25% (327 to 252 μmoles/kg in 10 sec). Evidently pyruvate is not decarboxylated or acetyl-CoA fails to reach the site of the condensing enzyme. Isocitrate levels (16 μmoles/kg) do not change during early ischemia. Although both its utilization and formation are blocked in the absence of NAD^+, α-ketoglutarate declines rapidly during the first 5 sec of ischemia (127 to 48 μmoles/kg). Goldberg *et al.* suggested that this may be related to the increase in NH_3 and NADH, and a shift of the glutamic dehydrogenase reaction toward glutamate. The level of glutamate (see amino acids, p.284) does not change during ischemia; but its concentration (9.5 mmoles/kg) is so much greater than that of the related keto acid that a small increment in glutamate might not be detected. Succinate increases initially, then declines, only to rise once again in a biphasic fashion (686 to 995 μmoles/kg at 30 sec). The late increase in succinate is unlikely to be due to α-ketoglutarate oxidation in the absence of oxygen, and Goldberg *et al.* suggested that it may be formed from lactate by way of a pathway (demonstrable in liver) involving fumarate and a reversal of the succinate dehydrogenase reaction. Fumarate (70 to 108 μmoles/kg) and malate (387 to 481 μmoles/kg) increase between 10 and 30 sec after the onset of ischemia. The levels of these substances[55] indicate that the reaction through fumarase is close to equilibrium in the steady state, possibly explaining the lack of significant change in the malate/-fumarate ratio during ischemia. Oxaloacetate, like α-ketoglutarate, declines during early ischemia (4 to 2 μmoles/kg in 5 sec), perhaps because of amination or transamination reactions. The small change in citrate mitigates against the concept that a change in this metabolite is a major factor in the disinhibition of phosphofructokinase and the facilitation of glycolysis, unless the glycolytic compartment experiences a more significant citrate change.

2. *High Energy Phosphate ($\sim P$) Intermediates*

Oxidative steps of carbohydrate metabolism normally contribute 36 of the 38 net high energy phosphate bonds (as ATP) generated by the aerobic metabolism of a single glucose molecule. During basal, steady-state, oxidative metabolism approximately 15% of brain glucose is converted to lactate and does not enter the citric acid cycle. Therefore, the total net gain of $\sim$P is only 33 equivalents per glucose molecule. Following failure of respiration, the cell becomes entirely dependent for its energy needs on a small reserve of creatine phosphate and the generation of 2 equivalents of $\sim$P for each molecule of glucose and 2.9 equivalents per glucosyl unit of glycogen metabolized anaerobically. This is not sufficient to meet metabolic demands, and energy reserves rapidly decline. The rates of change of creatine phosphate and ATP in anesthetized rabbit brain during ischemia were studied by Thorn *et al.*[22] and in both unanesthetized and anesthetized mouse brain by Lowry *et al.*,[1] Gatfield *et al.*,[27] Goldberg *et al.*,[55] and Maker *et al.*[24] Thorn produced ischemia by inflating a pressure cuff around the animal's neck; the other investigators used decapitation.

The earliest fluxes calculated from the changes during the first few seconds of ischemia are 11 to 17 mmoles/kg wet brain/min for creatine phosphate (CP), and 1.5 to 2 mmoles/kg/min for ATP.[1,55] The greater initial fluxes found by Goldberg *et al.* are apparently related to their use of *in situ* freezing to determine preischemic values. During the first 30 sec of ischemia the average rate of decline in CP is 2.8 to 6.0 mmoles/kg/min.[1,27,55] CP (initially 2.4 mmoles/kg) continues to decline, becoming undetectable after 60 sec of ischemia[1] (Fig. 1). During the first 30 sec of ischemia ATP (initially 2.4 mmoles/kg) declines at an average (as contrasted to initial) rate of 0.96 to 2.6 mmoles/kg/min.[1,27,55] Thus, in this period the decline in ATP is relatively slow. During the next 30 sec, following the exhaustion of most of the available CP and glucose, ATP levels decline more rapidly, paralleling the decline in glycogen[1] (Fig. 1). ATP levels are not depleted until the rate of delivery of $\sim$P from CP and glucose lags behind the rate of ATP utilization and ATP begins to fall, though carbohydrate reserves in the form of glycogen are still high.[1] After 60 sec the decline in ATP slows and it is still detectable at a very low level after 10 min of ischemia.[1,24,56] This portion of ATP may be in a compartment from which it is not readily mobilized or it may persist because metabolic demand has been reduced. The concentration of ATP remaining after 2 min of ischemia is similar to that of the membrane-associated ATP found after cell fractionation procedures,[57] and is apparently not available as an energy source for cell metabolism.

CP is used during ischemia to support ATP levels, and the changes in its concentration represent its flux during ischemia. On the other hand, ATP is both utilized and produced rapidly and its levels represent the net balance of these processes, averaged over various tissue compartments.

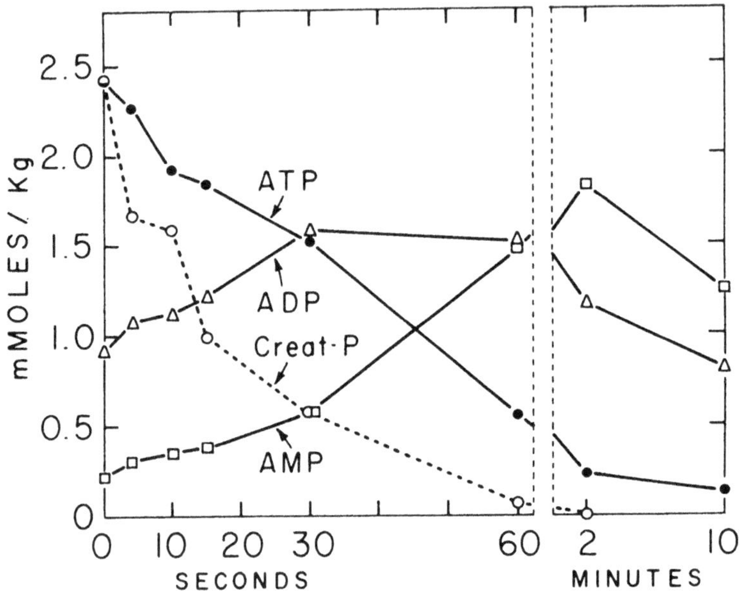

Fig. 1. Concentrations of creatine phosphate, ATP, ADP, and AMP in the brains of unanesthetized adult mice frozen at different intervals after decapitation. Based on the data of Lowry *et al.*[1]

To calculate the high energy flux ($\Delta \sim P$), not only are data required on the changes of ATP concentration, but the yield of $\sim P$ from all metabolites that contribute significantly must also be estimated (see Section II).

The "apparent" equilibrium constant of the ATP-creatine phosphotransferase (creatine phosphokinase, CK) reaction

$$K_{ap} = \frac{(ATP)\,(Cr)}{(ADP)\,(CP)}$$

(6.9 initially) rises during ischemia.[1,10] This has been attributed to an increase in the actual equilibrium constant of the reaction due to the fall in *p*H with ischemia.[58] However, the *p*H change does not entirely account for the equilibrium change in cat cortex.[10] It is therefore likely that the observed change in the K_{ap} is due to a departure from equilibrium conditions, possibly because of compartmentation of substrates.[59] Kuby and Nottman[60] and Wood[30] have thoroughly discussed the complexities of this reaction. Changes in the CP level parallel those of ATP during cation-modulated changes in cortical slice metabolism[61] and evidence has been found for a link between the activities of CK and the Na–K stimulated adenosine triphosphatase.[62] This direct association with the ion transport system, which has a high metabolic demand, may be partly responsible for the extremely rapid turnover of CP.

The initial rate of use of ~P during early ischemia was estimated by Lowry *et al.* from the rates of change in CP, ATP, glucose, and glycogen to be about 25 meq/kg/min.[1] The initial rate of ~P use is maintained only for about 15 sec (possibly because by that time the most active cells or cell parts have exhausted their energy supplies). It then falls to one-half the initial rate at which it is maintained until 60 sec after decapitation. There is little turnover of ~P after 2 min of ischemia and exhaustion of the major energy reserve (fasting state).[1]

During ischemia at 37°C the changes in ATP, ADP, and AMP reflect those expected from sequential reactions involving the dephosphorylation of ATP, followed after a slight lag by rather rapid equilibration via the adenylate kinase (AK) reaction. AMP, the final product, increases as ATP falls. ADP, the intermediate product (initially 910 μmoles/kg), increases rapidly during the first 30 sec of ischemia and then declines gradually despite the rapid ATP decline at this period, due to the adenylate kinase action (Fig. 1). There is a further slow fall in ADP between 2 and 10 min after decapitation.[1,22,56] AMP, very low under aerobic steady-state conditions (210 μmoles/kg), increases logarithmically during the first 60 sec of ischemia and more gradually for the next minute. It then decreases at a relatively low rate for the remainder of a 10 min ischemic period,[1] probably because of deamination to inosine monophosphate (IMP). AMP does not decline significantly between 2 and 10 min after decapitation if the brain temperature is allowed to fall.[56] The value of the K_{ap} for the adenylate kinase reaction [(ATP) (AMP)/(ADP)^2] declines slightly from the "true" equilibrium value of 0.6 after 6 min of ischemia in adult animals.[1]

During various metabolic stresses, changes in the K_{ap} of the creatine kinase and adenylate kinase reactions occur which favor ATP formation.[63] However, the decrease in the K_{ap} of the adenylate kinase reaction during ischemia found by Lowry *et al.* is not compatible with this concept.

Studies using anoxia rather than ischemia as the stimulus for glycolysis elucidate further the changes in the nucleoside phosphates. These indicate that the decline in ATP is paralleled by a decline in GTP but that UTP declines more slowly.[47] The biphasic change in ADP occurring in ischemia also is noted in anoxia and is reflected by UDP and possibly GDP. AMP increases more slowly in anoxia than during ischemia and is paralleled by changes in GMP and UMP. CMP is the only cytidine phosphate detected and shows no particular pattern of change.[47] Inosine monophosphate begins to increase after 30 sec of anoxia. This might explain the late decline in AMP and total adenylate phosphate (ATP, ADP and AMP) found by Lowry *et al.*,[1] since IMP may be the major product of AMP breakdown. Similar changes in nucleoside phosphates other than the adenylates during metabolic stress have also been noted by Minard and Davis.[63]

Schulz *et al.*[64] have discussed the difficulty of extracting and assaying inorganic phosphate (P_i) from tissue. In constructing a balance sheet for the changes in organic phosphate and the P_i released, the phosphate required in glycolysis must be included. In mouse brain P_i increases from 2.2 to 5.2 mmole/kg within 30 sec of decapitation and reaches 13.2 mmole/kg at 10 min.[64] Early in ischemia (30 sec) several investigators have found an approximate balance between organic phosphate disappearance and the appearance of P_i or a slight lag in the appearance of P_i, possibly because of its utilization for glycogenolysis.[1,64] At 10 min, however, P_i appears in excess of the drop in organic phosphate. This may be due to the release of phosphate from undetermined sources.[1,64]

3. *Glycolytic Pathway Intermediates*

Following the failure of the citric acid cycle early in ischemia, the continuing use of ~P soon results in local increases in ADP, AMP, and P_i, and decrease in ATP. These factors have been shown to modulate the activities of regulatory enzymes of several metabolic pathways. For glycolysis these enzymes are the phosphorylase system, hexokinase (HK), and phosphofructokinase (PFK)[28] and perhaps pyruvate kinase.[65] Local changes in metabolite levels result in an increased glycolytic flux in ischemic anoxia. Glucose (Fig. 2) declines rapidly in mouse brain at an initial flux of 6.5 mmoles/kg/min (approaching the rate of CP decline and considerably faster than the fall of ATP). By 30 sec glucose has fallen to 15% of its aerobic level (1.54 mmole/kg) and it slowly declines thereafter.[1]

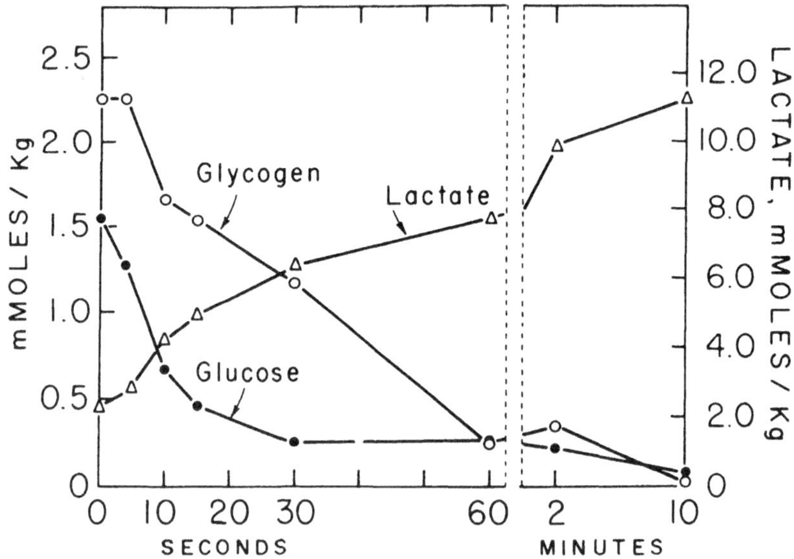

Fig. 2. Concentrations of glucose, glycogen and lactate in the brains of unanesthetized adult mice frozen at different intervals after decapitation. Based on data of Lowry *et al.*[1]

Since the brain glucose level depends on that in blood (within physiological limits), absolute levels will vary accordingly.[26]

Under steady-state conditions a metabolic control point between glucose and G6P is indicated by the observations that glucose, ATP, and hexokinase are present in concentrations higher than necessary to support the normal flux and that the equilibrium enormously favors G6P formation.[28] It has been postulated that HK may be regulatory but this could not be demonstrated experimentally.[29] During anoxia there is probably increased glucose uptake into brain from blood.[66] Merely slowing the circulation will increase the fraction of blood glucose extracted.[32] Compartmentation of substrates and enzyme may also be responsible for regulatory activity at the glucose phosphorylation step. It has even been proposed that there is no free intracellular glucose in brain, the determined levels being entirely extracellular.[67] The association of a significant portion of brain hexokinase with the mitochondria[68,69] may be of importance in separating substrate and enzyme. However, this association has been questioned.[70] G6P in brain, unless confined to a very small compartment, appears to be present in too small a concentration to inhibit hexokinase significantly. Although the Michaelis constants of hexokinase for each of its substrates are quite low, they tend to increase at low substrate levels. The low levels of ATP and glucose attained during ischemia result in the operation of hexokinase at substrate levels close to its K_m and may be partially responsible for the observed fall in flux before exhaustion of the substrates. Again, compartmentation of metabolites may also be a factor. The concentrations of F6P and G6P (initially 80 μmole/kg) decline roughly in parallel during the initial 30 sec of ischemia to about one-half their initial levels.[1] The G6P/F6P ratio slightly exceeds the equilibrium ratio of 3:1 for phosphohexose isomerase, but Lowry *et al.*[1] could not be certain of the accuracy of their F6P data. Even at peak glycolysis, equilibrium is approximated at this step.[1,29] The changes in ATP, glucose, and G6P in whole adult mouse brain found by Lowry *et al.* have been confirmed.[56] Studies of anesthetized rat brain, maintaining the severed head at 37°C, show similar changes in CP, ATP, ADP, and glucose.[71]

Fructose diphosphate (FDP) almost doubles within the first 4 sec of ischemia, then declines to initial levels by 30 sec and more slowly thereafter.[1] Decreases in G6P and F6P, while FDP and all subsequent intermediates between FDP and lactate increase during ischemia, indicate a significant facilitation of F6P phosphorylation.[29] PFK is inhibited by ATP and synergistically activated by a number of substances which increase during ischemia including: ADP, AMP, P_i, NH_4^+, and FDP.[66] Although a glycolytic control point beyond PFK has been postulated,[72] Lowry *et al.*[1] failed to find evidence for it. More recently, evidence for glycolytic control at pyruvate kinase in guinea pig cortex has been reported.[65]

During ischemia, the substrates at several steps in glycolysis are present at levels which suggest either compartmentation or that the enzymes at these points cannot handle the preceding flux and act as "dams" though not as "valves".[29] The levels of dihydroxyacetone phosphate (DHAP) during ischemia roughly parallel those of FDP.[1] The level of FDP is much higher initially and during peak glycolysis than the equilibrium ratio would indicate for the aldolase reaction. Lowry and Passonneau[29] suggested that this may be due to part of the FDP not being available for aldolase action rather than to inhibition of the enzyme. Data on intermediates between DHAP and pyruvate were obtained by these investigators only in 10-day-old animals. Concentrations of glyceraldehyde phosphate (GAP), 3-phosphogylcerate (3GP), 2-phosphoglycerate (2GP), phosphoenol pyruvate (PEP), and pyruvate all increase during ischemia, although less than FDP and DHAP.[1] During a 60-sec period the percentage increase in concentrations induced by ischemia is progressively less for successive glycolytic intermediates from FDP to PEP. Since the levels of intermediates between GAP and pyruvate do not increase as much as do their fluxes, the pyruvate kinase step may also be facilitated.[28] After 30 sec to 2 min of ischemia, levels of all glycolytic metabolites between FDP and pyruvate decline as the carbohydrate reserve is exhausted.[1]

Pyruvate (10-day-old anesthetized mice) increases at a rate proportional to that of lactate during the first 25 sec of ischemia (39 to 72μmole/kg), following which the lactate increase exceeds that of pyruvate. Pyruvate levels lag behind those of the preceding metabolites early in ischemia and then continue to rise while levels of its precursors are falling.[1] The products of the two anaerobic reactions in which NADH is oxidized (lactate dehydrogenase and α-glycerophosphate dehydrogenase) accumulate rapidly and proportionally during ischemia. In adult mouse brain lactate increases from an initial level of 2.26 mmole/kg to over 6 mmole/kg by 30 sec and reaches 11 mmole/kg after 10 min of ischemia[1] (Fig. 2). The initial lactate flux is 11.3 mmole/kg/min (3.5 mmole/kg/min in 10-day-old mice) while that of α-glycerophosphate (α-GP) is 0.1 mmole/kg/min in 10-day-old mice.[1] Although during the first 2 min they increase concomitantly, lactate levels are 24 times those of α-GP. Both increase almost linearly and proportionately for approximately 2 min and then level off with the exhaustion of the carbohydrate supply.[1] Maker *et al.*[24] found a similar lactate rise in mouse brain. Both Thorn *et al.*[22] and Schmahl *et al.*[10] in rabbit whole brain and cortex respectively, found similar early increases but higher ultimate levels of both these metabolites than those in mouse. Although this might reflect species differences, it may be related to differences in initial carbohydrate reserves, perhaps secondary to the anesthesia used in the rabbit studies.[26] Neither Thorn *et al.* nor Schmahl *et al.* measured carbohydrate levels. The utilization of high energy phosphate compounds was more rapid in mouse[1] and rat[71] brain than in rabbit brain[22,73] and it has been suggested that incomplete ischemia produced by neck compression in the rabbit studies might explain the differences.[71] Incomplete ischemia combined with anoxia (tracheal compression) would also account for the high final lactate levels in the rabbit brain since anoxia alone leads to higher brain lactate accumulation than does ischemia (see p. 289). In anesthetized rat brain[71] the lactate increase during the first minute of ischemia is similar to that of unanesthetized mouse brain[1] and was accounted for by the disappearance of glucose and glycogen. Final (15 min) lactate concentrations in fasted animals were similar to those in mice.[71]

Substrate concentrations and kinetic data make it appear doubtful[29] that pyruvate and lactate are maintained near equilibrium during peak glycolysis. The overall NADH levels measured were too low to support the observed lactate flux. However, in liver, Hohorst *et al.*[53] found that several extramitochondrial enzyme systems linked to NAD were apparently in equilibrium before and during a brief ischemia: yet the overall NAD^+/-NADH ratio, measured enzymatically, failed to change concomitantly with the apparent oxidation/reduction states of the NAD-dependent substrates. Apparently different metabolic compartments are involved. Both the lactic dehydrogenase and α-glycerophosphate dehydrogenase (soluble) reactions serve to oxidize NADH, a function essential for the continuation of glycolysis. Since during ischemia changes in the ratios pyruvate/lactate and DHAP/α-GP differ from each other in mouse brain[1] and cat cortex,[10] these reactions may not be in equilibrium with the same pool of NAD^+/-NADH in brain. This is unexpected since both ratios involve soluble cytoplasmic enzymes. Differences have been shown between LDH and α-GPDH in function, anatomic distribution and development.[74,75] α-GPDH has a role in lipid metabolism and in certain tissues in the transfer of reducing equivalents into mitochondria. Therefore, it is possible that these enzymes are associated with different pyridine nucleotide pools. Approximately 60% of NAD(H) may be intramitochondrial.[51]

A relatively small store of carbohydrate reserve in brain is in the form of glycogen.[1] Since no ATP is necessary for most of its initial phosphorylation, there is an approximately 45% greater equivalent of ~P generated in the glycolysis of a mole of glycogen (as glucosyl units) than of glucose. During ischemia (Fig. 2) glycogen declines in mouse brain at a rate of 2.2 mmole/kg/min for the first 30 sec and then more rapidly, at rates similar to those of ATP.[1] Low levels are still detectable after 10 min of ischemia.[1,56,71] The conversion of phosphorylase "b" to the "a" form occurs extremely rapidly in ischemic brain.[76] Thus, the delay in use of most of the glycogen until almost all the CP and glucose have disappeared cannot be attributed to delayed phosphorylase activation. It is likely that changes in the levels of P_i and other cofactors modifying the activity of this enzyme cause the delayed acceleration of glycogenolysis. Metabolite distribution factors may also be involved. There is no evidence that phosphoglucomutase is rate limiting in brain and insufficient evidence that the debrancher activity may be limiting.[77] Since PHRL "a" is reconverted to the AMP dependent "b" form during ischemia because of phosphatase action and low ATP level,[76] the continued active glycogenolysis is evidently dependent on the accumulation of AMP. It is of interest that delay in the use of glycogen during ischemia has been found in brain of all species studied, mammalian and nonmammalian, but not in peripheral nerve.[78] Variations in creatine phosphate levels at the time of glycogen decrease indicate that the concentration of this substance is not critical for the triggering of glycogen degradation.[1] Both the decline in G6P, an inhibitor of PHRL "a",[79] and the rise in FDP[80] have been suggested as mechanisms influencing the delayed activation of glycogenolysis.

4. Poikilothermic Species

McDougal *et al.*[81] measured the rates of change of creatine phosphate, ATP, glucose, glycogen, dihydroxyacetone phosphate, α-glycerophosphate, and lactate in poikilothermic vertebrates (fish, frog, and turtle) for 2 to 4 hr following decapitation. The rates of change of the metabolites were much lower than in mammalian brain. Although the lower ambient temperature accounted for some of this metabolic retardation, a major difference was an intrinsically lower rate of glycolysis in these species, perhaps related to a relative dearth of brain structures with high metabolic activity (neuropil). In these species the initial rates of use of high energy phosphates are only 1/25 to 1/50 of that in mouse brain.[1] The greatly expanded time scale of change in these species made clearer the patterns of changes in ischemia. However, there are significant differences from the mammalian pattern, possibly reflecting differences in metabolic regulation among the species. In somewhat simplified terms, it appeared that the lower rate of use of ATP results in a relatively stable level of this substance for a prolonged period during ischemia, which inhibits phosphofructokinase, particularly as the *p*H declines. The consequent increase in G6P may, in turn, inhibit hexokinase. Unlike the situation in mammalian brain, phosphofructokinase in these species appears to be inhibited during most of the ischemic period and metabolite levels and fluxes reflect this inhibition.[81] These differences limit the usefulness of these studies in evaluating mammalian brain metabolism; yet there may be compartments (e.g., glia) within mammalian brain in which metabolism more closely resembles that in lower vertebrates. Recent measurements suggest that the findings of McDougal *et al.*[81] may not be universally applicable. Rates of ~P utilization and lactate production during ischemia considerably higher than those in goldfish, frog, or turtle brain, though somewhat lower than in mammalian brain were found in the brainstem of the puffer fish.[82]

5. Human Brain

Kirsch and Leitner[83] measured several metabolites in anesthetized human cortex and white matter obtained by biopsy from patients during surgery for brain tumor and frozen immediately or incubated anaerobically up to 5 hr before study. The overall patterns of metabolism were similar to those of mouse brain. However, data for comparable early periods of ischemia were insufficient to allow comment on the significance of possible differences noted between the two species.

B. Other Metabolic Changes During Ischemia

1. *Hydrogen Ion*

The *p*H of the cortical surface (dog) declines from 7.2 to 6.8 during the first 2 min of ischemia and to 6.50 by 60 min.[84] Initially, this fall is apparently related to CO_2 accumulation and later to rise in lactate, since the CO_2 declines after oxygen is exhausted.[37] The minimum *p*H as well as the maximum lactate attained are dependent on the total available carbohydrate reserve initially present. Greater fluctuations in these parameters are seen during hypoxia with intact circulation and a concomitant maintained carbohydrate supply (*p*H 5.6)[85] than are seen with ischemia. Thus, cessation of glycolysis and the consequent exhaustion of $\sim$P in brain is more likely to be related to exhaustion of substrate rather than to a "toxic" acidity. Indeed, the levels of lactate reached are not in themselves incompatible with total recovery.[86–88] Defective carbohydrate metabolism and ionic concentrations have been demonstrated in brain slices maintained at *p*H 6.5 as compared to 7.4.[89] Many reactions either produce or absorb hydrogen ions and no single one is responsible for the overall *p*H change. The buffering capacity of rat brain toward lactic acid has been found to allow a decline of 0.024 *p*H unit for each millimole of lactic acid produced per kilogram of tissue water.[43a] The lactate level reached after 10 min of total ischemia[1] would, in itself, therefore produce a decline of 0.2 *p*H unit. However, higher lactate levels may be reached in certain brain regions or during partial ischemia. It should be noted that at the final *p*H actually recorded in ischemic brain, there are changes in the kinetics of certain enzymes. The maximal velocity of the creatine kinase reaction and its equilibrium constant increase in the direction favoring ATP formation while the ATP inhibition of phosphofructokinase declines.[90] Apparent cessation of *in vivo* brain glycolysis occurs if the *p*H is reduced to 6.5.[43a]

2. *Amines and Amino Acids*

Tews *et al.*[91] studied the changes in several amino acids in anesthetized dog brain before and after a 20-min ischemic period, evidently exposed to air and at room temperature. They found no significant changes in *N*-acetylaspartate, arginine, aspartate, glutamate, glycine, histidine, leucine, lysine, methionine plus cystathione, phenylalanine, serine, taurine, threonine, tyrosine, valine, urea, glutathione, glycerophosphoethanolamine or phosphoethanolamine. Increases were noted in alanine and γ-aminobutyric acid. In rabbit brain there is no change in glutamate and probably none in aspartate after 10 min of ischemia, while alanine increases.[92]

An increased incorporation of labeled glucose into brain alanine following a period of anoxia-ischemia was the only change in amino acid turnover found by Atkinson and Spector[93] and was attributed to the increased availability of pyruvate for transamination to alanine because of its failure to enter the citric acid cycle. Only a slight decrease in mouse brain aspartic acid and an increase in *N*-acetylaspartic acid was found after a 2-min ischemic period.[92] Studying the most abundant free amino acids in mouse brain, Young and Lowry[95] failed to find significant changes in glutamate, glutamine, or alanine during 10 min of ischemia at 37°C. Since the entrance of glutamate into the energy-yielding pathway occurs via the tricarboxylic acid cycle (after trans- or deamination), these results are consistent with the cessation of net flux by the latter path. However, Thorn *et al.*[92] reported a linear increase of ammonia at a rate of 0.337 mmole/liter/min during a 15-min ischemia period in rabbit brain, indicating catabolism of amine-containing substances during ischemia. AMP destruction is one such source and possibly a major one.

The level of γ-aminobutyric acid (GABA) increases following decapitation and any delay in tissue preparation leads to a further increment.[96,97] Hypoxia and ischemia also elevate GABA.[91,98] This has been attributed to continued functioning of glutamic decarboxylase while the oxidation-dependent GABA shunt pathway is inhibited.[98]

Acetylcholine concentrations decline rapidly to a maintained plateau unless special precautions are taken in tissue preparation.[18] This may be related to release of a portion of the bound acetylcholine early during ischemia and its prompt destruction by cholinesterase.

Several amines which may function as transmitters have been extensively studied in recent years (dihydroxyphenylethylamine, norepinephrine, tyramine, 5-hydroxytryptamine). Biochemical mechanisms exist in brain for their synthesis and modification, and there is evidence for relatively rapid turnover.[99] The turnover of these substances under steady-state conditions is much slower than that of acetylcholine and no special precautions have been taken to stop brain metabolism in their assay. However, we know of no studies determining their rate of change during ischemia.

3. *Enzymes*

Few studies have directly quantitated enzyme levels during ischemia. The early changes in phosphorylase have been mentioned with reference to glycogen metabolism. Smith *et al.*[25] and Robbins *et al.*[100] measured the stability of a number of enzymes in the layers of rabbit cerebellum up to 6 hr after death (at which time the body temperature was 30.5°C). Because of variability in the data, only changes over 20% were considered significant. For 6 hr postmortem no significant changes occurred in the activities of malic, lactic, glutamic, or glucose-6-phosphate dehydrogenases, fumarase, hexokinase, aldolase, phosphoglucoisomerase, β-glucuronidase, (unstimulated) adenosine triphosphatase, acid phosphatase, or purine nucleoside phosphorylase. The only change was a decline in phosphofructokinase activity in both cortical and white matter layers of the cerebellum over the 6 hr period studied. A similar loss of phosphofructokinase occurred in incubated brain homogenates. However, the assay conditions for this enzyme were not optimal. Thiamine pyrophosphatase, (unstimulated) adenosine triphosphatase, adenosine-5-phosphatase, and acid and alkaline phosphatases were stable in whole brain suspensions up to 24 hr.[101] Possible changes in enzyme distribution as found by slide histochemical reactions will be discussed later.

4. *Lipids*

Although long considered to be relatively inert metabolically, it is apparent that several lipids turn over rapidly and could be susceptible to changes early in ischemia.[102] Polyphosphoinositides of a rat myelin fraction declined during a 10-min postmortem period and more slowly up to 2 hr after death.[103] Similar reductions occur after *in vivo* ischemia.[104] A linear increase in free fatty acids at a rate of 12 mmole/kg fresh tissue/hr has been found in mature rat brain during the initial 4 min following decapitation and the possibility raised of their origin from membrane lipids.[104a] Lowden and Wolfe[105] found no changes in the chromatographic patterns of gangliosides up to 6 hr postmortem but did show lower levels of total ganglioside in patients with cerebrovascular disease and in brain asphyxiated for 5 to 8 min. However, Suzuki[106] could find no difference in total or individual ganglioside levels over postmortem periods from 30 min to 60 hr. Ganglioside loss in chronically ischemic brain probably results from neuronal loss. A decreased turnover in all phospholipid fractions of rat brain proportional to degree of hypoxia reported by Dvorkin[107] was later attributed entirely to a concomitant decline in body temperature in the earlier investigation.[108] With the failure of energy sources no significant synthesis of lipid can be expected and several studies based on the incorporation of labeled precursors confirm this expectation.[109]

V. ELECTROLYTE AND FUNCTIONAL CHANGES

The maintenance of electrochemical gradients across cell membranes is ultimately dependent on a supply of metabolic energy. If, as is currently hypothesized, the exchange of sodium and potassium is linked to a system involving sodium and potassium-stimulable magnesium-activated adenosine triphosphatase,[110] local availability of high energy phosphate would be critical for the maintenance of such gradients. During ischemia, ion shifts and conductivity changes are early occurrences; however, it is not yet possible to relate such changes directly to those in energy availability because of differences in experimental conditions and uncertainty as to the significance of the overall metabolite fluxes to those at cellular and subcellular levels. Experiments comparing electrolyte fluxes and ~P changes during stimulation and in the presence of specific blocking agents suggest a close link of energy supplies and the "sodium pump."[61]

Meyer *et al.*[37,38] have carried out a series of investigations on correlations between the changes in respiratory gases and electrical potentials and certain electrolytes occurring in exposed, anesthetized mammalian cerebral cortex during hypoxia and ischemia. They used surface electrodes sensitive to various ions and produced local ischemia by carotid embolization with plastic microspheres. When monkey cortical tissue pO_2 falls below 8 mm Hg (within 60 sec of artificial embolization), a net shift in sodium to the "intracellular" space occurs, followed within a few seconds by a net shift of potassium to the "extracellular" (recordable) space. These changes reach equilibrium by 4 min postembolization.[38] Preliminary data also indicated a chloride shift similar to that of sodium. Although these changes overlap in time with those in energy metabolites previously discussed, no precise correlation of these two metabolic parameters is now possible. "Metabolic recovery" after temporary interruptions of blood flow precedes reversal of the electrolyte shifts.[38] Van Harreveld and his collaborators[111] recorded electrical conductivity in exposed rabbit cortex during ischemia. A small, slow decrease in tissue conductivity early during ischemia was followed at 3.5 min by a sudden, rapid fall in conductivity over a 60-sec period. This coincided with increase in cortical surface electronegativity (asphyxial potential). The investigators attributed these electrical changes to a redistribution of ions, particularly a shift of sodium from an extracellular site into neuronal dendrites. The abrupt potential change suggested a threshold phenomenon (like the action potential). By contrast, Meyer *et al.*[38] found no such abrupt changes in ion distribution. To explain his findings, Van Harreveld postulated the release of a substance causing depolarization at a certain stage of ischemia. Using a histochemical procedure for microscopic localization of chloride, Van Harreveld[111] showed apparent shifts after several minutes of ischemia from extracellular sites into apical dendrites of cerebral cortical, Purkinje, and spinal neurons as well as into Bergmann astrocytes. Ultrastructural studies showed swelling of apical dendrites and Bergmann astrocytes over this same period.

Some correlations can be made between metabolic changes and neuronal function. Evidently by autoregulatory vascular mechanisms, the blood flow to a brain region increases at times of increased metabolic demand in relation to physiological or experimental stimuli, permitting normal electrical function until demand exceeds supply.[112-114] The previously noted changes in electrochemical gradients across cortical cell membranes together with the possible release of excitatory or depressant metabolites during ischemia, result in rapid changes in the electrocorticogram or electroencephalogram (EEG). The earliest electroencephalographic change in anesthetized rabbit brain occurs within 5 sec of the onset of total ischemia.[115] By 10 to 12 sec there is definite slowing of brain wave frequency followed by a decrease in amplitude. The EEG becomes isoelectric within 20 to 25 sec. In the unanesthetized, curare-paralyzed mouse, alpha rhythm is detectable for only 2 to 4 sec following decapitation.[23,116] Between 4 and 10 sec predominantly low voltage, fast activity at a slightly decreased amplitude is seen. Little recognizable brain activity is recordable after 10 sec postdecapitation and the EEG is completely isoelectric after 20 sec. The time course of these changes indicates the extreme dependence of neuronal function on high levels of oxidative metabolism. Changes begin while the overall levels of ~P are still quite high.[23,116] This suggests that much of the ~P is not readily available for such functions as maintaining synaptic ionic gradients. Since the ultrastructural location of the ion flux responsible for the EEG may be the dendritic trees of the neuropil, this may also be the location of the most rapid energy changes. This concept finds support in our own recent studies, showing a higher rate of creatine phosphate disappearance during ischemia in the adult molecular cerebellar layer, which is richest in neuropil, than in the other layers.[117]

VI. LATE CHEMICAL AND STRUCTURAL CHANGES DURING ISCHEMIA

As noted, the majority of enzymes determined quantitatively are stable for several hours following interruption of blood supply.[25,100] After the carbohydrate reserves have been exhausted (10–15 min) the major change in glycolytic metabolites is a slow decline in phosphorylated metabolites concomitant with an increase in inorganic phosphate.[83] Rats that survive an anoxic-ischemic episode may continue to have elevated lactate levels and low levels of adenosine phosphates up to 24 hr after the initial insult.[118] Late changes in water content and total electrolytes have also been reported.[119] The biochemical correlates of the ischemic structural changes in survivors include decreased capacity to incorporate carbon from glucose and leucine into proteins, nucleotides, acid-soluble substances and lipids.[120] Structural changes after anoxia,[121] ischemia,[111,122,123] and anoxic–ischemic lesions,[124] have been attributed, in part, to failure to maintain membrane-associated molecules (proteins, nucleotides, and lipids) and destruction by enzymes released from lysosomes (autolysis) followed in surviving animals by the action of local and circulation-borne macrophages and later by the proliferation of astrocytes and astrocytic processes. The durations of some of these changes are measured in weeks. The mechanism of release of lysosomal hydrolases is not clearly established. Some maintain that the *p*H decline due to lactate accumulation is responsible, but as discussed, this is not, in itself, a likely cause. Anderson[125] studied the time course of changes in postmortem (37°C) rat brain acid phosphatase from a bound to a soluble and electrophoretically changed form. If the solubilization of the acid hydrolase is a measure of lysosomal enzyme release, release is very slow; only a 50% increase in the soluble enzyme activity occurs over a 6 hr period.

Many enzyme studies utilizing slide histochemical techniques are open to methodological criticism with respect to specificity, localization, and particularly quantitation.[126] Becker[123] found some changes in enzymes (particularly lysosomal) relatively early in ischemia whereas

Zeman[127] found no significant enzyme changes prior to definite beginning necrosis. To explain this discrepancy Zeman differentiated between autolysis (liquefaction), in which lysosomal enzymes participate in the ischemic change, and coagulation, which occurs without these acid hydrolases. He speculated that without autolysis, enzyme changes are delayed. A similar differentiation between necrosis with and without liquefaction has been made by Lindenberg,[128] who attributed the structural dissolution to low *p*H, since it could be prevented by measures prior to ischemia which lowered the carbohydrate reserve, and therefore the eventual lactate production. As previously noted, high lactate levels per se do not appear to be toxic. The capacity of the brain to recover after anoxia was found to be independent of the degree of acidity reached[85] and extremely high lactate levels are compatible with recovery.[86–88] However, the role of lactate in metabolic and structural derangements remains to be defined. After administration of ammonium acetate, Schenker *et al.*[129] noted that, at the level of ammonia in rabbit brain corresponding to that attained after 10 min of ischemia,[92] rats exhibited drowsiness; yet cortical ATP and PC levels were normal though levels in brainstem ("base") were depressed. It is unlikely that ammonia accumulation contributes significantly to the severity of ischemic damage.

The concept that ischemic lesions are irreversible is often applied to the capacity to recover of the entire organism or to brain regions rather than to individual cells. Ultimately, the point of irreversibility depends on the level at which failure of energy metabolism and its consequences have become sufficiently disruptive. This point varies for different cell types in the nervous system and probably even for the same cell type in different locations or functional states. The temperature at the time of circulatory failure[22] and the functional state of the brain are also determining factors. The brains of certain inframammalian, poikilothermic species which have low requirements can survive prolonged periods of anoxia.[81] Variations in survival are apparently due to different methods of producing brain ischemia; thus the dogs of Neely *et al.*[130] survived 25 min of brain ischemia produced by elevating the intracranial pressure above arterial pressure. This is much longer than the 8–10 min survival times for most species.[88,131] Human brain function may fail to recover after a shorter time period (about 4 min) but the precise duration of clinical brain ischemia is difficult to determine and much longer survivals have been reported. The "allowable" ischemic period for the entire brain extends past the period when readily available energy stores are exhausted. Considerable electrolyte shifts occur early in ischemia and functional activity ceases well before full recovery has become impossible. Only minor ultrastructural changes have been noted at the time of greatly reduced energy flux and disturbed ion gradients.[121] The changes in the most vulnerable structures which occur between 2 and 10 min in the ischemic period and which lead to failure of recovery are unknown. The lactate increment in itself does not appear to be responsible and no enzyme changes are detectable quantitatively. Ames *et al.*[131] found that many closed precapillary and capillary vessels fail to reopen after several minutes of ischemia, apparently primarily because of endothelial changes. They postulate that this mechanism ("no-reflow") rather than a biochemical one is responsible for irreversibility. However, a longer delay before development of ultrastructural endothelial damage was reported by Hills.[132]

VII. HYPOGLYCEMIA AND HYPOXIA

Although hypoglycemia reproduces one of the events occurring with ischemia, the metabolic changes induced are very different from those occurring with circulatory failure; therefore, hypoglycemia cannot serve as a model for ischemia. During hypoglycemia, the onset of functional and metabolic change is quite slow. With the continued function of the tricarboxylic acid cycle (unlike ischemia), the endogeneous substrate is sufficient to maintain energy metabolism for 20–25 min.[133] Whereas some carbohydrate metabolites increase and some decrease in ischemia, the level of every intermediate metabolite from glucose to malate as well as lactate is diminished in hypoglycemia.[55] The continued function of the oxidative pathways is responsible for these differences while glucose, and later glycogen, diminish, as contrasted to the early block in the tricarboxylic acid cycle in ischemia because of the failure of respiration. The presence of oxygen also enables the brain to utilize some amino acids by transamination to keto acids of the TCA cycle and results in a decrease particularly of glutamate and glutamine. Alanine and its transamination partner, pyruvate, also decline.[133] Changes in creatine phosphate, the adenosine phosphates, inorganic phosphate, and ammonia are similar to those in ischemia, but since they do not occur until a stage of deep shock is reached,[133] circulatory factors may also be involved. Insulin is capable of inducing a convulsion at a normal total brain concentration of high energy phosphate,[23] suggesting possible differences in accessibility for brain function between overall ATP and that specifically generated by glycolytic–citric cycle activity.

In contrast to hypoglycemia, anoxia appears to reproduce the early changes occurring with ischemia.[22,47] Among the differences from ischemia are those related to lack of circulatory stasis. During early anoxia when most lactate can be removed by the bloodstream, brain levels are lower than during ischemia and the *p*H drops at only 0.05 units/min, one third the rate with ischemia; but eventually high lactate levels do occur.[22,116] Glucose utilization increases during anoxia, contributing to the increased lactate production.[33] The most significant departures of metabolite levels in anoxia from those of ischemia are at the earliest time intervals. By 6 sec of anoxia, mouse brain lactate triples, primarily derived from higher glucose uptake. At this time there is little change in ATP, indicating a greater capacity of the brain to sustain energy metabolism by glycolysis in anoxia with substrate available, than during ischemia.[116] Bakay and Lee[134] found significant structural differences in brain between simple anoxia and that accompanied by hypercapnia. These investigators believe that CO_2 accumulation is a major factor in making hypoxia a poor model for ischemia. If anoxia is sustained, the effect on other organs must be taken into account.

VIII. SEIZURE STUDIES AND ISCHEMIA

Most studies of brain metabolism during seizures have not been controlled for ventilation and muscle metabolic demand. During most convulsive seizures, the combination of increased neural and muscular activity and respiratory hypoxia along with decreased cardiac output and increased jugular venous pressure resulting from an increase in intrathoracic pressure during the tonic phase creates a condition in its early stages resembling ischemia. Later, a new steady state becomes established at some point and some energy reserves are usually maintained.[80] Overall, the metabolite and other biochemical changes resemble those of anoxia.[63,116,135] Consistent with an additional metabolic demand of massive neuronal discharge, the time course of metabolite changes is even more rapid than during simple ischemia when either intact mice or their severed heads are electrically stimulated.[23,116] As in anoxia, there is evidence for a greatly increased glucose uptake from blood during induced seizures with an intact circulation and a correspondingly high lactate production.[116] Blood glucose increases during seizures, partly accounting for the increased brain glucose.[77] If precautions are taken to maintain oxygenation and blood flow, i.e., in oxygenated, artifically respirated, curarized mice, no fall in $\sim$P and glucose or rise in lactate occurs with electroconvulsions despite a four- to sixfold increase in $\sim$P flux.[136] If the metabolic increment due to a seizure[23] is primarily a reflection of neuronal events, then the rapid and extensive changes in metabolic flux would suggest that the bulk of the brain energy reserves are so located that they are readily available to sustain neuronal and particularly synaptic processes. It is to be expected, however, that seizure stimuli also affect glial cells and the contribution of glial metabolism to the biochemical changes remains to be evaluated.

IX. RESPONSE TO ISCHEMIA BY BRAIN CELLS OF VARIOUS TYPES

In postmortem and experimental material it is well established that histological changes during ischemia generally occur in the order: neurons, glia, myelinated fibers, mesodermal elements. Furthermore, Purkinje and small neurons appear to degenerate before other large neurons[122] and oligodendroglia before astrocytes. However, ultrastructural changes may be seen relatively early in astrocytes.[124] This differential sensitivity of tissue elements has been variously attributed to differences among cell types in metabolic demand, peculiarities of local blood supply, the local aggregation of cells with high demand (cortex and nuclei), and neuronal stimulation during ischemia.[137] Indirect evidence suggesting that neurons and oligodendroglia have a relatively higher oxidative metabolism has been summarized.[137,138] This evidence is based on the correlation of respiration *in vitro* and respiratory enzyme activity of different brain layers, regions, or glial tumors, with the differential cell counts in areas or tissues studied. The high metabolic demand of the neuron might be related to the necessity of maintaining an extensive cell surface area (cell processes) and transmembrane, contraosmotic, ionic gradients constantly disturbed by local graded depolarizations and discharges,[137] and to requirements for transmitter synthesis and packaging. Some oligodendroglia are closely apposed to neuronal perikarya and their membranes form the large surfaces of the myelin layers.

A few metabolic studies have been made directly on single neurons isolated in the fresh state or dissected from freeze-dried sections. By necessity, these have been perikaryal samples. These are summarized in the chapter by Brand and Lehrer in Volume 5 of this Handbook. Of note is the finding that isolated rabbit lateral vestibular cells (Deiters') incubated anaerobically at 37°C lose potassium (as does the cortical "intracellular space") while this ion increases in the "glial" capsules surrounding these large neurons.[139] These and other findings suggest that, although both cell types maintain the potassium gradient by energy-dependent processes, those of the glia are less sensitive to anoxia. There is doubt, however, about the composition of the capsular material in these studies, since this tissue has a higher respiratory rate (with or without added glucose) than the neuron perikaryon itself.[141] It may well be that it contains a significant portion of dendritic neuropil which might make a large contribution to its oxidative metabolism. By contrast, the high level of MDH in a dorsal root ganglion cell body as compared to its capsule suggests a high respiration-related metabolism, at least in this neuron.[142]

No metabolic measurements have been reported differentiating areas rich in synaptic structures (neuropil) from other tissue elements of brain. Although neuropil might be expected to have both a high metabolic demand and high sensitivity to ischemia, indirect estimates by study of enzyme levels and substrate changes with ischemia in layers from which synapse-rich areas can be dissected (e.g. retina) have so far failed to substantiate this hypothesis.[68,143–145] The metabolism of retina, however, is complicated by the peculiarities of its blood supply. The activities of several enzymes (HK, MDH, LDH) are higher in areas of the central nervous system and peripheral ganglia rich in neuropil than in individual cell bodies or areas rich in perikarya.[142]

An important aspect of the problem of the response of the neuron to ischemia derives from the evidence that neuronal metabolism may adapt, to some extent, to local circulatory factors while still functioning.[66] The extensive data correlating the metabolism of retina with its blood supply obtained by Lowry and his collaborators[68,143–145] show that tissue functioning at a distance from a supply of nutrients and oxygen operates with enzyme and metabolite levels better suited to anaerobic metabolism than the homologous tissue closer to its supply. Similarly, enzyme levels in the giant supramedullary neurons of *Diodon* which are surrounded by individual capillary networks suggest that these neurons function under higher oxygen tensions[146] than the neurons studied by Lowry.[142] Lehrer and Bornstein[147] have shown that enzymatic gradients develop in long-term tissue cultures of rat cerebellum or mouse cerebrum which reveal adaptive trends in showing higher levels of lactic dehydrogenase in portions of the explants farthest removed from the nutrient medium. These adaptive changes occur during development but there is also evidence for metabolic adaptation in the adult. Chronic hypoxia not only increases the RNA content and cytochrome oxidase activity of Deiters' neurons (though not that of their capsules), but also increases the anaerobic glycolytic capacity of the neurons and perhaps of their capsules.[140] Such changes might occur in the tissue surrounding a necrotic area which is incompletely supplied by collateral circulation.

Despite the evidence for a greater metabolic demand by neurons, it has by no means been established why these cells degenerate early. Based on study of whole brain[1] and cerebellar layers for the first 30 sec of ischemia,[27] the energy reserves in the majority of brain cells appear to be nearly exhausted on a comparably short time scale compared to that of delayed differential structural changes. However, as discussed in Section X, some more slowly metabolizing compartments may be present. The answer seems to lie in the cellular events occurring after the cessation of detectable metabolism for whole brain and before the onset of visible degeneration. During this period possibly more rapid, poorly understood changes take place in the average neuron than in other cells.

X. REGIONAL EFFECTS OF ISCHEMIA

During generalized brain ischemia some regions show earlier and more significant damage than others.[148] Some regions receive a richer circulatory supply than others. Sokoloff,[149] using radioactive albumin, showed that blood flow to gray matter in general is significantly better than that to white matter and that there is little variation among various gray regions. Other studies found differences in the circulation among various gray matter regions.[150] During anesthesia the blood flow to gray matter declines compared to that in white.[149] Yet Gatfield *et al.*[27] found the metabolic rate of white matter to be the more reduced by anesthesia than that of gray. Sokoloff[151] used his data to postulate that circulation to a region was largely determined by its metabolism. Consistent with this hypothesis is the finding that succinic dehydrogenase and diaphorase distribution (by slide histochemistry) parallel capillary density in the cat brain.[152] Capillary density appears to be related not so much to the concentration of neuronal cell bodies as to that of synaptic terminals.[153] Thus, the trigeminal ganglion with many cell bodies but no neuronal interconnections has both a poor vascularity and low metabolic rate.[154,155]

The oxygen consumption of cerebral cortex (per wet weight) has been estimated to be four times that of whole brain.[156] When slices from various brain regions are incubated *in vitro*, the oxygen uptake per gram of caudate and thalamus exceeds even that of the cerebral cortex.[157] Under these conditions, the glycolytic rate (lactate formation) is also higher in these areas.[158] *In vitro* respiration and cytochrome oxidase activity per gram wet weight of tissue is highest in the cerebral and cerebellar cortices and declines in a rostrocaudal direction. However, when the data are expressed per cell (DNA), all neuron-rich areas of the central nervous system appear to respire similarly except for the densely cellular cerebellar cortex which shows relatively low oxygen consumption.[159,160] In slices stimulated by potassium, a rostrocaudal decremental gradient may be found in respiration per cell.[160] There is a correlation between respiration per cell and percent of total cells in a region which are neuronal if one uses Nurnberger's[161] differential cell counts in various areas. Although neurons in different regions may differ biochemically, regional differences in respiration and blood supply mainly appear to reflect differences in concentrations of neuropil.

A decreasing gradient of ATP and creatine phosphate concentrations (per wet weight) from outer to inner layers of cat and rat cortex could not be attributed simply to increasing myelin content since inorganic phosphate showed an inverse gradient and phospholipid remained level across the cortex.[162] However, the effect of freezing from the external surface inward would create a similar trend (see Section II). The changes in the lactate level during ischemia are parallel in both outer and inner cerebral cortex. Intermittent ischemia caused a decline in RNA concentration in cat motor and visual cortical areas but not in the auditory region.[162] Metabolite changes during ischemia in anesthetized cat cortex[10] suggest a somewhat greater initial ischemic metabolic flux in cortex than in whole brain.[22] But when the variability in the data and higher freezing rates for the external cortical tissue are considered, the significance of the flux difference is obscured. Gatfield *et al.*[27] compared the fluxes of several carbohydrate and high energy metabolites in several regions of mouse brain during a 20–30 sec period of ischemia (postdecapitation). Preischemic levels of energy reserve are similar in parietal cortex, Ammon's horn, and cerebellar cortex. Levels in these areas are slightly higher than in medulla but a higher lactate level in the latter structure suggested that it had taken longer to freeze. The decrease in energy reserve is greater in parietal cortex after 20 sec of ischemia than in whole brain after 30 sec. The total $\sim$P flux (per unit weight) appears similar in parietal and cerebellar cortices and medulla, and somewhat lower in Ammon's horn. Major discrepancies between the rates of disappearance of glycogen and glucose and the production of lactate make calculations of energy utilization uncertain. No significant differences in the rates of use of ~P (on a fat-free dry weight basis) were found among cerebellar molecular, granular, and white matter layers during 30 sec of ischemia.[27] This conclusion was based on the rate of lactate appearance. If the changes in measured carbohydrate levels are used in the calculation, the flux turns out less in white and granular layers than molecular.[27] Preliminary data obtained by Maker *et al.*[117] (Figs. 3 and 4) suggest a higher ischemic flux in cerebellar molecular than in granular layer which is more apparent after maturation of the brain is completed. The relatively high steady-state flux in cerebellar white matter has been confirmed.[117] Cerebellum of anesthetized rabbits (exclusive of basal nuclei) utilizes less ~P on a wet weight basis during early ischemia than does pons, but this difference may disappear if correction is made for lipid content.[163] The rate of decline of ATP is lower in cerebellar white matter than in the cortical layers[117] (Fig ,4). A rate of metabolism in ischemic white matter lower than that in whole brain is also suggested by the findings of Kirsch *et al.*[164] in mouse, but parallel studies of cortex

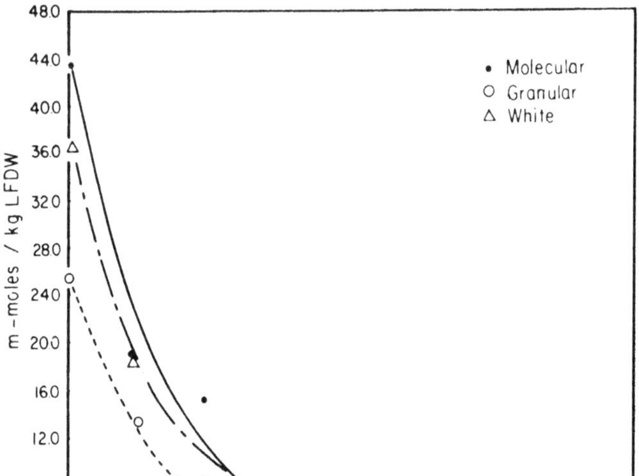

Fig. 3. Concentrations of creatine phosphate in the molecular, granular, and white matter layers of 8-month-old mouse cerebellum, expressed as mmole/kg lipid-free dry wt., at different intervals after decapitation. Based on data of Maker *et al.*[117]

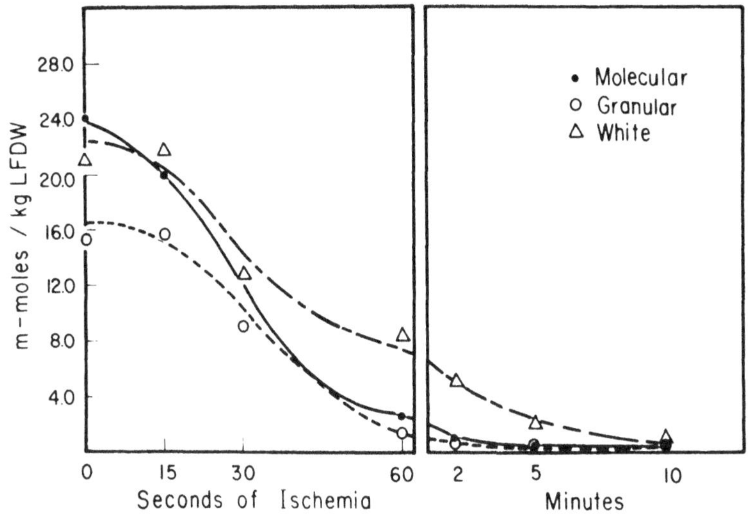

Fig. 4. Concentrations of ATP in the molecular, granular, and white matter layers of 8-month-old mouse cerebellum, expressed as mmole/kg lipid-free dry wt., at different intervals after decapitation. Based on data of Maker *et al.*[117]

were not reported. In the human, the high energy flux in early ischemia is higher in anesthetized parietal subcortical white matter than in gray, even when expressed on a wet weight basis.[83] However, the effect of anesthesia must be evaluated and discrepancies between the amounts of carbohydrate reserve change and lactate appearance throw the flux calculations into doubt.

XI. THE PROBLEM OF SELECTIVE VULNERABILITY

That some regions of the brain appear to be especially vulnerable to ischemia has been attributed to either or both of the following: (1) as first suggested by Spielmeyer, areas with more tenuous supply may be bereft of circulation more easily than generously vascularized areas; (2) there might be differences among various areas in dependence on oxidative metabolism. Evidence for both points of view was summarized in a recent symposium.[165] Against the purely vascular hypothesis would be the observation that total brain ischemia, completely depriving all areas of blood, results in differential effects on various regions. The finding that occasional lesions involve an entire homogeneous region (holotopistic) without conforming to its blood supply,[148] has also been quoted as contradicting the vascular hypothesis.

The difficulty in arriving at the causes of selective vulnerability can be illustrated by the extensive studies on the cerebral cortical laminae and on the hippocampus. Lamina III, the internal granular layer, appears to be particularly sensitive to hypoxia. This layer in certain areas (e.g., human frontal) is the thickest of the cortical layers and has less cell body density, a richer neuropil, and the same glial concentration as the other cortical layers.[137,166] The finding that it has a somewhat greater respiratory capacity (seemingly related to its neuropil content) than the other layers was suggested as contributing to its vulnerability to anoxia.[166] However, it appears that all cells, both neurons and glia, are affected in laminar lesions, leading Spector[167] to attribute the laminar pattern to vascular factors.

Uchimura[168] attributed the sensitivity to anoxia of certain parts of the hippocampus to peculiarities he found in their blood supply. Other studies, however, could not confirm that the vascular pattern Uchimura described was of common occurrence.[169] The area supposedly most vulnerable has had different and conflicting designations: Sommer's sector, Rose's sector, H_1,[170] H_2, CA_1.[171] It should be noted also that the hippocampus in the small rodent represents a major portion of the cerebrum and also differs in cell concentrations from the structure in higher mammals. By slide histochemical methods the "vulnerable" hippocampus region, as compared to its neighbors, shows a decreased reactivity for zinc, acid phosphatase, diaphorase, and acetylcholinesterase (human),[170] and in the mouse is more resistant to damage by acetylpyridine.[171] The zinc-containing material is actually contained in the mossy fibers which originate from the dentate gyrus.[172] These differences within the hippocampus do not appear to offer an easy explanation for the selective vulnerability seen in the human brain. Altered reactivity of various regions of the brain to ischemia may depend more on circulatory factors in one area and more on biochemical factors in another, but the question is far from settled.

XII. AGE AND RESPONSE TO ISCHEMIA

The brains of immature animals are better able to withstand anoxia and ischemia than those of mature animals. Although it has sometimes been stated that this difference is due to a higher anaerobic metabolic capacity in younger brain, this parameter is, in fact, lower in the young brain than in the adult.[2] More likely explanations may be: (1) because of a tendency of immature animals to be poikilothermic, their brain temperatures are frequently lower under experimental conditions, with consequent lower metabolic rates and $\sim$P use; (2) during maturation there is proliferation of structures such as synaptic apparatus which have high energy requirements; concurrently there must be increases in both brain glycolysis and respiration to supply this demand. Earlier evidence for this view has been summarized.[2] The so-called resistance of the young brain to anoxia or ischemia usually refers to the relative capacity of certain functions to survive oxygen deprivation and is directly related to the degree of immaturity at birth. This advantage is rapidly lost at a rate paralleling that of maturation. In the rat, the period during which the survival advantage declines also is concurrent with increasing vascularization of the brain,[173] maturation of neurons, growth of processes, and the great increase of neuronal interconnections, establishment of ionic gradients across cell membranes, increasing capacity to respond to electrical stimulation by changes in metabolic rate (McIlwain,[31] pp. 270–279) and development of spontaneous electrical activity.[174] These changes are associated with a great increase in metabolic demand, and, when they are complete, brain metabolism and resistance to anoxia are similar to those in the adult. Contrary to the earlier belief that a low fetal cerebral oxygen supply in mammals necessitated a significant anaerobic metabolism, the fetal cerebral and maternal arterial oxygen levels are equal.[175] There is ample evidence of the importance of lower body temperature in the survival of immature animals.[176] If the body temperature is maintained, the survival rate of perinatal rats in a nitrogen atmosphere declines more than that of the adult, although survival of both is adversely affected by temperature maintenance.[177]

That the functional survival advantage of immature rats during anoxia can be greatly reduced by agents such as iodoacetate which interfere with glycolysis[178] suggests that glycolysis can sustain the immature brain longer. But one cannot conclude that glycolysis is more active in the young brain. On the contrary, glycolysis is greater in mature rat and dog brain.[178,179] The immature animal does have a greater proportion of the M form of LDH which gives way to a predominance of the H forms in the mature brain, possibly showing better adaptation to anaerobiasis in the young. Lowry *et al.*[1] compared metabolite changes during ischemia in brains of 10-day-old animals (at which time resistance has significantly declined) to those of adult (actually adolescent, 17–22 g) mouse brain. Although the steady-state carbohydrate reserves in the brains of fasted young mice were lower than those of the adults, the metabolite concentration changes during ischemia followed the same patterns as in the adult but at a significantly lower rate. Regulation of carbohydrate flux during ischemia in 10-day-old animals is apparently similar to that of mature animals. Whether control in still younger or fetal brain would also follow the adult metabolic pattern is not known, but it appears unlikely. The brain of 10-day-old mice sustains $\sim$P levels (at approximately one-half of the initial flux) almost three times as long as does the adult (possibly by decreased utilization), but evidently not long enough to account, in itself, for the prolonged survival of the immature animal.[1] However, by 10 days of age anoxic survival time also resembles that of the adult (Himwich, p. 128[2,181]).

Thurston and McDougal[182] recently determined the ischemic metabolic flux in the brains of newborn mice. They found the ATP level sustained approximately four times longer and the ~P flux less than one-half that in 10-day-old mice. These investigators did not believe even this very low energy utilization would in itself account for the prolonged survival times of anoxic newborn mice. It is possible that a lesser capacity to accelerate glycolysis or other metabolic differences also favor survival of the immature brain. The vulnerability of other organs, particularly the heart, must be considered when simple survival is the criterion for resistance. Overall metabolic changes may not reflect those in certain more resistant brain regions when specialized functions such as gasping movements of the severed head are used as a measure of resistance.

Little information is available on the response to ischemia later in the life span. Preliminary data indicate that ~P disappearance during ischemia is greater in the cerebellar molecular layer of 8-month-old mice than in 30-day-old animals.[117] The rate of disappearance during ischemia of ATP and creatine phosphate is lower in the cerebellar cortex and white matter of 30-month-old mice (beyond the median life span) than at 8 months of age.[117] The significance of these age related differences remains to be determined.

XIII. PERIPHERAL NERVE

Peripheral nerve consists mainly of the myelinated axons of neurons whose cell bodies lie in the spinal cord or in dorsal root ganglia, along with Schwann cells and connective tissue. Since peripheral nerves are commonly studied by isolation *in vitro*, the relatively anoxic environment reproduces most of the conditions occurring in ischemia. Major differences are loss of metabolites to the medium and lower temperature. The extensive literature before 1957 on isolated nerve metabolism during anoxia has been reviewed.[183] There is still controversy as to the relative sensitivity to anoxia or ischemia of the different fiber types present in a mixed nerve.[184] The extremely long ischemic survival times reported for isolated peripheral nerves were derived from studies in invertebrates or amphibians. Mammalian nerve survival is much shorter, though still considerably longer than that of mammalian brain.[185] The rapid phase of electrophysiological changes in rat nerve at 24°C occurs after 16 min of ischemia,[185] corresponding to similar changes in rat brain after a shorter interval (Van Harreveld,[111] p. 69). As in brain, the greatest part of active axonal metabolism is probably devoted to ion transport.[186] There may be a change in metabolic pattern between the active and the resting state.[187] If the LDH isoenzyme pattern of a tissue reflects relative dependence on aerobic or anaerobic metabolic pathways, the preponderance of the M-type in peripheral nerve would suggest a greater metabolic capacity under hypoxic conditions than that of adult brain.[188]

The rate of change in overall pyridine nucleotide reduction by fluorescence measurement of anoxic, isolated nerve is lower than that in brain[183] and is particularly low in invertebrate nerve.[187] Carbohydrate metabolites and adenine and creatine phosphate cofactors in rabbit tibial nerve maintained anoxically at 35°C *in vitro* undergo much slower changes than in brain,[78] the metabolic rates being about 7% of those in mouse brain. These findings are consistent with the measured rates of oxygen consumption in mammalian peripheral nerve.[189] The pattern of change of metabolites during ischemia is similar to that in mouse brain, with the notable exception that glycogen is utilized faster than glucose in the peripheral nerve. As in brain, carbohydrate reserves are quite low in mammalian nerve.[78] During *in vitro* reoxygenation, the ~P level increases almost to its original value at a time when the lactate levels are still considerably elevated, indicating that in nerve as well as brain high lactate levels are not necessarily toxic. Recovery of metabolite levels is possible after 30 min of anoxia.[190]

The pattern of metabolism of mammalian nerve during ischemia appears to be more like that of mammalian brain than that of inframammalian brain, although both nerve and the latter tissue metabolize slowly.[81] Phosphofructokinase, for instance, is facilitated more strongly in both mammalian tissues than in the brains of several inframammalian species.[81] Isolated invertebrate nerve (crustacean stretch receptor), when stimulated aerobically (possibly relatively anoxic) undergoes metabolite and cofactor changes similar to those of mammalian nerve during ischemia,[191] including the evident utilization of glycogen before glucose. However, like amphibian, reptile, and fish brain, changes in G6P in the crustacean nerve indicate less facilitation of PFK than in mammalian nervous tissue. The metabolism of peripheral nerve thus reflects species differences as well as the structural and functional differentiation of central and peripheral neural tissue.

XIV. GLIAL TUMORS

Glial tumors may serve as models for glial metabolism. The response of glial tumors to variations in blood supply is also of interest for therapy, e.g., decrease in radiosensitivity with anoxia. Ischemia-induced changes in metabolite levels have been studied.[24,56,164,192]

Changes in the levels of several metabolites and cofactors of carbohydrate metabolism during ischemia were determined in a chemically induced mouse ependymoblastoma growing intracerebrally or subcutaneously, and compared to the levels of several enzymes involved in carbohydrate metabolism.[56,193] Steady-state levels of lactate, ADP, and AMP are higher and ATP and creatine phosphate lower in tumor than in brain. During an ischemic period of 10 min, progressive anaerobic metabolic changes in glucose, glycogen, lactate, G6P, creatine–phosphate, ATP, ADP, and AMP occur. However, the metabolic flux is lower in the subcutaneously growing tumor than in the other tissues studied. The steady-state levels of metabolites and cofactors in the tumors resemble those of ischemic brain, suggesting that even the small tumors tend to outgrow their blood supply. Evidence of circulatory insufficiency increased with the size of the tumors. Measurements of metabolite levels in various regions of tumor indicate a significant variation from area to area, apparently in relation to effectiveness of circulation.[56,164,194] The results suggest that environmental factors are an important determinant of the mode of metabolism of the tumor as a whole and of regions within it. Since the mode of metabolism of a neoplasm may be more clearly related to local factors than to the enzymatic potential of the tissue, the use of tumor tissue as a model for glia is limited. However, the metabolism of tumor tissue functioning under optimal conditions such as that dissected from histologically verified viable areas well supplied with blood may resemble that of the parent (glial) tissue as well as reflecting neoplastic change.

XV. SUMMARY

The changes taking place in whole mouse brain during ischemia have been described in approximately chronological sequence. Following the failure of vascular autoregulatory mechanisms which function to maintain a supply of oxygen sufficient to maintain oxidative metabolism, brain metabolism changes from a steady-state condition to a "closed-system" associated with a tendency of metabolic reactions toward equilibrium. Within 30 sec the O_2 level falls, mitochondrial respiration ceases, an increased proportion of the pyridine nucleotides becomes reduced, and the citric cycle reactants tend toward the levels present at equilibrium. The level of $\sim$P falls rapidly despite increased glycolysis and facilitation of regulatory enzymes. Within 60 sec both glycolytic substrates (glucose and glycogen) and $\sim$P are reduced to very low levels. During this period the LDH reaction is the major source of NAD^+ required for continued glycolysis and lactate rapidly accumulates until the carbohydrate supply is exhausted. However, the amount of lactate produced is not, in itself, sufficient to interfere with metabolism. During the first 30 sec of ischemia changes in Na^+ and K^+ occur in the direction of their ionic concentration gradients, probably because of insufficient $\sim$P necessary to maintain the gradients. Electrophysiological changes are associated with these ionic shifts. The levels of several substances which may function as transmitters in the central nervous system also change rapidly during ischemia.

Ischemic changes in cells of various types as well as in different brain regions have been compared, and the problem of selective tissue vulnerability discussed. Important variables are differences in circulation to various areas and the proportion of a given tissue volume occupied by structures with high energy requirement (neuropil). Hypoxia alone produces most of the changes seen with ischemia, but the metabolic changes in hypoglycemia differ markedly. The basis of "resistance" to ischemia shown by the brains of immature animals appears to be mainly their low energy requirement. The metabolic changes in mammalian peripheral nerve differ from those of brain in pattern of metabolite changes as well as in time course. The brains of poikilothermic species respond differently to ischemia than mammalian brain, because of both lesser energy requirement and different regulatory patterns of carbohydrate metabolism.

XVI. REFERENCES

1. O. H. Lowry, J. V. Passonneau, F. X. Hasselberger, and D. W. Schulz, Effect of ischemia on known substrates and cofactors of the glycolytic pathway in brain, *J. Biol. Chem.* **239**:18–30 (1964).
2. H. E. Himwich, *Brain Metabolism and Cerebral Disorders*, Williams and Wilkins Co., Baltimore (1951).
3. M. Schneider, *in Metabolism of the Nervous System* (D. Richter, ed.), pp. 238–244, Pergamon Press, New York (1957).
4. W. E. Stone, *in Neurochemistry* (K. A. C. Elliott, I. H. Page, J. H. Quastel, eds.), pp. 485-514, C. C. Thomas, Springfield, Ill. (1955).
5. S. S. Kety, Measurement of local circulation within the brain by means of inert, diffusible tracers, *Acta Neurol. Scand. Suppl.* **14**:20–23 (1965).
6. R. Rodnight, H. McIlwain, and M. A. Tresize, Analysis of arterial and cerebral venous blood from the rabbit, *J. Neurochem.* **3**:209–218 (1959).
7. Symposium, Regional cerebral blood flow, *Acta Neurol. Scand. Suppl.* **14**:111–197 (1965).
8. F. Gotoh, Y. Tazaki, K. Hamaguchi, and J. S. Meyer, Continuous recording of sodium and potassium ionic activity of blood and brain *in situ*, *J. Neurochem.* **9**:81–97 (1962).
9. S.-O. Brattgärd, H. Løvtrup-Rein, and M. L. Moss, Free nucleotides in nerve cells, *J. Neurochem.* **13**:1257–1260 (1966).
10. F. W. Schmahl, E. Betz, H. Talke, and H. J. Hohorst, Energiereiche Phosphate, und Metabolite des Energiestoffwechsels, *Biochem. Zeitschr.* **342**:518–531 (1965).
11. B. McL. Breckenridge and J. H. Norman, The conversion of phosphorylase b to phosphorylase a in brain, *J. Neurochem.* **12**:51–57 (1965).
12. N. Weiner, The content of adenine nucleotides and creatine phosphate in brain of normal and anaesthetized rats, *J. Neurochem.* **7**:241–250 (1961).
13. Symposium, Forms of water in biologic systems, *Ann. N.Y. Acad. Sci.* **125**:249–772 (1965).
14. Symposium, *Cryobiology*, *Fed. Proc.*, **24**:*Suppl.* 15 (1965).
15. R. Katzman and C. E. Wilson, Extraction of lipid and lipid cation from frozen brain tissue, *J. Neurochem.* **7**:110–127 (1960).
16. R. Katzman, personal communication to the authors.
17. D. Richter and R. M. C. Dawson, Brain metabolism in emotional excitement and in sleep, *Am. J. Physiol.* **154**:73–79 (1948).
18. R. Takahashi and M. H. Aprison, Acetylcholine content of discrete areas of the brain obtained by a near-freezing method, *J. Neurochem.* **11**:887–898 (1964).

19. R. A. Dale, Effect of sampling procedures on the contents of some intermediate metabolites of glycolysis in rat, *J. Physiol.* (*London*) **181**:701–711 (1965).
20. J. Crossland and K. J. Rogers, The glycogen, glucose and lactate content of the brain in experimental catatonia, *Biochem. Pharmacol.* **17**:1637–1645 (1968).
21. S. Levine, Anoxic-ischemic encephalopathy in rats, *Am. J. Path.* **36**:1–17 (1960).
22. W. Thorn, H. Scholl, G. Pfleiderer, and B. Müldener, Metabolic processes in the brain at normal and reduced temperatures and under anoxic and ischaemic conditions (Ger.), *J. Neurochem.* **2**:150–165 (1958).
23. L. J. King, G. M. Schoepfle, O. H. Lowry, J.V. Passonneau, and S. Wilson, Effects of electrical stimulation on metabolites in brain of decapitated mice, *J, Neurochem.* **14**:613–618 (1967).
24. H. S. Maker, G. M. Lehrer, C. Weiss, D. J. Silides, and L. C. Scheinberg, The quantitative histochemistry of a chemically induced ependymoblastoma. II. The effect of ischaemia on substrates of carbohydrate metabolism, *J. Neurochem.* **13**:1207–1212 (1966).
25. D. E. Smith, E. Robbins, K. M. Eydt, and G. E. Daesch, The validation of the quantitative histochemical method for use on post mortem material. I. The effect of time and temperature, *J. Lab. Invest.* **6**:447–457 (1957).
26. C. Mayman, P. D. Gatfield, and B. McL. Breckenridge, The glucose content of brain in anaesthesia, *J. Neurochem.* **11**:483–487 (1964).
27. P. D. Gatfield, O. H. Lowry, D. W. Schulz, and J. V. Passonneau, Regional energy reserves in mouse brain and changes with ischaemia and anaesthesia, *J. Neurochem.* **13**:185–195 (1966).
28. O. H. Lowry, Metabolite levels as indicators of control mechanisms, *Fed. Proc.* **25**:846–849 (1966).
29. O. H. Lowry and J. V. Passonneau, The relationships between substrates and enzymes of glycolysis in brain, *J. Biol. Chem.* **239**:31–42 (1964).
30. T. Wood, Adenosine triphosphate–creatine phosphotransferase from ox brain, 2. Properties and function, *Biochem. J.* **89**:210–220 (1963).
31. H. McIlwain, *Biochemistry and the Central Nervous System*, Little, Brown & Co., Boston (1966).
32. W. F. Agnew and C. Crone, Permeability of brain capillaries to hexoses and pentoses in the rabbit, *Acta Physiol. Scand.* **70**:168–175 (1967).
33. P. J. Cohen, S. C. Alexander, T. C. Smith, M. Reivich, and H. Wollman, Effects of hypoxia and normocarbia on cerebral blood flow and metabolism, *J. Appl. Physiol.* **23**:183–189 (1967).
34. P. C. Johnson, Review of previous studies and current theories of autoregulation, *Cir. Research. Suppl.* **15**:2–9 (1964).
35. M. N. Shalit, S. Shimojyo, and O. M. Reinmuth, Carbon dioxide and cerebral circulatory control: I. The extravascular effect, *Arch. Neurol.* **17**:298–330 (1967).
36. E. Häggendal, Blood flow autoregulation of the cerebral gray matter with comments on its mechanism, *Acta Neurol. Scand. Suppl.* **14**:104–109 (1965).
37. J. S. Meyer, F. Gotoh, and Y. Tazaki, Circulation and metabolism following experimental embolization *J. Neuropath. and Exptl. Neurol.* **21**:4–24 (1962).
38. J. S. Meyer, F. Gotoh, Y. Tazaki, K. Hamaguchi, S. Ishikawa, F. Novailhat, and L. Symon, Regional cerebral blood flow and metabolism *in vivo*. Effects of anoxia, hypoglycemia, ischemia, acidosis, alkalosis and alterations of blood pCO_2, *Arch. Neurol.* **7**:560–581 (1962).

39. F. Gotoh, J. S. Meyer, and Y. Takagi, Cerebral venous and arterial blood gas during Cheyne-Stokes respiration, *Neurology*, **19**:278 (1969).
40. J. W. Severinghaus, Role of cerebrospinal fluid pH in normalization of cerebral blood flow in chronic hypocapnia, *Acta Neurol. Scand. Suppl.* **14**:116–120 (1965).
41. U. Pontén and B. K. Siesjö, Brain tissue carbon dioxide changes and cerebral blood flow measurements, *Acta Neurol. Scand. Suppl.* **14**:129–134 (1965).
42. R. Cooper, H. J. Crow, W. Grey Walter, and A. L. Winter, Regional control of cerebral vascular reactivity and oxygen supply in man, *Brain Res.* **3**:174–191 (1966).
43. A. G. Waltz, Effect of blood pressure on blood flow in ischemic and in nonischemic cerebral cortex. The phenomena of autoregulation and luxury perfusion, *Neurology* **18**:613–621 (1968).
43a. A. Kaasile, L. Nilsson, and B. K. Siesjö, *Acta Physiol. Scand.* (in press) and personal communications Dr. Siesjö.
44. N. A. Lassen, The luxury-perfusion syndrome and its possible relation to acute metabolic acidosis localized within the brain, *Lancet* **2**:1113–1115 (1966).
45. K. Hoedt-Rasmussen, E. Skinhoj, O. Paulson, J. Ewald, J. K. Bjerrum, A. Fahrenkrug, and N. A. Lassen, Regional cerebral blood flow in acute apoplexy (The "luxury perfusion syndrome" of brain tissue), *Arch. Neurol.* **17**:271–281 (1967).
46. T. Hasagawa, J. R. Ravens, and J. F. Toole, Precapillary arteriovenous anastomoses "thoroughfare channels" in the brain, *Arch. Neurol.* **16**:217–224 (1967).
47. R. N. Lolley and F. E. Samson, Jr., Cerebral high-energy compounds: Changes in anoxia, *Am. J. Physiol.* **202**:77–79 (1962).
48. G. Thews, in *Selective Vulnerability of the Brain in Hypoxaemia* (J. P. Schade and W. H. McMenemey, eds.), pp. 27–35, F. A. Davis Co., Philadelphia (1963).
49. D. W. Lübbers, Capillary pattern and oxygen tension of the cerebral cortex, *Acta Neurol. Scand. Suppl.* **14**:92–93 (1965).
50. A. Bänder and M. Kiese, Die Wirkung des sauerstoffübertragenden Ferments in Mitochondrien aus Rattenlebern bei niedrigen Sauerstoffdrucken, *Naunyn-Schmiedebergs Arch. Exptl. Path. Pharmak.* **224**:312–321 (1955).
51. B. Chance, P. Cohen, F. Jobsis, and B. Schoener, Intracellular oxidation-reduction state *in vivo*, *Science* **137**:499–508 (1962).
52. H. A. Krebs and T. Gascoyne, The redox state of the nicotinic-adenine dinucleotides in rat liver homogenate, *Biochem. J.* **108**:513–520 (1968).
53. H. J. Hohorst, F. H. Kreutz, and M. Rein, Steady-state equilibria of DPN-linked reactions and the ox/red state of DPN/DPNH in the cytoplasmic compartment of liver cells *in vivo*, *Biochem. Biophys. Res. Commun.* **4**:159–162 (1961).
54. H. J. Hohorst, M. Reim, and H. Bartels, Equilibria of two partner reactions of energy metabolism in muscle, *Biochem. Biophys. Res. Commun.* **7**:137–141 (1962).
55. N. D. Goldberg, J. V. Passonneau, and O. H. Lowry, Effects of changes in brain metabolism on the levels of citric acid cycle intermediates, *J. Biol.. Chem.* **241**:3997–4003 (1966).
56. H. S. Maker, G. M. Lehrer, D. J. Silides, and C. Weiss, Circulatory factors in the carbohydrate metabolism of an experimental glial neoplasm, *Ann. N.Y. Acad. Sci.* (1969). (In press.)
57. M. Nyman and V. P. Whittaker, The distribution of adenosine triphosphates in subcellular fractions of brain tissue, *Biochem. J.* **87**:248–255 (1963).
58. T. Nihei, L. Noda, and M. F. Morales, Kinetic properties and equilibrium constant of the adenosine triphosphate-creatine transphosphorylase-catalyzed reaction, *J. Biol. Chem.* **236**:3203–3209 (1961).

59. H. J. Hohorst, M. Reim, and H. Bartels, Studies on the creatine kinase equilibrium in muscle and the significance of ATP and ADP levels, *Biochem. Biophys. Res. Comm.* **7**:142–146 (1962).
60. A. Kuby and E. A. Nottman, *in The Enzymes* (P. D. Boyer, H. Lardy, and K. M. Myrbäck, eds.), Vol. 6, 2nd ed., pp. 615–603, Academic Press, New York (1962).
61. P. D. Swanson, Effects of tetrodotoxin on electrically stimulated cerebral cortex slices, *Biochem. Pharm.* **17**:129–141 (1968).
62. T. Wood and P. D. Swanson, The occurrence of adenosine triphosphate-creatine transferase in particulate fractions from cerebral tissue and its coupling to the sodium-stimulated microsomal adenosine-triphosphatase, *J. Neurochem.* **11**:301–307 (1964).
63. F. N. Minard and R. V. Davis, The effects of electroshock on the acid-soluble phosphates of rat brain, *J. Biol. Chem.* **237**:1283–1289 (1962).
64. D. W. Schulz, J. V. Passonneau, and O. H. Lowry, An enzymic method for the measurement of inorganic phosphate, *Anal. Biochem.* **19**:300–314 (1967).
65. G. Takagi, Control of aerobic glycolysis and pyruvate kinase activity in cerebral cortex slices, *J. Neurochem.* **15**:903–916 (1968).
66. O. H. Lowry, *in Nerve As A Tissue* (K. Rodahl and B. Issekutz, eds.), pp. 163–174, Harper & Row, New York (1966).
67. H. S. Bachelard, The subcellular distribution and properties of hexokinases in the guinea pig cerebral cortex, *Biochem. J.* **104**:286–292 (1967).
68. O. H. Lowry, N. R. Roberts, D. W. Schulz, J. E. Clow, and J. R. Clark, Quantitative histochemistry of retina. II. Enzymes of glucose metabolism, *J. Biol. Chem.* **236**:2813–2820 (1961).
69. H. M. Katzen and K. T. Schimke, Multiple forms of hexokinase in the rat: distribution, age dependency, properties, *Proc. Nat. Acad. Sci.* **54**: 1218–1225 (1965).
70. R. Tanaka and L. G. Abood, Isolation from rat brain of mitochondria devoid of glycolytic activity, *J. Neurochem.* **10**:571–576 (1963).
71. G. Gercken and H. Preuss, The effect of breathing oxygen on the metabolism of the rat brain under normal and ischaemic conditions, *J. Neurochem.* **16**:761–767 (1969).
72. R. Balàzs and D. Richter, *in Regional Neurochemistry* (S. S. Kety and J. Elkes, eds.), pp. 49–56, Pergamon Press, New York (1961).
73. W. Thorn, Der Phosphocreatingehalt des unbelusten und des ischämischen Gehirns, *Pflüger's Arch. f.d. ges. Physiol.* **285**:331–334 (1965).
74. R. H. Laatsch, Glycerol phosphate dehydrogenase activity of developing rat central nervous system, *J. Neurochem.* **9**:487–492 (1962).
75. J. DeVellis and O. A. Schjeide, Time-dependence of the effect of *x*-irradiation on the formation of glycerol phosphate dehydrogenase and other dehydrogenases in the developing rat brain, *Biochem. J.* **107**:259–264 (1968).
76. B. M. Breckenridge and J. H. Norman, Glycogen phosphorylase in brain, *J. Neurochem.* **9**:383–392 (1962).
77. F. N. Minard, C. H. Kang, and I. K. Mushahwar, The effect of periodic convulsions induced by 1,1-dimethylhydrazine on the glycogen of rat brain, *J. Neurochem.* **12**:279–286 (1965).
78. M. A. Stewart, J. V. Passonneau, and O. H. Lowry, Substrate changes in peripheral nerve during ischemia and Wallerian degeneration, *J. Neurochem.* **12**:719–727 (1965).
79. V. N. Nigam, Studies on glycogen synthesis in pigeon liver homogenates. Glycogen synthesis from glucose monophosphates and UDP glucose, *Biochem. J.* **105**:515–519 (1967).

80. B. Sacktor, J. E. Wilson, and C. G. Tiekert, Regulation of glycolysis in brain, *in situ*, during convulsions, *J. Biol. Chem.* **241**:5071–5075 (1966).
81. D. B. McDougal, Jr., J. Holowach, M. C. Howe, E. M. Jones, and C. A. Thomas, The effects of anoxia upon energy sources and selected metabolic intermediates in the brains of fish, frog and turtle, *J. Neurochem.* **15**:577–588 (1968).
82. G. M. Lehrer, H. S. Maker, D. J. Silides, and C. Weiss, unpublished observations.
83. W. M. Kirsch and J. W. Leitner, Glycolytic metabolites and co-factors in human cerebral cortex and white matter during complete ischemia, *Brain Res.* **4**:358–368. (1967).
84. J. W. Crowell and B. W. Kaufman, Changes in tissue pH after circulatory arrest, *Am. J. Physiol.* **200**:743–745 (1961).
85. G. Gercken, P. v. Wichert, und C. Hintzen, pH an der Grosshirnrinde bei Künstlicher Perfusion, *Pflugers Arch. f. Ges. Physiol.* **278**:84 (1963).
86. F. Plum and J. B. Posner, Blood and cerebrospinal fluid lactate during hyperventilation, *Am. J. Physiol.* **212**:864–867 (1967).
87. J. B. Posner, personal communication to the authors.
88. B. S. Hurwitz and S. K. Wolfson, Jr., Brain lactate in anoxia and hypothermine relationship to brain viability, *Expt. Neurol.* **23**:426–434 (1969).
89. P. D. Swanson, pH Effects on the metabolic properties of isolated cerebral tissues, *Neurology* **19**:298 (1969).
90. O. H. Lowry and J. V. Passonneau, Kinetic evidence for multiple binding sites on phosphofructokinase, *J. Biol. Chem.* **241**:2268–2279 (1966).
91. J. K. Tews, S. H. Carter, P. D. Roa, and W. E. Stone, Free amino acids and related compounds in dog brain: Post-mortem and anoxic changes, effects of ammonium chloride infusion, and levels during seizures induced by picrotoxin and by phenylene tetrazol, *J. Neurochem.* **10**:641–653 (1963).
92. W. Thorn and J. Hermann, The effects of anoxia, ischemia, asphyxia and reduced temperature on the ammonia level in the brain and other organs (Germ.), *J. Neurochem.* **2**:166–177 (1958).
93. J. N. C. Atkinson and R. G. Spector, Metabolism of glucose in anoxic-ischaemic rat brain, *Brit. J. Exptl. Path.* **45**:393–397 (1964).
94. M. C. Fleming and O. H. Lowry, The measurement of free and *n*-acetylated aspartic acids in the nervous system, *J. Neurochem.* **13**:779–783 (1966).
95. R. L. Young and O. H. Lowry, Quantitative methods for measuring the histochemical distribution of alanine, glutamate and glutamine in brain, *J. Neurochem.* **13**:785–793 (1966).
96. V. Chimelař, I. M. Hais, and M. Hodāňovā, γ-Aminobutyric acid content in rat brain processed under different temperature conditions, *Acta Biochim.* **11**:327–335 (1964).
97. R. A. Lovell, S. J. and K. A. C. Elliott, The γ-aminobutyric acid and factor I content of brain, *J. Neurochem.* **10**:479–488 (1963).
98. J. D. Wood, J. Watson, and A. J. Drucker, The effect of hypoxia on brain γ-aminobutyric acid levels, *J. Neurochem.* **15**:603–608 (1968).
99. L. L. Iversen and J. Glowinski, Regional studies of catecholamines in the rat brain, I and II. Rate of turnover of catecholamines in various brain regions, *J. Neurochem.* **13**:671–682 (1966).
100. E. Robins, D. E. Smith, G. E. Daesch, and K. E. Payne, The validation of the quantitative histochemical method for use on post-mortem material. II. The effects of fever and uraemia, *J. Neurochem.* **3**:19–27 (1958).
101. D. Naidoo and O. E. Pratt, The validity of histochemical observations post-mortem on phosphatases in brain tissue, *Enzymologia: Acta Biocatalytica* **17**:1 (1954).

102. C. M. Redman and L. E. Hokin, Stimulation of the metabolism of phosphoinositol and phosphatidic acid in brain cytoplasmic fractions by low concentration of cholinergic agents, *J. Neurochem.* **11**:155–163 (1964).
103. J. Eichberg and G. Hauser, Concentrations and disappearance post-mortem of polyphosphoinositides in developing rat brain, *Biochim. Biophys. Acta* **144**:415–422 (1967).
104. A. C. Birnberger and S. G. Eliasson, Experimental ischemia and polyphosphoinositide metabolism, *Neurology* **19**:297 (1969).
104a. N. G. Bazan and M. Cummings, The turnover of brain fatty acids following decapitation or convulsions. *Second International Meetings of the International Society for Neurochemistry* (R. Paoletti, R. Fumagalli, and C. Galli, eds.), pp. 83–84, Tamburini editore, Milan (1969).
105. J. A. Lowden and L. S. Wolfe, Effect of hypoxia on brain gangliosides, *Nature* **197**:771–772 (1963).
106. K. Suzuki, The pattern of mammalian brain gangliosides. II. Evaluation of the extraction procedures, post-mortem changes and the effect of formalin preservation, *J. Neurochem.* **12**:629–638 (1965).
107. V. Ya Dvorkin, Turnover of individual phospholipid fractions in the rat brain during hypoxia, *Nature* **212**:1239 (1966).
108. D. A. Chetverikov and S. V. Gasteva, Possible mechanisms of depression of cerebral phospholipid metabolism during a deficiency of body oxygen, *Nature* **212**:1236–1237 (1966).
109. M. M. Cohen, The effect of anoxia on the chemistry and morphology of cerebral cortex slices *in vitro*. *J. Neurochem.* **9**:337–344 (1962).
110. P. Whittam, *in The Neurosciences* (G. C. Quarton, T. Melnechuk, and F. O. Schmitt, eds.), pp. 313–325, Rockefeller Press, New York (1967).
111. A. Van Harreveld, *Brain Tissue Electrolytes*, Butterworths, Washington (1966).
112. D. H. Ingvar, M. Baldy-Moulinier, I. Sulg, and S. Horman, Cerebral blood flow related to EEG, *Acta Neurol. Scand. Suppl.* **14**:179–182 (1965).
113. D. H. Ingvar and J. Risberg, Influence of mental activity upon cerebral blood flow in man, *Acta Neurol. Scand. Suppl.* **14**:183–186 (1965).
114. P. W. Davies and R. G. Grenell, Metabolism and function in the cerebral cortex under local perfusion, with the aid of an oxygen cathode for surface measurement of cortical oxygen consumption, *J. Neurophysiol.* **25**:651–683 (1962).
115. W. Thorn, G. Pfleiderer, R. A. Frowein, and I. Ross, Stoffwechselvorgänge im Gehirn bei akuter Anoxie, akuter Ischämie und in der Erholung, *Pfluger's Arch. Ges. Physiol.* **261**:334–360 (1955).
116. L. J. King, O. H. Lowry, J. V. Passonneau, and V. Venson, Effects of convulsants on energy reserves in the cerebral cortex, *J. Neurochem.* **14**:599–611 (1967).
117. H. S. Maker, G. M. Lehrer, D. J. Silides, and C. Weiss, Changes in ATP and creatine phosphate (CP) in mouse cerebellar layers during ischemia, *Neurology* **19**:297 (1969), and unpublished observations.
118. R. G. Spector, Content of lactic acid and adenosine mono-, di- and tri-phosphates in anoxic-ischaemic rat brain, *J. Path. Bact.* **90**:533–542 (1965).
119. R. G. Spector, Water content of the brain in anoxic-ischaemic encephalopathy in adult rats, *Brit. J. Exptl. Path.* **42**:623–630 (1961).
120. S.-L. Yap and R. G. Spector, Cerebral protein synthesis in anoxic-ischaemic brain injury in the rat, *J. Path. Bact.* **90**:543–549 (1965).
121. H. Hager, W. Hirschberger, and W. Scholz, Electron microscopy of Syrian hamster brain following acute hypoxia, *Aerospace Med.* **31**:379–387 (1960).

122. H. Jacob, *in Selective Vulnerability of the Brain in Hypoxaemia* (J. P. Schadé and W. H. McMenemey, eds.), pp. 153–163, F. A. Davis Co., Philadelphia (1963).
123. N. H. Becker, in *Selective Vulnerability of the Brain in Hypoxaemia* (J. P. Schadé and W. H. McMenemey, eds.), p. 317, F. A. Davis, Philadelphia (1963).
124. C. P. Hills, The ultrastructure of anoxic-ischaemic lesions in the cerebral cortex of the adult rat brain, *Guy's Hosp. Rep.* **113**:333–348 (1964).
125. P. J. Anderson, The effect of autolysis on the distribution of acid phosphatase in rat brain, *J. Neurochem.* **12**:919–925 (1965).
126. G. M. Lehrer, Discussion on enzyme localization (cholinergic mechanisms), *Ann. N.Y. Acad. Sci.* **144**:660–662 (1967).
127. W. Zeman, *in Selective Vulnerability of the Brain in Hypoxaemia* (J. P. Schadé and W. H. McMenemey, eds.), pp. 327–348, F. A. Davis Co., Philadelphia (1963).
128. R. Lindenberg, *in Selective Vulnerability of the Brain in Hypoxaemia* (J. P. Schadé and W. H. McMenemey, eds.), pp. 182–209, F. A. Davis Co., Philadelphia (1963).
129. S. Schenker, D. W. McCandless, E. Brophy, and M. S. Lewis, Studies on the intracerebral toxicity of ammonia, *J. Clin. Investigation* **46**:838–848 (1967).
130. W. A. Neely and J. R. Youmans, Anoxia of canine brain without damage, *JAMZ* **183**:1085–1087 (1963).
131. A. Ames III, R. L. Wright, M. Kowada, J. M. Thurston, and G. Majno, Cerebral ischemia II. The no-reflow phenomenen, *Am. J. Path.* **52**:437–447 (1968).
132. C. B. Hills, Ultrastructural changes in the capillary bed of the rat cerebral cortex in anoxic-ischaemic brain lesions. *Am. J. Path.* **44**:531–551 (1964).
133. T. K. Tews, S. H. Carter, and W. E. Stone, Chemical changes in the brain during insulin hypoglycemia and recovery, *J. Neurochem.* **12**:679–693 (1965).
134. L. Bakay and J. C. Lee, The effect of acute hypoxia and hypercapnia on the ultrastructure of the central nervous system, *Brain* **91**:697–706 (1968).
135. K. S. Warren and S. Schenker, Brain ammonia during convulsions in thiamine-deficient mice, *Nature* **193**:253–255 (1962).
136. R. C. Collins, J. B. Posner, and F. Plum, Cerebral metabolic response to electroconvulsions in paralyzed ventilated mouse, *Tr. Am. Neurol. A.* **94**:242–244 (1969).
137. H. H. Hess, *in Regional Neurochemistry* (S. S. Kety and J. Elkes, eds.), pp. 200–212, Pergamon Press, New York (1961).
138. A. Pope, *in Biology of Neuroglia* (W. F. Windel, ed.), pp. 211–222, C. C. Thomas, Springfield, Ill. (1958).
139. A. Hamberger and H. Röckert, Intracellular potassium in isolated nerve cells and glial cells, *J. Neurochem.* **11**:757–760 (1964).
140. A. Hamberger and H. Hydén, Inverse enzymatic changes in neurons and glia during increased function and hypoxia, *J. Cell Biol.* **16**:521–525 (1963).
141. A. Hamberger, Oxidation of tricarboxylic acid cycle intermediates, *J. Neurochem.* **8**:31–35 (1961).
142. O. H. Lowry, *in Metabolism of the Nervous System* (D. Richter, ed.), pp. 323–328, Pergamon Press, London (1957).
143. O. H. Lowry, N. R. Roberts, and C. Lewis, The quantitative histochemistry of the retina, *J. Biol. Chem.* **220**:879–892 (1956).
144. F. M. Matschinsky, Quantitative histochemistry of nicotinamide adenine nucleotides in retina of monkey and rabbit, *J. Neurochem.* **15**:643–657 (1968).
145. F. M. Matschinsky, J. V. Passonneau, and O. H. Lowry, Quantitative histochemical analysis of glycolytic intermediates and cofactors with an oil well technique, *J. Histochem. Cytochem.* **16**:29–39 (1968).

146. G. M. Lehrer, C. Weiss, D. J. Silides, C. Lichtman, and M. Furman, The quantitative histochemistry of supramedullary neurons of puffer fishes. I. Enzymes of glucose metabolism, *J. Cell. Biol.* **37**:575–579 (1968).
147. G. M. Lehrer, M. B. Bornstein, C. Weiss, and D. J. Silides, Enzymatic maturation of mouse cerebral neocortex *in vitro* and *in situ, Exptl. Neurol.* **26**:595–606 (1970).
148. W. Scholz, *in Selective Vulnerability of the Brain in Hypoxaemia* (J. P. Schade and W. H. McMenemey, eds.), pp. 257–267, F. A. Davis Co., Philadelphia (1963).
149. L. Sokoloff, *in Regional Neurochemistry* (S. S. Kety and J. Elkes, eds.), pp. 107–117, Pergamon Press, New York (1961).
150. V. Nair, D. Palm, and L. J. Roth, Relative vascularity of certain anatomical areas of the brain and other regions of the rat, *Nature* **188**:497–498 (1960).
151. L. Sokoloff, The action of drugs on the cerebral circulation, *Pharmacol. Rev.* **11**:1–85 (1959).
152. R. L. Friede, *in Regional Neurochemistry* (S. S. Kety and J. Elkes, eds.), pp. 151–159, Pergamon Press, New York (1961).
153. H. B. Hough and H. G. Wolff, The relative vascularity of subcortical ganglia of the cat's brain; the putamen, globus pallidus, substantia nigra, red nucleus and geniculate bodies, *J. Comp. Neurol.* **71**:427–436 (1939).
154. E. G. Holmes, Oxidation in central and peripheral nervous tissue, *Biochem. J.* **24**:914–925 (1930).
155. E. G. Holmes, The metabolic activity of the cells of the trigeminal ganglion, *Biochem. J.* **26**:2005–2009 (1932).
156. D. H. Ingvar, *in Regional Neurochemistry* (S. S. Kety and J. Elkes, eds.), pp. 55–61, Pergamon Press, New York (1961).
157. H. E. Himwich and J. F. Fazekas, Comparative studies of the metabolism of the brain of infant and adult dogs, *Am. J. Physiol.* **132**:454–459 (1941).
158. A. Chesler and H. E. Himwich, Glycolysis in the parts of the central nervous system of cats and dogs during growth, *Am. J. Physiol.* **142**:544–549 (1944).
159. J. W. Ridge, The distribution of cytochrome oxidase-activity in rabbit brain, *Biochem. J.* **102**:612–617 (1967).
160. J. W. Ridge, Resting and stimulated respiration *in vitro* in the CNS. Regional distribution and relation to cell density, *Biochem. J.* **105**:831–835 (1967).
161. J. I. Nurnberger *in Biology of Neuroglia* (W. Windle, ed.), pp. 193–202, C. C. Thomas, Springfield, Ill. (1958).
162. G. E. Vladimirov, M. N. Baranov, L. Z. Pevzner, and Wang Tsyn-Yan, *in Regional Neurochemistry* (S. S. Kety and J. Elkes, eds.), pp. 126–134, Pergamon Press, New York (1961).
163. W. W. Tourtellotte, O. H. Lowry, J. V. Passonneau, J. L. O'Leary, A. B. Harris, and M. J. Rowe, Carbohydrate metabolites in rabbit hereditary ataxia, *Arch. Neurol.* **15**:283–288 (1966).
164. W. M. Kirsch, D. Schulz, and J. W. Leitner, The effect of prolonged ischemia upon regional energy reserves in the experimental glioblastoma, *Cancer Research* **27**:2212–2219 (1967).
165. J. P. Schadé and W. H. McMenemey, eds., *Selective Vulnerability of the Brain in Hypoxaemia*, F. A. Davis Co., Philadelphia (1967).
166. H. H. Hess and A. Pope, Intralaminar distribution of cytochrome oxidase in human frontal isocortex, *J. Neurochem.* **5**:207–217 (1960).
167. R. G. Spector, Selective changes in dehydrogenase enzymes and pyridine nucleotides in rat brain anoxic-ischaemic encephalopathy, *Brit. J. Exptl. Path.* **44**:312–316 (1963).

168. Y. Uchimura, Über die Gefässversorgung des Ammonshornes, *Z. ges. Neurol. Psychiat.* **112**:1–10 (1928).
169. J. Muller and L. Shaw, Arterial vascularization of the human hippocampus. 1. Extracerebral relationships, *Arch. Neurol.* **13**:45–47 (1965).
170. R. L. Friede, The histochemical architecture of the Ammon's horn as related to its selective vulnerability, *Acta Neuropath.* (Berlin), **6**:1–13 (1966).
171. P. D. McLean, *in Selective Vulnerability of the Brain in Hypoxaemia* (J. P. Schadé and W. H. McMenemey, eds.), pp. 177–181, F. A. Davis Co., Philadelphia (1967).
172. T. McLardy, Neurosyncytial aspects of the hippocampal mossy fiber system, *Confin. Neurol.* **20**:1–17 (1960).
173. E. H. Craigie, The vascular supply of the archicortex of the rat. II. The albino rat at birth, *J. Comp. Neurol.* **52**:353–357 (1931).
174. S. M. Crain, Development of electrical activity in the cerebral cortex of the albino rat, *Proc. Exper. Biol. and Med.* **81**:49–51 (1952).
175. W. F. Lucas, T. Kirschbaum, and N. S. Assali, Cephalic circulation and oxygen consumption before and after birth, *Am. J. Physiol.* **210**:287–292 (1966).
176. J. F. Fazekas, F. A. D. Alexander, and H. E. Himwich, Tolerance of the newborn to anoxia, *Am. J. Physiol.* **134**:281–287 (1941).
177. F. E. Samson, Jr. and N. A. Dahl, Cerebral energy requirement of neonatal rats, *Am. J. Physiol.* **188**:277–280 (1957).
178. H. E. Himwich, A. O. Bernstein, H. Herrlich, A. Chesler, and J. F. Fazekas, Mechanisms for the maintenance of life in the newborn during anoxia, *Am. J. Physiol.* **135**:387–391 (1941–1942).
179. J. J. O'Neill and T. E. Duffy, Alternate pathways in newborn brain, *Life Sci.* **5**:1849–1857 (1966).
180. V. Bonavita, F. Ponte, and G. Amore, Lactate dehydrogenase isoenzymes IV: An ontogenetic study on the rat brain, *J. Neurochem.* **11**:39–47 (1964).
181. L. Jílek, J. Fischer, L. Krulich, and S. Trojan, The reaction of the brain to stagnant hypoxia and anoxia during ontogeny. *Progress in Brain Res.* **9**:113–131 (1964).
182. J. H. Thurston and D. B. McDougal, Jr., Effect of ischemia on metabolism of the brain of the newborn mouse. *Am. J. Physiol.* **216**:348–352 (1969).
183. F. Brink, *in Metabolism of the Nervous System* (D. Richter, ed.), pp. 187–207, Pergamon Press, New York (1957).
184. J. L. Fox and P. I. Kenmore, The effect of ischemia on nerve conduction, *Exptl. Neur.* **17**:403–417 (1967).
185. J. Maruhashi and E. B. Wright, Effect of oxygen lack on the single isolated mammalian (rat) nerve fiber, *J. Neurophysiol.* **30**:434–452 (1967).
186. J. M. Ritchie, The oxygen consumption of mammalian nonmyelinated nerve fibres at rest and during activity. *J. Physiol.* **188**:309–329 (1967).
187. M. Doane, Fluorometric measurement of pyridine nucleotide reduction in the giant axon of the squid, *J. Gen. Physiol.* **50**:2603–2632 (1967).
188. A. Lowenthal, D. Karcher, and M. Van Sande, Electrophoretic patterns of lactate dehydrogenase isoenzymes in nervous tissues, *J. Neurochem.* **11**:247–250 (1964).
189. P. F. Cranefield, F. Brink, and D. W. Bronk, The oxygen uptake of the peripheral nerve of the rat, *J. Neurochem.* **1**:245–249 (1957).
190. M. A. Stewart and G. I. Moonsammy, Substrate changes in peripheral nerve recovering from anoxia, *J. Neurochem.* **13**:1433–1439 (1966).
191. E. Giacobini and A. Grasso, Variations of glycolytic intermediates, phosphate compounds and pyridine nucleotides after prolonged stimulation of an isolated crustacean neurone, *Acta Physiol. Scandinav.* **66**:49–57 (1966).

192. W. M. Kirsch, Substrates of glycolysis in intracranial tumors during complete ischemia, *Cancer Research* **25**:433–439 (1965).
193. G. M. Lehrer, H. S. Maker, D. J. Silides, C. Weiss, and L. C. Scheinberg, The quantitative histochemistry of a chemically induced ependymoblastoma. I. Enzymes, *J. Neurochem.* **13**:1197–1206 (1966).
194. G. M. Lehrer, *in The Biology and Treatment of Intracranial Tumors* (W. S. Fields and P. C. Sharkey, eds.), pp. 140–154, C. C. Thomas, Springfield, Ill. (1962).

NOTE ADDED IN PROOF

Since submission of this review further studies pertinent to the chemistry of brain ischemia have appeared. Kauffman and Lowry (*J. Biol. Chem.* **244**:3687, 1969) assayed several intermediates of the pentose phosphate pathway in mouse brain after varying periods of ischemia. During ischemia there is a moderate decline in 6-phosphogluconate and all intermediates beyond 6-phosphogluconate dehydrogenase decline rapidly, apparently drained off into the glycolytic chain. The question of the role of hypoxia in producing the biochemical change occurring in convulsions remains unsettled. Sanders *et al.* (*Science* **169**:206, 1970) report a decline in ATP prior to the onset of overt metrazol induced seizure activity. However, the techniques used do not rule out a greater rate of change in ATP during freezing rather than drug induced lower steady state levels.

Folbergrova *et al.* (*J. Neurochem.* **17**:1155, 1970) examined the effect of 3- and 30-sec periods of ischemia on the levels of P-creatine, ATP, inorganic phosphorus, glucose and glycogen in several cortical layers and the white matter of mouse (20 g) cerebrum. Among their findings was a higher rate of energy utilization in white matter than in cortical layers I or V. As had Gatfield *et al.*,[27] they found a greater inhibition of energy utilization by phenobarbitone in white matter than in gray. Several explanations have been offered for this phenomenon which is not concordant with past studies on gray and white matter metabolism. One factor determining the rate of metabolism in white matter is apparently brain maturity. Kennedy *et al.* (*Neurology* **20**:614, 1970) found the regional cerebral blood flow of the white matter of immature dog brain to be greater than that in adult animals. This is consistent with the findings that the rate of energy utilization is greater in the white matter of maturing mice than in the adult (Maker and Lehrer, unpublished). These differences in metabolism may contribute to the differences in the distribution of hypoxic or ischemic lesions between the immature and mature human brain (Brand, *Drugs and Poisons in Relation to the Developing Nervous System*, p. 192, U.S. Public Health Service, 1967).

Chapter 11

BODY TEMPERATURE AND DRUG EFFECTS

Jill E. Cremer

Biochemical Mechanisms Section
Toxicology Research Unit, Medical Research Council
Woodmansterne Road, Carshalton, Surrey, England

I. INTRODUCTION

Many drugs and toxic substances bring about changes in body temperature. This is particularly true for small laboratory animals. Until quite recently the consequence of such temperature changes on intermediary metabolism was often ignored. It is the purpose of this chapter to consider and describe neurochemical changes in relation to body temperature. Such a consideration will have to take into account the interplay of environmental temperature and body temperature. These two factors can affect the distribution and metabolism of drugs acting on the CNS as well as the metabolism of the brain itself.

Although it is not the main purpose of this chapter to consider in detail the mechanism of thermoregulation and the possible sites of action of drugs on it, a brief outline does seem appropriate (for a more detailed discussion see Bligh and Cremer[1] and Bligh[2]). In physiological terms the essential elements of the thermoregulatory system are the anterior region of the hypothalamus which receives information from peripheral and central thermosensors and from which instructions are passed to the various effector organs to bring about heat loss or heat production. The balance between these latter two processes determines the core temperature of the body. The effector processes for heat production are shivering and non-shivering thermogenesis. Those for heat loss are peripheral vasomotor control, sweating, and panting. It is not well understood how the body is set at its particular temperature, but it appears that certain agents such as drugs and pyrogens can alter the set-point, so that the body maintains itself above or below the normal temperature.[2,3] Other drugs appear to knock out the hypothalamic centers so that the body temperature becomes particularly dependent on the environmental temperature.[4] Yet others can act on the peripheral effector organs, such that the body cannot maintain its normal temperature. (For a more detailed list see Bligh and Cremer.[1])

Heat production is the result of increased oxidation. For a new classification of mammalian thermogenesis, based on thermodynamic considerations expressed in biological terms see Prusiner and Poe.[5] These authors point out that the ultimate control of peripheral thermogenesis is at the level of mitochondrial oxidation. Shivering thermogenesis results from an increased rate of mitochondrial oxidation of skeletal muscle cells, due to the degradation of ATP during contraction. The main heat-producing organ of non-shivering thermogenesis is probably liver, although other organs, such as brown adipose tissue, can also respond to thermoregulation.[6] The brain in a small animal such as the rat contributes only a few per cent of the total heat production of resting basal metabolism. In many larger animals and certainly in man, the size of the brain (particularly of the cortex) in proportion to the total body weight is greater, so that it contributes a greater proportion of the total body heat. Even so, there is little indication in any homeothermic animal that the rate of oxidative metabolism in brain is varied in a direct response to thermoregulation. Therefore, although certain drugs may well be able to act directly on oxidative metabolic processes in brain tissue, other drugs may primarily affect the main thermoregulatory organs, and metabolic changes observed in the brain will be an indirect consequence of its altered temperature.

The unequivocal fact remains that a reduced rate of oxygen consumption by the whole animal is accompanied by a fall in body temperature and vice versa. Measurement of the rate of total oxygen consumption has proved of value in establishing the extent of hypothermia or hyperthermia caused by drugs.[7,8]

II. HYPOTHERMIA

A. Neurochemical Changes

1. *Absolute Amounts of Metabolites*

Data from the literature on absolute amounts of some metabolites in brains of animals which were hypothermic compared with the coenothermic animals have been assimilated in Table I. One of the changes appears to be a diminution in the amount of glutamic acid. This has always been observed following the administration of triethyltin[9,10] and during barbiturate anesthesia.[11,12] Changes observed following chlorpromazine have been variable[12,13] but although chlorpromazine can cause hypothermia, no clear indication of hypothermia is given in these studies. The "high energy" intermediates ATP and phosphocreatine fluctuate around the control level.[14–19] Glucose, on the other hand, is considerably increased in amount.[10,14,18,20,21] Factors such as a con-

TABLE I

Absolute Amounts of Metabolites in Brains of Hypothermic Animals

Cause of hypothermia	Metabolite measured	Percent of control (control = 100)	Literature reference
Triethyltinsulfate	Glutamic acid	85, 81	9, 10
Thiopentone		67	11
Phenobarbital		86	12
Chlorpromazine		94, 96	12, 13
Triethyltinsulfate	ATP + Phosphocreatine	110	19
Ether		107, 114, 118	15, 16, 17
Phenobarbital		96, 89, 125, 108	14, 15, 17, 18, 21
Pentobarbital		120	16
Amytal		123	17
Chlorpromazine		106, 94	14, 15
Triethyltinsulfate	Glucose	340	10
Ether		480, 304, 195	20, 17, 32
Phenobarbital		300, 340, 350, 380, 250	14, 21, 20, 17, 18, 33
		300	
Amytal		340	17
Chlorpromazine		700, 480	14, 20
Ether	Pyruvate	50	17
Phenobarbital		30, 75	17, 14
Amytal		35	17
Triethyltinsulfate	Lactic acid	109	10
Ether		58, 47	32, 20
Phenobarbital		40, 41, 47	18, 20, 21
Chlorpromazine		65	20

comitant increase in serum glucose have been fully discussed[20] and shown to only partially account for the increased brain glucose. Where simultaneous measurements have been made, pyruvate has been found to decrease when glucose increased.[14,17] Lactic acid values vary from being close to the control range,[10,20] to being significantly lower during deep anesthesia.[18] There are seemingly insuperable difficulties in obtaining values for glucose and lactic acid which unequivocally represent the amounts present in the brain of the live animal. The rapid glycolysis postmortem is exemplified in the work of Gatfield *et al.*[18] and of Mark *et al.*[22] In spite of the variable techniques used, an increase in brain glucose in animals which had become hypothermic as a result of the administration of a drug appears to be a consistent finding. A further consideration of this phenomenon will be given below.

2. *Dynamic Aspects*

Another approach to studying intermediary metabolism has been to attempt to assess the dynamic aspects. For this, use is usually made of radioactive precursors. A list of the type of experiment made and some of the results for brains of hypothermic animals are given in Table II. Some of the differences between the control and experimental groups are very striking, for example, the incorporation of ^{32}P into phospholipids,[23-25] of amino acids into protein,[26] and of ^{14}C from glucose into amino acids.[10,27,28] A fully quantitative interpretation of studies of this type, involving nonsteady states is complicated. In the instances cited here, the differences between the control and drug-treated groups appear great enough to be real. The percentage values given in Table II have been calculated from the data as presented in the appropriate references. It should be pointed out that sufficient account is not always taken of factors such as changing specific activities of precursors and size of pools of metabolites (see other chapters in this book for fuller details).

TABLE II

Dynamic Aspects of Brain Metabolism in Animals Made Hypothermic

Cause of hypothermia	Metabolic process measured	Percent of control (control = 100)	Literature reference
Triethyltinsulfate	^{32}P incorporation into phospholipids	69	25
Pentobarbital		37	24
Low environmental temperature		16	23
Pentobarbital	Amino acids into protein	12[a], 52	26, 26
Chlorpromazine		39	26
Reserpine		16	26
Triethyltinsulfate	^{14}C glucose into amino acids	39, 50	10, 28
Pentobarbital		62	27
Chlorpromazine		44	27

[a] Environmental temperature of 4°C.

A different approach to the dynamic aspect of metabolism has been to measure the quantities of metabolites in the brain at intervals following death, referred to as the "closed system" method.[18] This method has been used in relation to glucose metabolism and the data have been calculated in terms of production and utilization of high energy phosphate. In mice depressed with phenobarbitone[18,29] or with γ-hydroxybutyrate,[29] glucose metabolism appeared to be appreciably reduced and the "energy" metabolism was significantly less than in the controls. Since in neither case were attempts made to keep the animals warm, they were, presumably, hypothermic, although this needs to be established for γ-hydroxybutyrate. These results, taken in conjunction with others mentioned above, leave little doubt that glucose metabolism is reduced in the brains of animals made hypothermic. Since the absolute amounts of ATP and phosphocreatine in rapidly frozen brains do not differ significantly from control animals, their production and utilization would appear to be stopped equally.

For more detailed discussions on the increased absolute amounts of glucose in the brains of hypothermic animals, changes in glycolytic intermediates, and possible control sites, see Refs. 14, 17, 18, 20, 21.

B. Effect of Environmental Temperature

1. *Reversible Neurochemical Changes*

The extent and even direction of change in body temperature brought about by administration of a drug is closely dependent on the environmental temperature at which the recipient is maintained. For each drug there is a critical environmental temperature, at which a change in body temperature does not occur (see Section IV). In neurochemical studies it is now becoming possible to categorize those changes which are reversible and those which are irreversible as a result of maintaining the body temperature at normal levels by choosing a suitable environmental temperature. An attempt at such a classification is made here in Table III.

One of the earliest convincing studies that a neurochemical change was related to hypothermia was by Vladimirov *et al.*[23] They showed that the decreased incorporation of ^{32}P into cerebral phospholipids in brains of rats made hypothermic by either a low environmental temperature or by administration of amytal was completely reversed by maintaining the animals' body temperature close to 37°C. Some very similar results with pentobarbital had been reported earlier by Dawson and Richter.[24] In rats made hypothermic with triethyltin, there is a reduced incorporation of ^{32}P into several phospholipids of the brain, especially lecithin.[25] This reduction is completely overcome by maintaining rats at an environmental temperature of 33°C.[25] In the case of this toxic agent, it is of interest that the lethal dose was not changed, nor was the development of the characteristic edema in the central nervous system prevented by maintaining the animals in a warmed environment.

TABLE III

Neurochemical Changes in the Brains of Animals in Which Hypothermia Was Prevented by Raising the Environmental Temperature

Drug	Environmental temperature (°C)	Neurochemical test	Percent of control (control = 100)	Literature reference
Triethyltinsulfate	33	^{32}P into phospholipids	100	25
Pentobarbital	Body temp. 37		85	24
Amytal	Body temp. 37		100	23
Pentobarbital	37	Amino acids into protein	110	26
Chlorpromazine	37		101	26
Reserpine	37		83	26
Triethyltinsulfate	33	^{14}C-glucose into amino acids	52	10
Triethyltinsulfate	33	Glutamic acid	81	10
Thiopentone	37		88	11
Triethyltinsulfate	33	Glucose	240	10
Ether	Body temp. 38		153	32
Phenobarbital	Body temp. 37		147	33

There is well-documented evidence that the decreased entry of amino acids into the brains of animals treated with chlorpromazine, reserpine, or pentobarbital is largely prevented by maintaining the animal's body temperature at 37°C.[30,31] The inhibitory effect of the same three drugs on histidine incorporation into brain protein was shown to be prevented by keeping the animals warm.[26]

2. *Irreversible Neurochemical Changes*

Of the other neurochemical changes mentioned in Section II, A there remain the absolute amounts of glutamic acid and glucose and the rate of metabolism of glucose. From available data it can be tentatively stated that these changes are only partially reversed by preventing hypothermia. Studies of all the three factors were made on the brains of rats given triethyltin.[10] Some results are given in Table III. The differences from the controls were less in animals given triethyltin and maintained at an environmental temperature of 33°C than in those kept at 22°C, but were still statistically significant. The changes were accentuated by hypothermia, but were not a direct consequence of it. Only very limited data for other depressant drugs on the levels of metabolites in brains of animals at different environmental temperatures are available in the literature at the time of writing (see Table III). Comparative studies on brain glucose levels in coenothermic and hypothermic anesthetized animals have been made.[32,33] In guinea pigs anesthetized with ether, cerebral glucose levels were significantly elevated in both warm and cold animals, although the extent was greater in the latter.[32] In mice killed 8 hours after receiving phenobarbital, brain glucose was still higher than the controls in the coenothermic animals, but to a lesser extent than in hypothermic animals.[33] However, the rate of metabolism of phenobarbital in the two groups of animals may have been a complicating factor for a direct comparison (see Section IV). Nevertheless, their results show, in agreement with Mayman *et al.*[20] that the ratio of brain glucose to serum glucose is a better indicator than brain glucose alone. This ratio was still much higher in the coenothermic phenobarbital-treated animals.

Dynamic aspects of glucose metabolism in relation to hypothermia and depressant drugs await investigation. It seems worth noting at this point that recent work by the Japanese group[34] has shown that whereas the glucose uptake, lactate production, and oxygen consumption were very similar in perfused cat brains showing different degrees of nervous function recorded as EEG, the conversion of ^{14}C glucose into amino acids did appear to be directly related to function. Thus, in their A group of animals giving a fast wave pattern, there was a high rate of conversion of ^{14}C-glucose to amino acids. The B group in their experiments showing a lower level of brain function had a decreased rate of conversion very similar to that for hypothermic anesthetized[27] and hypothermic or coenothermic triethyltin poisoned animals *in vivo* (Table III). The question remains whether the altered pattern of glucose metabolism seen *in vivo* is a consequence of reduced nervous activity, or is due to a direct action on enzymes concerned with oxidative metabolism.

III. HYPERTHERMIA

A. Neurochemical Changes

The number of known drugs causing hyperthermia at the usual environmental temperature of around 20°C is far fewer than the number causing hypothermia. It is not surprising that little attention has been given to neurochemical changes related to hyperthermia. There is, however, a very interesting study in which hyperthermic animals were included.[17] The rectal temperature of mice was increased to 43–44°C by placing them in a heated chamber. Analyses of all the tricarboxylic acid cycle substrates, glucose-6-phosphate, and glucose were made on the brains of these animals.

There was a striking increase in pyruvate, α-oxoglutarate, fumarate, and malate, with a decrease in oxaloacetate. It is of considerable interest that the changes were the mirror image of those found in brains of mice anesthetized with either phenobarbital, amytal, or ether. The anesthetized mice were presumably hypothermic in these experiments. Glucose was only a few percent higher than controls in the hyperthermic mouse brain, whereas in the anesthetized animals it was increased three to four times. Measurements were also made of phosphocreatine and adenylates. Again the changes moved in opposite directions in the hyperthermic and anesthetized animals, although the extent of the change was small. There tended to be a decrease in phosphocreatine and ATP during hyperthermia. It would be interesting to compare these results with animals made hypothermic by placement in a cold environment.

2,4-Dinitrophenol and 3,5-dinitro-*o*-cresol (DNOC), well-known uncoupling agents of mitochondrial oxidative phosphorylation, cause hyperthermia *in vivo*.[35–37] There is strong evidence that the heat production resulting from the increased mitochondrial oxidation overwhelms the heat losing effector mechanisms in these animals. It is likely that these agents act directly on mitochondria in the brain, as well as those in other organs. Analyses made on rats killed following injections of DNOC showed the energy-rich phosphate compounds of the brain to be maximally depleted.[38]

Another drug well known for its hyperthermic effects is amphetamine. As yet there are very few neurochemical studies. In one such[27] in which ^{14}C-glucose was used, there was no effect on the incorporation of ^{14}C into carboxylic or amino acids in brains of rats given DL-amphetamine, except a slight increase in alanine radioactivity. The body temperature of these animals was not recorded.

IV. METABOLISM AND DISTRIBUTION OF DRUGS IN RELATION TO THE CNS

A. Body Temperature and Environmental Temperature

Since the time of the excellent review by Fuhrman and Fuhrman[39] on the effects of temperature on the action of drugs, our knowledge of the mechanisms of drug metabolism has increased considerably. An attempt to show the relevance of body temperature to drug metabolism is made in this section. The extent and duration of response to a drug is largely controlled by the rate at which the drug is metabolized and excreted. Many of the neurotropic drugs mentioned in this chapter are metabolized by the mixed function oxidase system which is found predominantly in the liver. The literature on these drug-metabolizing enzymes is so extensive and is increasing so rapidly that a reference to a recent review article only is given.[40] One of the complicating features of this system is that it can be quite rapidly altered quantitatively in the intact animal. One of the best-known agents for increasing the amount of these enzymes is phenobarbital. Since phenobarbital is itself metabolized by these enzymes, its induction property leads to its own increased destruction. In a list of compounds that stimulate their own metabolism[40] those acting on the CNS include hexobarbital, phenobarbital, pentobarbital, chlorpromazine, and meprobamate.

Since so many drugs interact with this system, they can also interfere with each other's rate of metabolism, often in a kinetically competitive manner. Although factors such as species, age, sex, and nutrition have been shown to greatly influence this drug-metabolizing system, the effect of temperature has been less well considered, although some indication of the influence of environmental temperature has been described.[41] Since hexobarbital is metabolized by the liver, a prolongation of hexobarbital sleeping time could be the result of a decrease in its rate of metabolism. A study of six adrenergic blocking agents showed that they all prolonged the hexobarbital sleeping time in mice maintained at room temperature. These agents also caused hypothermia. When this was prevented by maintaining the animals in a warmed environment, only four of the six agents prolonged the sleeping time. The same four could be shown by studies *in vitro* to block the drug-metabolizing enzymes of the liver, whereas the action of the other two was an indirect effect of the hypothermia which slowed the liver metabolizing enzymes.[41] There is slight controversy about the effect of environmental temperature on hexobarbital sleeping times.[42,43] The differences are probably related to dose. At higher doses, sleeping time is more prolonged the lower the environmental temperature,[42] at lower doses there was no significant difference between animals kept at 23° or 37°C.[43] In the latter experiments the rate of elimination of hexobarbital from the brain was exponential at each temperature. This was not so at higher doses where the variable degrees of hypothermia appeared to affect the rate of elimination.[42]

Another comprehensive study is the effect of temperature on the toxicity and distribution of pentobarbital and barbital.[44] The latter drug is not metabolized by the mixed function oxidases. An environmental temperature of 15°C increased the toxicity of both drugs over that of 30°C, although the animals died more slowly, especially after pentobarbital. The effect of variations in room temperature on the rate of elimination of barbital was small, whereas it markedly influenced the tissue concentrations of pentobarbital, including brain. It is interesting that at very early times after an intravenous injection there was more pentobarbital in the brains of animals kept at 30° than at 15°C. Within a few minutes this was reversed. Presumably the initial distribution of the drug was facilitated in the warmer animals and then its subsequent rate of metabolism was increased.

The relationship between body temperature, environmental temperature, and metabolism of a drug in different animal species is clearly brought out in studies on tremorine and oxotremorine.[45–49] The liver enzymes convert tremorine to the pharmacologically active oxotremorine and further metabolize oxotremorine to nonneurotropic products. The rate of elimination of oxotremorine is directly related to the hypothermia which it causes. During the time of maximal hypothermia in mice, the elimination both from the brain and the carcass virtually stops.[47] As the body temperature rises, the rate of elimination from the brain increases more quickly than from the carcass. In rats oxotremorine is normally rapidly metabolized and causes only a slight hypothermia, but if its metabolism in the liver is blocked by other drugs, then the degree of hypothermia is much greater.[46,48] Drugs which block the metabolism of oxotremorine influence the intensity and duration of its other effects, including tremor. On the other hand, administration of amphetamine or atropine sulfate to mice given oxotremorine not only protects against the tremor but also prevents the hypothermia and consequently increases the rate of elimination of oxotremorine.[45,49] A study by Cox and Potkonjak[50] carried out at different environmental temperatures helped to explain the conflicting reports on the effects of tremorine and oxotremorine on the levels of various amines and acetylcholine in the brain.

It is now well established that centrally acting cholinergic drugs can cause hypothermia[39,51] although the effect seems to be very species dependent.[51] In a study with various organophosphorus cholinesterase inhibitors, it was found that during hypothermia in the rat, the body temperature was abnormally dependent on the environmental temperature.[51] Hypothermia following an injection of DFP (diisopropyl phosphorofluoridate) could be prevented by placing the animals in a room kept at 31.5°C. The possibility of preventing a change in body temperature by choosing a particular environmental temperature led Shemano and Nickerson[35] to suggest that for most drugs there is a "critical ambient temperature" above which the body temperature would rise and below which it would fall. The critical ambient temperature for 2,4-dinitrophenol[35] is low compared with many drugs, being around 20°C, presumably because its predominant action is to increase heat production. Chlorpromazine,[35] on the other hand, which enhances heat loss, as well as reducing heat production, has a critical ambient temperature of 36°C. These values are for the rat. If avoidance of a change of body temperature is desired which would otherwise result from the administration of a drug, then the critical ambient temperature should be found for the particular set of circumstances, i.e., drug, time after administration, dose, species of animal, etc.

V. CONCLUDING REMARKS

A. Possible Use of Drugs in Understanding Thermoregulation

One of the standard methods of approach to biological research problems is to deliberately upset the system under study, in order to find out how it normally functions. Drugs by definition alter a biological system and their application to the problem of understanding thermoregulation is one which can be usefully made. In this connection, it would seem instructive to increase our knowledge of the distinction between neurochemical processes which are changed reversibly and irreversibly in a direct relation to changes in body temperature. Some of those which are irreversible might initiate temperature changes and possibly be part of the thermoregulatory mechanism. The few examples given in this chapter show the infancy of this subject.

The relationship between peripheral thermoregulatory effector organs and those of the CNS might be tackled by comparing drugs which can, and cannot enter the brain. An example of this is in the use of organophosphorus compounds. Meeter and Wolthuis[51] showed that two cholinesterase inhibitors which ionize did not cause hypothermia in rats when injected intravenously but did do so when injected into the subarachroidal space.

The pharmacological investigations on the drugs mentioned in this chapter outweigh the neurochemical. It is hoped that by drawing attention to the findings of these pharmacological studies in relation to temperature effects, subsequent neurochemical studies will take these parameters into account. Failure to do so might lead to greater confusion than is necessary in this complicated field.

On the other hand, in a study of neurochemical events use might sometimes be made of deliberately altering the duration of action of a drug by controlling its rate of metabolism. Indeed, it is only by intentionally changing the status quo that our understanding of regulatory mechanisms in the CNS will increase.

VI. REFERENCES

1. J. Bligh and J. E. Cremer, Body temperature and responses to drugs, *Brit. Med. Bulletin* **25**:299–306 (1969).
2. J. Bligh, The thermosensitivity of the hypothalamus and thermoregulation in mammals, *Biol. Rev.* **41**:317–367 (1966).
3. V. J. Lotti, P. Lomax, and R. George, Temperature responses in the rat following intracerebral micro-injection of morphine, *J. Pharmacol. Exptl. Therap.* **150**:135–139 (1965).
4. J. Kollias and R. W. Bullard, The influence of chlorpromazine on physical and chemical mechanisms of temperature regulation in the rat, *J. Pharmacol. Exptl. Therap.* **145**:373–381 (1964).
5. S. Prusiner and M. Poe, Thermodynamic considerations of mammalian thermogenesis, *Nature* **220**:235–237 (1968).
6. L. Jansky and J. S. Hart, Cardiac output and organ blood flow in warm and cold-acclimated rats exposed to cold, *Can. J. Physiol.* **46**:653–659 (1968).
7. V. J. Lotti, P. Lomax, and R. George, Heat production and heat loss in the rat following intracerebral and systemic administration of morphine, *Int. J. Neuropharmacol.* **5**:75–83 (1966).
8. P. Lomax, The hypothermic effect of pentobarbital in the rat: sites and mechanisms of action, *Brain Res.* **1**:296–302 (1966).

9. J. E. Cremer, Amino acid metabolism in rat brain studied with ^{14}C-labelled glucose, *J. Neurochem.* **11**:165–185 (1964).
10. J. E. Cremer, Selective inhibition of glucose oxidation by triethyltin in rat brain *in vivo, Biochem. J.* **119**:95–102 (1970).
11. R. M. C. Dawson, The metabolism and glutamic acid content of rat brain in relation to thiopentone anaesthesia, *Biochem. J.* **49**:138–144 (1951).
12. R. S. De Ropp and E. H. Snedeker, Effect of drugs on amino acid levels in brain: excitants and depressants, *Proc. Soc. Exp. Biol. Med.* **106**:696–700 (1961).
13. S. I. Singh and C. L. MalLotra, Amino acid content of monkey brain—IV effects of chlorpromazine on some amino acids of certain regions of monkey brain, *J. Neurochem.* **14**:135–140 (1967).
14. K. F. Gey, M. Rutishauser, and A. Pletscher, Suppression of glycolysis in rat brain *in vivo* by chlorpromazine, reserpine and phenobarbital, *Biochem. Pharmacol.* **14**:507–514 (1965).
15. F. N. Minard and R. V. Davis, Effect of chlorpromazine, ether and phenobarbital on the active-phosphate level of rat brain, *Nature* (*Lond.*) **193**:277–278 (1962).
16. N. Weiner, The content of adenine nucleotides and creatine phosphate in brain of normal and anesthetized rats, *J. Neurochem.* **7**:241–250 (1961).
17. N. D. Goldberg, J. V. Passonneau, and O. H. Lowry, *in Control of Energy Metabolism* (B. Chance, G. R. W. Estabrook, and J. R. Williamson, eds.), pp. 321–327, Academic Press, New York (1965).
18. P. D. Gatfield, D. H. Lowry, D. W. Schulz, and J. V. Passonneau, Regional energy reserves in mouse brain and changes with ischaemia and anaesthesia, *J. Neurochem.* **13**:185–195 (1966).
19. H. B. Stoner and C. J. Threlfall, The biochemistry of organotin compounds, *Biochem. J.* **69**:376–385 (1958).
20. C. I. Mayman, P. D. Gatfield, and B. McL. Breckenridge, The glucose content of brain in anaesthesia, *J. Neurochem.* **11**:483–487 (1964).
21. O. H. Lowry, J. V. Passonneau, F. X. Hasselberger, and D. W. Schulz, Effect of ischaemia on known substrates and co-factors of the glycolytic pathway in brain, *J. Biol. Chem.* **239**:18–30 (1964).
22. J. Mark, Y. Godin, and P. Mandel, Glucose and lactic acid content of the rat brain, *J. Neurochem.* **15**:141–143 (1968).
23. G. E. Vladimirov, T. N. Ivanova, N. I. Pravdina, and L. N. Rubel. The rate of turnover of phosphorus compounds in the brain in deep hypothermia, *Biokhimiya* **24**:818–824 (1959).
24. R. M. C. Dawson and D. Richter, The phosphorus metabolism of the brain, *Proc. Roy. Soc. B.* **137**:252–267 (1950).
25. M. S. Rose and W. N. Aldridge, Triethyltin and the incorporation of [^{32}P] phosphate into rat brain phospholipids, *J. Neurochem.* **13**:103–108 (1966).
26. L. Shuster and R. V. Hannam, The indirect inhibition of protein synthesis *in vivo* by chlorpromazine, *J. Biol. Chem.* **239**:3401–3406 (1964).
27. H. S. Bachelard and J. R. Lindsay, Effects of neurotropic drugs on glucose metabolism in rat brain *in vivo, Biochem. Pharmacol.* **15**:1053–1058 (1966).
28. J. E. Cremer, Some influences of triethyltin on brain glucose metabolism *in vivo* and *in vitro, Biochem. J.* **106**:8p–9p (1968).
29. M. C. Fleming and S. Lacourt, The comparative effect of γ-hydroxybutyrate and phenobarbital on brain energy metabolism, *Biochem. Pharmacol.* **14**:1905–1907 (1965).
30. A. Pletscher, E. Kunz, H. Staebler, and K. F. Gey, The uptake of tryptamine by brain *in vivo* and its alteration by drugs, *Biochem. Pharmacol.* **12**:1065–1070 (1963).

31. A. Lajtha and J. Toth, The effects of drugs on the uptake and exit of cerebral amino acids, *Biochem. Pharmacol.* **14**:729–738 (1965).
32. El-S. H. H. Hegab and J. A. Miller, Ether anesthesia, hypothermia and carbohydrate metabolism in adult guinea-pig, *J. Appl. Physiol.* **25**:130–133 (1968).
33. S. R. Nelson, D. W. Schulz, J. V. Passonneau, and O. H. Lowry, Control of glycogen levels in brain, *J. Neurochem.* **15**:1271–1279 (1968).
34. S. Otsuki, S. Watanabe, K. Ninomiya, T. Hoaki, and N. Okumura, Correlation between [U-^{14}C] glucose metabolism and function in perfused cat brain, *J. Neurochem.* **15**:859–865 (1968).
35. I. Shemano and M. Nickerson, Effect of ambient temperature on thermal responses to drugs, *Can. J. Physiol.* **36**:1243–1249 (1958).
36. J. A. Kaiser, Studies on the toxicity of diisophenol (2,6-diiodo-4-nitrophenol) to dogs and rodents plus some comparisons with 2,4-dinitrophenol, *Toxicol. Appl. Pharmacol.* **6**:232–244 (1964).
37. V. H. Parker, J. M. Barnes, and F. A. Denz, Some observations on the toxic properties of 3:5-dinitro-ortho-cresol. *Brit. J. Indust. Med.* **8**:226–235 (1951).
38. V. H. Parker, The effect of 3:5 Dinitro-orthocresol on phosphocreatine and the adenosine phosphate compounds of rat tissues, *Biochem. J.* 381–386 (1954).
39. G. J. Fuhrman and F. A. Fuhrman, Effects of temperature on the action of drugs, *Ann. Rev. Pharmacol.* **1**:65–78 (1961).
40. A. H. Conney, Pharmacological implications of microsomal enzyme induction, *Pharmacol. Rev.* **19**:317–366 (1967).
41. J. O. Mullen and J. R. Fouts, Prolongation of hexobarbital sleeping time and inhibition of hepatic drug metabolism by adrenergic blocking agents in mice, *Biochem. Pharmacol.* **14**:305–311 (1965).
42. D. Winne, Experimentelle und biometrische Untersuchungen über den Zusammenhang von Gehirnkonzentration, Eliminationsgeschwindigkeit, Schlafzeit und Körpertemperatur bei der Hexobarbitalnarkose der Maus, *Arch. Exptl. Pathol. Pharmakol.* **247**:278–294 (1964).
43. J. Noordhoek, Pharmacokinetics and dose sleeping time lines of hexobarbital in mice, *Europ. J. Pharmacol.* **3**:242–250 (1968).
44. I. Setnikar and O. Temelcou, Effect of temperature on toxicity and distribution of pentobarbital and barbital in rats and dogs, *J. Pharm. Exptl. Therap.* **135**:213–222 (1962).
45. P. S. J. Spencer, The antagonism of oxotremorine effects in the mouse by thymoleptics, *Life Sci.* **5**:1015–1023 (1966).
46. F. Svöqvist, W. Hammer, H. Schumacher, and J. R. Gillette, The effect of desmethylimipramine and other "anti-tremorine" drugs on the metabolism of tremorine and oxotremorine in rats and mice, *Biochem. Pharmacol.* **17**:915–934 (1968).
47. W. Hammer, B. Karlen, and F. Sjöqvist, The influence of body temperature on the elimination of oxotremorine in the mouse, *Biochem. Pharmacol.* **17**:935–944 (1968).
48. W. Hammer, B. Karlen, A. Rane, and F. Sjöqvist, Rate of metabolism of tremorine and oxotremorine in rats and mice, *Life Sci.* **7**:197–204 (1968).
49. W. Hammer, B. Karlen, and F. Sjöqvist, Drug-induced changes in brain levels of oxotremorine in mice, *Life Sci.* **7**:205–211 (1968).
50. B. Cox and D. Potkonjak, The effect of ambient temperature on the actions of tremorine on body temperature and on the concentration of noradrenaline, dopamine, 5-hydroxytryptamine and acetylcholine in rat brain, *Brit. J. Pharmac. Chemother.* **31**:356–366 (1967).
51. E. Meeter and O. L. Wolthuis, The effects of cholinesterase inhibitors on the body temperature of the rat, *Europ. J. Pharmacol.* **4**:18–24 (1968).

Chapter 12

PHENOTHIAZINES: NEUROCHEMICAL ASPECTS OF THEIR MODE OF ACTION

Sabit Gabay

Biochemical Research Laboratory
Veterans Administration Hospital
Brockton, Massachusetts
and
Harvard School of Dental Medicine
Boston, Massachusetts

I. INTRODUCTION AND PURPOSE

Since a term such as tranquilizer has neither specific nor useful neurochemical connotation, and since phenothiazines induce sedation without hypnosis, the term antipsychotic phenothiazines,* will be used throughout to distinguish these compounds from psychic sedatives (psycholeptics), neuroleptics, etc. This distinction, admittedly more relevant to the intended clinical use and perhaps not the most meaningful, would, however, be more descriptive as to the role of these agents affecting neurobiological processes by their presumed pharmacological mechanism. Under this heading are subsumed phenothiazines, particularly chlorpromazine and its congeners, which have proved most useful in treatment of psychotic patients with aggressive, disturbed, or overactive behavior—most of them suffering from schizophrenias, mania, or other functional psychoses.

One of the purposes of this review is to focus attention on a few selected investigations, particularly those which are neurochemically oriented, that have not been wholly considered in previous reviews on this subject. In the interest of brevity, ancillary literature and references on the effects of phenothiazines on peripheral tissues have been held to a strict minimum. Reviews and books of special interest to neuropsychopharmacology, biochemical pharmacology, clinical studies, etc. are outlined in the reference section preceding the actual bibliography.

* Abbreviation: phenothiazines.

II. EFFECTS OF PHENOTHIAZINES ON THE METABOLIC PROCESSES IN THE NERVOUS SYSTEM

Evidence concerning alterations in the metabolic processes induced by these agents, analogous in structure, will be reviewed critically and eclectically. Thus, a representative sample sufficient to serve as a basis for discussion of the background of this problem has been attempted.

A. Effects on Cerebral Respiration and Glycolysis

Since these compounds depressed nerve cell function *in vivo*, much of the early search for a biochemical explanation centered around the study of cellular respiration. Thus, chlorpromazine (CPZ) [10-(3-dimethylaminopropyl)-2-chlorophenothiazine], at very high concentrations (10^{-3} M), was found to inhibit respiration in brain slices and homogenates.[1-3] Kozak *et al.*[4] reported that CPZ inhibited the oxygen uptake of brain homogenates, in the presence of exogenous glucose, but not the endogenous respiration; however, the inhibition of endogenous respiration has also been reported.[5] Yanagawa[6] has shown that rat brain slice respiration is stimulated by low concentration of CPZ and inhibited by high concentrations. A biphasic effect has also been suggested by Wortiz and Jackim,[7] who found, after a single subcutaneous injection of CPZ, that there was a very slight inhibition of the respiration of brain homogenates and that this effect persisted for 16 hr, after which a very slight stimulation occurred.

Although one might expect a decreased blood sugar level with CPZ since it is antiadrenergic and the action of insulin would be unopposed,[8] CPZ has a marked hyperglycemic effect.[9,10] Máthé *et al.*[11] found an increased glucose level in the rat brain following the intraperitoneal administration of CPZ, and Larsson[12] demonstrated that the utilization of glucose was specifically depressed in the hypothalamus. Bernsohn *et al.*[13] and Moraczewski and Dubois[14] have shown, however, that high concentrations of CPZ do not affect anaerobic glycolysis in brain homogenates *in vitro*. The inability of CPZ to inhibit anaerobic glycolysis is further supported by the failure of this drug to inhibit hexokinase in brain slices and homogenates even in very high concentrations.[15,16] It also has been found to have no effect on cerebral glyceraldehyde phosphate dehydrogenase (GPD) and phosphoglycerokinase.[15]

Bernsohn *et al.*,[13] although unable to inhibit anaerobic glycolysis with CPZ, maintained that the compound did inhibit hexokinase activity in brain homogenates. However, they were unable to demonstrate inhibition of purified yeast hexokinase by CPZ. Löw[17] has confirmed this, but Masurat *et al.*[18] have shown a complete inhibition of this enzyme if the drug and enzyme were incubated and low concentrations of Mg^{2+} and adenosine triphosphate (ATP) were used. They also found that low concentrations of CPZ stimulated this enzyme in the presence of high concentrations of Mg^{2+} and ATP. Recently Gey *et al.*[19] measured the level of hexose phosphates in the rat brain and found it to be depressed at the time of maximum depression of motor activity following the injection of CPZ. They suggested, therefore, that CNS depressants suppress glycolysis by decreasing glucose phosphorylation.

Buzard[20] has suggested, after finding that anaerobic glycolysis in rat brain homogenates could be inhibited by CPZ if the reaction were run long enough, that CPZ was converted anaerobically to metabolites which were the actual inhibitors. Allenby and Collier had previously shown that 7.6×10^{-6}M phenothiazone would inhibit brain homogenate hexokinase after 1 hour of preincubation,[21] but the conversion of phenothiazine to phenothiazone by light was of insufficient magnitude to account for this inhibition.[22]

The inhibition of glycolysis might be associated with an activation of the pentose pathway,[23] but Gey *et al.*[19] have found no changes with CPZ in the relation between the pentose and glycolytic pathways in brain slices. In regard to specific enzymes involved in the pentose pathway, Kistner[24] has shown that glucose-6-phosphate dehydrogenase (G6PD) is inhibited by high concentrations and activated by low concentrations of CPZ under aerobic conditions, but is inhibited only anaerobically. Carver[25] has reported that G6PD from brain tissues is inhibited non-competitively by CPZ, but only after extended preincubation of the drug and enzyme. Brain 6-phosphogluconate dehydrogenase (6PGD) was not inhibited at all by CPZ. In contrast to brain tissue, however, both enzymes from the adrenal cortex were inhibited, but the inhibition of 6PGD was competitive. The activity of rat brain aldolase is not altered after CPZ injection,[15] but the dog plasma enzyme is inhibited.[26]

McIlwain and Greengard,[27] in view of the uncertain pertinence of the above-mentioned effects of CPZ on respiration, have suggested that the most significant effect of CPZ with regard to respiration is its inhibition of the increased oxygen uptake that occurs after electrically stimulating the guinea pig brain. Respiration and glycolysis were completely insensitive to CPZ unless the brain tissue was thus shocked. Lindan *et al.*[28] have supported this concept with their observation that CPZ inhibited the K^+ stimulated increase in the respiration of the isolated rat brain cortex.

B. Effects on Oxidative Phosphorylation and Related Processes

In further attempts to localize the site of action of CPZ with regard to the inhibition of respiration, many workers have focused their attention on some of the respiratory systems involved and on coupled phosphorylation. Thus, CPZ has been found to inhibit oxidation and to uncouple phosphorylation in brain mitochondria.[13,29–31] Century and Horwitt,[31] however, could not demonstrate any alteration in the phosphorylating capacity of brain homogenates after the injection of CPZ. Aghajanian[32] believed the effect of CPZ on brain mitochondria depended on their metabolic state, i.e., the mitochondria were very sensitive to CPZ during periods of slow electron transport and insensitive during periods of high rates of electron transport. Thus, he showed that CPZ (5×10^{-4}M) did not affect respiration in the presence of high concentrations of ADP and substrate, i.e., when mitochondria are most active. Lührs *et al.*(33) have reported that low concentrations of CPZ ($< 10^{-5}$M) increased the oxidation and phosphorylation of liver mitochondria, but Løvtrup(34) was unable to confirm this observation.

Although some have shown that the uncoupling in brain mitochondria was independent of the substrate used,(13,34,35) others have reported that there was no uncoupling with NAD-dependent substrates (glutamate, pyruvate, and α-ketoglutarate) even though their oxidation was inhibited.(30,36–39) The inhibition of the oxidation of NAD-dependent substrates could also be prevented by the addition of NAD to the reaction mixture.(40)

Berger *et al.*(41) have shown that in rat liver mitochondria CPZ has very little effect on oxygen consumption but uncouples phosphorylation with all substrates—especially that coupled to the oxidation of cytochrome *c*. Gallagher *et al.*(40) have confirmed this uncoupling action in liver mitochondria but found that succinate oxidation was not affected. Cerebral oxidation of succinate is apparently more sensitive to CPZ than liver, however, since it is inhibited even in brain homogenates.(42–44)

Although no inhibition of cytochrome oxidase has been found in brain slices and homogenates,(39,45) Abood(29) and Bernsohn *et al.*(13) have demonstrated that this enzyme was inhibited by CPZ in brain mitochondria. This has been confirmed by others,(35,43) and Dawkins *et al.*(46) have identified the inhibition as competitive. *In vivo* inhibition of cerebral cytochrome oxidase has been reported by Williams *et al.*(47) who also found that CPZ had no effect on the copper concentration of the brain. Kurokawa *et al.*(48) have shown a slight stimulatory effect of CPZ on cytochrome oxidase at concentrations $< 10^{-4}$M. Finally, although Berger[37] has demonstrated the inhibition by CPZ of the substrate phosphorylation accompanying the oxidation of α-ketoglutarate, the only step where uncoupling has been identified is in the oxidation of cytochrome *c* by cytochrome oxidase.(37,49)

C. Recapitulative Discussion

It appears unlikely, however, that these reported effects on oxidative metabolism and coupled phosphorylation form the biochemical basis for the action of phenothiazines as antipsychotic agents. As many authors have pointed out, the concentrations of drug required are quite in excess of the dosage needed for its antipsychotic effect.(13,39,41,44,46,50) The fact that brain mitochondria are generally less sensitive than liver mitochondria(37,43,49) also militates against this being the primary action of the drugs. Even more compelling evidence is the failure of acetylpromazine, a phenothiazine with definite antipsychotic properties, to inhibit these processes while demethylated CPZ, which has no efficacy as such, does inhibit them.[46] Bose *et al.*[51] found that promethazine, an antihistaminic phenothiazine, inhibited succinic dehydrogenase more effectively than CPZ, and Bernsohn *et al.*[13] have demonstrated that promazine, a very weak phenothiazine, was as effective as CPZ and promethazine in the inhibition of cytochrome oxidase. Finally, imipramine, which is an antidepressant, exhibited the same inhibitory power as CPZ in many of these reactions.[34,52]

If depressants acted by inhibiting respiration and/or uncoupling phosphorylation, decreased levels of high energy phosphate compounds in the brain should be evident. Therefore, it is surprising that Grenell *et al.*[53] have reported that the injection of CPZ caused a marked increase in brain ATP levels—especially in the hypothalamus. Weiner and Huls[54] and Minard and Davis,[55] however, have shown that there were no changes in ATP or phosphocreatine levels in the brain if the entire animal had been immersed in liquid nitrogen, which Grenell did not do. The increased levels of ATP and phosphocreatine in animals which were not thus frozen were explained by the inhibition by CPZ of the rapid degradation of ATP to AMP that occurred in control animals. Recently, however, Kaul *et al.*[56] have demonstrated biphasic changes in rat brain ATP levels which depended on the time of freezing the animals in liquid nitrogen after the injection of CPZ; ATP decreased in 3 hr and increased in 6 hr. They suggested that, since the CPZ concentration in the brain was low initially, phosphorylation was uncoupled, but that, as CPZ accumulated, ATPase was inhibited and the ATP levels were thus elevated. According to a subsequent report of Decsi,[57] there was a high correlation between the *in vivo* action of a wide variety of phenothiazines and other tranquilizers in their *in vitro* ability to inhibit oxidative phosphorylation and ATPase activity.

The inhibition of both NAD- and Mg^{2+}-activated ATPase in brain mitochondria by high concentrations of CPZ has been confirmed by many authors.[17,29,58–61] Löw[38] has also shown that NAD-ATPase is stimulated by low concentrations of CPZ and that FAD could partially overcome the inhibitory effects of high CPZ concentrations.

III. PHENOTHIAZINES AND ALTERATIONS IN MEMBRANE PERMEABILITY

A. Effects on Cellular and Subcellular Membranes

One area that has received much attention is the effect of psychotropic agents on various membranes. CPZ, in very high concentrations, has been shown to lyse bacterial cells[62,63] and to increase the permeability of various other microorganisms.[64–66] The latter effect could be prevented by Ca^{2+} and Mg^{2+}.[67] On the other hand, decreased permeability was proposed by Freeman and Spirtes[68,69] to explain the prevention by CPZ of the hemolysis of erythrocytes in water, urea, and glycerol. Guth and Spirtes[70] found that permeability to water was specifically suppressed and by concentrations of CPZ that were more likely to be within the pharmacological range of CPZ (10^{-6} to 10^{-5}M). These authors reviewed this literature extensively, as of 1965, and proposed that changes in membrane permeability may be the major underlying mechanism in phenothiazine action.[71]

As a cautionary note, it should be added that effects of this sort seem to be general rather than specific. Seeman and Weinstein[72] found that in addition to phenothiazines, antihistamines and local anesthetics in low concentration protect erythrocytes against hypotonic or mechanical hemolysis. At high concentrations, these compounds lysed the cells. The protective effect on membranes was enhanced at *p*H 8.0 and diminished at *p*H 6.0, compared with neutrality. Protection of erythrocyte membranes against lysis has also been observed with low concentrations of nonsteroidal antiinflammatory drugs,[73] acidic compounds which, in contrast to the bases mentioned in the previous sentence, were more effective as membrane stabilizers at *p*H 6.0 than at *p*H 8.0.

The ability of phenothiazines to affect subcellular membranes has been shown to occur in many biological systems. Thus, the effects of CPZ on mitochondrial membranes might be of even greater significance. Spirtes and Guth[74] have observed a preventive action by CPZ on the swelling of rat liver mitochondria in sucrose but not in KCl. Smith *et al.*[75] have found that CPZ and promethazine inhibited the glutathione- or ascorbate-induced swelling lysis and the hypotonic swelling of mitochondria. These compounds, in further contrast to the usual electron transport inhibitors and uncouplers, had no effect on electron transport-dependent swelling in isotonic media. In this regard, CPZ acted like an antioxidant. Decreased membrane permeability has also been implicated in the inhibition by CPZ ($>10^{-4}$M) of the release of acid phosphatase from rat liver lysosomes at *p*H 5.0.[76]

From a conceptual point of view, these studies, however interesting they may be, do not furnish fundamental information on the interrelationships between phenothiazine and permeability of cell membrane. As Sandberg and Piette[77] pointed out, all of these observations depend on the gross behavior of the whole cell or subcellular system. Their electron magnetic resonance studies of the interaction between CPZ and bovine erythrocyte were carried out in order to detect a direct molecular basis for the changes that might ensue. Since experiments of this sort would help greatly to identify the basic mechanism, it is hoped that their objective would be extended to purified brain organelles and their membranes.

It is amazing that virtually nothing is known about the alterations in membrane permeability brought about by phenothiazines on purified brain subcellular structural components or their membranes. Perhaps this has been largely due to the lack of preparative methods in purifying brain organelles, for they are still in the stage of development and new techniques are constantly being instituted.

Purified brain mitochondria may be a useful model system with which to initiate some studies on the effects of phenothiazines at a membrane level. These organelles, as well as the membranes thereof, may be isolated intact in a relatively pure, viable state, suitable for carrying out a number of systematic studies of this crucial problem. Methods for the study of permeability properties of mitochondrial membranes have been reviewed recently by Chappell.(78) As a cautionary note, it should be added that a great deal of complexity would be introduced into the understanding of the permeability if the membrane processes are not considered in conjunction with the enzymatic processes. It is worth noting that evidence exists that some membranes are made up of subunits, the arrangement of which (with respect to each other or with respect to ancillary enzyme systems) can be altered by the compound responsible for the permeability change.(79) A still more complicated situation exists in the apparent ease with which phenothiazines bind to protein. Thus, caution must be exercised in interpretation without complete knowledge of the distribution of phenothiazines across the membrane and their interactions with membrane lipids (see Section III).

B. Effects on Brain Amines

Alterations in membrane permeability have been suggested as the cause of the depletion of biogenic amines from brain cells that has been observed with CPZ.(80–83) In support of this theory, Giarman and Schanberg(84) found that this drug decreased the amount of bound serotonin (5-HT) and increased the free 5-HT in the rat brain without having any effect on the total amount present.

Even the increased monoamine levels that result from the administration of monoamine oxidase (MAO) inhibitors are decreased by CPZ.(81–83,85,86) This occurred without any apparent inhibition of MAO *in vivo*(80,81,87) or any interference with the penetration of MAO inhibitors into the brain.(80,81) It is possible, however, that the major effect of CPZ is to hinder the binding of the monoamines to their receptors.(82) Axelrod *et al.*(88) showed that CPZ decreased the *in vivo* uptake of labeled norepinephrine (NE) into various tissues of the cat, but could not demonstrate this effect in mice.(89) Bartlet(85,90,91) and West(92) believe that the decreased 5-HT level was due to the inhibition of 5-hydroxytryptophan decarboxylase *in vivo*; however, no inhibition of this enzyme has been demonstrated *in vitro*.(93)

Recently, in order to clarify the mechanism by which brain monoamines and neurochemically related drugs influence processes such as synthesis and storage of amines, affect their release, distribution, uptake and reuptake, etc., several experimental approaches have been devised.

CPZ, along with the antidepressant imipramine, or its derivative desmethylimipramine, have been most commonly employed for the study of uptake in the brain. Both drugs appear to be able to decrease the initial uptake and to prolong the disappearance of labeled amine.[94] These drugs can also influence the extent which brain slices take up or release NE when stimulated.[95] The fact that both CPZ[96,97] and desmethylimipramine[98,99] have been reported to increase the synthesis rate or turnover of NE,[100] and yet do not result in elevated levels, seems to confirm that these two drugs with diametrically opposite clinical effects do not affect endogenous brain amine concentration as reported by Sulser *et al.*[101] and Pletscher *et al.*[102] These findings, although consistent with the conclusion that CPZ diminishes membrane permeability, provide evidence of this effect with imipramine and desimipramine also. Nevertheless, the fact that the intracellular metabolism of NE in the brain is influenced by the latter compounds more than that of other biogenic amines is supported by a fairly large body of observations.[94,103–105] For example, the effects of desmethylimipramine appear to be limited to NE metabolism, since the drug does not retard the uptake of either 5-HT[103] or dopamine[104,105] in nerve endings. Furthermore, CPZ treatment results in enhanced fluorescence of diencephalic dopamine-containing neurons.[106]

Despite the wide variety of effects of phenothiazines on biogenic amines attributed to CPZ, it is also possible that the antipsychotic action produced by this compound may be fully explainable on the basis of its physiologic action in the reticular formation.[107] In this system, CPZ blocks neurons excited by NE and potentiates the inhibitory response of NE on the other responsive cells possibly by inhibiting reuptake.[108] Other authors,[109] however, reported that desimipramine and imipramine, but not CPZ, blocked uptake into brain NE injected into lateral ventricle of rats. These discrepancies in the observed effects of CPZ on NE have been discussed by Schanberg *et al.*[108]

Interested readers are referred to other reviews[110–113] that appear elsewhere in these volumes, whose authors have expertly treated this field of current great interest.

IV. PHENOTHIAZINES AND BRAIN LIPIDS

Alterations in lipid metabolism have also been suggested as the basis for the antipsychotic properties of the phenothiazines. CPZ (10^{-4}M) has been found to depress the incorporation of ^{32}P into brain phospholipids *in vitro*[114,115] and *in vivo*.[115–118] Magee *et al.*[114] stressed that, at this concentration, there was no change in oxygen consumption or in the concentrations of ATP or phosphocreatine. In contrast, nuclear, mitochondrial, and microsomal fractions of cerebral tissues show increased ^{32}P incorporation after stimulation.[119]

A biphasic effect has been reported by Grossi *et al.*[120] who found that low concentrations of CPZ (10^{-5}M) stimulated the incorporation of acetate, mevalonate, and glucose into rat brain phospholipids both *in vivo* and *in vitro* and that high concentrations (10^{-3}M) inhibited it. Glende and Cornatzer[(121)] have shown that ^{35}S incorporation into brain sulfalipids is also inhibited *in vivo*, but they believe this to be the result of a decreased availability of ATP to combine with the ^{35}S before incorporation. More recently, Century and Horwitt[(122)] reported on the effect of dietary lipids on CPZ depression of ^{32}P labeling of rat brain phospholipids.

From the analysis of the information collected in the preceding paragraphs, it seems evident that the pharmacological and phospholipid effects of phenothiazines are closely coupled. Since it is agreed that phenothiazine action brings about an alteration in the permeability of membrane (see Section II), and since it is established that membranes are mainly composed of protein and lipids, it would be of particular significance, therefore, to examine the interactions between these compounds and membrane lipids whose changes in organization and composition are most likely to be involved. This appears to remain unexplored by neurochemists and neuropsychopharmacologists. Admittedly, as Ansell[(123)] commented, "a direct relationship between the lipids of the nervous system and physiological-pharmacological activity has been rarely demonstrated."

Again, it must be remembered that membrane lipids are not homogeneous and they must be regarded as dynamic, so that particular structures and classes of lipids would modify various factors, such as the penetration, solubility, transport, and binding of the drugs participating in the interactions. Thus, without a keen appreciation of the part that lipid membranes play in membrane structure, attempts to correlate phenothiazine effects on brain lipids would not be meaningful.

V. PHENOTHIAZINES AND ENDOCRINE FUNCTIONS

Laborit and Huguenerd's[(124)] pioneering observation of CPZ-induced hypothermia greatly stimulated investigations on the interaction between phenothiazines and endocrine functions. As a result, numerous reports on the effect of CPZ and related compounds in relation to adenohypophysis functions have appeared.

In the pituitary–gonadal axis, CPZ, perphenazine, and promethazine inhibited the activity of gonadotrophic hormones.[(125–127)] These actions on reproductive functions resemble those of reserpine. Other findings[(128,129)] strongly supported the hypothesis that phenothiazines antagonize the normally operating inhibitory influence of the hypothalamus on the secretion of prolactin. Parenthetically, experiments on the effect of these compounds on gonadal function in the male are notably lacking. Conclusions drawn from various studies indicate that of all pituitary functions, the pituitary–gonadal axis is most susceptible to phenothiazines in animals and man. Clinically, the most frequently observed endocrine dysfunction lies in this axis, since female patients treated chronically with relatively high doses commonly developed amenorrhea and galactorrhea.

In the pituitary–adrenal axis, the action of the phenothiazines varies considerably. Some conclusions emerged from attempts to explain the interaction of nicotinamide with reserpine and CPZ. For example, after Burton *et al.*[130] found that nicotinamide greatly enhanced the anti-psychotic effect of CPZ, they reported[131] that the increased NAD levels in the liver which resulted from the injection of nicotinamide could be prolonged by CPZ. They suggested[132,133] that more NAD was produced as a result of the inhibition of the formation of nicotinamide metabolites since the rates of NAD synthesis and degradation were unaffected. Recently this group has presented evidence that high concentrations of CPZ (10^{-3}M) inhibited nicotinamide methyltransferase in mouse liver both *in vitro* and *in vivo*.[134] Shuster and Hannam,[135] however, believe the prolonged increase of NAD levels was produced secondarily to the hypothermic effect of CPZ. Another possibility, suggested by Greengard and Quinn,[136] was that CPZ suppressed the release of adrenocorticotrophic hormone of the pituitary gland (ACTH), since the administration of ACTH to hypophysectomized animals reversed the NAD elevation. The effect of CPZ on pituitary ACTH release has not been clearly established as yet, however. Various authors have reported that ACTH release is suppressed,[137,138] stimulated,[139–143] or completely unaffected.[144,145]

Recently this topic has been meticulously treated by de Wied[146] who, in a scholarly review of the literature, appraised in great detail the significance of CPZ and other phenothiazines in relation to endocrine functions. To interested readers seeking full information, his review is strongly recommended. To de Wied's comment that the "analysis of the literature indicates that phenothiazines affect pituitary activity only in amounts that exceed those necessary to induce psychodepressant effects in animal species," it may be appropriate to add that doses required to obtain consistent activity in some circumstances reach toxic levels.

It is difficult to evaluate the neurochemical significance of this section. However, evidence continues to accumulate suggesting that the hypothalamus as well as limbic structures (mesencephalic and rhinencephalic), which are known to be involved in pituitary activity, are affected by phenothiazines. Further understanding of this poorly delineated close relationship would undoubtedly be rewarding to neuroendocrinologists seeking the site of action of phenothiazines in the brain.

VI. SYNOPSIS OF INFORMATION ON PAST, CURRENT, AND FUTURE RESEARCH

This section was organized in an attempt to summarize the information gathered in this review and to present what appears to be the most significant interpretation of presently available data and to have the opportunity to express new viewpoints for future research work concerning the effects of phenothiazines on the various possibilities discussed above.

A. Collating Salient Points of Available Data

It is apparent from the preceding sections that phenothiazines have generally been assumed to produce their effects either by interacting with a particular enzyme system, or systems, or by bringing about an alteration in the permeability of biological membranes. In addition, investigations on (1) the effects of CPZ and other psychotropic agents on some free amino acids,[147,148] especially on glutamic acid and the "glutamate-group" (Glu, GABA, and Asp); and (2) the correlation between the structural organization of certain regions in the brain[149] have been reported. These observations represent only a quantitative data collection and conclusions drawn from these experiments are such that no common denominator has so far been discerned. As the authors agree, the alterations observed have failed to reveal a consistent picture of the drug effects on the brain amino acids.[148-150] Slight changes in the levels of amino acid by these agents have been attributed to membrane permeability which, as we know, remains an open question for the time being.

Nevertheless, evidence so far suggests that the overall effect of phenothiazines is to produce a general decrease in membrane permeability. But, as indicated previously (see Section II), several biophysical, biochemical, and neurochemical factors urgently need to be studied. It is well known that a wide range of types of drugs has pharmacological effects on the cell membranes.

Underlying much of the work in the field of enzyme interaction is the basic assumption that phenothiazines specifically inhibit some enzyme systems; thus, CPZ has been found to inhibit respiratory and glycolytic enzymes and cholinesterase, to uncouple oxidative phosphorylation, and to inhibit and stimulate ATPase.

B. Phenothiazines and FAD-Requiring Enzymes*

From the earlier literature available on the mechanism of enzyme-phenothiazine derivatives, it is evident that these antipsychotic agents inhibit some flavoenzyme systems. Thus, in addition to the effects of CPZ on succinic dehydrogenase,[42] other flavoenzymes reportedly inhibited by CPZ are $NADH_2$-cytochrome-*c* reductase[35,60] and D-amino acid oxidase (D-AAO) in dog brain.[151] Lasslo and Meyer,[152] however, reported that CPZ both inhibits and potentiates the activity of D-AAO. In addition, Kurokawa *et al.*[48] have found that succinic dehydrogenase is also both inhibited and potentiated by CPZ.

* This subsection concentrates primarily on the studies carried out in the author's laboratory.

One could visualize the structural analogy between the phenothiazine nucleus and the isoalloxazine moiety of FAD as the basis for an interaction between phenothiazines and FAD-requiring enzymes. Gabay and Harris[153] postulated that a systematic comparison of these interactions not only may elucidate metabolic reaction sequences and mechanisms, but also may lead us to the understanding of the mode of action of psychotropic drugs and may provide the ground rules for a systematic attempt to produce agents designed to change the functioning enzymes. They have been thoroughly investigating the effects of CPZ[153] and a variety of phenothiazines[154,155] on a purified D-AAO preparation, as a representative of the flavoenzymes, in isolated systems.

In their earlier work,[153] CPZ was found to inhibit D-AAO in competition with its coenzyme, FAD. The degree of inhibition varied with the method of exposure of the apooxidase to FAD and CPZ. In addition, this inhibition was found to be a function of the apooxidase protein concentration; that is, as the protein concentration increased, the extent of inhibition decreased. This was also confirmed by a decreased inhibition when albumin was added to the reaction mixture. Therefore, it was suggested that this nonspecific complexing between CPZ and protein might explain the large number of diverse effects attributed to CPZ which has undoubtedly complicated interpretation of inhibition (see Section I).

It thus became evident that the relative impurity of the enzyme preparation and the possible involvement of nonspecific binding of phenothiazines by foreign protein precluded the possibility of determining the mechanism of inhibition precisely. Since there appeared to be a definite relationship between the inhibitory potencies of CPZ, and its derivatives and their efficacy in antipsychotic therapy,[154] Gabay and Harris[155] considered it essential to study the kinetics of the reaction with a very highly purified D-AAO preparation.

Although their investigations have been documented *in extenso* in their publications,[153-156] a brief summary is provided as follows. The inhibition of a very highly purified D-AAO apoenzyme preparation by a variety of phenothiazine derivatives has been studied. Only derivatives possessing antipsychotic activity were found to be inhibitors. It should be mentioned that this is the first instance in which CPZ, promazine, and promethazine, an antihistaminic phenothiazine, have been shown to affect a biochemical system differently. Kinetic analysis demonstrated that the phenothiazines competed with FAD for the apoenzyme, and inhibitory constants (K_i) for several derivatives could thus be calculated. The relative inhibitory capacities of these compounds, as measured by their K_i, were nearly paralleled by their relative clinical efficacy and potency, side effects, certain pharmacologic factors, and antipsychotic daily therapeutic dosage, respectively. Phenothiazine derivatives with little or no clinical efficacy as antipsychotic agents (promazine, trimeprazine, promethazine, and the sulfoxides of CPZ and thioridazine) did not inhibit D-AAO even in concentrations as high as 4×10^{-4}M;

neither did imipramine, an antidepressant which is structurally similar to the phenothiazines. However, chlorprothixene, a thiaxanthene derivative with definite usefulness in antipsychotic management, was also a coenzyme-competitive inhibitor and had a K_i approximating that of CPZ. Since the K_i's were low enough to represent concentrations that might be encountered *in vivo*, the hypothesis that phenothiazines could act by inhibiting flavo-enzymes appears to gain further significance.

C. Viewpoints Projected Toward Future Research

The overall unsatisfactory state of understanding of the neurochemical behavior of phenothiazines is largely owing to the relative poverty of knowledge of their effect at the cellular or molecular level. The elucidation of drug action in neurochemical terms is complicated by the fact that such compounds may act by affecting different target sites.

Of special importance in regard to phenothiazines is the delineation of the diverse pharmacological effects of this large class of compounds[157] caused by subtle alterations of, or substitutions on, the basic phenothiazine nucleus. That is, the proposed mechanism must explain not only the reason some are primarily antipsychotics while others are basically anti-metics, antihistaminics, antihelminthics, antitussives, or even anti-depressants, but also why there are such large differences in clinical efficacy within a series of compounds which all exhibit the same major effect.

Furthermore, since data obtained with intact animals, or with non-purified *in vitro* systems of their organs, are so complex that they are of little use in a primary effort to interpret neurochemical or biochemical pharmacologic action at a molecular level, the study of the interaction of phenothiazines with much simpler systems is inevitable.

Still another suggestion for the action of CPZ is that it serves as a powerful electron donor.* Szent-Györgyi and his associates[158,159] reported that CPZ and 5-HT are relatively powerful electron donors capable of forming charge transfer complexes with acceptor molecules—such as riboflavin and its derivatives. Pullman and Pullman[160] in support of this theory pointed out that CPZ should be a very good electron donor. Moreover, Nash and Allison[161] have pointed out that hydrogen bonds can potentiate charge transfer, and that in this way the strength of interaction of CPZ (or any electron donor compound) with a membrane system can be increased. Thus, it is believed that this field warrants further systematic investigations using suitable neurochemical systems. The antipsychotic activity of CPZ and its congeners may very possibly be connected with their outstanding electron-donor properties.

In conjunction with the investigations of Gabay and Harris,[153-156] as described above, it is worth mentioning that their selection of D-AAO

* In the context of biochemistry, the topic of charge transfer has been most closely associated with the donor–acceptor complexes.

as an "enzyme approach" was only to serve as a model system. Although the physiological role of D-AAO remains doubtful, recently it has been demonstrated in the human brain stem,[162] and its distribution was reported in bovine and human nervous tissues.[163] At any rate, it would not be untoward to expect that much of the information obtained in regard to D-AAO might be equally valid with respect to flavoenzymes whose biochemical and physiological roles are definitely established in the nervous system; for example, succinic dehydrogenase,* $NADH_2$-cytochrome-*c* reductase, acyl-CoA dehydrogenase, etc.

Finally, these are some of the questions for which answers are still sought. Are the pharmacological and antipsychotic effects of CPZ, and its congeners, related in any neurochemical way to its remarkable action on membrane permeability? Since such an involvement seems to exist, how can one distinguish it from other effects, such as combination with active sites of membrane enzymes or binding to the so-called receptor sites? Extensive research and systematic studies of these more complex systems are desperately needed.

If this review, though incomplete and restricted in scope, has any stimulating influence on those readers who, through coordinated efforts, seek to understand some of the neurochemical mechanisms related to antipsychotic phenothiazines, the author considers it justified.

VII. REFERENCES

1. S. Courvoisier, J. Fournel, R. Ducrot, M. Kolsky, and P. Koetschet, Propriétés pharmacodynamiques du chlorhydrate de chloro-3 (dimethylamino-3′ propyl)-10 phenothiazine (4.560 R. P.): Étude expérimentale d'un nouveau corps utilisée dans l'anesthésie et dans l'hibernation artificielle, *Arch. Int. Pharmacodyn.* **92**:305–361 (1953).
2. L. Peruzzo and R.-B. Forni, Action de la chlorpromazine (Largactil) sur la respiration des tissus (consommation en oxygene), *Presse Medicale* **61**:1463 (1953).
3. M. Finkelstein, W. A. Spencer, and E. R. Ridgeway, Chlorpromazine and tissue metabolism, *Proc. Soc. Exptl. Biol.* **87**:343–344 (1954).
4. J. Kozak, N. Lang, and A. Zeleny, Some factors modifying the action of chlorpromazine upon the O_2 uptake of brain homogenates *in vitro*, *Experientia* **14**:454–455 (1958).
5. I. Yamamoto, A. Tsujimoto, Y. Tsujimura, M. Minami, and Y. Kurogochi, Effect of chlorpromazine on the cell metabolism, *Jap. J. Pharmac.* **6**:138–146 (1957).
6. M. Yanagawa, Chlorpromazine. 1. Effect of chlorpromazine on tissue respiration and carbohydrate metabolism of rat brain, *Nippon Yakurigaku Zasshi* **54**: 1141 (1958).
7. J. Wortiz and E. Jackim, Effects of chlorpromazine on brain tissue respiration, *Am. J. Psychiat.* **119**:363–366 (1962).
8. F. B. E. Charatan and N. G. Bartlett, The effect of chlorpromazine ("Largactil") on glucose tolerance, *J. Ment. Sci.* **101**:351–353 (1955).

* There is good evidence that this enzyme is located only in brain mitochondria, and it has been used as a marker for the biochemical homogeneity of purified mitochondrial preparation.[164]

9. A. B. Dobkin, R. G. B. Gilbert, and L. Lamoureux, Physiological effects of chlorpromazine, *Anaesthesia* **9**:157–174 (1954).
10. S. K. Gupta, M. A. Patel, and A. D. Joseph, Effects of chlorpromazine and epinephrine on blood-sugar of rabbits, *Arch. Int. Pharmacodyn.* **128**:82–88 (1960).
11. V. Máthé, Gy. Kassay, and K. Hunkár, Die Wirkung des Chlorpromazins auf den Kohlenhydratstoffwechsel des Rattengehirns, *Psychopharmacology* **2**, 334–341 (1961).
12. S. Larsson, The effect of chlorpromazine on the glucose metabolism in different parts of the goat brain, *Acta Physiol. Scand.* **53**:68–74 (1961).
13. J. Bernsohn, I. Namajuska, and L. S. G. Cochrane, The effect of chlorpromazine on respiration and glycolysis in rat brain, *Arch. Biochem. Biophys.* **62**:274–283 (1956).
14. A. S. Moraczewski and K. P. DuBois, The influence of chlorpromazine on intermediary carbohydrate metabolism, *Arch. Int. Pharmacodyn.* **120**, 201–221 (1959).
15. H. G. Albaum, The effect of CNS drugs on the enzyme activity of rat brain, School of Aviation Medicine, U.S. Air Force, Aerospace Medical Center, Brooks Air Force Base, Texas 60–86, 1–8 (1960).
16. H. J. Strecker, *in Psychopharmacology* (H. H. Pennes, ed.) pp. 23–46, Hoeber-Harper, New York (1958).
17. H. Löw, The effects of atebrine and chlorpromazine on some ATP-splitting enzymes, *Exptl. Cell Res.* **16**:456–458 (1959).
18. T. Masurat, S. M. Greenberg, E. G. Rice, J. F. Herndon, and E. J. Van Loon, The action of chlorpromazine on yeast hexokinase, *Biochem. Pharmac.* **5**:20–26 (1960).
19. K. F. Gey, M. Rutihauser, and A. Pletscher, Suppression of glycolysis in rat brain *in vivo* by chlorpromazine, reserpine, and phenobarbital, *Biochem. Pharmac.* **14**:507–514 (1965).
20. J. A. Buzard, *In vitro* inhibition of anaerobic glycolysis by pheno-thiazine ataractic drugs, *Experientia* **16**:153 (1960).
21. G. M. Allenby and H. B. Collier, Enzyme inhibition by derivatives of phenothiazine. V. Inhibition of rat brain hexokinase by phenothiazone, *Can. J. Med. Sci.* **30**:549–551 (1952).
22. H. B. Collier and G. M. Allenby, Enzyme inhibition by derivatives of phenothiazine. IV. Inhibition of succinoxidase activity of rat-liver mitochondria, *Can. J. Med. Sci.* **30**:443–446 (1952).
23. H. Laborit, Mecanismes d'orientation des voies metaboliques en fonction de l'environment. Proposition d'un principe, *Presse Med.* **69**:717–720 (1961).
24. S. Kistner, The effect of atabrine and promazine derivatives on the activity of the Old Yellow Enzyme, *Acta Chem. Scand.* **14**:1389–1402 (1960).
25. M. J. Carver, Differential effects of phenothiazines on hexose phosphate dehydrogenases, *Biochem. Pharmac.* **12**:19–24 (1963).
26. M. P. Schulman and W. H. Vogel, Inhibition of enzymes by aromatic compounds *in vitro* and *in vivo*, *Fed. Proc.* **24**:530 (1965).
27. H. McIlwain and O. Greengard, Excitants and depressants of the central nervous system on isolated electrically stimulated cerebral tissues, *J. Neurochem.* **1**:348–357 (1957).
28. O. Lindan, J. H. Quastel, and S. Sved, Biochemical studies on chlorpromazine. 1. The effect of chlorpromazine on respiratory activity of isolated rat brain cortex, *Can. J. Biochem.* **35**:1135–1144 (1957).
29. L. G. Abood, Effect of chlorpromazine on phosphorylation of brain mitochondria, *Proc. Soc. Exptl. Biol.* **88**:688–690 (1955).

30. A. Andrejew, G. Ducet, J. Louw, and A. J. Rosenberg, Influence de la chlorpromazine, du soneryl, et du phenthiobarbital sur les phosphorylations oxydatives, *Compt. Rend. Soc. Biol.* **150**:484–486 (1956).
31. B. Century and M. K. Horwitt, Actions of reserpine and chlorpromazine · HCl on rat brain oxidative phosphorylation and ATPase, *Proc. Soc. Expt. Biol.* **91**:493–497 (1956).
32. G. Aghajanian, The effect of chlorpromazine on respiration of brain mitochondria as a function of metabolic state, *Biochem. Pharmacol.* **12**:7–12 (1963).
33. W. Lührs, G. Bacigalupo, B. Kadenbach, and E. Heise, Der Einfluss von Chlorpromazin auf die oxydative Phosphorylierung von Tumormitochondrien, *Experientia* **15**: 376–377 (1959).
34. S. Løvtrup, A comparative study of the influence of chlorpromazine and imipramine on mitochondrial activity: oxidation and phosphorylation, *J. Neurochem.* **10**:471–477 (1963).
35. S. Løvtrup, A comparative study of the influence of chlorpromazine and imipramine on mitochondrial activity: Cytochrome oxidase and $NADH_2$-cytochrome c reductase, *J. Neurochem.* **11**:377–386 (1964).
36. A. Andrejew and A. J. Rosenberg, Action de la chlorpromazine sur la respiration des coupes et homogenats et sur les phosphorylations oxydatives des mitochondries du cerveau de rat, *Compt. Rend. Soc. Biol.* **152**:1366–1377 (1958).
37. M. Berger, Metabolic reactivity of brain and liver mitochondria towards chlorpromazine, *J. Neurochem.* **2**, 30–36 (1957).
38. H. Löw, The effects of promazines on mitochondrial ATP-ase reactions, *Biochim. Biophys. Acta* **32**:11–18 (1959).
39. L. Decsi, Wirkung von Largactil auf den Zellstoffwechsel, *Acta Physiol. Acad. Sci. Hung.* **10**:387–396 (1956).
40. C. H. Gallagher, J. H. Koch, and D. M. Mann, The effect of phenothiazine on the metabolism of liver mitochondria, *Biochem. Pharmacol.* **14**:789–797 (1965).
41. M. Berger, H. J. Strecker, and H. Waelsch, Action of chlorpromazine on oxidative phosphorylation of liver and brain mitochondria, *Nature*, **177**:1234–1235 (1956).
42. B. C. Bose and R. Vijayvargiva, Observations on the effect of tranquillizers on tissue respiration, succinic dehydrogenase system, and amino acid oxidation in rat tissues, *Arch. Int. Pharmacodyn.* **127**:27–32 (1960).
43. E. W. Helper, M. J. Carver, H. P. Jacobi, and J. A. Smith, The effect of tranquilizing agents and related compounds on the succinoxidase system, *Arch. Biochem. Biophys.* **76**:354–361 (1958).
44. E. E. Smith, C. Watanabe, J. Louie, W. J. Jones, H. Hoyt, and F. E. Hunter, Jr., The effect of chlorpromazine, promethazine, and diphenhydramine on swelling of isolated liver mitochondria, *Biochem. Pharmacol.* **13**:643–657 (1964).
45. J. Kozák, A. Zelený, and N. Lang, Effect of chlorpromazine on tissue metabolism *in vitro*, *Nature* **185**, 107–108 (1960).
46. M. J. R. Dawkins, J. D. Judah, and K. R. Rees, Action of some phenothiazine derivatives on the respiratory chain, *Biochem. Pharmacol.* **2**:112–120 (1959).
47. R. B. Williams, W. R. Humphries, and C. F. Mills, Influence of phenothiazine administration on cytochrome oxidase activity in lamb brain tissue, *Nature* **198**: 387–388 (1963).
48. M. Kurokawa, H. Naruse, M. Kato, and T. Yabe, Action of chlorpromazine on the respiratory system of guinea pig brain cell-free preparations, *Folia Psychiat. Neurol. Jap.* **10**:354–363 (1957).
49. M. J. R. Dawkins, J. D. Judah, and K. R. Rees, The effect of chlorpromazine on the respiratory chain, *Biochem. J.* **72**:204–209 (1959).

50. K. Kok, Investigation into the influence of chlorpromazine and 10-(3′-dimethylaminopropyl)-phenothiazine (RP 3276) on respiration and glycolysis of rat brain cortex slices, *Acta Physiol. Pharmacol. Neerl.* **5**:1–7 (1956).
51. B. C. Bose, R. Vijayvargiva, and A. Q. Saifi, Observations on the effect of CPZ and promethazine on tissue respiration and succinic dehydrogenase activity of brain and liver of rats, *Indian J. Med. Res.* **47**:36–39 (1959).
52. P. N. Abadom, K. Ahmed, and P. G. Scholefield, Biochemical studies on Tofranil, *Can. J. Biochem.* **39**:551–558 (1961).
53. R. G. Grenell, J. Mendelson, and W. D. McElroy, Effects of chlorpromazine on metabolism in the central nervous system, *Archs. Neurol. Psychiat.* **73**:347–351 (1955).
54. N. Weiner and H. N. Huls, Effect of chlorpromazine on levels of adenine nucleotides and creatine phosphate of brain, *J. Neurochem.* **7**:180–185 (1961).
55. F. N. Minard and R. V. Davis, Effect of chlorpromazine, ether, and phenobarbital on the active phosphate level of rat brain: An improved extraction technique for acid-soluble phosphates, *Nature* **193**:277–278 (1962).
56. C. L. Kaul, J. J. Lewis, and S. D. Livingstone, Influence of chlorpromazine on the levels of adenine nucleotides in the rat brain and hypothalamus *in vivo*, *Biochem. Pharmacol.* **14**:165–175 (1965).
57. L. Decsi, Further studies in metabolic background of tranquilizing drugs' action, *Psychopharmacologia* **2**:224–242 (1961).
58. A. Andrejew, G. Ducet, J. Louw, and A. J. Rosenberg, Influence de la chlorpromazine, du sonéryl, et du penthiobarbital sur l'activité ATP-asique, *Comp. Rend. Soc. Biol.* **150**:509–511 (1956).
59. A. Andrejew and A. J. Rosenberg, Action conjugeé de la chlorpromazine, du 2,4-dinitrophénol, du penthiobarbital, et du sonéryl sur l'activité ATP-asique, *Comp. Rend. Soc. Biol.* **150**:681–683 (1956).
60. M. J. R. Dawkins, J. D. Judah, and K. R. Rees, Action of chlorpromazine. 3. Mitochondrial ATP-ase and the ATP-ADP exchange, *Biochem. J.* **76**:200–205 (1960).
61. J. Bernsohn, I. Namajuska, and B. Boshes, The action of chlorpromazine and reserpine on brain enzyme systems, *J. Neurochem.* **1**:145–149 (1956).
62. R. P. Agrawal and A. Guha, Lytic action of chlorpromazine · HCl on *E. coli* G cells, *Brit. J. Pharmacol.* **24**:466–469 (1965).
63. E. D. Weinberg, J. H. Billman, and D. Borders, Lysis of *Bacillus subtilis* by amines, acridines, and phenothiazines, *Exptl. Cell Res.* **15**:625–628 (1958).
64. H. A. Nathan, Alteration of permeability of *Lactobacillus plantarum* caused by chlorpromazine, *Nature* **192**:471–472 (1961).
65. H. A. Nathan, Effect of nutritional deficiencies on synthesis of the inducible malic enzyme of *Lactobacillus plantarum*, *Arch. für Mikrobiol.* **38**:107–113 (1961).
66. H. A. Nathan and W. Friedman, Chlorpromazine affects permeability of resting cells of *Tetrahymena pyriformis*, *Science* **135**:793–794 (1962).
67. H. N. Guttman and W. Friedman, Protozoa as pharmacological tools: The phenothiazine tranquillizers, *Trans. N.Y. Acad. Sci.* **26**:75–89 (1963).
68. A. R. Freeman and M. A. Spirtes, Effect of some phenothiazine derivatives on the hemolysis of red blood cells *in vitro*, *Biochem. Pharmacol.* **11**:161–163 (1962).
69. A. R. Freeman and M. A. Spirtes, Effects of chlorpromazine on biological membranes. II. Chlorpromazine-induced changes in human erthyrocytes, *Biochem. Pharmacol.* **12**:47–53 (1963).
70. P. S. Guth and M. A. Spirtes, Mode of action of chlorpromazine, *Biochem. Pharmacol.* **8**:170 (1961).

71. P. S. Guth and M. A. Spirtes, *in International Review in Neurobiology* (C. C. Pfeiffer and J. R. Smithies, eds.) Vol. VII, pp. 231–278, Academic Press, New York (1965).
72. P. Seeman and J. Weinstein, I. Erythrocyte membrane stabilization by tranquilizers and antihistamines, *Biochem. Pharmacol.* **15**:1737–1752 (1966).
73. A. D. Inglot and E. Volna, Reactions of non-steroidal anti-inflammatory drugs with erythrocyte membrane, *Biochem. Pharmacol.* **17**:269–279 (1968).
74. M. A. Spirtes and P. S. Guth, Effects of chlorpromazine on biological membranes. I. Chlorpromazine-induced changes in liver mitochondria, *Biochem. Pharmacol.* **12**:37–46 (1963).
75. J. A. Smith, M. J. Carver, and E. W. Helper, The effect of tranquilizing drugs on enzyme systems, *Am. J. Psychiat.* **114**:1011–1014 (1958).
76. P. S. Guth, J. Amaro, O. Z. Sellinger, and L. Elmer, Studies *in vitro* and *in vivo* of the effects of chlorpromazine on rat liver lysosomes, *Biochem. Pharmacol.* **14**: 769–775 (1965).
77. H. E. Sandberg and L. H. Piette, EPR studies of psychotropic drug interaction at cell membrane, *Agressologie IX*(1), 59–65 (1968).
78. J. B. Chappell, Systems used for transport of substrates into mitochondria, *Brit. Med. Bull.* **24**:150–157 (1968).
79. D. E. Green, Membranes as expressions of repeating units, *Proc. Nat'l. Acad. Sci. U.S.A.* **55**:1295–1302 (1966).
80. K. F. Gey and A. Pletscher, Influence of chlorpromazine and chlorprothixene on the cerebral metabolism of 5-hydroxytryptamine, norepinephrine, and dopamine. *J. Pharmacol. Exptl. Therap.* **133**:18–24 (1961).
81. D. E. Schwartz, W. P. Burkard, M. Roth, K. F. Gey, and A. Pletscher, Effect of chlorpromazine on the penetration of monoamine oxidase inhibitors and monoamine oxidase releasers into rat brain, *Arch. Int. Pharmacodyn.* **141**:135–144 (1963).
82. O. Hornykiewicz, H. Ehringer, and K. Lechner, Beeinflussung der Iproniazidwirkung auf die Katecholamine und das 5-Hydroxytryptamin des Rattenhirnes durch Chlorpromazin, *Naunyn-Schmiedebergs Arch. Exptl. Pathol. Pharmakol.* **241**:198–199 (1961).
83. A. Pletscher and K. F. Gey, Wirkung von Chlorpromazin auf pharmakologische Veränderungen des 5-Hydroxytryptamin und Noradrenalin Gehaltes im Gehirn, *Med. Exp.* **2**:259–265 (1960).
84. N. J. Giarman and S. M. Schanberg, Drug-induced alterations of the sub-cellular distribution of 5-hydroxytryptamine (serotonin) in rat's brain. *Biochem. Pharmacol.* **9**:93–96 (1962).
85. A. L. Bartlet, The 5-hydroxytryptamine content of mouse brain and whole mice after treatment with some drugs affecting the CNS, *Brit. J. Pharmacol.* **15**:140–146 (1960).
86. C. Morpurgo, Influence of phenothiazine derivatives on the accumulation of brain amines induced by monoamine oxidase inhibitors, *Biochem. Pharmacol.* **11**:967–972 (1962).
87. H. Ehringer, O. Hornykiewicz, and K. Lechner, Die Wirkung des Chlorpromazin auf den Katecholamin- und 5-Hydroxytryptaminstoffwechsel im Gehirn der Ratte, *Naunyn-Schmeidebergs Arch. Exptl. Pathol. Pharmakol.* **239**:507–519 (1960).
88. J. Axelrod, L. G. Whitby, and G. Hertting, Effect of psychotropic drugs on the uptake of H^3-norepinephrine by tissues, *Science* **133**:383–384 (1961).
89. J. Axelrod and J. K. Inscoe, The uptake and binding of circulating serotonin and the effect of drugs, *J. Pharmacol. Exptl. Therap.* **141**:161–165 (1963).

90. A. Bartlet, The influence of chlorpromazine on the excretion of 5-HIAA from mice, *J. Physiol.* (London) **165**:25–26P (1963).
91. A. Bartlet, The influence of chlorpromazine on the metabolism of 5-hydroxytryptamine in the mouse, *Brit. J. Pharmacol.* **24**:497–509 (1965).
92. G. B. West, Studies on 5-hydroxytryptamine and 5-hydroxytryptophan, *J. Pharm. Pharmacol.* **10**:92–97T (1958).
93. K. F. Gey, W. P. Burkard, and A. Pletscher, Influence of chlorpromazine on decarboxylases of aromatic amino acids, *Biochem. Pharmacol.* 8:383–387 (1961).
94. J. Glowinski and R. J. Baldessarini, Metabolism of norepinephrine in the central nervous system, *Pharmacol. Revs.* **18**:1201–1238 (1966).
95. R. J. Baldessarini and I. J. Kopin, The effect of drugs on the release of norepinephrine-H^3 from central nervous system tissues by electrical stimulation *in vitro*, *J. Pharmacol. Exptl. Therap.* **156**:31–38 (1967).
96. H. Corrodi, K. Fuxe, and T. Hokfelt, The effect of neuroleptics on the activity of central catecholamine neurones, *Life Sci.* **6**:767–774 (1967).
97. W. P. Burkard, K. F. Gey, and A. Pletscher, Activation of tyrosine hydroxylation in rat brain *in vivo* by chlorpromazine, *Nature* **213**:732–733 (1967).
98. N. H. Neff and E. Costa, The influence of monoamine oxidase inhibition on catecholamine synthesis, *Life Sci.* **5**:951–959 (1966).
99. N. H. Neff and E. Costa, *in Antidepressant Drugs* (S. Garattini and M. N. G. Dukes, eds.) pp. 28–34, Excerpta Medica Foundation, New York (1967).
100. B. B. Brodie, E. Costa, A. Dlabac, N. H. Neff, and H. H. Smookler, Application of steady state kinetics to the estimation of synthesis rate and turnover time of tissue catecholamines, *J. Pharmacol. Exptl. Therap.* **154**:493–498 (1966).
101. F. Sulser, M. H. Bickel, and B. B. Brodie, The action of desmethylimipramine in counteracting sedation and cholinergic effects of reserpine-like drugs, *J. Pharmacol. (Kyoto)* **144**:321–328 (1964).
102. A. Pletscher, K. F. Gey, and E. Kunz, *in Progress in Brain Research* (H. E. and W. A. Himwich, eds.) Vol. 8, pp. 45–62, Elsevier Publishing Co., Amsterdam (1965).
103. K. Fuxe and U. Ungerstedt, Localization of 5-hydroxytryptamine uptake in rat brain after intraventricular injection, *J. Pharm. Pharmacol.* **19**:335–337 (1967).
104. J. Glowinski, J. Axelrod, and L. Iversen, Regional studies of catecholamines in the rat brain. IV. Effects of drugs on the disposition and metabolism of H^3-norepinephrine and H^3-dopamine, *J. Pharmacol. Exptl. Therap.* **153**:30–41 (1966).
105. J. Glowinski, and L. L. Iversen, Regional studies of catecholamines in the rat brain. I. The disposition of [^{3}H]norepinephrine, [^{3}H]dopamine, and [^{3}H]dopa in various regions of the brain, *J. Neurochem.* **13**:655–669 (1966).
106. N.-E. Anden, A. Dahlstrom, K. Fuxe, and T. Hokfelt, The effect of haloperidol and chlorpromazine on the amine levels of central monoamine neurons, *Acta Physiol. Scand.* **68**:419–420 (1966).
107. P. B. Bradley, H. H. Wolstencroft, L. Hoesli, and G. L. Avanzino, Neuronal basis for the central action of chlorpromazine, *Nature* **212**:1425–1427 (1966).
108. S. M. Schanberg, J. J. Schildkraut, and I. J. Kopin, The effect of psychoactive drugs on norepinephrine-^{3}H metabolism in brain, *Biochem. Pharmacol.* **16**:393–399 (1967).
109. J. Glowinski and J. Axelrod, Inhibition of uptake of tritiated-noradrenaline in the intact rat brain by imipramine and structurally related compounds, *Nature* **204**: 1318–1319 (1964).
110. J. Glowinski, *in Handbook of Neurochemistry* (A. Lajtha, ed.) Vol. IV, pp. 91–114, Plenum Press, New York (1970).

111. L. L. Iversen, *in Handbook of Neurochemistry* (A. Lajtha, ed.) Vol. IV, pp. 197–220, Plenum Press, New York (1970).
112. J. J. Schildkraut and S. Gershon, *in Handbook of Neurochemistry* (A. Lajtha, ed.) Vol. VI, pp. 357–386, Plenum Press, New York (1971).
113. H. Weil-Malherbe, *in Handbook of Neurochemistry* (A. Lajtha, ed.) Vol. VII, Plenum Press, New York. (In preparation.)
114. W. L. Magee, J. F. Berry, and R. J. Rossiter, Effect of chlorpromazine and azacyclonol on the labelling of phosphatides in brain slices, *Biochim. Biophys. Acta* **21**:408–409 (1956).
115. G. B. Ansell and E. F. Marshall, Some effects of chlorpromazine on brain phospholipid metabolism. II. Labelling of choline- and ethanolamine-containing phospholipids by radioactive precursors *in vitro*, *J. Neurochem.* **10**:883–888 (1963).
116. G. B. Ansell and E. F. Marshall, Some effects of chlorpromazine on brain phospholipid metabolism. I. Effects of the drug on phospholipids *in vivo* and *in vitro*, *J. Neurochem.* **10**:875–882 (1963).
117. G. B. Ansell and H. Dohman, The depression of phospholipid turnover in brain tissue by chlorpromazine, *J. Neurochem.* **1**:150–152 (1956).
118. B. Century and M. K. Horwitt, Effect of CPZ on P^{32} uptake into brain phospholipids as a function of the diet lipid, *Fed. Proc.* **24**:688 (1965).
119. L. E. Hokin and M. R. Hokin, Effects of acetylcholine on the turnover in phospholipids of brain cortex *in vitro*, *Biochim. Biophys. Acta* **16**:229–237 (1955).
120. E. Grossi, P. Paoletti, and R. Paoletti, The *in vitro* and *in vivo* effects of chlorpromazine on brain lipid synthesis, *J. Neurochem.* **6**:73–78 (1960).
121. E. A. Glende and W. E. Cornatzer, The effect of chlorpromazine, trifluoperazine, and promethazine on brain sulfalipids, *J. Pharmacol. Exptl. Therap.* **139**:377–382 (1963).
122. B. Century and M. K. Horwitt, Effect of dietary lipid on chlorpromazine depression of ^{32}P labeling of rat brain phospholipids, *Life Sci.* **8**:215–223 (1969).
123. G. B. Ansell, Cerebral lipids; their chemistry, metabolism and contribution to membrane phenomena, *Biochem. J.* **96**:43P–44P (1965).
124. H. Laborit and P. Huguenerd, L'hibernation artificielle par moyens pharmacodynamique et physiques, *Press Méd.* **64**:1329 (1951).
125. F. Meidinger, Action compareé sur l'hypophysis d'antihistaminiques, de neuroplégique et d'anti-inflammatoires, *Compt. Rend. Soc. Biol.* **148**:1086–1090 (1954).
126. M. Y. Khan and E. C. Bernstorf, Effect of chlorpromazine and reserpine upon pituitary function, *Exptl. Med. Surg.* **22**:363–378 (1964).
127. M. X. Zarrow and K. Brown-Grant, Inhibition of ovulation in the gonadotrophin-treated immature rat by chlorpromazine, *J. Endocrinol.* **30**:87–95 (1964).
128. A. Danon, S. Dihstein, and F. G. Sulman, Stimulation of prolactin secretion by perphenazine in pituitary-hypothalamus organ culture, *Proc. Soc. Exptl. Biol. Med.* **144**:366–368 (1963).
129. J. Meites, C. S. Nicoll, and P. K. Talwalker, *in Advances in Neuroendocrinology* (A. V. Nalbandov, ed.), pp. 238–277, University of Illinois Press, Urbana (1963).
130. R. M. Burton, R. A. Salvador, A. Goldin, and S. R. Humphries, Interaction of nicotinamide with reserpine and chlorpromazine. II. Some effects on the central nervous system of the mouse, *Arch. Int. Pharmacodyn.* **128**:253–259 (1960).
131. R. M. Burton, R. A. Salvador, K. Smith, and R. E. Howard, The effect of chlorpromazine, nicotinamide, and nicotinic acid on pyridine nucleotide levels of human blood. *Ann. N.Y. Acad. Sci.* **96**:353–355 (1962).

132. R. M. Burton and R. A. Salvador, The effect of chlorpromazine on nicotinamide methylpherase: An example of drug-enzyme interaction at the molecular level. *Ann. N.Y. Acad. Sci.* **96**:353–355 (1962).
133. R. M. Burton, N. O. Kaplan, A. Goldin, M. Leitenberg, and S. R. Humphries, Interaction of nicotinamide with reserpine and chlorpromazine. III. Some effects on the DPN content in liver, *Arch. Int. Pharmacodyn.* **128**:260–275 (1960).
134. R. A. Salvador and R. M. Burton, Inhibition of the methylation of nicotinamide by chlorpromazine, *Biochem. Pharmacol.* **14**:1185–1196 (1965).
135. L. Shuster and R. V. Hannam, The indirect inhibition of protein synthesis *in vivo* by chlorpromazine, *J. Biol. Chem.* **239**:3401–3406 (1964).
136. P. Greengard and G. P. Quinn, Metabolic effects of tranquillizers and hypophysectomy, *Ann. N.Y. Acad. Sci.* **96**:179–184 (1962).
137. C. C. J. Olling and D. de Wied, Inhibition of the release of corticotrophin from the hypophysis by chlorpromazine, *Acta Endocrinol.* **22**:283–292 (1956).
138. F. G. Sulman and H. Z. Winnik, Hormonal depression due to treatment of animals with chlorpromazine, *Nature* **178**, 365 (1956).
139. A. Yuwiler, E. Geller, S. Shapiro, and G. G. Slater, Adrenocortical and enzymic effects of imipramine and chlorpromazine, *Biochem. Pharmacol.* **14**:621–623 (1965).
140. R. H. Egdahl and J. B. Richards, Effect of chlorpromazine on pituitary ACTH secretion in the dog, *Am. J. Physiol.* **185**:235–238 (1956).
141. R. J. Jarrett, Some endocrine effects of two phenothiazine derivatives, chlorpromazine and perphenazine, in the female mouse, *Brit. J. Pharmacol. and Chem.* **20**:497–506 (1963).
142. P. K. Talwalker, J. Meites, C. S. Nicoll, and T. F. Hopkins, Effects of chlorpromazine on mammary glands of rats, *Am. J. Physiol.* **199**:1073–1076 (1960).
143. A. Ashford and M. Shapero, Effect of chlorpromazine, reserpine, benactyzine, and phenobarbitone on the release of corticotrophin in the rat, *Brit. J. Pharmacol.* **19**:458–463 (1962).
144. D. Betz and W. F. Ganong, Effect of chlorpromazine on pituitary-adrenal function in the dog, *Acta Endocrinol.* **43**:264–270 (1963).
145. W. F. Ganong, *in Advances in Neuroendocrinology* (A. V. Nalbandov, ed.), pp. 92–149, University of Illinois Press, Urbana (1963).
146. D. de Wied, Chlorpromazine and endocrine function, *Pharmacol. Revs.* **19**:251–288 (1967).
147. M. J. E. Ernsting, W. F. Kafoe, W. Th. Nauta, H. K. Oosterhuis, and P. A. Roukema, *in Amino Acid Pools* (J. T. Holden, ed.) pp. 493–498, Elsevier Publishing Co., Amsterdam (1963).
148. H. H. Tallan, *in Amino Acid Pools* (J. T. Holden, ed.) pp. 465–470, Elsevier Publishing Co., Amsterdam (1963).
149. S. I. Singh and C. L. Malhotra, Effects of chlorpromazine on some amino acids of certain regions of monkey brain, *J. Neurochem.* **14**:135–140 (1967).
150. E. Mussini and F. Marcucci, *in Amino Acid Pools* (J. T. Holden, ed.) pp. 486–492, Elsevier Publishing Co., Amsterdam (1963).
151. K. Yagi, T. Nagatsu, and T. Ozawa, Inhibitory action of chlorpromazine on the oxidation of D-amino-acid in the diencephalon part of the brain, *Nature* **177**:891–892 (1956).
152. A. Lasslo and A. Meyer, Effect of chlorpromazine on isolated *d*-amino acid systems, *Nature* **184**:1922–1929 (1959).
153. S. Gabay and S. R. Harris, Studies of FAD-requiring enzymes and phenothiazines. I. Interactions of chlorpromazine and D-amino acid oxidase, *Biochem. Pharmacol.* **14**: 17–26 (1965).

154. S. Gabay and S. R. Harris, Studies of FAD-requiring enzymes and phenothiazines. II. Structural requirements for D-amino acid oxidase inhibition, *Biochem. Pharmacol.* **15**:317–322 (1966).
155. S. Gabay and S. R. Harris, Studies of FAD-requiring enzymes and phenothiazines. III. Inhibition kinetics with highly-purified D-amino acid oxidase, *Biochem. Pharmacol.* **16**: 803–812 (1967).
156. S. Gabay and S. R. Harris, Inhibition of flavoenzymes by phenothiazines, *Agressologie* **IX**(1):79–89 (1968).
157. M. Gordon, P. N. Craig, and C. L. Zirkle, Molecular modification in the development of phenothiazine drugs, *Advances in Chem. Series* **45**:140–147 (1964).
158. A. Szent-Györgyi, *Introduction to Submolecular Biology*, Academic Press, New York (1960).
159. G. Karreman, I. Isenberg, and A. Szent-Györgyi, On the mechanism of action of chlorpromazine, *Science* **30**:1191–1192 (1959).
160. B. Pullman, Aspects de la structure électronique des phenothiazines, *Agressologie* **IX**(1):19–26 (1968).
161. T. Nash and A. C. Allison, Charge transfer, hydrogen bonding and drug action, *Biochem. Pharmacol.* **12**:601–602 (1963).
162. J. T. Dunn and G. T. Perkoff, D-amino acid oxidase in human tissues, *Biochim. Biophys. Acta* **73**:327–331 (1963).
163. A. H. Neims, W. D. Zieverink, and J. D. Smilack, Distribution of D-amino acid oxidase in bovine and human nervous tissues, *J. Neurochem.* **13**:163–168 (1966).
164. S. Gabay and N. N. Valanju, *in The Present Status of Psychotropic Drugs* (A. Cerletti and F. J. Bové, eds.) pp. 307–309, Excerpta Medica Foundation, New York (1968).

VIII. APPENDIX—SOME RECENT REVIEW ARTICLES OF SPECIAL INTEREST

1. *General:*

M. W. Parkes, Tranquillizers, *in Progress in Medicinal Chemistry* (G. P. Ellis and G. B. West, eds.) pp. 72–131, Butterworths, London (1961).

2. *Chemistry:*

E. Schenker and H. Herbst, Phenothiazine und Azophenothiazine als Heilmittel, *in Forschritte der Arzneimittelforschung* (E. Jucker, ed.) Vol. V, pp. 269–627, Birkhauser Verlag, Basel, Stuttgart (1963). (A complete survey of the phenothiazine literature—it contains 102 tables with chemical structure and 6800 literature references.)

3. *Pharmacology:*

E. F. Domino, R. D. Hudson, and G. Zografi, Substituted phenothiazines: Pharmacological and chemical structure, *in Drugs Affecting the Central Nervous System* (A. Burger, ed.) pp. 387–397, Dekker, New York (1968).

G. Curzon, Biochemical pharmacology of phenothiazines and related substances, *in Biochemical Aspects of Neurological Disorders* (J. N. Cumings and M. Kremer, eds.) pp. 82–98, Blackwell Scientific Publications, Oxford, England (1968).

4. *Neurochemistry:*

P. B. Bradley, Synaptic transmission in the central nervous system and its relevance for drug action, *Internat'l. Rev. Neurobiol.* **11**, 1–56 (1968).

5. *Congress Proceedings:*

Second International Symposium on Action Mechanism and Metabolism of Psychoactive Drugs Derived from Phenothiazine and Structurally Related Compounds, October 17–19, 1967 Paris, France: *In Agressologie* (I. S. Forrest and B. Weber, eds.) Vol. IX, Nos. 1 and 2, l'Imprimerie Cooperative Arpajonaise, Essonne, France (1968).

Chapter 13

RESERPINE: A SURVEY OF ITS PHARMACOLOGY

P. A. Shore

Department of Pharmacology
University of Texas Southwestern Medical School
Dallas, Texas

I. INTRODUCTION

Studies on the biochemical and pharmacological actions of reserpine and related agents have led to considerable expansion of our knowledge of functional mechanisms in the nervous system. Several chapters in this *Handbook of Neurochemistry* deal with specific areas of investigation in which studies with reserpine have been intimately involved. Such chapters include those on storage, release, and metabolism of catecholamines and serotonin; antidepressants; the biochemistry of affective disorders; Parkinsonism; and others. This discussion, accordingly, will be restricted to a brief description of the essentials of reserpine's pharmacology. It must be stressed that references cited are often only representative of the many studies performed on certain aspects of the research described.

Reserpine, in crude or purified form, was introduced into Western psychiatric practice almost simultaneously with chlorpromazine. Although the use of reserpine as a tranquilizer has declined greatly in favor of the phenothiazines, reserpine still finds considerable use as an antihypertensive agent, and the drug has been of great importance as a pharmacological tool, being instrumental in the pharmacological dissection of amine storage mechanisms in the adrenergic neuron, insight into the mechanism of action of numerous other drugs, and for use in producing experimental models of chemical sympathectomy, Parkinsonism, and mental depression.

II. HISTORY

Extracts of the plant *Rauwolfia serpentina* have found use for centuries in India for treatment of a variety of disorders. In more recent times, the plant root was described in Indian medical journals as a treatment for psychoses and hypertension. Introduction to Western medicine came only in the 1950's, at which time several of the alkaloids of the plant extracts were isolated, identified, and investigated for pharmacological activity. It was soon apparent that one of the alkaloids, reserpine, possesses, in high potency, the pharmacological activity of the crude plant extract. Other similarly active *Rauwolfia* alkaloids include the close congeners, deserpidine and rescinnamine (Fig. 1), but reserpine remains the most important of the *Rauwolfia* alkaloids. Information regarding the many alkaloids isolated from *Rauwolfia* may be found in the review by Bein.[1]

III. PHARMACOLOGY OF RESERPINE

A. General Effects

Injection of reserpine in sufficient dosage elicits a sedative or calming effect quite different from the sedative action of barbiturates. Unlike the latter substances, reserpine is not a hypnotic, even with high doses, and does not produce anesthesia. The depression of locomotor activity caused by reserpine, unlike the action of barbiturates, is readily reversible by external stimulation. Upon removal of the stimulus, the reserpine-treated animal reverts to a somnolent state. Other notable effects of higher doses of reserpine include a Parkinsonismlike tremor, marked potentiation of the

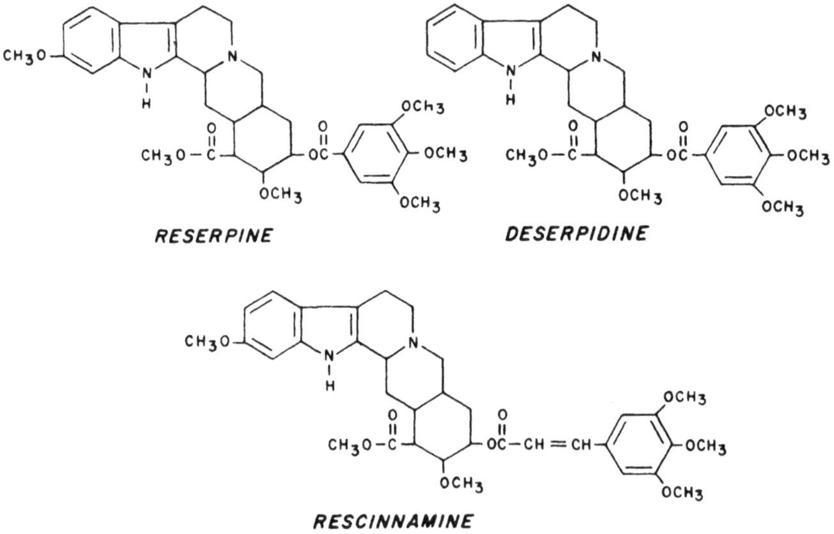

Fig. 1. Structural formulae of reserpine and similarly active *Rauwolfia* alkaloids.

effects of other central nervous system depressants such as barbiturates or ethanol, hypothermia, bilateral ptosis, miosis, and diarrhea. A profound relaxation of the nictating membrane is seen in several species. Another marked difference between reserpine and the barbiturates is that, unlike the latter drugs, reserpine lowers the threshold to convulsants such as pentylenetetrazol or electroshock. Parasympathetic predominance is seen in a lowered blood pressure coupled with slowing of the heart rate.

In general, higher doses of reserpine are required to elicit a major tranquilizing (antipsychotic) effect than are required to alter peripheral adrenergic function. When the drug is given to man for antipsychotic effects, side reactions include, among others, extrapyramidal signs, nightmares, psychotic depressions, and signs of parasympathetic predominance such as hypersecretion of gastric hydrochloric acid with possible exacerbation of peptic ulcer, nasal stuffiness, and bradycardia. The latter effects are seen also with the lower doses employed in the treatment of hypertension. Higher doses also induce endocrinological changes such as stimulation of prolactin secretion and ACTH release.

B. Absorption, Localization, and Disposition

Being a very lipid-soluble compound, reserpine is readily absorbed from the gastrointestinal tract and readily crosses the blood–brain barrier. It is rapidly metabolized in the body, and it was initially thought that the pharmacological effects of reserpine persisted long after the disappearance of the drug from the body,[2] but it is now known that minute amounts of reserpine are retained in association with some aspects, at least, of the pharmacological effects of the drug.[3] This point is discussed in more detail in a later section of this article.

The major routes of metabolism of reserpine occur by hydrolysis to form methyl reserpate, reserpic acid, and trimethoxybenzoic acid (Fig. 2). In some species, an additional route of metabolism, *O*-demethylation, leads to the syringoyl derivative of reserpine with subsequent hydrolysis to syringic acid.[4]

C. Actions on Monoamine Storage and Metabolism

The most interesting scientific aspect of reserpine, by far, is its profound effect on monoamine storage and metabolism. The first evidence of this was the action of reserpine in causing the release of serotonin (5-hydroxytryptamine) from its storage sites in brain, gastrointestinal tract, and blood platelets.[5,6] Only those *Rauwolfia* alkaloids exerting a tranquilizing effect cause amine depletion.[7] Subsequent to this, it was found that the catecholamines, norepinephrine, epinephrine, and dopamine, are also released from their storage sites centrally and peripherally by the same alkaloids.[8–10] The depletion of norepinephrine stores in peripheral adrenergic neurons plays a major role in the antihypertensive action of the drug,[8] but

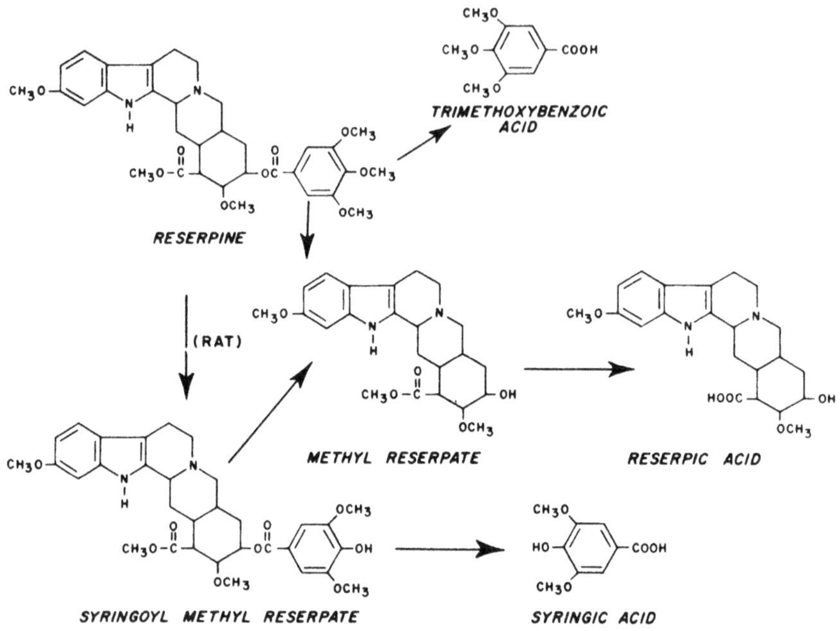

Fig. 2. Pathways of metabolism of reserpine in the body.

the role of brain amine depletion in the antipsychotic action of reserpine is not so clear despite years of concerted investigation.

As there was evidence that reserpine does not inhibit the biosynthetic enzymes involved in the formation of the monoamines, investigation centered on the mechanism of the direct release of these amines by reserpine. It was found that the addition of reserpine in very low concentrations to blood platelets *in vitro* caused a release of serotonin from the platelets, one molecule of reserpine effecting the release of hundreds of serotonin molecules.[11] Furthermore, while normal platelets rapidly accumulated serotonin from the medium, small quantities of reserpine added to normal platelets (or with platelets from reserpine-treated animals) caused a marked blockade of the accumulation process.[12] Similar experiments demonstrated that brain and heart slices accumulate norepinephrine from their incubation medium and that reserpine also blocks the accumulation of catecholamines.[13,14] It was thus clear that reserpine inhibits a monoamine uptake process in various tissues, and it was postulated that the drug blocks the "pump" of an amine "pump and leak" mechanism, much as cardiac glycosides block the Na^+ pump.

More recent research in the special area of catecholamine storage allowed the action of reserpine to be further pinpointed. Thanks to application of differential centrifugal techniques[15] and the development of fluorescent visualization techniques for catecholamine stores,[16] it became clear that in the adrenergic neurone, norepinephrine (NE) is stored in granulated vesicles of about 500 Å in diameter, little if any existing in the axoplasm. As shown by fluorescence microscopy and also by biochemical techniques, it was soon apparent that there exist two NE uptake and/or storage mechanisms, one at the axonal membrane and one at the level of the intraneuronal storage granules. Reserpine, it was found, did not seem to interfere with amine transport at the neuronal membrane, but rather inhibited the intraneuronal granule amine storage mechanism.[17,18] Thus, it could be demonstrated that after incubation with norepinephrine, although reserpine inhibited the retention of norepinephrine, the drug did not prevent passage of the amine into the neuron, but instead promoted the intraneuronal metabolism of norepinephrine by monoamine oxidase. After blockade of this enzyme by any of several inhibitors, amine uptake was little affected by reserpine although most of the amine now accumulated in free form in the axoplasm instead of in the granules. Similar findings and conclusions were obtained by using the norepinephrine congener, metaraminol, which is not a substrate for monoamine oxidase. Thus, the accumulation of metaraminol by heart slices is not affected by reserpine under normal circumstances, emphasizing that the major norepinephrine-depleting and accumulation-blocking action of reserpine must be at the site of the intraneuronal catecholamine storage granules.[18]

The precise nature of the action of reserpine on the granules is not yet fully understood. Utilizing the large catecholamine-containing adrenal medullary granules, it has been shown that amine uptake is promoted by the presence of Mg^{2+} and ATP, while reserpine in low concentration competitively inhibits amine uptake.[19] This finding is attractive in explaining reserpine's action, but the competitive nature of the inhibition is puzzling, since a single dose of reserpine inhibits amine storage for days, even when most of the drug has disappeared, suggesting an irreversible or noncompetitive mechanism. Furthermore, as recent investigations have shown, reserpine is persistently bound only in certain organs such as brain, heart, and spleen,[3,4,20] but even in these organs only minute amounts of reserpine persist, bound, it seems, essentially irreversibly to some cell constituent.[3] Thus, after the injection of ^{3}H-reserpine into rats, the labeled reserpine bound persistently in heart does not mix with subsequently added reserpine.[3] Even adrenergically innervated tissues have a limited ability to persistently bind reserpine, In rats, for example, the injection of 25 μg/kg of ^{3}H-reserpine resulted in the binding in heart 18 hours later of about 5 ng/g, while injection of 100 μg/kg led to the binding of only about 7 ng/g. The degree of norepinephrine depletion in rat heart correlated well at this time with the concentration of bound reserpine, and it could be calculated that a single reserpine molecule was associated with the depletion of about 500 norepinephrine molecules, with about 20 molecules of reserpine being persistently bound for each affected amine storage granule in rat heart.[3] Whether or not reserpine binding is actually at the storage granule has not yet been established.

Quite recently it was observed in experiments *in vitro* utilizing the nonmetabolizable catecholamine congener, metaraminol, that although reserpine does not, in a normal medium, significantly inhibit amine uptake by the Na^+-dependent membrane amine transport system, the drug in low concentrations potentiates the inhibitory effect of a lowered $[Na^+]$ on the rate of uptake of *l*-metaraminol but not *d*-metaraminol.[21] As only the former is stereochemically acceptable to the storage granules,[22] and as the potentiating effect of reserpine is seen most strikingly at low $[Na^+]$, it has been suggested that amine storage in the granules may occur by a carrier dependent on the low $[Na^+]$ present intraneuronally, a carrier blocked noncompetitively by reserpine.[21]

An additional effect of reserpine has been noted. Again using the nonmetabolizable amines *l*- and *d*-metaraminol, it was observed[23] that after injection of high doses of reserpine into rabbits, the efflux rates of both *l*- and *d*-metaraminol are greatly enhanced despite the fact that only the levo isomer is bound intraneuronally. This observation, and the finding that this type of efflux is promoted only after high doses of reserpine (0.5 mg/kg or higher) brings up the possibility of an effect of high doses of the drug on the permeability of the adrenergic neuronal membrane. It is interesting to note also that high doses of reserpine produce structural changes in heart mitochondria[24] and appear to alter the permeability of the mitochondrial membrane to norepinephrine.[25] It is tempting to consider that such effects may occur on other types of cell as well, suggesting that higher doses of reserpine may enhance the permeability of a variety of cells, thus, perhaps, leading to the nonspecific leakage of still other substances from certain cells. Thus, for example, the release of ACTH following high doses of reserpine[26] could in fact be caused by a nonspecific leakage of the hormone or its releasing factor.

Knowledge of the biochemical actions of reserpine and related compounds on monoamines has aided investigation into the pharmacotherapy of Parkinsonism and mental depression. Thus the association of brain dopamine depletion by reserpine with the extrapyramidal effects produced clinically by the drug played a role in the initiation of DOPA as an anti-Parkinsonism drug.[27] Also, the action of the tricyclic antidepressants in inhibiting the adrenergic neuron membrane amine carrier mechanism, and the observation that pretreatment of animals with the tricyclics alleviated or even reversed the central depressant action of reserpine[28] enhanced the proposal that the tricyclics may act by blocking the reuptake of norepinephrine released in the brain.

IV. CONCLUSION

Despite the greatly diminished use of reserpine in psychiatric practice, and the introduction of newer drugs for the treatment of hypertension, reserpine remains an agent of great and continuing scientific interest. Not only does it remain of great importance in investigations into the intricacies of the adrenergic neuron, but the model central depression and the Parkinsonianlike syndrome induced by the drug permit further laboratory investigations into these disease entities.

As stated in the introduction, this chapter has presented a brief survey of the essentials of reserpine actions. For a more detailed discussion of biochemical and pharmacological aspects of reserpine, the interested reader is referred to other chapters in these volumes as well as to several comprehensive review articles.[29-31]

V. REFERENCES

1. H. J. Bein, The pharmacology of *Rauwolfia*, *Pharmacol. Rev.* **8**:435–483 (1956).
2. S. M. Hess, P. A. Shore, and B. B. Brodie, Persistence of reserpine action after the disappearance of drug from brain: effect on serotonin, *J. Pharmacol. Exptl. Therap.* **118**:84–89 (1956).
3. H. S. Alpers and P. A. Shore, Specific binding of reserpine: association with norepinephrine depletion, *Biochem. Pharmacol.* **18**:1363–1372 (1969).
4. A. J. Plummer, H. Sheppard, and A. R. Schulert, *in Psychotropic Drugs* (S. Garattini and V. Ghetti, eds.) pp. 350–362, Elsevier Publ. Co., Amsterdam (1957).
5. P. A. Shore, S. L. Silver, and B. B. Brodie, Interaction of reserpine, serotonin, and lysergic acid diethylamide in brain, *Science* **122**:284–285 (1955).
6. A. Pletscher, P. A. Shore, and B. B. Brodie, Serotonin as a mediator of reserpine action in brain, *J. Pharmacol. Exptl. Therap.* **116**:84–89 (1956).
7. B. B. Brodie, P. A. Shore, and A. Pletscher, Serotonin-releasing activity limited to *Rauwolfia* alkaloids with tranquilizing action, *Science* **123**:992–993 (1956).
8. A. Carlsson, E. Rosengren, A. Bertler, and J. Nilsson, *in Psychotropic Drugs* (S. Garattini and V. Ghetti, eds.) pp. 363–372, Elsevier Publ. Co., Amsterdam (1957).
9. P. A. Shore and B. B. Brodie, *in Psychotropic Drugs* (S. Garattini and V. Ghetti, eds.) pp. 423–427, Elsevier Publ. Co., Amsterdam (1957).
10. M. Holzbauer and M. Vogt, Depression by reserpine of the noradrenaline concentration in the hypothalamus of the cat, *J. Neurochem.* **1**:8–11 (1956).
11. A. Carlsson, P. A. Shore, and B. B. Brodie, Release of serotonin from blood platelets by reserpine *in vitro*, *J. Pharmacol. Exptl. Therap.* **120**:334–339 (1957).
12. B. B. Brodie, E. G. Tomich, R. Kuntzman, and P. A. Shore, On the mechanism of action of reserpine: effect of reserpine on the capacity of tissues to bind serotonin, *J. Pharmacol. Exptl. Therap.* **119**:461–467 (1957).
13. H. J. Dengler, I. A. Michaelson, H. E. Spiegel, and E. O. Titus, Effects of drugs on uptake of isotopic norepinephrine by cat tissues, *Nature* **191**:816–817 (1961).
14. G. Hertting, J. Axelrod, and L. G. Whitby, Effect of drugs on the uptake and metabolism of ^{3}H-norepinephrine, *J. Pharmacol. Exptl. Therap.* **134**:146–153 (1961).
15. L. T. Potter and J. Axelrod, Studies on the storage of norepinephrine and the effect of drugs, *J. Pharmacol. Exptl. Therap.* **140**:199–206 (1963).
16. A. Carlsson, B. Falck, and N.-A. Hillarp, Cellular localization of monoamines, *Acta Physiol. Scand. 56* (*Suppl. 196*): 1–27 (1962).
17. A. Carlsson, *in Mechanisms of Release of Biogenic Amines* (U. S. von Euler, S. Rosell, and B. Uvnas, eds.) pp. 331–345, Pergamon Press, Oxford (1966).
18. A. Giachetti and P. A. Shore, Studies *in vitro* of amine uptake mechanisms in heart, *Biochem. Pharmacol.* **15**:607–614 (1966).

19. A. Carlsson, N.-A. Hillarp, and B. Waldeck, A Mg^{++}-ATP-dependent storage mechanism in the amine granules of the adrenal medulla, *Med. Exp.* **6**:47–53 (1962).
20. L. Manara and S. Garattini, Time course of H^3-reserpine levels in brain of normal and tetrabenazine-pretreated rats, *European J. Pharmacol.* **2**:139–141 (1967).
21. M. F. Sugrue and P. A. Shore, The mode of sodium-dependency of the adrenergic neuron amine carrier: evidence for a second, sodium-dependent, optically-specific, and reserpine-sensitive system, *J. Pharmacol. Exptl. Therap.* **170**:239–245 (1969).
22. P. A. Shore, D. Busfield, and H. S. Alpers, Binding and release of metaraminol, *J. Pharmacol. Exp. Ther.* **146**:194–199 (1964).
23. A. Giachetti and P. A. Shore, Permeability changes induced in the adrenergic neurone by reserpine, *Biochem. Pharmacol.* **19**:1621–1626 (1970).
24. S.-C. Sun, R. S. Sohal, H. L. Colcolough, and G. E. Burch, Histochemical and electron microscopic studies of the effects of reserpine on the heart muscle of mice, *J. Pharmacol. Exp. Ther.* **161**:210–221 (1968).
25. F. Izumi, M. Oka, H. Yoshida, and R. Imaizumi, Stimulatory effect of reserpine on monoamine oxidase in guinea pig heart, *Biochem. Pharmacol.* **18**:1739–1748 (1969).
26. E. Westermann, R. P. Maickel, and B. B. Brodie, On the mechanism of pituitary-adrenal stimulation by reserpine, *J. Pharmacol. Exp. Ther.* **138**:208–217 (1962).
27. O. Hornykiewicz, *in Biochemistry and Pharmacology of the Basal Ganglia* (E. Costa, L. J. Cote, and M. D. Yahr, eds.) pp. 171–186, Raven Press, Hewlett, N.Y. (1966).
28. F. Sulser, M. H. Bickel, and B. B. Brodie, The action of desmethylimipramine in counteracting sedation and cholinergic effects of reserpine-like drugs, *J. Pharmacol. Exptl. Therap.* **144**:321–330 (1964).
29. P. A. Shore, Release of serotonin and catecholamines by drugs, *Pharmacol. Rev.* **14**:531–550 (1962).
30. A. Carlsson, *in Handbook of Experimental Pharmacology*, Vol. XIX; "5-Hydroxytryptamine and Related Indolealkylamines" (O. Eichler and A. Farah, eds.) pp. 529–592 (1965).
31. F. E. Bloom and N. J. Giarman, Physiologic and pharmacologic considerations of biogenic amines in the nervous system, *Ann. Rev. Pharmacol.* **8**:229-258 (1968).

Chapter 14

ANTIDEPRESSANTS AND RELATED DRUGS*

Joseph J. Schildkraut
and
Elliot S. Gershon

Neuropsychopharmacology Laboratory
Massachusetts Mental Health Center
Department of Psychiatry
Harvard Medical School
Boston, Massachusetts

I. INTRODUCTION

The clinical efficacy of various drugs used in the treatment of affective disorders has been well established during the past decade.[1-7] The biological actions of these drugs, including their neurochemical and behavioral effects, have been fairly extensively investigated[3-18] and aspects of this literature have been reviewed recently. This chapter will focus primarily on the effects of these drugs on the physiology and metabolism of the biogenic amines in the nervous system. The effects of these drugs on behavior in animals, as well as in man, will also be reviewed, since the correlation of alterations in behavior with neurochemical changes has been of considerable interest in this field of research.

II. CHARACTERISTICS OF THE AFFECTIVE DISORDERS

A brief summary of the clinical characteristics of the affective disorders, i.e., the depressions and manias, will provide an essential background for considering the neuropharmacology of the drugs used in the treatment of these disorders. The term depression as it is commonly used clinically, may refer to the symptom of sadness of affect or to one of the various psychiatric syndromes in which depressed mood is usually a prominent feature. Depressed mood states do, of course, also occur normally. The

* This work was supported in part by U.S. Public Health Service grant M.H. 15413 from the National Institute of Mental Health.

clinical syndromes of depression include various clusterings of the following symptoms: depressed mood (of greater severity and duration than may be regarded as normal), crying, feelings of hopelessness, helplessness, worthlessness and guilt, suicidal preoccupations, loss of drive and ambition, fatigue, psychic retardation, motor retardation or agitation, anxiety, sleep disturbances, loss of appetite, weight loss, constipation, and diurnal fluctuations of symptoms. Distinguishing and classifying the separate syndromes of depression on the basis of clinical phenomena has been a long-standing problem in psychiatry; a number of overlapping systems for classifying these disorders are currently used by various investigators and clinicians (e.g., Refs. 18, 19).

The depressions for which antidepressant drugs (or electroconvulsive therapy) are of most value clinically are those disorders which have been designated "the endogenous depressions" in several systems of classification.[1,18,20–22] These depressions characteristically show autonomy of the depressive symptoms once the illness is established and lack of reactivity of the symptoms either to day-to-day changes in the patient's environment or to interpersonal interactions. The manic-depressive disorders and involutional melancholias are included among the endogenous depressions. Depressive syndromes which are excluded from this category include schizoaffective depressions, chronic characterological depressions or situational depressions, i.e., the depressive reactions. The problems of diagnosis and classification of the depressions are discussed more fully elsewhere.[18,19] It should be emphasized that these diagnostic classifications are based largely on clinical history and phenomenology. As yet, there is no set of biochemical or physiological criteria for distinguishing among the various depressive disorders, although it appears that at least some of these may be genetically distinct entities.[23–26]

Manias and hypomanias (similar but milder manic states) are psychiatric syndromes which include elation alternating with irritability, grandiosity, pressed and rapid speech, flights of ideas, and increased motor activity. There is loss of discretion and judgment; grandiose unrealistic plans and commitments are often undertaken while in this state. In many cases there is a cycling between depressive and manic states, e.g, the patient may present with a severe depression, which may abruptly (often overnight) change to a manic episode. The term manic-depressive illness strictly refers to these cycling disorders, but is often also used to include episodes of mania alone or certain types of recurrent depressions occurring without episodes of mania.

III. EFFECTS OF ANTIDEPRESSANT AND RELATED DRUGS ON CLINICAL STATE AND BEHAVIOR

The two classes of drugs most commonly used in the treatment of the depressive disorders are the monoamine oxidase inhibitors and the tricyclic antidepressants, i.e., imipramine and its congeners. While the psychomotor stimulants (of which amphetamine is the prototype) have been used clinically for a number of decades, these drugs have been of only limited value in the treatment of major depressive disorders.[1,2,5,7] Effective pharmacotherapy of the depressions was not possible prior to the late 1950s, when the two classes of major antidepressants, i.e., the monoamine oxidase inhibitors and the tricyclic antidepressants, were introduced.[20,27,28] For many years prior to the introduction of the major antidepressants, electroconvulsive therapy (ECT) was the only truly effective somatic treatment for many depressions, and ECT still remains the most effective treatment for some depressed patients.[22]

There is no truly adequate animal analogue of the clinical depressions. It is obvious, but, nonetheless, important to note that clinical depression is different from sedation, or decreased behavioral activity, or depression of conditioned behavioral responses, or inhibition of neuronal membrane depolarizations. While the effects of antidepressants and related drugs have been studied in animals, on one or more of these parameters of behavior or neurophysiological output, the relationship of these pharmacological actions to the clinical effects of these drugs remains obscure.

A number of recent reviews of the clinical effectiveness of the monoamine oxidase inhibitors in the treatment of depressive disorders have concluded that some but not all of these compounds were clinically effective.[1,2] The spectrum of depressive disorders which respond to monoamine oxidase inhibitors is not as broad as the spectrum which may be treated successfully with tricyclic antidepressants, but it has been suggested that one or another group of depressed patients may be particularly responsive to treatment with the monoamine oxidase inhibitors.[29,30]

The clinical antidepressant actions of imipramine and amitriptyline, the two tricyclic antidepressants which have been studied most extensively, have been well documented in double blind controlled studies.[1] Other tricyclic antidepressants including desmethylimipramine (a demethylated derivative of imipramine), nortriptyline (demethylated amitriptyline) and protriptyline also appear to be clinically effective drugs.[1] Clinical improvement is generally not observed in patients with depressions until 1–4 weeks after the beginning of treatment with tricyclic antidepressants or monoamine oxidase inhibitors. Mild transient self-limited hypomanic symptomatology is commonly observed in patients with endogenous depressions during treatment with imipramine.[20,31]

Monoamine oxidase inhibitors produce increased spontaneous motor activity in a number of animal species. A comparable increase in spontaneous motor activity is not produced in mammals by imipramine or other tricyclic antidepressants, although enhanced behavioral output is seen in pigeons in certain operant conditioning situations.[32]

Various phenothiazines (e.g., chlorpromazine, thioridazine) which are chemically related to the tricyclic antidepressants have been reported by some, but not all investigators, to be effective in the treatment of certain depressive disorders.[33–36] Phenothiazines however, do not seem to be effective in the treatment of retarded depressions, whereas the tricyclic antidepressants are most effective in this group of disorders.[34]

The psychomotor stimulants (e.g., amphetamine) cause mood elevation, wakefulness, and increased motor and mental activity in normal human subjects.[37] It has generally been found, however, that the psychomotor stimulants are not effective in the treatment of the major (endogenous) depressions.[1,2,5,7] Some symptomatic improvement may occur after amphetamine is administered to certain depressed patients, but such effects are usually transient and all too often are accompanied by various side effects. Tachyphylaxis, the diminished effect with repeated dosage, may be observed in human subjects after amphetamine administration. The psychomotor stimulation which occurs in man shortly after amphetamine administration may be subsequently followed by a "rebound" depression and fatigue.[38]

Certain depressed patients who fail to respond to pharmacological treatment with one of the major antidepressant drugs may be successfully treated with ECT. The clinical response to ECT occurs fairly rapidly (often within several days, and after only several treatments); this is to be compared with the period of chronic administration (usually several weeks) which is required to obtain a clinical response with the major antidepressants. ECT is also used in the treatment of manias, but manic patients often require more frequent treatments and a longer total course of such treatments than depressed patients.

The effectiveness of lithium salts in the treatment of mania has been fairly well documented, since its initial discovery over 20 years ago.[39] Recently there has been much interest in the use of lithium salts as prophylactic agents in patients with recurrent manic-depressive disorders or recurrent depressions,[3,40] but this has not been definitively established.[41] Lithium salts do not appear to be generally effective as antidepressants when used in the treatment of severe major depressive disorders.[3,42] Several investigators have suggested, however, largely on the basis of uncontrolled observations, that lithium salts may be of some use in the treatment of certain depressed patients.[3] Further controlled clinical studies will be needed to verify this.

Reserpine is of relatively little value in the clinical treatment of the affective disorders, except possibly when used in combination with other drugs as discussed below. Considerable neuropharmacological interest has been focused upon reserpine, in part because of its effects on human mood and animal behavior.[43] Sedation or tranquilization is generally observed in man after administration of reserpine. In a small but significant fraction (up to 15%) of patients treated with reserpine, this drug may produce severe depressive disorders which are difficult to distinguish from naturally occurring endogenous depressions.[44–47] In animals, reserpine produces alterations in behavior including sedation, decreased motor activity and disruption of the conditioned avoidance response. The characteristic syndrome of reserpine-induced sedation may be prevented by prior administration of various monoamine oxidase inhibitors,[48] or tricyclic antidepressants,[49] and may be antagonized by amphetamine.[50,51] The behavioral syndrome of reserpine induced sedation, as well as the neurochemical changes produced by reserpine have been studied as possible models of the clinical depressive disorders.[43] While of some heuristic value, the validity of such models has not yet been confirmed.

α-Methylparatyrosine (an inhibitor of the synthesis of catecholamines) has been observed to produce sedation in human subjects treated with this compound. A small number of patients have become depressed during treatment with this agent, but further observations will be needed to determine whether this is a regularly occurring clinical effect of α-methylparatyrosine. After cessation of α-methylparatyrosine, transient insomnia and hypomanic-like states have been observed.[52,53] In animals, α-methylparatyrosine produces sedation, as well as decreases in both spontaneous activity and conditioned avoidance responses.[54–59]

IV. CHEMICAL EFFECTS OF DRUGS USED IN THE TREATMENT OF THE AFFECTIVE DISORDERS

Of the many and varied neurochemical changes which may be produced by drugs used in the treatment of the affective disorders, their effects on the metabolism and physiology of the biogenic amines have been most extensively studied during the past decade.[60] The biogenic amines of particular pharmacological interest include the catecholamines norepinephrine and dopamine, and the indoleamine serotonin. The physiology and metabolism of these monoamines within the central nervous system, which is discussed in greater detail elsewhere in this series as well as in many other references (8, 16, 61–66) will be summarized only briefly in this chapter.

The biogenic amines are present in many areas of the brain, but there is a marked variability in the regional concentrations of one or another of these amines; e.g., norepinephrine and serotonin are most highly concentrated in the hypothalamus, while high concentrations of dopamine are found in the basal ganglia.[67–73] The distribution of the biogenic amines in brain has been extensively studied by means of histochemical fluorescence techniques, and neuronal systems containing one or another of these monoamines have been described.[74,75] Membrane depolarization has been observed locally when one or another of these monoamines has been applied within the brain, by various microtechniques.[66] It has been suggested that these monoamines may serve as neurotransmitters or neuromodulators within the central nervous system, but this has not yet been definitively established.[16,66,76,77] Norepinephrine, which is highly concentrated in peripheral sympathetic nerves, has been shown to be the neurotransmitter at peripheral sympathetic nerve terminals.[68]

The synthesis and metabolism of norepinephrine has been studied rather more extensively in the peripheral nervous system than in the brain, where similar although not identical biochemical pathways appear to occur.[62,64] Norepinephrine is synthesized in the neuron from the amino acid tyrosine, which is hydroxylated to form 3,4-dihydroxyphenylalanine (DOPA). This amino acid is then decarboxylated, forming dopamine which may be converted to norepinephrine by β-hydroxylation. The synthesized catecholamines are stored within the neuron in intraneuronal storage granules. Under normal physiological conditions, tyrosine hydroxylase appears to be the rate-limiting enzyme, involved in the synthesis of the catecholamines.[79,80]

Serotonin is synthesized within neurons from the precursor amino acid tryptophan, which is hydroxylated by the enzyme tryptophan hydroxylase to form 5-hydroxytryptophan. Decarboxylation of 5-hydroxytryptophan results in the formation of serotonin which is stored within the neuron.[63,65]

In the peripheral sympathetic nervous system, there is evidence to suggest that, after neuronal discharge, norepinephrine is removed from the synaptic space by reuptake of the released transmitter into the presynaptic neuron of origin. Enzymatic inactivation of norepinephrine may not be of comparable importance in removing norepinephrine from the synapse after neuronal discharge. When the reuptake of norepinephrine into presynaptic neurons is inhibited, marked potentiation of the effects of norepinephrine is observed. In contrast, after inhibition of monoamine oxidase or catechol *O*-methyl transferase, the enzymes that metabolize norepinephrine, potentiation of the effects of this amine is less pronounced, if present at all. Uptake of norepinephrine from the synaptic cleft is thought to be of major importance in the inactivation of norepinephrine centrally, but it has not been possible to study directly the physiological effects of impaired norepinephrine uptake in the brain.[64,81]

Recent studies suggest a similar mechanism for the neuronal uptake of serotonin in the brain.[82,83] The physiological significance of the processes of neuronal reuptake of biogenic amines (norepinephrine, dopamine, and serotonin) in brain remains to be determined. It cannot be assumed that reuptake is comparably important for the inactivation of each of these biogenic amines, nor has it been demonstrated that inactivation by neuronal reuptake of a given biogenic amine necessarily occurs to the same extent in different regions of the brain.

Within the presynaptic neuron, norepinephrine may be inactivated by mitochondrial monoamine oxidase, forming the deaminated catechol metabolites, 3,4-dihydroxymandelic acid and 3,4-dihydroxyphenyl glycol; deamination of norepinephrine may also occur extraneuronally. Most norepinephrine discharged extraneuronally which escapes reuptake into the presynaptic neuron, however, is thought to be enzymatically inactivated by catechol *O*-methyl transferase to form normetanephrine. The level of normetanephrine may thus reflect the level of norepinephrine present at receptors.[81,84,85] This has not been directly demonstrated, however, and further study will be required to determine whether normetanephrine levels necessarily reflect the level of noradrenergic activity under any or all conditions.

Deaminated catechol metabolites may be *O*-methylated or normetanephrine may be deaminated to form the deaminated *O*-methylated metabolites, 3-methoxy-4-hydroxymandelic acid (VMA) or 3-methoxy-4-hydroxyphenyl glycol (MHPG), depending upon species or tissue.[85,86] Such secondary *O*-methylation or deamination can occur in the central nervous system, as well as in peripheral tissue.[87] Most 3-methoxy-4-hydroxyphenyl glycol present in urine is conjugated with sulfate.[85,88] The sulfate conjugate of 3-methoxy-4-hydroxyphenyl glycol has recently been identified in rat brain as the major metabolite of norepinephrine and normetanephrine[89] and this compound has been found in human cerebrospinal fluid.[90] Studies of the origin of the urinary metabolites of norepinephrine in dog support this finding and suggest that most norepinephrine which comes from brain is metabolized to 3-methoxy-4-hydroxyphenyl glycol.[91] This metabolite, however, may also derive from norepinephrine originating in peripheral sympathetic nerves. Further studies will be required to determine whether measurements of urinary MHPG sulfate will be of value in separating the turnover and metabolism of norepinephrine in the central nervous system from norepinephrine turnover and metabolism in peripheral sympathetic tissues.

The metabolism of dopamine in the central nervous system has been studied less extensively. In general, it appears that the pathways for the metabolism of dopamine are analogous to those for norepinephrine.[64]

Serotonin is principally metabolized by monoamine oxidase, forming the deaminated metabolites, 5-hydroxyindoleacetic acid (5-HIAA) or 5-hydroxytryptophol. A pathway for serotonin metabolism which is analogous to the *O*-methylation of norepinephrine does not appear to be present in mammalian brain, although the *O*-methylated indole melatonin may be formed in the pineal gland.[63,65,85,92,93]

A. Tricyclic Antidepressant Drugs

Tricyclic antidepressants have been found to potentiate the physiological effects of norepinephrine and serotonin outside of the central nervous system.[6,12,94–101] This has been demonstrated *in vivo* as well as in various isolated organ preparations. While it has generally been assumed that tricyclic antidepressants might produce similar potentiation of the physiological effects of these biogenic amines within the central nervous system, direct evidence for such potentiation in the brain is presently unavailable. At extremely high doses, tricyclic antidepressants may block, rather than potentiate the effects of these amines in various physiological systems outside of the central nervous system; this too has not been demonstrated in the brain.

A possible mechanism to account for the potentiation of norepinephrine by tricyclic antidepressants was provided by the findings that imipramine inhibited uptake of norepinephrine both in peripheral tissues and in the brain.[102-104] Recent studies suggest that tricyclic antidepressants may act at the cell membrane to inhibit entry of norepinephrine into the neuron,[105] although a further action of these drugs on storage granules cannot be excluded.

Inhibition of norepinephrine uptake in brain is greater after acute administration of desmethylimipramine or protriptyline than after imipramine or nortriptyline.[106] Acute administration of amitriptyline, however, was not found to inhibit norepinephrine uptake in brain in two recent studies performed *in vivo*[106,107] but other investigators have reported inhibition of norepinephrine uptake with this drug.[104]

The uptake of dopamine into presumed dopamine containing neurons (i.e., regions of the brain such as the striatum known to contain high levels of endogenous dopamine) did not appear to be inhibited by tricyclic antidepressant drugs in several recent studies.[108,109] The uptake of dopamine into cortical brain slices, however, has been found to be inhibited by desmethylimipramine.[110] Further investigation, however, will be needed to confirm these observations.

The effects of tricyclic antidepressants on the neuronal uptake of serotonin are less well established. Recent evidence from studies *in vitro* suggests that certain tricyclic antidepressants may inhibit serotonin uptake in brain, but comparable findings have not been consistently obtained in studies performed *in vivo*. Imipramine and desmethylimipramine have been reported to inhibit the uptake of serotonin into brain slices[82,83] and isolated nerve endings.[111-113] The uptake of serotonin into brain after injection of this amine into the lateral ventricle has been reported to be inhibited by imipramine, but not by desmethylimipramine.[114,115] The uptake of radioactive serotonin in the brain, however, did not appear to be inhibited by desmethylimipramine in a study in which the ventricles were perfused with serotonin.[116] Neither imipramine nor desmethylimipramine decreased the uptake into brain of intracisternally administered radioactive serotonin in another recent study.[117]

It is not possible to resolve the apparent discrepancies in these findings at the present time, since many factors may be involved. Higher concentrations of serotonin were generally used in studies performed *in vivo* than in those studies done *in vitro*, and it is possible that the effects of these tricyclic antidepressants on specific transport mechanisms may have been masked by the entry of serotonin into the brain by nonspecific diffusion. It has been suggested recently that serotonin administered by intraventricular injection may be found in the brain at sites different from those which contained endogenously formed serotonin.[118] Uptake and metabolism of serotonin under conditions achieved *in vitro* may differ from that which occurs *in vivo*, and such differences might further account for the discrepancies in these various findings.

The levels of norepinephrine or serotonin in brain do not appear to be altered after acute administration of tricyclic antidepressants. Turnover of these amines, however, may be altered by tricyclic antidepressants, and this has been studied by means of one or another of the techniques currently available for assaying monoamine turnover in the brain. The disappearance of intracisternally administered ^{3}H-norepinephrine from brain is slowed after acute administration of imipramine or desmethylimipramine[119,120]; this suggests a decrease in turnover of norepinephrine. It has been reported, however, that, after inhibition of catecholamine synthesis, the disappearance of endogenous norepinephrine from brain is accelerated in animals treated with desmethylimipramine[121]; this would suggest an increase in norepinephrine turnover under these conditions. Further studies are indicated to resolve these apparent discrepancies.

The disappearance of endogenous serotonin from brain after inhibition of serotonin synthesis has been reported to be slowed in animals treated with imipramine, but not with desmethylimipramine.[122,123] Similarly, imipramine, but not desmethylimipramine has been found to slow the disappearance of intracisternally administered radioactive serotonin from the brain.[117]

The effects of tricyclic antidepressants on the metabolism of biogenic amines have been studied by a number of investigators. Various tricyclic antidepressants including imipramine, desmethylimipramine, protriptyline, nortriptyline, and amitriptyline have been found to increase levels of tritiated normetanephrine present in brain after intraventricular or intracisternal injection of ^{3}H-norepinephrine, and to decrease tritiated deaminated catechol metabolites.[106,109,120] Tricyclic antidepressants have been observed to produce similar changes in the metabolism of norepinephrine by brain tissue when studied *in vitro*.[124] In clinical studies of depressed patients, decreased urinary excretion of 3-methoxy-4-hydroxymandelic acid (VMA), the major deaminated *O*-methylated metabolite of norepinephrine,* has been observed with imipramine as well as with phenelzine (a monoamine oxidase inhibitor).[125,126] These findings suggest that tricyclic antidepressants may decrease the deamination of norepinephrine. Other interpretations of these findings, however, are possible, since the rate of formation of deaminated metabolites of norepinephrine in brain was not measured in any of the studies performed *in vivo*. It cannot be ruled out, therefore, that alterations in the rate of clearance of deaminated catechol metabolites from brain or alterations in the turnover of norepinephrine as well as inhibition of neuronal uptake of norepinephrine might possibly contribute to or account for one or another of these findings. Taken together, however, these data seem to suggest that tricyclic antidepressants prevent the deamination of norepinephrine by mitochondrial monoamine oxidase.

Tricyclic antidepressants in high concentrations have been reported to inhibit monoamine oxidase *in vitro*.[127,128] These findings, however, need

* Most urinary VMA probably derives from norepinephrine originating in peripheral sympathetic nerves, rather than from norepinephrine coming from the brain.[91]

not necessarily reflect direct chemical inhibition of the enzyme, since these experiments were not performed using a chemically pure monoamine oxidase. The possibility must also be considered that tricyclic antidepressants might act at intraneuronal membranes (e.g., mitochondrial membranes) to prevent norepinephrine from reacting with monoamine oxidase.[125]

The effects of tricyclic antidepressants on the metabolism of serotonin in brain have been studied less extensively, and the findings appear less coherent at the present time. Imipramine has been reported to decrease the levels in brain of the deaminated metabolite, 5-hydroxyindoleacetic acid, formed from endogenous serotonin.[129] Neither imipramine nor desmethylimipramine, however, caused a significant change in the levels of deaminated metabolites of intracisternally administered radioactive serotonin in the brain.[117] When radioactive serotonin was perfused through the lateral ventricle in rat brain, desmethylimipramine was observed to lower levels of endogenous, but not radioactive 5-hydroxyindoleacetic acid in the brain.[116] Desmethylimipramine, moreover, may alter the rate of clearance of 5-HIAA from the brain, since this drug has been observed to slow the release from brain of radioactive 5-HIAA administered by intraventricular injection.[130] Decreased urinary excretion of 5-HIAA has been observed after treatment with imipramine in clinical studies[131]; only a small fraction of urinary 5-HIAA, however, is thought to be derived from the brain.

It has been suggested that tricyclic antidepressants may exert actions at various neuronal membranes—presynaptic or postsynaptic, as well as intraneuronal—and that the effect of any given drug (or dose of a drug) may depend upon the relative potencies of action at these various membrane sites.[125,132,133] For example, the potentiation of norepinephrine produced by moderate doses of tricyclic antidepressants may result from the inhibition of inactivation of norepinephrine by presynaptic neuronal uptake, while the adrenolytic effect observed with high doses of tricyclic antidepressants may be explained by an overriding block of the postsynaptic receptor at the higher dose. The changes in the metabolism of norepinephrine produced by these drugs may also account for some of their physiological and behavioral effects. Similar alterations in the disposition or metabolism of serotonin or other biogenic amines may also be of importance in determining the physiological and behavioral changes observed with one or another of the tricyclic drugs.

B. Monoamine Oxidase Inhibitors

Monoamine oxidase inhibitors, as their name implies, inhibit the activity of the enzyme monoamine oxidase, thereby preventing deamination of catecholamines and indoleamines.[11,134] After administration of monoamine oxidase inhibitors, levels of norepinephrine, dopamine, and serotonin in brain are increased, and the deaminated metabolites of these amines are decreased.[11] In clinical studies monoamine oxidase inhibitors have been shown to decrease the urinary excretion of the deaminated metabolites of both norepinephrine and serotonin.[125,135,136] Levels of normetanephrine in brain are markedly increased after inhibition of monoamine oxidase.[11] Since deamination appears to be necessary for normetanephrine to be cleared from the brain,[89] these increased levels of normetanephrine may reflect decreased metabolism of normetanephrine, as well as increased production of this metabolite. The turnover or synthesis of various monoamines in brain have been reported to be decreased after administration of monoamine oxidase inhibitors, possibly as a result of end product inhibition of monoamine synthesis.[11,80,137,138] A large number of studies have examined the effects of monoamine oxidase inhibitors on brain levels of biogenic amines in relation to the increases in behavioral activity produced by these drugs. This literature has been reviewed elsewhere,[9,11,18] and will not be considered in any detail in this chapter. Attempts to relate the behavioral changes produced by monoamine oxidase inhibitors to alterations in the metabolism of any one specific amine have led to some discrepancies in findings[11,139] and the behavioral excitation observed with monoamine oxidase inhibitors has, at one time or another, been attributed to increased levels of norepinephrine, dopamine, or serotonin.[140-144]

Monoamine oxidase inhibitors counteract or prevent reserpine-induced sedation in animals. After administration of reserpine to animals previously treated with a monoamine oxidase inhibitor, increased behavioral activity is observed.[48,140,141,143,145-147]

After administration of monoamine oxidase inhibitors, derivatives of tyrosine such as octopamine may accumulate in the brain. These monoamines are normally present in very much lower concentrations, if at all.[11,148,149] In the peripheral sympathetic nervous system, such monoamines have been shown to accumulate during treatment with monoamine oxidase inhibitors and to be released by nerve stimulation. These compounds are generally less active than norepinephrine at adrenergic receptors and the term false transmitter has been applied. It has been suggested that accumulation and release of these "false transmitters" may account for the hypotension during treatment with monoamine oxidase inhibitors.[148] The physiological effects of these compounds within the central nervous system remain to be investigated; their possible role in the clinical antidepressant action of the monoamine oxidase inhibitors will require further study.

C. Electroconvulsive Therapy

Electroconvulsive therapy is the most effective treatment for certain depressed patients[22]; dramatic clinical response is often seen after a relatively small number of treatments. Electroconvulsive therapy has also been used in the treatment of mania, but this often requires a larger number of more closely spaced treatments. Alterations in the levels of various biogenic amines in brain have been observed in animals after electroconvulsive shock. Most studies have reported a decrease in levels of norepinephrine or an increase in serotonin levels after repeated electroconvulsive shock[150–154] but this has not been observed in all studies.[155] Similar findings have also been reported to occur after various forms of prolonged stress.[156,157] In a recent study, alterations in biogenic amine levels observed in "pseudo-shocked" animals (i.e., unshocked animals handled like the shocked animals) were similar to (but slightly smaller than) the changes observed after electroconvulsive shock; this further suggests that the stress of the shock situation may contribute, in part, to the biochemical alterations which have been observed with electroconvulsive shock.[158]

The effects of electroconvulsive shock on the turnover of biogenic amines in animal brain have been studied by a number of techniques. Levels of dopamine and its deaminated *O*-methylated metabolite, homovanillic acid, have been reported to be elevated in brain after electroconvulsive shock.[159] In another study, levels of 5-hydroxyindoleacetic acid, as well as homovanillic acid, in the cerebrospinal fluid were increased after chronically administered electroconvulsive shocks.[160] On the basis of these findings, it has been suggested that the turnover of dopamine and serotonin in brain may be increased after electroconvulsive shock, but further studies will be needed to establish this.

Electroconvulsive shock increases the rate of disappearance of intracisternally administered tritiated norepinephrine from the brain.[161,162] Levels of ^{3}H-normetanephrine in brain are increased under this condition, suggesting that electroconvulsive shock causes norepinephrine to be discharged extraneuronally.[161] After chronic administration of electroconvulsive shocks, the increase in norepinephrine turnover is maintained for some time after the final electroconvulsion.[162] Norepinephrine turnover, however, has also been found to be increased when electrical shocks are administered to the feet.[163]

From the presently available findings on the effects of electroconvulsive shock on the turnover and metabolism of biogenic amines, one cannot determine whether any or all of these changes may be necessary for the clinical effects of electroconvulsive therapy in the affective disorders. Further studies will be required to clarify this and to explore the possible differences between effects specific to electroconvulsion and the more general effects of stress on the turnover and metabolism of the various biogenic amines.

D. Amphetamine

Amphetamine (or its congener methamphetamine) produces stimulant and euphoriant effects in man after acute administration. This drug, however, is of little value in the treatment of the major depressive disorders.

Amphetamine appears to cause the release of norepinephrine and dopamine from neurons and may also inhibit the neuronal uptake of norepinephrine in the brain.[103,108,109,164,165] The uptake of serotonin into brain slices *in vitro* also has been reported to be decreased by amphetamine.[83] Amphetamine or methamphetamine has been reported by many investigators to alter the levels of various biogenic amines in brain,[166–169] but the magnitude and even the direction of change in levels (i.e., increase or decrease) may vary with the dose of the drug and the time after drug administration.[170] Amphetamine has been found to alter the metabolism of norepinephrine in brain. After intraventricular injection of ^{3}H-norepinephrine in animals treated with amphetamine, levels of ^{3}H-normetanephrine in brain have been found to be increased while tritiated deaminated catechol metabolites were found to be decreased.[165] From these and other findings, it has been suggested that amphetamine may exert its central action by releasing and potentiating one or another of the biogenic amines in the brain.

Amphetamine has been observed to antagonize various effects of reserpine on behavior, both in man and in animals.[50,51] Since reserpine depletes biogenic amines, it has been suggested that amphetamine might exert a direct action at neuronal receptors in the brain.[166,171,172] It has recently been observed, however, that amphetamine does not produce characteristic stimulation of animals when catecholamine synthesis is inhibited by α-methylparatyrosine.[173–175] This action of amphetamine is restored after administration of dihydroxyphenylalanine, a catecholamine precursor,[176] suggesting that at least some aspects of the psychomotor stimulation produced by amphetamine in animals may depend upon the presence of newly synthesized catecholamines in the brain.[173,177] It has been observed, however, that under certain conditions, amphetamine may counteract the behavioral depression and the depletion of norepinephrine produced by α-methylparatyrosine.[178,179] Further studies will be required for clarification of this finding. The behavioral stimulation produced by amphetamine does not appear to be decreased after inhibition of serotonin synthesis with parachlorophenylalanine.[175]

Tachyphylaxis, i.e., a diminished pharmacological effect with repeated dosage, is observed clinically with amphetamine. The acute psychomotor stimulation is often followed by a "rebound" depression and fatigue, particularly after large doses of amphetamine.[38] This may reflect the temporary depletion of catecholamines available for continued release.

Many of the behavioral effects of amphetamine are potentiated by imipramine.[177,180,181,182] While recent studies have shown that imipramine slows the metabolic inactivation of amphetamine,[183,184] the possibility cannot be excluded that this potentiation may also, in part, depend upon the synergistic effects of imipramine and amphetamine on biogenic amine metabolism.

E. Cocaine

Cocaine, which is a stimulant and euphoriant in man, has been observed to potentiate many of the physiological effects of norepinephrine and serotonin in animals.[185] Cocaine inhibits the neuronal uptake of norepinephrine in brain[103,161] just as it does in peripheral sympathetic nerves.[102] It has recently been reported that cocaine inhibits the uptake of serotonin in brain slices and isolated synaptosomes.[82,83,111] Cocaine produces alterations in norepinephrine metabolism similar to those observed with amphetamine or tricyclic antidepressants. Increased levels of ^{3}H-normetanephrine and decreased levels of tritiated deaminated catechol metabolites of intracisternally administered ^{3}H-norepinephrine are found in brain when animals are previously treated with cocaine.[161]

F. Reserpine and Reserpinelike Drugs

The observation that reserpine induces depressions in some patients has made the pharmacology of reserpine a subject of considerable psychiatric interest. Reserpine depletes neuronal stores of the biogenic amines, norepinephrine, dopamine and serotonin, both in the brain and in peripheral tissue.[186-189] It is generally agreed that reserpine interferes with the process by which monoamines are stored within neuronal storage granules. Some investigators have suggested that reserpine may also act at the cell membrane, but this possibility is less well documented. The mechanism by which reserpine may impair storage of monoamines has been extensively studied, and the possible effects of reserpine on the magnesium and ATP dependent process which may be of importance in the binding of norepinephrine have been of particular interest. This literature is reviewed elsewhere in this volume.[61,190]

It was recently reported by a number of investigators that norepinephrine synthesis appears to be decreased after administration of reserpine.[191-193] Synthesis of dopamine does not appear to be impaired by reserpine. It has been suggested that reserpine may inhibit the synthesis of norepinephrine by preventing the accumulation of dopamine in the storage granules (which contain the enzyme dopamine β-hydroxylase), where conversion to norepinephrine normally occurs. Alterations in the turnover of norepinephrine have also been observed with reserpine,[138] but these data are difficult to interpret in light of the apparent decrease in norepinephrine synthesis.

Reserpine also appears to alter the metabolism of catecholamines. Increased intraneuronal deamination is thought to occur after treatment with reserpine, as a result of the impaired binding of these amines in storage granules.[194,195]

Reserpine has been reported to increase synthesis of serotonin; and deamination of serotonin may be increased after administration of reserpine.[196] Reserpine may also alter storage and metabolism of other putative neuronal transmitters in brain, e.g., histamine, γ-aminobutyric acid or acetylcholine[16]; this is discussed elsewhere in this volume.[190]

Considerable research effort has been devoted to attempts to establish causal connections between the neurochemical effect of reserpine and the behavioral effects of this drug. Many research strategies have been employed in such studies. Correlations of behavioral effects with levels of one or another of the monoamines in brain have been reported, but only limited conclusions may be drawn from such measurements of total monoamine levels in brain, since physiological function may depend upon the content in one or another small and specifically localized functional pool, e.g., this may be the newly synthesized monoamines.[139,197] It is of some interest to note that after treatment with reserpine, restoration of the capacity of central neurons to accumulate biogenic amines has been reported to be correlated in time with the disappearance of reserpine-induced sedation.[198,199]

Most investigators have assumed the behavioral syndrome produced by reserpine to result from the decreased availability of one or another of the biogenic amines. Since the amino acid precursors of the biogenic amines, unlike the amines, can cross the blood–brain barrier, administration of such precursor amino acids has been used to increase the content of the derivative monoamines in the brain of the reserpine-treated animal.[200] The catecholamine precursor, 3,4-dihydroxyphenylalanine, has been reported to reverse reserpine-induced sedation in animals and to restore behavior to a near-normal state. The indoleamine precursor, 5-hydroxytryptophan, in contrast, does not effectively restore normal behavior in animals treated with reserpine, although administration of this amino acid may increase motor activity.[147,200–203] Dihydroxyphenylalanine, in one study, has been reported to counteract the psychological effects of reserpine in human subjects.[204] On the basis of these observations, it has been suggested that reserpine-induced sedation and possibly also reserpine-induced depression in man may result from depletion of catecholamines. In an attempt to determine the relative importance of dopamine or norepinephrine depletion on reserpine-induced sedation, dihydroxyphenylserine, which is decarboxylated to form norepinephrine without dopamine as an intermediate, has been administered to animals previously treated with reserpine. The behavioral depression induced by reserpine was not reversed by dihydroxyphenylserine in a recent study,[205] but not all investigators agree.[206] Only limited conclusions may be drawn from such studies, since the administration of these precursor amino acids cannot be assumed to result simply in a replacement of the derivative biogenic amines. These precursor amino acids themselves may have pharmacological effects, and since the distribution of amines formed from such exogenously administered precursors is not necessarily identical with the distribution of the endogenous amines[202,207,208] the derivative monoamines may be exerting actions or effects not normally observed under usual physiological conditions.

When indoleamine synthesis is interrupted by administration of parachlorophenylalanine (an inhibitor of tryptophan hydroxylase) levels of serotonin in brain are markedly reduced (without changes in levels of norepinephrine); under these conditions, sedation has not been observed.

Following reserpine administration, however, such animals previously treated with parachlorophenylalanine become sedated.[209] When catecholamine synthesis is interrupted by administration of α-methylparatyrosine, sedation or impairment of behavior has been observed.[54,173] These findings have been interpreted to suggest that decreased availability of catecholamines, rather than indoleamines is of major importance in the syndrome of reserpine-induced sedation; but other interpretations of these data are possible, since parachlorophenylalanine may not totally inhibit indoleamine synthesis nor does it totally deplete serotonin from the brain.[210]

While most investigators have assumed reserpine-induced sedation to result from decreased levels of one or another biogenic amine in the brain, it has also been suggested that impairment of amine binding by reserpine, with continued indoleamine synthesis, results not in a decrease, but rather in an increase in the free (unbound) serotonin available to interact with receptor sites in brain. Such increase in serotonin at receptors has been suggested to account for reserpine-induced sedation,[198,210] but this formulation has been questioned by some investigators.[211]

Prior treatment with either monoamine oxidase inhibitors or tricyclic compounds can prevent the sedation which is normally observed after administration of reserpine. Animals previously treated with monoamine oxidase inhibitors may show marked behavioral excitation after administration of reserpine. Under such conditions, monoamines released by reserpine cannot be inactivated by monoamine oxidase. It has been suggested that the spillover of one or another of the monoamines onto receptors might account for the excited behavior observed with this combination of drugs.[48,140,141,143,144,146,147]

Similarly, animals previously treated with a tricyclic antidepressant drug (particularly desmethylimipramine) show behavioral excitation after administration of reserpinelike drugs.[49] It has been suggested that this results from the spillover of free catecholamines onto receptors, since excitation is not observed with this combination of drugs, if catecholamines are previously depleted with α-methylmetatyrosine. It has recently been reported that the deamination of norepinephrine released by reserpine in the brain is relatively decreased after prior treatment with desmethylimipramine, whereas extraneuronal norepinephrine and normetanephrine (levels found in perfusion fluid) are relatively increased[212]; these observations are compatible with the decrease in deamination of norepinephrine in brain produced by tricyclic antidepressants under other conditions.[106]

On the basis of these findings in animals, reserpine or tetrabenazine has been administered to depressed patients who had failed to respond to tricyclic antidepressants alone. Clinical improvement has been reported to have occurred in some patients treated with this combination of drugs.[213–215] In one study, the degree of clinical improvement appeared to be correlated with the increase in 5-HIAA excretion after reserpine[215]; this, however, could be only a nonspecific indicator of the pharmacological activity of reserpine.

G. Lithium Salts

The effectiveness of lithium salts in the treatment of mania stimulated investigation of the effects of lithium ion on the metabolism of norepinephrine and serotonin. Since the initial report that lithium salts could alter the metabolism of norepinephrine in animal brain,(216) a number of studies have shown lithium salts to effect the turnover and metabolism of norepinephrine and serotonin.

Lithium salts have been found to increase the rate of depletion of norepinephrine from brain after inhibition of catecholamine synthesis(217,218) and to increase the rate of disappearance of intracisternally administered ^{3}H-norepinephrine from brain.(219,220) These findings suggest that lithium salts may increase the turnover of norepinephrine in brain. In contrast, lithium salts did not appear to alter the rate of disappearance of serotonin from brain after inhibition of the synthesis of this amine,(217) but lithium salts were found to slow the disappearance of intracisternally administered radioactive serotonin from brain.(117)

The uptake of ^{3}H-norepinephrine into synaptosomes isolated from rat brain has been reported to be increased after chronic administration of lithium carbonate.(221) The uptake of intracisternally administered ^{3}H-norepinephrine into rat brain (*in vivo*), however, was not significantly changed after the acute administration of lithium chloride,(120) nor was a significant change in ^{3}H-norepinephrine uptake into brain (*in vivo*) seen after chronic treatment with lithium chloride.(219) Further studies will be needed to explore this apparent discrepancy.

The metabolism of biogenic amines in brain may be altered by lithium salts. After intracisternal injection of ^{3}H-norepinephrine, levels of tritiated deaminated catechol metabolites in brain are increased, and levels of ^{3}H-normetanephrine are decreased in animals treated with lithium salts.(210,219) On the basis of these findings, it was suggested that lithium salts might increase the deamination of norepinephrine in the brain and decrease the norepinephrine available to receptors, but other interpretations of these findings are also possible, as discussed below. In a recent clinical study, patients treated with lithium salts were reported to have an increase in the urinary excretion of VMA (the major deaminated *O*-methylated metabolite of norepinephrine in man) and decreased urinary excretion of normetanephrine and metanephrine.(222) After intracisternal administration of radioactive serotonin, increased levels of radioactive deaminated metabolites were found in brains of animals treated with lithium salts.(117)

The alterations in levels of metabolites of norepinephrine or serotonin in brain after administration of lithium salts could result from changes in the turnover or metabolism of these amines or alterations in the transport of one or another of these metabolites from the brain. It has recently been suggested that lithium salts may alter the transport of acids in kidney and brain.(223) Such an action may possibly account, in part, for the increased levels of deaminated metabolites of intracisternally administered radioactive norepinephrine or serotonin found in brain after the administration of lithium salts.

V. CONCLUSION

The known clinical effects of drugs or electroconvulsive shock on affective states in man seem to correlate with the effects of these drugs on the metabolism of biogenic amines (particularly norepinephrine). Antidepressants such as the monoamine oxidase inhibitors or the tricyclic compounds, stimulants such as amphetamine or cocaine and electroconvulsive shock, all cause changes in norepinephrine metabolism which suggest an increase in levels of norepinephrine at receptors in brain. In contrast, reserpinelike drugs which can produce clinical depressions or lithium salts which are effective in the treatment of manias may decrease norepinephrine available to receptors in the brain. Many of these drugs cause similar changes in serotonin metabolism, and the importance of serotonin as well as other monoamines in the clinical actions of these drugs is suggested by the presently available data.

Definitive interpretations of these findings are limited by many methodological problems. Moreover, the precise physiological functions which norepinephrine or other biogenic amines serve in the brain have yet to be determined. Until further information is obtained on the role of the biogenic amines in the central nervous system, it will not be possible to ascertain whether the changes in biogenic amine metabolism produced by psychoactive drugs are necessary or sufficient conditions for their clinical actions.

While recognizing the limitations on the interpretations of the pharmacological data, these findings have nonetheless prompted studies of biogenic amine metabolism in patients with affective disorders. Direct biochemical assay of brain tissue in living man is, of course, not feasible and most research has therefore involved assay of the monoamines or their metabolites in body fluids under various clinical conditions. The results of these clinical studies have been reviewed in detail elsewhere.[9,13,18,224] The heterogeneity of patient populations under investigation and the apparent inconsistencies in some studies make a brief summary of the conclusions of these clinical studies rather difficult. Patients with retarded or manic-depressive depressions appear to have lower urinary excretion of norepinephrine and normetanephrine during depression than after recovery, whereas manic patients have increased urinary excretion of these amines.[31,225,226] Moreover, recent studies of the urinary excretion of the sulfate conjugate of 3-methoxy-4-hydroxyphenylglycol (MHPG sulfate), the metabolite which may best reflect the level of noradrenergic activity in the brain, suggest that urinary MHPG excretion may be decreased in depressions and increased during hypomania.[224,227,228] Indoleamine metabolism has also been studied in patients with affective disorders. Decreased levels of 5-hydroxyindoleacetic have been observed in the cerebrospinal fluid of depressed patients[229,230]; and decreased urinary tryptamine excretion has been reported to occur in depressions.[231,232]

The findings of the clinical studies are thus compatible with, but do not definitively confirm, the hypothesis that antidepressants and related drugs may exert their clinical effects by altering biogenic amine metabolism in the brain. Considerable further investigation including the exploration of the normal physiological functions of biogenic amines in the brain will be required in order to test this hypothesis.

VI. REFERENCES

1. J. M. Davis, G. L. Klerman, and J. J. Schildkraut, *in Psychopharmacology: A Review of Progress, 1957–1967* (D. H. Efron, ed.), pp. 719–747, U.S. Government Printing Office, Washington (1968).
2. M. Shepherd, M. Lader, and R. Rodnight, *Clinical Psychopharmacology.* The English Universities Press, Ltd., London (1968).
3. M. Schou, Lithium in psychiatric therapy and prophylaxis, *J. Psychiat. Res.* **6**:67–95 (1968).
4. H. E. Lehmann, Clinical perspectives on antidepressant therapy, *Amer. J. Psychiat.* (May Suppl.):12–21 (1968).
5. J. O. Cole and J. M. Davis, *in Comprehensive Textbook of Psychiatry* (A. M. Freedman and H. I. Kaplan, eds.), pp. 1263–1275, Williams and Wilkins Co., Baltimore (1967).
6. G. L. Klerman and J. O. Cole, Clinical pharmacology of imipramine and related antidepressant compounds, *Pharmacol. Rev.* **17**:101–141 (1965).
7. A. Hordern, The antidepressant drugs, *New Eng. J. Med.* **272**:1159–1169 (1965).
8. F. E. Bloom and N. J. Giarman, Physiologic and pharmacologic considerations of biogenic amines in the nervous system, *Ann. Rev. Pharm.* **8**:229–258 (1968).
9. J. J. Schildkraut and S. S. Kety, Biogenic amines and emotion, *Science* **156**:21–30 (1967).
10. D. X. Freedman, *in Psychiatric Drugs* (P. Solomon, ed.), pp. 32–57, Grune and Stratton, New York (1966).
11. A. Pletscher, *in Psychopharmacology: A Review of Progress, 1957–1967* (D. H. Efron, ed.), pp. 649–654, U.S. Government Printing Office, Washington (1968).
12. E. B. Sigg, *in Psychopharmacology: A Review of Progress, 1957–1967* (D. H. Efron, ed.), pp. 655–669, U.S. Government Printing Office, Washington (1968).
13. J. J. Schildkraut, The catecholamine hypothesis of affective disorders: A review of supporting evidence, *Am. J. Psychiat.* **122**:509–522 (1965).
14. W. E. Bunney, Jr. and J. M. Davis, Norepinephrine in depressive reactions, *Arch. Gen. Psychiat.* **13**:483–494 (1965).
15. J. Durell and J. J. Schildkraut, *in American Handbook of Psychiatry III* (S. Arieti, ed.), pp. 423–457, Basic Books, New York (1966).
16. J. Crossland, Psychotropic drugs and neurohumoral substances in the central nervous system, *Progr. Med. Chem.* **5**:251–319 (1967).
17. L. Gyermek, The pharmacology of imipramine and related antidepressants, *Int. Rev. Neurobiol.* **9**:95–143 (1966).

18. J. J. Schildkraut, J. M. Davis, and G. L. Klerman, *in Psychopharmacology: A Review of Progress, 1957–1967* (D. H. Efron, ed.), pp. 625–648, U.S. Government Printing Office, Washington, D.C. (1968).
19. Committee on Nomenclature and Statistics of the American Psychiatric Association. *Diagnostic and Statistical Manual of Mental Disorders*. 2nd ed. American Psychiatric Association, Washington (1968).
20. R. Kuhn, The treatment of depressive states with G22355 (imipramine hydrochloride), *Am. J. Psychiat.* **115**:459–464 (1958).
21. L. G. Kiloh, J. R. B. Ball, and R. F. Garside, Prognostic factors in treatment of depressive states with imipramine, *Brit. Med. J.* **1**:1225–1227 (1962).
22. M. Greenblatt, G. H. Grosser, and H. Wechsler, Differential response of hospitalized depressed patients to somatic therapy, *Am. J. Psychiat.* **120**:935–943 (1964).
23. G. Winokur and P. Clayton, Family history studies I. Two types of affective disorders separated according to genetic and clinical factors, *Rec. Adv. Biol. Psychiat.* **9**:35–50 (1967).
24. C. Perris, A study of bipolar (manic-depressive) and unipolar recurrent depressive psychoses, *Acta Psychiat. Scand. Suppl.* **194**, 42 (1966).
25. J. Angst, Zür Ätiologie und Nosologie endogener depressiver Psychosen, *Monogr. Gesamtgeb. Neurol. Psychiatrie*, Heft 112, (1966).
26. J. D. Rainer, *in Proceedings of the Research Conference on The Depressive Group of Illnesses* (G. J. Sarwer-Foner, ed.) *Can. Psychiat. Assoc. J.* 11 (GWAN Suppl.), S29–S33 (1966).
27. G. E. Crane, Iproniazid (marsilid) phosphate, a therapeutic agent for mental disorders and debilitating diseases, *Psychiat. Res. Rep. Am. Psychiat. Assn.* **8**:142–152 (1957).
28. H. P. Loomer, J. C. Saunders, and N. S. Kline, A clinical and pharmacodynamic evaluation of iproniazid as a psychic energizer, *Psychiat. Res. Rep. Am. Psychiat. Assn.* **8**:129–141 (1957).
29. E. D. West and P. J. Dally, Effects of iproniazid in depressive syndromes, *Brit. Med. J.* **1**:1491–1494 (1959).
30. C. M. B. Pare, L. Rees, and M. J. Sainsbury, Differentiation of two genetically specific types of depression by the response to antidepressants, *Lancet* **11**:1340–1343 (1962).
31. J. J. Schildkraut, R. Green, E. K. Gordon, and J. Durell, Normetanephrine excretion and affective state in depressed patients treated with imipramine, *Am. J. Psychiat.* **123**:690–700 (1966).
32. P. B. Dews, A behavioral output enhancing effect of imipramine in pigeons, *Int. J. Neuropharmacol.* **1**:265–272 (1962).
33. M. Fink, D. F. Klein, and J. C. Kramer, Clinical efficacy of chlorpromazine-procyclidine combination, imipramine and placebo in depressive disorders, *Psychopharmacologia* **7**:27–36 (1965).
34. L. E. Hollister and J. E. Overall, Phenothiazine derivatives as antidepressants, *Agressologie* **4**:289–292 (1968).
35. E. S. Paykel, J. S. Price, R. U. Gillan, G. Palmai, and E. S. Chessner, A comparative trial of imipramine and chlorpromazine in depressed patients, *Brit. J. Psychiat.* **114**:1281–1297 (1968).
36. A. Raskin, High dosage chlorpromazine alone and in combination with an antiparkinsonian agent (procyclidine) in the treatment of hospitalized depressions, *J. Nerv. Ment. Dis.* **147**:184–195 (1968).
37. B. Weiss and V. G. Laties, Enhancement of human performance by caffeine and the amphetamines, *Pharmacol. Rev.* **14**:1–36 (1962).

38. L. S. Goodman and A. Gilman, *The Pharmacological Basis of Therapeutics*. 2nd ed., pp. 361, 518, Macmillan, New York (1955).
39. J. F. J. Cade, Lithium salts in the treatment of psychotic excitement, *Med. J. Australia* **2**:349–352 (1949).
40. P. C. Baastrup and M. Schou, Lithium as a prophylactic agent, *Arch. Gen. Psychiat.* **16**:162–172 (1967).
41. B. Blackwell and M. Shepherd, Prophylactic lithium: Another therapeutic myth? An examination of the evidence to date, *Lancet*: **1**:968–971 (1968).
42. R. R. Fieve, S. R. Platman, and R. R. Plutchik, The use of lithium in affective disorders: I. Acute endogenous depression. II. Prophylaxis of depression in chronic recurrent affective disorder, *Am. J. Psychiat.* **125**:487–498 (1968).
43. B. B. Brodie, *in The Scientific Basis of Drug Therapy in Psychiatry* (J. Marks and C. M. B. Pare, eds.), pp. 127–154, Pergamon Press, Oxford (1965).
44. R. W. P. Achor, N. O. Hanson, and R. W. Gifford, Jr., Hypertension treated with rauwolfia serpentina (whole root) and with reserpine, *JAMA* **159**:841–845 (1955).
45. J. C. Muller, W. W. Pryor, J. E. Gibbons, and E. S. Orgain, Depression and anxiety occurring during rauwolfia therapy, *JAMA* **159**:836–839 (1955).
46. G. Lemieux, A. Davignon, and J. Genest, Depressive states during rauwolfia therapy for arterial hypertension: A report of 30 cases, *Can. Med. Assn. J.* **74**:522–526 (1956).
47. T. H. Harris, Depression induced by rauwolfia compounds, *Am. J. Psychiat.* **113**:950 (1957).
48. M. Chessin, E. R. Kramer, and C. T. Scott, Modification of the pharmacology of reserpine and serotonin by iproniazid, *J. Pharmacol. Exptl. Therap.* **119**:453–460 (1957).
49. F. Sulser, M. H. Bickel, and B. B. Brodie, The action of desmethylimipramine in counteracting sedation and cholinergic effects of reserpine-like drugs, *J. Pharmacol. Exptl. Therap.* **144**:321–330 (1964).
50. A. Randrup and W. Jonas, Brain dopamine and the amphetaminereserpine interaction, *J. Pharm. Pharmacol.* **19**:483–484 (1967).
51. M. G. Gelder and J. R. Vane, Interaction of the effects of tyramine, amphetamine and reserpine in man, *Psychopharmacologia* **3**:231–241 (1962).
52. K. Engleman and A. Sjoerdsma, Inhibition of catecholamine biosynthesis in man, *Circulat. Res.* **18**:1104–1108 (1966).
53. K. Engelman, D. Horwitz, E. Jequier, and A. Sjoerdsma, Biochemical and pharmacologic effects of α-methyltyrosine in man, *J. Clin. Invest.* **47**:577–594 (1968).
54. S. Spector, A. Sjoerdsma, and S. Udenfriend, Blockade of endogenous norepinephrine synthesis by α-methyltyrosine, an inhibitor of tyrosine hydroxylase, *J. Pharmacol. Exptl. Therap.* **147**:86–95 (1965).
55. K. E. Moore and R. H. Rech, Antagonism by monoamine oxidase inhibitors of α-methyltyrosine-induced catecholamine depletion and behavioral depression, *J. Pharmacol. Exp. Ther.* **156**:70–75 (1967).
56. H. Corrodi and L. C. F. Hanson, Central effects of an inhibitor of tyrosine hydroxylation, *Psychopharmacologia* **10**:116–125 (1966).
57. L. C. F. Hanson, The disruption of conditioned avoidance response following selective depletion of brain catechol amines, *Psychopharmacologia* **8**:100–110 (1965).
58. K. E. Moore, Effects of α-methyltyrosine on brain catecholamines and conditioned behavior in guinea pigs, *Life Sci.* **5**:55–65 (1966).
59. R. H. Rech, H. E. Borys, and K. E. Moore, Alterations in behavior and brain catecholamine levels in rats treated with α-methyltyrosine, *J. Pharmacol. Exptl. Therap.* **153**:412–419 (1966).

60. D. H. Efron, ed. *Psychopharmacology: A Review of Progress, 1957–1967.* U.S. Government Printing Office, Washington, D.C. (1968).
61. A. Lajtha, ed. *Handbook of Neurochemistry*, 7 Vols., Plenum Press, New York (1969–1971).
62. J. Glowinski and R. J. Baldessarini, Metabolism of norepinephrine in the central nervous system, *Pharmacol. Rev.* **18**:1201–1238 (1966).
63. S. Garattini and L. Valzelli, *Serotonin.* Elsevier, Amsterdam (1965).
64. G. H. Acheson, *Second Symposium on Catecholamines.* Reprinted from *Pharmacol. Rev.* **18**, March, 1966. Williams and Wilkins, Co., Baltimore (1966).
65. E. Costa and M. Sandler, eds. *Advances in Pharmacology*, Vol. 6A & 6B, Academic Press, New York (1968).
66. G. C. Salmoiraghi and C. N. Stefanis, A critique of iontophoretic studies of central nervous system neurons, *Int. Rev. Neurobiol.* **10**:1–30 (1967).
67. M. Vogt, The concentration of sympathin in different parts of the central nervous system under normal conditions and after the administration of drugs, *J. Physiol.* **123**:451–481 (1954).
68. U. S. von Euler, *Noradrenaline.* Charles C Thomas, Springfield, Ill. (1956).
69. M. Vogt, Catecholamines in the brain, *Pharmacol. Rev.* **11**:483–489 (1959).
70. A. Carlsson, M. Lindqvist, and T. Magnusson, *in Adrenergic Mechanisms* (J. R. Vane, G. E. W. Wolstenholme, and M. O'Connor, eds.), pp. 432–439, Little, Brown and Co., Boston (1960).
71. T. B. B. Crawford, *in 5-Hydroxytryptamine* (G. P. Lewis, ed.), pp. 20–25, Pergamon Press, New York (1958).
72. J. Crossland, *in The Clinical Chemistry of Monoamines* (H. Varley and A. H. Gowenlock, eds.), pp. 175–190, Elsevier, Amsterdam (1963),
73. H. McLennan, *Synaptic Transmission*, Saunders, Philadelphia, 1963.
74. B. Falck, *in Progress in Brain Research, VIII*, "Biogenic Amines." (H. E. Himwich and W. A. Himwich, eds.), pp. 28–44, Elsevier, Amsterdam (1964).
75. N. A. Hillarp, K. Fuxe, and A. Dahlstrom, Demonstration and mapping of central neurons containing dopamine, noradrenaline and 5-hydroxytryptamine and their reactions to psychopharmaca, *Pharmacol. Rev.* **18**:727–741 (1966).
76. B. B. Brodie and P. A. Shore, A concept for a role of serotonin and norepinephrine as chemical mediators in the brain, *Ann. N.Y. Acad. Sci.* **66**:631–642 (1957)
77. N. J. Giarman, Neurohumors in the brain, *Yale J. Biol. Med.* **32**:73–92 (1959).
78. S. Udenfriend, *in Second Symposium on Catecholamines* (G. H. Acheson, ed.), pp. 43–51 (*Pharmacol. Rev.* **18**), Williams and Wilkins Co., Baltimore (1966).
79. S. Spector, *in Second Symposium on Catecholamines* (G. H. Acheson, ed.), pp. 599–609 (*Pharmacol. Rev.* **18**), Williams and Wilkins, Co., Baltimore (1966).
80. S. Spector, R. Gordon, A. Sjoerdsma, and S. Udenfriend, Endproduct inhibition of tyrosine hydroxylase as a possible mechanism for regulation of norepinephrine synthesis, *Molec. Pharmacol.* **3**:549–555 (1967).
81. I. J. Kopin, *in Second Symposium on Catecholamines* (G. H. Acheson, ed.), pp. 513–523 (*Pharmacol. Rev.* **18**), Williams and Wilkins Co., Baltimore (1966).
82. K. J. Blackburn, P. C. French, and R. J. Merrills, 5-Hydroxytryptamine uptake in rat brain *in vitro*, *Life Sci.* **6**:1653–1663 (1967).
83. S. B. Ross and A. L. Renyi, Accumulation of tritiated 5-hydroxytryptamine in brain slices, *Life Sci.* **6**:1407–1415 (1967).
84. I. J. Kopin, Storage and metabolism of catecholamines: The role of monoamine-oxidase. *Pharmacol. Rev.* **16**:179–191 (1964).
85. J. Axelrod, *in Second Symposium on Catecholamines* (G. H. Acheson, ed.), pp. 95–113 (*Pharmacol. Rev.* **18**), Williams and Wilkins Co., Baltimore (1966).

86. M. D. Armstrong, A. McMillan, and K. N. F. Shaw, 3-Methoxy-4-hydroxy-O-mandelic acid: A urinary metabolite of norepinephrine, *Biochim. Biophys. Acta* **25**:422–423 (1957).
87. E. Mannarino, N. Kirschner, and B. S. Nashold, Jr. The metabolism of (C-14) noradrenaline by cat brain *in vivo*, *J. Neurochem.* **10**:373–379 (1963).
88. J. Axelrod, I. J. Kopin, and J. D. Mann, 3-Methoxy-4-hydroxyphenylglycol sulfate, a new metabolite of epinephrine and norepinephrine, *Biochim. Biophys. Acta* **36**:576–577 (1959).
89. S. M. Schanberg, J. J. Schildkraut, G. R. Breese, and I. J. Kopin, Metabolism of normetanephrine-H^3 in rat brain: Identification of conjugated 3-methoxy-4-hydroxyphenylglycol as the major metabolite, *Biochem. Pharmacol.* **17**:247–254 (1968).
90. S. M. Schanberg, G. R. Breese, J. J. Schildkraut, E. K. Gordon, and I. J. Kopin, 3-Methoxy-4-hydroxyphenylglycol sulfate in brain and cerebrospinal fluid, *Biochem. Pharmacol.* **17**:2005–2008 (1968).
91. J. W. Maas and D. H. Landis, *in vivo* studies of the metabolism of norepinephrine in the central nervous system, *J. Pharmacol. Exptl. Therap.* **163**:147–162 (1968).
92. H. Blaschko, *in 5-Hydroxytryptamine* (G. P. Lewis, ed.), pp. 50–57, Pergamon Press, Oxford (1958).
93. S. Udenfriend, *in 5-Hydroxytryptamine* (G. P. Lewis, ed.), pp. 43–49, Pergamon Press, Oxford, 1958.
94. E. B. Sigg, Pharmacological studies with tofranil, *Can. Psychiat. Ass. J.* **4**:S75–S85 (1959).
95. R. W. Ryall, Effects of cocaine and anti-depressant drugs on nictitating membrane of the cat, *Brit. J. Pharmacol.* **17**:339–357 (1961).
96. E. B. Sigg, L. Soffer, and L. Gyermek, Influence of imipramine and related psychoactive agents on the effect of 5-hydroxytryptamine and catecholamines on the cat nictitating membrane, *J. Pharmacol. Exptl. Therap.* **142**:13–20 (1963).
97. H. Thoenen, A. Hürlimann, and W. Haefely, Mode of action of imipramine and 5-(3′methylaminopropyliden)dibenzo(a,e) cyclohepta (1,3,5)trien hydrochloride (Ro 4-6011), a new antidepressant drug, on peripheral adrenergic mechanisms, *J, Pharmacol. Exptl. Therap.* **144**:405–414 (1964).
98. L. Gyermek, *in Antidepressant Drugs of Non-MAO Inhibitor Type* (D. H. Efron and S. S. Kety, eds.), pp. 41–62, Workshop of Pharmacology Unit, NIMH, No. 1 (1968).
99. S. Gershon, G. Holmberg, E. Mattsson, N. Mattsson, and A. Marshall, Imipramine hydrochloride, *Arch. Gen. Psychiat.* **6**:96–102 (1962).
100. A. J. Prange, Jr., E. Postrom, and C. M. Cochrane, Imipramine enhancement of norepinephrine in normal humans, *Psychiat. Digest* **125** (1964).
101. N. Svednyr, The influence of a tricyclic antidepressive agent (protriptyline) on some of the circulatory effects of noradrenaline and adrenaline in man, *Life Sci.* **7**:77–88 (1968).
102. G. Hertting, J. Axelrod, and L. G. Whitby, Effect of drugs on the uptake and metabolism of ^{3}H-norepinephrine, *J. Pharmacol. Exp. Ther.* **134**:146–153 (1961).
103. H. G. Dengler, H. E. Spiegel, and E. O. Titus, Effect of drugs on uptake of isotopic norepinephrine by cat tissues, *Nature* **191**:816–817 (1961).
104. J. Glowinski and J. Axelrod, Inhibition of uptake of tritiated noradrenaline in the intact rat brain by imipramine and related compounds, *Nature* **204**:1318–1319 (1964).
105. A. Carlsson and B. Waldeck, Mechanism of amine transport in the cell membranes of the adrenergic nerves, *Acta Pharmacol.* **22**:293–300 (1965).

106. J. J. Schildkraut, G. A. Dodge, and M. A. Logue, Effects of tricyclic antidepressants on the uptake and metabolism of intracisternally administered norepinephrine-H^3 in rat brain, *J. Psychiat. Res.* (In press.)
107. G. Stille, Pharmacological investigation of antidepressant compounds, *Pharmakopsychiatrie Neuro-Psychopharmakologie* **1**:92–106 (1968).
108. A. Carlsson, K. Fuxe, B. Hamberger, and M. Lindqvist, Biochemical and histochemical studies on the effects of imipramine-like drugs and (+)-amphetamine on central and peripheral catecholamine neurons, *Acta Physiol. Scand.* **67**:481–497 (1966).
109. J. Glowinski, J. Axelrod, and L. Iversen, Regional studies of catecholamines in the rat brain IV. Effects of drugs on the disposition and metabolism of H^3-norepinephrine and H^3-dopamine, *J. Pharmacol. Exp. Ther.* **153**:30–41 (1966).
110. S. B. Ross and A. L. Renyi, Uptake of some tritiated sympathomimetic amines by mouse brain cortex slices in vitro. *Acta Pharmacol. (Kobenhavn)* **24**:297–309 (1966).
111. T. Segawa and I. Kuruma, The influence of drugs on the uptake of 5-hydroxytryptamine by nerve-ending particles of rabbit brain stem, *J. Pharm. Pharmac.* **20**:320–322 (1968).
112. C. D. Wise and H. W. Ruelius, The binding of serotonin in brain: A study *in vitro* of the influence of physicochemical factors and drugs, *Biochem. Pharmacol.* **17**:617–631 (1968).
113. J. M. Davis, R. Colburn, D. L. Murphy, and W. E. Bunney, Jr., *in Scientific Proceedings*, pp. 228–229, American Psychiatric Association (1968).
114. K. Fuxe and U. Ungerstedt, Localization of 5-hydroxytryptamine uptake in rat brain after intraventricular injection, *J. Pharm. Pharmac.* **19**:335–337 (1967).
115. A. Carlsson, K. Fuxe, and U. Understedt, The effect of imipramine on central 5-hydroxytryptamine neurons, *J. Pharm. Pharmac.* **20**:150–151 (1968).
116. D. Palaic, I. H. Page, and P. A. Khairallah, Uptake and metabolism of (^{14}C) serotonin in rat brain, *J. Neurochem.* **14**:63–69 (1967).
117. J. J. Schildkraut, S. M. Schanberg, G. R. Breese, and I. J. Kopin, Effects of psychoactive drugs on the metabolism of intracisternally administered serotonin in rat brain. *Biochem. Pharmacol.* (In press.)
118. E. Sanders-Bush and F. Sulser, Selective effect of drugs on brain serotonin at different functional "sites," *Pharmacologist* **9**:210 (1968).
119. J. Glowinski and J. Axelrod, Effects of drugs on the disposition of H^3-norepinephrine in the rat brain, *Pharmacol. Rev.* **18**:775–785 (1966).
120. S. M. Schanberg, J. J. Schildkraut, and I. J. Kopin, The effects of psychoactive drugs on norepinephrine-H^3 metabolism in brain, *Biochem. Pharmacol.* **16**:393–399 (1967).
121. N. H. Neff and E. Costa, *in Antidepressant Drugs* (S. Garattini and M. N. G. Dukes, eds.), pp. 28–34, Excerpta Medica, Amsterdam (1967).
122. H. Corrodi and K. Fuxe, The effect of imipramine on central monoamine neurons. *J. Pharm. Pharmac.* **20**:230–231 (1968).
123. H. Corrodi, K. Fuxe, and T. Hökfelt, The effect of some psychoactive drugs on central monoamine neurons. *Europ. J. Pharmacol.* **1**:363–368 (1967).
124. J. Jonason and C. O. Rutledge, The effect of protriptyline on the metabolism of dopamine and noradrenaline in rabbit brain *in vitro*, *Acta Physiol. Scand.* **73**:161–175 (1968).
125. J. J. Schildkraut, G. L. Klerman, R. Hammond, and D. G. Friend, Excretion of 3-methoxy-4-hydroxymandelic acid (VMA) in depressed patients treated with anti-depressant drugs, *J. Psychiat. Res.* **2**:257–266 (1964).

126. J. J. Schildkraut, E. K. Gordon, and J. Durell, Catecholamine metabolism in affective disorders: I. Normetanephrine and VMA excretion in depressed patients treated with imipramine, *J. Psychiat. Res.* **3**:213–228 (1965).
127. R. Pulver, B. Exer, and B. Herrmann, Einige Wirkungen des N-(γ-dimethylamino-propyl)-iminodibenzyl-HCl und seiner Metabolite auf den Stoffwechsel von Neurohormonen. *Arzneimittel-Forschung* **10**:530–533 (1960).
128. S. Gabay and A. J. Valcourt, *in Recent Advances in Biological Psychiatry* (J. Wortis, ed.), Vol. X, pp. 29–41, Plenum Press, New York (1968).
129. M. DaPrada and A. Pletscher, On the mechanism of chlorpromazine-induced changes of cerebral homovanillic acid levels, *J. Pharm. Pharmac.* **18**:628–630 (1966).
130. D. Eccleston, Personal communication.
131. L. Haskovec and K. Rysanek, Excretion of 3-methoxy-4-hydroxymandelic acid and 5-hydroxyindoleacetic acid in depressed patients treated with imipramine, *J. Psychiat. Res.* **5**:213–220 (1967).
132. W. v. Haefely, A. Hürlimann, and H. Thoenen, Scheinbar paradoxe Beeinflussung von peripheren Noradrenalinwirkungen durch einige Thymoleptica, *Helv. Physiol. Acta* **22**:15–33 (1964).
133. A. Pletscher, *in Neuropsychopharmacology* (H. Brill, J. O. Cole, P. Deniker, H. Hippius and P. B. Bradley, eds.), pp. 571–577, Excerpta Medica, Amsterdam (1967).
134. E. A. Zeller and J. Barsky, *In vivo* inhibition of liver and brain monoamine oxidase by 1-isonicotinyl-2-isopropyl hydrazine, *Proc. Soc. Exptl. Biol. Med.* **81**:459–461 (1952).
135. A. Sjoerdsma, L. Gillespie, Jr., and S. Udenfriend, A simple method for the measurement of monoamine-oxidase inhibitors in man, *Lancet* **11**:159–160 (1958).
136. W. v. Studnitz, Effect of marsilid on excretion of 3-methoxy-4-hydroxymandelic acid in man, *Scand. J. Clin. Lab. Invest.* **11**:224–225 (1959).
137. S. Spector, *in Psychopharmacology: A Review of Progress, 1957–1967* (D. H. Efron, ed.), pp. 13–16, U.S. Government Printing Office, Washington, D.C. (1968).
138. N. H. Neff and E. Costa, Application of steady-state kinetics to the study of catecholamine turnover after monoamine oxidase inhibition or reserpine administration, *J. Pharmacol. Exptl. Therap.* **160**:40–47 (1968).
139. S. S. Kety, *in Ultrastructure and Metabolism of the Nervous System* (S. R. Korey, A. Pope and E. Robbins, eds.), pp. 311–324, Williams and Wilkins Co., Baltimore (1962).
140. S. Spector, P. A. Shore, and B. B. Brodie, Biochemical and pharmacological effects of the monoamine oxidase inhibitors, iproniazid, 1-phenyl-2-hydrazinopropane (JB516) and 1-phenyl-3-hydrazinobutane (JB835), *J. Pharmacol. Exptl. Therap.* **128**:15–21 (1960).
141. S. Spector, C. W. Hirsch, and B. B. Brodie, Association of behavioral effects of pargyline, a non-hydrazide MAO inhibitor with increase in brain norepinephrine, *Int. J. Neuropharmacol.* **2**:81–93 (1963).
142. G. M. Everett and R. G. Wiegand, *in Proceedings of the First International Pharmacological Meeting* (W. D. M. Paton and P. Lindgren, eds.), Vol. 8, pp. 85–95, Pergamon Press, New York (1962).
143. H. Green and R. W. Erickson, Further studies with tranylcypromine (monoamine oxidase inhibitor) and its interaction with reserpine in rat brain, *Arch. Int. Pharmacodynam.* **135**:407–425 (1962).
144. H. Corrodi, Blockade of the psychotic syndrome caused by nialamide in mice, *J. Pharm. Pharmacol.* **18**:197–199 (1966).

145. L. Stein and O. S. Ray, Accelerated recovery from reserpine depression by monoamine oxidase inhibitors, *Nature* **188**:1199–1200 (1960).
146. A. Carlsson, *in Proceedings of the Second Meeting of the Collegium Internationale Neuro-psychopharmacologicum* (E. Rothlin, ed.), pp. 417–421, Elsevier, New York (1961).
147. F. G. Graeff, J. G. Leme, and M. Rocha e Silva, Role played by catechol and indoleamines in the central actions of reserpine after monoamine oxidase inhibition, *Int. J. Neuropharmacol.* **4**:17–26 (1965).
148. I. J. Kopin, *in Psychopharmacology: A Review of Progress, 1957–1967* (D. H. Efron, ed.), pp. 57–60, U.S. Government Printing Office, Washington, D.C. (1968).
149. Y. Kakimoto and M. Armstrong, On the identification of octopamine in mammals, *J. Biol. Chem.* **237**:422–427 (1962).
150. A. A. Shatalova and E. K. Antonov, Content of adrenaline and noradrenaline in adrenal and brain tissues and in the blood of rabbits in convulsive states, *Psychopharmacol. Abstr.* **1**:341 (1961).
151. C. Breitner, A. Picchioni, and L. Chin, Neurohormone levels in brain after CNS stimulation including electrotherapy. *J. Neuropsychiat.* **5**:153–158 (1964).
152. S. Garattini and L. Valzelli, *in Psychotropic Drugs* (S. Garattini and V. Ghetti, eds.), pp. 428–436, Elsevier, Amsterdam (1957).
153. S. Garattini, R. Kato, L. Lamesta, and L. Valzelli, Electroshock, brain serotonin and barbiturate narcosis, *Experientia* **16**:156 (1960).
154. C. Breitner, A. Picchioni, L. Chin, and L. E. Burton, The effect of electrostimulation on brain 5-hydroxytryptamine concentration, *Dis. Nerv. Syst.* **22** (April Suppl.): 93–96 (1961).
155. D. D. Bonneycastle, N. J. Giarman, and M. K. Paasonen, Anticonvulsant compounds and 5-hydroxytryptamine in rat brain, *Brit. J. Pharmacol.* **12**:228–231 (1957).
156. D. X. Freedman, Psychotomimetic drugs and brain biogenic amines, *Am. J. Psychiat.* **119**:843–850 (1963).
157. E. W. Maynert and R. Levi, Stress-induced release of brain norepinephrine and its inhibition by drugs, *J. Pharmacol. Exptl. Therap.* **143**:90–95 (1964).
158. R. K. Hinesley, J. A. Norton, and M. H. Aprison, Serotonin, norepinephrine and 3,4-dihydroxyphenylethylamine in rat brain parts following electroconvulsive shock, *J. Psychiat. Res.* **6**:143–152 (1968).
159. J. Engel, L. C. F. Hanson, B. E. Roos, and L. E. Strombergsson, Effect of electroshock on dopamine metabolism in rat brain, *Psychopharmacologia* **13**:140–144 (1968).
160. A. J. Cooper, A. T. B. Moir, and H. C. Guldberg, The effect of electroconvulsive shock on the cerebral metabolism of dopamine and 5-hydroxytryptamine, *J. Pharm. Pharmac.* **20**:729–730 (1968).
161. J. J. Schildkraut, S. M. Schanberg, G. R. Breese, and I. J. Kopin, Norepinephrine metabolism and drugs used in the affective disorders: A possible mechanism of action, *Am. J. Psychiat.* **124**:600–608 (1967).
162. S. S. Kety, F. Javoy, A. M. Thierry, L. Julou, and J. Glowinski, A sustained effect of electroconvulsive shock on the turnover of norepinephrine in the central nervous system of the rat, *Proc. Nat. Acad. Sci.* **58**:1249–1254 (1967).
163. A. M. Thierry, F. Javoy, J. Glowinski, and S. S. Kety, Effects of stress on the metabolism of norepinephrine, dopamine and serotonin in the central nervous system of the rat, I. Modifications of norepinephrine turnover, *J. Pharmacol. Exptl. Therap.* **163**:159–171 (1968).

164. L. Stein, Self-stimulation of the brain and the central stimulant action of amphetamine, *Fed. Proc.* **23**:836–850 (1964).
165. J. Glowinski and J. Axelrod, The effect of drugs on the uptake, release and metabolism of H^3-norepinephrine in the rat brain, *J. Pharmacol. Exptl. Therap.* **149**:43–49 (1965).
166. C. B. Smith, Effects of *d*-amphetamine upon brain amine content and locomotor activity of mice, *J. Pharmacol. Exptl. Therap.* **147**:96–102 (1965).
167. J. McLean and M. McCartney, Effect of *d*-amphetamine on rat brain noradrenaline and serotonin, *Proc. Soc. Exptl. Biol. Med.* **107**:77–79 (1961).
168. S. Sanan and M. Vogt, Effects of drugs on the noradrenaline content of brain and peripheral tissues and its significance. *Brit. J. Pharmacol.* **18**:109–127 (1962).
169. K. E. Moore and E. W. Lariviere, Effects of *d*-amphetamine and restraint on the content of norepinephrine and dopamine in rat brain, *Biochem. Pharmacol.* **12**:1283–1288 (1963).
170. J. Cook and S. Schanberg, Effect of methamphetamine on norepinephrine metabolism in brain, *Pharmacologist* **10**:195 (1968).
171. J. R. Vane, *in Adrenergic Mechanisms* (J. R. Vane, G. E. W. Wolstenholme and M. O'Connor, eds.), pp. 356–372, Little, Brown and Co., Boston (1960).
172. J. M. VanRossum, J. B. van der Schoot, and J. A. Th. M. Horkmans, Mechanism of action of cocaine and amphetamine in the brain, *Experientia* **18**:229–231 (1962).
173. A. Weissman, B. K. Koe, and S. S. Tenen, Antiamphetamine effects following inhibition of tyrosine hydroxylase, *J. Pharmacol. Exptl. Therap.* **151**:339–352 (1966).
174. L. C. F. Hanson, Evidence that the central action of (+)-amphetamine is mediated via catecholamines, *Psychopharmacologia* **10**:289–297 (1967).
175. F. Sulser, M. L. Owens, M. R. Norvich, and J. V. Dingell, The relative role of storage and synthesis of brain norepinephrine in the psychomotor stimulation evoked by amphetamine or by desipramine and tetrabenazine, *Psychopharmacologia* **12**:322–332 (1968).
176. A. Randrup and J. Munkvad, Role of catecholamines in the amphetamine excitatory response, *Nature* **211**:540 (1966).
177. L. Stein, *in Antidepressant Drugs* (S. Garattini and M. N. G. Dukes, eds.), pp. 130–140, Excerpta Medica, Amsterdam (1967).
178. K. E. Moore and R. H. Rech, Reversal of α-methyltyrosine-induced behavioural depression with dihydroxyphenylalanine and amphetamine, *J. Pharm. Pharmac.* **19**:405–407 (1967).
179. B. L. Welch and A. S. Welch, Stimulus-dependent antagonism of the α-methyltyrosine-induced lowering of brain catecholamines by (+)-amphetamine in intact mice. *J. Pharm. Pharmacol.* **19**:841–843 (1967).
180. P. L. Carlton, Potentiation of behavioral effects of amphetamine by imipramine, *Psychopharmacologia* **2**:364–376 (1961).
181. A. Weissman, Interaction effects of imipramine and d-amphetamine on non-discriminated avoidance, *Pharmacologist* **3**:60 (1961).
182. C. L. Scheckel and E. Boff, Behavioral effects of interacting imipramine and other drugs with *d*-amphetamine, cocaine and tetrabenazine, *Psychopharmacologia* **5**:198–208 (1964).
183. F. Sulser, M. L. Owens, and J. V. Dingell, On the mechanism of amphetamine potentiation by desipramine (DMI), *Life Sci.* **5**:2005–2010 (1966).
184. L. Valzelli, S. Consolo, and C. Morpurgo, *in Antidepressant Drugs* (S. Garattini and M. N. G. Dukes, eds.), pp. 61–69, Excerpta Medica, Amsterdam (1967).

185. U. Trendelenburg, The supersensitivity caused by cocaine. *J. Pharmacol. Exptl. Therap.* **125**:55–65 (1961).
186. A. Pletscher, P. A. Shore, and B. B. Brodie, Serotonin release as a possible mechanism of reserpine action, *Science* **122**:374–375 (1955).
187. A. Bertler, A. Carlsson, and E. Rosengren, Release by reserpine of catechol amines from rabbits' hearts, *Naturwissenschaften* **43**:521 (1956).
188. M. Holzbauer and M. Vogt, Depression by reserpine of the noradrenaline concentration in the hypothalamus of the cat, *J. Neurochem.* **1**:8–11 (1956).
189. P. A. Shore, Release of serotonin and catecholamines by drugs, *Pharmacol. Rev.* **14**:531–550 (1962).
190. P. A. Shore, *in Handbook of Neurochemistry*, Vol. VI (A. Lajtha, ed.), Plenum Press, New York. (In preparation.)
191. C. O. Rutledge and N. Weiner, The effect of reserpine upon synthesis of norepinephrine in the isolated rabbit heart, *J. Pharmacol. Exptl. Therap.* **157**:290–302 (1967).
192. I. J. Kopin and V. K. Weise, Effect of reserpine and metarminol on excretion of homovanillic acid and 3-methoxy-4-hydroxyphenylglycol in the rat, *Biochem. Pharmacol*, **17**:1461–1464 (1968).
193. R. H. Roth and E. A. Stone, The action of reserpine on noradrenaline biosynthesis in sympathetic nerve tissue. *Biochem. Pharmacol*, **17**:1581–1590 (1968).
194. I. J. Kopin and E. K. Gordon, Metabolism of norepinephrine-H^3 released by tyramine and reserpine, *J. Pharmacol. Exptl. Therap.* **138**:351–359 (1962).
195. I. J. Kopin and E. K. Gordon, Metabolism of administered and drug-released norepinephrine-7-H^3 in the rat, *J. Pharmacol. Exptl. Therap.* **140**:207–216 (1963).
196. J. N. Tozer, N. H. Neff, and B. B. Brodie, Application of steady state kinetics to the synthesis rate and turnover time of serotonin in the brain of normal and reserpine-treated rats, *J. Pharmacol. Exptl. Therap.* **153**:177–182 (1966).
197. J. Häggendal and M. Lindqvist, Disclosure of labile monoamine fractions in brain and their correlation to behavior, *Acta Physiol. Scand.* **60**:351–357 (1964).
198. B. B. Brodie, M. S. Comer, E. Costa, and A. Dlabac, The role of brain serotonin in the mechanism of a central action of reserpine, *J. Pharmacol. Exptl. Therap.* **152**: 340–349 (1966).
199. J. Glowinski, L. Iversen, and J. Axelrod, Storage and synthesis of norepinephrine in the reserpine-treated rat brain. *J. Pharmacol. Exptl. Therap.* **151**:385–399 (1966).
200. A. Carlsson, M. Lindqvist, and T. Magnusson, 3,4-Dihydroxyphenylalanine and 5-hydroxytryptophan as reserpine antagonists, *Nature* **180**:1200 (1957).
201. G. M. Everett and J. E. P. Toman, *in Biological Psychiatry* (J. H. Masserman, ed.), pp. 78–81, Grune and Stratton, New York (1959).
202. P. L. McGeer, E. G. McGeer, and J. A. Wada, Central aromatic amine levels in behavior, II: Serotonin and catecholamine levels in various cat brain areas following administration of psychoactive drugs or amine precursors, *Arch. Neurol.* **9**:81–89 (1963).
203. J. A. Wada, J. Wrinch, D. Hill, P. L. McGeer, and E. G. McGeer, Central aromatic amine levels and behavior. I. Conditioned avoidance response in cats, following administration of psychoactive drugs or precursors, *Arch. Neurol.* **9**:69–80 (1963).
204. R. Degkwitz, R. Frowein, C. Kulenkampff, and U. Mohs, Uber die Wirkungen des L-DOPA beim Menschen und deren Beeinflussung durch Reserpin, Chlorpromazin, Iproniazid und Vitamin B_6, *Klin. Wschr.* **38**:120–123 (1960).
205. C. R. Creveling, J. Daly, T. Tokuyama, and B. Witkop, The combined use of α-methyltyrosine and threo-dihydroxyphenylserine-selective reductions of dopamine levels in the central nervous system, *Biochem. Pharmacol.* **17**:65–70 (1968).

206. A. Carlsson, Functional significance of drug-induced change in brain monoamine levels, *in Biogenic Amines* (H. E. Himwich and W. A. Himwich, eds.), pp. 9–27, Elsevier, Amsterdam (1964).
207. H. Green and J. L. Sawyer, *in Biogenic Amines* (H. E. Himwich and W. A. Himwich, eds.), pp. 150–167, Elsevier, Amsterdam (1964).
208. A. Bertler, B. Falck, C. Owman, and E. Rosengren, The localization of monoaminergic blood–brain barrier mechanisms, *Pharmacol. Rev.* **18**:369–385 (1966).
209. B. K. Koe and A. Weissman, Marked depletion of brain serotonin by *p*-chlorophenylalanine, *Fed. Proc.* **25**:452 (1966).
210. B. B. Brodie and W. D. Reid, Serotonin in brain: Functional considerations, *Advances in Pharmacology* **6B**:97–113 (1968).
211. A. Carlsson, Reporter's remarks. *Advances in Pharmacology* **6B**:115–119 (1968).
212. F. Sulser, M. L. Owens, and J. V. Dingell, *in vivo* modification of biochemical effects of reserpine by desipramine in the hypothalamus of the rat, *Pharmacologist* **9**:213 (1967).
213. W. Poldinger, Combined administration of desipramine and reserpine or tetrabenazine in depressive patients, *Psychopharmacologia* **4**:308–310 (1963).
214. P. Dick and P. Roch, *in Antidepressant Drugs* (S. Garattini and M. N. G. Dukes, eds.), pp. 311–315, Excerpta Medica, Amsterdam (1967).
215. L. Haskovec and K. Rysanek, The action of reserpine in imipramine resistant depressive patients, *Psychopharmacologia* **11**:18–30 (1967).
216. J. J. Schildkraut, S. M. Schanberg, and I. J. Kopin, The effects of lithium ion on H^3-norepinephrine metabolism in brain, *Life Sci.* **5**:1479–1483 (1966).
217. H. Corrodi, K. Fuxe, T. Hökfelt, and M. Schou, The effect of lithium on cerebral monoamine neurons, *Psychopharmacologia* **11**:345–353 (1967).
218. D. M. Stern, R. Fieve, N. Neff, and E. Costa, The effect of lithium on the turnover of brain and heart catecholamines, *Pharmacologist* **9**:210 (1967).
219. J. J. Schildkraut, M. A. Logue, and G. A. Dodge, The effects of lithium salts on the turnover and metabolism of norepinephrine in rat brain, *Psychopharmacologia* **14**:135–141 (1969).
220. K. Greenspan, M. Aronoff, and D. F. Bogdanski, Personal communication.
221. R. W. Colburn, F. K. Goodwin, W. E. Bunney, Jr., and J. M. Davis, Effect of lithium treatment on the uptake of norepinephrine by synaptosomes, *Nature* **215**:1395–1397 (1967).
222. L. Haskovec and K. Rysanek, Die Wirkung von Lithium auf den Metabolismus der Katecholamine und Indolalkylamine beim Menschen, *Arzneimittel-Forschung.* (In press.)
223. A. Anumonye, H. W. Reading, F. Knight, and G. W. Ashcroft, Uric acid metabolism in manic-depressive illness and during lithium therapy, *Lancet* **11**:1290–1293 (1968).
224. J. J. Schildkraut, *in Recent Advances in the Psychobiology of Depressive Illnesses—Proceedings of NIMH Workshop, April 30–May 2, 1969* (T. Williams, ed.), U.S. Government Printing Office, Washington, D.C. (In press.)
225. R. Strom-Olsen and H. Weil-Malherbe, Humoral changes in manic-depressive psychosis with particular reference to the excretion of catechol amines in urine, *J. Ment. Sci.* **104**:696–704 (1958).
226. K. Greenspan, J. J. Schildkraut, E. K. Gordon, B. Levy, and J. Durell, Catecholamine metabolism in affective disorders II. Norepinephrine, normetanephrine, epinephrine, metanephrine and VMA excretion in hypomanic patients, *Arch. Gen. Psychiat.* (In press.)

227. J. W. Maas, J. Fawcett, and H. Dekirmenjian, 3-Methoxy-4-hydroxyphenylglycol (MHPG) excretion in depressive states: A pilot study, *Arch. Gen. Psychiat.* **19**:129–134 (1968).
228. K. Greenspan, J. J. Schildkraut, E. K. Gordon, L. Baer, M. S. Aronoff, and J. Durell, Catecholamine metabolism in affective disorders III. MHPG and other catecholamine metabolites in patients treated with lithium carbonate. *J. Psychiat. Res.* (In press.)
229. G. W. Ashcroft, T. B. B. Crawford, D. Eccleston, D. F. Sharman, E. J. MacDougall, J. B. Stanton, and J. K. Binns, 5-Hydroxyindole compounds in the cerebrospinal fluid of patients with psychiatric or neurological diseases, *Lancet* **11**:1049–1054 (1966).
230. S. J. Dencker, U. Malm, B. E. Roos, and B. Werdinius, Acid monoamine metabolites of cerebrospinal fluid in mental depression and mania. *J. Neurochem.* **13**:1545–1548 (1966).
231. R. Rodnight, Body fluid indoles in mental illness. *Int. Rev. Neurobiol.* **3**:251–292 (1961).
232. A. Coppen, D. M. Shaw, A. Malleson, E. Eccleston, and G. Gundy, Tryptamine metabolism in depression, *Brit. J. Psychiat.* **111**:993–998 (1965).

Chapter 15

LITHIUM*

Mogens Schou

The Psychopharmacology Research Unit
Aarhus University Psychiatric Institute
Risskov, Denmark

I. INTRODUCTION—THERAPEUTIC AND PROPHYLACTIC PROPERTIES

Lithium differs in many respects from other agents used in psychiatric therapy.[1,2] It is therapeutically active against the manic phase of manic-depressive psychosis, and the effect is more specific than that of conventional neuroleptic drugs. Furthermore, when lithium is given as maintenance treatment, it may prevent further relapses in recurrent affective disorders of bipolar (manic-depressive) and monopolar (recurrent depressive) type.[3] Lithium is the only drug for which prophylactic action against one of the major psychoses has been demonstrated.

II. DISTRIBUTION IN THE ORGANISM

Chemically, lithium stands apart from the usual neuroleptic and antidepressive drugs. It is a monovalent cation, which belongs to the group of alkali metals that includes sodium, potassium, rubidium, cesium, and francium. That the lithium ion itself is the active agent makes monitoring of treatment possible through flame photometric determination of lithium concentrations in the blood. It is also an advantage for studying the biochemistry of this drug, which is not metabolized in the organism.

Like sodium and potassium, lithium passes slowly from blood into brain, but at equilibrium it is distributed differently from either of these ions. The concentration in brain tissue is about the same as in blood plasma, and there are only small concentration differences between the various regions of the brain. Lithium concentration in spinal fluid is about one-half that in brain and blood plasma.

* The manuscript was completed in September 1968.

During therapeutic and prophylactic administration of lithium to psychiatric patients the serum lithium concentration is maintained at about 1 meq/liter. Extrapolation from rat data would indicate that lithium concentration in the patient's brain is about 1 meq/kg wet weight and in spinal fluid about 0.5 meq/liter.

III. EFFECT ON MONOAMINES

Much speculation has centered around a possible connection between affective disorders and cerebral monoamines. Lithium interferes with monoamine metabolism in the brain, but it is not known whether this has any relation to the psychiatric effects of the drug.

Corrodi *et al.*[4] gave single intraperitoneal injections of lithium to rats. At various intervals after the injection the rats were killed and their brains removed. Lithium concentration in the tissue was determined by flame photometry, and the concentrations of norepinephrine, dopamine, and serotonin by chemical and histochemical (fluorimetric) methods. Lithium administered alone did not alter the monoamine content of the brain, but when given with an inhibitor of tyrosine hydroxylase it produced a fall of brain norepinephrine that was significantly more rapid than that seen when the inhibitor was given alone. Histochemical studies showed that this effect could be seen cranial, but not caudal, to a spinal cord transection, indicating that lithium produced an increase of the impulse flow in cerebral norepinephrine neurons. The effect was correlated to the lithium concentration of the brain tissue. Corresponding results have been obtained by Stern *et al.*[5] with brain and heart.

In these experiments lithium did not interfere with either dopamine or serotonin metabolism. Later experiments have indicated that when lithium is administered with the food in small doses over long periods the activity of cerebral serotonin neurons is inhibited while that of norepinephrine neurons remains largely unaffected.[6] This effect was observed in brains with a lithium content of about 1 meq/kg wet weight.

Schildkraut and his associates[7,8] injected tritiated norepinephrine intracisternally into rats either before or after lithium administration and determined the labeling of norepinephrine and its metabolites in the brain after varying periods. Lithium produced a shift of norepinephrine metabolism from *o*-methylation to intraneuronal deamination; there were signs of increase in norepinephrine turnover but it was not statistically significant.

A third approach was followed by Colburn *et al.*,[9] who studied the *in vitro* uptake of tritium-labeled norepinephrine in a system containing isolated nerve endings from rat brain. Synaptosomes from animals pretreated with lithium for a week showed a 30% higher uptake of labeled norepinephrine than did synaptosomes from control animals.

According to current hypotheses, neuroleptic drugs act on mania and antidepressive drugs on depression by lowering and raising, respectively, the level of catecholamines or indole amines at postganglionic receptor sites. Lithium differs from these drugs in its effect on monoamine metabolism and by its "normothymic," mood-stabilizing action. Attempts have been made to correlate these phenomena,[10] but further observations are needed to put speculation on a firm basis.

IV. EFFECT ON ELECTROLYTES

Lithium is closely related to sodium and potassium, which are of fundamental importance for almost all nervous activity. Its mechanism of action might therefore be sought in an interference with the metabolism and function of these ions.

According to studies by Coppen,[11] the transfer of radioactive sodium from blood to cerebrospinal fluid is significantly slower in depressed patients than in the same persons after recovery and in healthy subjects. Since it appeared possible that this process might be affected by lithium; the passage rate of radioactive sodium from blood into brain tissue was studied in rats after lithium administration.[12] Single high lithium doses and repeated small lithium doses were, however, found equally ineffective.

Komissarova[13] studied the electrolyte composition of rat brain after lithium administration and was unable to detect any changes. When brain cortex slices are incubated in media with a high lithium concentration, lithium enters the tissue at the expense of sodium and potassium.[14]

V. EFFECT ON INTERMEDIARY METABOLISM

The metabolism of brain cortex slices is influenced by the surrounding ionic milieu. When sodium is present in the medium, addition of potassium leads to an increase of the respiratory rate. Hertz and Schou[15] showed that lithium exerts a twofold action in this system: like potassium it produces a high oxygen uptake; unlike potassium it does so even in the absence of sodium. Presumably lithium acts as its own cofactor.

Lithium has been compared with sodium and potassium in other preparations of nerve and brain tissue. The following systems show a specific requirement for sodium, and replacement by lithium leads to partial or total loss of activity: active uptake of choline,[15] amino acids,[17–19] and other nonelectrolytes[20]; fixation of carbon dioxide[21]; increase of phosphoprotein turnover in response to electrical stimulation[22]; and phosphorus metabolism of stimulated crab nerve.[23] On the other hand, an increase in the uptake of calcium ions by crab nerve could be noted in artificial sea water with lithium instead of sodium.[24]

The Na–K-stimulated adenosine triphosphatase of crab nerve requires the presence of magnesium, sodium, and potassium for full activity. When magnesium alone is present, lithium activates the enzyme slightly; the effect is intermediate between those produced by sodium and by potassium. When magnesium and sodium are present, lithium activates more, but the activity is lower than that obtained with ammonium, potassium, rubidium, and cesium.(25) Similar properties are shown by adenosine triphosphatase in microsomal fractions from rat brain and human brain.(26,27) In a synaptosome preparation from rat brain, lithium was without effect on *p*-nitrophenylphosphatase activity, but stimulatory to acylphosphatase activity.(28)

VI. EFFECT ON NERVOUS TRANSMISSION

It has long been known that lithium, as the only metal cation, can be substituted for sodium in the medium surrounding nerves without loss of conductivity.(29–32) Sodium has two roles in the neuron. It carries the inward current of the action potential, and it stimulates the active transport system; lithium seems to take part in only the first of these.(33) According to Meves(34) the active nerve membrane discriminates between the alkali metal ions in the ratio:

$$\text{Li}:\text{Na}:\text{K}:\text{Rb}:\text{Cs} = 1.1:1:1/12:1/40:1/61.$$

Lithium ions are extruded from spinal motoneurons by the sodium pump at about half the rate for sodium extrusion(35); this is in contrast to the ineffective extrusion of lithium from other tissues.(36) Substitution of lithium for sodium does not appreciably affect the resting potential of nerve; resting permeability of mammalian nonmyelinated fibers to lithium is about 70% of that to sodium.(37)

Pappano and Volle(38) substituted lithium for sodium in perfusion experiments with mammalian sympathetic ganglia. Ganglionic transmission and the response to acetylcholine were initially supported and then abolished; development of block was accelerated by repetitive preganglionic stimulation. Klingman(39) used a similar system with varying lithium replacements. With 6% lithium replacement the active potential decreased just perceptibly; when 70% of the sodium was replaced by lithium, there was an irreversible loss of action potential. The experiments indicated that ganglionic transmission is more sensitive to the lack of sodium or the presence of lithium than is transmission in preganglionic fibers. Substitution of lithium for sodium also led to changes in phospholipid and amino acid metabolism of the ganglia.(40,41)

Observations of lithium effects on the metabolism of nerve and brain tissue preparations have mostly been incidental to studies on the effects of sodium and potassium. Lithium was usually added in high concentration and often substituted for sodium. These circumstances are hardly ideal for obtaining information about metabolic responses of human brain to the low lithium concentrations prevailing during psychiatric use of the drug. It is therefore of interest that experiments have been reported recently which showed metabolic effects of lithium in a concentration of a few milliequivalents per liter medium: inhibition of brain mitochondrial respiration,[42] inhibition of potassium uptake into brain cortex slices,[43] and interference with amino acid metabolism of superior cervical ganglion.[41]

Further studies will presumably disclose interrelations between electrolytes, monoamines, and intermediary metabolism in the central nervous system. It seems possible that lithium may counteract pathological mood swings by exerting a stabilizing action on cerebral functions that are critically dependent on particular protein–electrolyte–water patterns.

VII. REFERENCES

1. M. Schou, Biology and pharmacology of the lithium ion, *Pharmacol. Rev.* **9**:17–58 (1957).
2. M. Schou, Lithium in psychiatric therapy and prophylaxis, *J. Psychiat. Res.* **6**:67–95 (1968).
3. P. C. Baastrup and M. Schou, Lithium as a prophylactic agent. Its effect against recurrent depressions and manic-depressive psychosis, *Arch. Gen. Psychiat.* **16**:162–172 (1967).
4. H. Corrodi, K. Fuxe, T. Hökfelt, and M. Schou, The effect of lithium on cerebral monoamine neurons, *Psychopharmacologia (Berl.)* **11**:345–353 (1967).
5. D. N. Stern, R. Fieve, N. Neff, and E. Costa, The effect of lithium on the turnover of brain and heart catecholamines, *Pharmacologist* **9**:210–220 (1967).
6. H. Corrodi, K. Fuxe, and M. Schou, The effect of prolonged lithium administration on cerebral monoamine neurons in the rat, *Life Sci.* **8**:643–652 (1969).
7. J. J. Schildkraut, S. M. Schanberg, and I. J. Kopin, The effects of lithium ion on H^3-norepinephrine metabolism in brain, *Life Sci.* **5**:1479–1483 (1966).
8. S. M. Schanberg, J. J. Schildkraut, and I. J. Kopin, The effects of psychoactive drugs on norepinephrine-^3H metabolism in brain, *Biochem. Pharmacol.* **16**:393–399 (1967).
9. R. W. Colburn, F. G. Goodwin, W. E. Bunney, Jr., and J. M. Davis, Effect of lithium on the uptake of noradrenaline by synaptosomes, *Nature* **215**:1395–1397 (1967).
10. M. Schou, Biochemie der Depressionen; mögliche Wirkungsmechanismen des Lithiums, *in Das depressive Syndrom* (H. Hippius and H. Selbach, eds.), pp. 9–15, Berlin, 1969.
11. A. Coppen, Abnormality of the blood–cerebrospinal-fluid barrier of patients suffering from a depressive illness, *J. Neurol. Neurosurg. Psychiat.* **23**:156–161 (1960).
12. A. Amdisen and M. Schou, Lithium and the transfer rate of sodium across the blood–brain barrier, *Psychopharmacologia (Berl.)* **12**:236–238 (1968).
13. R. A. Komissarova, Contribution á l'étude du mécanisme de l'action sédative du lithium carbonique, *Zh. Nevropat. Psikhiat.* **66**:917–921 (1966).
14. R. S. Bourke and D. B. Tower, Fluid compartmentation and electrolytes of cat cerebral cortex *in vitro*—II. Sodium, potassium and chloride of mature cerebral cortex, *J. Neurochem.* **13**:1099–1117 (1966).
15. L. Hertz and M. Schou, Univalent cations and the respiration of brain-cortex slices, *Biochem. J.* **85**:93–104 (1962).

16. J. Schuberth, A. Sundwall, and B. Sörbo, Relation between Na^+-K^+ transport and the uptake of choline by brain slices, *Life Sci.* **6**:293–295 (1967).
17. S. Lahiri and A. Lajtha, Cerebral amino acid transport *in vitro*—I. Some requirements and properties of uptake, *J. Neurochem.* **11**:77–86 (1964).
18. A. Cherayil, J. Kandera, and A. Lajtha, Cerebral amino acid transport *in vitro*—IV. The effect of inhibitors on exit from brain slices, *J. Neurochem.* **14**:105–115 (1968).
19. D. F. de Almeida, E. B. Chain, and F. Pocchiari, Effects of ammonium and other ions on the glucose-dependent active transport of *l*-histidine in slices of rat-brain cortex, *Biochem. J.* **95**:793–796 (1965).
20. R. Villegas, G. M. Villegas, M. Blei, F. C. Herrera, and J. Villegas, Nonelectrolyte penetration and sodium fluxes through the axolemma of resting and stimulated medium sized axons of the squid *Doryteuthis plei*, *J. Gen. Physiol.* **50**:43–59 (1966).
21. S.-C. Cheng and P. Mela, CO_2 fixation in the nervous system. II. Environmental effects on CO_2 fixation in lobster nerve, *J. Neurochemistry* **13**:281–287 (1966).
22. K. Ahmed, J. D. Judah, and H. Wallgren, Phosphoproteins and ion transport of cerebral cortex slices, *Biochim. Biophys. Acta (Amst.)* **69**:428–430 (1963).
23. P. F. Baker, The relationship between phosphorus metabolism and the sodium pump in intact crab nerve. *Biochim. Biophys. Acta (Amst.)* **75**:287–289 (1963).
24. P. F. Baker and M. P. Blaustein, Sodium-dependent uptake of calcium by crab nerve, *Biochim. Biophys. Acta (Amst.)* **150**:167–170 (1968).
25. J. C. Skou, Enzymatic basis for active transport of Na^+ and K^+ across cell membrane, *Physiol. Rev.* **45**:596–617 (1965).
26. W. N. Aldridge, Adenosine triphosphatase in the microsomal fraction from rat brain, *Biochem. J.* **83**:527–533 (1962).
27. F. Samaha, Studies on Na^+-K^+-stimulated ATPase of human brain, *J. Neurochem.* **14**:333–341 (1967).
28. B. Formby and J. Clausen, Comparative studies of K^+-p-nitrophenylphosphatase, K^+-acylphosphatase and (Na^+ + K^+)-adenosinetriphosphatase in synaptosomes of rat brain, *Hoppe-Seylers Z. Physiol. Chem.* **349**:909–919 (1968).
29. R. Höber, Über den Einfluss der Salze auf den Ruhestrom des Froschmuskels, *Pflügers Arch. Ges. Physiol.* **106**:599–635 (1905).
30. A. L. Hodgkin and B. Katz, The effect of sodium ions on the electrical activity of the giant axon of the squid, *J. Physiol.* **108**:37–77 (1949).
31. A. F. Huxley and R. Stämpfli, Effect of potassium and sodium on resting and action potentials of single myelinated nerve fibres, *J. Physiol.* **112**:496–508 (1951).
32. J. W. Moore, N. Anderson, M. Blaustein, M. Takata, J. Y. Lettvin, W. F. Pickard, T. Bernstein, and J. Pooler, Alkali cation selectivity of squid axon membrane. *Ann. N.Y. Acad. Sci.* **137**:818–829 (1966).
33. D. R. Gardner and G. A. Kerkut, A comparison of the effects of sodium and lithium ions on action potentials from *Helix aspersa* neurones. *Comp. Biochem. Physiol.* **25**:33–48 (1968).
34. H. Meves, Experiments on internally perfused squid giant axons. *Ann. N.Y. Acad. Sci.* **137**:807–817 (1966).
35. T. Araki, M. Ito, P. G. Kostyuk, O. Oscarsson, and T. Oshima, The effects of alkaline cations on the responses of cat spinal motoneurons, and their removal from the cells. *Proc. Roy. Soc. B* **162**:319–332 (1965).
36. R. D. Keynes and R. C. Swan, The permeability of frog muscle fibres to lithium ions. *J. Physiol.* **147**:626–638 (1959).

37. C. J. Armett and J. M. Ritchie, On the permeability of mammalian non-myelinated fibres to sodium and to lithium ions, *J. Physiol.* **165**:130–140 (1963).
38. A. J. Pappano and R. L. Volle, Actions of lithium ions in mammalian sympathetic ganglia, *J. Pharmacol. Exptl. Therap.* **157**:346–355 (1967).
39. J. D. Klingman, Effects of lithium ions on the rat superior cervical ganglion, *Life Sci.* **5**:365–373 (1966).
40. J. D. Klingman, Ionic influences on rat superior cervical ganglion phospholipid metabolism, *Life Sci.* **5**:1397–1407 (1966).
41. J. D. Klingman and W. McBride, Amino acid metabolism in rat superior cervical ganglion. Paper read at the First Meeting of *Internat. Soc. Neurochemistry*, Strasbourg, July, 1967. (In press.)
42. A. R. Krall, Potassium and sodium-like effects of lithium on brain mitochondrial phosphorylation, *Life Sci.* **6**:1339–1344 (1967).
43. Y. Israel, H. Kalant, and A. E. LeBlanc, Effects of lower alcohols on potassium transport and microsomal adenosinetriphosphatase activity of rat cerebral cortex, *Biochem. J.* **100**:27–33 (1966).

Chapter 16

ANESTHETICS

L. J. Mullins

Department of Biophysics
University of Maryland School of Medicine
Baltimore, Maryland

I. INTRODUCTION

The study of anesthetic action is an extremely broad field of endeavor, especially if one defines anesthesia as the reversible inhibition of many sorts of physiological parameters. To the experimentalist, the field offers the possibility of studying the action of chemically inert compounds in controlling insects, killing weeds, or depressing the function of the higher centers of the central nervous system. To the theoretician, the field presents a relatively compact and firmly established body of data in need of explanation.

In discussing the subject, it seems useful at the outset to summarize some of the facts on which there is general agreement and to set forth some of the reasons why other aspects of the subject remain obscure. It is clear that no chemical reaction takes place between an anesthetic molecule and the organism and it appears further that in depressing electrical excitability anesthetics act on the cell membrane rather than on any intracellular phase. There are no specific chemical groupings that are required if a molecule is to be an anesthetic and a knowledge of the physical rather than the chemical properties of a molecule is sufficient in many instances to enable one to predict anesthetic action.

From the statements above, it can be inferred that one is dealing with the solution of molecules in a membrane phase of the cell; hence a study of the physical chemistry of solutions is essential to an understanding of anesthetic action. It is unfortunate that the liquid state (as distinct from the gaseous or solid state) is the most difficult to treat in a physically rigorous manner; the reasons for this are set forth below.

II. SOLUTIONS

A. Nonideal Solutions

Nonideal solutions are the rule rather than the exception. The reasons for solution nonideality can be dealt with under two headings: (1) those in which the heat of mixing is not zero or (2) those in which the entropy of mixing is nonideal. A heat of mixing arises because of interaction between molecules of various sorts (an extreme example is chemical compound formation) while a nonideal entropy of mixing occurs because of the tendency of some molecules to order other molecules around themselves. Both of these factors are likely to operate in the mixing of fluorochloroparaffins with oil and because the nature of the interactions is likely to be complex, theoretical calculations of the deviations to be expected may often be only rough approximations.

There are a number of theoretical treatments of nonideal solutions of a relatively sophisticated sort but these are limited to rather simple cases. A treatment by Hildebrand,[1] although semiempirical, is probably the most generally useful one for the problems of anesthesia. The Hildebrand treatment assumes that one can correct for nonideal behavior by measuring solvent-solvent interaction in the pure solvent, solute-solute interaction in the pure liquid solute, and thereby predict solvent-solute interaction. The heat of vaporization per unit molar volume ($\Delta H_v/V_m$) is taken as a measure of the energy of interaction for both the pure solvent and the pure solute, and is a measure of the forces holding the liquid molecules together.

Hildebrand's solubility theory has been applied to anesthesia previously[2] and the results will be summarized only briefly here. The solubility parameter for a liquid, δ, is defined as $\delta^2 = (\Delta H/V_m)$. The advantage of δ in predicting solution nonideality is that, for paraffins, for example, it is constant for all members of the series while other indices, such as boiling point, rise as the number of carbon atoms in a molecule increases.

1. *Experimental Measurement of Solution Nonideality*

Instead of attempting to calculate nonideal behavior of solutions, one can measure the nonideal behavior experimentally provided that (1) the proper solvent for such measurements is known, and (2) the solvent has all the properties exhibited by the anesthetic receptor. Proper experimental measurements would then consist of a plot of the vapor pressure of the anesthetic *vs* its mole fraction in the solvent, over a concentration range appropriate to that encountered in anesthesia. Such measurements should employ compounds of extreme purity and would involve the measurement of partial pressures of vapor of only a fraction of a millimeter of mercury. Unfortunately, many measurements which have been made are of oil/water partition coefficient—a needlessly complicated index and one that does not correlate well with anesthetic data (SF_6 has a coefficient of 200, while ethyl ether is 3.2; yet SF_6 has no physiological action at pressures of the order of 1 atm).

In reporting experiments with anesthetics, current practice is to correlate the results with the so-called oil/gas partition coefficient (the Bunsen coefficient). This is a satisfactory procedure if the oil used (usually olive oil) has solvent properties sufficiently similar to the phase where anesthetics act in the nervous system so that nonidealities of the resulting solution are the same in both cases (for example, in the case of highly fluorinated molecules). However, taking the rather extensive data that have accumulated on anesthetic thresholds, there is no clear indication of a correlation between the experimentally measured oil/gas coefficient and the theoretically derived thermodynamic activity of the substance when this has been corrected for the expected nonideality of the solution by employing calculated activity coefficients.

As as example of the sorts of nonideality to be expected from perfluoroparaffins, the extrapolated vapor pressures of both methane and perfluoromethane are about 300 atm. The partial pressures for anesthesia are 3.7 atm for methane and 26 atm for CF_4.[3] For methane, the thermodynamic activity is $3.7/300 = 0.012$, a value very similar to that observed for other anesthetics, while for CF_4 it would be $26/300 = 0.09$, a value that is certainly high. Taking the measured oil/gas coefficient of 0.07 and converting this Bunsen coefficient to p^0 as indicated above, one gets $p^0 = 850$ atm or an activity of $26/850 = 0.031$ which is well within the range of values for anesthesia. Some improvement in numerical accuracy could be obtained by making fugacity corrections for the pressures involved, as these are very high. The point of the calculation, however, is to show that it is possible to obtain thermodynamic activities from oil/gas partition coefficients. An alternate test is to calculate the activity coefficient of CF_4 using Hildebrand solubility theory. Here δ for CH_4 is 7 and for CF_4 it is 5 and the activity coefficient calculated is about 4, so that p^0 would be effectively increased to 4×300 atm or 1200 atm and the activity would be $26/1200 = 0.022$, again a value in agreement with data for other anesthetics.

B. Ideal Solutions

1. *Raoult's Law*

The most fundamental relationship between the quantity of material that dissolves in a solvent and the partial pressure of the substance involved is given by

$$p_1/p^0 = X_1$$

where p_1 is the partial pressures of the solute, p^0 the vapor pressures of the pure liquid solute, and X_1 is the mole fraction of the solute. This relationship is valid only if the mixing of solute and solvent is ideal (i.e., there is no heat of mixing and the entropy of mixing is $-R \ln X$). Ideal solutions are formed by mixing solutes with solvents of very similar chemical structures (e.g., paraffins mix almost ideally with each other) but this is not always true if the substances in question form hydrogen bonds, or if there are highly electron-dense substituents in the molecule.

In spite of the nonideality of many solutions, it will be convenient later in the discussion to refer to Raoult's law as defining the limiting concentrations necessary for anesthesia.

2. *Henry's Law*

This is a practical, rather than a theoretical, statement of Raoult's law. It is usually stated as

$$p = C/\alpha$$

where p is the partial pressure of the gas in equilibrium with the solution, C is the concentration of the gas in the solvent in (volume gas)/(volume solvent) and α is the Bunsen solubility coefficient. The Bunsen solubility coefficient α is the volume of gas dissolved per atmosphere of gas partial pressure; it has the dimensions of atmosphere^{-1}; and it can be readily shown that for ideal solutions it is simply the reciprocal of the vapor pressure of the pure liquid solute, i.e., $p^0 = 1/\alpha$. Thus, for an ideal case, Henry's law can be written

$$p/p^0 = C'$$

where C' is the concentration of dissolved gas expressed in mole fraction notation.

The foregoing is relevant to the usual method of expressing anesthetic potency by relating it to oil/gas partition coefficients. Since many anesthetics can be expected to form ideal solutions with olive oil, the Bunsen coefficient as measured experimentally and the vapor pressure of the pure liquid should be related; stated another way, the oil/gas coefficient should be readily calculated if the anesthetic forms an ideal and dilute solution in oil. The Bunsen coefficient α is defined as

$$\alpha = \text{liter gas dissolved/liter oil} \cdot \text{atm} = \text{atm}^{-1}$$

if we transform liters of gas to moles and liters of oil to moles by assuming a density of 800 g/liter and a molecular weight for oil of 300 g/mole, and take 22.4 liters/mole as the conversion factor for the dissolved gas, then

$$\frac{\text{liter gas}}{22.4\ \text{liter/mole}} \Bigg/ \frac{\text{liter oil} \cdot 800\ \text{g/liter} \cdot \text{atm}}{300\ \text{g/mole}} = \alpha' = \frac{\alpha}{58}$$

As $1/\alpha$ (or $1/\alpha'$) has the same dimensions as p^0 (atmospheres) the vapor pressure of the pure liquid, the relationship can be tested by comparing values of p^0 calculated using α, with literature values.

TABLE I

The Relationship between Oil/Gas Coefficient and Vapor Pressure of Anesthetics (37°C)

Substance	α (atm^{-1})	$1/\alpha'$ (atm)	p^0 (atm)
Diethyl ether	65	0.90	1.0
Chloroform	205	0.22	0.4
Cyclopropane	11	5.2	9.8
Halothane	224	0.26	0.63
Fluroxene	48	1.20	0.84

Table I then suggests that reasonable agreement can be obtained between measured values of α and values calculated from vapor pressure measurements. There is a suggestion that the agreement is better for paraffins and ethers, as compared with halogenated paraffins, and indeed some nonideality is to be expected from such compounds.

3. *The Ferguson Principle*

If anesthesia depends upon the existence, in some phase of the cell, of a definite concentration of anesthetic molecules, then for an ideal solution this concentration will be some particular fraction of the saturated vapor of the anesthetic liquid. This fact was first appreciated by Ferguson[(4)] and his hypothesis has proved to be a quite durable contribution to an understanding of anesthetic action. To a pharmacologist accustomed to expressing anesthetic concentrations as millimeters of partial pressure of a compound necessary for anesthesia, it would appear that strong anesthetics are those with a low partial pressure required, and this is the basis for the often repeated statement that N_2O is a weak anesthetic; it is actually a very strong anesthetic but the partial pressure required for its action is greater than 1 atm. The Ferguson principle rests on the empirical observation that about 3% of the saturated vapor of many substances produces general anesthesia, and on the theoretical concept that for ideal solutions, Raoult's law would predict that at 3% of saturation, the mole fraction of the substance in solution would also be 3%. From a thermodynamic point of view, if the standard state of a solute is defined as the pure liquid, then the Ferguson principle suggests that the same amount of work is done in bringing about anesthesia no matter what the molecule involved. This is so because if p is the partial pressure for anesthesia and p^0 the vapor pressure of the pure liquid, then $p/p^0 = a$, the thermodynamic activity of the solute (this is identical to mole fraction for ideal solutions). The chemical potential of the solute is $\mu = \mu^0 - RT \ln a$, where μ^0 is the chemical potential in the standard state.

Exceptions to the Ferguson principle can be expected (1) if the anesthetic can be expected to form nonideal solutions with the cell phase where it acts (perfluoroparaffins are a suitable example of such nonideal solutions) or (2) if the molecule is larger than a certain critical size. This point will be discussed below, but it can be noted here that in all homologous series of compounds, the higher members of the series require higher and higher thermodynamic activities until finally the series produces inert compounds.

4. *Effect of Molecular Volume on Solution Ideality*

Many physical properties of anesthetics have been cited by various authors as correlating well with anesthetic action. Some of these, such as surface activity or interfacial activity, are not general requirements since many anesthetics are devoid of these characteristics. Other properties such as polarizability, boiling point, molecular weight, and oil/gas coefficient, are related parameters and it is largely a matter of preference which of these is chosen to correlate with anesthetic action. It is to be emphasized that none of the physical parameters suggested will suffice to explain the anesthetic action of large or highly fluorinated molecules. As an example, the oil/gas partition coefficient can be expected to increase about threefold for each addition of CH_2 to a paraffin chain, yet a point is reached where paraffins are inert. A similar experience has been noted with halogenation of paraffins (especially with fluoro-substitution). This limit on anesthetic action cannot be explained on the basis of either a solution nonideality or a lipid solubility; hence the need to introduce another consideration into discussion.

As more substituents are added to a molecule, its molecular weight and volume increase (as does its boiling point). Therefore it seems more logical to consider molar volume (molar weight/density) rather than molecular weight as a measure of the change produced since the substituents added are often heavy ones such as Br or Cl. The actual molecular volume of the anesthetic molecule together with information about its shape (spherical, linear, branched, etc.) will probably prove to be more important, but such information is generally only available when the substance is in a crystal.

C. Hydrate vs Lipid Solubility Correlations with Anesthetic Action

While most correlations between anesthetic potency and physical properties of a molecule have envisaged a solution of the molecule in some nonaqueous phase, a quite different sort of correlation was developed independently by Pauling[5] and Miller.[6] These authors supposed that it was the ability of a molecule to surround itself with a cage of water molecules and thus to perturb water structure at the cell surface which was important.

Because the same sorts of molecular parameters which lead to a high lipid solubility also are responsible for the formation of hydrate structure in water, a correlation between hydrate dissociation pressure and lipid solubility is to be expected. It was suggested by Miller, Patton, and Smith[7] that fluorocarbons might offer a means of discriminating between the Pauling–Miller theory and lipid solubility correlations. Perhaps the best data for this comparison are that of Miller, Eger, and Lundgren.[8] They find the results shown in Table II below, where it can be noted that the effective concentration of the anesthetic in oil (MAC × oil/gas) varies from 1.4 to 2.6, while the hydrate concentration varies from 0.12 to 6.4; even if the extreme deviation shown by SF_6 is eliminated, the variation is from 0.12 to 0.80, or a distinctly poorer correlation than that given by lipid solubility considerations. Even more impressive is the range of variation of oil/gas partition coefficient CF_4 = 0.073 to $CHCl_3$ = 265 or a 3600-fold change while (MAC × oil/gas) for these substances is from 1.9 to 2.08, a change that is within the error of measurement.

Another difficulty with the hydrate hypothesis is that it offers no real explanation for the *mechanism* of anesthetic action. Very extensive alterations of water structure near the cell surface are to be expected from strong salt solutions (such as a hypertonic seawater with 1 M NaCl) or strong solutions of molecules such as sucrose which engage in extensive hydrogen bonding; yet neither sorts of solution materially affect the electrical excitability of nerve.

III. PROCESSES AFFECTED BY ANESTHETICS

Since it is clear that anesthetics do affect such diverse cellular processes as electrical excitability, cell division, and substrate transfer into

TABLE II

Correlation between Minimum Anesthetic Concentration (MAC) and Oil/Gas and Hydrate Dissociation Pressure[8]

Gas	MAC (atm)	Oil/gas partition coefficient (atm^{-1})	MAC × oil/gas	f Hydrate dissociation pressure at 0°(atm)	$\frac{MAC}{f_{diss}}$
CF_4	26	0.073	1.90	41.5	0.63
SF_6	4.9	0.293	1.44	0.77	6.4
N_2O	1.9	1.40	2.63	10.0	0.19
Xe	1.2	1.90	2.26	1.50	0.80
C_3H_6	0.18	11.8	2.06	0.63	0.29
$CHCl_3$	0.0077	265	2.08	0.062	0.12

mitochondria, a question naturally arises as to whether these processes are all connected somehow or whether it is possible to rule out some of them as unrelated to a loss of electrical excitability.

Fortunately, in the case of axon conduction it is possible to eliminate from consideration most processes other than electrical excitability as being affected by the anesthetic agent supplied. This statement is possible because of the demonstration by Baker, Hodgkin, and Shaw[9] that 95% of the axoplasm from squid giant axons can be removed and replaced by a flowing salt solution such as potassium fluoride. Such a treatment can be expected to remove from the inside of the nerve fiber many sorts of substrates, salts, transmitters, and other water-soluble molecules normally present. The presence of 0.5 M fluoride can, additionally, be expected to inhibit many enzymes which might be present in the remaining thin rim of axoplasm immediately adjacent to the excitable membrane and in addition to remove divalent cations such as Ca and Mg which form insoluble salts with fluoride. In spite of this extensive alteration of the internal environment of the fiber, normal action potentials can be generated over periods of many hours. The results of experiments such as that cited above appear to justify the conclusions that the only requirements for the maintenance of electrical excitability in axons are the presence of the normal concentration gradients for Na and K across the membrane, and for the presence of some Ca^{2+} outside the membrane. A further requirement for any kind of biochemical activity in the cytoplasm would appear to be clearly excluded by the results obtained.

A. Anesthesia in Nerve Fibers

Because the work of Hodgkin and Huxley has given such a complete and quantitative picture of the processes involved in nerve excitation, it is possible to analyze the effect that general anesthetics have on peripheral nerve fibers with far greater precision than is possible for any other bioelectric phenomena. Normally, excitation in nerve results from the interaction between membrane potential of the nerve fiber and the molecular mechanism in the excitable membrane which initiates a large and highly specific change in the permeability of the fiber membrane for Na^+. This sodium permeability increase is a transient which soon decays and is followed by a large and specific increase in the K^+ permeability of the fiber membrane. While the detailed molecular mechanisms involved in the transient increase in Na^+ permeability of the fiber and the sustained increase in K^+ permeability of the fiber are not known, there are a number of reasons for thinking that the ion flows involved in nerve excitation take place through aqueous channels or pores in the cell membrane. It also seems unlikely that there are large numbers of such channels, so that most ion movement takes place at discrete sites in the membrane and these sites are rather far separated from each other. The effects of both general and local anesthetics on nerve excitation have been studied by a number of investigators including Taylor,[10] Moore *et al.*,[11] and Armstrong and Binstock.[12] The results obtained show that anesthetics interfere with the movement of both Na^+ and K^+ as they cross the membrane during an action potential. Using the voltage clamp technique, it is possible to measure not only the effects of anesthetics on the Na and K currents of the nerve fiber, but also to measure the kinetic parameters that turn on the Na conductance, turn it off, and turn on the K conductance. It has been found that the kinetic parameters involving rates of turning-on of ion conductance are not affected by anesthetics nor is the extent to which the Na conductance mechanism is activated by membrane potential. The two Hodgkin–Huxley parameters which are affected by anesthetics applied to nerve are the maximum Na conductance and the maximum K conductance that the fiber shows. Such experimental findings can be interpreted in two ways: either the anesthetic blocks some of the passageways that would carry Na^+ and K^+, or it uniformly lowers the rate at which Na^+ and K^+ can be carried through the existing channels.

From experimental information given above, we can synthesize a picture of what seems to be happening when anesthetics are applied to the nerve membrane. The turn-on of Na conductance requires that some kind of a gate be present in the membrane; this opens when the potential across the membrane decreases but the rate of opening of the gate is not anesthetic-sensitive so that it need not concern us here. When the channel is open, sodium ions begin to flow. If the interior of the channel is mostly water molecules (as is normal), sodium ions flow through the channel with their maximum velocity; if anesthetic molecules can lodge in the channel and associate with the channel walls, they will then provide a barrier to ion flow, presumably because they associate more strongly with the channel walls than do water molecules. Hence they cannot be swept out of the way as easily and the mobility of the Na^+ in the channel is reduced. This decreased mobility will reduce the quantity of Na^+ that can be carried in a given time before the mechanism responsible for turning off Na flow operates to terminate this ionic current and substitute for it a K^+ flow.

A second result which can be inferred from the experimental data is that although the membrane is lipid in nature, and anesthetics undoubtedly dissolve in it, the channels that carry the ionic currents are relatively far apart; thus it seems most unlikely that anesthetic molecules dissolving in the membrane lipids per se will have any influence on the capacity of the channels to carry ions. As one gets closer to the region where ionic currents are carried (in the channel) there is more and more of a possibility for lipid-soluble molecules to act on the channel. If we note that anesthetic thresholds are of the order of 0.03, this means that a 3% change in the volume of the membrane might be expected when it was equilibrated with an anesthetic dose. This seems a very small volume change in the membrane as a whole and one not likely to lead to interference with the Na^+ currents;

on the other hand, if the anesthetic molecules are in the channels themselves, a direct interference with ionic current flow seems much more probable.

B. Anesthesia at Synapses

A much more adequate model of anesthetic action as it applies to the central nervous system could be obtained if we had more information available on the action of anesthetics at synapses. It is known[13] that in the superior cervical ganglion the fibers synapsing in the ganglion are considerably more sensitive to anesthetics than the axons that run straight through the ganglion.

Enough is known about synaptic excitation to draw up the following scheme:

Metabolism → Transmitter synthesis in nerve terminals → Packaging of transmitter in vesicles →

Transmitter release on nerve excitation → Combination of transmitter with postsynaptic membrane receptor → Induction of permeability change in postsynaptic membrane

There are several suggestions in the literature that it is the last step in the above sequence (the induction of permeability change) which is most sensitive to anesthetic action. If this proves to be the case, it would make synaptic excitation somewhat analogous to nerve excitation in that in both cases it is flow of ionic current which is affected by the anesthetic agent. Fairly complex behavior of a single postsynaptic membrane is possible if, for example, the regions around inhibitory postsynaptic receptors are more sensitive to anesthetics than similar regions around excitatory receptors. In such a case, very low concentrations of anesthetic will have the action of a convulsant. Some of the fluorinated ethers[14] have indeed been shown to have such an action. These substances, as noted elsewhere in this chapter, are often just at the threshold of size where anesthetic action is possible so that minor changes in molecular shape may have a decisive effect on the actual concentration developed in the membrane. It is somewhat surprising that very little work has been done on the effect of subthreshold concentrations of anesthetics on synaptic transmission since, according to the hypothesis advanced, a differential effect on inhibitory transmission might result from the administration of low concentrations of any anesthetic.

C. Effects on the Central Nervous System

Anesthesia in man is defined by certain clinical signs of depression of higher cortical function; this, however, does not mean that there are no other effects of the agent administered. The depression in CNS function undoubtedly arises from the inhibition of certain synaptic pathways so that the number of neurons firing per unit time is less than normal. It is, therefore, not surprising to find that oxygen consumption and substrate utilization by the brain are depressed by anesthesia. These effects are a result of neuronal depression, not a mechanism for anesthesia. The brain normally consumes an appreciable fraction of the total oxygen uptake; as neurons are not dividing cells, and their synthetic activities appear modest, it is not unreasonable to assume that a large fraction of the oxygen consumed is used for the production of ATP by oxidative phosphorylation.

In turn, bioelectric activity that leads to the generation of action potentials involves Na entry into the cells, and a consequence of such activity is the need to pump out Na to maintain the normal concentration gradient. The substrate for Na extrusion is ATP and we can infer that very large quantities of this substance will be required for the operation of the Na pump. The following sequence would appear to summarize the indirect connection between anesthesia and metabolism.

anesthesia → fewer neurons firing → less Na entry into neurons
→ less Na pumping → lowered ATP consumption
→ lowered oxygen consumption

The foregoing is not meant to imply that anesthetics have no effect on metabolism; this is a separate topic. It is intended to suggest that observable changes in energy consumption by the central nervous system are to be expected when neuronal function is depressed. Similarly, studies of the distribution of anesthetics in internal organs (including brain) are likely to be reasonable measures of the fat content of the tissue but are unlikely to yield any useful information about the distribution of anesthetics at their site of action.

D. Effects of Anesthetics on Unicellular Organisms

Chemical agents which are commonly thought of as anesthetics not only produce conduction block in peripheral nerve and synaptic block in more complex nervous connections, but also affect a variety of biological processes, among which cell division figures prominently. Such observations make it clear that phases in the cell where anesthetics may act are ubiquitous and this complicates the analysis of the site of anesthetic action. Thus, while the action of anesthetics on the excitation process in nerve is beyond question, it is also clear that such substances also act on the mitochondrial membrane[15] and on cell division. In this connection, the work of Allison and Nunn[16] is of great interest. They observed that anesthetics in pharmacological doses inhibit the formation of microtubules which are the structural basis for the mitotic spindle; hence anesthetic action can be related to the inhibition of cell division. In the adult central nervous system where cell division is nonexistent, it is clearly the action of the anesthetic on the excitable membrane which is responsible for the phenomena of anesthesia, but in other tissues of the body, alterations in function may result from an inhibition of cell division induced by anesthetics.

A very valuable discussion of the general biological effects of anesthetics is given by Schreiner.[17] In particular, he has noted that inhibition of growth of *Neurospora* or HeLa cells can be inhibited by N_2, A, Kr, or Xe as well as an inhibition of acetylcholine esterase. For enzyme inhibition, however, the partial pressures required are more than ten times those required for cell division inhibition. For the series, He–Xe, the polarizability and pressure for 50% inhibition of *Neurospora* growth show a linear relationship.

IV. BIOCHEMICAL VERSUS BIOPHYSICAL MODES OF ANESTHETIC ACTION

A. The Depression of Na Transport Activity by Anesthetics

The maintenance of a low internal Na^+ concentration in excitable cells (and indeed in many other types of cell) has been shown to be dependent on a mechanism called the Na pump. This system performs the following functions:

$$Na^+(\mathrm{in}) + ATP + E \rightarrow E{\sim}P + ADP + Na^+(\mathrm{out})$$

$$E{\sim}P + K^+(\mathrm{out}) \rightarrow E + P_i + K^+(\mathrm{in})$$

Thus the hydrolysis of ATP → ADP and P_i is able to move Na^+ to the outside of the cell and to bring K^+ inside. The amounts of Na moved are generally much larger than the amounts of K so that the mechanism also generates an electric potential in connection with its operation. An inhibition of the Na pump can be expected to allow Na^+ to accumulate in the cell and thus to abolish the ionic gradient on which the action potential depends. It can also be expected to cause a loss of K^+ inside the cell and thereby to abolish the K^+ gradient upon which the resting potential of the cell depends. Additionally, if the electrogenic action of the Na pump is holding the membrane potential at a value higher than that given by the K^+ concentration ratio across the membrane, then a cessation of pumping will lead to a decrease in resting potential.

It has been shown that diethyl ether[18] and butanol[19] depress the Na pump of squid giant axons at concentrations identical with those necessary to block conduction. The question then is: Has conduction block resulted from an inhibition of the biochemical mechanisms underlying the Na pump? Fortunately, in this instance it is possible to be quite sure that this is not the case because: (1) the Na pump in squid axons can be blocked by a variety of means with no effect on electrical excitability; (2) there is no measurable change in the membrane potential of this fiber when the Na pump is inhibited; and (3) the gradients of Na and K across the membrane change very slowly when the pump is inhibited, owing to the large size of the nerve fiber.

It would appear then that although anesthetics do have a powerful inhibitory action on the biochemical system responsible for maintaining the electrical excitability of nerve, their primary action is on the passive ion currents which flow during the action potential.

Very small nerve fibers are much more dependent on metabolism for the maintenance of their ion gradients; if these fibers also had electrogenic Na pumps, then it is conceivable that blocking might result from the cessation of pumping. A test for such an hypothesis would be to show that the application of an anesthetic depolarized the membrane potential before blocking conduction.

B. Chemical Transformation of Anesthetic Agents

The diversity of chemical structures capable of producing anesthesia (xenon to halothane) is so great that it has always seemed unlikely that there was any common chemical reaction in which such substances would participate. Nevertheless, studies have been made to see if there were any chemical transformation of anesthetics by organisms, and these have shown that while many substances in common use are metabolized to some small degree, there is no common end product of such degradation. Further, anesthetics such as Xe clearly are not metabolized.

An alternate suggestion has been that anesthetic agents are inhibitors of metabolism but act without the necessity of chemical transformation. In this sense, the inhibitory action of anesthetics on ionic flows during excitation of nerve is established, as is the inhibitory action of such agents on active transport. It seems likely, however, that such inhibition is related to the changes in a lipid phase produced by the anesthetics rather than by there being a specific chemical reaction that is inhibited by the agents.

C. Interactions of Anesthetics with Proteins

While the foregoing discussion has emphasized that anesthetic molecules will tend to accumulate in any hydrophobic environment and the obvious sort of environment in the cell is a lipid phase, it is to be emphasized that many proteins have parts of their structure which are lipidlike and such structures cannot be ignored in considering sites for the action of anesthetics.

Experimental work on the interaction of proteins with paraffins is of rather recent origin but enough has been done to make it clear that there are strong interactions between ordinary globular proteins and paraffins. Demonstrations of a binding between proteins and simple paraffins were made by Wishnia[20] and by Wetlaufer and Lovrien.[21] In a more recent study,[22] it has been possible to show that there is a conformational change in both β-lactoglobulin and bovine plasma albumin when these are brought in contact with a variety of anesthetics. These measurements were made by polarimetric methods and in the case of the interaction of butane with β-lactoglobulin, the optical rotatory dispersion was measured. The results of the latter measurements suggest that the helical content of the protein molecule is unchanged but that side chains are rearranged. The quantitative aspects of this study to evaluate anesthetic potency were unfortunately based on a standard state of 1 atm. In turn, this led to the conclusion that methoxyflurane, chloroform, and halothane were powerful agents in affecting optical rotation, while N_2O and acetylene were weak in such interactions. If the Ferguson principle of the pure liquid as a standard state had been chosen, the conclusions of these authors regarding dose vs. optical rotation of the proteins might have been reversed. However, the main point of the studies cited above is that they lead to the conclusion that proteins do have the ability to bind substantial quantities of pure hydrocarbons such as butane, as well as more polar anesthetic agents, and that upon binding there are conformational changes in the protein molecule.

An extreme example of binding between an anesthetic agent and a protein has been demonstrated by Schoenborn,[23] who discovered an exceptionally strong binding between xenon and myoglobin. These X-ray diffraction studies located the site of Xe binding as a natural cavity in the myoglobin molecule. This cavity allows the surrounding atoms (representing amino acid chains) to develop a van der Waals' type of binding which is estimated to be of the order of 10 kcal/mole. An interesting feature in this Xe binding is that it apparently does not affect the catalytic ability of myoglobin to take up and discharge O_2. However, it is quite possible to suppose that other catalytic macromolecules may well have cavities in their structure which are somewhat closer to their active sites and that in such cases there might well be a conformational change in the molecule which would affect catalytic activity. Further exploration of this cavity in hemoglobin has shown that cyclopropane will bind in it[24] although N_2O will not. Such a finding is understandable because the cavity is spherical and the linear N_2O molecule would not be expected to fit in the manner of spherical molecules. In connection with models of the site for anesthetic action, it is interesting that the Xe site in myoglobin will not accept molecules much larger than Xe so that binding is limited to relatively small molecules. While the site for anesthetic action clearly accepts molecules much larger than Xe, there is evidence that there is a size limit for molecules that are to produce anesthesia, as well as for molecules that are to fit into protein structures. Further information is most desirable with respect to both the extent to which protein interactions with hydrocarbons lead to conformational changes and the extent to which anesthetic binding by proteins is limited by site size.

V. MECHANISM OF ANESTHETIC ACTION

A. Enzyme Inhibition

One obvious way to control cellular activity is by affecting the rate of chemical catalysis of intracellular enzymes. It is not surprising, therefore, that enzyme inhibition should have been postulated as a possible way by which anesthetics could affect chemical reactions occurring in cells. The fact that, as discussed earlier in this chapter, the site for the action of anesthetics appears to be hydrophobic in nature need not conflict with the known hydrophilic properties of many enzymes; in a molecule as large as enzyme or other protein, hydrophobic regions can exist in the protein interior while hydrophilic groupings (which confer water solubility on the molecule) are on the outside.

The preceding discussion has suggested that the action of anesthetics may be a pluralistic one in affecting many sorts of cellular processes, quite possibly by different sorts of mechanisms. The possibility that proteins as well as lipids can take up anesthetics, and thereby undergo structural changes, vastly increases the number of mechanisms which might be thought to operate in cellular control.

As an example, one might consider simple proteolytic enzymes and the possibility that anesthetics inhibit such action. However, experiments show that concentrations of anesthetics tenfold greater than threshold for anesthetic action fail to affect rates of hydrolysis.

On the other hand, although mitochondria can, in part, be regarded as arrays of enzymes embedded in a lipid matrix, anesthetics profoundly affect substrate and divalent cation transfer across the mitochondrial membrane(15) so that in this sense anesthetics affect enzyme action. Because of the demonstrated interaction of anesthetics with protein, the possibility must be kept open that enzyme inhibition occurs when the enzyme involved is not associated with a membrane phase.

That protein side chains can present the equivalent of an "oil" phase has been emphasized by Featherstone and Muehlbacher(25) in their review of anesthetic action. This suggestion is perhaps one of the most important changes in our conceptual framework regarding the site of action of anesthetics because it allows a direct connection between the site of action of anesthetics (admittedly a hydrophobic one) and a protein, which is a more plausible structure for controlling the electrical excitation in nerve.

In a nuclear magnetic resonance study of the interaction of benzyl alcohol with erythrocyte membranes,(26) it was noted that at low concentrations (up to 50 mM) the molecules of the alcohol were strongly immobilized by the membrane structure although at higher concentrations this immobilization was lost. The partition coefficient was also measured as a function of concentration and this appeared reasonably constant in the range of 2–10 mM, but rose rapidly at higher concentrations, suggesting that perhaps dimerization was taking place.

B. Interference with Neuronal Transmission

Since synaptic transmission is more sensitive to anesthetic action[13] than is conduction in axons, attempts to sort out the mechanism responsible are more likely to be directly comparable to anesthesia in the central nervous system than to conduction in peripheral nerve. One should therefore examine some of the effects likely to be encountered in synaptic transmission and see what plausible role can be ascribed to anesthetics. To do this, an idealized scheme for synaptic excitation must be assumed and this is shown in Fig. 1. A nerve cell (N) is assumed to have a single excitatory (E) and a single inhibitory fiber (I) synapsing on its surface. Excitatory action is assumed to decrease membrane resistance for Na, K, and Cl while inhibitory action is assumed to selectively decrease only R_K and R_{Cl} and not to change the membrane potential. Let us see what effect an anesthetic would have if it doubled membrane resistance (R_m). At rest $1/R_m \cong 1/(R_K + R_{Cl})$ so that for a volley of inhibitory impulses previously sufficient to reduce R_m to half its normal value in the unanesthetized synapse, membrane resistance would go from $2 \cdot R_m$ to R_m in the anesthetized postsynaptic membrane. In other words, the inhibitory transmitter action has been abolished by the change in membrane resistance postulated to follow anesthetic action. The failure of R_m to decrease below normal levels means, of course, that the postsynaptic membrane is more vulnerable to excitatory action.

If we consider the action of an excitatory transmitter under circumstances similar to those given above, it is important to note that its action will be to cause a nonselective increase in ion permeability; the principal ion carrying current will be Na (because K and Cl are assumed to be in electrochemical equilibrium across the membrane). In turn, there will be some depolarization of the postsynaptic membrane even though the

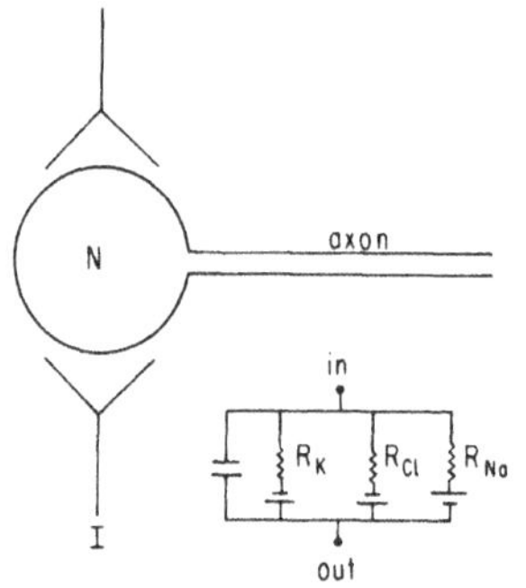

Fig. 1. A simple synapse with inhibitory (I) and excitatory (E) fibers. The circuit diagram relates R_{Cl} and R_K to inhibitory action, and total membrane resistance to excitatory action.

membrane resistance is unchanged. This is because under excitatory transmitter action a greater fraction of the membrane current is carried by Na.

Pursuing this oversimplified version of synaptic action, a purely excitatory volley in presynaptic fibers would in unanesthetized synapses, reduce R_m to $\frac{1}{2}$ of its normal value and practically all this change would be a change in R_{Na} because R_K and R_{Cl} are so much larger at rest than is R_{Na}. The results of this volley are that a Na^+ current flows and the postsynaptic membrane potential decreases. According to the hypothesis advanced, low concentrations of anesthetics do not affect excitatory synaptic action (so that the same Na^+ current will flow in both cases). The potential change, which determines whether the axon of the postsynaptic neuron will fire, will therefore be given by $i_{Na}R_m$. Inhibitory action opposes this potential change by making R_m smaller. If, as suggested earlier, low concentrations of anesthetics increase postsynaptic membrane resistance, this change by itself makes the synapse more sensitive to excitatory action because less Na^+ current is necessary for the same potential change.

The mechanism is probably too speculative to be pressed further but it does illustrate the notion that a simple membrane resistance change in the postsynaptic membrane induced by anesthetics is sufficient to explain an excitatory action of anesthetics at low concentrations (because of an inhibition of inhibitory action). A block in excitatory synaptic transmission at larger anesthetic concentrations is to be expected when sensitivity to transmitter action is abolished.

C. Interference with Ion Interchange in Axons

For the axon it is possible to give a reasonably complete picture of the effect of anesthetics on excitation. It is known that what is affected by the anesthetic is $\bar{g}_{Na}$, the maximum conductance that the axon can have for Na^+ (the K^+ conductance is also affected but this is not directly involved in the initial events of excitation). The Hodgkin–Huxley equation for Na conductance is $g_{Na} = \bar{g}_{Na}m^3h$. The parameters m and h are not affected by anesthetics nor are their time derivatives. The diagram shown in Fig. 2 is intended to translate excitation into mechanisms. The membrane is shown as an oily phase with occasional aqueous channels or pores through which ions pass. The pore at the top has a cover which is closed and water molecules inside the channel—it is intended to represent the unanesthetized state. When the membrane potential is reduced, the cover opens, sodium ions flow through the channel, the membrane potential is reduced, and excitation takes place.

The lower channel has its cover open, anesthetic molecules in the channel, and Na^+ is unable to flow because the interaction of the anesthetic with the channel walls is strong enough to prevent Na^+ flow. It might be thought that these channels could also be blocked by aqueous cages containing anesthetic molecules; however, it is difficult to see how the water

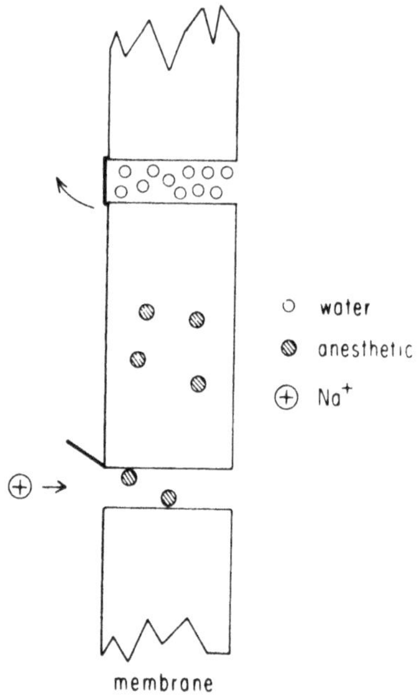

Fig. 2. A diagram of the excitable membrane of an axon. Two channels are shown: at the top a closed channel containing water; at the bottom an open channel containing anesthetic molecules.

of the cage, in its interaction with the channel walls, would be different from ordinary water. In turn, it seems less probable that such cages would impede the ion flow that is required if excitability is to be abolished.

Anesthetic molecules are also shown in the lipid phase of the membrane, but remote from the channels. They do not affect the excitation process, but are included in the diagram to stress the difference between lipid solubility in the membrane which is not limited by molecular size, and penetration into membrane channels which does have molecule size limitation. There are reasons for believing that the channel radius is about 4 Å.

D. Irreversible Actions of Anesthetics

It is customary to stress the reversible nature of the anesthetic molecule–cell membrane interaction but in fact reversible anesthesia is only possible within a quite restricted range of concentrations. Irreversible loss of excitability can take place at concentrations of anesthetic only twice the effective concentration. This narrow range of pharmacological action may be of some help in understanding mechanisms of anesthetic action.

One thought is that too much anesthetic in the lipid phase of the membrane may make for instability of the structure, and the membrane simply collapses. Against this explanation are the following: (1) in work with artificial bimolecular lipid membranes, large quantities of solvent can be incorporated without membrane instability and the incorporation and removal of solvent are quite reversible; while (2) when axon conduction is irreversibly blocked by substances such as ether (Mullins, unpublished), the membrane potential is not affected so that clearly the membrane has not collapsed. What is needed as an explanation for the irreversible loss of excitability is a mechanism by which the binding of anesthetic molecules enhances the binding of the anesthetic molecules already present—in other words an autocatalytic process. Since we have earlier ruled out chemical reactivity, we are left with the idea that as more and more anesthetic molecules enter a channel, the interaction between these molecules and the channel wall increases because they deform the channel. Ultimately, one can think of a phase transition or "freezing" of this part of the membrane with a great increase in binding of the anesthetic molecules. While there is no experimental information of a dose-response nature, the suggestion mentioned above can be given more explicit form by saying that if an ion channel has, on the average, one anesthetic molecule, then there is some interference with nerve excitation, while 2 molecules/channel will block and 4 molecules/channel will increase the binding of all 4 molecules such that recovery of excitation will not be measurable over the usual experimental period of an hour or so. This mechanism ought to be temperature sensitive and a study of anesthetic binding *vs* temperature might well contribute to our better understanding of irreversible loss of excitability.

VI. THE DESIGN OF PRACTICAL ANESTHETICS

A. Molecular Size

For small molecules such as Xe, N_2O, or C_2H_4, anesthetic potencies can be predicted in a quite straightforward way using thermodynamic activity considerations. As mentioned earlier in this chapter, N_2O, which is often characterized as a "weak" anesthetic, is in fact as powerful an agent as one might wish although anesthesia will require a pressure of something like 1.4 atm. Small molecules which are totally fluorinated, such as CF_4, will show marked deviations from ideal behavior in paraffinic liquids so that activity coefficient corrections will be necessary to accurately predict anesthetic thresholds. As one goes to larger molecules, further considerations (whose nature is not yet understood) come into play so that an exceedingly intriguing problem is posed by Table III.

The first five compounds of Table III all have virtually identical boiling points and therefore, by inference, similar intramolecular attractive forces. Yet, there is the fact that the first three compounds have identical

TABLE III

Anesthesia of Mice[a]

Compound number	C			C			*b.p.*	Anesthetic threshold
1	F	F	F	H	Br	Cl	50	0.03
2	F	F	F	H	Br	CH_3	49	0.07
3	F	F	F	Cl	Cl	Cl	47	0.13
4	F	F	Br	F	F	Br	46	0.11
5	F	F	Cl	F	Cl	Cl	48	0.16
6	F	F	Cl	H	Cl	Cl	72	0.05
7	F	F	F	H	Br	Br	73	0.05

[a]Adapted from Suckling.(27)

F-substitution on one carbon of ethane, and differ in anesthetic threshold by more than a factor of 4. The similarity of boiling point among these three compounds is relatively easy to understand as compounds 1 and 2 have a Br,Cl *vs* a Br,CH_3 substitution while compound 3 has a Cl_3 substitution and these changes ought to be roughly comparable. Note, however, that compounds 5 and 6 differ greatly in boiling point although they differ only in a single substituent on the 2 carbon, F *vs* H. This is a case in which empirical notions of the effect of substituents on boiling point simply fail; the anesthetic thresholds are also greatly different, a fact that cannot be readily explained.

There is a suggestion by Suckling(27) that the presence of an H substitution on the 2 carbon is necessary if a low anesthetic threshold is to be obtained. The data of this table support such a notion, but it is hardly a helpful suggestion in reality if one cannot explain the boiling point anomaly mentioned above. The data in the table are important in the sense that they give a detailed description of the shape of the anesthetic receptor; this is essential since small differences in molecular shape and local attractive force are important in dictating anesthetic binding. It will be profitable to leave further consideration of this point for the moment and return to it after considering large molecules.

Relatively little work has been done on paraffin anesthetics of molecular weight greater than about C_5, and the literature suggests that the studies which have been done may have used impure compounds. Compound purity is far from a trivial factor, as the following will show. Simple distillation of hydrocarbons, even in columns with a large number of effective plates, can be expected to lead to impurities in the distillates of the order of a few percent. Since the lower hydrocarbons are effective anesthetics, a small contamination in higher hydrocarbons will lead to spurious results in the sense that higher hydrocarbons will be suspected of anesthetic action. Fortunately, preparative gas chromatography is highly developed and can separate hydrocarbons in a high state of purity. It is therefore possible to state that *n*-decane (supplied as National Bureau of Standards reference sample) has no anesthetic action even when animals are caused to breath its saturated vapors for prolonged periods of time. Clearly the animal is being supplied with the substance at a thermodynamic activity of 1.0 (*vs* the usual thermodynamic threshold of 0.02). The precise point at which physiological inactivity of the paraffins takes place is not clear since high purity compounds below C_{10} have not been studied. At any rate, *n*-decane has a molar volume of 194 ml/mole and we may expect that a somewhat lower volume actually represents the point at which inert behavior of paraffins appears.

The conventional explanation of why higher paraffins are physiologically inert is that the substances are insufficiently water soluble to allow their distribution to the central nervous system through the blood. This is a kinetic argument which is easily disposed of by allowing longer equilibration times with the anesthetic agent. It is also not quite a precise analysis because a lower aqueous solubility means that blood will equilibrate more rapidly with the agent involved and so less time will be consumed in this process (compare diethylether equilibration with halothane, for example). At any rate, one can also argue that since hexane has anesthetic properties which appear promptly, and water solubility of paraffins declines about threefold for each carbon added to the chain, then decane should require only 81 times as long as hexane for a similar equilibration.

A second test of whether molar volume (or molecular size) is a relevant consideration in limiting anesthetic action is to look at other sorts of compounds. Perfluoropentane (C_5F_{12}) is also an inert substance for anesthesia; it has a molar volume of 214 ml/mole or a value similar to that for decane. Because this is a perfluoro compound, an allowance of as much as a factor of 10 might be made for nonidealities of mixing, but even so, the compound is supplied at an activity of 1.0 so that 0.1 would be the applied activity corrected for nonidealities of solution.

An interesting series of ethers is also helpful in the analysis of the problem of molar volume *vs* anesthetic action. The substances shown below differ greatly in their action; the perfluoroether is inert when applied at 0.8 atm or an activity of $0.8/3.2 = 0.25$ while the hexafluoroether is a convulsant and the trifluoroether a weak anesthetic. While there are undoubtedly activity coefficient corrections necessary for all the compounds cited above, it is reasonable to expect that these would not materially change the conclusion that a large molar volume is incompatible with anesthesia even though all other indices of anesthesia (such as oil/gas partition coefficient)

TABLE IV

The Size and Vapor Pressure of Some Fluorine-Substituted Ethers

Compound	V_m	p° atm
Di(perfluoroethyl)ether	180	3.2
Hexafluorodiethylether	131	0.5
Trifluorodiethylether	100	0.6

would predict anesthetic action. The hexafluoroether would appear to be a compound of great interest in the sense that it is too large to be an anesthetic but is able to develop sufficient binding in the central nervous system so that it can differentially depress inhibitory postsynaptic action. This in turn would be expected to lead to convulsant action much like that encountered with strychnine which appears to be a specific chemical inhibitor of postsynaptic inhibitory action. It is important to note that what is being suggested is not that there is any chemical specificity to hexafluorodiethylether, but that this substance is unable to reach concentrations adequate for anesthesia although it is able to reach lower levels of concentration such that it can exert a differential action in depressing the sensitivity of the postsynaptic membrane to inhibitory transmitter action.

The most commonly used anesthetic with the highest boiling point is methoxyflurane (CH_3-O-CF_2-CCl_2H), boiling point 105°C. The molar volume of this substance is only 116 ml/mole, a value well below the postulated threshold for anesthesia; halothane, on the other hand, with a V_m of 138 ml/mole is just at the threshold of molecular size one supposes to be the limit for anesthetic action. This fact may explain the critical distribution of substituents that has been noted for molecules related to halothane.

Reasons for a limitation in molecular size related to membrane excitation have already been given. The matter can be approached another way by seeing how molecules of variable size might interact with a site of fixed size when the forces involved are dispersion forces. To simplify the discussion we may take the molecule as spherical, the site as hemispherical, and consider that the binding is proportional to a specified fraction of the spherical surface which is in effective contact with the site. Figure 3 shows the situation for a series of molecules with diameters from 0.5 to 1.25, the diameter of the site. The plot is of the area in contact *vs* molecular radius and as would be expected, binding is maximal for a molecule radius which just fits the site and then falls rapidly to zero.

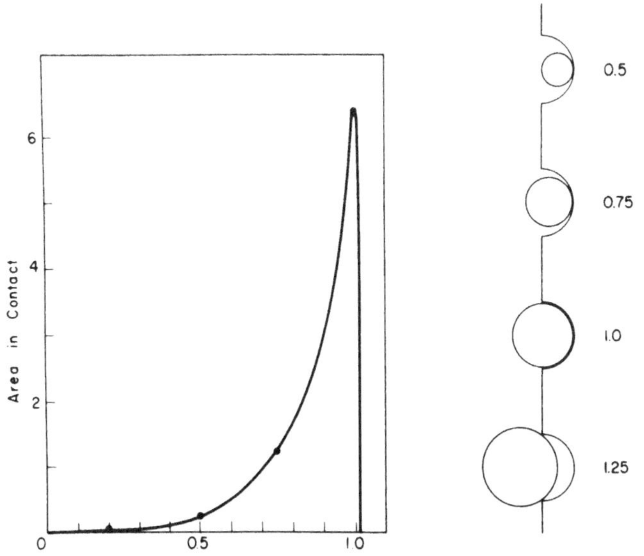

Fig. 3. The binding of molecules of various sizes to a hemispherical site is given in purely geometrical terms of an "area in contact". The model is introduced to explain a limitation in site size for anesthetic action.

A more realistic model would include a mean site size $s \pm d$, where d is a dispersion in mean site size. For this arrangement, binding energy would fall less abruptly than indicated in the figure.

B. The Distribution of Electron-Dense Substituents

It has been suggested[(2)] that inasmuch as paraffins, with a low solubility parameter, often seemed to be satisfactory practical anesthetics while heavy halogen-substituted paraffins (Cl, Br, I), with very high solubility parameters, seemed to have unsatisfactory effects on cardiac excitability, it might be possible to take advantage of the extremely low solubility parameter of fluorocarbons and by combining F and Cl substitutions in a paraffin, to produce a practical anesthetic with a solubility parameter between that of the fluoro- and chloroparaffins.

This suggestion was made purely on the grounds of designing a molecule which would partition to the nervous system in the way that paraffins do and yet would have a higher boiling point than the corresponding paraffin, and of course would be nonflammable. A number of such molecules have been synthesized and the interesting (and unforeseen) fact which has emerged is that structural requirements for such molecules are extremely important. It seems likely that this steric effect is related to the suggestion that the anesthetic receptor is of finite size and that therefore even the substituted ethanes (of which halothane is the prototype) are molecules with a size close to the limit that can enter the receptor. If this is so, then a small change in the profile of the molecule may make the difference between penetration and nonpenetration. Since substitution of various atoms in the molecule may not be the sole cause of such changes in profile, it seems worth speculating that restrictions on free rotation about the C–C bond in the molecules of Table III may affect the profile of the molecule and hence the penetration. This point could easily be examined by infrared spectroscopy of the compounds involved. There is reason to guess that compounds 3, 4, and 5 might have restrictions on free rotation because of the way that halogen substitution is distributed in the molecule.

A second mechanism that may change the effective profile of the anesthetic molecule occurs when all the substituents on one carbon are of low electron density, while those on the other carbon are of high electron density; under these conditions there may be a tendency for interaction between molecules of the anesthetic in the pure liquid such that electron-dense ends of the molecule associate. This tendency to dimerize would then affect the boiling point and might help to explain some of the anomalies found with these compounds. It is interesting that although the perfluoroparaffins and paraffins each boil at about the same temperature for the same number of carbon atoms, Krantz and Rudo[28] point out that compounds such as CH_3Cl and CF_3Cl differ in boiling point by 60°C. This holds true for Br or I substitution as well. The inference is then that the presence of F in the molecule vastly attenuates the Cl–Cl interaction which is mainly responsible for setting the boiling point of CH_3Cl.

C. Structural Requirements for Anesthetic Action

If the potency of an anesthetic is judged by how small the thermodynamic activity for anesthesia (*a*) is, then small molecules such as N_2O and Xe are extremely potent (although applied partial pressures may be more than 1 atm) and there is no common structural requirement for an anesthetic action. As molecular size increases, a complex and as yet incompletely understood set of structural requirements appears. Anesthetic action ceases when molecular size reaches some particular limit (apparently decane for the hydrocarbon series). This general problem of structural requirements for anesthetic action has been rather extensively explored in a search for compounds that might be useful as practical anesthetics. Unfortunately, a survey of the literature shows that the purity of many compounds tested was questionable, and that often there is no physicochemical data such as partition coefficients which would allow one to compare thermodynamic properties with pharmacological action.

Initial considerations in the search for new anesthetics were to improve on paraffins such as cyclopropane by making a noncombustible anesthetic. Halogenated hydrocarbons have been extensively investigated, especially fluorine-substituted hydrocarbons. A definitive review of the properties of these substances has been made by Krantz and Rudo[28] and the following points are relevant to the present discussion. Perfluoroparaffins can be expected to show large deviations from ideal solution behavior even in oil; as the molar volume of a perfluoroparaffin is about twice that of the corresponding hydrocarbon, such substances can be expected to reach the critical volume beyond which anesthesia is impossible at a smaller number of carbon atoms in a paraffin chain. The behavior of such substances follows in a rather straightforward way from differences in the solubility parameter between paraffin and perfluoroparaffin.

The boiling point of a liquid is a useful index of the intramolecular forces in a liquid, measuring as it does the temperature at which the thermal energy of the molecules is sufficient to overcome the attractive forces in the liquid state. The liquid with the lowest boiling point (and hence lowest intramolecular attractive force) is He. Yet there is every reason to believe that He is an anesthetic although the required partial pressure for an activity of 0.01 would be very high. Very low intramolecular attractive forces such as those in CH_4 (b.p. $-162°C$) or the rather high forces in CCl_4 (b.p. 77°C) do not influence anesthesia as both substances are anesthetics at about the same activity. On the other hand, decane (b.p. 174°C) is not anesthetic while there are a variety of fluoro-chloro-substituted paraffins with relatively high boiling points which are anesthetics. Clearly, anesthetic action does not correlate with boiling point when a great variety of compounds is considered although, as indicated above, boiling point does give important clues to the nature of the intramolecular attractive forces of the molecule under consideration.

VII. FUTURE RESEARCH

With the great improvements currently being made in methods of isolating and purifying cell membranes, it is reasonable to look forward to an improved description of the site or sites of action of anesthetics with such structures. It also seems likely that the method used so successfully to characterize the xenon site in myoglobin may be very helpful (when applied to membrane proteins) in similarly characterizing the site responsible for anesthetic-membrane binding.

In clinical anesthesia it seems likely that improvements in the control of anesthesia may come about by the use of mixtures of anesthetics. While it is already current practice to use N_2O-halothane mixtures in order to reduce the partial pressure of halothane required, the possibilities inherent in the use of mixtures are much greater than this. One may expect simple additivity of small and large molecules insofar as their effects on a particular physiological process are concerned. This is implicit in a constant thermodynamic activity for anesthesia. It is, therefore, the possibility of a differential distribution of molecules to various parts of the nervous system as well as nonnervous cells that may be of considerable interest. There are numerous possibilities for work along these lines. Some of the studies of interest are:

1. Mixtures of anesthetics and convulsants that might extend the range between anesthetic and toxic concentrations of these agents.
2. Mixtures of two or more anesthetics that might have the same action as (1) because the threshold for irreversible loss of excitability might be higher than for a single agent.
3. Mixtures especially designed to suppress certain types of nonneuronal responses in tissues (such as cardiac arythmias). This suggestion is based on evidence that many simple chloroparaffins such as ethyl chloride seem to exert a convulsant action on cardiac tissue at doses where there is CNS anesthesia.

ACKNOWLEDGMENTS

I am indebted to Drs. J. P. Bunker, E. I. Eger, II, and R. M. Featherstone for having read this manuscript and for much useful advice concerning the presentation.

VIII. REFERENCES

1. J. H. Hildebrand and R. Scott, *The Solubility of Non-Electrolytes*, Reinhold Publishing Co., New York (1950).
2. L. J. Mullins, Some physical mechanisms in narcosis, *Chem. Rev.* **54**:289–323 (1954).
3. E. I. Eger, II, C. Lundgren, S. L. Miller, and W. C. Stevens, Anesthetic potencies of sulphur hexafluoride, carbon tetrafluoride, chloroform, and ethane in dogs: Correlation with the hydrate and lipid theories of anesthetic action, *Anesthesiology* **30**:129–135 (1969).
4. J. Ferguson, The use of chemical potentials as indices of toxicity, *Proc. Roy. Soc. (London)* **B127**:387–404 (1939).
5. L. Pauling, Molecular theory of general anesthesia, *Science* **134**:15–21 (1961).
6. S. L. Miller, Theory of gaseous anesthetics, *Proc. Nat. Acad. Sci. (Wash.)* **47**:1515–1524 (1961).
7. K. W. Miller, W. D. M. Patton, and E. B. Smith, Site of action of general anesthetics, *Nature* **206**:574–577 (1965).
8. S. L. Miller, E. I. Eger, II, and C. Lundgren, Anaesthetic potency of CF_6 and SF_6 in dogs, *Nature* **221**:468–469 (1969).
9. P. F. Baker, A. L. Hodgkin, and T. I. Shaw, Replacement of the axoplasm of giant nerve fibres with artificial solutions, *J. Physiol. (London)* **164**:330–354 (1962).
10. R. E. Taylor, Effect of procaine on electrical properties of squid axon membrane, *Am. J. Physiol.* **196**:1071–1078 (1959).

11. J. W. Moore, W. Ulbricht, and M. Takata, Effect of ethanol on the sodium and potassium conductances of the squid axon membrane, *J. Gen. Physiol.* **48**:279–295 (1964).
12. C. M. Armstrong and L. Binstock, The effects of several alcohols on the properties of the squid giant axon, *J. Gen. Physiol.* **48**:265–277 (1964).
13. M. G. Larrabee and J. M. Posternak, Selective action of anesthetics on synapses and axons in mammalian sympathetic ganglia, *J. Neurophysiol.* **15**:91–114 (1952).
14. J. C. Krantz, Jr., E. B. Truitt, Jr., A. S. C. Ling, and L. Speers, Anesthesia LV. The pharmacologic response to hexafluorodiethyl ether, *J. Pharmacol. Exp. Therap.* **121**:362–368 (1957).
15. B. Chance, M. Pring, A. Azzi, C. P. Lee, and L. Mela, Kinetics of membrane transitions, *Biophys. J.* **9**:A90 (1969).
16. A. C. Allison and J. F. Nunn, Effects of general anesthetics on microtubules, *The Lancet*, 1326–1329 (1968).
17. H. R. Schreiner, General biological effects of the helium–xenon series of elements, *Fed. Proc.* **27**:872–877 (1968).
18. E. A. Schwartz, Effect of diethylether on sodium efflux from squid axons, *Currents Mod. Biol.* **2**:1–3 (1968).
19. P. F. Baker, M. P. Blaustein, R. D. Keynes, J. Manil, T. I. Shaw, and R. A. Steinhardt, The ouabain-sensitive fluxes of sodium and potassium in squid giant axons, *J. Physiol.* (*London*) **200**:459–496 (1969).
20. A. Wishnia, The solubility of hydrocarbon gases in protein solutions, *Proc. Nat. Acad. Sci.* **48**:2200–2204 (1962).
21. D. B. Wetlaufer and R. Lovrien, Induction of reversible structural changes in proteins by nonpolar substances, *J. Biol. Chem.* **239**:596–603 (1964).
22. D. Balasubramanian and D. B. Wetlaufer, Reversible alteration of the structure of globular proteins by anesthetic agents, *Proc. Nat. Acad. Sci.* **55**:762–765 (1966).
23. B. P. Schoenborn, H. C. Watson, and J. C. Kendrew, Binding of xenon to sperm whale myoglobin, *Nature* **207**:28–30 (1965).
24. G. A. Gregory and E. I. Eger, II, Partition coefficients in blood and blood fractions at various concentrations of cyclopropane, *Fed. Proc.* **27**:705 (1968).
25. R. M. Featherstone and C. A. Muehlbacher, The current role of inert gases in the search for anesthesia mechanisms, *Pharmacol. Rev.* **15**:97–121 (1963).
26. J. C. Metcalf, P. Seeman, and A. S. V. Burgen, The proton relaxation of benzyl alcohol in erythrocyte membranes. *Molec. Pharmacol.* **4**:87–95 (1968).
27. C. W. Suckling, Some chemical and physical factors in the development of fluothane, *Brit. J. Anesth.* **29**:466–472 (1957).
28. J. C. Krantz, Jr., and F. G. Rudo, *in Handbook of Experimental Pharmacology* (O. Eichler, A. Farah, H. Herken, A. D. Welch, eds.), Vol. XX/I pp. 501–564, Springer-Verlag, New York (1966).

Chapter 17

LOCAL ANESTHETICS

Wolf-Dietrich Dettbarn

Department of Pharmacology
Vanderbilt University School of Medicine
Nashville, Tennessee

I. INTRODUCTION

It is the intention in this chapter to establish a possible common basis for further discussions on the mechanism of action of local anesthetics. The large variety of compounds having local anesthetic action has made it difficult to clearly establish a common mode of action. This is complicated even more by the fact that local anesthetic activity is often found in combination with other pharmacological actions: agents having antihistaminic, sympathomimetic, parasympathomimetic, antiseptic or antibiotic action in one form or other exhibit a local anesthetic quality.[1-3] Attempts to explain the mechanism of action of such compounds through interference with normal function of enzyme systems or with the general biochemistry of cells have not been too rewarding as there is a considerable discrepancy between electrophysiological and biochemical changes which they produce. The common denominator of all local anesthetic action is the reversible block of electrical activity of nerve fibers without affecting the membrane potential or causing any general toxic action. The pharmacological and neurophysiological mechanisms by which this block is produced may not be the same for the various groups of local anesthetics. To simplify the discussion, one group of local anesthetics used clinically will be discussed in some detail as it seems reasonable that in addition to their close structural relationship their mode of action is probably similar. Mainly these are esters of benzoic acid and tertiary or secondary amino alcohols. These structural characteristics are clearly favorable for local anesthetic action; however, there is no one specific chemical configuration for a local anesthetic. An additional difficulty in the understanding of their action is the limited knowledge of the molecular mechanisms that control the permeability changes during conduction of nerve impulses. While the ionic basis of the generation of the nerve impulse is well known, the underlying mechanism by which changes in ion permeability are brought about is poorly understood. Thus, at the present time an attempt to explain inhibition of nerve function by the action of specific local anesthetics is difficult. However, certain hypotheses have been proposed from studies relating chemical structure of anesthetics to the modification of conduction in the nerve membrane.

Before starting a discussion of the mechanism of local anesthetic action, it may be helpful to say something about the electrophysiology and the nature of the nerve impulse. The interested reader is referred to the large number of review articles that have appeared during the last few years dealing with the various aspects of local anesthetic action.[4-7]

II. THE NATURE OF THE NERVE IMPULSE

One of the characteristics of living cells is their ability to maintain an uneven ion distribution between the internal and the external environment. This is accomplished by a Na–K pump. Sodium ions that do enter the fiber are quickly expelled by a sodium pump which is coupled with an uptake of potassium ions. Owing to this ionic imbalance, a potential difference exists across the membrane. Like other cells, nerve fibers maintain a potential difference across the membrane with a negative charge on the inside. This membrane potential is referred to as the resting potential to distinguish it from the action potential that occurs during stimulation. The size of the resting potential depends mainly on the relative permeability of the membrane to potassium and sodium ions. The greater the ratio of permeability of potassium ions to that of sodium ions, the greater the resting potential. Of the two factors, the higher diffusion rate for potassium and the sodium–potassium pump, it is the potassium ion concentration gradient that primarily determines the resting potential. If the internal potassium ion concentration of the fiber is increased, a greater resting potential (hyperpolarization) will result. The opposite is true when the potassium ion concentration of the external medium is increased; the resting potential will be decreased (depolarization).

While membrane potentials are common to all cells, only a few types of cells have the property of being excitable, namely nerve and muscle. On stimulation the permeability characteristics of the membrane are changed and the mobility of the sodium ions is increased; thus sodium ions move inward more rapidly. Because of this sudden influx of sodium ions, the inside becomes positive as compared to the outside and the potential is reversed from negative to positive. It is this sudden change in permeability to sodium ions that makes the action potential possible. The rising phase of the action potential is caused by a brief and highly specific increase in the membrane permeability to sodium ions which tends to drive the membrane charge toward the sodium equilibrium potential. The return to the

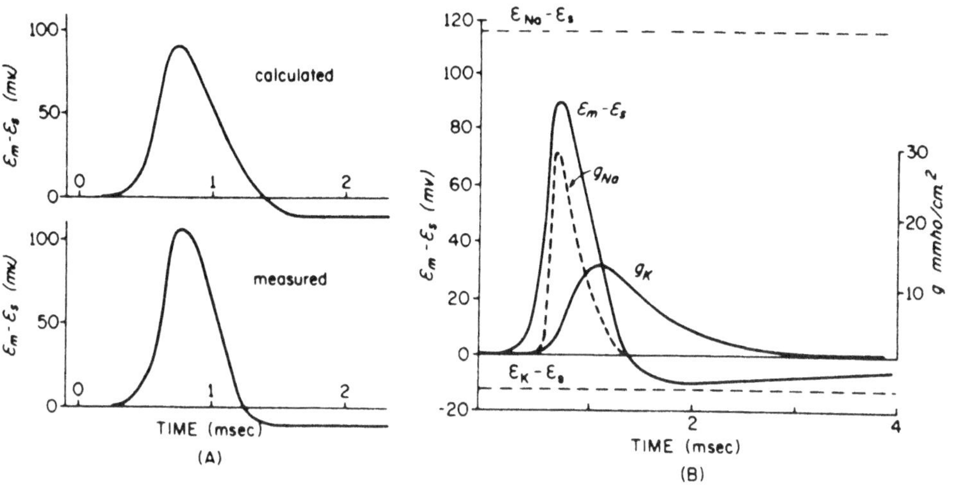

Fig. 1. (A) Top. Action potential calculated from measurements on voltage clamped axon. Calculated conduction speed, 18.8 m/second. Bottom. Propagated action potential in an axon. Measured conduction speed 21.2 m/second. Calculated and measured action potentials differ somewhat, but both show same general features. (From Hodgkin and Huxley.[7a])

(B) Calculated time courses of membrane voltage ($\epsilon_m - \epsilon_s$), sodium conductance (g_{Na}) and potassium conductance (g_K) in squid giant axon. Note time relationships between upstroke of action potential and g_{Na} and between g_K and downstroke and after hyperpolarization. (From Hodgkin.[7b])

resting state or the repolarization is brought about by a decrease in sodium ion permeability coupled with an increase in potassium permeability that exceeds that of the sodium ion (Fig. 1).[7a,b] In order to initiate the action potential there must be a strong enough stimulus to permit the initial change in sodium permeability to occur. Once an action potential has been initiated, the sodium permeability is increased in the excited region and the inside of the membrane becomes positive. An adjacent region not yet reached by the stimulus has a negative charge inside, since the sodium permeability is still low. However, there is a potential difference between these regions, and current will flow from the active to the inactive region through the intracellular fluid and back on the outside in the interstitial fluid. This local circuit reduces the membrane charge and potential in the inactive region. When a critical threshold is reached, the sodium permeability increases rapidly and the action potential is initiated (Fig. 2).[7c] While these phenomena have been widely accepted, the molecular mechanism underlying this permeability change is still under discussion. The majority of investigators accept the view that the initial depolarization causes the described changes in permeability because charged particles move to new

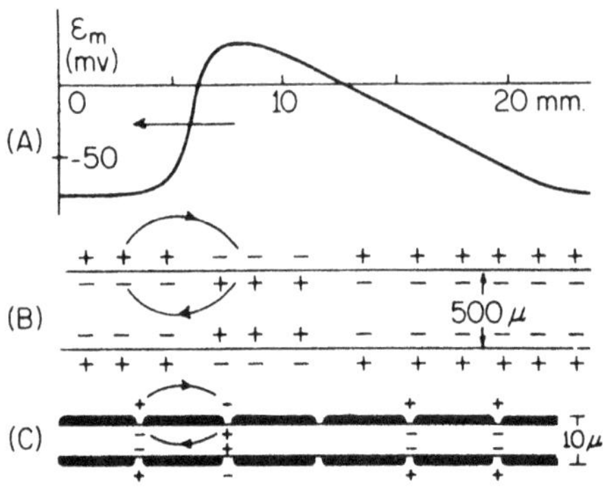

Fig. 2. Propagation by local circuit stimulation. (A) Spatial variation of an action potential at a fixed time. Ordinate, transmembrane potential, ϵ_m; abscissa, distance along fibers shown in (B) and (C). Action potential is propagated at a speed of 20 m/second to the left, as shown by the arrow crossing its upstroke. Note that the upstroke is much steeper than the down stroke. (B) Unmyelinated nerve fiber. Plus and minus signs represent approximately the transmembrane voltage given accurately in (A). Distance scale of (A) applies. Diameter of cell is grossly exaggerated with respect to length as shown. Arrows represent current flow in a local circuit or loop due to the differences in transmembrane potential caused by the different properties of the membrane: highly permeable to K^+ on left and right, even more highly permeable to Na in central regions. Local circuit flow acts to reduce charge in inactive regions but has much less effect on charge in active regions because of high Na permeability. Propagation is achieved by depolarizing action of local current flow. (C) Same as (B) except myelinated. For clarity, distance between nodes is shown as 4 mm or twice actual distance. Because of low capacity of sheath, charge is shown only at nodes (amount of charge in whole internodal region is about one half that at the node). Local circuit flow is thus largely from node to node as shown by arrows. (From Woodbury.[7c])

positions on the membrane under the influence of the electric field. When the membrane repolarizes, charged particles return to their previous positions and the only alteration that has taken place in the system is that some sodium and potassium ions have moved down their concentration gradients.[8] Another view involves a specific chemical reaction which triggers the change in permeability. The main features of this theory have been developed by Nachmansohn.[9] He has suggested that in the excitable membranes ACh controls permeability to ions associated with the action potential. During rest ACh is bound to a storage protein and is released by a stimulus reaching the membrane. The released ester reacts then with the ACh receptor, presumably a protein. In the reaction with the receptor a conformational change of the protein is produced which leads to a shift of charges. It is this reaction that changes the permeability of the membrane. The ACh and the receptor complex are in a dynamic equilibrium with the free ester and receptor. The free ester is rapidly hydrolyzed by acetylcholinesterase (AChE) and the receptor freed from ACh returns to its original condition. The membrane then returns to its resting state. The main factors supporting this view are as follows: ACh and the enzymes which hydrolyze and synthesize the ester, i.e., AChE and choline acetylase (CHA) are present in many nerve fibers. AChE which is localized in the excitable membrane or close to it hydrolyzes ACh in a few microseconds. An ACh receptor with which ACh reacts is present in some nerves. ACh causes spontaneous activity, depolarization of the membrane, and block of conduction. Furthermore, some nerve fibers release ACh continuously during the resting state. On stimulation, a significant increase in the release of ACh has been demonstrated.[10] The possible role of the ACh system in conduction is of particular interest since the hypothesis has been proposed that certain local anesthetics act by competing with ACh at the receptor site and thus cause the block of conduction.[10,11]

III. THE EFFECT OF LOCAL ANESTHETICS ON CONDUCTION

Local anesthetics are defined as compounds which block the nerve impulse without causing a change in the resting potential or any irreversible toxic effect. A small degree of hyperpolarization has been reported in a variety of nerve preparations. However, this only occurs when minimum blocking concentrations are used. Most local anesthetics will depolarize the membrane in concentrations higher than are needed for blocking action. An additional action seen is a slight reduction of potassium permeability. However, the main effect of local anesthetics is to prevent the Na permeability. This has been shown by applying procaine to the voltage clamped giant axon of the squid (Fig. 3) and with other local anesthetics on a variety of nerve preparations. Without affecting the resting potential, the increase to sodium ions is depressed while that to potassium ions is only slightly changed.[13,10] This effect on Na permeability can be counteracted by increasing the Na ion concentrations outside of the nerve. Conversely, a local anesthetic action is increased when the outside Na ion concentration is lowered.[14] This effect of local anesthetics, the reduced permeability of the membrane, also called stabilization, is similar to that caused by Ca ions. Since there is some similarity in the action of Ca and local anesthetics, insofar as both "stabilize" this excitable membrane, it has been proposed

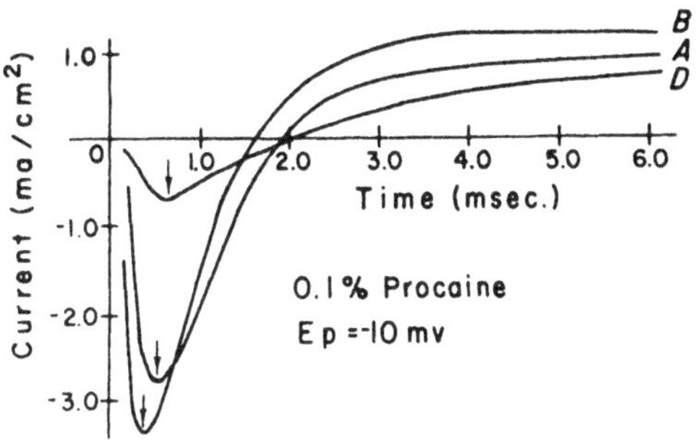

Fig. 3. Current through the membrane of the squid giant axon following a depolarizing voltage step (outward current upward). D. During application of 0.1% procaine. B, A. Before and after. (From Taylor.[13])

that both act on the same system. Both Ca and local anesthetics increase the threshold required for the initiation of an action potential, abolish spontaneous activity and block conduction without depolarization. It has also been shown that under certain experimental conditions Ca and

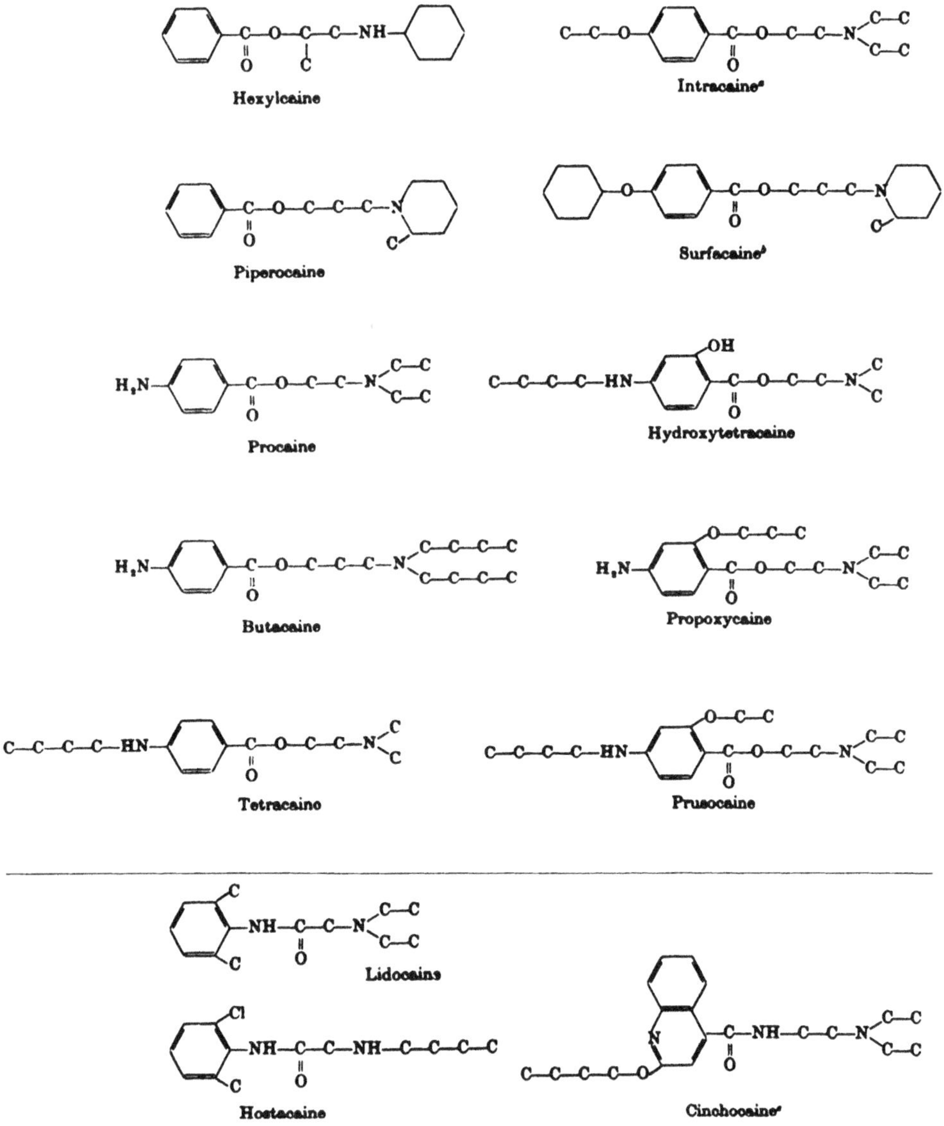

[a] Parathoxycaine. [b] Cyclomethycaine. [c] Dibucaine.

Fig. 4. Local anesthetics used clinically. (From Ariens.[4])

procaine act competitively on the nerve activity. An increase in extracellular Ca reduces the effect of procaine whereas a decrease in Ca increases the action of the local anesthetic. Complete removal of extracellular Ca causes a depolarization which can be completely reversed by procaine.[15–17]

IV. MECHANISM OF ACTION

A. Structure–Activity Relationship

There is no specific chemical configuration which confers local anesthetic activity on a compound. This discussion will be limited to the mechanism of action of the more generally used local anesthetics, a prototype of which is procaine (Fig. 4). Most local anesthetics in clinical use are related in structure and pharmacological action to procaine. The basic structure consists of a lipophilic group linked through an intermediate chain with a hydrophilic group. The distance between these two groups varies from five to two interatomic distances. The lipophilic group consists of a carbonyl group usually linked to an aromatic ring. Both the aromatic ring and the oxygen atom of the carbonyl group will attract electrons and increase the positive charge on the carbonyl carbon atom. In the case of the benzoic acid esters, the proximity of the carbonyl double bond with double bonds in the aromatic ring can result in an electron shift increasing the negative charge of the carbonyl oxygen atom (Fig. 5). Any increase in the negativity of the carbonyl oxygen by introducing NH_2, NRH, or OH groups into the aromatic ring in either *ortho* or *para* position may increase the local anesthetic action of the compound. Substitutions in the ring which attract electrons will reduce the local anesthetic action. This is seen when NO_2 is introduced into the ring in a *para* position to the carbonyl group (Fig. 6).[17a] The role of the aromatic component of the local anesthetic may thus be characterized as follows: it provides lipid solubility, it facilitates

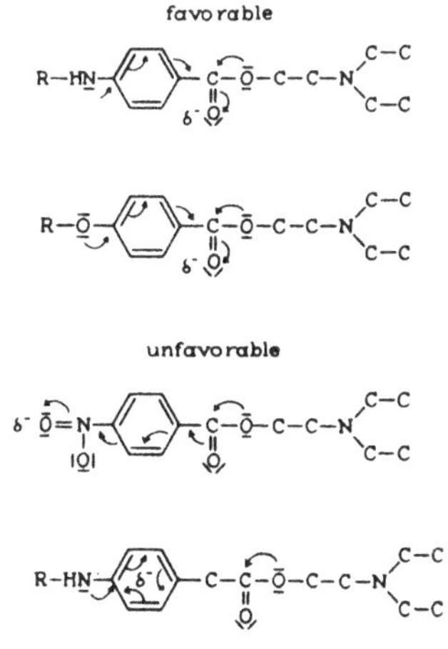

Fig. 5. Electromeric effects on local anesthetic action. (From Ariens.[4])

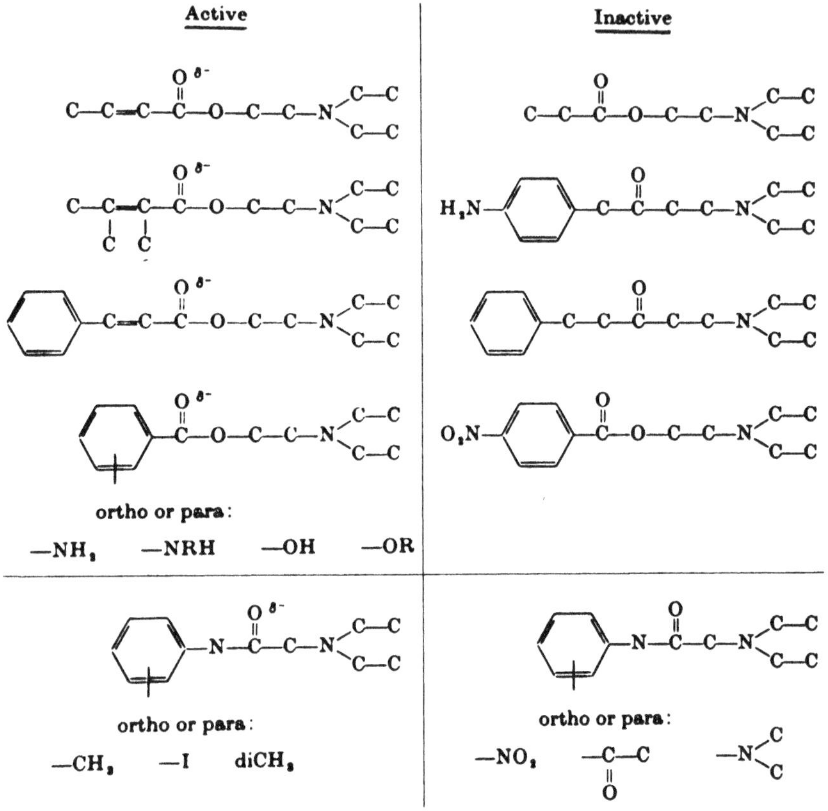

Fig. 6. Influence of various acyl and anilide groups on local anesthetic activity. (From Büchi and Perlia.[17a])

the receptor binding by van der Waals' forces, and it is part of a strong anionic reaction site. The basic amino groups probably contribute also to the action of local anesthetics; however, not as much in determining the anesthetic action as they contribute to the binding properties as a cationic hydrophilic factor in facilitating the transport mechanisms to the receptor sites of the membrane. A variety of other groups may be substituted for the amino group. However, quaternization of the nitrogen leads to either a loss or a reduction in the local anesthetic effect. Quaternization reduces the lipid solubility of the compounds which may explain the reduction in action. The main factor influencing the anesthetic action seems to be a favorable relation between lipophilic and hydrophilic properties of the two groups in the molecule. An increase in the hydrophilic nature can cause loss of anesthetic action. Dominance in the lipophilic character reduces its water solubility. Local anesthetic action of antihistaminics can be explained by their structural characteristics. Although they lack a carbonyl group, they do have a general charge distribution pattern comparable to that of local anesthetics of the ester type, i.e., an ionizable amino group and a reinforceable dipole linked by an intermediate chain (Fig. 7). It is not surprising that amines known to have local anesthetic action have a similar structure. For a more detailed discussion see Refs. 4, 18.

B. The Active Form of Local Anesthetics

The majority of the more commonly used local anesthetics are tertiary or secondary amines At the *p*H of biological solutions they will be present in either the dissociated (ionized) or undissociated (unionized) form, depending on the pK_a values. A controversy over whether the charged or uncharged form is the active form of the local anesthetic has continued for a number of years.[6] The ionized form is important for the interaction with charged groups of a possible receptor in the membrane. However, in order to reach this receptor site within or close to the membrane, the drug has to penetrate through an interposing barrier of connective tissue and Schwann cells. The demonstration of penetration through these barriers and the membrane itself suggests that the uncharged moiety is the active form of the molecule. However, most recent evidence seems to indicate that local anesthetics penetrate the tissue as uncharged molecules and then interact with the "receptor" in their cationic form.[6,19–21] In most cases the nerve membrane will allow only the nonionic species to penetrate.

C. Interaction with the ACh System

The structural similarity between the more commonly used local anesthetics and the cholinergic drugs has lent support to the hypothesis that local anesthetics compete with ACh at the receptor site. Nachmansohn has proposed that local anesthetics may act as competitive antagonists of ACh. In contrast to ACh, these compounds do not depolarize the membrane in concentrations large enough to block conduction. Quaternary compounds like ACh depolarize the membrane; however, by modifying the ACh structure it is possible to analyze the molecular changes related to increased local anesthetic potency.[22] This relationship between ACh and local anesthetics is summarized in Fig. 8. A substitution of a methyl group on the nitrogen makes this compound more lipid soluble. Lipid solubility

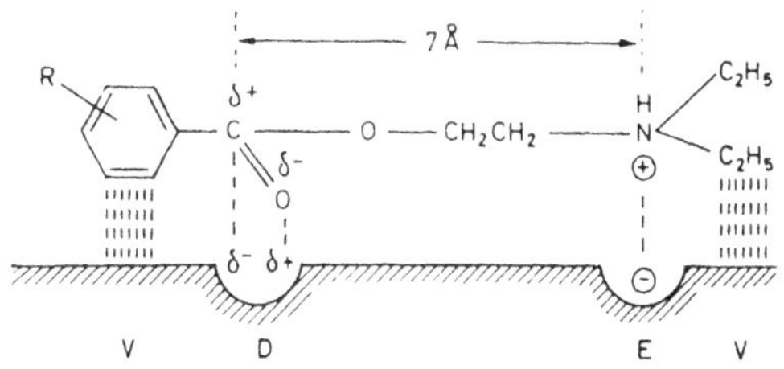

Fig. 7. Model of binding of local anesthetic molecule to the receptor by V, van der Waals forces; D, dipole-dipole interaction; E, electrostatic forces. (From Büchi and Perlia.[17a])

Compound	Action on synapses: Activator		Action on synapses: Inhibitor		Action on conducting membrane	
	Mconc.		Mconc.		Mconc.	
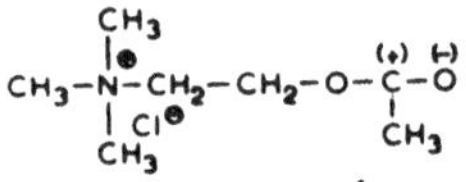 Acetylcholine	2.5×10^{-6}	+		0		0
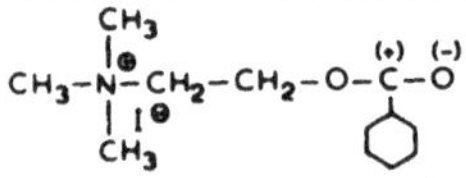 Hexahydrobenzoylcholine	5×10^{-4}	+		0		0
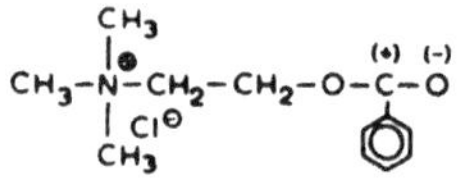 Benzoylcholine	5×10^{-4}	+	5×10^{-4}	+	1×10^{-3}	+
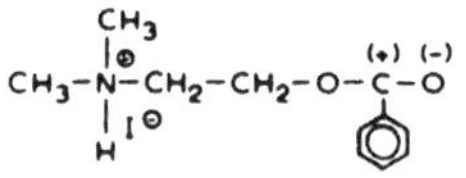 2-dimethylamino-ethylbenzoate		0	1×10^{-3}	+	(5×10^{-3})	(+)
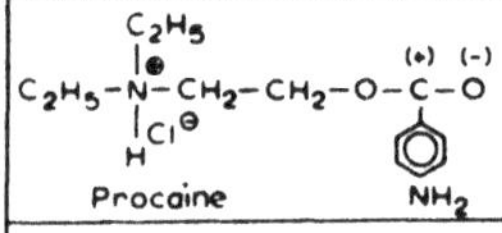 Procaine		0			5×10^{-4}	+
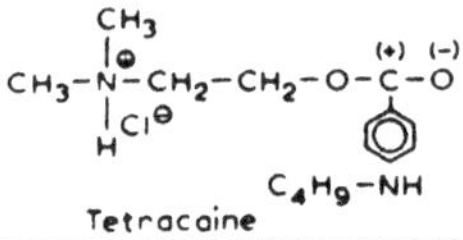 Tetracaine		0	5×10^{-5}	+	3.3×10^{-5}	+
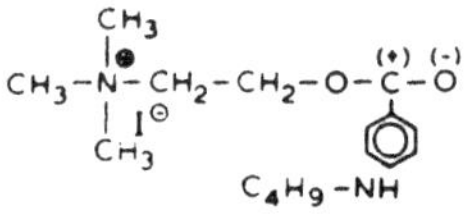 Tetracaine-methiodide		0	2×10^{-5}	+	1×10^{-5}	+

Fig. 8. Effect of ACh, local anesthetics, and intermediary forms on the excitable membrane of the monocellular electroplax. The concentrations listed produce comparable effects. (From Bartels.[22])

is further enhanced by the substitution of an aniline ring for the methyl group on the carbonyl carbon. The change from a quaternary to a tertiary nitrogen may contribute to a reduction in the depolarizing action of ACh. The suggestion that these compounds (the local anesthetics) act as competitive antagonists to ACh is an interesting one. However, this is difficult to demonstrate on peripheral nerve fibers since the action of ACh itself is poorly understood. Most of the nerves studied are insensitive to externally applied ACh due to its low lipid solubility which prevents it from reaching the active site of the receptor. However, even in the cases where ACh can be shown to act upon the peripheral nerve, no true competition can be established. The concentration of ACh needed to change the membrane potential is already so high that reliable studies with local anesthetics are impossible. Even in preparations such as the neuromuscular junction where it is possible to use low concentrations of ACh, no typical competitive antagonism between local anesthetics and ACh has been established.[23] The most that can be concluded from such experiments is that ACh and local anesthetics are possibly bound to separate sites on the same "receptor" molecule so that binding to one site influences the properties of the other.

D. The Membrane

All of the information available at this time indicates that the primary action of local anesthetics is limited to the membrane of the nerve fiber. The local anesthetic combines only loosely with the receptors and is easily removed by washing Although considerable information is available on the electrical characteristics of the excitable nerve membrane, very little is known about its molecular composition and structure. The nerve, whether myelinated or unmyelinated, is surrounded by Schwann cells which are in close contact with the plasma membrane of the axon. Around each nerve is an axolemma, an intercellular space, and the Schwann cell. However, the only real permeability barrier is the plasma membrane which surrounds the axon, a membrane which has been studied extensively as the component responsible for the electrical characteristics of nerve fibers. The most familiar model of the nerve membrane is that of Danielli and Davson[24] with modifications by Robertson.[25,26] It consists of a bilayer of lipids with opposed hydrocarbon chains to form a continuous nonpolar phase. The outer as well as the inner surface of this lipid layer is covered by a monolayer of proteins. Depending upon the nature of a given cell with specific functions, this unit membrane has variations in the molecular composition of its lipids or proteins. Variations in the chemical structure of the phospholipids with their associated physical-chemical characteristics may account at least in part for the numerous variations in the structure and properties of cell membranes. It may be assumed that each membrane consists of a certain mixture of lipids in a specific uniform molar ratio. Presumably each type of lipoprotein makes its characteristic contribution to the specific properties of the membrane. The polar side chains of the phospholipids differ in size and charge and may thus be responsible for not only the membrane specificity but also the nature of the interaction with drugs. The membrane also contains cholesterol which forms complexes with the phosphoglycerides and leads to a certain rigidity of the membrane. Not much is known about the membrane proteins, as it is difficult to obtain lipid-free protein fractions.

E. Mode of Local Anesthetic Action

There are several theories as to the nature of the mechanism of the action of local anesthetics which will not be discussed here. Experiments have demonstrated that the blocking effects of local anesthetics are influenced by factors such as fiber size and the presence or absence of Ca or Na ions.[15,16] This indicates that the main action probably takes place in the membrane. Furthermore, the effect of local anesthetics on the Na activation mechanisms during the generation of action potentials limits the action to the membrane, and in order to find a reasonable explanation for the local anesthetic action, more needs to be known about the membrane. Structurally unrelated nonspecific agents such as alcohols seem to act by accumulating inside the cell and may thus disorganize metabolic processes. However, they owe this action to some favorable chemical constitution, namely a favorable balance between hydrophobic and hydrophilic groups in the molecule. There may be an accumulation in the axolemma causing swelling which could be responsible for a spatial separation of structures or enzymes needed for the complex reactions underlying the process of excitation. It has been shown that small cations can penetrate the membrane. Whether there are pores, permanent or functional ones, remains to be determined. It is possible that the lipophilic part of the local anesthetic is bound to the membrane surface or parts of the molecule are interpositioned with the hydrophobic part of the lipid structure. By attachment of the anesthetics to the surface, the area of permeability (pores) for ions could be covered up or the local anesthetic might even enter the pores and thus reduce the size of the passage for charged ions. Another possibility is that the lipid phase of the membrane may expand. Lipid-soluble compounds may accumulate in lipid structures and thus increase their volume. These structures may be surrounding the pores or form large areas between the channels. The increased volume due to arrival and solvation of local anesthetics in these areas would increase lateral pressure which then could interfere with change in diameter of the pore.[16] The local anesthetic would be more likely to interfere with a possible enlargement of the pores during processes leading to increased permeability. This sort of nonspecific mechanism certainly would help to explain the anesthetic effect of uncharged molecules such as inert gases or alcohols.

There is no unifying concept for the mechanism of local anesthetic action, the reason for this being that the basic process of permeability changes is still unsolved. However, some speculation may be appropriate. It appears now that the initial step in the excitation cycle is the removal of calcium from receptor sites in the membrane.[8] The flow of ions, mainly Na and K, is controlled by the membrane potential and a protein-lined double layer of lipids and phospholipids. The phosphate groups of certain phospholipids may act as ion exchange sites whose affinity for cations depends on the configuration of the dipolar complexes which in turn depend on the electric field strengths and the presence of appropriate cations. The ion transfer which occurs mainly through these sites is controlled by electrical and ionic gradients. Of all the physiologically known cations, Ca is most intimately involved with the excitation process in nerve fibers.[8] Calcium may act as the trigger of electrical events. However, little is known about the movement and localization of Ca within the membrane. Upon stimulation it may be released from one locus and react with another site in the membrane where it triggers the events causing the rapid permeability changes.

A favorable candidate for these Ca^{2+}-binding sites are the polar heads of the phospholipids. Charged compounds such as local anesthetics would tend to utilize the negative charges of the phospholipids with the rest of the molecule in the lipids (Fig. 9). This is supported by experimental data showing that the local anesthetics and Ca have similar actions on the resting potential of the fiber.[15,16] If the polar heads of the phospholipids are the sites which control the permeability changes of the membrane, any binding of local anesthetics to these sites would tend to decrease the facility to change the configuration during stimulation.[9] Ions which can form strong bonds with polar groups such as phospholipids have in general a local anesthetic action. The inhibition of electrically induced changes in the phospholipid part of the membrane or an increase of the pressure in the lipid layer by lipid-soluble material would thus interfere with the excitability. Local anesthetics compete with Ca for binding sites of phosphatidyl serine and inhibit the Ca uptake in a direct relation to their local anesthetic potency (Table I). Conversely, it can be shown that the binding of ^{14}C-labeled local anesthetics to phosphatidyl serine is antagonized by Ca ions which thus will interfere with the binding of local anesthetics to the phospholipid. Local anesthetics which in their active forms are cations may interact within ionic sites of the phospholipids or proteins and compete with Ca ions for binding sites. In fact, it seems that the procaine calcium ratio is a major factor determining the effectiveness of a local anesthetic. This is evident by the fact that vertebrate nerves which have a low outside concentration of Ca require less procaine in order to block electrical activity than lobster or squid nerves, which have a fivefold higher Ca ion concentration. That Ca ions and local anesthetics compete for the same sites in the membrane, sites which are directly involved with a

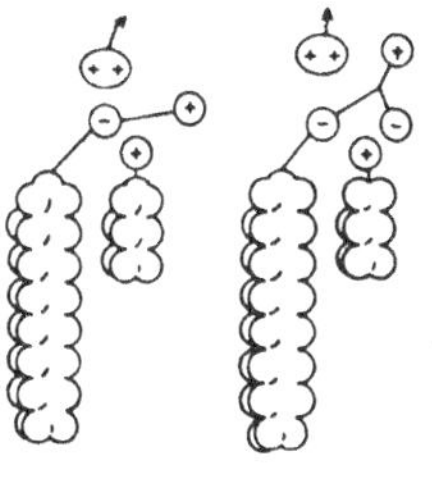

Fig. 9. Molecular models of interaction between divalent cations and cationic drugs. *Left*, Phosphatidyl ethanolamine. *Right*, Phosphatidyl serine. The nonpolar portion of the local anesthetic lies between the fatty acid chains of the lipid molecules, while the polar group is in close proximity to the polar head of the phospholipid. Cationic drugs like procaine could neutralize the negatively charged sites on the phospholipid so that fewer sites would be available for the binding of divalent cations. (From Blaustein and Goldman.[29])

TABLE I

Effect of Cationic Drugs on Calcium Binding by Phosphatidyl-L-serine

Drug	Relative inhibition of Ca binding[a]	Relative anesthetic potency
Procaine	1.0	1.0
Mepivacaine	1.6	2.1
Lidocaine	3.5	3.8
Hexylcaine	4.4	2.7
Piperocaine	4.7	3.5
Tetracaine	8.7	36.5
Dibucaine	10.1	53.8

[a] The relative inhibition of calcium binding was determined from the reciprocal of the concentration of drug necessary to inhibit calcium binding to phosphatidyl-L-serine by 20%.
The relative anesthetic potency was determined on frog sciatic nerve trunk at *p*H 7.18 to 7.20. (From Blaustein and Goldman.[29])

control mechanism for Na and K permeabilities, appears to be a reasonable hypothesis.[27–30]

F. Models

At this moment explanations based on studies relating local anesthetic structure to actions on the membrane are limited mainly by our scant knowledge of the origin of the impulse generation and the membrane. Model systems are frequently useful in such studies. Some correlations have been found between local anesthetic action on nerve conduction and their effects on models. The design of models has been influenced by the assumption that the bimolecular leaflet is the most probable form assumed by phospholipids in solution. This observation has triggered the introduction of a number of systems and investigations attempting to correlate the findings on models with the findings on conducting fibers. For a review on model systems see Refs. 6, 16, 17, and 31.

It is relatively easy to select a model and to predict the effect of ions, enzymes, and drugs on such a model. However, any interpretation of such studies should be approached with caution. No model yet described is able to imitate all the properties of a living cell membrane. However, a sectional approach to examining one of the properties of the cell membrane is facilitated by studying model systems. Most of the models used contain a single lipid or a mixture of lipids which are arranged as monolayers or bilayers. Without exception, all model systems are inadequate replicas of living cells and nerve membranes. Membranes are living entities containing enzymes and continuous regenerating processes. Model systems in order to exhibit some aspects of the living cell must take into consideration the physical chemical characteristics of the axolemma. The high resistance across the nerve membrane is primarily due to the lipids. This resistance makes possible a transmembrane potential modified by the permeability properties and the metabolic activities of the cell.

Another promising approach is the study of molecular complexes of local anesthetics with bis(*p*-nitrophenyl phosphate). Many proteins containing phosphodiesters will react with local anesthetics, so it may be of help to isolate and characterize the complexes formed by local anesthetics with tissues in order to gather more information about the nature of the receptor for local anesthetics.[31]

One avenue of approach which has not been explored too widely as yet is the effect of anesthetics on the modification of the composition of the cell membrane. Such a modification could impose a change in the normally appearing concentrations of naturally occurring phospholipids. Another approach would be to use abnormal lipids for incorporation into the cell membrane. Such modifications are possible.[31]

V. CONCLUSION

The block of conduction caused by local anesthetics is well understood and explained in terms of nerve membrane conductance. Knowledge as to the precise mechanism of local anesthetic action is limited by our still incomplete understanding of the molecular basis of the action potential. Experimental data support the assumption that the reactive sites for local anesthetics in the nerve membrane are either anionic, cationic, or possibly lipophilic groups of proteins and lipids The physicochemical properties of the local anesthetics in use clinically are almost exclusively cationic by nature. In addition, support derives from a favorable combination of lipophilic and hydrophilic characteristics. All these factors determine the transport, anionic binding, and local anesthetic action. Since there is no uniform theory as to the mechanism of action, a number of theories have been proposed. For most of them the supporting evidence is still inadequate. This is true for the competition of local anesthetics for the ACh receptor or any metabolic interaction of local anesthetics with organic phosphates as well as other metabolic processes. Most of the metabolic actions seen are with concentrations higher than needed to block the action potential. The inhibition of the conductance change for sodium and potassium caused by local anesthetics is still under discussion. The favored view is the interaction of local anesthetics with anionic charged sites of the phospholipids and/or proteins, thus preventing a movement of calcium, sodium, or potassium ions. This is supported by the observed conductance changes. Additional support may come from further study of the interaction with lipophilic groups of the membrane lipids and proteins.

VI. REFERENCES

1. H. Haas, *Histamin und Antihistamin*, Vols. 1 and 2, Cantor, Aulendorf, Württemberg, Germany (1951–1952).
2. L. S. Goodman and A. Gilman, *The Pharmacological Basis of Therapeutics*, Macmillan, New York (1955).
3. N. Ludwigs and K. Wiemers, *Arch. Exptl. Pathol. Pharmacol.* **213**:355 (1951).
4. E. J. Ariens, ed., *Molecular Pharmacology*, Academic Press, New York (1964).
5. P. G. Watson, The mode of action of local anesthetics, *J. Pharm. Pharmacol.* **12**:257 (1960).
6. J. M. Ritchie and P. Greengard, On the mode of action of local anesthetics, *Ann. Rev. Pharmacol.* **6**:405 (1966).
7. H. F. Zipf, Die Wirkungsmechanismen der Localanasthetica, *Pharmaceutica Acta Helvetica* **42**:480 (1967).

7a. A. L. Hodgkin and H. E. Huxley, *J. Physiol.* **117**:500 (1952).

7b. A. L. Hodgkin, *Proc. Roy. Soc. Ser. B* **148**:1 (1958).

7c. J. W. Woodbury, Physiology and Biophysics, Ruch and Patton, eds., Saunders, Philadelphia (1965).

8. A. L. Hodgkin, *The Nature of the Nerve Impulse*, p. 108, Univ. Press, Liverpool (1966).
9. D. Nachmansohn, *Chemical and Molecular Basis of Nerve Activity*, p. 235, Academic Press, New York (1959).
10. W-D. Dettbarn, The ACh system in peripheral nerve, *Am. N.Y. Acad. Sci.* **144**, Art. 2, p. 483 (1967).
11. E. Bartels and D. Nachmansohn, Molecular structure determining the action of local anesthetics on the ACh receptor, *Biochem. Z.* **342**:359 (1965).
12. A. M. Shanes, W. H. Freygang, H. Grundfest, and E. Amatniek, Anesthetic and calcium action in the voltage-clamped squid giant axon, *J. Gen. Physiol.* **42**:793 (1959).
13. R. E. Taylor, Effect of procaine on electrical properties of squid axon membrane, *Am. J. Physiol.* **196**:1071 (1958).
14. G. A. Condouris, A study on the mechanism of action of cocaine on amphibian peripheral nerve, *J. Pharmacol. Exptl. Therap.* **131**:243 (1963).
15. F. A. Davis and W-D. Dettbarn, Depolarizing action of calcium ion depletion on frog nerve and its inhibition by compounds acting on the ACh system, *Biochim. Biophys. Acta* **63**:349 (1962).
16. A. M. Shanes, Electrochemical aspects of physiological and pharmacological action in excitable cells, I, *Pharmacol. Rev.* **10**:59 (1958).
17. A. M. Shanes, Electrochemical aspects of physiological and pharmacological action in excitable cells, II, *Pharmacol. Rev.* **10**:104 (1958).

17a. J. Büchi and X. Perlia, *Arzneimittel-Forsch.* **10**:1 (1960).

18. N. Löfgren, *Studies on Local Anesthetics, Xilocaine, A New Synthetic Drug*, p. 152, Ivor Haeggström, Stockholm (1948).

19. J. M. Ritchie and P. Greengard, On the active structure of local anesthetics, *J. Pharmacol. Exptl. Therap.* **133**:241 (1961).
20. W-D. Dettbarn, Active form of local anesthetics, *Biochim. Biophys. Acta* **57**:73 (1962).
21. T. Narahashi, J. W. Moore, and R. N. Roshin, Anesthetic blocking of nerve membrane conductances by internal and external applications, *J. Neurobiology.* (In press.)
22. E. Bartels, Relationship between ACh and local anesthetics, *Biochim. Biophys. Acta* **109**:194 (1965).
23. A. Steinbach, A kinetic model for the action of xilocaine on receptors for ACh, *J. Gen. Physiol.* **52**:162 (1968).
24. J. F. Danielli and H. Davson, A contribution to the theory of permeability of thin films, *J. Cell Comp. Physiol.* **5**:495 (1935).
25. J. D. Robertson, The ultrastructure of cell membranes and their derivatives, *Biochem. Soc. Symp.* **16**:3 (1959).
26. A. L. Lehninger, The neuronal membrane, *Proc. Nat Acad. Sci.* **60**:4 (1968).
27. M. B. Feinstein, Reaction of local anesthetics with phospholipids, *J. Gen. Physiol.* **48**:357 (1964).
28. M. B. Feinstein and M. Paimre, Specific reaction of local anesthetics with phosphodiester groups, *Biochim. Biophys. Acta* **115**:33 (1965).
29. M. P. Blaustein and D. Goldman, Action of anionic and cationic nerve-blocking agents: Experiment and interpretations, *Science* **153**:429 (1966).
30. M. P. Blaustein, Phospholipids as ion exchangers: Implications for a possible role in biological membrane excitability and anesthesia, *Biochim. Biophys. Acta* **135**:653 (1967).
31. A. W. Cuthbert, Membrane lipids and drug action, *Pharmacol. Rev.* **19**:59 (1967).

Chapter 18

HALLUCINOGENS

Stephen I. Szara

Section on Psychopharmacology
Division of Special Mental Health Research
National Institute of Mental Health
St. Elizabeths Hospital, Washington, D.C.

I. INTRODUCTION: SCOPE AND EXTENT

The discovery that a chemical such as lysergic acid diethylamide (LSD-25 or LSD) in extremely small quantities can profoundly alter the normal functions of the CNS[1] has focused interest on the biochemical basis of behavior and stimulated a tremendous amount of research in chemistry, pharmacology, neurophysiology, psychiatry, and psychology. A quarter of a century after its discovery and thousands of scientific publications later, we still do not have the answer to the basic questions: how can a few micrograms of a chemical produce the spectacular psychological changes observed after taking LSD? Yet, while we were searching for the answer, we became aware of the multiplicity of the levels of relationships between the biochemical mechanisms and the particular substrata for perception and behavior. On all of these levels the causal relationships have to be clarified before we can hope to tie them together for a final understanding of the mechanism of action of these drugs.

It was suspected quite early[2,3] that the basic link in the mechanism of action of these drugs is at the biochemical level, probably owing to interference with neurohumoral transmission mechanisms. In a recent review on the basic actions of psychoactive drugs, Carlsson pointed out that "Mescaline, LSD and related hallucinogens probably also act as monoamine analogs, although the mechanisms involved are obscure."[4] It is now realized that the action of these drugs is so complex that we cannot explain even the biochemical effects by a single mechanism, but we have to consider both direct and indirect, as well as multiple sites of action.[5]

The situation is further complicated by the fact, that, even if it is possible to pinpoint certain processes in the metabolism or receptor interaction of the amines in the brain which are interfered with by the hallucinogens, a truly satisfactory understanding of the mechanism of action must await the demonstration of the physiological functions subserved by these amines.[6]

The literature on the biochemical aspects of the actions of psychotomimetic drugs has been exhaustively and critically reviewed up to 1965 by Giarman and Freedman[7]; therefore the present review will cover only the recent developments in this field, referring to earlier data only when it is necessary for understanding certain significant trends in research today.

We would also restrict our review to those hallucinogens which reliably and consistently produce periods of altered perceptions, mood, and experience without clouded consciousness and without marked physiological changes. Further criteria for selection of drugs were similar to those used by Giarman and Freedman in their 1965 review[7] i.e., availability of clinical data in man which appears to be at least minimally reliable, and availability of laboratory data in regard to their metabolism and to their effect on some biochemical parameter in the brain. The piperidyl glycollates, bufotenin and adrenochrome, are excluded mainly on clinical grounds,[7] while Δ^9-tetrahydrocannabinol (THC) although shown recently to be one of the clinically effective ingredients of marihuana,[8] is excluded because no data on its metabolism or biochemical effects in the brain have been published as yet.

This leaves us with two major classes of hallucinogens: certain indole derivatives (for example LSD and DMT) and some phenylethylamine derivatives (for example mescaline and STP).

II. METABOLISM OF HALLUCINOGENS RELEVANT TO MECHANISMS OF ACTION

As with any other drug, the intensity and the duration of action of the hallucinogenic drugs depends on the dose given and on the tissue levels of the drug which in turn is a function of the rate of elimination by enzymatic breakdown, by binding, and by active or passive transport.

The problem of the formation of pharmacologically active metabolites has remained a controversial issue. Although some indirect evidence is suggestive of the possibility that active metabolites may play a role in the psychotropic action of these drugs, direct evidence for this effect is still lacking.

During the past few years an increasing awareness could be detected in the published literature toward the importance of regional economy of the brain, and the interaction of drugs with regional biochemical processes underlying the normal function of the brain. Perhaps a better understanding of the role of these regional processes in behavior may clarify not only the behavioral action of these drugs but also the controversies over the issue of toxic metabolites.

A. Lysergic Acid Diethylamide (LSD) and Related Indole Derivatives

LSD, the most potent hallucinogenic drug known today, was shown by Snyder and Reivich to be unequally distributed in the brain of squirrel monkeys.[9] Highest concentration of the drug was found in the pituitary and pineal glands, reaching seven to eight times the amount found in the cortex. Structures of the limbic system and certain subcortical areas thought to be related to visual and auditory reflex functions contained two to three times higher concentrations than in the cortex. The authors argue that this cannot be explained by regional differences in blood flow or lipid solubility, and suggest that the selective concentration of LSD might be related to the perceptual and emotional effects this drug is able to produce.

An additional support for the notion that some of the psychological responses to LSD are related to "tissue concentration" of the drug comes from high correlations of performance test scores with estimated "tissue concentration" of LSD in human subjects.[10] The biological half-life of LSD for loss of the drug from the plasma compartment was estimated to be 103 min using a two-compartment open model in the mathematical analysis of the data. The psychological performance test shown to be disturbed by LSD in the first hours after administration of the drug to normal volunteers was a simple arithmetic test, probably sensitive to attention processes.

Many of the psychological effects, however, last for 8 to 10 hr, which necessitates the assumption that some additional factors may also play a role in the effect of the drug.

A longer lasting disturbance of the chemical and/or physiological equilibrium may be one possible factor which we shall discuss later. Another possibility is the formation of some active metabolite(s) which might have a longer half-life than that of the unchanged LSD itself. The oxidation of LSD in the 2-position and the hydroxylation probably in the 13-position by microsomal liver enzymes have been suggested some time ago as possible pathways of metabolism, but, since Giarman and Freedman's review in 1965 no new data have been published.

The short-acting hallucinogen *N,N*-dimethyltryptamine (DMT) has also been shown to be distributed unevenly in the brain.[11] The compound, injected intraperitoneally, enters the brain of mice quickly and reaches a maximum level in the neocortex in 10 min, at which time its concentration in the hippocampus is about half as high as in the cortex. Since the concentration of the drug in the cortex declines with time more sharply than in the hippocampus, the hippocampal level of the drug 20 to 40 min after injection is higher than in other areas.

The role of metabolism in the effects of hallucinogenic tryptamine derivatives has also remained controversial in spite of some new data published recently.

In a clinical study[11-13] involving the comparison of three compounds: *N,N*-diethyltryptamine (DET), *N,N*-dipropyltryptamine (DPT), and 6-fluoro *N,N*-diethyltryptamine (6-FDET) it was found that DET and DPT were about equally active as hallucinogens, but the third compound (6-FDET) was found to produce autonomic changes only and was completely void of hallucinogenic action in the same subjects. The authors noted conspicuous differences in the metabolism of these compounds in terms of 6-hydroxylation and dealkylation and suggested that both metabolic pathways might be necessary in producing a psychologically active metabolite.[13,14] Since these metabolic steps may lead to the formation of biologically active β-carbolines (harmine type compounds), this remains an attractive hypothesis to be tested.[15]

The hallucinogenic activity of several β-carboline derivatives has been compared in man, and found to be varied with small substitution changes in the basic ring structure.[16]

B. Mescaline and the Hallucinogenic Phenylethylamine Derivatives

In the case of mescaline, the question of the role of psychoactive metabolites is even more pertinent than it is in case of the indole-type hallucinogens, primarily because of the relatively high dose needed to elicit psychological effects in man, and because the behavioral effects of mescaline do not coincide with the time of its maximum concentration in brain of animals.[7]

In the cat, using ^{14}C-labeled mescaline, Neff *et al.*[17] found a biological half-life of 90–120 min on the basis of disappearance from plasma, and a distribution in the brain which did not show the pattern described with LSD but followed the more general pattern of distribution of ^{82}Br-ion, ^{14}C-labeled thiopental or urea, indicating no special characteristics which could be related to the hallucinogenic action of mescaline. They failed to find any metabolites of mescaline in cat brain except labeled trimethoxyphenylacetic acid.

The metabolism of ^{14}C-mescaline in man was studied by Charalampous *et al.*[18] They found a half-life for the drug about twice as long (6 hr) as that reported in the cat. During the first 2 hr after administration, about 87% of the dose was excreted of which more than half (55–60%) was the unchanged form and a considerable portion (27–30%) was the corresponding 3,4,5-trimethoxyphenylacetic acid. Five percent of the excreted radioactivity was found in the form of *N*-acetyl-β-hydroxyphenyl-ethylamine, indicating that demethylation and *N*-acetylation has also taken place in man. This same metabolite was also found in the rat[19] and suggestive evidence was presented that *N*-acetylation probably precedes *O*-demethylation to form the metabolite. Although *N*-acetylation sometimes leads to active metabolites (examples: acetylcholine, melatonin) in most cases, probably including mescaline, this metabolic pathway represents an inactivation process.[18,19]

A compound called "STP" on the black market is chemically related to both mescaline and amphetamine, having the structure 2,5-dimethoxy-4-methyl-amphetamine (DOM). It has been shown to be about 100 times more potent as a hallucinogen than mescaline.[20] Very little is known about its metabolism. Regardless of the dose given, about 20% of "STP" is excreted unchanged in 24 hr, and the peak of excretion was found to be between 3 and 6 hr. Snyder *et al.*[20] suggest that this might be the result of slow absorption of the drug into the circulation and might be related to the relatively long-lasting psychotropic effects produced by STP.

Amphetamine itself, and methamphetamine are drugs widely abused by certain fringe elements of society, but are not considered as typical hallucinogens; therefore their metabolism and biochemical effects are not discussed in this chapter.

3,4-Dimethoxyphenylethylamine (DIMPEA), another chemical relative of mescaline, has been suspected to be the long sought-after psychotoxic metabolite identical with a "pink spot" found more often in schizophrenic than in normal urine.[21,22]

Closer examination, however, revealed that the "pink spot" was not identical with DIMPEA and that it most likely represented a compound of dietary or bacterial origin.[23,24]

III. ALTERATION OF NEUROCHEMICAL PARAMETERS AND SUGGESTED MECHANISMS OF ACTION

The problem of the mechanism of action of hallucinogenic drugs at the biochemical level has been hampered by the fact that such studies are feasible only in animals and that we are not certain what kind of animal behavior can be used as a reasonable model for the complex psychological effects—perceptual, emotional, and thinking disturbances—produced by these drugs in man.

Most of the hallucinogenic drugs would affect the behavior of animals in a rather nonspecific way in a simple test which uses the spontaneous exploratory behavior, or some simple conditioned avoidance response as tests.[25,26]

More complex conditioning tests showed more selectivity and appeared to have more relevance to the symptoms hallucinogenic drugs produce in humans.[27,28]

On the basis of their data, Key and Bradley[27] have postulated two sites of action: (1) direct action on the brainstem reticular formation; and, (2) a more indirect action on afferent collaterals impinging upon the reticular formation.

Most of the subsequently published behavioral data seems to be consistent with this idea, and it appears to provide an anatomical and physiological anchoring point for biochemical approaches to the problem.

A. Interrelationships with Serotonin

The controversy about the role of interaction with serotonin in the mechanism of action of hallucinogenic drugs has been somewhat clarified since the review of Giarman and Freedman.(7)

The simplistic notion that the psychological effect of LSD and similar drugs can be explained on the basis of the synergistic(29) or antagonizing(2) action they show *in vitro* when 5-HT is used in certain peripheral preparations has been replaced by the view that these drugs do not simply *act* on certain tryptamine receptors, but *interact* with a neural system in which a dynamic process of constant synthesis and elimination of functionally important amines, including serotonin, serves some biological function.(30,31) Most of the experimental data leading to this view have been published recently.

Jouvet has presented an impressive mass of data obtained in cats, strongly suggesting that the medullary and midbrain raphe system and its serotonin-containing neurons may be a major sleep-inducing structure, and that cerebral 5-HT seems to play a determinant role in the active process underlying sleep.(32)

Aghajanian and Sheard challenged this view on the basis of their experiments with awake unrestrained rats using electrodes implanted in the midbrain raphe.(33) No somnolence was found during periods of stimulation through the electrode but a specific behavioral effect was observed characterized by a failure of habituation to repeated and not reinforced sensory stimuli.

This stimulation of the raphe was accompanied by a release of 5-HT in forebrain cortical structures and experiments with *p*-chlorophenylalanine (PCPA) and 5-HTP suggest that these behavioral effects are dependent upon the availablity of serotonin in the brain.(34)

These authors further suggest that, since both LSD(27) and electrical stimulation of the raphe produce a failure of habituation to repetitive sensory stimulation, the underlying biochemical process may be the same; these results seem to support Woolley's more recent hypothesis that LSD mimics the action of serotonin in the brain.(35)

Some other data could be quoted which are consistent with this hypothesis. Freedman and Giarman have shown that LSD induces a small but significant increase of brain serotonin in the rat(36) and Szara has shown that DET similarly increases brain serotonin levels in the rat and rabbit.(37) Rosecrans *et al.* later showed that LSD influences 5-HT metabolism in brain in a complex way.(38) Three phases were distinguished:

1. In the first 30 min after administration of the drug brain 5HT increases and 5-HIAA decreases.

2. From 30 to 60 min 5-HT still increases, while the 5-HIAA level is returning to normal.
3. After 60 min both 5-HT and 5-HIAA levels decrease at a parallel rate.

These results are interpreted as an effect of LSD on control mechanisms that regulate 5-HT turnover in the rat brain.[38]

Two other groups draw similar conclusions from their work on the effect of LSD on 5-HT metabolism. Diaz *et al.*[39] and Anden *et al.*[40] have shown that LSD decreases the turnover rate of 5-HT in the brain. Diaz *et al.* suggested that LSD might exert a bretyliumlike action on central 5-HT neurons.[39] However, Anden has found that LSD has a 5-HT-like, apparently direct action on central neurons.[40] A negative feedback mechanism is invoked by these investigators to account for the retarded 5-HT turnover.

This negative feedback effect is assumed to be biochemical in nature and operating either at the tryptophan 5-hydroxylase step[38,41] or at the decarboxylase step[42] in the 5-HT synthetic pathway.

Although the biochemical negative feedback mechanism elicited by the increased levels of 5-HT explains the decreased turnover rate of 5-HT after administration of LSD, the mechanism for increasing the *level* of 5-HT is still unknown. It might be a direct biochemical effect of LSD on the binding mechanism[38] or it might be indirectly produced by a neural feedback inhibition of the presynaptic central 5-HT neurons.[43]

Of interest in this respect is the finding by Marchbanks[44] that LSD at low concentrations (10^{-6}M) inhibited the binding of 5-HT to nerve ending particles prepared from certain areas of the rat brain. Similar findings were reported in guinea pig brain by another group of investigators.[45] The apparent discrepancy between these[44,45] and other[38–40] findings is usually explained on the basis that the former results were obtained *in vitro* while the others were *in vivo* experiments.

There is, however, another possible explanation, namely that sharp regional differences may be obscured by the whole brain level approach, giving a slightly increased 5-HT level when the differences are averaged out, while the selection of certain areas for the *in vitro* experiment may have favored those areas in which LSD indeed inhibited the binding of 5-HT. The sharp regional differences in the levels of 5-HT after the administration of LSD or other hallucinogenic drugs have been repeatedly shown by Szara and his collaborators in the brains of rabbits[46] and of mice.[47,48]

The importance of these regional differences is further supported by experiments with stimulation-induced release of serotonin from brain.[49–51]

Yet a further possibility to account for the decreased activity of some 5-HT neurons in the raphe was suggested by Koella.[52] He has shown that the area postrema of the brainstem contains receptors sensitive to both LSD and peripherally injected 5-HT, producing EEG changes in cats. These data suggested the possibility that from the area postrema certain neural pathways may lead to the serotonin-containing medullary and mesencephalic raphe structures, thus accounting for the lowered activity of these 5-HT neurons.[52]

It was indeed found that LSD in very low concentrations reversibly blocks the spontaneous activity of raphe neurons[53] and that the primary site for the electroencephalographic action of hallucinogenic drugs is in the medullary rather than midbrain level.[54,55]

Since interconnecting neural pathways may contain some other potential neurotransmitters, the possibility that hallucinogenic drugs may act on the serotonin-containing neurons indirectly via catecholamine-containing neurons or other pathways, cannot be excluded.

B. Interrelationships with Catecholamines

Freedman's early observation[56] that LSD produces not only an increase in brain 5-HT levels but also a reduction in brain norepinephrine levels has been confirmed and extended by Diaz *et al.*[39] These authors have found that LSD decreases both norepinephrine and dopamine levels in the brain which they interpret as probably a result of persistent activation of catecholaminergic neurons. Werdinius[43] in discussing the results of Diaz *et al.* presented data confirming the effect of LSD on norepinephrine depletion rate, but in his experiments the dopamine metabolism was unaffected.

Bloom and his collaborators working with LSD applied by microelectrophoresis to single neurons in the olfactory bulb have found that LSD occasionally mimicked the action of 5-HT but it never blocked responses to 5-HT without also blocking the response to norepinephrine.[57]

Dixon has used LSD-induced aberrant behavior in the rat to detect the role of biogenic amines in the mediation of this behavior.[58] Various drugs that influence catechol or 5-HT metabolism were given prior to LSD and the resulting change in the aberrant behavior pattern was observed. Pretreatment of the animals with *p*-chlorophenylalanine in doses known to reduce serotonin concentration in the brain, without affecting catecholamine levels, failed to exert any influence on the LSD-induced aberrant behavior. On the other hand, pretreatment of the animals with α-methylparatyrosine, a selective inhibitor of catecholamine biosynthesis in the brain, produced a strong inhibition of LSD-induced behavior. These data were interpreted by Dixon in favor of the hypothesis that catecholamine mediation is responsible for the excitement observed with LSD in the testing situation. The author, however, admits that it is questionable whether this behavioral disturbance in animals is representative of the psychotomimetic action in man.

Conflicting with these results are those of Menon *et al.*[59] who found that the hypermotility and hyperthermia caused by injection of LSD in mice, remained unaffected by pretreating the animals with α-methyl-*p*-tyrosine in doses sufficient to deplete brain norepinephrine by 65%. Similar pretreatment of mice significantly reduced the central stimulant effect of mescaline or amphetamine.

It is difficult to fit these conflicting data on interactions with catecholamine metabolism into a comprehensive picture of mechanism of action. The fact that catecholamine levels are changed after administration of LSD and similar drugs cannot be denied, but whether they are secondary to the stressful experience[56] or have a more primary role in the mechanism of action cannot be established on the basis of available evidence. The *in vitro* data on the effect of drugs on norepinephrine uptake by subcellular fractions of brain[60] indicate no obvious correlation between activity in this system and psychotomimetic activity. Unfortunately, regional data on the effect of LSD on catecholamine metabolism are not available, although it is now well established that the distribution of the catecholamines, especially dopamine, is highly localized in certain regions and pathways of the brain.[61]

C. Interrelationships with Miscellaneous Neurochemical Substances

LSD and the other hallucinogens may interact with various biogenic substances in the brain, causing a change in their normal level.

Lewis *et al.*[62] have shown that several hallucinogenic drugs (LSD, psilocybine, mescaline, bufotenin) raised ATP and phosphocreatine levels and the ATP/ADP ratio in the rat brain. There appears to be some correlation between the ability of these drugs to cause behavioral stimulation, and the brain level changes in high energy phosphates. These changes, however, are similar to those produced by amphetaminelike compounds and the authors' experiments suggest that this may reflect an improved cerebral blood flow secondary to peripheral sympathomimetic action of the drugs. Since DOPA and 5-HTP were ineffective in producing similar biochemical changes, the role of NE or 5-HT as a mediator can probably be excluded.

Appelt *et al.*[63] have demonstrated that mescaline can stimulate the synthesis of nicotinamide adenine dinucleotide (NAD) in the central nervous system. Mice which received mescaline 90 min prior to an intracerebral injection of nicotinamide had higher levels of brain NAD at the second and third-hour time intervals than a comparable control group.

Varma *et al.*[64] have tested the effect of LSD and mescaline on heat stress changes in the glutathione levels of brain and blood of rats. The decrease in the level of this tripeptide after heat stress was further aggravated by pretreating the animals with LSD or mescaline. The results are interpreted as an increased stimulation of the hypothalamico-hypophyseo-adrenocortical axis due to the stress produced by these drugs.

It is of interest to mention here, that in spite of the effects of hallucinogenic drugs on levels of biogenic amines which are formed by decarboxylation of the corresponding amino acids, and on peptides such as glutathione, these drugs do not seem to affect the uptake of amino acids into the brain. Lajtha and Toth[65] have demonstrated that LSD and mescaline had no significant effect on cycloleucine uptake *in vivo* in mouse brain, while other drugs, such as pentobarbital, chlorpromazine, or reserpine had significant inhibitory effect on this process.

Mescaline applied locally on the cortex of the cat was used by Mison-Crighel *et al.*[66,67] to produce convulsions in the animals and to study the role of glutamine metabolites in the provoked phenomena. They found, that except for GABA levels in the focus, the treatment produced significant changes in GABA as well as in the glutamine, glutamic acid, and ammonia level in the whole cortex of the cat. These changes showed no relation to the appearance and development of the convulsive focus but rather seemed to be related to a functional modification of the whole cortex.

Penicillin, another convulsant drug when applied topically, also produced changes in ammonia and glutamic acid levels, but the regional pattern of these changes was distinctly different from those produced by mescaline.[68]

The effect of the locally applied mescaline treatment on the ATP level of the cortex of the cat was studied by Dosseva *et al.*[69]

At the time the electrical signs of epileptic activity appeared (5–8 min after topical application of mescaline), the ATP level was 26% lower in focal tissues than in contralateral tissues. When spiking activity reached a maximum (at 8–12 min) the ATP in the focal tissue was 92% higher than control. In the author's interpretation the ATP changes were not the direct cause of the appearance of epileptogenic activity but rather were regarded as an adjustment of the cerebral tissue energy supply for the increased bioelectrical activity in the epileptogenic focus.

Another interesting bit of information was reported on the effect of LSD on the iron content of the brain of rat by Hadzovic *et al.*[70] LSD is known to produce tremors in rat that cannot be inhibited by anticholinergic drugs. LSD, parallel with its tremor-producing activity, decreased the iron content of the brain. Another tremor-producing agent, tremorine, caused a temporary decrease of iron content of the brain lasting only as long as the clinical signs of tremor were present. The significance of these findings for the mechanism of action of LSD remains to be elucidated.

IV. SUMMARY

The effect of hallucinogenic drugs on the biochemical machinery of the brain has remained a lively interest for basic psychopharmacological research during the recent years. This interest is based on the hope that insights into the interaction between these drugs and functionally important biochemical processes may shed some light on brain mechanisms which play a part in the perceptual, emotional, and thinking processes these drugs are able to alter.

Another motivating factor for an interest in this area was that progress in the area of metabolism and biochemical action of hallucinogenic drugs may lead to the eventual discovery of a metabolic disorder which is hypothesized to be a possible etiologic factor in endogenous mental disorders such as schizophrenia.

Although this expected breakthrough did not materialize, significant progress was made in exploring the biochemical factors related to the action of hallucinogenic drugs.

We can look at these biochemical processes from various points of views. The drugs are metabolized by the biochemical machinery of the body, thus playing an important part in determining the length of time the drug remains in the body. But drugs can also be metabolized so that pharmacologically active metabolites may be formed. No direct evidence has been reported for the existence of such a metabolite in case of hallucinogens but some indirect evidence stands out as probably relevant to this problem. One is the finding that the specific psychological effects of the drug occur at the time when the concentration of the unchanged drug in the brain has declined. The other is that the psychological effectiveness of a drug depends on a particular configuration of the molecule and certain substitutions would abolish its activity. This last point could, of course, be related to receptor specificity in the pharmacological sense, but the possibility that it reflects enzyme specificity involved either in the breakdown of the molecule or in the formation of an active metabolite has not been ruled out.

The most consistent biochemical effect of LSD and similar drugs is the decrease of turnover rate of 5-HT in the brain. Since the nonhallucinogenic drug imipramine reportedly produces the same effect,[71] although probably by a different mechanism, the turnover data must be interpreted cautiously. Selective regional factors probably play an important role and would have to be considered in relating biochemical findings to functional behavioral parameters. It remains a major problem that we are not certain which behavioral effect in animals is relevant to the specific hallucinogenic action in man. The effect on the process of habituation to repeated nonreinforced stimuli seems to be a good candidate for such a parameter,[72] but much more research is needed before all the biochemical and behavioral data can be fitted into a comprehensive theory on the mechanism of action of hallucinogenic drugs.

V. REFERENCES

1. W. A. Stoll, Lysergsäure-diäthylamide, ein Phantasticum aus der Mutterkorngruppe, *Schweiz. Arch. Neurol. Psychiat.* **60**:279 (1947).
2. J. H. Gaddum, *in Ciba Found. Symp. on Hypertension*, Little, Brown and Co. Boston, Mass. (1954).
3. D. W. Woolley and E. Shaw, A biochemical and pharmacological suggestion about certain mental disorders, *Proc. Nat. Acad. Sci.* **40**:228–231 (1954).
4. A. Carlsson, Basic actions of psychoactive drugs, *Intern. J. Neurology.* **6**:27–45 (1967).

5. D. X. Freedman, Psychodysleptics and cerebral monoamines. Paper given at the 2nd Laval Pharmacological Conferences, Quebec, Canada, September 16–20, 1968.
6. F. E. Bloom and N. J. Giarman, Physiologic and pharmacologic considerations of biogenic amines in the nervous system, *Ann. Rev. Pharmacol.* **8**:229–258 (1968).
7. N. J. Giarman and D. X. Freedman, Biochemical aspects of the actions of psychotomimetic drugs, *Pharmacol. Rev.* **17**:1–25 (1965).
8. H. Isbell, C. W. Gorodetzky, and D. Jasinsky, Effects of delta-9-trans-tetrahydrocannabinol in man, *Psychopharmacol.* **11**:184–188 (1967).
9. S. H. Snyder and M. Reivich, Regional localization of lysergic acid diethylamide in monkey brain, *Nature* **209**:1093–1095 (1966).
10. J. G. Wagner, G. K. Aghajanian, and O. H. L. Bing, Correlation of performance test scores with "tissue concentration" of lysergic acid diethylamide in human subjects. *Clin. Pharmacol. Therap.* **9**:635–638 (1968).
11. S. Szara, *in Amines and Schizophrenia* (H. E. Himwich *et al.*, eds.), pp. 181–198, Pergamon Press, Oxford (1967).
12. L. A. Faillace, A. Vourlekis, and S. Szara, Clinical evaluation of some hallucinogenic tryptamine derivatives, *J. Nerv. Ment. Dis.* **145**:306–313 (1967).
13. S. Szara, L. A. Faillace, and L. B. Speck, Metabolic and physiological reaction to three short-acting tryptamine derivatives, Proc. Vth Int. Congress of C.I.N.P., *Excerpta Med.* **129**:1115 (1967).
14. S. Szara, Discussion of fate and metabolism of some hallucinogenic indolealkylamines, *Adv. Pharmacol.* **6/B**:230–231 (1968).
15. W. M. McIssac and R. T. Harris, Discussion of the pharmacologic studies on the structure-activity relationships of indolealkylamines, *Adv. Pharmacol.* **6**:247–248 (1968).
16. C. Naranjo, *in Ethnopharmacol. Search for Psychoactive Drugs* (D. H. Efron, ed.) No. 1645, pp. 385–391, PHS Publication, Washington, D.C. (1967).
17. N. Neff, G. V. Rossi, G. D. Chase, and J. L. Rabinowitz, Distribution and metabolism of mescaline-C^{14} in the cat brain, *J. Pharm. Exptl. Therap.* **144**:1–7 (1964).
18. K. D. Charalampous, K. E. Waller, and J. Kinross-Wright, Metabolic fate of mescaline in man, *Psychopharmacol.* **9**:48–63 (1966).
19. J. M. Musacchio and M. Goldstein, The metabolism of mescaline C^{14} in rats, *Biochem. Pharmacol.* **16**:963–970 (1967).
20. S. H. Snyder, L. A. Faillace, and L. Hollister, 2,5-Dimethoxy-4-methyl-amphetamine (STP): A new hallucinogenic drug, *Science* **158**:669–670 (1967).
21. A. J. Friedhoff and E. Van Winkle, Isolation and characterization of a compound from the urine of schizophrenics, *Nature* **194**:897–898 (1962).
22. M. Takesada, Y. Kakimoto, I. Sano, and Z. Kaneko, 3,4-Dimethoxyphenylethylamine and other amines in the urine of schizophrenic patients, *Nature* **199**:203–204 (1963).
23. W. von Studnitz, G. E. Nyman, Excretion of 3,4-dimethoxy-phenylethylamine in schizophrenia, *Acta Psychiat. Scand.* **41**:117–121 (1965).
24. T. L. Perry, S. Hansen, L. MacDougall, and C. J. Schwartz, *in Amines and Schizophrenia* (H. E. Himwich *et al.*, eds.), pp. 31–41, Pergamon Press, Oxford (1967).
25. S. S. Kety, The pharmacology of psychotomimetic and psychotherapeutic drugs, *Ann. N.Y. Acad. Sci.* **66/3**:417–840 (1957).
26. M. Ya. Mikhalson and V. G. Longo, *Pharmacology of Conditioning, Learning and Retention*, Pergamon Press, New York (1965).
27. B. F. Key and P. B. Bradley, The effect of drugs on conditioning and habituation to arousal stimuli in animals, *Psychopharmacol.* **1**:450–462 (1960).

28. J. R. Smythies, R. J. Bradley, V. S. Johnston, F. Benington, R. D. Morin, and L. C. Clark, Jr., Structure-activity relationship studies on mescaline, III. The influence of the methoxy groups, *Psychopharmacol.* **10**:379–387 (1967).
29. L. Gyermek and T. Sumi, Potentiating actions of indolealkylamines and lysergic acid diethylamide on reflexes elicited by 5-hydroxytryptamine, *Proc. Soc. Exp. Biol. Med.* **114**:436–439 (1963).
30. M. Vogt, Cerebral tryptamine receptors: functional considerations, *Adv, Pharmacol.* **6/B**:19–28 (1968).
31. E. Costa and N. H. Neff, Estimation of turnover rates to study the metabolic regulation of the steady-state level of neuronal monoamines, *Handbook of Neurochemistry* (A. Lajtha, ed.), Vol. 4, pp. 45–90, Plenum Press, New York (1970).
32. M. Jouvet, Insomnia and decrease of cerebral 5-hydroxytryptamine after destruction of the raphe system in the cat, *Adv. Pharmacol.* **6/B**:265–279 (1968).
33. G. K. Aghajanian and M. H. Sheard, Behavioral effects of midbrain raphe stimulation—dependence on serotonin, *Communications in Behav. Biol.* **1**:37–41 (1968).
34. M. H. Sheard and G. K. Aghajanian, Stimulation of the midbrain raphe: effect on serotonin metabolism, *J. Pharmacol. Exptl. Therap.* **163**:425–430 (1968).
35. M. H. Sheard and G. K. Aghajanian, Stimulation of midbrain raphe neurons: behavioral effects of serotonin release, *Life Sci.* **7**:19–25 (1968).
36. D. X. Freedman and N. J. Giarman, LSD-25 and the status and level of brain serotonin, *Ann. N.Y. Acad. Sci.* **96**:98–107 (1962).
37. S. Szara, *in Comparative Neurochemistry* (D. Richter, ed.), pp. 425–432, Pergamon Press, Oxford (1964).
38. J. A. Rosecrans, R. A. Lovell, and D. X. Freedman, Effects of lysergic acid diethylamide on the metabolism of brain 5-hydroxy-tryptamine, *Biochem. Pharmacol.* **16**:2011–2021 (1967).
39. P. M. Diaz, S. H. Ngai, and E. Costa, Factors modulating brain serotonin turnover, *Adv. Pharmacol.* **6/B**:75–92 (1968).
40. N. E. Anden, H. Corrodi, K. Fuxe, and T. Hökfelt, Evidence for a central 5-hydroxytryptamine receptor stimulation by lysergic acid diethylamide, *Brit. J. Pharmacol.* **34**:1–7 (1968).
41. S. Eiduson, 5-Hydroxytryptamine in the developing chick brain: its normal and altered development and possible control by end-product repression, *J. Neurochem.* **13**:923–932 (1966).
42. S. F. Contractor and M. K. Jeacock, A possible feedback mechanism controlling the biosynthesis of 5-HT, *Biochem. Pharmacol.* **16**:1981–1987 (1967).
43. B. Werdinius, Discussion of factors modulating brain serotonin turnover, *Adv. Pharmacol.* **6/B**:93–95 (1968).
44. R. M. Marchbanks, Serotonin binding to nerve-ending particles and other preparations from rat brain, *J. Neurochem.* **13**:1481–1493 (1966).
45. C. D. Wise and H. W. Ruelius, The binding of serotonin in brain: A study *in vitro* of the influence of physicochemical factors and drugs, *Biochem. Pharmacol.* **17**:617–632 (1968).
46. S. Szara, *in Neuropsychopharmacology* (P. B. Bradley, F. Flugel, and P. Hoch, eds.), Vol. 3, pp. 412–416, Elsevier, Amsterdam (1964).
47. S. Szara, D. M. Morton, and A. Aikens, Comparison of hallucinogenic congeners on regional serotonin metabolism in brain, *The Pharmacologist*, **9**:250 (1967).
48. S. Szara, Brain serotonin, dehabituation and the hallucinogenic drugs (Discussion of Dr. Aghajanian's Paper). Proceedings of the 1967 Annual Meeting of the American College of Neuropsychopharmacology. (In press.)

49. T. N. Chase, G. R. Breese, and I. J. Kopin, Serotonin release from brain slices by electrical stimulation: regional differences and effect of LSD, *Science* **157**:1461–1463 (1967).
50. T. N. Chase, G. R. Breese, and D. O. Carpenter, Stimulation-induced release of serotonin, *Adv. Pharmacol.* **6**:351–364 (1968).
51. W. Feldberg and R. D. Myers, Appearance of 5-hydroxytryptamine and an unidentified pharmacologically active lipid acid in effluent from perfused cerebral ventricles, *J. Physiol.* **184**:837–855 (1966).
52. W. P. Koella, Discussion of insomnia and decrease of cerebral 5-hydroxytryptamine after destruction of the raphe system in the cat, *Adv. Pharmacol.* **6/B**:280–282 (1968).
53. G. K. Aghajanian, E. W. Foote, and M. H. Sheard, Lysergic acid diethylamide: sensitive neuronal units in the midbrain raphe, *Science* **161**:706–708 (1968).
54. Y. Takeo and H. E. Himwich, The significance of methyl groups in the electroencephalographic effects of indolealkylamines in the rabbit, *Biochem. Pharmacol.* **16**: 1013–1022 (1967).
55. M. Fujimosi and H. E. Himwich, Electroencephalographic alerting sites of *d*-amphetamine and 2,5-dimethoxy-4-methyl-amphetamine, *Nature* **220**:491–494 (1968).
56. D. X. Freedman, Psychotomimetic drugs and brain biogenic amines, *Am. J. Psychiat.* **119**:843–850 (1963).
57. F. E. Bloom, E. Costa, and G. C. Salmoiraghi, Analysis of individual rabbit olfactory bulb neuron responses to the microelectrophoresis of acetylcholine, norepinephrine and serotonin synergists and antagonists, *J. Pharmacol.* **146**:16–33 (1964).
58. A. K. Dixon, Evidence of catecholamine mediation in the aberrant behavior induced by lysergic acid diethylamide (LSD) in the rat, *Experientia* **24**:743–747 (1968).
59. M. K. Menon, D. C. Dendiya, and J. S. Bapna, Modification of the effect of some central stimulants in mice pretreated with alphamethyl-l-tyrosine, *Psychopharmacol.* **10**:437–444 (1967).
60. W. F. Herblin and R. D. O'Brien, Interaction of norepinephrine with subcellular fractions of rat brain. I. Characteristics of norepinephrine uptake, *Brain Research* **8**:298–309 (1968).
61. N. A. Hillarp, K. Fuxe, and A. Dahlstrom, Demonstration and mapping of central neurons containing dopamine, adrenaline, and 5-hydroxy-tryptamine and their reactions to psychopharmaca, *Pharmacol. Rev.* **18**:727–742 (1966).
62. J. J. Lewis, A. P. Ritchie, and G. R. van Petten, The influence of hallucinogenic drugs upon *in vivo* brain levels of adenine nucleotides, phosphocreatine and inorganic phosphate in the rat, *Brit. J. Pharmacol.* **25**:631–637 (1965).
63. G. D. Appelt, N. O. Walker, and R. G. Brown, Effect of mescaline on nicotinamide adenine dinucleotide synthesis in the central nervous system, *J. Pharm. Sci.* **57**:527–528 (1968).
64. R. R. Varma, K. P. Khuteta, and P. C. Dandiya, The effect of some psychopharmacological agents on heat-stressed changes in the glutathione levels of brain and blood in rats, *Psychopharmacol.* **12**:170–175 (1968).
65. A. Lajtha and J. Toth, The effects of drugs on uptake and exit of cerebral amino acids, *Biochem. Pharmacol.* **14**:729–738 (1965).
66. N. Mison-Crighel, N. Luca, and E. Crighel, The effect of an epileptogenic focus, induced by topical application of mescaline, on glutamic acid, glutamine and GABA in the neocortex of the cat, *J. Neurochem.* **11**:333–340 (1964).
67. G. Badin and N. Mison-Crighel, The effect of an ectosylvian epileptogenic focus induced by topical application of mescaline on nitrogenous compounds in the neocortex of cat, *J. Neurochem.* **13**:1217–1222 (1966).

68. C. Pintillie, N. Mison-Crighel, and G. Badin, Convulsive effect elicited by topical application of penicillin on glutamate–glutamine system of brain, *Nature* **214**:1131 (1967).
69. I. Dosseva, Z. Tencheva, and S. Dimov, Correlation between bioelectric activity and adenosin-5′-triphosphate metabolism in cat brain mescaline induced epileptogenic focus, *Compt. Rend. Acad. Bulgare Sci.* **19**:313–316 (1966).
70. S. Hadzovic, B. Nicolin, and P. Stern, The effect of tremorine and lysergic acid diethylamide on the iron content of the brain, *J. Neurochem.* **12**:908–909 (1965).
71. H. Corrodi and K. Fuxe, The effect of imipramine on central monoamine neurons, *J. Pharm. Pharmacol.* **20**:230–231 (1968).
72. S. Szara, The hallucinogenic drugs' controversy, Paper presented on the First International Symposium on Psychodysleptics (Hallucinogens) held at Laval University, Quebec, Canada, on September 16–19, 1968.

Chapter 19

DEMYELINATING CHOLINESTERASE INHIBITORS: LIPID AND PROTEIN METABOLISM

Giuseppe Porcellati

Department of Biological Chemistry
University of Pavia
Pavia, Italy

I. INTRODUCTION

Organophosphorus compounds (OPC) are known to cause a specific secondary type of demyelination in nervous tissues: lesions in the axis cylinder and myelin sheath, both in the central nervous system and the peripheral nervous system, have been described, as a result of their neurotoxic action.(1–5) To our knowledge, no review or recent monograph has appeared, dealing with the effects brought about by demyelinating OPC on protein and lipid metabolism of CNS and PNS. Some contributions in this field have, however, appeared in these last years, partially covering this subject matter.(4,6–10) Therefore, I have concentrated mostly on this problem and have tried to collect the most relevant and recent literature in the specific field. As is known, the economic importance of the OPC as insecticides is remarkable.(11,12) A detailed and adequate study of their toxicity and metabolism also is of interest, but this aspect will not be particularly reviewed in this paper.

A. The Demyelinating Organophosphorus Compounds

Certain of the anticholinesterase OPC can cause chronic, irreversible, demyelinating lesions in the nervous system. The OPC that has been most extensively studied, in this connection, is tri-*ortho*-cresyl phosphate (TOCP), which was shown to be the causative agent in man of an outbreak of motor neuritis in the United States in 1930 and in Morocco in 1958–1959 (see Ref. 5). The clinical picture, first described by Harris(13) and Smith and Lillie,(1) was characterized by a peripheral and central paresis, with a predominant distally appearing flaccid paresis of the lower extremities.

Paralysis with TOCP has been subsequently produced in experimental animals,[1–3,6,7,14] and recently also with a mixture of *o-p-m*, *o-p-p* and *p-m-m* esters of tricresylphosphate (TCP).[15]

Other OPC, notably the alkyl phosphates, have been found to exert a direct action upon the nervous system, being more neurotoxic than the above-mentioned aryl phosphates. The OPC most thoroughly examined are diisopropylfluorophosphonate (DFP)[2,16,17] and bis(isopropylamino)-fluorophosphine oxide (mipafox).[2,4,16] Although TOCP differs chemically from DFP and mipafox, histological examination of the nervous tissues of the hen after administration of the three organophosphorus compounds has revealed similar demyelination of axon sheaths and degeneration of the axon in the sciatic nerve and spinal cord, with paralytic signs.[2,3,11,18] More precisely, repeated small doses of TOCP exert, as compared to DFP and mipafox, more slowly damaging effects upon the nervous system of the cat and hen, whereas poisoning with single doses of TOCP in the same animals brings about the same effect produced by mipafox.[18]

Other OPC have been shown in recent years to act as neurotoxic demyelinating agents: a metabolite of TOCP,[19,20] the tributyl phosphorotrithiolate (DEF),[21] the tributyl phosphorotrithioite (merphos),[21] and the tri-*p*-ethylphenyl phosphate (see Ref. 5). The central and peripheral lesions, accompanied by clinical signs of ataxia, are similar to those seen following administration of TOCP. Recently, fourteen other OPC, mostly ethyl-*p*-nitrophenyl derivatives, have been found to produce in adult hens neurotoxic effects like those produced by TOCP.[22] Also, the well-known *O*,*O*-dimethyl-*S*-(1,2-dicarbethoxyethyl)dithiophosphate (malathion) and the *O*-ethyl *O*-*p*-nitrophenyl phenylphosphonothioate (EPN) produce paralysis in chickens, as do TOCP, DFP, and mipafox.[23] The main difference, however, between the two groups of paralytic compounds is that the syndrome does not develop before 12–14 days after DFP, TOCP, or mipafox administration,[2,16] whereas with EPN and malathion the syndrome usually develops within 1 or 2 days (see Ref. 23).

B. Toxicology

Organophosphorus compounds produce chronic lesions and acute effects on the nervous system. With the exception of malathion and EPN, they all show a 10–12-day delay period before the chronic neurotoxic symptoms develop. It has never been possible to shorten this period by altering the route or manner of administering the OPC. As will be discussed later, the manner in which OPC exert these late effects is largely unknown. Older neuropharmacological studies have indicated that the acute effects of these compounds are due to the inhibition of true cholinesterase (E.C.3.1.1.7), an enzyme particularly concentrated in the nervous system. Liver esterases also seem to be acutely inhibited.[24] The acute poisoning, caused by the accumulation at cholinergic synapses of unhydrolyzed acetylcholine, develops rapidly and may be lethal. Normally, however, the acute symptoms pass off in a few hours.

C. Animal Species Susceptibility

The cat and the adult hen are most sensitive to OPC.[2,3,18,25] The young chicken, on the other hand, is completely resistant, and becomes sensitive only between 8 and 12 weeks of age. The guinea pig, whose intoxication by DFP has been recently examined,[26] appears to be unaffected by administered OPC, whereas the rabbit[2] and the rat[2,4] need more repeated doses of the compounds to achieve some demyelination in the peripheral nerve and in the fasciculus gracilis, though often no permanent paralysis develops. Mipafox induces in rats, within 3–4 months of repeated doses, a typical peripheral neuropathy, consisting of a subtotal degeneration of the branches of the sciatic nerve, while its proximal portion is histologically unaffected.[4] Man also is susceptible to paralysis caused by TOCP and mipafox.[1,13,23,25] It has not been possible as yet to decide whether the differences arise from special properties of the nerve cells of different animals or whether insensitivity depends on other factors. Interestingly, it was in rats and rabbits that Myers *et al.*[27] demonstrated that a potent cholinesterase inhibitor is formed from TOCP. The compound, identified as the 2-*o*-cresyl-4H-1,3,2-benzodioxaphosphoran-2-one,[20] has been recently found to act as a neurotoxic agent in the cat.[25]

II. A BRIEF DESCRIPTION OF THE NEUROPATHOLOGICAL STUDIES

The histopathology of the paralysis, both in man and experimental animals, first described by Smith and Lillie,[1] has been studied also by other workers.[2,3] In birds, nerve-fiber degeneration, both ascending and descending, has been demonstrated.[28,29] The distribution of the lesions is the same after poisoning with either DFP, TOCP, or mipafox, and long spinal tracts with primary sensory fibers are constantly damaged. In the spinal cord, the fasciculus gracilis in the posterior columns, the spinocerebellar tract in the lateral columns, and long association fibers in the anterior columns are principally involved. In the bird, the ventral tract is affected mostly in the lumbar region, the spinocerebellar tracts and to a lesser degree the posterior columns are damaged in the cervical region, whereas in the brainstem the large decussating fibers of the vestibulocochlear system may also be involved. With hens treated with DEF and merphos,[21] the regions of the spinal cord most evidently affected are the ventral tracts near the ventromedian fissure, dorsolateral tracts of the sacral area, and dorsal and ventral tracts in the cervical area.

Various histological findings[2,4,18,30] have demonstrated that not only the long tracts of the spinal cord and certain cerebellar systems, but also the peripheral nerve fibers, display degenerative lesions. Moreover, in OPC neurotoxicity, degeneration is always distal first; it is rarely found in the cat sciatic nerve above the knee,[18] from which it spreads centrally with the course of time. The peripheral nerve fibers are strikingly affected as early as one day after the onset of clinical signs in the TOCP-poisoned hen, and the histological alterations include myelinated as well as unmyelinated nerve fibers. Fibers of large diameter and greater length appear more susceptible than smaller-sized fibers[18]; the spindle primary sensory fibers are therefore always damaged first and most extensively.

The pathological process caused by intoxication by OPC has long been considered predominantly motor in type and essentially a peripheral neuropathy, but it has been demonstrated to be more of a general systemic nature, taking place not only at the level of the peripheral nerve and long spinal tracts in the chicken, but also in the CNS, at least in the cat. Recently, detailed lesions in the cat CNS have been reported[31]; they were shown, by means of preterminal studies (Nauta), to occur in the projection areas of most of the long ascending and descending spinal tracts, and no other additional areas of degeneration were found elsewhere in the brain. Fiber diameter seems to have less influence in the CNS than in peripheral nerve lesions.[31]

On the basis of biochemical studies, it has been postulated that the peripheral demyelination produced by OPC is a type of secondary demyelination,[6,32–34] since evidence suggests that initially the whole neuronal process suffers first. Histological and biochemical evidence is being accumulated in this connection.[4,5,7,8,15,18] Concurrently, the axons are sufficiently damaged to account for the degenerative changes in the myelin sheath, and to indicate that the myelin changes are of the Wallerian or secondary type. Moreover, the degenerative changes in the myelin sheath of peripheral nerves are not segmental,[3] and are matched by an equal amount of axonal break-up.[3] Schwann cell and connective tissue proliferation and infiltration of macrophages have also been reported.[3] No inflammatory or reactive cell changes have been described. In treated animals, counts of nuclei in the gracile tracts and peripheral nerves show an increase in the number of nuclei, which vary according to the duration of survival after poisoning.[18,31] While various workers draw parallels between the alterations of the nerve fibers to those of the typical Wallerian degeneration, Cavanagh, in a later report,[18] interprets the process as a disease of the whole neuron, more precisely as a type of retrogression (dying back), where the longest and thickest peripheral nerve fibers are mainly affected.[18]

The distal neuropathy produced by OPC poisoning affects both motor and sensory nerve fibers and occurs in the presence of apparently normal nerve cells, which do not show any visible concomitant morphological changes.[18] Recently, a Wallerian-type chromatolytic change in the morphology of neurons of the anterior and posterior horns of chick spinal cord has been detected[15] following application of 0.22 ml/kg of TCP. The neuronal changes were accompanied by ascending fiber degeneration.

According to these studies, TCP alters not only the distal portion of long-fiber tracts, but the neuron also, by affecting first some portion of the neuronal cytoplasm (perinuclear or axonal).

With electron microscopic studies, it has been shown that TOCP poisoning in chickens produces alterations which primarily affect the axoplasm, with subsequent secondary deterioration of the myelin sheath.(30) A pathological pattern of the axon-Schwann cell compound membrane has been visualized,(30) since the normal gap existing between the paired unit membranes appears to be discontinuously closed, suggesting an alteration at the level of the axon membrane. Boutons terminaux and preterminal degeneration in cats after TOCP poisoning have also been examined.(35)

III. LIPID METABOLISM

A. Lipid Class Concentration and Constitution

Webster(36) observed no change in lipid phosphorus concentration of the sciatic nerves and spinal cord of TOCP-poisoned hens. It has been also demonstrated(34) that the concentration of the individual phospholipid classes does not vary in the hen peripheral nerves during administration of DFP and TOCP, although a slight fall of some phospholipid concentration occurs at the thirtieth day after the TOCP treatment.(34) The total phospholipid concentration is similarly unchanged.(34) Comparable results have been obtained later, in the hen, after poisoning with TOCP and mipafox.(10,37) Berry and Cevallos(37) have only found an increase of cholesterol esters and phosphatidyl choline and a marked decrease of cerebrosides and triglyceride, accompanied by an increase in mono- and diglycerides. Lysolecithin concentration is also decreased.(37) The increase of the peripheral nerve cholesterol esters caused by the TOCP treatment(37) has not been confirmed in hen brain, spinal cord, and peripheral nerves after TOCP(38) and mipafox(10) administration.

The amount of P in the total acid-soluble fraction of the hen spinal cord and peripheral nerve does not vary during treatment with DFP(34) and TOCP.(36) The levels of free phosphoryl ethanolamine, phosphoryl choline, and probably glycerophosphate are significantly increased in the hen sciatic nerve 2 days after the administration of DFP, i.e., in the early stages of the demyelination,(33,34) and those of phosphoryl choline in the rat brain during DFP poisoning.(39) The amounts of glycerylphosphorylcholine (GPC) and glycerylphosphorylethanolamine (GPE) are stable in the hen peripheral nerves during DFP and TOCP poisoning, increasing only at the thirtieth day after TOCP administration by a value of about 70%.(33)

B. Lipid Synthesis

No significant change of ^{32}P-phosphate uptake into the total phospholipid and acid-soluble P fractions in the peripheral nerves and spinal cord of TOCP-poisoned hens was found by Webster in 1954.[36] The result refers to the whole phospholipid fraction and does not preclude the possibility that the metabolism of the individual phospholipid fractions may have been interfered with. A 20% decrease of lipogenesis from ^{14}C-acetate in hen peripheral nerves was observed 24 hours after the DFP treatment (approximate concentration in the tissue of 5.5×10^{-6}M, if evenly distributed),[40] although it was no longer observed at the fifth and seventeenth day. Partially comparable findings are those of Majno and Karnovsky,[4,41] who found a depression of lipid biosynthesis from labeled acetate in the rat and hen peripheral nerve and spinal cord after mipafox poisoning.

A decrease in ^{32}P-phosphate uptake into brain lipids of DFP-poisoned mice, (2.5 mg/kg body wt.) was found by Nelson and Barnum.[39] These authors showed, 3 hours after the DFP administration, a marked inhibition of the ^{32}P-phosphate incorporation rate into phosphatidyl choline, a normal concentration of the phosphatidyl choline, and an increase of the phosphoryl choline levels. The inhibition appears to follow the phosphorylation of choline and seems to be due to a reduction in the rate of transfer of phosphoryl choline to phosphatidyl choline by the Kennedy pathway, a finding which may be substantiated by the fact that the level of phosphoryl choline in the brain is raised by the DFP.[39] In contrast to these findings, Ansell and Chojnacki[43] have found that DFP does not interfere with the formation of phosphatidyl choline by inhibiting either the cholinephosphate cytidylyltransferase (E.C.2.7.7.15) or the cholinephosphotransferase (E.C.2.7.8.2).

^{32}P-phosphate incorporation into individual phospholipid classes is significantly depressed both *in vivo* and *in vitro* in the hen peripheral nerve, at the early stages of DFP and TOCP administration.[34] The uptake into phosphatidyl choline, phosphatidyl ethanolamine, phosphatidyl inositol and cardiolipin is depressed, whereas that into phosphatidyl serine and phosphatidic acid is unaffected, as shown in Table I.

The peripheral nerves of DFP- or TOCP-treated hens normally synthesize glycerophosphate *in vivo* and normally produce through the Kennedy pathway phosphoryl choline and phosphatidic acid* *in vitro* from the appropriate precursors.[7,44] On the contrary, the *in vitro* synthesis of phospholipid from labeled phosphoryl choline is significantly depressed in the peripheral nerves (Table II), whereas it is not affected when ^{32}P-labeled-cytidinediphosphatecholine (CDPC) and D-2,3-diglyceride are used as precursors.[7,44] No experimental studies have been carried out to test whether the phosphatidate phosphohydrolase activity (E.C.3.1.3.4) is affected in the peripheral nerves by treating with the OPC. The hypothesis has been put forward that treating with DFP and TOCP produces early biochemical defects at the last stages of the phospholipid biosynthesis,[34]

* Synthesis of phosphatidic acid through the diglyceride kinase-catalyzed mechanism is irreversibly inhibited in a particulate fraction from guinea pig cerebrum on incubation with DFP.[45]

TABLE I

The Incorporation of ^{32}P-Orthophosphate into the Alkali-Labile Phospholipids of the Hen Peripheral Nerve Preparations During TOCP Intoxication[a,b]

Fraction[c]	Intervals from TOCP administration (hours)						
	0	3	48	144	288	480	720
I	621 ± 26	401 ± 28	451 ± 31	467 ± 35	420 ± 31	430 ± 27	321 ± 38
II	48 ± 4	30 ± 2	22 ± 2	31 ± 2	—	30 ± 2	31 ± 3
III	59 ± 3	51 ± 4	60 ± 4	52 ± 3	51 ± 4	48 ± 5	58 ± 6
IV	2221 ± 165	2010 ± 134	1491 ± 121	1560 ± 103	1521 ± 123	1700 ± 107	1401 ± 147
V	4150 ± 213	4020 ± 145	3920 ± 123	4370 ± 178	4199 ± 156	4070 ± 201	4021 ± 286
VI	175 ± 16	166 ± 13	79 ± 7	101 ± 9	—	109 ± 9	—

[a] Taken from Porcellati and Mastrantonio.(34)
[b] Values are reported in counts/min/μg of P for each fraction ±S.E. (7 experiments for each value).
[c] I: PhC; II: PhE; III: PhS; IV: PhI; V: PA; VI: CA.

TABLE II

The Formation of Phosphatidyl Choline from ^{32}P-Labeled Phosphoryl Choline in the Sciatic Nerves of Normal and of DFP-Treated Chickens[a b]

	Hours after administration of DFP					
	0	3	48	144	384	720
Total lipid P fraction	7.1	6.1	2.0	3.5	2.5	4.8
Phosphatidyl choline fraction	29 ± 2.1	22 ± 1.8	9 ± 0.8	13 ± 0.9	12 ± 1.0	15 ± 1.2
P (Fisher's *t*-test)		>0.001	>0.01	>0.001	>0.01	< 0.3

[a] Taken from Porcellati.(7,44)
[b] Data expressed as counts/min/μg of phosphate ±S.E.M. (mean of 8 experiments).

probably those related to the conversion of the phospholipid phosphoric esters to their cytidine diphosphate derivatives.[7,44] The phosphoryl choline to cytidine triphosphate cytidylyltransferase activity (E.C.2.7.7.15) of the hen peripheral nerve is noticeably decreased in fact during the various stages of the DFP treatment,[8] and is not restored by adding lysolecithin or lysophosphatidyl ethanolamine or by increasing the amounts of the cytidine triphosphate added.[8]

The calcium-stimulated conversion of ser to phosphatidyl serine through endogenous phospholipid in DFP-treated hen peripheral nerve preparations is unaltered.[8] This finding may explain why no decrease of the phosphatidyl serine labeling takes place in demyelinating peripheral nerve[34].

^{3}H-inositol incorporation into polyphosphoinositides of the hen peripheral nerve is partially inhibited by administering DFP or TOCP[46]; no inhibition occurs if the ^{3}H-inositol is supplemented with cytidine-diphosphatediglyceride. An indication has been reported that the OPC affect the step related to the conversion of the phosphatidic acid moiety to CDP-diglyceride.[46] The addition of lysolecithin lysophosphatidyl ethanolamine or surface-active agents, or of various amounts of CTP, does not restore the CDP-diglyceride-dependent synthesis of the polyphosphoinositides in the DFP-treated hen peripheral nerves preparations.

Simultaneously with a decreased uptake of labeled precursors into some phospholipid of peripheral nerve at early stages of OPC intoxication, an increased ^{32}P-phosphate incorporation in the phosphoryl choline and phosphoryl ethanolamine takes place.[34] This result may signify the establishment of a preparative stage for *de novo* phospholipid synthesis in the peripheral nerves.

A somewhat increased specific activity of nerve lipid has been described in the more pronounced stages of intoxication by OPC,[4] a finding which is balanced by the slight decrease of phospholipid concentration[34] and by the increase of the GPC and GPE levels[33] (see Section III, E). Similarly, other results have indicated that in TOCP-poisoned cats PhE is increasingly synthesized in the white substance of spinal cord.[47]

C. Lipid Degradation

OPC do not act *in vitro* on phospholipases or lipases (see Ref. 5). It has been shown that the phospholipid concentration of peripheral nerve during intoxication by OPC is stable at least for the first 30 days after poisoning,[10,34,37] thus indicating that lipid degradation is not very active. A decrease of the peripheral nerve content of lysolecithin, however, has been reported.[37] In the advanced stages of the OPC treatment, a moderate increase of phospholipid breakdown takes place.[34] Simultaneously, a slight increase in the GPC and GPE concentration of hen peripheral nerve occurs,[33] together with a moderate increase of the phosphoryl choline and phosphoryl ethanolamine phosphohydrolase activity,[7,44] which are all signs of enhanced degradative events.

D. Fatty Acid Metabolism

Fatty acid composition of individual lipid classes of peripheral nerve and spinal cord from TOCP- and mipafox-treated hens has been examined.[10,37] Conflicting results have been obtained. During the TOCP intoxication, changes in fatty acid composition occur, and appear in some lipid classes before the level of that lipid class is affected.[37] Free fatty acid composition also changes, particularly with a decrease of 16:0, 16:1, and an increase of 18:1 and 18:2.[37] During mipafox treatment, on the contrary, the polyunsaturated fatty acid patterns of brain, spinal cord, and peripheral nerve remain strikingly constant[10]; only in brain tissue is there a barely significant (6%) increase in the amount of tetraenoic acid present and a highly significant (20%) decrease in the amount of pentaenoic acid present.[10] It is not clear whether these last changes are really due to the mipafox intoxication or are merely side effects of the starvation which often accompanies mipafox treatment.[10] Differences in the results, which have been summarized,[10,37] may be due to the different chemical structure of the OPC used or to the fact that experimental animals may have been at different stages of the neuropathy when examined. According to Hove (see Ref. 10), the administration of large doses of TOCP to rats produces a reduction in the content of "arachidonic acid" in the brain.

E. Discussion

The importance of the role of starvation or the poor nutritional state and the relative deterioration, which accompany the TOCP and mipafox treatment, has been stressed.[4,10,33,34] The hen is particularly sensitive to the malnutrition state.[4,34] Careful controls and parallel studies with force-fed-animals have been carried out for most of the experiments on lipid metabolism.

Administration of DFP, TOCP, or mipafox to hens and other animals produces no changes in the amounts and compositions of several lipid classes in the brain, spinal cord, and peripheral nerve, despite the production of severe, functional, histological, and metabolic lesions, especially in the peripheral nerve tissues. Most probably, the demyelination of the PNS is much weaker than the classical Wallerian degeneration, thus influencing the lipid breakdown in a less dramatic sense.

Nevertheless, DFP, TOCP, and mipafox affect lipid metabolism in the nervous system, and particularly its synthetic pathways (see Section III, B). The incorporation of acetate into lipids is easily depressed and seems to represent a more sensitive indicator of regressive changes in the myelin and Schwann cell. The incorporation of P is also early, although not consistently, depressed, and then rises with the late stages of the neuropathy, probably indicating, in this last case, progressive changes in the Schwann cells and possible infiltration. The early decreased incorporation of acetate has been found in every type of neuropathy in which it has been sought[4]; the early decrease of the ^{32}P-phosphate uptake into phospholipid, on the contrary, seems to be more specific for this type of neurotoxic alteration. The biochemical damages are, however, not consistent, and vary considerably with the type of OPC, with the stage of the neuropathy, and with the animal species. This may be obvious, if related quantitatively to the morphological changes that can be found in the nerve fibers and spinal tracts following administration, and to their correspondent stages.

The main biochemical damage of phospholipid biosynthesis, produced at the early stages of the DFP and TOCP poisoning, is a depression of the CTP-dependent conversion of phosphoryl choline (and probably phosphoryl ethanolamine) and phosphatidic acid to their CDP-derivatives (CDPC, CDPE, and CDP-diglyceride). The inhibition of phospholipid biosynthesis, at a point subsequent to the phosphorylation of the bases, is a feature exhibited by some other drugs which affect phospholipid metabolism.[9,42] The decrease of the phosphoryl choline to CTP-cytidylyltransferase activity[8] does not seem to be due to an interference with the CTP formation,[8] although the results obtained *in vitro* do not necessarily mean that the CTP is efficiently supplied *in vivo* during the DFP and TOCP poisoning. The enzymatic synthesis of the various CDP compounds is not restored by the addition of lysolecithin, lysophosphatidyl ethanolamine, or surface-active agents.[8] An inefficient *in vivo* production of suitable amounts of D-2,3-diglyceride acceptor, presumably caused by the inhibition of the phosphatidate phosphohydrolase, may also occur, and may explain the decreased synthesis of phosphatidyl choline and phosphatidyl ethanolamine at a point subsequent to the formation of phosphoryl choline and phosphoryl ethanolamine.[43] This point, however, has not been proved. Molecular changes of the phosphoryl choline to CTP-cytidylyltransferase (E.C.2.7.7.15) and of the CTP to 1,2-diacylglycerophosphate cytidylyltransferase (E.C.2.7.7), brought about by OPC, may also be responsible for the damage of the CDPC and CDP-diglyceride production,[8] as well as variations in the control of the permeability of nerve cell and organelle membrane to precursors and metabolites.[7,8,48]

The enzymatic reaction, which is affected by administering OPC, is to be considered the rate-limiting step of the overall synthetic process.[8] As a consequence, the process of renewal of some phospholipid appears to be partially blocked by treating with DFP, mipafox, or TOCP.[4,34,40] Formation of phosphatidyl serine and phosphatidic acid is unaffected by the injection of DFP and TOCP.[34] Strictly speaking, the synthesis of these two lipids does not involve the participation of the CDP-compounds.

The biochemical defect of lipid biosynthesis is not related, however, to the capability of the OPC in producing myelin degeneration.[7,8] Rather similar damage has been found, in fact, to occur also in the brain of mice,[39] in which no demyelination normally takes place. Further, the addition of mipafox *in vitro* to rat and hen nervous tissues[4] inhibits the uptake of labeled precursors into phospholipid, but in amounts quite similar to those used also for the "non-demyelinating" compounds.[4] Porcellati[7,8] and Ansell and Chojnacki[43] have been unable to obtain similar results. A decreased uptake of labeled acetate into lipids has been obtained by the direct addition of DFP to the incubated peripheral nerve preparation,[40] although high concentrations of the drug were used (10^{-3}M).

Two stages of alteration of phospholipid metabolism take place in the demyelinating peripheral nerve structures. The first occurs at early periods and reflects the action of the drugs on the lipid metabolism. It is characterized by a general depression of the synthesis of phosphatidyl choline, phosphatidyl ethanolamine, and phosphatidyl inositol,[34] by an evident biochemical damage at the level of the formation of the CDP-derivatives,[7,8] by an increase of the levels of free serine, phosphoryl ethanolamine and phosphoryl choline.[33,34] The content of the individual phospholipid is unchanged.[10,34,37] The second stage takes place at about the thirtieth day after administration of TOCP, and represents the effects of the demyelination process upon lipid metabolism. In this stage, the phospholipid content is somewhat decreased,[33,34] the GPC and GPE levels slightly increased,[33] the phosphoryl choline and phosphoryl ethanolamine-phosphohydrolase activity moderately increased,[7,44] and the specific labeling of lipids from radioactive acetate increased.[4]

IV. PROTEIN METABOLISM

A. The Concentration of Protein, Amino Acids, and Related Compounds

Determination has been carried out of the more abundant amino acids in the spinal cord proteolipid-protein from normal samples, and from samples of DFP-treated cats.[6] No significant differences have been found between the concentration and the amino acid patterns obtained from the two different sources.[6]

The amounts of free Ser, Thr, and Glu of the hen peripheral nerve increase 60–80% at the sixteenth day after DFP and TOCP poisoning[33]; DFP produces slightly higher variations. The amounts of GPC and GPE increase by 70% at the thirtieth day after TOCP administration.[33] The concentration of hen sciatic nerve phosphoryl choline, phosphoryl ethanolamine, and glycerophosphate also increases during OPC treatment,[34] but in a different way. The amounts of the free amino acids increase essentially at 10–12 days after treatment, maximal values being reached at the sixteenth day and control levels at about the thirtieth. The phosphoryl choline and phosphoryl ethanolamine concentrations, on the other hand, increase at the beginning of the administration and then remain stable for long periods.

B. Protein Synthesis

Results on the incorporation of ^{14}C-leucine and ^{14}C-lysine into the protein of the spinal nerve and spinal ganglia of the cat during DFP poisoning have been reported.[6] While incorporation into the spinal nerve protein increases after one week, a significant decrease of proteosynthesis occurs in the spinal ganglia within the first periods after administration of the demyelinating-DFP (Table III). On the contrary, in rat brain an increase in protein turnover occurs during the first two days after the administration in the spinal fluid of the "nondemyelinating" OPC 217AO (2-diethoxyphosphinylthioethyldimethylamine acid oxalate).[50] Results on the leucine or lysine incorporation into the proteolipid-protein of the cat spinal cord during the DFP administration show a rather profound depression of the incorporation rates,[49] probably related to the more delayed response of the central white matter to noxious stimuli.

C. Protein Degradation

The neutral proteinase activity increases in the hen spinal cord and peripheral nerve during demyelination by DFP and TOCP,[32] probably as an effect of the neurotoxic process. The increase starts after the first week, reaches a maximal value on the sixteenth to twentieth day, and does not occur in the brain tissue. Free amino acid concentration increases in the peripheral nerve by the sixth day after the DFP and TOCP administration,[33] as a consequence of the increase of the enzymatic activity. No change has been reported in hen nervous tissues after the administration of the "non-demyelinating" tetraethylpyrophosphate (TEPP).[51]

TABLE III

The Incorporation of ^{14}C-Labeled Amino Acids into the Protein of the Cat Spinal Ganglia (SG) and Spinal Nerve (SN) During Poisoning by DFP[a,b]

Labeled amino acid	Sample	Days after treatment 1	4	18	30
^{14}C-leucine	SG	39	61	161	114
^{14}C-leucine	SN	107	131	226	212
^{14}C-lysine	SG	50	47	107	93
^{14}C-lysine	SN	86	78	264	240

[a] Taken from Porcellati.[6]
[b] Data are obtained by equating the activities of the normal controls to 100. Values for normal samples (expressed as counts/min/mg protein-bound leucine or lysine ±S.E.M.): 1421 ± 122 and 1101 ± 109, respectively, for normal spinal nerve, and 801 ± 77 and 581 ± 49, respectively, for normal spinal ganglia.

TABLE IV

The Proteinase Activity of Cat Spinal Cord (SC), Spinal Ganglia (SG), and Spinal Nerve (SN) During DFP Poisoning[a,b]

Proteinase activity	Sample	Days after treatment					
		1	2	4	12	18	30
"Neutral"	SC	112	104	116	101	131	167
"Neutral"	SG	87	103	91	144	191	170
"Neutral"	SN	109	117	100	139	167	161
"Acidic"	SC	179	191	104	101	92	107
"Acidic"	SG	—	167	125	93	111	97
"Acidic"	SN	140	155	143	106	106	102

[a] From Porcellati.[6]

[b] Data are obtained by equating the activities of the normal controls to 100. Values of 9 experiments. "Neutral" proteinase activity of samples of normal SC, SG, and SN: 7.4 ± 0.66, 5.1 ± 0.49, and 8.3 ± 0.56 μmoles of amino-N/g of wet wt./hr, respectively; "acidic" proteinase: 4.3 ± 0.38, 9.4 ± 0.87, and 5.7 ± 0.42, respectively.

After DFP administration to cats, neutral proteinase also increases in the spinal ganglia, spinal nerve and cord, whereas acid cathepsin activity shows an earlier increase in the same structures (Table IV).[6] The change in acidic proteinase activity is probably connected with the first biochemical events of the Wallerian-like changes.[6,8] Increased rates of acidic proteinase activity also occur in frog brain homogenates soon after injection of a "nondemyelinating" OPC,[50] but results of inhibition of the same enzyme were obtained with rat brain homogenates.[50]

The addition of a number of OPC to normal nervous tissues of the hen produces inhibition of the neutral proteinase enzymes *in vitro*.[52] Peripheral nerve acidic proteinase is resistant to DFP, while brain and peripheral nerve aminopeptidase is inactivated by the same compound (see Ref. 5).

D. Discussion

The time course of the changes of neutral and acidic proteinases in peripheral nerve during OPC poisoning is different.[6] The increase of the acidic enzyme, which represents an early biochemical phenomenon, may be related to the initial changes in the mitochondrial apparatus, which has been found to occur almost immediately after OPC treatment.[18] The early increase seems not to be related to the capability of producing myelin degeneration, because a similar increase in brain has also been found following administration of a "nondemyelinating" compound.[50] The later increase of the neutral proteinase, on the other hand, follows the first definite signs of true demyelination and does not precede the structural and physiological nerve damage.[32] It might signify a chemical response to cellular proliferation, and appears to be an effect of the demyelination produced by DFP and TOCP.[32]

The "liberation" or activation of proteinase within the myelin sheath during demyelination by OPC might result in the disruption of protein-bound lipid and in modification of proteolipid structures.[6,33] The possible participation of the lysosomal particles and of the lysosomal enzymes in the events of the secondary demyelinations has been recently discussed.[6,53,55] Neutral proteinase and amino peptidase enzymes have been located in the peripheral nerve myelin.[5,56] Both enzymes are inhibited *in vitro* by DFP,[5,52] whereas acidic catheptic activity is unaffected.[5] However, there does not seem to be any relation between the neutral proteinase inhibition brought about by OPC *in vitro* and the demyelination produced *in vivo* by the same compounds.[52]

According to Clouet and Waelsch,[50] OPC have an effect on protein metabolism, in addition to their specific inhibition of cholinesterase enzymes. Accordingly, the demyelinating-DFP is capable of decreasing the rate of protein synthesis of the nerve cell *in vivo* almost at the commencement of the treatment,[6] in spite of the fact that the increase of the proteinase activity,[6,32] taking place at the same time, should bring about an increased need of proteosynthesis in order to replace degraded protein, as it really does in the case of a "nondemyelinating" OPC.[50] The decrease of the protein synthesis may represent therefore an important primary cause of the lesions in the motor paralysis produced by DFP and TOCP,[6-8] although no visible concomitant morphological alteration has been observed in nerve cells during the poisoning.[18]

Interference of OPC with protein metabolism, oxidative processes, and energy utilization and transference, as possible primary causes of the neurotoxic process, have been recently discussed.[6-8] It has also been reported that DFP^{32} binds slowly but consistently with the brain proteins of the mouse with slow recovery thereafter.[57] Activation or "liberation" of cellular hydrolases, such as acidic proteinase[6] or ribonuclease,[44] may also play an hypothetic important role, in connection with the decreased rate of protein synthesis found during administration of OPC.

V. A DISCUSSION OF THE POSSIBLE DIFFERENCES BETWEEN DEMYELINATION BY ORGANOPHOSPHATES AND OTHER SECONDARY DEGENERATIONS

A. Introduction

Little is known about the similarity of the histopathological features of peripheral nerve degeneration by section and by OPC poisoning. While some workers[15] draw parallels between the alterations of the nerve fibers and those of the classical Wallerian degeneration, others interpret the process as a type of retrogression.[18] According to Janzik and Gless,[15] who have found some retrograde cytological changes in the parent cell of the degenerated nerve fibers, the neuronal degeneration and the tract degeneration are causally related, with a Wallerian-like mechanism. On the other hand, it has been found that during TOCP poisoning the terminal sections of the large-calibered, long nerve fibers are primarily attacked,[18] and that Wallerian degeneration, i.e., fragmentation of axon and myelin sheath, with reactive modifications of Schwann cells and macrophages, is rarely present in the isolated fibers of the cat sciatic peripheral nerves above the knee. Wallerian changes are present principally in the large nerve trunk below the knee, and below the elbow in the forelimb.

Electronmicroscopic findings of the hen PNS during TOCP poisoning have been reported[30]; some differences have been found with the changes of Wallerian degeneration. Nevertheless, in the neuropathy caused by OPC certainly axon and myelin sheath appear to break down simultaneously,[3] and this suggests an analogy with Wallerian degeneration, in which axon and myelin break down because they are severed from the perikaryon. In the neuropathies caused by DFP or TOCP a similar mechanism may occur, in that the fibers may demyelinate because the perikaryon cannot sustain them. Majno and Karnovsky[4] have shown a suggestive histopathological similarity between the distal segment of the sciatic nerve from mipafox-treated rats and the same nerve during degeneration by section.

Similarities of enzyme changes in Wallerian degeneration and in the neuropathy by OPC have been reported,[6,32,51,58] thus indicating that the enzyme organization of the peripheral nerve shows the same rearrangements both after transection and after OPC poisoning.

B. Lipid Metabolism

The stability of the spinal cord and peripheral nerve lipid concentration (see Section III, A), found during treatment with OPC, appears to contrast with the studies on Wallerian degeneration,[59-64] which have revealed diminutions in the amounts of phospholipid, total cholesterol, and cerebrosides; the appearance of cholesterol esters; and changes in the fatty acid patterns. Moreover, lysophosphatides of peripheral nerve increase during Wallerian degeneration,[60,65] and decrease after TOCP poisoning.[37] However, no detectable changes of phospholipid concentration and decomposition occur in peripheral nerve during the first week after nerve section,[63,65] even though the nerve had lost the ability to conduct an action potential. In the study of mipafox intoxication,[4] the histopathological appearance of demyelinating fibers is largely similar to the early stage, i.e., the first week, of the Wallerian degeneration. It has been postulated, therefore, that the changes in lipid composition do not appear yet, during the first weeks of mipafox administration, owing to the weaker histopathological changes.[4] This assumption bears some relationship to results which indicate that detectable phospholipid decomposition takes place in the hen peripheral nerve only after the thirtieth day from TOCP poisoning,[33] when the Wallerian-like changes certainly become more evident.

During Wallerian degeneration by section and during nerve regeneration after crush, a noticeable increase of the incorporation rates of various labeled precursors in the peripheral nerve lipids, both *in vivo* and *in vitro*, has been described,[59,64,66–70] as well as a marked stimulation of the corresponding enzymatic activities.[53,71–73] These metabolic changes do not occur during demyelination by OPC,[4,7,8,44] and a greater labeling of lipid *in vitro* occurs only in later stages of the intoxication process.[4] In addition, most of the enzymes related to phospholipid biosynthesis are neither decreased nor increased.[7,8] The disparity of these two different biochemical situations may be explained in terms of metabolic controls. Since the enzymes responsible for biosynthesis of most lipids are under the regulation of a "feedback" mechanism which controls their synthesis, the stability of the phospholipid concentration, observed during demyelination by OPC, may bring about the observed constancy of the enzyme activity, whereas the increased enzymatic activity of demyelinating peripheral nerve after transection may be explained in terms of reduced content of phospholipid in that tissue. When some higher lipid labeling occurs *in vitro* in the later stages of mipafox administration,[4] then some degradation of phospholipid also occurs.[33,34] Similarly, the noticeable increase of hydrolytic enzymes acting upon the phospholipid phosphoric esters, phosphoryl ethanolamine, phosphoryl choline, GPE, and GPC during Wallerian degeneration[66,74] is not evident during demyelination by OPC and becomes detectable only in the latest stages of the intoxication (see Section III, C, E).

Similar findings have been reported on lipogenesis from ^{14}C-acetate, in the sciatic nerve of mipafox-treated animals and in transected sciatic nerve and spinal column.[4,75] In the rat, lipid biosynthesis from acetate is depressed in the proximal portion of the sciatic nerve (where no lesions are evident) and increased in the degenerating branches.[4] These changes are similar to those found in early and late stages of Wallerian degeneration, respectively.[75] In the hen, incorporation of acetate is generally depressed as in early stages of Wallerian degeneration or during chordotomy.[75]

The main alteration of the lipid metabolism, which takes place during the early stages of demyelination by OPC and which does not occur during degeneration following transection, is the decreased rate of synthesis of phospholipid from ^{32}P-phosphate in the peripheral nerve.[7,8,34] This damage is due to an enzymatic defect at the level of the phosphoryl choline: CTP-cytidylyltransferase, which has been largely discussed (see Section III, B). The enzymatic activity is not stimulated by adding lysolecithin, lysophosphatidyl ethanolamine, or surface-active agents, as it occurs in normal peripheral nerve,[53] and this result might signify that rather complex effects may be brought about by DFP and TOCP on the transferase enzyme, probably altering its conformational architecture.

C. Protein Metabolism

The similarity of the time course of changes of acidic and neutral proteinases in the DFP-dependent demyelination and in Wallerian degeneration[6] indicate close relationships between the two processes. A similar mechanism for the appearance of early biochemical events, in both types of demyelination, has in fact been proposed,[6] and the possible participation of lysosomal enzymes in both events has recently been discussed.[6,53–55]

The amino acid incorporation into nerve protein and proteolipid-protein is similarly increased during nerve degeneration by section and by OPC administration,[6] thus indicating similar histopathological sequences in both events. Moreover, the results of the incorporation of leucine and lysine into the proteolipid-protein of the cat spinal cord during DFP administration show a rather profound depression of the incorporation rates,[6] also noticed during chordotomy.[49] This last finding indicates that the response of the central white matter to the two types of noxious stimuli is probably similar.

The decreased rate of protein synthesis found in spinal ganglia, at the early stages of DFP poisoning,[6] is probably comparable to the decrease in proteosynthesis found in peripheral nerves during nerve degeneration following administration of diphtheric toxin.[76]

VI. A BRIEF DISCUSSION ON THE MECHANISMS OF ACTION OF THE NEUROTOXIC DEMYELINATING ORGANOPHOSPHORUS COMPOUNDS

The primary site of action of the neurotoxic, demyelinating OPC is not clearly known. The inhibition of the pseudocholinesterase (E.C.3.1.1.8), which for a long time has been considered to be involved in the production of neurological lesions in domestic fowls, does not play a primary role in the development of the lesions.[77] The inhibition by neurotoxic OPC of some esterases of brain and spinal cord tissues is not very likely to be involved in the production of the lesions,[22,77] because these enzymes are also inhibited by compounds which do not cause ataxia in chickens. Some not-easily-detectable esterases may be involved.[77] Preliminary studies have been carried out on the binding of labeled DFP with brain protein.[78]

The early decrease of the protein synthesis found in some nervous tissues may represent an important primary cause of the lesions produced by OPC.[6] Interference with oxidative processes, at the level of the neuronal cytoplasm and organelles, as well as with energy transference and utilization mechanisms, may be indicated as other possible primary causes of the neurotoxic process. "Liberation" of cellular hydrolases, such as "acidic proteinase" or ribonuclease, or complex changes of the molecular architecture of membrane-bound enzymes or structural components may also be of outstanding importance. Interactions with proteinases of lysosomal origin have been visualized.[6] OPC may affect the integrity of lysosomal membranes, and more easily "evidentiate" proteinase activity in the DFP- and TOCP-treated peripheral nerves.[6,8] This consideration is partially supported by studies with liver systems.[48] Further suggestions are reported elsewhere.[8]

The biochemical defect of the phospholipid metabolism, although representing an early change caused by the intoxication process, does not signify primary damage and is not linked directly to the pathogenesis of the characteristic lesions produced by the demyelinating OPC.

VII. REFERENCES

1. M. I. Smith and R. D. Lillie, The histopathology of tri-orthocresyl phosphate poisoning, *Arch. Neurol. Psychiatr.* (Chicago) **26**:976–992 (1931).
2. J. M. Barnes and F. A. Denz, Experimental demyelination with organo-phosphorus compounds, *J. Path Bact.* **65**:597–605 (1953).
3. J. B. Cavanagh, The toxic effects of tri-ortho-cresylphosphate on the nervous system. An experimental study in hens, *J. Neurol. Neurosurg. Psychiatr.* **17**:163–172 (1954).
4. G. Majno and M. L. Karnovsky, A biochemical and morphological study of myelination and demyelination. III. Effect of an organophosphorus compound (mipafox) on the biosynthesis of lipids by nervous tissue of rats and hens, *J. Neurochem.* **8**:1–16 (1961).
5. C. W. M. Adams (ed.), *Neurohistochemistry*, Elsevier, Amsterdam (1965).
6. G. Porcellati, in *Protides of the Biological Fluids* (H. Peeters, ed.), Vol. 13, pp. 115–126, Elsevier, Amsterdam (1966).
7. G. Porcellati, The effect of organophosphorus compounds on nerve phospholipid metabolism, *Progr. Biochem. Pharmacol.* **3**:49–58 (1967).
8. G. Porcellati, in *Third International Symposium on Drugs Affecting Lipid Metabolism* (W. H. Holmes, L. A. Carlson, and R. Paoletti, eds.), *Advances in Experimental Medicine and Biology* Vol. 4, pp. 295–306, Plenum Press, New York (1969).
9. G. B. Ansell, *in Advances in Lipid Research* (R. Paoletti and D. Kritchevsky, eds.), Vol. 3, pp. 139–171, Academic Press, New York (1965).
10. C. D. Joel, H. W. Moser, G. Majno, and M. L. Karnovsky, Effects of bis(mono-*iso*propylamino)fluoro-phosphine oxide (mipafox) and of starvation on the lipids in the nervous system of the hen, *J. Neurochem.* **14**:479–488 (1967).
11. D. F. Heath, *Organophosphorus Poisons*, Pergamon Press, Oxford (1961).
12. R. D. O'Brien, *Toxic Phosphorus Esters*, Academic Press, New York (1960).
13. S. Harris, Jr., Jamaica ginger paralysis (peripheral polyneuritis), *Sth. med. J.* (Bgham, Ala) **23**:375–380 (1930).
14. P. Glees, Experimentelle Markscheidendegeneration durch Tri-ortho-kresyl Phosphat und ihre Verhütung durch Cortisonacetat, *Dtsch. Med. Wschr.* **86**:1175–1178 (1961).
15. H. H. Janzik and P. Glees, Chromatolysing spinal neurons in the chick following tricresylphosphate (TCP) intoxication, *Acta Neuropath.* **6**:303–306 (1966).

16. A. N. Davison, Some observations on the cholinesterases of the central nervous system after the administration of the organophosphorus compounds, *Brit. J. Pharmacol.* **8**:212–216 (1953).
17. M. C. Lancaster, A note on the demyelination produced in hens by dialkyfluoridates, *Brit. J. Pharmacol.* **15**:279–281 (1960).
18. J. B. Cavanagh, Peripheral nerve changes in ortho-cresyl phosphate poisoning in the cat, *J. Path. Bact.* **87**:365–383 (1964).
19. R. L. Baron and J. E. Casida, Enzymatic and antidotal studies on the neurotoxic effect of certain organophosphates, *Biochem. Pharmacol.* **11**:1129–1136 (1962).
20. M. Eto, J. E. Casida, and T. Eto, Hydroxylation and cyclization reactions involved in the metabolism of TOCP, *Biochem. Pharmacol.* **11**:337–342 (1962).
21. R. L. Baron and H. Johnson, Neurological disruption produced in hens by two organophosphate esters, *Brit. J. Pharmacol.* **23**:295–304 (1964).
22. W. N. Aldridge and J. M. Barnes, Further observations on the neurotoxicity of organophosphorus compounds, *Biochem. Pharmacol.* **15**:541–548 (1966).
23. R. F. Witter and T. B. Graines, Relationship between depression of brain or plasma cholinesterase and paralysis in chickens caused by certain organic phosphorus compounds, *Biochem. Phamacol.* **12**:1377–1386 (1963).
24. D. J. Ecobichon and W. Kalow, Action of organophosphorus compounds upon esterases of human liver, *Can. J. Biochem. Physiol.* **41**:1537–1546 (1963).
25. J. D. Taylor, A neurotoxic syndrome produced in cats by a cyclic phosphate metabolite of tri-*o*-cresyl phosphate, *Toxicol. Appl. Pharmacol.* **11**:538–546 (1967).
26. D. Hausen, E. Schaum, and O. Wassermann, Distribution and metabolism of DFP in the guinea-pig, *Arch. für Toxikol.* **23**:73–81 (1968).
27. D. K. Myers, J. B. J. Rebel, C. Veeger, A. Kemp, and E. G. L. Simons, Metabolism of triaryl phosphates in rodents, *Nature* (London) **176**:260–262 (1955).
28. P. Glees and W. G. White, A study of skin absorption effects of tri-ortho-cresyl phosphate (TOCP) in hens, *J. Physiol. (London)* **153**:20P–22P (1960).
29. P. Glees and H. H. Janzik, Chemically (TCP) induced fibre degeneration in the central nervous system, with reference to clinical and neuropharmacological aspects, *Progr. Brain Res.* **14**:97–121 (1965).
30. A. Bischoff, The ultrastructure of tri-*ortho*-cresyl phosphate-poisoning. I. Studies on myelin and axonal alterations in the sciatic nerve, *Acta Neuropath.* **9**:158–174 (1967).
31. J. B. Cavanagh and G. N. Patangia, Changes in the central nervous system in the cat as the result of tri-*o*-cresyl phosphate poisoning, *Brain* **88**:165–180 (1965).
32. G. Porcellàti, A. Millo, and I. Manocchio, Proteinase activity of nervous tissues in organophosphorus compound poisoning, *J. Neurochem.* **7**:317–320 (1961).
33. I. Montanini and G. Porcellati, Protein metabolism of peripheral nerves during demyelination by organophosphorus compounds, *Ital. J. Biochem.* **13**:230–239 (1964).
34. G. Porcellati and M. A. Mastrantonio, Phospholipid metabolism of peripheral nerves during demyelination by organophosphorus compounds, *Ital. J. Biochem.* **13**:332–352 (1964).
35. L. Illis, G. N. Patangia, and J. B. Cavanagh, Boutons terminaux and tri-*ortho*-cresyl phosphate neurotoxicity, *Exp. Neurol.* **14**:160–174 (1966).
36. G. R. Webster, The distribution and metabolism of phosphorus compounds in normal and demyelinating nervous tissue of the chicken, *Biochem. J.* **57**:153–158 (1954).
37. J. F. Berry and W. H. Cevallos, Lipid class and fatty acid composition of peripheral nerve from normal and organophosphorus-poisoned chickens, *J. Neurochem.* **13**:117–124 (1966).

38. C. H. Williams, N. J. Johnson, and J. L. Casterline, Cholesterol content of spinal cord and sciatic nerve of hens after organo-phosphate and carbamate administration, *J. Neurochem.* **13**:471–474 (1966).
39. W. L. Nelson and C. P. Barnum, The effect of di*iso*propylphosphorofluoridate (DFP) on mouse brain phosphorus metabolism, *J. Neurochem.* **6**:43–49 (1960).
40. L. Austin, Lipid biosynthesis and chemically induced paralysis in the chicken, *Brit. J. Pharmacol.* **12**:356-360 (1957).
41. G. Majno and M. L. Karnovsky, Respiration and lipid biosynthesis by rat nerves *in vitro*, after treatment with a cholinesterase inhibitor (mipafox), *Fed. Proc.* **14**: 250 (1955).
42. G. B. Ansell, *in Metabolism and Physiological Significance of Lipids* (R. M. C. Dawson and D. N. Rhodes, eds.), pp. 481–499, John Wiley, London (1963).
43. G. B. Ansell and T. Chojnacki, The effect of diisopropylfluorophosphonate on enzymes concerned in brain phospholipid synthesis, *Progr. Biochem. Pharmacol.* **3**:189–195 (1967).
44. G. Porcellati, in *Convegno Internazionale di Studi sulla Sclerosi Multipla* (C. L. Cazzullo, ed.), pp. 195–206, La Tipografica Varese, Gallarate (1966).
45. L. E. Hokin, A. Yoda, and R. Sandhu, Irreversible inhibition of adenosine triphosphatases, diglyceride kinase and phosvitin kinase of brain by diisopropylphosphorofluoridate, *Biochim. Biophys. Acta* **126**:100–116 (1966).
46. G. Porcellati, Organofosforici demielinizzanti e lipogenesi del nervo periferico, *Acta neurologica* **22**:168–172 (1967).
47. J. D. Taylor, The effect of tri-ortho-cresyl phosphate intoxication on phospholipid synthesis in cat spinal cord, *Can. J. Physiol Pharmacol.* **43**:715–721 (1965).
48. C. Ntiforo and M. Stein, Labilization of lysosomes as an aspect of the biochemical toxicology of anticholinesterase pesticides, *Biochem. J.* **102**:44P (1967).
49. G. Porcellati, Il metabolismo proteico nella degenerazione e rigenerazione del tessuto nervoso periferico, *Farmaco*, **20**: 586–605 (1965).
50. D. H. Clouet and H. Waelsch, Amino acid and protein metabolism of the brain. IX. The effect of an organophosphorus inhibitor on the incorporation of ^{14}C-lysine into the proteins of rat brain, *J. Neurochem.* **10**:51–63 (1963).
51. G. Porcellati, Proteinase activity of nervous tissues during experimental demyelination, *Proc. VIIth Intern. Congr. Neurol.* 815–822 (1961).
52. A. Millo and G. Porcellati, The action of some organo-phosphorus compounds on "neutral proteinase" *in vitro*, *Ital. J. Biochem.* **10**: 333–343 (1961).
53. G. Porcellati, La reazione fosforilcolina:citidililtrasferasica nella degenerazione mielinica secondaria, *Acta Neurol.* **24**:119–124 (1969).
54. G. Porcellati, in *Protein Metabolism of the Nervous System* (A. Lajtha, ed.), 601–620, Plenum Press, New York (1970).
55. G. Porcellati, in *Handbook of Neurochemistry* (A. Lajtha, ed.) Vol. 7, Plenum Press, New York (1971), in press.
56. G. Porcellati, in *Handbook of Neurochemistry* (A. Lajtha, ed.) Vol. 2, pp. 393–422, Plenum Press, New York (1969).
57. B. V. Ramachandran, Distribution of DF^{32}P in mouse organs. III. Incorporation in the brain tissue, *Biochem. Pharmacol.* **16**:1381–1383 (1967).
58. V. Bonavita and R. Guarneri, NAD- and NADP-linked dehydrogenases in the sciatic nerve of rats injected with di-isopropylfluorophosphonate, *Brain Res.* **2**:200–207 (1966).
59. R. J. Rossiter, in *Chemical Pathology of the Nervous System* (J. Folch-Pi, ed.), pp. 207–230, Pergamon Press, London (1961).

60. J. F. Berry, W. H. Cevallos, and R. R. Wade, Lipid class and fatty acid composition of intact peripheral nerve and during Wallerian degeneration, *J. Amer. Oil Chem. Soc.* **42**:492–500 (1965).
61. R. E. McCaman, in *Ultrastructure and Metabolism of the Nervous System* (S. R. Korey, A. Pope, and E. Robins, eds.), pp. 169–181, Williams and Wilkins, Baltimore (1962).
62. A. C. Johnson, A. R. McNabb, and R. J. Rossiter, Chemical studies of peripheral nerve during Wallerian degeneration. I. Lipids, *Biochem. J.* **45**:500–508 (1949).
63. A. C. Johnson, A. R. McNabb, and R. J. Rossiter, Chemistry of Wallerian degeneration. A review of recent studies, *Arch. Neurol. Psychiatr.* (Chicago) **64**:105–121 (1950).
64. N. Miani, The relationship between axon and Schwann cell. Phospholipid metabolism of degenerating and regenerating peroneal–tibial nerves of the rabbit *in vitro*, *J. Neurochem.* **9**:525–536 (1962).
65. J. Domonkos and L. Heiner, On the possibility of lysophosphatide formation during Wallerian degeneration, *J. Neurochem.* **15**:93–98 (1968).
66. J. F. Berry and J. D. Coonrad, Hydrolysis of nucleoside diphosphate esters in peripheral nerve during Wallerian degeneration, *J. Neurochem.* **14**:245–255 (1967).
67. N. Miani, Metabolism and chemical changes in regenerating neurons. III. The rate of incorporation of radioactive phosphate into individual phospholipids of the nerve-cell perikaryon of the C.8 spinal ganglion *in vitro*, *J. Neurochem.* **9**:537–541 (1962).
68. D. Nicholls and R. J. Rossiter, Metabolism of lipids of peripheral nerve regenerating after crush, *J. Neurochem.* **11**: 813–818 (1964).
69. R. J. Rossiter, E. T. Pritchard, and K. P. Strickland, Metabolism of nucleic acids and lipids in brain and in peripheral nerve during experimental demyelination. *Neurol.* **8**; Suppl. 1, 34–40 (1958).
70. G. Majno and M. L. Karnovsky, A biochemical and morphological study of myelination and demyelination. II. Lipogenesis *in vitro* by cat nerve following transection. *J. Exp. Med.* **108**:197–214 (1958).
71. R. E. McCaman, Intermediary metabolism of phospholipids in brain tissue. Microdetermination of choline phosphokinase, *J. Biol. Chem.* **237**:672–676 (1962).
72. R. E. McCaman, M. Smith, and K. Cook, Intermediary metabolism of phospholipids in brain tissue. II. Phosphatidic acid phosphatase, *J. Biol Chem.* **240**:3513–3517 (1965).
73. R. E. McCaman and K. Cook, Intermediary metabolism of phospholipids in brain tissue. III. Phosphocholine-glyceride transferase, *J. Biol. Chem.* **241**:3390–3394 (1966).
74. G. Porcellati and B. Curti, L'effetto della demielinizzazione sperimentale sulla attività di alcune fosfatasi dei nervi periferici, *Bioch. Appl.* **7**:42–56 (1960).
75. M. L. Karnovsky and G. Majno, *in Chemical Pathology of the Nervous System* (J. Folch-Pi, ed.), pp. 261–266, Pergamon Press, London (1961).
76. D. F. Matheson and J. B. Cavanagh, Protein synthesis in peripheral nerve and susceptibility to diphtheric neuropathy, *Nature* (*London*) **214**:721–722 (1967).
77. W. N. Aldridge and J. M. Barnes, Esterases and neurotoxicity of some organophosphorus compounds, *Biochem. Pharmacol.* **15**:549–554 (1966).
78. M. K. Johnson, Brain proteins labelled by DFP^{32} and their connection with the delayed neurotoxic effect of some organophosphorus compounds, First International Congress of Neurochemistry (Strassbourg, 23–28th July, 1967), p. 111.

Chapter 20

THE ALTERATION OF BRAIN METABOLISM BY NARCOTIC ANALGESIC DRUGS

Doris H. Clouet*

New York State Research Institute for Neurochemistry and Drug Addiction Ward's Island, New York, New York

I. INTRODUCTION

The term, narcotic analgesic drug, as used in this chapter includes the naturally occurring opiates and their derivatives and other compounds of similar pharmacological activity and structural relationships. It has been suggested that the pharmacological activity resides in a backbone structure of a tertiary nitrogen linked through an ethylene bridge and a quaternary carbon to a planar phenyl ring[1–3]; this structure may be discerned in the common narcotic drugs and antagonists (Fig. 1), as well as in the more complex highly active synthetic narcotic agents.[4] The synthesis of the latter group of drugs has forced a revision of the structural requirements for agonist or antagonist activity. The *N*-allyl derivatives of morphine and levorphanol (nalorphine and levallorphan) and the *N*-methylcyclopropyl derivatives of the benzomorphan series (cyclazocine) have mainly antagonist with some agonist activity, while the *N*-allyl derivative of oxymorphone, naloxone, is considered a pure antagonist.[5,6] However, the nature of the alkyl group on the tertiary nitrogen is not the sole determinant of agonist or antagonist activity.[4,5]

The administration of a single dose of morphine or other narcotic drugs leads to a spectrum of pharmacological responses including depressant effects: a decrease in respiratory rate, blood pressure, awareness of pain, gastrointestinal motility, anxiety, etc., and stimulant effects: euphoria and convulsions (particularly at high doses of drug). Quantitation of the drug effect is accomplished by measuring one or more of the pharmacological responses, such as the period of analgesia (by lack of response to a painful stimulus) or hypothermia (by oral or core temperature measure-

* Now at New York State Narcotic Addiction Control Commission Testing and Research Laboratory, Brooklyn, New York.

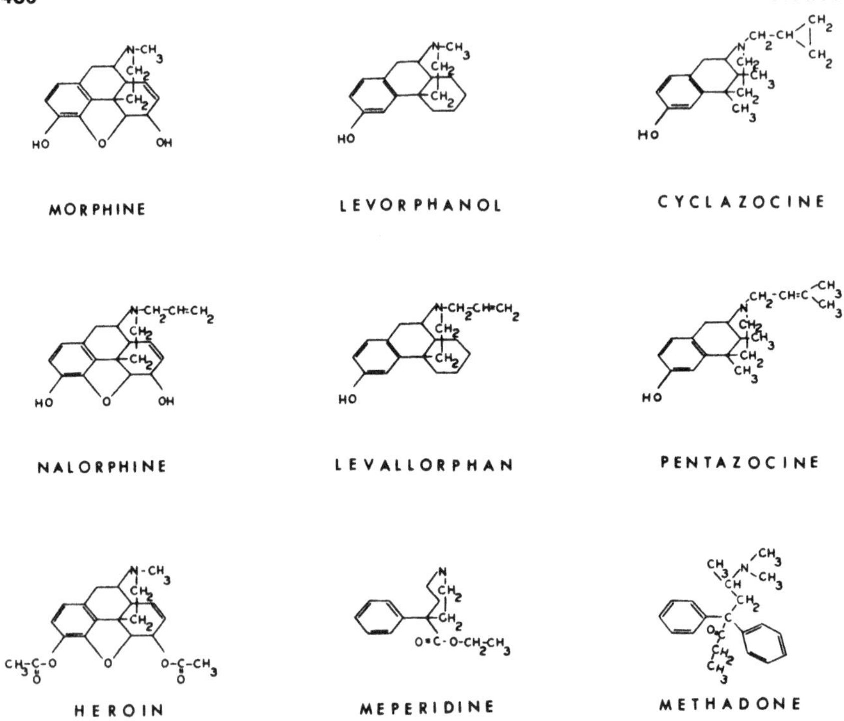

Fig. 1. Structures of some narcotic analgesic drugs and antagonists. Nalorphine, levallorphan, cyclazocine, and pentazocine are antagonists, the rest are agonists.

ments). The chronic administration of any agonist in this category of drugs results in a diminished pharmacological response to succeeding doses, i.e., tolerance. Tolerance is measured quantitatively by relating the response after multiple doses to that after the initial dose. Another response characteristic of narcotic analgesic drugs is the withdrawal syndrome, found when drug use is discontinued either through elimination of drug consumption or by the administration of an antagonist, which precipitates immediate withdrawal symptoms. Drug dependence, a term found preferable to addiction, is defined by the phenomena of tolerance and habituation (drug-seeking behavior) with chronic drug use and withdrawal symptoms upon discontinuance of drug use.[(7)] In this chapter, the pharmacological state of experimental animals will be described in the terms defined above whenever attempts have been made to relate biochemical changes to pharmacological effects. In addition, in assessing the effects of drug administration, both species differences and dose effects are relevant in relating studies from various laboratories, particularly since the dose necessary to produce a stimulant or depressant effect is species-specific.

Overwhelming evidence from physiological and biochemical studies indicates that most of the responses to the administration of narcotic analgesic drugs are reactions of the central nervous system. It is the aim of this contributor to review the experimental studies concerning the biochemical responses in the nervous system to the administration of narcotic drugs, and to offer suggestions for the ordering of the responses into modes of chemical control in the nervous system.

II. METABOLIC DISPOSITION OF NARCOTIC DRUGS IN NERVOUS TISSUE

A. Uptake and Transport of Narcotic Drugs in the Nervous System

When narcotic analgesic drugs are administered to experimental animals, the drugs enter most tissues rapidly, with no particular concentration in the nervous system.(8) However, with time, as the drugs are cleared from plasma, they are retained against a concentration gradient both in brain and in nerve.(9,10) There are variations in the levels of drug in the nervous system and the time necessary in order to reach maximal concentration after each route of administration; both parameters are dependent on the drug used and the dose of drug. The highest level of ^{14}C-morphine in the nervous system is found at 35 min to 4 hr after injection by the intraperitoneal route in both tolerant and nontolerant dogs.(11) In general, after systemic administration, the drugs reach peak levels in blood in less than 30 min, and in nervous tissue, at the same time or slightly later.(8,12) The route of administration is important both for attaining adequate drug levels in brain, as will be discussed later, and for bypassing catabolism in the liver.(10) For example, morphine-*N*-oxide, which is about one-twentieth as active as morphine when introduced subcutaneously, is much less active when introduced intraperitoneally because the *N*-oxide is converted to normorphine during passage through the liver.(14) For chronic drug administration, pellets of morphine may be implanted subcutaneously with drug release for several weeks, during which tolerance develops.(15)

Studies of the regional distribution of ^{14}C-morphine in the brains of dogs reveal that there is a fairly even distribution of drug from area to area throughout the brain, with higher levels in gray than in white matter for each area.(16,11)

Differential centrifugation of homogenates of whole brain from narcotic-treated rats reveals that most of the injected drug reaching the brain after the injection of tritiated dihydromorphine remains in the soluble fraction of the brain, and that the centrifugal distribution is quite similar when the drug is added to brain homogenates *in vitro*.(17,18) After the intracisternal injection of tritiated dihydromorphine, the labeled drug has been found both in ribonucleic acid fractions and in synaptosomal fractions prepared from the brains of the injected animals.(19,20) Both nucleic acid and synaptosomal fractions are also labeled *in vitro*, when tritiated dihydromorphine is incubated with brain slices or homogenates before fractionation.(19–21)

There seems to be a consensus of evidence that the rate of entrance and the distribution of narcotic drug are not related to the previous drug history of the experimental animal, although minor differences have been found between tolerant and nontolerant animals.[8,11] The reported differences are in opposite directions: labeled cyclazocine is excreted faster in nontolerant dogs than in tolerant dogs,[22] while plasma levels of morphine decrease more rapidly in tolerant than in nontolerant animals.[11] The most recent studies of drug levels in tolerant and nontolerant animals uphold the conclusions of earlier studies that the pharmacological response to narcotic drugs is not directly proportional to the concentration of drug in whole brain.[11,23-25] Thus, a tenfold difference in the brain level/dose ratio for etorphine over morphine cannot account for the thousandfold difference in pharmacological activity of etorphine over morphine.[26] In nontolerant animals, the simultaneous administration of the antagonist nalorphine with the agonist has been reported to increase the brain level of morphine, and to decrease the brain level of levorphanol from the level in animals receiving the agonist alone.[16,27] It is quite clear that the simultaneous administration of etorphine and its structural antagonist, cyprenorphine, to rats results in decreased levels of the agonist in the brain, muscle, blood, and particularly, stomach, suggesting that there is a decrease in drug uptake into tissue and its excretion into bile in the presence of the antagonist.[26]

Evidence is equivocal concerning the existence of a blood–brain barrier to narcotic drugs. Both the ED_{50} and the LD_{50} of morphine in rats increase with the age of the rat, as does the dose necessary to obtain a certain level of drug in the brain,[28-30] suggesting the development of a barrier system. In the human infant, too, there seems to be a barrier effect since the dose of morphine or meperidine in infants is about one-third that in adults.[31] In adult rats, there seems to be a partial barrier to the entrance of quaternary congener, *N*-methyl morphine, into brain, since the injection systemically produces respiratory failure but no analgesia or hypothermia, although these two latter responses are found when the drug is injected intracerebrally.[32] On the other hand, the LD_{50} of codeine does not change with age in the same way as morphine in rats and no dependence of dose effect with age is seen with heroin or meperidine in the rat, and no age effect at all in the guinea pig.[29,30] The comparable effective dose of morphine administered systemically or into the cerebrospinal fluid is about 60:1, which would be expected on the basis of the relative tissue space available for the initial drug distribution, so that if there is a blood–brain barrier to morphine, it is not formidable.

There is a placental barrier to both dihydromorphine and etorphine, since maternal blood levels of these drugs remain higher than fetal levels up to 2 hr after drug administration.[33,34] Etorphine crosses into fetal brain more rapidly than dihydromorphine and is accumulated in fetal brain at a higher brain/plasma level than dihydromorphine.[34] The lack of a brain barrier in the fetus is shown by the higher brain/blood ratios of both dihydromorphine and etorphine in the fetus after the administration of the drugs to pregnant animals.[34]

Narcotic drugs are accumulated against a concentration gradient in isolated preparations from the nervous system such as choroid plexus and cerebral cortex slices.[35-37] The accumulation of morphine, dihydromorphine, nalorphine, codeine, and the *d*- and *l*-isomers levorphan and dextrophan, by pieces of choroid plexus has been shown to be inhibited by the addition of metabolic inhibitors, and competitively, by other narcotic drugs, as is the transport of dihydromorphine into cerebral cortex slices.[35,36] The transport is specific for tertiary amines since hexamethonium, decamethonium, and carbachol do not inhibit the uptake of dihydromorphine, nor does morphine inhibit the uptake of decamethonium.[36,38]

B. Metabolic Conversions of Narcotic Drugs

The biotransformations of narcotic drugs in man and in experimental animals include: *N*-dealkylation (morphine, methadone, nalorphine, etc.), *O*-dealkylation (codeine), deacetylation (heroin), dealkylation followed by cyclization (methadone), hydrolysis (meperidine), hydroxylation of the methyl group followed by oxidation to carboxylic acid (pentazocine), and conjugation as glucuronides or ethereal sulfates (morphine).[39-45] Both morphine 3-glucuronide and morphine ethereal sulfate have been isolated from the urine of man and experimental animals with the major metabolites of morphine differing in various species.[44,46-48] In addition, the dealkylated derivatives may be remethylated using *S*-adenosyl methionine as a methyl donor.[49,50] Most of these conversions take place in the liver in reactions involving the liver microsomal drug metabolizing enzymes and cytochrome P-450.[51] Only dealkylation, deacetylation, and remethylation have been shown to occur in the nervous system.[49,52,53] The importance of the metabolic conversions of narcotic drugs in nonnervous tissue lies in the ability of such reactions to alter effective drug levels in plasma and in tissues, and, consequently, the pharmacological and biochemical responses to the drug. The administration of nonnarcotic drugs[54,55] and of hormones[56,57] does alter the rate of metabolism of narcotic drugs by inducing the activity of the drug-metabolizing enzymes in liver. In rats, the chronic administration of morphine or other narcotic drugs results in changes in the activity of liver enzyme activity catalyzing the metabolism of morphine, meperidine, codeine, dihydromorphine, levorphan and nalorphine, including conjugation.[54,55,58-62] The activities of these enzymes decrease as the pharmacological responses decrease.[61,63] However, this phenomenon is most likely peculiar to male rats. In rats there is a sex difference in the activity of liver drug-metabolizing enzymes which is responsive to sex hormone treatment,[57] while in mice and guinea pigs neither sex differences in nor decreases during chronic morphine treatment of demethylase activity are found.[64] Even in male rats, the alterations in the activity of liver enzymes metabolizing morphine have very slight effect on the development of tolerance in morphine-treated animals.[65] The lack of specificity in the drug-metabolizing enzyme response to the administration of inducer drugs has been related to a reaction common to all enzyme activities involving the microsomal cytochrome system in liver, for which drugs undergoing oxidative degradation are substrates.[51] However, it is possible to induce *N*-dealkylating activity in liver by administering the inducer drug, phenobarbital, without at the same time inducing *O*-dealkylating activity for codeine, so that factors other than the common cytochrome system are involved.[66]

III. EFFECTS OF NARCOTIC DRUGS ON GENERAL METABOLIC SYSTEMS

A. Energy and Intermediary Metabolism

One of the depressant effects of narcotic drugs in man and animals is the depression of respiration. After morphine treatment, oxygen consumption decreases as does both the tidal volume and respiratory rate.[67-69] However, the reflection of these physiological changes in biochemical effects has been difficult to demonstrate. The oxygen consumption of cerebral cortex slices isolated from morphine-treated rats does not differ from those from untreated animals, nor does the presence of morphine in the incubation medium affect the uptake of oxygen. It is only when the stimulatory effect of potassium ions on oxygen consumption of cortex slices is examined, that a difference emerges between slices from tolerant or control animals. In the presence of 10^{-3} M morphine in the medium, there is a decreased stimulation of oxygen consumption on the addition of potassium ions.[70] This effect is dependent on a low concentration of calcium ions in the medium, and is not found when the cortex slices are isolated from tolerant rats.[70,71] In addition to morphine, levorphanol, meperidine, and methadone also prevent the full stimulation of respiration by potassium ions in cortex slices from nontolerant animals, but have no effect on slices from morphine-tolerant animals.[72,73] In *in vivo* experiments, it has been shown that less radioactivity from ^{14}C-glucose is found in citric acid cycle intermediates of brain 30 min after the administration of a single dose of morphine to rats than in untreated animals.[74] In an isolated muscle preparation, the rat diaphragm, the uptake of glucose is influenced both by the presence of morphine in the medium, and by the prior treatment of the rat with morphine. The addition of morphine *in vitro* increases the uptake of glucose in muscles of untreated rats, and decreases the uptake in diaphragms from tolerant animals.[75]

The effect of the administration of morphine to rodents on the levels of ATP and phosphocreatine in brain has been most variable, and is probably related to the time of sampling and the relative excitant/depressant ratios of drug effects.[76,77] One of the early effects of narcotic drugs on *Escherichia coli* or HeLa cell cultures is a slow loss of ATP through leakage out of the cells,[78] so that the increases of ATP in brain found 1 hr after the administration of drug may be counter-responses to an initial loss of ATP. In this regard, studies of the fate of ^{14}C-leucine injected into the cerebrospinal fluid reveal that the oxidation of leucine to glutamate, glutamine, aspartate and other compounds related to the citric acid cycle in the brain was the same in control and morphine-treated animals killed 2 hr after the injection.[79]

B. Protein Metabolism

The rate of protein synthesis in rat brain is inhibited transiently when the brains are derived from morphine-treated animals, whether protein synthesis is measured by the incorporation of a labeled precursor into brain protein *in vivo*, or in isolated brain ribosomal preparations.[80,81] The addition of morphine to an amino acid incorporating system with brain ribosomes *in vitro* has no effect on the rate of amino acid incorporation into protein. However, when intact cells of *E. coli* or HeLa cultures are incubated in the presence of another narcotic drug, levorphanol, or its congener antagonist, levallorphan, protein synthesis is inhibited.[82,83] This inhibition has been attributed to an initial decrease in ATP level, followed by a decreased incorporation of ^{14}C-amino acids into proteins in both types of cells.[78] It has not been possible to establish such a relationship of the inhibition of protein synthesis to a decrease in ATP levels in the brains of morphine-treated rats, since one would then expect ribosomes isolated from the brains of morphine-treated rats to have normal incorporating activity in the presence of an optimal energy supply, instead of the decreased activity found experimentally.[81] In HeLa cells in culture, another mechanism for the inhibition of protein synthesis has been found in the presence of a large number of narcotic agonists and antagonists: the reversible dissociation of ribosomes during the incubation of the cells with the drugs.[84]

The discussion of whether protein synthesis in brain is necessary for the development of tolerance to narcotic drugs will be deferred until the end of the next section, so that the action of inhibitors both of protein and RNA synthesis on the development of tolerance may be considered at the same time.

In addition to indirect effects of the administration of morphine on the synthesis and release of peptide hormones of the pituitary and hypothalamus (described in Section IV, B), the injection of morphine into isolated cervical ganglia, or intravenously, has been shown to inhibit the action of the peptide hormones, angiotensin and bradykin, in stimulating ganglionic transmission.[85] It is suggested that the drug and hormone act on a common site in the postreceptor pathway.

C. Nucleic Acid Metabolism

In *E. coli*, the addition of 10^{-3} M levorphanol inhibits the incorporation of ^{32}P into RNA.[82,86] That this is a specific inhibition of ribosomal RNA has been deduced from a number of experiments in which the turnover of messenger and transfer RNA is not affected by the addition of drug, while the synthesis of 5S and ribosomal RNA is inhibited.[86,87] In the presence of 10^{-3} M levorphanol or levallorphan, the rate of incorporation of labeled guanine into the RNA of HeLa cells is similarly reduced in an inhibition involving all species of RNA.[83] Greene and Magasanik have made a detailed study in both types of single cell cultures.[78] In *E. coli*, the addition of 3 mM levallorphan produces a rapid inhibition of both protein and RNA synthesis, as well as an inhibition of the incorporation of ^{14}C-thymidine into DNA. In HeLa cells, in addition to an inhibition of RNA synthesis, a loss of ATP with a decrease in dephosphorylated metabolites is seen 30 sec after the addition of drug. The effect of the narcotic drug on the cells is interpreted as an initial action resembling an ATPase activation with subsequent events related to permeability changes and loss of ATP.[78]

The administration of morphine to rats produces a similar inhibition of the synthesis of ribosomal RNA in the liver,[88] and also alters the rate of synthesis of RNA in a biphasic way, using the incorporation of ^{14}C-orotic acid into brain RNA, with an initial inhibition of synthesis followed by an increased incorporation into heavy RNA.[89] When tritiated dihydromorphine is injected into the CSF, some of the drug is recovered loosely bound to RNA isolated from brain nuclear or microsomal–supernatant fractions.[19]

The development of tolerance to narcotic drugs could involve the synthesis of a "memory" molecule (be it RNA, protein, or polysaccharide) or a change in the rate of synthesis of an enzyme (or the messenger RNA for the enzyme) necessary for the synthesis of an additional amount of any component produced during the development of tolerance. A large number of studies have been undertaken to explore this possibility.[90–97] The administration of an inhibitor of RNA synthesis, actinomycin D, results in a delay in the development of tolerance to the analgesic effects of chronic morphine treatment in mice.[90] Another inhibitor of RNA synthesis, 8-azaguanine, produces the same effect: tolerance to chronic morphine administration is less in mice also receiving the inhibitor.[91,92] Tolerance may be divided into short-term tolerance, a hyposensitivity to the drug occurring 2–5 hr after the administration of the drug, and long-term tolerance occurring 6 hr or later. The administration of actinomycin D blocks the development of long-term tolerance but not that found immediately after drug use.[93] However, when morphine, meperidine, or heroin is infused intravenously into rats, the development of tolerance during the infusion is prevented by the simultaneous infusion of actinomycin D, although established tolerance is not reversed by the inhibitor.[94] Other inhibitors of RNA synthesis, 6-mercaptopurine and 5-fluorouracil, also reduce the level of tolerance developed during morphine infusion.[98,99] Thus, the evidence tends to indicate that synthesis of RNA is required for the development of tolerance to narcotic drugs. Inhibition of RNA synthesis affects the synthesis of new messenger RNA and preribosomal units, so that protein synthesis is also inhibited in time. Whether the inhibition of protein synthesis in brain without interference with RNA turnover also affects the rate of development of tolerance is the subject of several experimental studies. The development of tolerance to the production of lenticular opacity in mice during chronic levorphanol treatment is significantly depressed when an inhibitor of protein synthesis, puromycin, is administered simultaneously with levorphanol.[93,95] The administration of another inhibitor, cycloheximide,[97,100] or of a suppressor of the immune response, cyclophosphoramide,[96] also decreases the level of tolerance to morphine arising during multiple injection schedules. When protein synthesis in brain is reduced by 40% by the administration of cycloheximide or puromycin, the development of acute tolerance to morphine infusion is also reduced.[98] These latter studies suggest that the action of inhibitors of RNA synthesis on the development of tolerance is affected through the subsequent inhibition of protein synthesis. However, further studies correlating biochemical effects in the nervous system with pharmacological responses are necessary before a firm conclusion about the requirement of protein synthesis in brain for the development of tolerance may be reached.

D. Lipid Metabolism

In order to distinguish between direct effects of the administration, or the addition in *in vitro* experiments, of narcotic drugs on lipid metabolism in the nervous system and indirect effects consequent to changes in energy metabolism, oxygen uptake, or ATP pool size, these parameters have been examined at the same time as phospholipid synthesis. As already described, the oxygen uptake in a brain slice does not change in the presence of morphine. In the same experiments, the incorporation of ^{32}P into phospholipid is increased 40% with the largest increases in phosphoinositides and phosphatidic acids.[101] That these changes in the rate of synthesis of phospholipids reflect relative changes in the rate of synthesis of individual phospholipid classes is shown by the studies of Mulé, who examined the time curves for the incorporation of ^{32}P, ^{14}C-glycerol or ^{14}C-choline into the phospholipids of guinea pig cerebral cortex slices.[102] In the presence of morphine there is more synthesis of phosphatidic acids and phosphatidyl inositide and less synthesis of phosphorylcholine. Mulé suggests that the 1,2-diglyceride becomes limiting in brain slice experiments. When morphine and ^{32}P are given to guinea pigs *in vivo*, all of the classes of phospholipids in brain show an increase in new synthesis, suggesting that in the whole animal, the diglyceride level is not limiting.[103] In a model system consisting of an organic and an aqueous phase with the movement of morphine from the organic to the aqueous phase as an indicator of binding and transport of the drug, both calcium ions and phospholipids are required, with phosphatidic acids isolated from brain the most effective in transporting the drug into the aqueous phase.[104]

In the isolated guinea pig ileum preparation, morphine in very low doses (0.01 μg/ml) inhibits the effect of prostaglandin PGE_1.[105] Nalorphine and etoniazine have similar effects.

IV. EFFECTS OF NARCOTIC DRUGS ON SPECIFIC METABOLIC MECHANISMS

A. Amines

1. *Catecholamines*

The administration of a single dose of a narcotic drug to animals or man results in a release of catecholamines from the adrenal gland.[106] A decrease is found in the number of granular catecholamine storage vesicles in adrenergic nerves of morphine-dependent rats at the same time that the content of agranular vesicles increases, suggesting that the terminal adrenergic nerves are also depleted of catecholamines by the narcotic drug.[107] In brain, the release and depletion of norepinephrine and dopamine has been demonstrated following the acute administration of morphine in various mammalian species.[108–111] After multiple doses of morphine, the level of norepinephrine either shows no change or is actually increased.[108,109,112,113] In the hypothalamus, norepinephrine levels are lower 1 hr after the introduction of morphine into the cerebrospinal fluid.[114] In this area of brain the decrease in norepinephrine level has been related to a depressed synthesis of dopamine in the hypothalamus after the administration of morphine.[115] In whole mouse brain, the depletion of dopamine precedes the depletion of norepinephrine, implicating the precursor dopamine in the lower levels of norepinephrine found after treatment with morphine.[111] The administration of an inhibitor of dopamine β-hydroxylase, sodium diethyldithiocarbamate, to morphine-treated rats results in increased level of brain dopamine and potentiation of analgesia.[116] In rats treated chronically with morphine, there is no change in the turnover rate of norepinephrine in the brain if the turnover is measured by the disappearance of injected ^{3}H-norepinephrine.[117] However, the biosynthesis of ^{14}C-dopamine and ^{14}C-norepinephrine from ^{14}C-tyrosine in the intact brain of the rat, as well as in the hypothalamus and striatum, is increased after a single injection of morphine, and further increased upon subsequent injections.[118] It has been suggested that morphine alters the levels of catecholamines only in those areas of brain in which there is a daily monamine rhythm: substantia nigra, anterior hypothalamus, caudate nucleus, and cervical cord.[119]

The administration of nonnarcotic drugs which affect the levels of brain catecholamines also alters the biochemical response to morphine. Pretreatment of animals with chlorpromazine prevents the depletion of brain norepinephrine and pretreatment with reserpine, α-methyl tyrosine, or tetrabenazine results in smaller changes in catecholamine levels since the prior action of these drugs is to deplete the supply of brain amines.[108,112,120–122] The pretreatment with reserpine 24 hr prior to an injection of morphine seems to abolish the analgesic response, although the measurement of analgesia is complicated by the potentiation of analgesia by reserpine observed when the drugs are introduced simultaneously.[123,124] Monoamine oxidase inhibitors such as nialamide or iproniazid increase the pharmacological activity of morphine and other narcotic drugs by inhibiting the catabolism of the amines in the nervous system,[125,126] while β-adrenergic blockers such as propranolol block some of the responses to morphine and antagonize the synergistic effects of norepinephrine in prolonging the analgesic response to morphine.[127] Both norepinephrine and epinephrine potentiate some of the pharmacological effects ofmorphine, particularly when introduced into the CSF.[127–129] Further confusion concerning which catecholamine is the most important in the action of morphine and other narcotic drugs is introduced by the results of a study showing that when the level of dopamine is normal and the level of norepinephrine is reduced by an inhibitor of dopamine β-hydroxylase, diethyldithiocarbamate, the analgesic response to morphine is not affected.[116]

Morphine and other narcotic drugs are species-specific in the nature of the predominant pharmacological responses: either excitation or depression, or both. The depletion of catecholamines from nervous tissue seems to be a general response in all species. The comparative alterations of amine metabolism in the brains of various species of morphine-treated animals are discussed by Maynert.[130,131]

The major conclusion of the studies described in this section is the need for a functioning sympathetic nervous system in order for the pharmacological effects of morphine to occur. This conclusion is reinforced by a study in which unilateral sympathectomy is shown to abolish the effects of morphine unilaterally.[132]

2. *Serotonin*

The administration of narcotic drugs to animals has no effect on the levels of serotonin in whole brain.[133] There are, however, several studies in which the turnover of brain serotonin responds to the narcotic drug. An increased catabolism of serotonin is shown by an increased concentration of a metabolite, 5-hydroxyindoleacetic acid, in the brains of rats receiving a single dose of morphine.[134] If the catabolism of serotonin and other monoamines is inhibited by monoamine oxidase inhibitors such as pargyline or tranylcypromine, then serotonin piles up in the brain after an injection of morphine or meperidine.[135,136] In tolerant animals, the increase in brain serotonin levels induced by the injection of pargyline is greater than in nontolerant animals.[135]

p-Chlorophenylalanine depletes brain serotonin by competing with tryptophan as a substrate for tryptophan hydroxylase in the biosynthesis of serotonin. If this analog is administered to mice, the development of analgesic tolerance to morphine or methadone is impeded, and tolerance in the hypothermic effect is abolished, suggesting that the development of tolerance depends on an adequate store of serotonin in the nervous system.[135,137,138] Further evidence on this point stems from studies in which animals are pretreated with reserpine before the administration of narcotic drugs. Reserpine, which acts as an amine depletor in the central nervous system, antagonizes some of the pharmacological responses to morphine. The antagonism is reversed by the administration of a precursor of serotonin, 5-hydroxytryptophan, but not by serotonin given peripherally.[139] If the blood–brain barrier is bypassed by introducing serotonin intraventricularly, then the effect of reserpine in abolishing analgesia is reversed.[129] The administration of pargyline also reverses the antagonism of reserpine to morphine, presumably by maintaining high brain serotonin levels through a decreased catabolism.[137,140]

In studies conducted *in vitro*, many narcotic drugs inhibit the effect of serotonin on ileum contraction at a low drug concentration (10^{-7} M), while nonnarcotic depressants do not have this effect.[141]

3. *Acetylcholine*

The levels of acetylcholine and its synthesizing and hydrolyzing enzyme systems, choline acetylase and acetylcholinesterase, have been examined in the nervous system after the administration of narcotic drugs. At very low doses, there is no change in the level of acetylcholine in the brain, but at higher doses (100 mg/kg) a transient rise in acetylcholine is observed in the brains of nontolerant mice, with a maximal increase 30 min after the injection of the drug.[131,142–144] During the development of tolerance to chronic drug administration, the increase in acetylcholine is no longer seen. In withdrawal induced by either discontinuing the drug, or by nalorphine injection, the level of acetylcholine in brain sharply increases, peaking 39–46 hr after the last injection of morphine.[144] The activity of choline acetylase in one area of brain, the caudate nucleus, reflects the changes in acetylcholine level: 1 hr after the administration of a single dose of morphine, the activity of the enzyme is lower, returning to normal levels upon subsequent injections of morphine.[145] No change has been seen in the level of a precursor of acetylcholine, acetyl CoA, at the time that the changes in acetylcholine and choline acetylase are found.[146–148]

It has been suggested that the rise in bound acetylcholine in brain seen following the acute administration of morphine is due to a decreased liberation of acetylcholine from its storage site.[142] The inhibitory effect of morphine on acetylcholine release has also been seen in isolated preparations of preganglionic and superior cervical ganglion, and in perfused whole brain of the cat.[147] In brain cortex slices, the addition of morphine inhibits the release of preloaded acetylcholine from the slice; a reaction dependent on an adequate level of calcium ions.[149,150] It is likely that morphine competes with acetylcholine for the releasing mechanism, since the binding of acetylcholine has been shown to be competitively inhibited by morphine and other narcotic drugs both in brain cortex slices, and in synaptic-vesicle-rich fractions of brain.[151,152]

The hydrolysis of acetylcholine by brain acetylcholinesterase is inhibited *in vitro* in the presence of morphine and other agonists and antagonists.[153,154] Erythrocyte cholinesterase is also inhibited by narcotic drugs, including the newly introduced highly active compounds.[155,156] The competitive nature of the interaction of narcotic drugs and acetylcholine for sites on serum cholinesterase has been shown to be limited to a single site, the hydrolytic site.[157,158] In spite of this strong inhibitory effect of narcotic drugs on the activity of acetylcholinesterase *in vitro*, no effect of morphine administration on the activity of brain acetylcholinesterase *in vivo* has been shown in either tolerant or nontolerant animals.[142,153] It is possible that cholinesterase is present in the nervous system in such excess that local inhibition is masked.

Thus, it has been demonstrated that morphine and other narcotic drugs compete with acetylcholine at two metabolic sites: the binding and the hydrolytic sites. The relative importance of the interference in acetylcholine metabolism to the spectrum of pharmacological responses is difficult to assess. Maynert suggests the interesting hypothesis that acetylcholine is involved in the excitatory responses, while catecholamines are related to the depressant effects of morphine.[130,131]

B. Hormones

1. *Corticosteroids*

The response of the adrenal gland to the administration of narcotic drugs is not limited to the release of catecholamines, as adrenal corticosteroid hormones are also released. In man and in experimental animals, the injection of a single dose of a narcotic drug such as morphine is followed by a transient increase in the plasma level of corticosteroids and an increased excretion of urinary steroids.[159–163] In chronically addicted man, the plasma level of corticosteroids remains at about one-quarter that in nonaddicted man, although the adrenal gland is completely responsive to the administration of ACTH.[160] Conversely, the presence or absence of the adrenal gland influences the pharmacological responses to morphine, which are intensified in adrenalectomized animals, and decreased upon the administration of cortisone.[164,165] The uptake of methadone or morphine in brain and spinal cord corresponds to the pharmacological responses:

lower brain levels of ^{14}C-methadone after the administration of cortisone and higher brain levels of ^{14}C-morphine in adrenalectomized animals.[165,166]

2. *ACTH and Other Pituitary Hormones*

That some of the secondary effects of narcotic drugs are mediated through responses of the pituitary hormones is shown by several kinds of evidence, in addition to that indicated by the release of adrenal corticosteroids and catecholamines as described in previous sections. ACTH is released from the anterior pituitary gland after a single injection of morphine, but is no longer released in chronic drug treatment.[167,168] However, chronic morphine treatment does not block hormone release in the anterior pituitary gland, but instead interferes with pituitary-releasing factors. Morphine blocks stress-induced and vasopressin-induced ACTH release but does not block the effect of corticotropin-releasing factor in the release of ACTH.[169] The hypothalamic pituitary hormone-releasing factors have also been implicated in two other hormone responses. The administration of morphine is followed by a release of the antidiuretic hormone from the posterior pituitary gland with the expected consequent changes in urinary excretion patterns.[170–174b] Tolerance develops to this effect, which is also blocked by the simultaneous administration of narcotic agonist and antagonist.[172,173] The opposite response to acute drug administration is seen in the release of thyroid-stimulating hormone: an inhibition of thyroid activity. However, this action of morphine is considered an increase in the inhibitory activity in the hypothalamus.[174a,175] Morphological studies show that the neurosecretory material in the posterior pituitary increases after a single injection of morphine and fails to do so on chronic drug treatment.[176] It is clear that even the response of the hypothalamic-releasing factors to the administration of morphine is most likely a secondary response, since morphine depletes the level of catecholamines in the hypothalamus, as already described.[114,115]

C. Inorganic Ions

The one persistent alteration in ion concentration in the central nervous system found in various kinds of studies of narcotic analgesic treatment involves calcium. After a single injection of morphine to mice, there is a transient fall in ionic calcium levels in the brain, with a return to normal values in 24 hr.[177] This effect no longer occurs in the tolerant mouse. It has been suggested that calcium plays a role in analgesia, since the administration of ionic calcium into the cerebrospinal fluid decreases the analgesia produced by morphine, and the administration of calcium chelators such as citrate or EDTA enhances the analgesic response to the injection of morphine.[178] The presence or absence of calcium also alters other biochemical responses to morphine: the depression of potassium-stimulated respiration by brain cortex slices is greater in Ca^{2+}-free medium;

the release of acetylcholine from brain slices is no longer inhibited by morphine in Ca^{2+}-free medium.[71,179] No effect is seen *in vitro* when morphine is added to brain microsomal preparations exhibiting ATP-dependent calcium accumulation.[180] However, in a two-phase system (water–chloroform), the phospholipid-facilitated transport of calcium ions from the aqueous to the organic phase is inhibited in the presence of various narcotic analgesic drugs.[104] The interaction of narcotic analgesic drugs and calcium may not be specific for narcotic drugs, but may instead be related to a general situation in the nervous system, in which low levels of calcium are concomitant with excitation, and high levels with depression.[181]

D. Electropharmacological Responses

The application of morphine or levorphanol to the surface of rabbit brain produces a high voltage slow wave in an electroencephalographic recording.[182] This type of activity is also seen in man.[183] The administration of morphine or other narcotic analgesic drugs to several species of experimental animals produces a dose-dependent effect: a blocked arousal and elimination of REM sleep at low doses, the restoration of arousal at high doses, and convulsive seizure activity at very high doses.[184,185] Similar changes are seen in EEG patterns of fetal guinea pigs when the pregnant female is injected with meperidine.[186] In the rat, a complex response to the acute administration of morphine includes the almost complete loss of paradoxical sleep and prominent "slow bursts." Tolerance develops rapidly to the decrease in paradoxical sleep.[187] Tolerance to the shift to higher voltage in EEG recording from the hypothalamus has been seen for 7 days after a single injection of a fairly low dose of morphine in the rat (8 mg/kg).[188] The use of EEG recordings as an extremely sensitive method of measuring narcotic drug response is limited only by the relative lack of specificity of the response and the technical difficulty of measuring EEGs in large numbers of experimental animals. In man, EEG recording is a most valuable technique for examining the effects of narcotic drugs in an objective way.[183,189]

V. TRANSFER OF TOLERANCE

A. Brain Factors

The hypothesis that tolerance to the pharmacological activity of narcotic analgesic drugs occurs through the synthesis of new components, or an increased synthesis of existing substances, in the nervous system during chronic drug treatment leads logically to attempts to isolate such substances from the nervous system. In one such study, it has been reported that mice which are injected with homogenates prepared from the brains of rats tolerant to morphine are more resistant to a test dose of morphine than animals which receive brain homogenates prepared from untreated animals.[190,191] In other laboratories, however, no differences have been found in the pharmacological responses to a test dose of morphine among animals receiving brain homogenates from either tolerant or untreated rats, or saline.[192,193] In another study, a substance has been isolated from brain homogenates of mice implanted with pellets of morphine subcutaneously, which increases the analgesic effect of morphine and decreases the abstinence symptoms after nalorphine when injected into nontolerant mice.[194] In this latter study, the substance has been found to be soluble in water, acid, acetone, and alcohol, but it is not morphine, which suggests that an active metabolite may be accumulating during the chronic absorption of large amounts of the narcotic drug from the implanted pellet.

B. Serum Factors

The comparison of the development of tolerance to the chronic consumption of narcotic drugs to the immune response has led to two kinds of studies, in one of which the length of time that sensitivity (tolerance) persists after the administration of morphine has been measured, and in the other, a search for serum factors which can transfer immunity to the effects of a subsequent injection of the narcotic drug. In the first kind of study, a single dose of morphine has a demonstrable effect on the response to a second injection of morphine, if the doses are as far apart as 180 days.[194,196] However, results from the second type of study are opposite to that anticipated from the immune response model: the administration of serum from morphine-tolerant animals to naïve mice or rats potentiates the effect of an injection of a test dose of morphine rather than causes the transfer of tolerance.[197,198] The whole question of hypo- and hypersensitivity to the effect of a second injection of a narcotic drug needs reexamination in the light of newer information on the importance of dose and time between doses on the rate of development of tolerance, and to the periods of increased or decreased responses to the drug.[199]

VI. SITES OF BIOCHEMICAL ACTION

A. Central Nervous System

The localization of the action of morphine in the CNS is dependent to some extent on which parameter of response to morphine is examined. The effect of morphine and other narcotic analgesic agents on electrophysiological responses, presumably related to the release of acetylcholine (see Section IV, A, 3), has been used to identify cholinergic fibers in central and peripheral nervous system, and the effect on various pharmacological responses has served to identify the various structures which are required for the pharmacological activity.[200] Two techniques have been particularly useful in exploring the CNS for possible anatomical sites for the action of narcotics as related to biochemical activity: (1) a microinjection technique by which a drug is introduced into a small area of brain, and (2) a microablation technique by which a small discrete area of brain tissue is removed.

Introducing morphine into the CSF of the IV° ventricle by way of the cisterna magna of rats reduces the dose required for equivalent analgesic or hypothermic response to 1/60th that necessary by the subcutaneous route (Clouet, unpublished). When the drug is introduced into the hypothalamic nuclei, the equivalent dose is further reduced to 1/400th of that required by intravenous injection.[199] However, there is a segregation of responses when the area into which the drug penetrates is limited. In the cat, for example, morphine introduced into the IV° ventricle or subarachnoid space produces respiratory depression, while injection into the III° ventricle produces excitatory as well as depressant responses.[201] The presence of a blood–brain barrier to narcotic drugs (see Section II, A) may require that the drug be transported from the CSF to brain tissue by way of the extra-blood–brain barrier structures such as the choroid plexus, the area postrema, or the subfornical organ.[202] The latter two structures differ in response to morphine: there is no decrease in norepinephrine levels in the area postrema while the amine levels fall in the subfornical organ after morphine administration.[203] Similarly, a low dose of morphine introduced into the III° ventricle of the cat has a greater hyperglycemic effect than the same dose in a lateral ventricle.[204,205]

The microinjection of morphine directly into regions of the anterior hypothalamus in the rat has indicated that the hypothermic response occurs only when the drug is injected in the area of the preoptic or anterior hypothalamic nuclei.[206] An injection of morphine into the medial hypothalamus produces an increase in plasma corticosteroids and a decrease of adrenal ascorbic acid, evidence of ACTH release mediated by this area of the hypothalamus.[207]

The inhibition of the release of thyrotropic hormone (TSH) produced by the administration of morphine has been used as an indicator in an exploration of the hypothalamus with microablation techniques. There is no effect on the TSH release when lesions are made in the anterior hypothalamus. In the posterior hypothalamus, a site whose loss prevents morphine activity is the medial mammillary nucleus.[175] The distinction between probable sites of action of morphine in the hypothalamus and sites of transport such as the subfornical organ is shown in studies in which lesions of the subfornical organ reduced the effectiveness of morphine injected intraventricularly, but not of morphine injected intravenously.[205]

Actual changes in neurosecretory material in the hypothalamus and posterior pituitary, presumably representing the hypothalamic pituitary-hormone-releasing factors, have been observed histochemically during the treatment of rats with morphine.[176] After a single injection of morphine, the amount of neurosecretory material increases in the posterior pituitary, and also in the axons of the hypothalamic supraoptic and periventricular nuclei. During chronic morphine treatment, the neurosecretory material is almost absent from the posterior pituitary, but much higher in the perikarya of the neurons of the hypothalamus rather than in the axons.[176]

Such experiments as these suggest that the sites of action of narcotic drugs are multiple both anatomically and biochemically, but have not, as yet, identified the initial biochemical lesion produced by the drug. This lesion may well precede temporally some of the secondary effects measured in the experiments described above.

B. Peripheral Tissue

It is difficult to separate secondary effects of narcotic drugs from primary ones in peripheral tissue. For example, the depressant effects of narcotic drugs on the respiratory center are well documented, although various drugs have different effects on the respiratory rate and on the tidal volume.[67] On the other hand, the effects of the administration of morphine on the secretion of adrenal hormones or on kidney water balance function are secondary effects requiring an intact pituitary for drug action, and the development of lens opacity in levorphanol-treated mice requires an adequate level of norepinephrine in the CNS.[93,95]

In isolated diaphragm muscle preparations, the addition of morphine results in an increased glucose uptake which is reversed when the muscle is derived from tolerant animals.[208] The sensitivity to insulin *in vitro* is also morphine-dependent in these preparations.[209] The isolated preparation of ileum also shows a direct response to the application of morphine, with a decrease in acetylcholine release[210,211] or a decrease in the release of serotonin in the presence of calcium.[212]

C. Sites in Single Cells

Attempts to develop bacterial variants with a requirement for a narcotic analgesic drug have not been successful.[86] However, the addition of fairly high concentrations of a narcotic drug to the culture medium has produced a number of metabolic effects. At 1.35×10^{-3} M levorphanol, the growth of *E. coli* is inhibited.[86] The synthesis of RNA is also inhibited at this drug concentration, while protein synthesis is inhibited to a lesser extent.[82] The narcotic antagonist, levallorphan, has a similar effect on the growth of *E. coli*.[78] The inhibition of RNA synthesis may be selective, only involving ribosomal RNA synthesis,[86] or may be complete, involving all species of RNA,[78] perhaps depending on drug concentration. One of the first biochemical changes noted when levallorphan is added to *E. coli* cultures is a loss of ATP both by leakage from the cells and by breakdown to AMP.[78]

The addition of narcotic drugs to single cell preparations from mammalian tissue also results in a change in cellular activity.[83,213] HeLa cells in culture with either levorphanol or levallorphan added have an inhibition of the synthesis of RNA and protein.[83] There is a partially selective loss of ATP by leakage from the cells.[78]

The early occurrence of selective effects on the transport of small molecular weight compounds both in mammalian and bacterial systems suggests that membrane components may be vulnerable to the action of narcotic drugs.

VII. CONCLUSIONS

The administration of narcotic analgesic drugs to experimental animals has a profound effect on metabolism in the central nervous system. Alterations have been observed in protein, nucleic acid, lipid, and energy metabolism. In addition, the biogenic amines and the pituitary–adrenal hormonal mechanisms are implicated in the biochemical response to narcotic drugs. Because all of the metabolic systems in brain are in a delicate equilibrium, responsive to the slightest change in chemical environment, it is likely that most of the observed effects are secondary shifts consequent to the initial biochemical lesion. Such factors as the competition between acetylcholine and morphine for hydrolysis and transport, the release *in vivo* of serotonin and norepinephrine by morphine, the inhibition of protein and ribonucleic acid synthesis by narcotic drugs, the effects of the drugs on membrane phospholipids and on the loss of ATP in single cells, all suggest possible sites of action for the drugs. The possibility that there is more than one biochemical site of action should be considered, both because of the many responses to drug administration in biochemical and pharmacological parameters, and because of the structural similarities between narcotic analgesic drugs and tissue components such as steroid hormones and the biogenic amines.

Whatever the site of the biochemical lesion in the central nervous system produced by a single injection of a narcotic drug, it is apparent that there are two general responses to chronic drug administration: (1) less response, or the absence of any response during chronic drug treatment in a system which shows change after a single dose, or (2) an increased response to chronic treatment. In the first category of responses are the effects on K^+-stimulated oxygen and glucose uptake by brain slices, the release of Ca^{2+} in the brain *in vivo*, the release of hypothalamic pituitary-hormone-releasing factors, trophic hormones from the pituitary and the changes in biogenic amines in the central nervous system. These may be considered adaptive responses related to pharmacological tolerance since the effect of the initial injection is reversed by chronic drug use. In the second category of response are the effects on the synthesis of brain proteins and of biogenic amines, which show larger changes with chronic drug administration. An attractive hypothesis, for which the evidence is not conclusive, is that these metabolic systems are part of the control mechanism for attaining new equilibria in the biochemical systems of the first category, i.e., the mechanism for the development of tolerance.

An important question remains unanswered: whether the sum total of the biochemical responses in the central nervous system to the administration of narcotic drugs represent a major disturbance of brain metabolism, or instead, a minor transient effect qualitatively, and quantitatively, similar to events occurring often in the CNS as a result of other modes of stimulation.

VIII. REFERENCES

1. N. B. Eddy, The relation of chemical structure to analgesic action, *J. Am. Pharm. Assn.* **39**:245–251 (1950).
2. A. H. Beckett and A. F. Casy, Synthetic analgesics: stereochemical considerations, *J. Pharm. Pharmacol.* **6**:986–1001 (1954).
3. A. H. Beckett, A. F. Casy, and N. J. Harper, Analgesics and their antagonists: Some steric and chemical considerations, *J. Pharm. Pharmacol.* **8**:874–884 (1956).
4. K. W. Bentley, A. L. A. Boura, A. E. Fitzgerald, D. G. Hardy, A. McCoubrey, M. L. Aikman, and R. E. Lister, Compound possessing morphine-antagonizing or powerful analgesic properties, *Nature.* **206**:102–103 (1965).
5. W. R. Martin, Opioid antagonists, *Pharmacol. Rev.* **19**:463–521 (1967).
6. H. Blumberg, H. B. Dayton, M. George, and D. N. Rappaport, N-Allylnoroxymorphone: A potent narcotic antagonist, *Fed. Proc.* **20**:311 (1961).
7. N. B. Eddy, H. Halbach, H. Isbell, and M. H. Seevers, Drug dependence; Its significance and characteristics, *Psychopharmacol. Bull.* **3**:1–12 (1966).
8. E. L. Way and T. K. Adler, The pharmacologic implications of the fate of morphine and its surrogates, *Pharmacol. Rev.* **12**:383–446 (1960).
9. L. A. Woods, Comparative distribution of morphine and nalorphine in dog brain, *J. Pharmacol. Exptl. Therap.* **120**:58–62 (1959).
10. H. W. Kosterlitz, J. W. Thompson, and D. I. Wallis, Dihydromorphine in peripheral nervous tissue, *Nature*, **201**:719 (1964).
11. S. J. Mulé and L. A. Woods, Distribution of N-C^{14} methyl labeled morphine: In the central nervous system of non-tolerant and tolerant dogs, *J. Pharmacol. Exptl. Therap.* **136**:232–241 (1962).
12. S. Y. Yeh and L. A. Woods, Physiologic disposition of N-C^{14} methyl codeine in the rat, *J. Pharmacol. Exptl. Therap.* **166**:86–95 (1969).
13. T. Johannesson and K. Milthers, Morphine and normorphine in the brain of rats: A comparison of subcutaneous, intraperitoneal and intravenous administration, *Acta Pharmacol. Toxicol.* **19**:241–246 (1962).
14. M. R. Fennessy, The analgesic action of morphine-*N*-oxide, *Brit. J. Pharmacol.* **34**:337–344 (1968).
15. C. Maggiolo and F. Huidobro, Administration of pellets of morphine to mice; abstinence syndrome, *Acta Physiol. Lat. Am.* **11**:70–78 (1961).
16. S. J. Mulé, L. A. Woods, and L. B. Mellett, The effect of nalorphine in the central nervous system of non-tolerant dogs, *J. Pharmacol. Exptl. Therap.* **136**:242–249 (1962).

17. H. Kaneto and L. B. Mellett, The intracellular binding of N-methyl C^{14}-morphine in brain tissue of the rat, *The Pharmacologist* **2**:98(1960).
18. D. Van Praag and E. J. Simon, Studies on the intracellular distribution and tissue binding of dihydromorphine-7,8-H^3 in the rat, *Proc. Soc. Exptl. Biol. Med.* **122**:6 (1966).
19. D. H. Clouet and N. Williams, Effect of the administration of narcotic drugs on the RNA of rat brain, *Fed. Proc.* **27**:753 (1969).
20. J. T. Scrafani, N. Williams, and D. H. Clouet, Binding of dihydromorphine to subcellular fractions of rat brain, *The Pharmacologist* **11**:256 (1969).
21. C. C. Hug and T. Oka, Uptake of dihydromorphine-H^3 by synaptosomes of brain, *The Pharmacologist* **11**:293 (1969).
22. S. J. Mulé and C. W. Gorodetsky, Physiologic disposition of H^3-cyclazocine in non-tolerant, tolerant and abstinent dogs, *J. Pharmacol. Exptl. Therap.* **154**:632–645 (1966).
23. J. W. Miller and H. W. Elliott, Rat tissue levels of carbon-14 labeled analgetics as related to pharmacological activity, *J. Pharmacol. Exptl. Therap.* **113**:283–291 (1955).
24. J. C. Szerb and D. H. McCurdy, Concentration of morphine in blood and brain after intravenous injection of morphine in non-tolerant, tolerant and neostigmine-treated rats, *J. Pharmacol. Exptl. Therap.* **118**:446–450 (1956).
25. S. J. Mulé, C. M. Redman, and J. W. Flesher, Intracellular disposition of H^3-morphine in the brain and liver of non-tolerant and tolerant guinea pigs, *J. Pharmacol. Exptl. Therap.* **157**:459–471 (1967).
26. H. E. Dobbs, Effect of cyprenorphine (M285), a morphine antagonist, on the distribution and excretion of etorphine (M99), a potent morphine-like drug, *J. Pharmacol. Exptl. Therap.* **160**:407–414 (1968).
27. K. D. Wuepper, S. Y. Yeh, and L. A. Woods, Effect of nalorphine and levallorphan on brain concentration of levorphanol in the dog, *Proc. Soc. Exptl. Biol. Med.* **124**:1146–1150 (1967).
28. H. J. Kupferberg and E. L. Way, Pharmacologic basis for the increased sensitivity of the newborn rat to morphine, *J. Pharmacol. Exptl. Therap.* **141**:105–112 (1963).
29. H. Braeunlich, Age-dependent effects of codeine and morphine in rats, *Acta Biol. Med. Ger.* **16**:178–186 (1966).
30. E. L. Way, Brain uptake of morphine: Pharmacologic implications, *Fed. Proc.* **26**:1115–1118 (1967).
31. W. L. Way, E. C. Costley, and E. L. Way, Respiratory sensitivity of the newborn infant to meperidine and morphine, *Clin. Pharmacol. Therap.* **6**:454–461 (1965).
32. R. S. Foster, D. J. Jenden, and P. Lomax, A comparison of the pharmacologic effects of morphine and CH_3-N-morphine, *J. Pharmacol. Exptl. Therap.* **157**:185–195 (1967).
33. J. H. Sanner and L. A. Woods, Comparative distribution of tritium-labeled dihydromorphine between maternal and fetal rats, *J. Pharmacol. Exptl. Therap.* **148**:176–184 (1965).
34. G. F. Blane and H. E. Dobbs, Distribution of tritium-labeled etorphine (M99) and dihydromorphine in pregnant rats at term, *Brit. J. Pharmacol.* **30**:166–172 (1967).
35. C. C. Hug, Transport of narcotic analgesics by choroid plexus and kidney tissue *in vitro*, *Biochem. Pharmacol.* **16**:345–359 (1967).
36. J. T. Scrafani and C. C. Hug, Jr, Active uptake of dihydromorphine and other narcotic analgesics by cerebral cortex slices, *Biochem. Pharmacol.* **17**:1557–1566 (1968).

37. A. E. Takemori and M. W. Stenwick, Studies on the uptake of morphine by the choroid plexus *in vitro*, *J. Pharmacol. Exptl. Therap.* **154**:586–594 (1966).
38. D. B. Taylor, R. Creese, and T. C. Lu, The mode of action of morphine on the uptake of carbachol and decamethonium by slices of rat cerebral cortex, *J. Pharmacol. Exptl. Therap.* **165**:310–319 (1969).
39. J. Axelrod, The enzymatic N-demethylation of narcotic drugs, *J. Pharmacol. Exptl. Therap.* **117**:322–330 (1956).
40. T. K. Adler, A newly identified metabolic product of codeine, *N*-demethylated codeine, *J. Pharmacol. Exptl. Therap.* **106**:371 (1952).
41. A. H. Beckett, J. F. Taylor, A. F. Casy, and M. M. A. Hassan, Biotransformation of methadone in man: Synthesis and identification of a major metabolite, *J. Pharm. Pharmacol.* **20**:754–762 (1968).
42. J. Kemp and E. L. Way, *In vitro* studies on the metabolism of heroin, *Fed. Proc.* **18**:409 (1959).
43. N. P. Plotnikoff, E. L. Way, and H. W. Elliott, Biotransformation of meperidine excreted in the urine of man. *J. Pharmacol. Exptl. Therap.* **117**:414–419 (1956).
44. H. Yoshimura, K. Oguri, and H. Tsukamoto, Isolation and identification of morphine glucuronides in urine and bile of rabbits, *Biochem. Pharmacol.* **18**:279–286 (1969).
45. K. A. Pittman, D. Rosi, R. Cherniak, A. J. Merola, and W. D. Conway, Metabolism *in vitro* and *in vivo* of pentazocine, *Biochem. Pharmacol.* **18**:1673–1678 (1969).
46. J. M. Fujimoto and V. B. Haarstad, The isolation of morphine ethereal sulfate from the urine of the chicken and cat, *J. Pharmacol. Exptl. Therap.* **165**:45–51 (1969).
47. J. M. Fujimoto, W. M. Watrous, and V. B. Haarstad, Isolation of nalorphine ethereal sulfate and nalorphine glucuronide from urine of cat and rabbit, *Proc. Soc. Exp. Biol. Med.* **130**:546–549 (1969).
48. J. M. Fujimoto, Isolation of two different glucuronide metabolites of naloxone from the urine of rabbit and chicken, *J. Pharmacol. Exptl. Therap.* **168**:180–186 (1969).
49. D. H. Clouet, The methylation of normorphine in rat brain *in vivo*, *Biochem. Pharmacol.* **12**:967–972 (1963).
50. J. Axelrod, The enzymatic *N*-methylation of serotonin and other amines, *J. Pharmacol. Exptl. Therap.* **138**:28–33 (1962).
51. P. L. Gigon, T. E. Gram, and J. R. Gillette, Effect of drug substrates on the reduction of hepatic microsomal cytochrome P-450 by NADH, *Biochem. Biophys. Res. Comm.* **31**:558–562 (1968).
52. K. Milthers, The *in vivo* transformation of morphine and nalorphine in the brain of rats, *Acta Pharmacol. Toxicol.* **19**:235–240 (1962).
53. E. L. Way, J. W. Kemp, J. M. Young, and D. R. Grassetti, The pharmacological effects of heroin in relationship to its rate of biotransformation, *J. Pharmacol. Therap.* **129**:144–154 (1960).
54. L. Leadbeater and D. R. Davies, A comparison of 6-14-endoetheno-7-1-hydroxy-1-cyclohexyl-tetrahydrooripavine with morphine, *Biochem. Pharmacol.* **17**:219–226 (1968).
55. G. J. Mannering and A. E. Takemori, Effect of repeated doses of levorphan-dextrophan and morphine on the capacity of rat liver preparations to demethylate, *J. Pharmacol. Exptl. Therap.* **127**:187–190 (1959).
56. J. Cochin and L. Sokoloff, Effect of administration of L-thryroxin on liver *N*-demethylating activity in normal and morphine-treated rats, *Proc. Soc. Exptl. Biol. Med.* **104**:504–506 (1960).

57. J. Axelrod, The enzymatic *N*-demethylation of narcotic drugs, *J. Pharmacol. Therap.* **117**:322–330 (1956).
58. T. K. Adler, Studies on morphine tolerance in mice. I. *In vivo N*-demethylation of morphine and *N*- and *O*-demethylation of codeine, *J. Pharmacol. Exptl. Therap.* **156**:585–590 (1967).
59. A. E. Takemori, Certain enzymic responses to morphine administration in rats, *J. Pharmacol. Exptl. Therap.* **130**:360–374 (1960).
60. S. Y. Yeh and L. A. Woods, The metabolism of C^{14}-methyl dihydromorphine in the rat, *The Pharmacologist* **9**:218 (1967).
61. J. Cochin and J. Axelrod, Biochemical and pharmacological changes in the rat following chronic administration of morphine, nalorphine and normorphine, *J. Pharmacol. Exp. Ther.* **125**:105–110 (1959).
62. J. Axelrod and J. Cochin, The inhibitory action of nalorphine on the enzymatic *N*-demethylation of narcotic drugs, *J. Pharmacol. Exptl. Therap.* **121**:107–112 (1957).
63. H. Remmer and B. Alseben, The activation of detoxication in liver microsomes during habituation, *Klin. Wochenschr.* **36**:332–333 (1958).
64. R. Kato and K. Onodo, Effect of morphine administration on the activities of microsomal drug metabolizing enzymes in liver of different species, *Japan. J. Pharmacol.* **16**:217–218 (1966).
65. D. H. Clouet and M. Ratner, The effect of altering liver microsomal-*N*-demethylase activity on the development of tolerance to morphine in rats, *J. Pharmacol. Exptl. Therap.* **144**:362–372 (1964).
66. W. J. George and T. R. Tephly, Studies on hepatic microsomal *N*- and *O*-dealkylation of morphine analogues, *Mol. Pharmacol.* **4**:502–509 (1968).
67. R. B. Nelson and H. W. Elliott, A comparison of some central effects of morphine, morphinone, and thebaine on rats and mice, *J. Pharmacol. Exptl. Therap.* **155**:516–520 (1967).
68. N. Kokka, H. W. Elliott, and E. L. Way, Some effects of morphine on respiration and metabolism of rats, *J. Pharmacol. Exptl. Therap.* **148**:386–392 (1965).
69. E. T. Zlenko, Effect of morphine on oxidative processes in the brain, *Farmakol. Toksikol.* **4**:9–11 (1968).
70. A. E. Takemori, Cellular adaptation to morphine in rats, *Science.* **133**:1018–1019 (1961).
71. H. W. Elliott, N. Kokka, and E. L. Way, Influence of calcium-deficit on morphine inhibition of Q_{O_2} of rat cerebral cortex slices, *Proc. Soc. Exptl. Biol. Med.* **113**:1049–1052 (1963).
72. A. E. Takemori, Cross-cellular adaptation to methadone and meperidine in cerebral cortex slices from morphinized rats, *J. Pharmacol. Exptl. Therap.* **135**:252–255 (1962).
73. J. A. Richter and R. M. Marchbanks, Effects of potassium and levorphanol in brain slices on the release of AcCh and its synthesis from radioactive choline, *Abs. Eleventh Int. Meeting, Int. Soc. for Neurochem.* p. 338 (1969).
74. H. S. Bachelard and J. R. Lindsay, Effects of neurotrophic drugs on glucose metabolism in rat brain *in vivo*, *Biochem. Pharmacol.* **15**:1053–1058 (1966).
75. M. W. Poon, E. O'F. Walsh, and M. L. Ng, Morphine-induced changes in sensitivity of the glucose uptake system of muscle to extracellular Mg^{++}, *Biochem. Pharmacol.* **17**:1575–1580 (1968).
76. C. J. Estler, The influence of morphine and levallorphan on motility, oxygen consumption, rectal temperature and on the creatine-phosphate, ATP, ADP, lactic acid, glycogen and CoA content of the mouse brain, *Biochem. Pharmacol.* **8**:68 (1961).

77. P. W. Dodge and A. E. Takemori, Changes in rat cerebral glycolytic intermediates *in vivo* after treatment with morphine, nalorphine, or pentobarbitol, *The Pharmacologist*. **9**:218 (1967).
78. R. Greene and B. Magasanik, The mode of action of levallorphan as an inhibitor of cell growth, *Mol. Pharmacol.* 3:453–472 (1967).
79. D. H. Clouet and A. Neidle, The effect of morphine administration on the transport and metabolism of intracisternally-injected leucine in the rat, *J. Neurochem.* **17**:1069–1074 (1970).
80. D. H. Clouet and M. Ratner, The effect of the administration of morphine on the incorporation of C^{14}-leucine into the proteins of rat brain *in vivo*, *Brain Res.* **4**:33–43 (1967).
81. D. H. Clouet and M. Ratner, The effect of morphine administration on the incorporation of C^{14}-leucine into protein of cell-free systems from rat liver and brain, *J. Neurochem.* **15**:17–23 (1969).
82. E. J. Simon and D. Van Praag, Inhibition of RNA synthesis in *Escherichia coli* by levorphanol, *Proc. Nat. Acad. Sci. U.S.* **51**:877–883 (1964).
83. W. D. Noteboom and G. C. Mueller, Inhibition of protein and RNA synthesis in HeLa cells by levallorphan and levorphanol, *Mol. Pharmacol.* **2**:534–542 (1966).
84. W. D. Noteboom and G. C. Mueller, Inhibition of cell growth and the synthesis of RNA and protein in HeLa cells by morphinans and related compounds, *Mol. Pharmacol.* **5**:38–48 (1969).
85. G. P. Lewis and E. Reit, Further studies on the action of peptides on the superior cervical ganglion and suprarenal medulla, *Brit. J. Pharmacol.* **26**:444–460 (1966).
86. E. J. Simon, Inhibition of synthesis of RNA in *E. coli* by the narcotic drug, levorphanol, *Nature* **198**:794–795 (1963).
87. R. Roschenthaler, M. A. Devynck, P. Fromageot, and E. J. Simon, Inhibition of the synthesis of 5S and RNA in *E. coli* by levallorphan, *Biochim. Biophys. Acta* **182**:481–490 (1969).
88. S. Sakiyama, Selective inhibition of ribosomal RNA synthesis by levorphanol in cells, *Seikagaku* **40**:214–219 (1968).
89. D. H. Clouet, *in Drug Abuse: Social and Psychopharmacological Aspects* (J. O. Cole and J. R. Wittenborn, eds.) C. C. Thomas, Springfield, Ill. (1969).
90. M. Cohen, A. S. Keats, W. Krivoy, and G. Ungar, Effect of actinomycin-D on morphine tolerance, *Proc. Soc. Exptl. Biol. Med.* **119**:381–384 (1965).
91. J. Yamamoto, R. Inoki, Y. Tamari, and K. Iwatsubo, Inhibitory effect of 8-azaguanine on the development of tolerance in the analgesic action of morphine, *Japan. J. Pharmacol.* **17**:140–142 (1967).
92. M. T. Spoerlein and J. Scrafani, Effects of time and 8-azaguanine on the development of morphine tolerance, *Life Sci.* **6**:1549–1564 (1967).
93. A. A. Smith, M. Karmin, and J. Gavitt, Tolerance to the lenticular effects of opiates, *J. Pharmacol. Exptl. Therap.* **156**:85–91 (1967).
94. B. M. Cox, M. Ginsburg, and O. H. Osman, Acute tolerance to narcotic analgesic drugs in rats, *Brit. J. Pharmacol.* **33**:245–256 (1968).
95. A. A. Smith, M. Karmin, and J. Gavitt, Blocking effect of puromycin, ethanol and chloroform on the development of tolerance to an opiate, *Biochem. Pharmacol.* **15**:1877–1879 (1966).
96. M. P. Feinberg and J. Cochin, Effect of cyclophosphoramide (Cytoxan) on tolerance to morphine, *The Pharmacologist* **10**:188 (1968).
97. H. H. Loh, F. Shen, and E. L. Way, Effects of cycloheximide on the development of morphine tolerance and physical dependence, *The Pharmacologist* **10**:188 (1968).

98. B. M. Cox and O. H. Osman, The role of protein synthesis inhibition in the prevention of morphine tolerance, *Brit. J. Pharmacol.* **35**:373 (1969).
99. B. M. Cox and M. Ginsburg, *in Scientific Basis of Drug Dependence* (H. Steinberg, ed.), J. and A. Churchill, London (1969).
100. M. P. Feinberg and J. Cochin, Effect of weekly doses of cycloheximide on tolerance to morphine in the rat, *The Pharmacologist* **11**:256 (1969).
101. M. Brossard and J. H. Quastel, Effect of morphine and tofranil on the incorporation of P^{32} into phospholipids of slices, *Biochem. Pharmacol.* **12**:766–768 (1963).
102. S. J. Mulé, Effect of morphine and nalorphine on the metabolism of phospholipids in guinea-pig cerebral cortex slices, *J. Pharmacol. Exptl. Therap.* **154**:370–383 (1966).
103. S. J. Mulé, Morphine and the incorporation of P^{32} into brain phospholipids of non-tolerant, tolerant and abstinent guinea-pigs, *J. Pharmacol. Exptl. Therap.* **156**: 92–100 (1967).
104. S. J. Mulé, Inhibition of phospholipid-facilitated Ca^{++} transport by CNS-acting drugs, *Biochem. Pharmacol.* **18**:339–346 (1969).
105. R. Jaques, Morphine as an inhibitor of prostaglandin in isolated guinea-pig intestine, *Experientia* **25**:1059–1060 (1969).
106. A. K. Ray, M. Mukherj, and J. S. Ghosh, Adrenal catecholamines and related changes during different phases of morphine administration-A histochemical study, *J. Neurochem.* **15**:875–881 (1968).
107. J. D. P. Graham, J. D. Lever, and T. L. B. Spriggs, The effect of morphine dependence on the vesicular content of adrenergic nerves in relation to arteriolar smooth muscle in the pancreas of the rat, *Brit. J. Pharmacol.* **37**:19–23 (1969).
108. G. P. Quinn and B. B. Brodie, Effect of chlorpromazine and reserpine on the central actions of morphine in the cat, *Med. Exptl.* **4**:349–355 (1961).
109. E. W. Maynert and G. I. Klingman, Tolerance to morphine I. Effects on catecholamines in the brain and adrenal gland, *J. Pharmacol. Exptl. Therap.* **135**:285–295 (1962).
110. E. W. Maynert and R. Levi, Stress-induced release of brain noradrenalin and its inhibition by drugs, *J. Pharmacol. Exptl. Therap.* **143**:90–95 (1964).
111. H. Tagaki and M. Nakama, Effect of morphine and nalorphine on the content of dopamine in mouse brain, *Japan. J. Pharmacol.* **16**:483–484 (1966).
112. L. M. Gunne, Catecholamines and 5-hydroxytryptamine in morphine tolerance and withdrawal, *Nature* **195**:815–816 (1962).
113. J. W. Sloan, J. W. Brooks, A. J. Eisenman, and W. R. Martin, The effect of addiction to and abstinence from morphine on rat tissue catecholamines and serotonin levels, *Psychopharmacologia* **4**:261–270 (1963).
114. K. E. Moore, L. E. McCarthy, and H. L. Borison, Blood glucose and brain catecholamine levels in the cat following the injection of morphine into the cerebrospinal fluid, *J. Pharmacol. Exptl. Therap.* **148**:169–175 (1965).
115. R. Laverty and D. F. Sharman, Modification by drugs of the metabolism of 3,4-dihydroxyphenylethylamine in tissues, *Brit. J. Pharmacol.* **24**:759–772 (1965).
116. K. Watanabe, Y. Matsui, and H. Iwata, Enhancement of the analgesic effect of morphine by sodium diethyldithiocarbamate in rats, *Experientia* **25**:950–951 (1969).
117. M. J. Neal, Failure of morphine dependence in rats to influence brain noradrenaline turnover, *J. Pharm. Pharmacol.* **20**:950–953 (1968).
118. D. H. Clouet and M. Ratner, *Science* **168**:854–856 (1970).
119. D. J. Reis, M. Rifkin, and A. Corvelli, Effects of morphine on cat brain norepinephrine in regions with daily monamine rhythms, *Eur. J. Pharmacol.* **8**:149–152 (1969).

120. L. M. Gunne and J. Jonsson, Effects of morphine intoxication in brain catecholamine neurons, *Eur. J. Pharmacol.* **5**:338–342 (1969).
121. T. Takagi, T. Kuruma, and K. Kimura, Mechanism of action of tetrabenazine as a morphine antagonist. I. Relation between the amine-depleting effect of tetrabenazine and its morphine antagonism, *Nippon Yakurigku Zasshi* **60**:556–562 (1964).
122. C. Munoz and C. Paeile, Changes in morphine analgesia induced by drugs which modify catecholamine content of the brain, *Arch. Biol. Med. Exp.* **4**:63–68 (1967).
123. R. A. Verri, F. G. Graeff, and A. P. Corrado, Antagonism of morphine analgesia by reserpine and α-methyl tyrosine and the role played by catecholamines in morphine analgesic action, *J. Pharm. Pharmacol.* **19**:264–265 (1967).
124. I. M. Mazurkiewicz and F. C. Lu, Potentiation of analgesia by reserpine, chlorpromazine, *J. Pharm. Pharmacol.* **12**:103–108 (1960).
125. S. K. Gupta and H. J. Kulkarni, Modification of morphine analgesia in rats by monoamine oxidase inhibitors, *J. India Med. Assoc.* **46**:197–198 (1966).
126. E. Contreras, L. Tamayo, and L. Quijeda, Effect of nialamide and methyldopa on the analgesic action of morphine in rats and mice, *J. Pharm. Pharmacol.* **21**:391–393 (1969).
127. B. Heller, J. M. Saavedra, and E. Fischer, Influence of adrenergic blocking agents upon morphine and catecholamine analgesic effect, *Experientia* **24**:804–805 (1968).
128. S. S. Chang and C. I. Feng, Analgesic action of intracisternal injection of adrenalin in relation to that of morphine, *Sheng Li Hsueh Pao* **28**:352–357 (1965).
129. C. G. Sparkes and P. S. J. Spencer, Modification of morphine analgesia in the rat by biogenic amines administered intraventricularly, *Brit. J. Pharmacol.* **35**:362p. (1969).
130. E. W. Maynert, Some aspects of the comparative pharmacology of morphine, *Fed. Proc.* **26**:1111–1114 (1967).
131. E. W. Maynert, Analgesic drugs and brain neurotransmitters. 1. Effects of morphine on acetylcholine and certain other neurotransmitters, *Arch. Biol. Med. Exp.* **4**:136–137 (1967).
132. M. Weinstock and A. S. Marshall, The influence of the sympathetic nervous system on the action of drugs on the lens, *J. Pharmacol. Exptl. Therap.* **166**:8–13 (1969).
133. E. W. Maynert, G. I. Klingman, and H. K. Kaki, Tolerance to morphine. II. Lack of effects on 5-OH tryptamine and GABA, *J. Pharmacol. Exptl. Therap.* **135**:296–299 (1962).
134. D. R. Haubrich and D. E. Blake, Effect of acute and chronic administration of morphine on the metabolism of brain serotonin in rats, *Fed. Proc.* **28**:793 (1969).
135. E. L. Way, H. H. Loh, and F-H. Shen, Morphine tolerance, physical dependence and the synthesis of brain serotonin, *Science*, **162**:1290–1292 (1968).
136. K. J. Rogers and J. A. Thornton, The interaction between monoamine oxidase inhibitors and narcotic analgesics in mice, *Brit. J. Pharmacol.* **36**:470–480 (1969).
137. S. S. Tenen, Antagonism of the analgesic effect of morphine and other drugs by *p*-chlorophenylalanine, a serotonin depletor, *Psychopharmacologia* **12**:278–285 (1968).
138. P. J. Medon, D. R. Haubrick, and D. E. Blake, Modification of the hypothermic action of morphine after the depletion of brain serotonin or catecholamines, *The Pharmacologist* **11**:258 (1969).
139. E. Contreras and L. Tamayo, Influence of changes in brain 5-OH tryptamine on morphine analgesia, *Arch. Biol. Med. Exptl.* **4**:69–71 (1967).
140. F. H. Shen, H. H. Loh, and E. L. Way, Reserpine antagonism of morphine analgesia in tolerant mice, *Fed. Proc.* **28**:793 (1969).

141. V. Petkov and P. Statkov, Nature of serotonin receptors. I. Analgesics and serotonin, *Eksp. Med. Morfol.* **7**:193–201 (1968).
142. K. Hano, H. Kaneto, T. Kakunaga, and N. Moribayashi, Pharmacological studies of analgesics VI. The administration of morphine and changes in acetylcholine metabolism in mouse brain, *Biochem. Pharmacol.* **13**:441–447 (1964).
143. J. F. Howes, L. S. Harris, W. L. Dewey, and C. A. Voyda, Brain acetylcholine levels and inhibition of the tail-flick reflex in mice, *J. Pharmacol. Exptl. Therap.* **169**:23–28 (1969).
144. W. A. Large and A. S. Milton, The effect of acute and chronic morphine administration on brain acetylcholine levels in the rat, *Proc. Brit. Pharmacol. Soc.* **January** (1970).
145. L. Thal and I. Wajda, The effect of morphine on choline acetyltransferase levels in the caudate nucleus of the rat, *Fed. Proc.* **28**:261 (1969).
146. R. W. Morris, Effects of drugs on the biosynthesis of acetylcholinesterase: pentobarbital, morphine and morphinan derivatives, *Arch. Int. Pharmacodyn. Therap.* **133**:236–243 (1961).
147. D. Beleslin and R. L. Polak, Depression by morphine and chloralose of acetylcholinesterase release from the cat's brain, *J. Physiol. (London)* **177**:411–419 (1965).
148. J. Schuberth, J. Sollenberg, A. Sundwall, and B. Sörbo, Acetyl CoA in brain: The effect of centrally active drugs, insulin coma and hypoxia, *J. Neurochem.* **13**:819–822 (1966).
149. T. Shikimi, H. Kaneto, and K. Hano, Effect of morphine on the liberation of acetylcholine from the mouse cerebral cortical slices in relation to Ca^{++} concentration in the medium, *Japan J. Pharmacol.* **17**:136–137 (1967).
150. M. Sharkawi and M. P. Schulman, Inhibition by morphine of the release of acetylcholine-C^{14} from rat brain cortex slices, *J. Pharm. Pharmacol.* **21**:546–547 (1969).
151. J. Schuberth and A. Sundwall, Effects of some drugs on the uptake of acetylcholine in cortex slices of mouse brain, *J. Neurochem.* **14**:807–812 (1967).
152. K. Kuriyama, E. Roberts, and J. Vos, Some characteristics of the binding of GABA and AcCh to a synaptic vesicle fraction from mouse brain, *Brain Res.* **9**:231–252 (1968).
153. W. Schaumann, A hypothesis of a cholinergic mechanism for the action of morphine, *Naunyn-Schmiedebergs Arch. Pharmakol. Exptl. Pathol.* **237**:229–240 (1959).
154. W. L. Dewey and L. S. Harris, Narcotic-antagonist analgesics, effects on brain cholinesterases. *The Pharmacologist* **9**:230 (1967).
155. A. C. Lane, I. R. Macfarlane, and A. McCoubrey, Inhibition of cholinesterases by complex derivatives of morphine, *Biochem. Pharmacol.* **15**:122–123 (1966).
156. G. E. Hein and K. Powell, Evaluation of kinetic constants for mixed inhibitors of cholinesterase, *Biochem. Pharmacol.* **16**:567-573 (1967)
157. M. J. Ettinger and A. Gero, Interactions of narcotics and their antagonists with human serum esterase I. Sites of action and nature of the inhibitory effect, *Arch. Int. Pharmacodyn. Therap.* **164**:96–110 (1966).
158. M. J. Ettinger and A. Gero, Interactions of narcotics, and their antagonists with human serum esterase II. Nature of the antagonism, *Arch. Int. Pharmacodyn. Therap.* **164**:111–119 (1966).
159. M. G. Slusher and B. Browning, Morphine inhibition of plasma corticosteroid levels in chronic venous catheterized rats, *Am. J. Physiol.* **200**:1032–1034 (1961).
160. A. J. Eisenman, H. T. Fraser, and J. W. Brooks, Urinary excretion and plasma levels of 17-hydroxycorticosteroids during a cycle of addiction to morphine, *J. Pharmacol. Exptl. Therap.* **132**:226–231 (1961).

161. R. K. McDonald, F. T. Evans, V. K. Weise, and R. W. Patrick, Effect of morphine and nalorphine on plasma hydroxycorticosteroid levels in man, *J. Pharmacol. Exptl. Therap.* **125**:241–247 (1959).
162. E. Paroli and P. Melchiorri, Urinary excretion of hydroxysteroids, 17-keto steroids and aldosterone in rats during a cycle of treatment with morphine, *Biochem. Pharmacol.* **6**:1–17 (1961).
163. O. Nikodijevic and R. P. Maickel, Some effects of morphine in pituitary–adrenocortical function in the rat, *Biochem. Pharmacol.* **16**:2137–2142 (1967).
164. C. A. Winter and L. Flataker, The effect of cortisone, desoxycortisone and ACTH on the responses of animals to analgesic drugs, *J. Pharmacol. Exptl. Therap.* **103**:93–105 (1951).
165. T. K. Adler, H. W. Elliott, and R. George, Some factors affecting the biological disposition of small doses of morphine in rats, *J. Pharmacol. Exptl. Therap.* **120**:475–485 (1967).
166. H. W. Elliott and C. Elison, The influence of some adrenal and pituitary hormones on the absorption and biological disposition of methadone, *J. Pharmacol. Exp. Therap.* **131**:31–37 (1961).
167. F. N. Briggs and P. L. Munson, Studies on the mechanism of stimulation of ACTH secretion with the aid of morphine as a blocking agent, *Endocrinology* **157**:205–219 (1955).
168. R. Guillemin, W. E. Dear, B. Nicholas, Jr., and H. S. Lipscomb, ACTH releasing activity *in vivo* of a CRF preparation and lysine vasopressin, *Proc. Soc. Exptl. Biol. Med.* **101**:107–111 (1969).
169. A. V. Schally, W. H. Carter, I. C. Hearn, and C. Y. Bowers, Determination of CRF activity in rats treated with monase, desamethasone and morphine, *Am. J. Physiol.* **209**:1169–1173 (1965).
170. H. Rodeck and R. Braukmann, Effect of morphine sulfate oxidation on the neurosecretor system, *Z. Zellforsch. Mikroskop. Anat.* **75**:517–526 (1966).
171. C. R. Marchand and J. M. Fujimoto, Na^{+} and K^{+} excretion in rats chronically treated with morphine, *Proc. Soc. Exptl. Biol. Med.* **123**:600–603 (1966).
172. C. Marchand and G. Denis, Diuretic effect of chronic morphine treatment in rats, *J. Pharmacol. Exptl. Therap.* **162**:331–337 (1968).
173. C. E. Inturrisi, D. G. May, and J. M. Fujimoto, The diuretic effect of morphine in the chicken, *Eur. J. Pharmacol.* **5**:79–84 (1968).
174a. T. W. Redding, C. Y. Bowers, and A. V. Schally, Effects of morphine and other narcotics on thyroid function in mice, *Acta Endocrinol.* **51**:391–399 (1966).
174b. C. Marchand and G. Davis, Diuretic effect of chronic morphine treatment in rats, *J. Pharmacol. Exptl. Therap.* **162**:331–337 (1968).
175. P. Lomax and R. George, Thyroid activity following the administration of morphine in rats with hypothalamic lesions, *Brain Res.* **2**:361–367 (1966).
176. A. K. Ray and J. J. Ghosh, Changes in the hypothalamo–neurohypophysial neurosecretory materials of rats during different phases of morphine administration, *J. Neurochem.* **16**:1–5 (1969).
177. T. Shikimi, H. Kaneto, and K. Hano, Changes in brain Ca^{++} content in normal and morphinized mice after administration of morphine, *Japan. J. Pharmacol.* **17**:135–136 (1967).
178. T. Kakunaga, H. Kaneto, and K. Hano, Pharmacologic studies on analgesics VIII. Significance of the calcium ion in morphine analgesia, *J. Pharmacol. Exptl. Therap.* **153**:134–141 (1966).

179. T. Shikimi, H. Kaneto, and K. Hano, Effect of morphine on the liberation of acetylcholine from mouse cerebral cortex slices in relation to the Ca^{++} concentration, *Japan, J. Pharmacol.* **17**:136–137 (1967).
180. J. D. Robinson and W. D. Lust, ATP-dependent Ca^{++} accumulation of brain microsomes, *Arch. Biochem. Biophys.* **125**:286–294 (1968).
181. L. J. Graziani, R. K. Kaplan, A. Escriva, and R. Katzman, Ca^{++} flux into CSF during ventricular and ventriculocisternal perfusion, *Am. J. Physiol.* **213**:629–636 (1967).
182. H. Gangloff and M. Monnier, The topical action of morphine, levorphanol, and the morphine antagonist levallorphan on the unanesthetized rabbit's brain, *J. Pharmacol. Exptl. Therap.* **121**:78–95 (1957).
183. A. Zaks, A. Bruner, M. Fink, and A. M. Freedman, Intravenous diacetylmorphine (heroin) in studies of opiate dependence, *Dis. Nerv. Sys.* **30**:Suppl. 2, 89–96 (1969).
184. A. S. DeCarolis and V. G. Longo, The effects of morphine and related drugs on the electrical activity of the brain: their relationships with analgesic action, *Arch. Biol. Med. Exptl.* **4**:24–28 (1967).
185. S. D. Echols and R. E. Jewett, Effects of morphine on the sleep of cats, *The Pharmacologist* **11**:254 (1969).
186. W. A. Bleyer and M. G. Rosen, Meperidine-induced changes in the maternal and fetal EEGs of the guinea-pig, *Electroencephogr. Clin. Neurophys.* **24**:249–258 (1968).
187. N. Khazan, J. R. Weeks, and L. A. Schroeder, Electroencephalographic, electromyographic and behavioral correlates during a cycle of self-maintained morphine addiction in the rat, *J. Pharmacol. Exp. Therap.* **155**:521–531 (1967).
188. J. M. Nelson and C. Kornetsky, Single dose tolerance to morphine sulfate: EEG changes, *The Pharmacologist*, **10**:188 (1968).
189. M. Fink, EEG and human psychopharmacology, *Ann. Rev. Pharmacol.* **9**:241–258 (1969).
190. G. Ungar, Transfer of morphine tolerance by material extracted from the brain, *Fed. Proc.* **24**:548 (1965).
191. G. Ungar and L. Galvan, Conditions of transfer of morphine tolerance by brain extracts, *Proc. Soc. Exptl. Biol. Med.* **130**:287–291 (1969).
192. R. Tirri, Transfer of induced tolerance to morphine and promazine by brain homogenates, *Experientia* **23**:278–279 (1967).
193. S. E. Smits and A. E. Takemori, Lack of transfer of morphine tolerance by administration of rat cerebral homogenates, *Proc. Soc. Exptl. Biol. Med.* **127**:1167–1171 (1968).
194. F. Huidobro and H. Miranda, The presence in mice chronically treated with morphine of a substance that modifies morphine responses, *Biochem. Pharmacol.* **17**:1099–1105 (1968).
195. J. Cochin and C. Kornetsky, Development and loss of tolerance in the rat after single and multiple injections of morphine, *J. Pharmacol. Exptl. Therap.* **145**:1–10 (1964).
196. C. Kornetsky and G. Bain, Single dose tolerance to morphine, *The Pharmacologist* **9**:219 (1967).
197. G. F. Kiplinger and J. W. Clift, Pharmacological properties of morphine potentiating serum obtained from morphine tolerant dogs and men, *J. Pharmacol. Exptl. Therap.* **146**:139–146 (1964).
198. C. Kornetsky and G. G. Kiplinger, Potentiation of an effect of morphine in the rat by sera from morphine-tolerant and abstinent dogs and monkeys, *Psychopharmacologia* **4**:66–71 (1963).

199. V. J. Lotti, P. Lomax, and R. George, Acute tolerance to morphine following systemic and intracerebral injection in the rat, *Int. J. Neuropharmacol.* **5**:35–42 (1966).
200. E. F. Domino, *in The Addictive States* (A. Wikler, ed.) Williams and Wilkins, Baltimore (1968).
201. J. Florez, L. E. McCarthy, and H. L. Borison, A comparative study in the cat of the respiratory effects of morphine injected intravenously and into the cerebrospinal fluid, *J. Pharmacol. Exptl. Therap.* **163**:448–455 (1968).
202. W. P. Koella and J. Sutin, Extra blood brain barrier brain structures, *Int. Rev. Neurobiol.* **10**:31–55 (1968).
203. M. Vogt, Concentration of sympathin in different parts of the CNS under normal conditions and after the administration of drugs, *J. Physiol.* **123**:451–481 (1954).
204. H. L. Borison, B. R. Fishburn, N. K. Bhide, and L. E. McCarthy, Morphine-induced hyperglycemia in the cat, *J. Pharmacol. Exptl. Therap.* **138**:229–234 (1962).
205. H. L. Borison, B. R. Fishburn, and L. E. McCarthy, A possible receptor role of the subfornical organ in morphine-induced hyperglycemia, *Neurology* **10**:1049–1055 (1964).
206. V. J. Lotti, P. Lomax, and R. George, Temperature response in the rat following intracerebral microinjection of morphine, *J. Pharmacol. Exptl. Therap.* **150**:135–139 (1965).
207. V. J. Lotti, N. Kokka, and R. George, Pituitary-adrenal activation following intrahypothalamic micro-injection of morphine, *Neuroendocrinol.* **4**:326–332 (1969).
208. E. O'F. Walsh, C. H. Lee Pang, and M. L. Ng, Reversal of hormonal effects as a result of chronic morphinization, *Nature* **204**:698–699 (1964).
209. E. O'F. Walsh, C. H. Lee Pang, and S. C. Wong, Sensitivity to insulin *in vitro* is morphine-dependent in muscles of chronically morphinized rats, *Biochem. Pharmacol.* **18**:1529–1530 (1969).
210. W. D. M. Paton, The action of morphine and related substances on contraction and acetylcholine output, *Brit. J. Pharmacol.* **12**:119–124 (1957).
211. W. Schaumann, The inhibition by morphine of the release of acetylcholine from the intestine of the guinea-pig, *Brit. J. Pharmacol.* **12**:115–118 (1957).
212. T. F. Burks and J. P. Long, The release of intestinal 5-hydroxytryptamine by morphine and related agents, *J. Pharmacol. Exp. Ther.* **156**:267–276 (1967).
213. N. E. Ruffin, B. L. Reed, and B. C. Finnin, Effects of morphine withdrawal on cells in continuous culture, *Life Sci.* **8**:671–675 (1969).

Chapter 21

ALCOHOL

Henrik Wallgren

Research Laboratories of the State Alcohol Monopoly (Alko)
Helsinki, Finland

I. INTRODUCTION

Ethyl alcohol[1] is a general depressant capable of reversibly suppressing a number of biological activities. The organ most susceptible to this type of alcohol action is the brain. Alcohol therefore is closely related to the general anesthetics. Among drugs acting primarily on the central nervous system, and widely used because of these effects, ethanol is unique in also being a source of calories. Ethanol is almost completely oxidized to carbon dioxide and water, in humans at a rate of about 1 g per 10 kg of body weight per hour. Of the chemical energy available to the body, amounting to 7 kcal per g ethanol, one-third is released in oxidation of ethanol to acetaldehyde and acetate in the liver. Most of the acetate formed is oxidized extrahepatically. The process of ethanol oxidation induces marked changes in the metabolic patterns of the liver, and covers nearly the entire energy requirement of the liver and about half the energy need of moderately active human subjects.

As is the case with almost any external agent, the effects of ethanol are dosage—(or, rather) concentration-dependent. Effects of ethanol on biochemical processes in the central nervous system include three categories of phenomena: those underlying or immediately linked with the inebriating and narcotic action of ethanol; those arising through interaction with peripheral metabolic alterations; and those arising through the unspecific stressing actions of large doses of ethanol. It is by no means always easy to distinguish between these categories.

Biochemical aspects of the effects of ethanol on the central nervous system were extensively reviewed by Himwich.[1] Kalant[2] has reviewed and discussed selected aspects of neurochemical work with an emphasis

* This review deals with the effects of ethanol on the central nervous system. Effects of other alcohols are cited for comparison, but not considered in any detail. The terms ethanol and alcohol are used interchangeably.

on interpretation. Quastel[3,4] has contributed brief reviews, mainly on work with cerebral tissue *in vitro*, and Wallgren[5] has given fairly detailed coverage of most of the aspects considered in the present review, with one major exception, viz., the effects of prolonged administration of alcohol.

II. RESPIRATORY METABOLISM

The interest in this problem area has its roots in Verworn's[6] asphyxial theory of anesthesia. The classical studies of Quastel and Wheatly in the 1930s led to the modified concept attributing anesthetic action to inhibition of key enzymes in the respiratory metabolism of the neurons.

A. Observations *in Vivo*

Oxygen consumption and glucose uptake of the intact human brain are depressed in heavy alcohol intoxication with blood alcohol levels of about 0.3% or more, while the cerebral blood flow increases.[7–9] Blood alcohol levels up to 0.2% do not significantly change the overall cerebral metabolic rate or blood flow.

A number of reports deal with changes in the concentrations of metabolites in the brains of experimental animals, killed by immersion in liquid nitrogen at various intervals after administration of ethanol. In mice, 4.1 g/kg raised the cerebral concentration of glucose, fructose-1,6-phosphate, creatine phosphate, and adenosine monophosphate (AMP), and depressed the concentrations of glycogen, pyruvate, lactate, coenzyme A, and adenosine diphosphate (ADP).[10] Coenzyme A did not return to normal until 20 hours after injection, whereas lactate was lowered for more than 40 hours, or long after the disappearance of ethanol from the body.[10] An alcohol dose of 1.5 g/kg given intravenously to mice raised the concentrations of glucose and creatine phosphate, whereas glycogen, pyruvate lactate, and coenzyme A decreased.[11] Increase in the concentration of ATP and of the ratio ATP/ADP has been found also in rats.[12,13]

The observations cited are consistent with the view that the functional depression caused by ethanol secondarily leads to decreased activity of glycolytic reactions and the citric acid cycle, as a result of diminished energy demand. An activation of the glucose-6-phosphate pathway of direct oxidation of glucose has been suggested,[11] but hardly has any functional significance since this pathway has a very low activity in the brain. The increase in glycogenolysis is augmented by injection of epinephrine[14] and inhibited by sympathetic blockage,[15] and thus seems to be due to activation of phosphorylase by epinephrine released in response to the stress of intoxication. The decrease in activity of coenzyme A has been attributed to the binding of acetaldehyde to the —SH group of the coenzyme.[16]

B. Studies *in Vitro*

1. *Cell-free Preparations*

In relation to the lethal concentration of ethanol, which is about 0.5% (0.11 M) in humans and 0.9% (0.2 M) in rats, the concentrations of ethanol affecting the respiration of cell-free preparations are high. In rat brain homogenate, Grenell[17] found only insignificant decrease in oxygen consumption with 0.1 M ethanol. The respiration of rat brain mitochondria with several substrates is not affected until the concentration of ethanol markedly exceeds that lethal *in vivo*.[18-20] The increase in respiration of cerebral mitochondria caused by dinitrophenol is partly inhibited by 0.3 M ethanol.[21] This action on the mitochondria presumably explains the finding[22,23] that ethanol depresses the respiration of dinitrophenol-stimulated cerebral cortex slices.

2. *Tissue Slices*

Thin sections of cerebral tissue prepared for studies of metabolism *in vitro* show a comparatively low level of respiratory activity unless stimulated by manipulation of the composition of the medium or application of suitable electrical pulses. Experiments with unstimulated tissue slices reviewed by Wallgren[5] almost all have shown at least transient stimulation of the respiration with concentrations of ethanol compatible with reversible action *in vivo*. Such observations have been made also with human cerebral tissue.[24] In rat tissue, glucose consumption was found to increase in proportion to the oxygen consumption whereas formation of lactate remained unchanged.[25] One explanation of the rise in oxygen consumption might be a change in membrane permeability, increasing the need for ion transport. Evidence that such a mechanism stimulates glycolysis of erythrocytes incubated in the presence of ethanol has been presented.[26]

Several studies with cerebral cortex tissue stimulated by means of high concentrations of potassium in the medium have shown depression of the respiration with concentrations of ethanol beginning from about 0.05 M (for review, see Ref. 5). At about 0.8 M concentration of ethanol, the respiratory response to ethanol is abolished.[27] An unexplained peculiarity is that lactate production, already greatly enhanced by potassium stimulation, is further increased by concentrations of ethanol depressing respiration by less than 10%.[24,25] Formation of carbon dioxide from uniformly labeled glucose is depressed by ethanol.[27]

Electrically stimulated cerebral cortex tissue is more sensitive to ethanol than the potassium-stimulated preparation. In rat tissue, respiration and glucose consumption decrease by about 10% at an ethanol concentration of 0.087 M (0.4%),[25] whereas the concentration lethal *in vivo* (approximately 0.2 M) diminishes the response to stimulation by about 50%.[28] Wallgren[28,5] has studied the interaction of metabolic inhibitors and pharmacological agents with ethanol. Quaternary analogues of acetylcholine as well as atropine with its tertiary nitrogen grouping enhanced the effects of alcohol, presumably by interaction with membrane components, whereas inhibitors such as malonate and an inhibitor of active transport, ouabain, did not modify the effects of ethanol. Another finding important for the interpretation of the mode of action of ethanol in electrically stimulated tissue is that 0.4% ethanol strongly diminished the rapid breakdown of creatine phosphate and ATP which occurs on application of electrical pulses.[29] There was no apparent effect on the synthesis of energy-rich phosphate. This result supports the view of a primary depressant effect on the excitable membranes which is reflected in decreased rates of energy metabolism. It also agrees with other evidence, discussed by Wallgren,[5] which indicates that ethanol does not interfere with oxidative phosphorylation.

3. *Comparisons Between Alcohols*

An important generalization from a large number of experiments in many types of biological preparations is that the depressant potency of aliphatic alcohols increases with increasing length of the carbon chain. The change in efficacy parallels changes in physicochemical properties. Similar relationships are found also among other types of anesthetic agents.[30-32] In unstimulated cerebral tissue, alcohols depress respiration in accordance with this rule, but only in concentrations markedly exceeding those lethal *in vivo*.[17] In electrically stimulated tissue, depression of respiration closely correlates with the chemical potentials of aliphatic alcohols, added in concentrations compatible with reversible action *in vivo*.[33] The importance of this observation lies in the fact that it links the effects of ethanol with processes involved in the response of the neurons to electrical stimulation, and thus presumably with the excitation cycle of the responding membrane structures. The rule also holds true in potassium-stimulated cerebral cortex tissue.[34,35] It is evident that any theory of the mode of action of ethanol on neuronal functioning should be able to account for the correlation between the thermodynamic activity and the anesthetic potency of alcohols.

III. TRANSMITTERS AND RELATED COMPOUNDS

A. Acetylcholine

Choline acetyltransferase of a cerebral mitochondrial fraction is not affected by ethanol.[36] The suggestion[10] was already mentioned that acetaldehyde produced in the oxidation of ethanol might decrease the availability of coenzyme A for transacetylations, but there is no evidence for a functional significance of such a process. In unstimulated slices of cerebral cortex tissue, ethanol diminished the release of acetylcholine into the medium, whereas in potassium-stimulated tissue, no change was seen.[36,37] The activity of acetylcholinesterase in the brains of rats undergoes a slight, functionally significant reduction during ethanol intoxication.[40] In the brains of rabbits[39] and in cerebral cortex slices *in vitro*,[36] the activity of acetylcholinesterase is unaffected. Electrophysiological studies on peripheral preparations have shown that ethanol enhances the depolarizing effect of acetylcholine on the postsynaptic membrane.[40]

B. Biogenic Amines

Changes in the metabolism and concentration of monamines in the brain correlate with many drug-induced behavioral changes. However, little is known about effects of ethanol on cerebral monoamines. Ethanol has been reported to deplete 5-HT and norepinephrine in the brainstems of rabbits[41] and to raise the concentration of 5-HT in the brains of rats.[42] A number of studies[43–46] did not reveal changes in the concentrations of these amines. Tyce, Flock, and Owen[47] found that in rat brains ethanol influenced neither concentration, turnover, nor metabolites of 5-HT. This stands in contrast to the pronounced shift in the liver to degradation of monoamines over reductive pathways during oxidation of ethanol, reported for instance by Davis *et al.*[48] This is due to the redox shift caused by ethanol in the liver. The finding that the changes in hepatic monoamine metabolism are not reflected in the brain is consistent with the poor penetration of monoamines and their metabolites from the blood into the brain. Still, there is more indirect evidence which suggests that ethanol may interact with cerebral monoamines. Histochemical studies[49] suggest that ethanol accelerates depletion of norepinephrine from most of the adrenergic nerve terminals in the brains of rats. Rosenfeld[50] showed that injected 5-HT, tryptamine, dopamine, and tyramine augmented the intoxicating effect of ethanol in mice. In this case, ethanol may have facilitated penetration of the large doses of amines given, since Estler and Ammon[14] demonstrated that such a mechanism operates for epinephrine.

C. Metabolism and Transport of Amino Acids

The assumed role of GABA as an inhibitory transmitter has elicited interest in possible effects of ethanol on levels and metabolism of GABA. Studies on this amino acid are complicated by the fact that in the brain it is very actively metabolized by a system closely associated with the citric acid cycle. It is therefore difficult to make distinction between changes arising through an effect on metabolic pathways and effects directly involved in an as yet hypothetical functional alteration.

Ethanol has been reported to raise the concentrations of GABA, glutamate, and aspartate, and to decrease glutamine, in the brains of rats *in vivo*[51] and in electrically stimulated cerebral cortex tissue *in vitro*.[52] These changes apparently have no obligatory relation to the process of ethanol intoxication since they do not occur consistently.[5,53–55] The observations on amino acid metabolism are of interest, however, in relation to the regulation of metabolic reaction rates in the brain. Häkkinen, Kulonen, and Wallgren[52] suggested that ethanol may cause changes in the concentrations of intermediates of the GABA shunt by its depressing effect on reaction rates. Häkkinen and Kulonen[56,57] have also shown that the effects of ethanol persist in subcellular fractions of cerebral tissue from rats killed during intoxication. In homogenates as well as particulate and soluble fractions of brain tissue, gycolysis was also increased after treatment of the animals with ethanol *in vivo*, and it was suggested than an activation of glycolysis relative to gluconeogenesis might lead to an accumulation of certain amino acids.[58]

Ethanol in concentrations of 0.5–4% diminished both influx and efflux of dihydroxyphenylalanine in cerebral cortex slices, but net accumulation was caused by stronger inhibition of the efflux.[59] On the other hand, neither ethanol (1–4%) nor butanol (0.1%) affected exit of α-aminoisobutyrate from cerebral cortex slices loaded with this nonmetabolizable amino acid.[60]

IV. ACTIVE ION TRANSPORT

The information concerning effects of ethanol on active ion transport (transport of ions against electrochemical gradients) is restricted to observations on transport-related ATPase activity and net movements of sodium and potassium.

The Na^+, K^+, Mg^{2+}-stimulated microsomal ATPase, first described by Skou,[61] has been extensively used as a biochemical model for studies of ion transport (cf. reviews by Skou[62] and Albers[63]). Inhibition of rat cerebral microsomal ATPase by ethanol was first reported by Järnefelt.[64] Israel, Kalant, and Laufer[65] found that the inhibitory effect of ethanol was competitive with potassium, but not with sodium. It has been reported that competition with potassium occurs also in a bovine cerebral ATPase preparation, as well as inhibition by a series of aliphatic alcohols in parallel with their anesthetic potency.[66]

Net accumulation of potassium in cerebral cortex slices depleted of the ion by standing in the cold or by exposure to anoxia has been found to be inhibited by ethanol,[67] and in this preparation, competition with potassium was also observed.[67,68] Lithium ions had actions rather similar to those of ethanol.[68] In unpublished experiments, the present author has confirmed the inhibitory effect of ethanol on net accumulation of potassium in rat cerebral cortex slices subjected to a period of anoxia whereas distribution of ions as measured by means of inulin was not affected when the slices had reached equilibrium. No effect was found when the slices were incubated throughout in aerobic conditions.

The effects so far reported thus do not indicate any dramatic effects of ethanol on ion transport: moreover, effects on ion transport have been found only in selected tissues.[69] However, some indirect evidence concerning interaction between calcium ions and alcohols at membrane sites illustrates how desirable it is to obtain more information on effects of ethanol on the movements and action of inorganic ions. Observations on peripheral nerve and muscle preparations[70,71] suggest antagonism between the effect of calcium and ethanol. Ehrenpreis[72] isolated from rabbit sciatic nerve a protein fraction, presumably originating in the membrane, which combined and formed precipitates with calcium ions. The precipitate formation was markedly augmented by isopropyl alcohol.

V. THE POSSIBLE ROLE OF METABOLISM OF ETHANOL

If ethanol were metabolized by brain tissue, changes might be induced in the metabolic patterns which could contribute to the effects of ethanol. There are a few reports indicating that cerebral tissue indeed oxidizes ethanol,[73-75] and Sutherland, Hine, and Burbridge[24] discussed the implications of such a process for the anesthetic action of ethanol. However, in a number of studies, metabolism of ethanol could not be detected in cerebral tissue. The evidence includes failure to detect utilization of ethanol as a substrate by cerebral tissue, lack of formation by brain tissue of labeled carbon dioxide from radioactive ethanol, and absence of alcohol dehydrogenase in the brain (references in review by Wallgren[5]). Also, alcohol dehydrogenase could not be demonstrated histochemically in brain tissue.[76]

Acetaldehyde, formed in the hepatic oxidation of ethanol, has repeatedly been suggested as an agent contributing to the intoxicating effect of ethanol.[77,78] Normally, however, such a relationship is highly unlikely because even in the liver the concentrations formed from ethanol remain low. In the systemic circulation, acetaldehyde is further diluted, and since it is also oxidized by most tissues, only minute quantities reach the brain.[79] It is also evident that the symptoms of alcoholic intoxication are proportional to the concentration of ethanol and not to that of acetaldehyde or other metabolites. Thus, we may conclude that ethanol as such causes the disturbance of nerve function.

An exception to the conclusion just made is provided by the reaction to the combined administration of disulfiram and alcohol, in which acetaldehyde presumably effectively contributes to the unpleasant sensations. Results of some recent work also suggests that acetaldehyde in some situations may contribute to the effects of alcohol in chronic alcoholic patients. Scholz[80] has shown that in the liver, ethanol and acetaldehyde form a redox-pair which is dependent on the ratio NADH/NAD in the same manner as for instance lactate and pyruvate. The ratio NADH/NAD normally increases markedly during oxidation of ethanol and is accompanied by a large increase in the hepatic lactate/pyruvate ratio, but this change does not occur in fatty livers from choline-deficient animals.[81]

Since the normal increase in the ratio of lactate/pyruvate is primarily due to decrease in the concentration of pyruvate,[82] failure of ethanol to change the ratio implies a higher concentration of pyruvate and presumably also of acetaldehyde. Thus, liver disease of alcoholic patients may be accompanied by raised concentrations of acetaldehyde in the liver, and this could possibly result in transfer of sufficient amounts of the highly toxic acetaldehyde to the brain to affect its functioning.

VI. EFFECTS OF CHRONIC ALCOHOL CONSUMPTION

Possible neurochemical consequences of prolonged ingestion of alcohol have been largely neglected in both experimental and clinical work. A crude symptom-based classification of the conditions encountered includes increased tolerance to alcohol, the withdrawal syndrome, and encephalopathies of chronic alcoholic patients. Of these conditions, only increased tolerance is readily induced in animals.[83,84] A full-blown withdrawal syndrome has not been produced in animals, with the exception of monkeys (Seevers, personal communication). With respect to the encephalopathies, it has not been possible to induce in animals a condition clearly relevant to the complex human symptomatology.

Kinard and Hay[38] demonstrated a decrease in brain acetylcholinesterase activity of rats behaviorally tolerant to alcohol, but concluded that the change was too small to have functional significance. Kalant and Grose[37] found that cerebral cortex slices of rats tolerant to alcohol became resistant to the suppression of acetylcholine release normally caused by ethanol. The respiration of electrically stimulated cerebral cortex slices from tolerant rats was as sensitive to ethanol-induced suppression as that from control animals.[83] Studies on enzyme activities of cerebral cortex[84] and concentration of acetylcholine[85] showed increases for the former and decrease for the latter after prolonged feeding of alcohol in the drinking water, but the implications were not clear because functional changes in the central nervous system were not reported. Kiessling and Tilander[86] found no change in respiration and capacity of oxidative phosphorylation of rat cerebral mitochondria with pyruvate or succinate as substrate when the animals had received alcohol for prolonged periods. In the brains of rats given 15% alcohol as sole fluid for 23 days, bound GABA was slightly lowered whereas free GABA was not changed.[87] As the change in bound GABA correlated with the level of blood alcohol at killing, it probably arose from a direct action of ethanol rather than through some more permanent change. In one study on humans which involved 11 chronic alcoholic patients, Sutherland *et al.*[88] reported a change in cerebral metabolism of glutamic acid and glutamine, estimated from arteriovenous differences, and elevation of the blood levels of glutamate and glutamine. Administration of alcohol seemed to normalize the cerebral metabolism of glutamate, whereas it raised the blood level of glutamate.

The slight changes and lack of appropriate controls render interpretation of the findings difficult.

VII. DISCUSSION

With respect to acute effects of ethanol, the biochemical information so far obtained is consistent with a primary effect on the functioning of the neuronal membrane. Observations on respiratory metabolism, levels and metabolism of energy-rich phosphates, and changes in the concentrations of amino acids metabolically closely associated with the citric acid cycle provide data which generally support the view that changes occur as a consequence of inhibition of the functional activity of the nerve cells. Moreover, it is clear that the effect arises through a direct action of ethanol molecules, and not through an action of metabolites derived from ethanol. The problem then to be solved is of how ethanol interferes with membrane functioning. Suggestive data have been obtained, such as those indicating interaction of membrane components presumably involving fixed negative charges with ethanol, divalent cations, and acetylcholine and related compounds. Observations on the effects of alcohols in aqueous micellar systems of electrolytes with long hydrocarbon chains cited by Wallgren[(89)] seem to provide another lead to mechanisms by which ethanol may act.

As yet, however, it is not clear which aspect of membrane functioning is most sensitive to ethanol. Although the evidence for inhibition of ion transport is in keeping with the concept of a direct action on neuronal membranes, it would be rash to assume that this effect is directly responsible for the depression of nerve function caused by ethanol. One objection to such a conclusion is the rather slow and slight effect on ion transport in contrast to the prompt and readily reversed action of ethanol; another is the fact pointed out by Kalant and Israel[(69)] that appreciable effects on ion transport are found only in selected tissues. Neurochemical workers in this field should take notice of the important finding that in squid giant axons[(90)] and in frog sartorius muscle[(40)] ethanol has a direct action on the excitation cycle, and particularly seems to suppress selectively the increase in sodium conductance which constitutes the rising phase of the action potential. These phenomena are not readily accessible for study in the central nervous system, but experiments on ion fluxes in cerebral tissue stimulated electrically should yield pertinent information. The experiment on the comparative efficacy of various aliphatic alcohols in suppressing the respiratory response of cerebral cortex slices to electrical stimulation[(33)] was already cited as giving indirect evidence for an effect on the immediate response to stimulation. Evidently, progress in this field will in part depend on advances in the knowledge of membrane structure and function.

With respect to the study of known or assumed transmitter and moderator substances, there is more information about changes in the liver during ethanol intoxication than about the situation in the central nervous system. Most of these compounds do not readily pass through the blood–brain barrier. Since the brain contains complete systems for their synthesis, storage, release, and removal or breakdown, it is obvious that effects of ethanol should be studied in the brain. So far, most of the meager information available concerns cerebral concentrations of monoamines. The results generally indicate little or no effects of ethanol. Attention should be given to the possibility that ethanol might alter storage, release, and action on the membranes of the monoamines. There is already evidence which shows that ethanol interacts with acetylcholine at membrane sites.

From the knowledge of the acute effects of ethanol, it is reasonable to expect that the increase in tolerance and the abstinence syndrome which result from prolonged and heavy consumption of alcohol originate in changes of membrane structure and function. No information is available in this area. Progress has in part been impeded by the lack of a suitable animal preparation in which these phenomena can be reproduced. However, increased tolerance is readily induced in animals. McQuarrie and Fingl[(91)] demonstrated in mice that when chronic alcohol administration was abruptly terminated, the nervous system was left in a state of hyperexcitability. This state presumably embodies elements corresponding to the human withdrawal syndrome and may provide a suitable basis for experiments.

VIII. SUMMARY

Studies on the effects of ethanol on the respiratory metabolism in intact human and animal subjects suggest that ethanol primarily affects nervous functioning with a consequent decrease in the metabolic rates. *In vitro*, only cerebral cortex tissue stimulated by potassium ions or electrical pulses shows metabolic changes similar to those found *in vivo*; thus preservation of the cell structure and of neuronal activity are required for a response. Ethanol causes parallel decreases in oxygen and substrate utilization. Changes in the metabolite levels and the lowered rates of utilization of ATP and creatine phosphate further corroborate the tentative conclusion from observations *in vivo* that the metabolic changes arise through a decreased requirement for metabolic energy. Inhibition of active ion transport has been demonstrated, but it seems likely that the effect on the excitation cycle, known from electrophysiological studies, has greater immediate importance for the anesthetic action of ethanol. Interactions of ethanol with divalent cations and acetylcholine and analogous substances provide further evidence for effects on membrane functioning. In connection with chronic alcohol administration, only a few studies of enzyme activities and substrate utilization have been performed, and the results are rather inconclusive. Studies are needed of possible changes in membrane composition and function in the conditions of increased tolerance and abstinence illness resulting from chronic alcohol ingestion and withdrawal. A prerequisite for such studies is that presence of the condition at which the study is aimed is properly verified. Both in relation to acute and chronic alcohol intoxication, more information should be obtained concerning storage, release, and metabolism of acetylcholine, biogenic amines, and other possible transmitter and moderator substances.

IX. REFERENCES

1. H. E. Himwich, in *Alcoholism* (G. N. Thompson, ed.), pp. 291–408, Charles C Thomas, Springfield, Ill. (1956).
2. H. Kalant, Some recent physiological and biochemical investigations on alcohol and alcoholism. A review, *Quart. J. Studies Alc.* **23**:52–93 (1962).
3. J. H. Quastel, in *Neurochemistry* (K. A. C. Elliott, I. H. Page, and J. H. Quastel, ed.) 2nd ed., pp. 790–812, Charles C Thomas, Springfield, Ill. (1962).
4. J. H. Quastel, Effects of drugs on metabolism of the brain *in vitro*, *Brit. Med. Bull.* **21**:49–56 (1965).
5. H. Wallgren, Effects of alcohol on biochemical processes in the central nervous system, *Psychosomat. Med.* **28**:431–442 (1966).
6. M. Verworn, Narcosis, *Harvey Lectures* 1911–1912, Lippincott, Philadelphia (1912).
7. C. H. Hine, A. F. Shick, L. Margolis, T. N. Burbridge, and A. Simon, Effects of alcohol in small doses and tetraethylthiuramdisulphide (Antabus) on the cerebral blood flow and cerebral metabolism, *J. Pharmacol. Exptl. Therap.* **106**:253–260 (1952).
8. L. L. Battey, A. Heyman, and J. L. Patterson, Effects of ethyl alcohol on cerebral blood flow and metabolism, *J. Am. Med. Assoc.* **152**:6–10 (1953).
9. J. F. Fazekas, S. N. Albert, and R. W. Alman, Influence of chlorpromazine and alcohol on cerebral hemodynamics and metabolism, *Am. J. Med. Sci.* **230**:128–132 (1955).
10. H. P. T. Ammon, C.-J. Estler, and F. Heim, Der Einfluss von Äthylakohol auf den Kohlenhydrat- und Energiestoffwechsel des Gehirns weisser Mäuse, *Arch. Intern. Pharmacodyn.* **154**:108–121 (1965).
11. F. Heim, H. P. T. Ammon, C.-J. Estler, and D. Mikschiczek, Funktion und Stoffwechsel des Gehirns unter Einwirkung niedriger Alkoholkonzentrationen, *Med. Pharmacol. Exptl.* **13**:361–370 (1965).
12. H. Saito, Study of the effect of central nervous depressants on the acid-soluble nucleotides of the brain (Jap.), *Tohoku Med. J.* **56**:113–126 (1962). (Abstr. *Quart. J. Studies Alc.* **25**:759, 1964).
13. H. M. Redetzki, Effects of alcohol on adenine nucleotide levels of mouse brain, *Quart. J. Studies Alc.* **28**:225–230 (1967).
14. C.-J. Estler and H. P. T. Ammon, Phosphorylaseaktivität und Glykogengehalt des Gehirns unter dem Einfluss von Äthanol und Adrenalin, *J. Neurochem.* **12**:871–876 (1965).
15. H. P. T. Ammon and C.-J. Estler, Inhibition of ethanol-induced glycogenolysis in brain and liver by adrenergic β-blockade, *J. Pharm. Pharmacol.* **20**:164–165 (1968).

16. H. P. T. Ammon, F. Heim, C.-J. Estler, G. Fickeis, and M. Wagner, The influence of aliphatic alcohols and their halogen derivatives on the coenzyme A in the liver of mice, *Biochem. Pharmacol.* **16**:1533–1537 (1967).
17. R. G. Grenell, in *Alcoholism. Basic Aspects and Treatment* (H. E. Himwich, ed.), pp. 7–17, Publ. No. 47 of Am. Assoc. Advan. Sci., Washington, D.C. (1957).
18. C. T. Beer and J. H. Quastel, The effects of aliphatic alcohols on the respiration of rat brain cortex slices and rat brain mitochondria, *Can. J. Biochem. Physiol.* **36**:543–546 (1958).
19. E. B. Truitt, F. K. Bell, and J. C. Krantz, Anesthesia. LIII. Effects of alcohols and acetaldehyde on oxidative phosphorylation in brain. *Quart. J. Studies Alc.* **17**:594–600 (1956).
20. G. De Gregorio, N. E. Lofrumento, A. Alifano, M. L. Manno, C. Serra, and R. Logoluso, Effetto dell'alcool sulla respirazione dei mitocondrio di cervello, *Boll. Soc. It. Biol. Sper.* **41**:425–426 (1965).
21. T. Nukada and N. Andoh, Ethyl alcohol inhibition of brain mitochondrial respiration stimulated by dinitrophenol, *Japan. J. Pharmacol.* **17**:325–326 (1967).
22. J. J. Ghosh and J. H. Quastel, Narcotics and brain respiration, *Nature* **174**:28–31 (1954).
23. E. Fischer, in *Alcoholism. Basic Aspects and Treatment* (H. E. Himwich, ed.), pp. 19–27, Publ. No. 47 of Am. Assoc. Adv. Sci., Washington, D.C. (1957).
24. V. C. Sutherland, C. H. Hine, and T. N. Burbridge, The effects of ethanol on cerebral cortex metabolism *in vitro*, *J. Pharmacol. Exptl. Therap.* **116**:469–479 (1956).
25. H. Wallgren and E. Kulonen, Effect of ethanol on respiration of rat-brain-cortex slices, *Biochem. J.* **75**:150–158 (1960).
26. H. M. Redetzki and J. E. Redetzki, Activation of glycolysis and decrease of sodium-potassium concentration gradients of erythrocytes by ethanol, *Fed. Proc.* **27**, No. 2, 708 (1968).
27. J. H. Quastel, Enzymatic mechanisms of the brain and the effects of some neurotropic agents, *Proc. Intern. Congr. Biochem.*, *4th Vienna*, 1958, Vol. 3, 90–114 (1959).
28. H. Wallgren, Effects of acetylcholine analogues and ethanol on the respiration of brain cortex tissue *in vitro*, *Biochem. Pharmacol.* **6**:195–204 (1961).
29. H. Wallgren, Rapid changes in creatine and adenosine phosphates of cerebral cortex slices on electrical stimulation with special reference to the effect of ethanol, *J. Neurochem.* **10**:349–362 (1963).
30. F. Brink and J. M. Posternak, Thermodynamic analysis of the relative effectiveness of narcotics, *J. Cellular Comp. Physiol.* **32**:211–233 (1948).
31. J. Ferguson, *in Mécanisme de la narcose*, pp. 25–39, Centre National de la Récherche Scientifique, Paris (1951).
32. K. W. Miller, W. D. M. Paton, and E. B. Smith, Site of action of general anesthetics, *Nature* **206**:574–577 (1965).
33. R. Lindbohm and H. Wallgren, Changes in respiration of rat brain cortex slices induced by some aliphatic alcohols, *Acta Pharmacol. Toxicol.* **19**:53–58 (1962).
34. E. Majchrowicz, Effects of aliphatic alcohols and aldehydes on the metabolism of potassium-stimulated rat brain cortex slices, *Can. J. Biochem. Physiol.* **43**:1041–1051 (1965).
35. C. T. Beer and J. H. Quastel, The effects of aliphatic aldehydes on the respiration of rat brain cortex slices and rat brain mitochondria, *Can. J. Biochem. Physiol.* **36**:531–541 (1958).
36. H. Kalant, Y. Israel, and M. A. Mahon, The effects of ethanol on acetylcholine synthesis, release, and degradation in brain, *Can. J. Physiol. Pharmacol.* **45**:172–176 (1967).

37. H. Kalant and W. Grose, Effects of ethanol and pentobarbital on release of acetylcholine from cerebral cortex slices, *J. Pharmacol. Exptl. Therap.* **158**:386–393 (1967).
38. F. W. Kinard and M. G. Hay, Effect of ethanol administration on brain and liver enzyme activities, *Am. J. Physiol.* **198**:657–658 (1960).
39. F. Fujita, Experimental studies on habituation to alcohols, *Folia Pharmacol. Japon.* **50**:258–263 (1954). (Abstr. *Excerpta Med., Sect. II*, **8**:No. 2460, 1955).
40. F. Inoue and G. B. Frank, Effects of ethyl alcohol on excitability and on neuromuscular transmission in frog skeletal muscle, *Brit. J. Pharmacol.* **30**:186–193 (1967).
41. D. Gursey and R. E. Olson, Depression of serotonin and norepinephrine levels in brain stem of rabbit by ethanol, *Proc. Soc. Exptl. Biol. Med.* **104**:280–281 (1960).
42. D. D. Bonnycastle, M. F. Bonnycastle, and E. G. Anderson, The effect of a number of central depressant drugs upon 5-hydroxytryptamine levels in the rat, *J. Pharmacol.* **135**:17–20 (1962).
43. J. Häggendal and M. Lindqvist, Ineffectiveness of ethanol on noradrenaline, dopamine or 5-hydroxytryptamine levels in brain, *Acta Pharmacol. Toxicol.* **18**:278–280 (1961).
44. G. R. Pscheidt, B. Issekutz, and H. E. Himwich, Failure of ethanol to lower brain stem concentration of biogenic amines, *Quart. J. Studies Alc.* **22**:550–553 (1961).
45. D. H. Efron and G. L. Gessa, Failure of ethanol and barbiturates to alter brain monoamine content, *Arch. Intern. Pharmacodyn.* **142**:111–116 (1963).
46. N. Rudas and L. Vacca, Influenza dell'alcool sul contenuto di monoamine del nervoso centrale, *Acta Neurol.* (Napoli) **19**:848–849 (1964).
47. G. M. Tyce, E. V. Flock, and C. A. Owen, Effect of ethanol, reserpine, monoamine-oxidase-inhibition, or glucuronyl transferase deficiency on metabolism of 5-hydroxytryptamine (5HT) by liver, *Fed. Proc.* **26**:729 (1967).
48. V. E. Davis, H. Brown, J. A. Huff, and J. L. Cashaw, Ethanol-induced alterations of norepinephrine metabolism in man, *J. Lab. Clin. Med.* **69**:787–799 (1967).
49. H. Corrodi, K. Fuxe, and T. Hökfelt, The effect of ethanol on the activity of central catecholamine neurones in rat brain, *J. Pharm. Pharmacol.* **18**:821–823 (1966).
50. G. Rosenfeld, Potentiation of the narcotic action and acute toxicity of alcohol by primary aromatic monoamines, *Quart. J. Studies Alc.* **21**:584–596 (1960).
51. H.-M. Häkkinen and E. Kulonen, The effect of alcohol on the amino acids of the rat brain with a reference to the administration of glutamine, *Biochem. J.* **78**:588–593 (1961).
52. H.-M. Häkkinen, E. Kulonen, and H. Wallgren, The effect of ethanol and electrical stimulation on the amino acid metabolism of rat brain-cortex slices *in vitro*, *Biochem. J.* **88**:488–498 (1963).
53. V. C. Sutherland and M. Rikimaru, The regional affects of adrenalectomy and ethanol on cerebral amino acids in the rat, *Intern. J. Neuropharmacol.* **3**: 135–139 (1964).
54. E. R. Gordon, The effect of ethanol on the concentration of γ-aminobutyric acid in the rat brain, *Can. J. Physiol. Pharmacol.* **45**:915–918 (1967).
55. M. Mouton, C. Lefournier-Contensou, and H. Chalopin, Incidence de l'intoxication alcoolique sur la teneur en acide γ-aminobutyrique du cerveau de la Souris, *Compt. Rend. Acad. Sci. (Paris)* 264:2649–2650 (1965).
56. H.-M. Häkkinen and E. Kulonen, The amino acid metabolism in fractions of rat-brain homogenates with reference to the effect of ethanol, *Biochem. J.* **96**:65P (1965).
57. H.-M. Häkkinen and E. Kulonen, Amino acid metabolism in various fractions of rat-brain homogenates with special reference to the effect of ethanol, *Biochem. J.* **105**:261–269 (1967).

58. H.-M. Häkkinen and E. Kulonen, Effect of ethanol on the metabolism of glucose in homogenate fractions of rat brain, *Scand. J. Clin. Lab. Invest.* **21**:Suppl. 101, 18 (1968).
59. K. Kaniike and H. Yoshida, Increase in accumulation of L-dopa (3,4-dihydroxyphenylalanine) in brain slices by alcohol, *Japan. J. Pharmacol.* **13**:292–296 (1963).
60. A. Cherayil, J. Kandera, and A. Lajtha, Cerebral amino acid transport *in vitro.*—IV. The effects of inhibitors on exit from brain slices. *J. Neurochem.* **14**:105–115 (1967).
61. J. C. Skou, The influence of some cations on an adenosine triphosphatase from peripheral nerves, *Biochim. Biophys. Acta* **23**:394–401 (1957).
62. J. C. Skou, Enzymatic basis for active transport of Na^+ and K^+ across cell membrane, *Physiol. Rev.* **45**:596–617 (1965).
63. R. W. Albers, Biochemical aspects of active transport, *Ann. Rev. Biochem.* **36**:727–756 (1967).
64. J. Järnefelt, Inhibition of the brain microsomal adenosine-triphosphatase by depolarizing agents, *Biochim. Biophys. Acta* **48**:111–116 (1961).
65. Y. Israel, H. Kalant, and I. Laufer, Effects of ethanol on Na, K, Mg-stimulated microsomal ATPase activity, *Biochem. Pharmacol.* **14**:1803–1814 (1965).
66. Y. Israel and I. Salazar, Inhibition of brain microsomal adenosine triphosphatases by general depressants, *Arch. Biochem. Biophys.* **122**:310–317 (1967).
67. Y. Israel-Jacard and H. Kalant, Effect of ethanol on electrolyte transport and electrogenesis in animal tissues, *J. Cellular Comp. Physiol.* **65**:127–132 (1965).
68. Y. Israel, H. Kalant, and A. E. LeBlanc, Effects of lower alcohols on potassium transport and microsomal adenosinetriphosphatase activity of rat cerebral cortex, *Biochem. J.* **100**:27–33 (1965).
69. H. Kalant and Y. Israel, in *Biochemical Factors in Alcoholism* (R. P. Maickel, ed.), pp. 25–37, Pergamon Press, Oxford (1967).
70. L. Hurwitz, S. v. Hagen, and P. D. Joiner, Acetylcholine and calcium on membrane permeability and contraction of intestinal smooth muscle, *J. Gen. Physiol.* **50**:1157–1172 (1967).
71. P. Rosenberg and E. Bartels, Drug effects on the spontaneous electrical activity of the squid giant axon, *J. Pharmacol. Exptl. Therap.* **155**:532–544 (1967).
72. S. Ehrenpreis, An approach to the molecular basis of nerve activity, *J. Cellular Comp. Physiol.* **66**:Suppl. 2, 159–164 (1966).
73. J. G. Dewan, Chemical steps in the metabolism of alcohol by brain *in vitro*, *Quart. J. Studies Alc.* **4**:357–361 (1943).
74. V. C. Sutherland, T. N. Burbridge, and A. Simon, Metabolism of $C\text{–}1^{14}$ ethanol to $C^{14}O_2$ by cerebral cortex *in vitro*, *Fed. Proc.* **17**:413(1958).
75. T. N. Burbridge, V. C. Sutherland, C. H. Hine, and A. Simon, Some aspects of the metabolism of alcohol *in vitro*, *J. Pharmacol. Exptl. Therap.* **126**:70–75 (1959).
76. M. M. Ferguson, Observations on the histochemical distribution of alcohol dehydrogenase, *Quart. J. Microscop. Sci.* **106**:289–297 (1965).
77. J. Akabane, Pharmacological aspects of manifestation of the acute after-effects of alcoholic beverages: a role of acetaldehyde in alcoholism, *Med. J. Shinshu U.* **5**:113–122(1960).
78. E. B. Truitt and G. Duritz, in *Biochemical Factors in Alcoholism* (R. P. Maickel, ed.) pp. 61–69, Pergamon Press, Oxford (1967).
79. J. W. Ridge, The metabolism of acetaldehyde by the brain *in vivo*, *Biochem. J.* **88**: 95–100 (1963).
80. R. Scholz, in *Stoffwechsel der isoliert perfundierten Leber* (W. Staib and R. Scholz, eds.), pp. 25–47, Springer, Berlin–Heidelberg–New York (1968).

81. M. P. Salaspuro and P. H. Mäenpää, Influence of ethanol on the metabolism of perfused normal, fatty and cirrhotic rat liver, *Biochem. J.* **100**:768–774 (1966).
82. O. Forsander, N. Räihä, M. Salaspuro, and P. Mäenpää, Influence of ethanol on the liver metabolism of fed and starved rats, *Biochem. J.* **94**:259–265 (1965).
83. H. Wallgren and R. Lindbohm, Adaptation to ethanol in rats with special reference to brain tissue respiration, *Biochem. Pharmacol.* **8**:423–424 (1961).
84. N. Abe, Dehydrogenase activity in alcohol-habituated rats (Jap.), *Tohoku Med. J.* **64**:267–269 (1961). (Abstr. *Quart. J. Studies. Alc.* **25**:581, 1964).
85. J. E. Moss, R. D. Smyth, H. Beck, and G. J. Martin, Ethanol impact on brain acetylcholine and its modification by cysteine, *Arch. Intern. Pharmacodyn.* **168**:235–238 (1967).
86. K.-H. Kiessling and K. Tilander, The effect of prolonged alcohol treatment on the respiration of liver and brain mitochondria from male and female rats, *Exptl. Cell Res.* **30**:476–480 (1963).
87. D. Hagen, GABA levels in rat brain after prolonged alcohol intake, *Quart. J. Studies Alc.* **28**:613–617 (1967).
88. V. C. Sutherland, T. N. Burbridge, J. E. Adams, and A. Simon, Cerebral metabolism in problem drinkers under the influence of alcohol and chlorpromazine hydrochloride, *J. Appl. Physiol.* **15**:189–196 (1960).
89. H. Wallgren, Biochemical aspects of the effects of ethanol on the central nervous system, XXXVI Congrès International de la Chimie Industrielle, Bruxelles 1966, *Comptes Rendues* **3**:812–815 (1967).
90. J. W. Moore, Effects of ethanol on ionic conductances in the squid axon membrane, *Psychosomat. Med.* **28**:450–457 (1966).
91. D. G. McQuarrie and E. Fingl, Effects of single doses and chronic administration of ethanol on experimental seizures in mice, *J. Pharmacol. Exptl. Therap.* **124**:264–271(1958).

Chapter 22

EFFECTS OF IONIZING RADIATION*

Enrique Egaña

Laboratory of Neurochemistry-Institute of Experimental Medicine
School of Medicine
University of Chile
Santiago, Chile

I. INTRODUCTION

The study of the effects of ionizing radiations on both growing and adult CNS appears as an aspect directly related to the general theme of this Handbook, since it refers to the "radiobiochemical lesion" of the neuroaxis. Such effects can resemble those produced by other noxious agents. Because we deal here with the intrinsic mechanism through which the ionizing radiation affects the nervous tissue, there is a radiobiological review at the beginning of the chapter. Neuroradiobiochemistry, one of the least known and cultivated frontiers of neurochemistry, does not represent more than 1 to 2% of the published papers of specialized journals, congresses, or symposia, in spite of the fact that the CNS is structurally the best system of animal economy and biologically the one most apt to register any energy change of the external medium; it could be supposed that the absorption of radiant energy would be rapidly transformed into physicochemical, biochemical, and functional alterations of the system. In this chapter we shall refer to neuroradiochemical facts of major significance, attempting to explain, as far as possible, the eventual mechanisms which may have bearing on such processes. In the past 8 years numerous books, reviews, and articles have been published on the effect of ionizing radiation on CNS. They contain excellent references to the radioneurobiochemical studies.[1-22]

* Most of the experimental work reported by E. Egaña *et al.* in this chapter has been carried out under a grant from the International Atomic Energy Agency, Vienna. Contracts No. 95, RB/1 and RB/2, and, at present, No. 887, also under a grant from the Comisión de la Investigación Científica, Faculty of Medicine, University of Chile, 1967–1970.

II. RADIATION ACTION ON LIVING MATTER WITH SPECIAL REFERENCE TO CNS

A. Radiation Action

The absorption of high energy particles or photons in living matter presents a sequence of observable events at atomic level of complex characteristics, which results in the instantaneous (less than 10^{-12} sec) formation of electronically excited or ionized molecules. The effects of radiation may be studied from diverse points of view, two of which are discussed here:

1. *General Conditions on Which Radiation Effects Depend*

Among the conditions on which radiation effects depend, the following are worth mentioning:

The physical quality of radiation (α and β particles, X, gamma and δ-rays, fast and slow neutrons, protons, etc.), refers to the reaction of radiation with electrons in matter, except neutrons, which may react with the atom nuclei with much greater probability than other charged particles; to the fact that β-particles have a variable in the energy distribution in opposition to α-particles (helium nucleus) emitted with equal energetic characteristics; that π meson (pion) interacts, causing several recoil particles of high LET (linear energy transfer); that X- and γ-rays have identical interaction depending on the energetic charge of the incident photon (photoelectric and Compton effects; ion pair formation).

The intensity of irradiation refers to total dose, the total of imparted (and absorbed) energy; to dose rate, which represents the rate or velocity at which energy is deposited; to LD_{50}, LD_{100}, which relate the amount of energy absorbed and the deleterious results on the organ, e.g., the central nervous system (CNS), or the organism *in toto* and its time of survival.

The method of administration can be external (the most usual experimentally or clinically) or internal, in which radioactive substances are administered (^{32}P, ^{131}I, ^{14}C, ^{24}Na, etc.), in which case the physical and biological decay must be considered. It is thought that CNS would be more vulnerable to internal irradiation at the same total absorbed dose. As regards the CNS, the external irradiation can be effected as whole body radiation, or localized to the head. The authors agree that the former method is far more noxious for the CNS because of the action of the extra-neuroaxis factors (abscopal effects), since the CNS tolerates far greater doses in the head irradiation.

Biological media and quality of tissue are mentioned in radiobiology, but from the neuroradiochemical point of view the following are worth analyzing: (1). The dynamic physicochemical and biochemical structure of tissue, which in the CNS shows different structure of the neuron and of glia cells, not only in thermodynamic characteristics but also in capacity for repair and growth which make the glia more radiosensitive; the active pulsation of the neuron in general is another condition worth considering. (2). The regenerative capacity. We do not have to insist that the neuron is a nonregenerative cell once the CNS is mature and that all the nonneuronal cells of CNS have a reparative capacity. (3). The local homeostasis of the CNS within which the cells work metabolically and energetically, is a determining condition: thus the p_{O_2} (oxygen effect), the degree of tissue hydration (effect through radiolysis of water), the pH ambient, the local content of SH groups (supposedly protective), and the local temperature (hypothermy diminishes the effects). These characteristics may help to explain why the hypothalamus should be far more radiosensitive than brain cortex and spinal cord under β-internal irradiation, as we shall see further. This could also possibly explain the greater radiosensitivity of fetal and newborn CNS, not only owing to its active DNA and RNA metabolism but also to its proportionally high H_2O content in tissue.

2. *Phases of Radiation Process*

Various authors,[23–29] with only minor differences, have proposed schemes regarding the temporary and consecutive stages which occur during the irradiation action.

a. Physical Stage. This is instantaneous, of a duration of about 10^{-13} to 10^{-15} sec, in which the incident energy diminishes at atomic level with the formation of a great number, depending on the energy of the particle ($\leq$32–34 eV for excitation), of activated molecules. Each mode of activation depends on an effective cross section, Q_n for excitation and Q_i for ionization. We can differentiate between these two processes because in case of ionization the residual charge and/or excited level may alter in various ways the conduct of the molecule,[26] this last event having great importance in radiobiology. Excitation and ionization are primary processes, but are not the only ones; others, proportionally far less important, are the ejection of an atom from its normal position, and the Auger effect, in which an inner orbital electron becomes lost.[26] It is necessary to point out that, since excitation and ionization are primary processes, one of the most decisive facts within the dynamics of radiation effects, they are the least known and measurable, because, as is expressed by Platzmann[29] in the measurement of a physical effect of radiation such as emission of light, formation of ions, or chemical effects, both the appearance of a new optical absorption spectrum or the ESR spectrum, the excited molecules or the radicals or the ions observed are usually not the primary products themselves but later products, having been formed perhaps 10^{-6} sec later.

b. Physical Chemical Stage (Secondary Products). This stage of radiation covers chemical or chemicophysical processes between 10^{-14} sec and 0.1 sec. It deals, from the initial creation of a dry electron through its solvation, fixation, and transfer, through excited singlet states and final triplet states up to when it eventually settles in thermodynamically stable entities. This enormous field has been seldom studied. There is a strong need for further work in the elucidation of the key processes of transient intermediates affecting the eventual reactivity of biochemically important entities which react at perhaps 1 sec. We have to bridge the gap between the sophistication of studies at 10^{-14} sec and life processes, which are relatively slow.

c. Chemical (Biochemical) Stage. This is the time at which the process has reached a thermic equilibrium and in which the recently formed ions or free radicals react with themselves or with the medium. This stage encompasses a time interval of about 10^{-8} to 10^{-6} sec and operates perhaps 10^{-10} longer than the primary effects. Within this stage, there appear varied biochemical alterations which can be *in situ* (within the irradiated volume) or at a distance; and, as regards time, "prompt" or "delayed." The latter require a certain lapse to develop. Such alterations can also be chronic, arising from a lethal dose and at a delay of one or more months.[(30)] These biochemical lesions will be analyzed in a more detailed manner in the section on neurochemical studies.

d. The Biological Stage. This is the reaction of the affected organisms in which subcellular and cellular aspects with special interest in their genetic reference [true gene mutation or intergenetic effect, according to Wolff[(31)]] are analyzed, as are also the functional and structural changes. The hematopoietic system, the CNS, the gastrointestinal system, and the masculine gonad are the most affected.

e. The Syndrome Irradiation (Clinical Significance). This refers to the signs and symptoms deriving from the irradiant action, which we merely mention.

3. *"Direct" (Target Theory) and "Indirect" (Radiolysis of Solvent) Effects*

From a radiobiological point of view, there are two fundamental mechanisms of radiation action, (1) The "direct effect" (target theory),[(32–34)] in which the excitation or ionization liberates locally a quantity of energy sufficient to disrupt a small surrounding area. An ample variety of biological molecules, enzymes, DNA and RNA, antigens, virus, bacteria, etc., could be inactivated by the passage of a single particle or photon; the inactivation is the product mainly of ionization rather than of excitation, although it cannot be fully ascertained. The ESR spectrum produced by this effect has been especially studied by Gordy *et al.*[(35–37)] in nucleic acids and proteins, some of which have a spectrum which is identical to that of irradiated cystine (an example of energy transfer to S-S bonds). (2) The "indirect effect" (radiolysis of water, "free radical" theory) is the second mechanism of radiation action. It can be safely said that the most frequent mechanism of radiation within aqueous cells is through the radiolysis of water, which represents, at least in the CNS, from 95 of 75% of the cellular contents, depending on age. The principal primary species of the irradiation of water are: OH, H, e_{aq}, O, H_2O_{aq}, as per the following reactions:

$$\rightsquigarrow H_2O \rightarrow H_2O^+ + e \qquad (1)$$

the ion H_2O rapidly undergoes a proton transfer reaction with neighboring water molecules:

$$H_2O^+ + H_2O \rightarrow H_3O^+ + OH \qquad (2)$$

and becomes solvated into $H_3O^+_{aq}$.

The electron ejected in the ionizing process has a low energy and would become solvated into e_{aq}. Other minor processes occur such as the dissociation of excited water:

$$H_2O \rightarrow H + OH \qquad (3)$$

Allen[38] and Daniel *et al.*[39] have suggested the formation of oxygen atoms from the dissociation of

$$H_2O^+ : H_2O^- \rightarrow OH^+ + H \qquad (4)$$

$$OH^+ \rightarrow O + H \qquad (5)$$

The free radicals originating from ionization or excitation are produced in groups or spurs and pairwise reactions which lead to hydrogen and hydrogen peroxide formation:

$$H + H \rightarrow H_2 \qquad (6)$$

$$OH + OH \rightarrow H_2O_2 \qquad (7)$$

Reactions (6) and (7) are typical of the exposure of aerated or oxygenated water; the powerful reducing hydrogen radical converts to the relatively inert hydroperoxy radical:

$$H + O_2 \rightarrow HO_2 \qquad (8)$$

The H_2O_2 yield decreases in oxygenated water continually until reaching an equilibrium; it decomposes in irradiated water in accordance with the general reaction:

$$H_2O_2 \rightarrow H_2O + \tfrac{1}{2}O_2 \qquad (9)$$

In turn, the hydrogen and hydroxyl radicals destroy H_2O_2 according to:

$$H + H_2O_2 \rightarrow H_2O + OH \qquad (10)$$

$$OH + H_2O_2 \rightarrow H_2O + H_2O \qquad (11)$$

We shall refer to the "H_2O_2 effect" later.

B. General Criterion for the Study of CNS Irradiation Effects

The criterion of such study in general is marked by the techniques with which the researcher examines the CNS altered by the irradiation.

1. *Morphopathological Studies*

Morphopathological studies are perhaps the most abundant. In these the macroscopic and the microscopic alterations are analyzed, also the subcellular constituents, not only of the neuron and neuroglial cells but also of the microcirculation of the neuroaxis. There exist numerous reports on the subject, the papers of Hicks *et al.*[41–44] and of Haymaker *et al.*[45–48] being perhaps the most complete ones. The most relevant facts, according to the authors quoted, are as follows: A schematic pattern of CNS response to irradiation is accepted—early reaction, lasting several weeks, is characterized by exudative phenomena, cell hypertrophy, and necrosis, and a delayed damage with necrosis, especially in the white substance.[49] In the early reactions, the radiolesion of the CNS depends on various factors: dose, form of exposure, and the specific radiosensitivity of the neuroaxis of every species (monkey: 1500 r or more, rabbit: 2440–2850 r and rat 2500 r to obtain the same type of radionecrosis).

The radiovulnerability of the neurons is different. Very radiosensitive are granule cells of cerebellum, neurons of the olfactory bulbs, microneurons of the dentates gyrus. Purkinje cells are resistant. Even within the same type, there are differences: the astrocytes of the cortex are more sensitive than those of the white substance,[50,51] a diversity related to DNA and RNA eventual differences or the cytoplasmic membrane system.[52]

Andersson *et al.*[53] and Larssen[54] believe that the radiolesion of cortex is vascular-dependent based on their experiments with high dose, 10,000–40,000 r, which could also be valid for low doses, of 250 r.[55,56] Others, such as Schummelfeder[57] state that the neuronal damage, independent of circulatory alterations, does exist per se. It would seem that the quantity of irradiation and the field of exposure are dominant factors (with doses as large as 400,000 r and 40 μ of incidence, the damage is neuronal, primarily); the more the area of radio absorption is enlarged, the more the damage of the microcirculation is determined. As to radiosensitivity, the oligodendroglia would be the most vulnerable, with acute swelling, hypertrophy, accumulation of glycogen[58] and the disintegration of myelin; the astrocyte requires greater doses to exhibit this type of lesion, which is equally circulation-dependent. The cortical neurons are the most radioresistant. The chronical radiogenic lesions are probably effects of microcirculatory alterations (swelling and consecutive fibrosis of the walls), endothelial cell hypertrophy, etc.; the white matter is most affected.

2. *Neurophysiological Studies*

Studies of the register of potential at rest and in action (EEG, ECG, ERG), the analysis of specific functions of CNS, etc., have been made. This subject is mentioned in passing: for more information, see references 59–66.

3. *Neurobiochemical Studies*

These are discussed in this chapter.

4. *Behavioral Studies*

Behavioral studies include tests of learning, motivation, avoidance, and other conditioned reflexes; effects on perceptual and aversive stimuli; sleep rhythm, especially in humans; maze performance, visual discrimination, etc. For more information, see references 67–68.

5. *Clinical Studies*

Clinical studies are self-explanatory.

C. The CNS, a Radiobiological Sensitive System

Some factors to be considered in studying the CNS are (1) Age: The greater radiovulnerability of embryo and fetus as compared to adult has already been pointed out; for the CNS advancing maturity signifies progressive acquisition of radioresistance. (2) The dose of exposure: arbitrarily the author has indicated 20% of the dose (LD_{50})[30] for the human being and laboratory animals, estimated at 65–160 r or rad (mean 100 approx.), as the dose capable of altering in a significant manner any index of adult CNS. The fact is that the mammalian CNS, even at lower than the 100 r threshold mentioned, appears to be a radiovulnerable system. This would indicate that it may be considered a definitely radiosensitive system.

Does the CNS have a sensory receptor which specifically registers, or recognizes, the radiant stimulus? This would, *a priori*, imply a specific receptor (specific depolarization of the receptor membrane) through an energetic or chemical mechanism (stimulated through the radiolysis of water *in situ* or target effect of an important molecule of the receptor). We are unable to answer this question.[79–83] On the other hand, the capacity of the CNS to register energy signals as low as 0.1–0.5 rad (in thermodynamic terms 10^{-4} to 10^{-5} g/cal-sec^{-1} g^{-1} tissue fresh weight), lower than voice level intensity,[84] is a fact which is hard to accept, although there exists in the animal economy a nervous structure, the retina, in which the rhodopsin registers in transient chemical alterations upon 2 eV photon absorption. There exist, however, experimental results that indicate that the CNS is a highly radiosensitive structure. We shall quote some examples, and in Table I, some neurobiochemical data on this aspect are given.

TABLE I

Some Neurobiochemical Responses of CNS to Radiation

Dose as total dose, r or rad	Species	CNS process or content	Characteristics of the exposure	References and remarks[a]
0.005 r	Human	Excitation of rhodopsin	X-local irradiation on the eye (visual sensation)	(87)
0.2–0.6 rad	Adult rat	Decrease of encephalon ATP	β-internal (^{32}P)	Fig. 2
0.5 rad	Adult rat	Increase of *in vitro* respiration	β-internal (^{32}P)	Table II
5 r	Newborn rat	RCR (whole encephalon)	gamma-whole body-(^{60}Co)	Table IV
5 r	Adult rat	Glutamate decrease and GABA increase in brain cortex	gamma-whole body-(^{60}Co)	Table VII
14 rad	Adult rat	Synthesis of ACh in hypothalamus, increase	β-internal (^{32}P)	Table IX
50 rad	Adult rat	1-^{14}C-glucose conversion into free CNS amino acids, increase and decrease according to the amino acid	gamma-whole body-(^{60}Co)	Table VI
5–22 rad	Adult rat	Increase of 5-HT content hypothalamus and "diencephalon"	β-internal (^{32}P)	Table X
50 rad	Adult rat	Decrease of aerobic and anaerobic glycolysis in diverse CNS levels	β-internal (^{32}P)	Table V (anaerobic glycolysis)

[a] Tables and figures refer to this chapter.

Kudritsky[85] finds, as do Livanov *et al.*,[86] changes in the central time of cerebrospinal reflexes, the former at doses as low as 0.1–0.5 and 10 r; Pape,[87] quoted by Livshits, refers to visual excitation at a dose of 0,005 r; Sherkov[88] reports on EGG changes in patients receiving 2–4 mC of ^{32}P, with an initial increase of the amplitude of α-waves, which is followed by a decrease. Grigor'yev *et al.*[89,90] report an EEC depression in rabbits with 0.05–1.3 r and changes in the cortical excitability with 1 r. Hicks *et al.*[91] presented excellent morphopathological data on the CNS with 10, 20, 30, or 40 r exposures, the threshold being 10 r in newborn and below 10 r in late fetal life. Prenatally, the most immature neurons of the second and fifth cortex layers are the most sensitive.

III. NEUROBIOCHEMICAL STUDIES

A. The Respiratory Process and Oxidative Phosphorylation

The CNS in its development gradually changes its preferential anaerobic into an aerobic metabolism; in the adult CNS there exists a gradient from the cortex to the spinal cord. This is especially true when it metabolized glucose.[92–96] The effect of irradiation on the CNS respiratory process has been studied using diverse techniques. The most relevant results are as follows: (1) The free O_2 content after head, abdomen, and chest irradiation of rats and rabbits with 600 to 1000 r has variable changes.[97,98] Snezhko[99,100] using polarographic methods measured the O_2 content of cortex and subcortical areas of rabbit with head exposure (1000–3000 r) and found the p_{O_2} diminished in both areas. (2) The CNS respiration studied *in vitro* has been analyzed by various authors. Florsheim *et al.*[101] do not report significant changes in the O_2 uptake in homogenized mouse brain irradiated with 500 to 19000 r and in lapses from 0 to 190 hr. Timiras *et al.*[102] find in 75-day experiments in rats irradiated with 500 r at 2 days old and killed at diverse lapses after exposure, a significant increase of Q_{O_2} of the cortex and a diminution in the hypothalamus; the Q_{CO_2} appears decreased in the cortex and remains invariable in the hypothalamus. These experiments are carried out with "endogenous" substrate; no exogenous glucose is added during the incubation.[102,103] Lott *et al.*[94] irradiated *in vitro* slices of brain cortex, thalamus, midbrain, and pons/medulla of adult rats with 10000 to 20000 r; the Q_{O_2} does not change in the first hour; however, in the second the respiration decreases in cortex slices remains unchanged in thalamus and midbrain, and increases in pons/medulla. Egaña[104,105] studied the effects of β-internal radiation (^{32}P) at variable doses from 0.1, 0.3, 0.4, 0.6 to 1.0 rad as very low, from 12 to 118 rad as low, from 130 to 357 rad as infra-LD_{50}[30] dose and from 504 to 2400 rad as high, on the oxygen uptake *in vitro* of all CNS levels (from brain cortex to spinal cord) in Ringer solutions of diverse cation composition. The general results are as follows: there are structures in

TABLE II

Oxygen Consumption of Slices of Some CNS Areas, Internally β-Irradiated (^{32}P) $\frac{\mu\text{atom oxygen}}{\text{mg protein}}$

Condition[a]	Exposure total dose, rad			CNS area[b]		
	Brain cortex	Hypo-thalamus	Spinal cord	Brain cortex	Hypothalamus	Spinal cord
Normal				1.40 ± 0.12 (45)	1.42 ± 0.18 (40)	0.61 ± 0.07 (36)
Irradiated						
Single dose	0.6	0.3	0.7	1.76 ± 0.38 (20)	1.41 ± 0.05 (20)	0.61 ± 0.19 (18)
	5	2.3	5.2	1.37 ± 0.18 (20)	1.37 ± 0.16 (20)	0.54 ± 0.1 (20)
	12	5.0	12	2.23[c] ± 0.40 (20)	1.94[c] ± 0.30 (20)	0.80 ± 0.2 (20)
	19	11	22	1.54 ± 0.4 (24)	2.19[c] ± 0.5 (22)	0.61 ± 0.14 (20)
	41	25	48	1.29 ± 0.24 (24)	1.87[c] ± 0.22 (20)	0.57 ± 0.16 (18)
	67	38	79	1.96[c] ± 0.40 (23)	2.01[c] ± 0.32 (23)	0.82 ± 0.22 (18)
	531	296	631	1.1 ± 0.16 (18)	1.32 ± 0.3 (20)	0.58 ± 0.13 (18)
One-month course	892	542	1111	1.1 ± 0.18 (37)	1.34 ± 0.2 (35)	0.44 ± 0.16 (30)
Four doses	1950	1108	2214	1.17 ± 0.21 (25)	1.04 ± 0.12 (22)	0.60 ± 0.25 (20)

[a] β-irradiated (^{32}P). Single dose 125 μC ^{32}P/100 g rat and the animals sacrificed at different times after irradiation: 2, 6, 12, 24, 48, and 72 hr, and 90–100 days. One month course given twice a week and totaling 320–350 μC/100 g rat. Four doses totaling 500 μC/100 g, the days on which the injection were given at 1, 7, 30, and 37 days, being sacrificed at 90–100 days after the first injection. Ringer–Elliott–bicarbonate–10 mM glucose–2.5 mM K^+-pyruvate; 60 min, gas-phase O_2 95%–CO_2 5%; Warburg vessel.

[b] Mean, S.D. and number of experiments (in parentheses).

[c] $P < 0.01$ between normal and irradiated. For "diencephalon," olfactory bulbs, cerebellum, midbrain, and pons/medulla studies see Ref. 104.

which the respiration increases with doses as low as 0.1, 0.3, 0.7 rad (midbrain, cerebellum, and spinal cord), the brain cortex at 12 rad, the hypothalamus at 5 rad, and the olfactory bulbs at 13 rad. The descent of O_2 uptake occurs at diverse doses, e.g., the olfactory bulbs at 675 rad, the spinal cord at 1110 rad.

In summary, the response of the CNS respiratory process *in vitro* under internal β-irradiation appears to be a variable phenomenon, from an increase at very low dose, passing to a stage of return to normal values at low dose, and ending in a final progressive inhibition at high dose (Table II).

1. *Mitochondrial Respiration in Irradiated CNS*

It is known that cytoplasmic organelles contain the electron transport system (ETS), the enzymes for ATP synthesis, the dehydrogenases of the tricarboxylic acids, the enzymes for protein synthesis, etc. The electron transfer system consists of at least nine oxidoreduction steps, and its entirety is made up of four distinct complexes[106–118] The constancy of mitochondria of adult and newborn CNS has been studied with the electron microscope[119] and by the protein mitochondria content/100 mg fresh tissue. From these studies, Egaña *et al.*[120] obtained the following results: (1) Adult CNS. β-internal irradiated mitochondria shows a slight increase in respiration at low dose and a definite decrease at high doses. This has been corroborated by experiments with whole-body gamma irradiation (50 rad)[121] in which a discrete oxygen consumption enhancement has also been observed. There would exist consequently a concordance of β-internal and gamma external radiations in mitochondria, and β-internal in slices. (2) Growing CNS. First of all, it must be remembered that the normal process of maturing is accompanied by a progressive increase of mitochondrial respiration[122–125] which attains its highest values in the adult (Table III), a statement which has been contradicted by the results of Dahl *et al.*[126] The whole-body gamma-exposure with 5, 50, and 500 r does not usually produce a great alteration in the oxygen consumption in mitochondria of newborn CNS when studied after 8–10 min of the irradiation. Nevertheless, at 21 days after exposure, respiration decreases; in Table III, data have been taken from experiments carried out with 50 r.[121,125] In the experiments shown in Fig. 1, the action of substrate (source of e), both endogenous and exogenous, has been studied; the model published here corresponds to succinate, but studies have also been carried out with ATP, ADP, glutamate, glucose/hexokinase, Ca^{2+}, etc. The most evident action has been obtained with succinate.

Although oxygen consumption is the basic metabolic index for the CNS and the incidence of irradiation on it is of importance, it represents nevertheless an overall process the fluctuation of which do not supply information on the finer metabolic steps. More information has been sought through the study of irradiation effects on the RCR (respiratory

TABLE III

Oxygen Consumption of Normal and Irradiated Mitochondria of Growing and Adult CNS[a]

$$\frac{\text{m}\mu\text{atom O/min}}{\text{mg mitochondria protein}}$$

	Normal	Irradiated[b]
Newborn (< 48 hr)	28 ± 1.6 (30)	22[c] ± 1.9 (30)[d]
7 days	44 ± 3.5 (25)	—
21 days	86 ± 6 (28)	43[c] ± 5 (20)[e]
Adult	104 ± 7.2 (16)	109 ± 8.2 (22)

[a] Both sexes; incubation time 75 min; Warburg vessel; gas-phase O_2; succinate 40 mM, ATP and ADP 2 mM, glutamate 10 mM; Ringer mitochondria K^+–Mg^{2+}–phosphate, 0.5 mM EDTA, *p*H 7.4. In the case of normals the *P* values (< 0.01) indicate that the oxygen uptake increases significantly with growth.

[b] Exposure: 50 r whole-body-gamma.

[c] $*P < 0.01$ between normal and irradiated.[121,125]

[d] Prompt: the preparation of mitochondria begins at 8–10 min after exposure.

[e] Delayed: after 21 days, after being irradiated at the same time as "prompt."

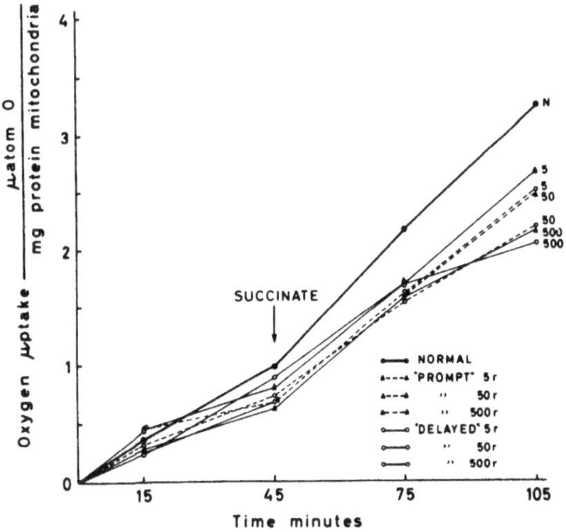

Fig. 1. Effect of "endogenous" and "exogenous" succinate on mitochondrial respiration. Incubation in Ringer–mitochondria; succinate 40 mM; the incubation medium: ATP and ADP 2.0 mM; β-NAD 1.5 mM; glutamate 10 mM; hexokinase 0.16 mg/ml and glucose 10 mM; EDTA 0.5 mM. At ↓ succinate is added from the side-vessel. Newborn for normal and "prompt" effects; for "delayed": 5 and 50 r 21 days and 500 r 7 days after the exposure, respectively.[125]

control rate) of mitochondria of growing CNS exposed to gamma whole-body (^{60}Co) irradiation. The RCR is defined by Chance *et al.*[127] as "the ratio of the oxidation rate in the presence of excess substrate and ADP (state 3) to the oxidation rate after ADP has been phosphorylated to a steady-state concentration (state 4)." In practice, it is obtained by means of oxigraphic determination[124,127] or manometrically[125,128] or with both methods[129] from the ratio of

$$\frac{\mu\text{atoms oxygen consumed in presence of ADP} + \text{substrate}}{\mu\text{atoms oxygen consumed in presence of substrate only}}$$

Egaña *et al.*[125] studied RCR in mitochondria of growing CNS of whole-body rat that has been gamma-irradiated. In Table IV some results are given.

The RCR studies could make one presume *a priori* that irradiation may have diverse effects on respiration, increasing it in the controlled state, and depressing it in the active state. This idea requires further investigation.

Experiments quite different from those mentioned have been carried out by Smith *et al.*[130]; they used a macrodose of irradiation (90,000 r)

TABLE IV

RCR of Irradiated Growing CNS Mitochondria[a]

Condition	RCR
Normal (< 48 hr)	2.0 ± 0.05 (20)
"Prompt" effects	
Irradiated	
5 r (< 48 hr)	1.64[b] ± 0.04 (15)
50 r (< 48 hr)	1.51[b] ± 0.06 (14)
500 r (< 48 hr)	1.50[b] ± 0.06 (15)
"Delayed" effects	
5 r (21 days)	1.9[b] ± 0.07 (14)
50 r (21 days)	1.61 ± 0.06 (14)
500 r (7 days)	1.42 ± 0.06 (14)

[a] Whole-body gamma-irradiation; all groups are exposed at the age of < 48 hr, i.e., newborn. RCR = μatom oxygen uptake in presence of succinate + ADP/μatom oxygen uptake in presence of succinate. Manometric method (Warburg vessel); gas phase: O_2. The main components of mitochondria physiological medium are succinate 40 mM, or succinate plus ADP 2 mM. For further technical details, see Table III. "Prompt": the newborn rats are killed 8–10 min after exposure; "delayed": 5 and 50 r at 21 days and 500 r at 7 days after exposure.[125]

[b] $P \leqslant 0.01$.

and found that the respiration of rat brain mitochondria with succinate as substrate is considerably depressed, the same as the oxidative phosphorylation. They conclude that these results are probably due more to a "mitochondrial disorganization" than to the destruction of respiratory enzymes.

As a tentative corollary to these findings as a whole it could be said that the respiratory process of both growing and adult CNS is radiosensitive when studied in slices and adult and growing mitochondria. With low doses, the respiration increases in the adult CNS, and decreases in the growing CNS. At high doses both groups respond with an inhibition of the process. We do not know what mechanism is involved in these processes (increase, decrease, or no change). One could suspect a change in the permeability of the cellular or mitochondrial membrane with a loss of K^+; a pronounced increase of ADP, as we shall see later; radiosensitivity of certain ETP components, etc. In the case of CNS-tissue which is glucose-dependent the respiratory process affected by irradiation could be in part elucidated by studying the oxidative phosphorylation and aerobic and anaerobic glycolysis.

2. *Oxidative Phosphorylation*

It is well known that in the CNS the capacity to catalyze oxidative phosphorylation is in the mitochondria and that the electron transport chain is directly coupled to the phosphorylation. We do not know, however, what the exact mechanism is through which both processes are tied.

Chance and his group[106] propose the formation of an energy-rich E~I which would represent a common pool of energy proceeding from various energetic reactions. Mitchell *et al.*[131,132] suggest the proton-gradient scheme for oxidation of a carrier, C_{red}, localized at one side of the membrane by means of a second C_{ox} bound to the other side. According to Mahler *et al.*[133] the evidence would then favor chemical rather than physical *p*H gradients and it is quite possible that in the future both models will be merged into one scheme. From the experimental data covering the effects of irradiation on oxidative phosphorylation in other tissues (not CNS), one could conclude that CNS mitochondria had a reduced capacity for ATP formation (see Ref. 30). It seems, nevertheless, that the more recent experimental findings do not corroborate this assertion. The effects of radiation on this process have been investigated using diverse techniques: Uptake in the CNS of ^{32}P, to a great extent is incorporated into ATP. Florsheim *et al.*[134] do not find modification after 18–20 kr. Correlated studies of ^{32}P uptake, its location in some phosphorus compounds of CNS, the ATP, Cr-P (creatine phosphate), P_i, etc., and $AV^{32}PD$ (carotid-jugular) have also been made. Egaña *et al.*[135] have undertaken studies of this type in adult rats subjected to β-internal radiation at greatly varied doses (already mentioned). The principal results indicate that the ^{32}P

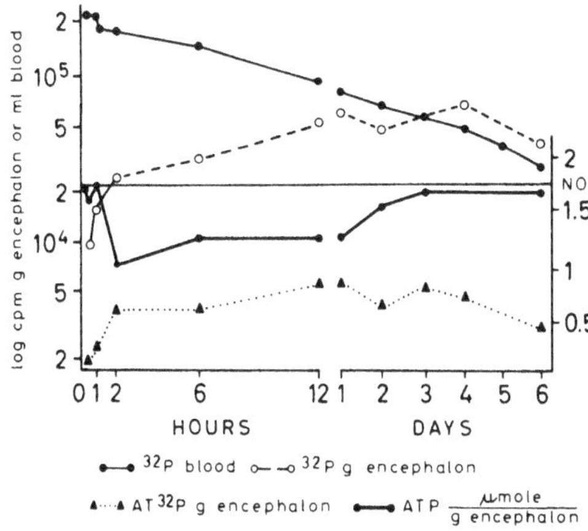

Fig. 2. ^{32}P Incorporation into ATP and total phosphorus compounds of the irradiated whole encephalon. Exposure: β-irradiation (internal) single dose experiments 125 μC/100 g rat. ^{32}P blood values determined in a mixture of carotid–jugular sample and expressed as cpm/ml. AT^{32}P and ^{32}P of whole encephalon as cpm/g fresh tissue. ATP content: μmole/g fresh tissue. Normal refers to the ATP content. Left ordinate: log cpm per g encephalon or ml blood. Right ordinate: μmole ATP/g fresh tissue. Abscissa: time in hours or days.[135]

incorporation rate into ATP does not alter; the ATP content decreases significantly; Cr-^{32}P remains normal; ^{32}P increases whenever ATP diminishes; the ratio

$$\frac{\text{sa } ^{32}\text{P-insoluble compounds (nucleic acids, phospholipids, etc.)}}{\text{sa } ^{32}\text{P-soluble compounds (ATP, ADP, Cr-P, } P_i\text{, etc.)}}$$

changes because of the increase of the soluble fraction during the first hours; later it becomes normal, finally to change again because of the increase of the insoluble fraction, Figure 2 shows these findings, which occur at low doses, 14, 31, and 56 rad. P/O and ADP/O ratios have been made. We shall only refer to the latter, which according to Chance *et al.*[106] and Milstein *et al.*[124] is identical to the former, and which according to Ozawa *et al.*[136,137] would be, in a certain sense, parallel to the RCR. Egaña *et al.*[125] studied ADP/O ratio in newborn rats, CNS mitochondria, whole body, 5, 50, and 500 r gamma-irradiated. An increase ($P < 0.01$) is observed in irradiated mitochondria; succinate and/or glutamate were used in these experiments. Gorodissakaya *et al.*[138] studied in rat CNS the immediate effects (2 hr) of β-internal irradiation, from 0.1 to 10 μC ^{32}P, on the radiophosphorus incorporation; the sa increases at 2 hr, which coincides with a high organic/inorganic ratio of brain phosphate.

Cherkasov *et al.*[20] find a light increment of CNS ATP under whole body 600–700 r gamma rays during 48 hr after the exposure, which decreases afterward. The ^{32}P incorporation into ATP and Cr-P decreases proportionately to the respective lowering of content. The author is not acquainted with other PO studies in irradiated CNS, nor knows of studies on ATPase, Mg^{2+}-, K^{+}-, Na^{+}-, or Ca^{2+}-activated irradiated CNS.

B. Carbohydrate Metabolism

The CNS is the most glucose-dependent tissue of the organism, since through its degradation it obtains 80% or more of its energy for structural and functional purposes. Many papers have been published on the subject of the effect of irradiation on CNS glucose metabolism. Grayevskaya *et al.*[139] found in dogs irradiated with 300–500 and 1000 r at 2 to 40 days a decrease of the glucose consumption by CNS; paradoxically, in later lapses the CNS delivers glucose toward the venous circulation. Egaña *et al.*[140] in whole-body gamma-exposed (50 r) rats found a greater supply of glucose to the encephalon as against normal (0.86 μmole/ml more), although the AVglucose remains equal to normal (1.18, normal being 1.15) during the first 15 min. It increases to 1.88 at 60 mm after exposure. This fact, i.e., the greater supply of glucose to the encephalon, has to be examined closely through the several factors: CNS glucose content, permeability of the neuronal and glial membranes to this metabolite, etc. (as far as the author knows, only the former has been studied in irradiated CNS).[140,141] On the other hand, several authors have verified an accumulation of glycogen in astroglial and oligodendrous glial cells after 1200–6000 rad (α-particles in surface dose) as early as 24 hr post-irradiation with a peak at 48 hr.[16,48] Haymaker *et al.* have interpreted this as an unspecific reaction of the irradiation on CNS which would result in a diminished degradation of glucose and a subsequent cerebral accumulation of glycogen. Cherkasova *et al.*[18,20] reported in 1957 an increase of total CNS glycogen in whole-body gamma-irradiated (600–700 r) rats conjugated with protein and free proteins, until the fourth day after exposure.

1. *Anaerobic and Aerobic Glucose Metabolism*

Egaña *et al.*[141] studied the glucose content of total encephalon of whole-body gamma-irradiated rat, 50 r, with the following preliminary results (μmole glucose/g fresh tissue): a decrease of 50% after 30 min of the exposure and of 30% after 60 min.

In the *in vitro* studies of anaerobic glycolysis of all the adult CNS areas (from cortex to spinal cord) a decrease in the formation of lactic acid has been found, measured in Warburg vessels[104,105] (Table V). On the other hand, Kay *et al.*[142] by means of glucose-U-^{14}C conversion into lactate ^{14}C report a lowering of the net synthesis of lactate. These authors have different interpretations regarding this point. Apart from

TABLE V

Anaerobic Glycolysis $\frac{\mu\text{mole lactic acid}}{\text{mg dry weight}}$

Condition	Exposure, total dose (rad)				Ratio			
	Cerebellum	Midbrain	Pons/medulla	Spinal cord	Cerebellum	Midbrain	Pons/Medulla	Spinal cord
Normal					0.28 ± 0.1 (50)	0.30 ± 0.04 (50)	0.15 ± 0.05 (50)	0.12 ± 0.04 (50)
Irradiated								
Single dose	0.3	0.1	0.5	0.7	0.17 ± 0.06 (20)	0.16 ± 0.009 (22)	0.14 ± 0.06 (21)	0.13 ± 0.05 (21)
	2.2	0.7	2.7	5.2	0.18 ± 0.009 (20)	0.18 ± 0.007 (21)	0.14 ± 0.06 (21)	0.15 ± 0.04 (20±)
	5.3	1.7	6.2	12	0.15 ± 0.04 (21)	0.17 ± 0.06 (21)	0.14 ± 0.03 (21)	0.14 ± 0.04 (21)
	10	4	12	22	0.18 ± 0.05 (19)	0.15 ± 0.03 (18)	0.14 ± 0.03 (19)	0.13 ± 0.03 (21)
	23	9	25	48	0.32 ± 0.1 (21)	0.18 ± 0.005 (20)	0.17 ± 0.03 (20)	0.09 ± 0.003 (20)
	38	16	43	79	0.22 ± 0.01 (16)	0.14 ± 0.01 (18)	0.16 ± 0.03 (18)	0.18 ± 0.09 (18)
	303	132	354	631	0.29 ± 0.1 (26)	0.18 ± 0.05 (28)	0.16 ± 0.05 (28)	0.16 ± 0.007 (28)
One month	510	233	651	1111	0.13 ± 0.06 (22)	0.15 ± 0.04 (24)	0.16 ± 0.07 (23)	0.12 ± 0.005 (20)
Four doses	1124	510	1340	2214	0.16 ± 0.05 (16)	0.10 ± 0.05 (16)	0.10 ± 0.05 (16)	0.11 ± 0.002 (16)

[a] β-irradiation (internal exposure) (^{32}P). Ringer–Elliott–bicarbonate–30 mM glucose; gas-phase N_2 95%–CO_2 5%, manometric method. For other CNS areas (brain cortex, "diencephalon." hypothalamus, olfactory bulbs), see Ref. 104.

the different conditions of exposure (β-internal in the first case and X-rays in the second), there also exist other variables. However, Kay *et al.*[142] believe that anaerobic glycolysis is not affected and admit the possibility that the lactate synthesis from substrates other than glucose may be diminished. Timiras *et al.*[102] in rats at 75 days, 500 r X-rays irradiated at 2 days, obtain a decrease of anaerobic glycolysis in brain cortex and a lowering tendency in hypothalamus (the slices are incubated in physiological medium without glucose). Other authors have also studied anaerobic glycolysis in the developing CNS[125] in homogenized whole encephalon of newborn CNS whole-body gamma-irradiated (5, 50, and 500 r), and in "prompt" and "delayed" time (7 and 21 days after newborn irradiation). The results would indicate that there are no immediate changes after irradiation except at 50 r, and an increase in "delayed" lapses, except at 500 r.

There are also some results in aerobic glycolysis. Egaña[104,105] finds in *in vitro* experiments with slices, using all levels of CNS of adult rats, β-internally exposed, 0.1–2400 rad, a lowering of the glucose consumption in diverse CNS areas, principally in brain cortex and hypothalamus; pons/medulla and spinal cord are radioresistant in this respect. In newborn rats that have been whole-body X-irradiated, Nakazawa *et al.*[143] reported that respiration and ^{14}C-carbon dioxide from ^{14}C-glucose in whole brain slices from 4 to 11 day old rats, do not change. De Vellis[144,145] in 2-day-old rats, head-exposed 750 r, X-rayed, investigated the aerobic lactate formation in the presence or absence of chloral. It increases in medulla slices but does not change in brain cortex and midbrain slices.

2. *Enzymes Related to Glucose Metabolism*

According to de Vellis *et al.*[144,146] the LDH does not alter in brain cortex and midbrain of newborn rats under the same conditions; the ratio of lactate dehydrogenase to glycerolphosphate dehydrogenase is found slightly increased in brain cortex and midbrain and more than doubled in medulla. The authors connect the two findings, expressing that the increase of lactate production in irradiated medulla would depend on the increase of LDH CPDH ratio. Bonavita *et al.*[141] in newborn and adult rats, head X-ray exposed (300, 400, and 500 r), found in the newborn an accelerated process of maturity of CNS LDH isoenzymes pattern with LDH-pyruvate affinity diminished. In the adult, the content is not affected, but the indicated affinity is.

C. Protein and Amino Acid Metabolism

The actual protein and amino acid content of CNS, dynamically considered, both in the growing and adult animal, depends on the active synthesis and breakdown, the migration of protein through the axon, the in-and-out passage of amino acids of the brain, etc. Studies with ^{14}C, ^{35}S labeled amino acids confirm these indications.[148] In this section, we shall cover two topics of fundamental importance: (1) the irradiation of amino acids and proteins in general, and (2) amino acids and proteins in irradiated CNS.

1. *Irradiation of Amino Acids and Protein*

At present, we are acquainted with certain important facts: in the irradiation of amino acids and proteins, free radicals are formed by the breakage of $C_\alpha H$ bonds; there remain electrons the spin concentration of which takes place mainly on C_α or by breakage of S—S bonds and concentration of spin density on sulfur atoms, as also by means of a secondary formation process of radicals due to the addition of hydrogen on the aromatic ringed groups of phenylalamine, tyrosine, and tryptophan.[149–151] Duchesne *et al.*[152,153] discovered a certain kind of protein radioresistance which would be more related to the aromatic side chain amino acids than to the aliphatic ones. Furthermore, they established that the Fe and Cu ions have a radioprotective action. In this regard, Shimazu *et al.*[154] believe that the radiolability of amino acid varies according to the protein of which it forms part and it could be due essentially to the nature of the lateral R group in amino acids, e.g., phenyl of phenylalamine, the SH groups of cysteine, and SCH_3 of methionine are very radiovulnerable. In the aliphatics, it depends on the length of the chain. We know from the first papers of Dale and his group[155,156] that irradiation of amino acids yields, through decomposition, CO_2 and NH_3.

Another fact of importance is based on the findings of Hasselstrom *et al.*[157] who obtain in β-irradiated ammonium acetate in water solution, the formation in addition of di- and tricarboxylic acids, a small amount of glycine, aspartic acids, etc. Also Singh *et al.*[158] studied the ESR of gamma-irradiated proteins, particularly those containing sulfur ions; they point out the importance of these ions and the increase of unpaired electrons (SH and S—S groups) in respect to the radioresistance of the protein molecule.

2. *Amino Acids and Protein of Irradiated CNS*

According to some authors, e.g., Chain and his group,[159] Balazs *et al.*[160,161] Gaitonde *et al.*,[162] Vrba *et al.*,[163] Geiger,[164] and Cremer *et al.*,[165] the glucose carbon metabolized by the CNS enters the amino acid pool via the tricarboxylic acid cycle, which acts so that the oxidation of pyruvate brings an accumulation of α-oxoglutarate, which through a transamination reaction converts rapidly into glutamate. This reaction operates in mitochondria to a great extent. Even though one may assume that one of the major actions which radiation could affect in CNS would be at the metabolic level of amino acids and proteins, not much research work has been carried out in this respect. Egaña *et al.*[166] have done studies showing the 1-^{14}C-glucose conversion into CNS free amino acids after whole-body gamma-50 r exposure in adult rats receiving 5 μC of radioglucose per 100 g. Table VI shows certain results which refer to aspartic

TABLE VI

Free Amino Acids from 1-^{14}C-Glucose: Whole-Encephalon Specific Activity[a] $\frac{\text{cpm}}{\mu\text{mole g fresh weight}}$

Time after 1-^{14}C-glucose injection (min)	Aspartic[b]		Glutamic[b]		Glutamine[b]		GABA[b]	
	Normal	Irrad.	Normal	Irrad.	Normal	Irrad.	Normal	Irrad.
15	126 ± 5 (5)	158 ± 18 (5)	251 ± 19 (5)	286 ± 49 (4)	261 ± 28 (4)	326 ± 24 (5)	134 ± 19 (5)	395 ± 64 (5)
30	591 ± 128 (6)	380 ± 81 (8)	847 ± 90 (6)	614 ± 39 (7)	624 ± 82 (5)	380 ± 49 (9)[c]	1.520 ± 239 (5)	464 ± 74 (8)[c]
45	361 ± 65 (7)	560 ± 79 (6)	673 ± 65 (7)	788 ± 118 (6)	512 ± 88 (8)	523 ± 64 (7)	686 ± 93 (8)	345 ± 60 (7)
60	172 ± 11 (5)	250 ± 31 (6)[c]	384 ± 39 (7)	407 ± 52 (7)	408 ± 62 (7)	488 ± 48 (7)	487 ± 42 (6)	430 ± 40 (4)

[a] From E. Egaña *et al.*(140,166)
[b] N, normal; Irrad. 50 r (whole-body) gamma-irradiation (^{60}Co). Specific activity (cpm/μM g fresh tissue); radiochromatographic measurement.(166,281)
[c] *$P < 0.01$ between normal and irradiated.

acid, glutamic acid, glutamine, and GABA, although the work also covered taurine, glycine, serine, threonine, alanine, and reduced glutathione. As can be seen, the conversion of 1-^{14}C-glucose is not affected to a great extent by 50 r exposure. There are further possibilities which can be explored: the change of dose and the different labeling of glucose (U-^{14}C or 1-^{14}C) which represent location, quantity, and velocity differences as far as the conversion into amino acids and also into acetyl CoA is concerned. To advance somewhat in this problem of metabolism of the amino acids in irradiated CNS, glutamic acid and GABA in adult encephalon of rats after varied doses of 5, 50, 500 r, whole-body exposure, have been determined,(167) Table VII.

Alexandrov(168) finds that ^{35}S-methionine is incorporated into the brain of adult mice, whole-body X-irradiated, at great speed up to 2 hr, then declines sharply toward the third hour. De Vellis *et al.*(169) studied in 2-day-old rats, 750 r X-head-irradiated at diverse lapses the ^{14}C-leucine incorporation into the protein of brain stem slices. They find a decrease of protein from the eighth day and of proteolipid protein from the twelfth. The problem of irradiation and CNS protein and amino acid metabolism is connected with the response to exterior stimulus. Richter *et al.*(170,171) reported results on electroshock seizures and incorporation of U-^{14}C-glucose in CNS protein and nucleic acid fractions; Orrego *et al.*(172) reported the effect of electric stimulation in brain slices *in vitro*; Talwar *et al.*(173) reported light stimulations produced an increase in the incorporation of 1-^{14}C-valine and U-^{14}C-glucose in the sensory-visual area of rabbits, previously maintained in darkness. The CNS irradiation, as a typical energy absorption, should act on the CNS in a similar manner as other kinds of energy.

As a brief summary, after considering the main findings communicated in this section, apparently the results of irradiation of protein and amino acids are different *in vitro* and *in vivo*. As regards proteins it is not clear, from another point of view, which process is more radiovulnerable: whether the synthesis of the proteic molecule, its stability as an already-formed molecule, or its catabolism. It is not advisable to extend the range of this short statement, as more extensive experimental information on the subject must be awaited.

D. Lipid Metabolism

Lipid metabolism has been little explored, as far as the effects of radiation are concerned. Years ago, Florsheim *et al.*(174) had established that doses of 500 to 19,000 r X-rays do not alter ^{32}P incorporation into CNS phospholipids; this was confirmed later by Casteva *et al.*(175) who declared that a high irradiation does not alter the phospholipid exchange in the CNS. Vinogradov(176) with 2-^{14}C-acetate established that the whole-body irradiation does not greatly affect the turnover of brain phospholipid; Smirnova(177) found alterations in lecithin and cephalin in mice CNS

TABLE VII

Glutamic and GABA Content of Some CNS Areas: Normal and Gamma Irradiated (Whole-Body)[a] $\frac{\mu\text{mole}}{\text{g fresh weight}}$

	Brain cortex[b]		Hypothalamus[b]		Midbrain[b]	
	Glutamic	GABA	Glutamic	GABA	Glutamic	GABA
Normal	9.0 ± 0.7 (12)	1.60 ± 0.1 (12)	10.4 ± 0.8 (12)	6.1 ± 0.7 (12)	8.7 ± 1 (12)	5.8 ± 0.4 (12)
Irradiated						
5 r	4.5[c] ± 0.6 (10)	2.7[c] ± 0.5 (12)	8.8 ± 0.9 (11)	7 ± 0.9 (12)	8.7 ± 0.8 (12)	3.8[c] ± 0.3 (12)
50 r	6.2[c] ± 0.5 (11)	0.97[c] ± 0.1 (11)	8.7 ± 0.7 (12)	2.8[c] ± 0.3 (12)	7.3 ± 0.9 (10)	2.9[c] ± 0.3 (10)
500 r	6.8[c] ± 0.5 (10)	1.1[c] ± 0.2 (10)	7.6[c] ± 0.8 (12)	4.6[c] ± 0.5 (12)	6.7[c] ± 0.7 (10)	4.8[c] ± 0.4 (11)

[a] From Egaña *et al.*[167]
[b] Exposure: whole-body gamma-irradiation. Chromatographic method[281] with modifications.[167]
[c] $P < 0.01$ between normal and irradiated; there are also several differences among irradiated.

after sublethal whole-body X-irradiation. Grossi *et al.*[178] found the incorporation of 1-^{14}C-acetate diminished in brain cholesterol and fatty acids in 10-day-old rat, 3500 r gamma irradiated; this lowering does not occur when 2-^{14}C-mevalonate is the tracer; on the contrary, there is an increase of cholesterol sa. On the other hand, CNS cholesterol and fatty acid content diminishes prematurely in irradiated mice. Vernadakis *et al.*[179] exposed 14-day fetus to 100 r; measuring total lipids and cerebrosides of cortex and diencephalon shows an increase in the exposed groups. Scheijde *et al.*[180] in 700 r head-irradiated rat 2 days old, measured the cholesterol and total phospholipids in brain stem. The former appears increased and the latter diminished. The results do not show major changes in the protein, total sterol, sphingomeylin, phosphotidylcholine, and other CNS fat components. However, the myelin is altered in its content of unsaturated fatty acids and unsaturated aldehydes. The authors conclude that the alteration takes place in the glial cells, producers of myelin. Lee[181] in brain and spinal cord mitochondria from 100–3000 r irradiated rat, found a heightening of synthesis of sphingomyelin and related compounds in cerebellum and spinal cord, in addition to an alteration of the carbonyl functional group, especially of aldehydes.

E. DNA and RNA

The study of the metabolism of CNS nucleic acids has fundamental implications for the developing and adult neuroaxis, hence the interest in exploring the effects of radiation on DNA and RNA metabolism. One could affirm that CNS DNA is more radiosensitive than RNA.

1. *DNA*

Carter *et al.*[182,183] consider that the CNS has an intermediate radiosensitivity within a classification of the organs and systems. They found in CNS rat 700 r whole-body irradiated, a significant decrease of DNA at 1–2 hr, a return to normal at 3–4 hr to decline progressively until death. De Vellis *et al.*[169] established at 24 hr postirradiation a 12–15% lowering of the brain stem DNA content in 750 r whole-body 2-day-old irradiated rats; the same occurs with the adult. The RNA, protein, and GPDH contents are also altered; there are also other changes in malate dehydrogenase, isocitric dehydrogenase, LDH, etc.; facts which, as a whole, make the authors believe that the DNA loss is not due to cellular death but to an altered synthesis which, they believe, is confirmed by the fast and immediate diminution in the incorporation of 2-^{14}C-thymidine with 375 r, an incorporation which increases progressively in function of time.

2. *RNA*

Aleksandrovskaya[184,185] reported years ago an increase in RNA in the Betz cells, after whole-body 50 r X-ray; in basal ganglia, a decrease;

with 150 r there was a loss of RNA in the neurons of II and V cortical layers; and, with 250 r, the whole brain decrement becomes general. Artyukhina[186,187] with 650 r whole-body X-ray finds no alteration in CNS RNA; however, with 1300 to 1750 r, the RNA increases. Yamamoto *et al.*[188] carried out experiments in two groups of rats, whole-body and head-irradiation, 500 r X-rays, which received ^{3}H-cytidine; their results indicate more alterations in whole-body irradiated animals. A cytoplasmatic RNA decrease in cortex, hypocampus, Purkinje cells, etc., is established; the nuclear RNA does not change. De Vellis finds, in rats in which DNA was also studied, a lowering of RNA in midbrain and a depressed incorporation (25–30%) of 5-^{3}H-uridine into RNA on the ninth day.

General comments: the CNS RNA, possibly synthesized in greater quantity in neurons, is more radioresistant than DNA (G, the chemical yield in molecules/100 eV, 200% less). The most important mechanism of the irradiation is discussed: rupture of hydrogen ion pair between the base[189–191] and denaturation by loss of its helicoidal structure, which according to some authors,[192,193] would be due to the interaction between the bases stacked parallel on top of each other. One of the mechanisms of radiogenic alteration of DNA and RNA functions through the products of water radiolysis, the H atoms which act on the component bases and the OH radicals which affect sugar components (20–25% of total yield).[194] It is worth mentioning that authors such as Singh *et al.*[195] consider that an energy transfer does take place *per se*, even before the formation of radicals, thus the DNA molecules would act as e scavengers. It is known that these molecules have a great affinity with the e_{aq} mechanism in which the primary sulfur radicals act (disulfide anions formed by trapping of e). It has been assumed that the primary radicals found in irradiated DNA (and possibly in RNA) could be caused by the intramolecular migration of unpaired electrons to the purine bases.[195,196] The role of oxygen in the radiation breakage of the ribonucleic acids, at least in DNA, is debatable.[197] As to the effect of irradiation on synthesis, there are authors, such as Walwick *et al.*[198] who believe that the DNA primer and the enzyme system for DNA synthesis (preformed kinases, polymerases, glycolytic enzymes, etc.) would not be significantly affected in the dose range 250–10,000 rad. This contention is strongly contradicted by experiments of de Vellis *et al.*[169] and of Merits *et al.*[199]

Finally, the physicochemical mechanism which would explain the radioresistance or radiovulnerability of the nucleic acids is of a difficult formulation. Pullman[200–204] states that these compounds are conjugated systems which have mobile electrons (π electrons); radiation may interfere with the electronic structure of these molecules, thus culminating in the formation of free radicals derived from the constituent bases. The action of radiation on the proteins could have a similar mechanism, which caused Szent-Györgyi[205,206] to suggest that the total supramolecular structure (be it nucleic acids or proteins) implies a certain degree of "general electronic delocalization" bound to ligaments of hydrogen and π electrons. This seems so evident that for Duchesne[207,208] the degree of radioresistance (or radiosensitivity) is tied to π electrons and to their delocalization, which, together with other expressions, is shown in the form of resonance energy. It is known that Gordy and his group[149,209,210] in 1956 made the first study of ESR in order to explore the effect of radiations on powdered samples of nucleic acids. Starting from this pioneering stage, in which the fact has been discovered that the free radicals produced are sufficiently long-lived for detection with ESR, Duchesne,[207,208] Muller,[211] and Pullman[200,204] have contributed to our understanding of the intimate mechanism of radiation on nucleic acids. The writer is not acquainted with the paper on the "DNA enzymatic repair system",[212] which could deal with CNS ionizing radiation exposure.

As a corollary and based on the analysis of the experimental facts communicated, it could be stated that the CNS DNA is more radio sensitive than RNA, particularly in the growing stage; the mechanisms could be both "indirect" (products of water radiolysis) and "direct" (impact on the macromolecules themselves) on one or various stages of the synthesis of nucleic acids. It is possible that in both the electronic structure of these molecules undergoes a change from their constituent bases. This postulate needs to be translated into terms of biological functions of the nucleic acids. In summary, the mechanism of the alterations stated requires further elaboration, at least from the radioneurobiochemical point of view.

F. Water and Ions

There exist few sufficiently informative papers on the effects of radiation on water and ions. Two groups of representative papers have been selected. Gregersen and his group[213-216] reported that 1000 r whole-body exposure in dog induces an H_2O increase in the gray and white matter and in the caudate nucleus; contrarywise, K^+ diminishes in gray and white substances. After 2000–5000 r head-irradiation, observed in the animal at 4–10 days, there are no alterations in the water content; Na^+ is augmented. With 20,000 r head-irradiation Na^+ and Cl^- increase in the nucleus. It is curious to note that CNS edema appears in whole-body irradiation and not in head-irradiation; there is a difference in species, since the pattern of dose effects described does not occur in rabbit. Volcana *et al.*[217] studied the growing CNS irradiated during gestation with 100 r (whole-body) and analyzed at diverse postnatal lapses after birth. Results: K^+ is enhanced as in normal development; Na^+ and Cl^- decrease, making irradiated equal to normal. It is difficult to interpret for the moment these alterations. Eventually it may be thought, as suggested by Gregersen *et al.*[213-216] that the H_2O accumulation could well be intracellular and could be explained in terms of endothelial permeability of the microcirculation in the first place and also by that of the cellular and intracellular compartmental membranes. On this point Willis[218] in *in vitro* studies shows an alteration of the permeability of irradiated mitochondria, measured through transport of K^+, Na^+, and Ca^{2+}, together with ATPase activity, and concludes that the alterations of transport were caused by the loss of permeability of the mitochondrial membrane, without the occurrence of a mitochondrial ATPase or of oxidative phosphorylation inhibition. There is a loss of K^+ when the mitochondria is in a low K^+ medium and a gain in a medium of high K^+.

G. Hydrogen Peroxide and Catalase Studies

1. *Hydrogen Peroxide*

There have been attempts to investigate the participation of the "H_2O_2 effect" in CNS irradiation. Experiments have been published particularly as regards the *in vitro* respiration with glucose as substrate in all the areas of CNS of β-internal exposed adult rats (with doses already mentioned). The H_2O_2 concentration has been chosen on the basis of two factors: the dosimetry expressed in total dose,[219] and the H_2O_2 yield, calculated according to Le Fort (0.288 μmole/100 rad absorbed)[220] and to Aebi 0.3–0.4 μmole/100 r of 250 kV X-rays.[221] Thus H_2O_2 has been used in concentrations ranging from 0.0001, 0.001, 0.01 to 1, 5, 30, and 50 mmole. The results are as follows: the lowest H_2O_2 concentrations increase the uptake of O_2 in normal CNS, particularly brain cortex, cerebellum, and spinal cord, by 25, 70, and 100%, respectively. Higher concentrations inhibit the respiratory process in all areas. In contrast, in the CNS-irradiated areas, the H_2O_2 inhibits even at the lowest concentrations in all areas, with the exception of olfactory bulbs, which tend to increase.[222]

One cannot invoke reasons of greater availability of O_2 to the tissue in the case of the increased respiratory process because it works with overoxygenated medium; nor can it be accepted that H_2O_2 should be acting as an eventual electron acceptor and carrier.[222]

2. *Catalase Studies*

It is well known that catalase reacts with H_2O_2 in two principal ways:

as catalase *per se*:

$$H_2O_2 + H_2O_2 \rightarrow H_2O + \tfrac{1}{2}O \tag{12}$$

peroxidase reaction:

$$H_2O_2 + RH_2 \rightarrow H_2O + R \tag{13}$$

The catalase reaction in which there is O_2 evolution can be considered to be a special type of peroxidase reaction in which H_2O_2 acts as a substrate and acceptor; in a peroxidase reaction oxygen is not given off but an acceptor is oxidized.

In view of the results obtained with H_2O_2 on normal and irradiated CNS areas, Egaña[223] has studied *in vitro* the action of nearly physiological catalase concentration (0.02, 0.002, and 0.0002 mM) on the respiratory effects of hydrogen peroxide. The general results are: in normal CNS slices, in the presence of diverse H_2O_2 concentrations, the catalase inhibits the stimulant H_2O_2 effect. In irradiated tissue the catalase conveys a respiration rate comparable to that obtained without H_2O_2 effect (equal to what is shown in Table II) (Table VIII). We are unable to explain the main results communicated by Egaña *et al.*[222,223] i.e.: (1) H_2O_2 enhancement of respiratory effects in some normal CNS areas and inhibition in irradiated ones. (2) Decrease of normal tissue respiration by H_2O_2/catalase association and comparative enhancement of irradiated CNS levels. Some direct and collateral facts about CNS may be remembered: (1) The brain is, perhaps, one of the organs lowest in catalase content; thus, in normal mice, Feinstein *et al.*[224] find values as low as 0.65 PU/g (♂ and ♀ mean). It is strange, however, that in acetalasemic and hypocatalasemic mutant mice, the CNS content is 37–48 PU/g, which is the opposite of what is found in blood, liver, and kidney. (2) In catalase-negative bacteria, the addition of catalase to the medium neutralizes the delayed effects of exposure (of "H_2O_2 effect", among others).[225,226] (3) The administration of liver catalase before 750–800 r whole-body X-ray irradiation in mice does not produce a greater lapse of survival, nor does it influence the postirradiation survival time.[227] Le Van *et al.*[228,229] suggest that H_2O_2 could be one of the factors involved in the aversion reaction in mice. Feinstein[230] and his group find that in 25–50 whole-body X-irradiated acatalasemic mice, the exposure accentuates the diminished saccharine aversion as against normal.

It is rather difficult to formulate a prior conclusion on these catalase studies, due, mainly, to two reasons: (1) There exists only one type of study on the subject, i.e., the effect on the overall process of CNS respiration. Finer points of this process still have to be clarified. (2) We do not know the significance for the normal CNS (at least *in vitro*) of placing at its disposal a quantity of catalase (exogen) which is greater than normal *in vivo*. In any case, it appears impossible to translate these results *in vitro* to what could eventually occur *in vivo*. As regards this latter aspect, there are no experimental results known to the author.

IV. CNS NEUROTRANSMITTERS

A. Acetylcholine

There is strong evidence that ACh plays the role of a central neurotransmitter, the enhanced liberation and synthesis of which has been found in the CNS for the first time during the excitation of the optical nerve.[231] Actually, the ACh is found in the presynaptic terminals and is liberated during nerve stimulation; it produces in the presynaptic membrane

TABLE VIII

The Effect of H_2O_2 and Catalase on Respiration (Cerebellum)[a] $\frac{\mu\text{atom oxygen}}{\text{mg protein}}$

Condition	Without H_2O_2	H_2O_2 (1 mM)	H_2O_2 (1 mM) and Catalase		
			0.2 mM	0.02 mM	0.002 mM
Normal	1.1 ± 0.25 (42)	1.65[b] ± 0.30 (33)	0.88[b] ± 0.18 (17)	1.34 ± 0.12 (24)	0.66[b] ± 0.18 (30)
Irradiated total dose (rad) One single dose					
0.3	1.12 ± 0.15 (40)	1.24 ± 0.18 (33)	1.15 ± 0.03 (21)	1.20 ± 0.25 (16)	0.90 ± 0.03 (16)
10	1.25 ± 0.05 (40)	1.52 ± 0.16 (32)	0.96 ± 0.11 (22)	0.89 ± 0.12 (17)	0.84 ± 0.09 (16)
307	0.90 ± 0.09 (29)	0.90 ± 0.06 (28)	0.66 ± 0.1 (18)	0.75 ± 0.03 (24)	0.78 ± 0.10 (20)
1124	1.08 ± 0.18 (37)	0.80 ± 0.17 (26)	0.90 ± 0.03 (17)	0.90 ± 0.04 (20)	0.90 ± 0.07 (17)

[a] From Egaña *et al.*[222,223] β-radiation (internal) (^{32}P). *In vitro* respiration, slices; Ringer–Elliott–bicarbonate–pyruvate as in Table II. "Blank reagents" (H_2O_2 and H_2O_2 plus catalase) were considered.

[b] $P < 0.01$ between normal with and without H_2O_2 and between H_2O_2 alone and H_2O_2 plus catalase. In the irradiated tissue, both the H_2O_2 and the H_2O_2 plus catalase effects are different according to the dose. The other CNS areas and other H_2O_2 and catalase concentrations have been also studied.

a local depolarization (excitatory) or hyperpolarization (inhibitory), if locally applied. It produces nerve stimulation; furthermore, it has active mechanisms of inactivation as well as of synthesis.[232] Mammalian CNS contains a fair quantity of ACh in three forms: free, not tied to a cellular or subcellular fraction which diffuses easily; labile bound, located in the synaptic vesicles of the mitochondrial fraction which, if treated with a hypoosmotic agent, diffuses; and stable bound, which represents the fraction also localized in the monosynaptic vesicles and in the membrane of nerve endings.[233-241]

The CNS has high quantities of enzyme which catalyze the ACh synthesis, i.e., the reaction AcCoA + choline $\rightarrow$ ACh, found in the detached nerve endings. It localizes the same as ACh, possibly in two compartments.[234,241] The CNS possesses the enzyme which hydrolyzes ACh, the acetylcholine hydrolase (choline esterase) AChE, located as ACh in the cholinergic nerve endings (for complete study, see Ref. 242).

As early as 1954, Florsheim *et al.*[134] observed that radiation diminishes the AChE in the CNS; Petrovnina[243] found that in rat, 550–600 r whole-body X-irradiated, an increase of AChE in the CNS took place, which reverted to normal at 3–7 days; the ACh increased, with an ulterior decrement; ACh_{ac} showed changes similar to those of ACh. Egaña[104,105] has studied the ACh synthesis *in vitro* in the different areas of CNS β-internally irradiated, by means of the measurement of its free (diffusible), labile bound, and stable bound forms. He found a progressive increase at low exposures (from about 25 to 40 rad) and a lowering at greater doses (300–500 rad). The free and labile bound do not change significantly. The hypothalamus shows the greatest increase (Table IX). The choline acetyltransferase activity, studied in aero- and anaerobiosis, in optimum medium (CoA, ATP, β-NAD, sodium acetate, *l*-cystine, choline-HCl, eserine, K^+, Mg^{2+}, NaF) shows no change at very low doses; it increases in anaerobiosis at 56 rad. High doses alter the enzyme activity.[244] Maletta *et al.*[245,246] investigated the AChE activity in 2-day-old rat, X-irradiated (450 r) at various times postexposure, which appeared lowered in cortex and hypothalamus, and showed no change in brain stem and cerebellum. They extended their studies to butyrylcholine esterase. Further, they published experiments at 14 days pregnancy and 100 r exposure; the AChE diminishes, especially in brain cortex: the spinal cord appeared as a radioresistant structure.

It is extremely difficult to formulate an overall conclusion on the effect of ionizing radiation on the complex ACh–AChE–AC_{ac} system in the CNS. *A priori* it could be assumed that there would exist a difference between the adult and growing brain. For the adult, and in the light of the experiments with β-irradiation, one could suppose that the synthesis suffers no changes (acetylcholinetransferase mechanism), nor does the permeability of cellular and intracellular membranes (the diffusion remains approximately equal), and that the stable bound form increases most, at least in some CNS areas. It remains to establish the AChE activity, as the

TABLE IX

Acetylcholine Synthesis by Hypothalamus $\frac{\mu g\ ACh}{g\ fresh\ weight}$

Condition		Acetylcholine synthesis				Total synthesized
			In the tissue			
		"Free" (diffusible)	Labile bound	Stable bound	Total	
Normal		13.9 ± 2.3 (24)	11.4 ± 1.9 (24)	32.3 ± 2.3 (24)	43.7 ± 7.4	57.6 ± 7.6
Irradiated total-dose (rad)						
Single doses	0.3	13.3 ± 1.9 (16)	11.5 ± 2.2 (16)	42.9 ± 10 (16)	54.4 ± 10	67.7 ± 11
	2.3	13.5 ± 1.5 (16)	13 ± 1.5 (16)	32.2 ± 8.3 (16)	45.2 ± 8.6	58.7 ± 8.9
	5	15.3 ± 3.7 (16)	12.4 ± 1.1 (16)	37.9 ± 7.6 (16)	50.3 ± 7.6	65.6 ± 7.2
	11	14.6 ± 2.2 (15)	13.1 ± 1.5 (15)	29.5 ± 6.8 (15)	42.6 ± 6.8	57.2 ± 7.5
	25	15.9[b] ± 4 (16)	15.3[b] ± 1.8 (16)	49.3[b] ± 5.9 (16)	64.6[b] ± 6.2	80.5[b] ± 8.4
	38	9.2[b] ± 1.2 (18)	9.5[b] ± 1.1 (18)	11.3[b] ± 2.1 (18)	20.8[b] ± 1.7	30[b] ± 2.1
	296	12.5 ± 2.3 (25)	10.9 ± 2.3 (25)	19.8[b] ± 5 (25)	30.7[b] ± 5.4	43[b] ± 6.8
One month	542	11.3 ± 1.8 (22)	10.5 ± 1.8 (22)	16.6[b] ± 1.6 (22)	27.1[b] ± 3.6	38.4[b] ± 4.7
Four doses	1108	9.8[b] ± 2 (20)	11.3 ± 2.2 (20)	19.2[b] ± 5.9 (20)	30.5[b] ± 6.5	40.3[b] ± 5

[a] From Egaña.[104,105] β-irradiated (internal) (^{32}P). Strips; incubation in Warburg vessel, 60 min, 38°; Ringer–Elliott-bicarbonate, *p*H initial 7.6, gas phase O_2 95%/CO_2 5%; pyruvate 2.5 mM, choline-HCl 9×10^{-3} M; eserine sulfate 3.85×10^{-4} M. ACh: "free" measured at the end of the incubation in the Ringer solution; "labile bound" obtained by grinding the tissue finely in presence of eserine (5 mg); "stable bound" in the same manner, in presence of eserine, plus 0.33 *N*-HCl; both represent the "total" in the tissue; "total synthesized" means "free" plus "total bound." ACh determination; eserinized frog's rectus muscle, in triplicate to sextuplicate. Mean, S.D., and number of experiments.

[b] $P < 0.01$ between normal and irradiated.

work has been carried out in eserinized media. There is also the great task of the study of these postirradiation processes, with marked molecules, e.g., AcCoA precursors and choline, and at a subcellular level or in brain slices from normal CNS.[247] In the growing CNS, the answer to the problems posed is more difficult, owing to the complexity of the maturation process.

B. Catecholamines (*l*-Noradrenaline and Adrenaline)

The writer is acquainted with only a few papers on CNS catecholamine and irradiation, although an important nexus seem quite evident: the changes in behavior produced by irradiation and the possible participation of catecholamines in them. Three papers were selected on this subject. Palaic *et al.*[248] reported that whole-body (900 r) X-irradiation produces a significant decrease of NA of CNS; this occurs more promptly in the newborn than in the adult. It is suggested that the NA pool is more radiosensitive than the 5-HT pool. The chlorpromazine and isopromazide do not inhibit the NA-induced radiation liberation. Prasad *et al.*[249] stated that in young mice whole-body (700–800 r) X-irradiated, dopamine injected prior to exposure protects survival. Finally, it is justified to recall that years ago J. L. Gray *et al.*[250] reported the protective action of adrenaline in whole-body X-irradiation, a characteristic not shared by NA.

C. γ-Aminobutyric Acid

Several problems can be stated. In the first place, is GABA a true central neurotransmitter? There is no definite criterion on this point, although a high content of GABA in the CNS is an established fact.[251–255] Further, the brain has the enzymes which catalyze its formation (GAD) and its reversible transamination with α-ketoglutarate (GABA-T). Its major distribution takes place in the neurons and the synapses represented in the nerve ending fraction; also, GABA was shown to have inhibitory physiological effects on varied types of nervous structures, from crayfish stretch receptor to monkey brain cortex.[253] In vertebrates, exogenously administered GABA produces no changes in the neuronal membrane, although an action on conductance required to produce inhibition is ascribed to it. Na^+ is necessary for GABA binding in subcellular vesicles. Na^+ inhibits ACh binding.[255] Nevertheless, the problem is complicated because GABA has been found in kidney, liver, spleen, diaphragm, muscle, and lung of mice in concentrations of 10^{-8}–10^{-10} mole/g fresh tissue.[256] Egaña has found GABA in the physiological medium in which liver slices had been incubated.[257] Obviously, in these organs GABA would not act as a neurotransmitter. At the same time, it has not been proven that GABA would be liberated in the specialized presynaptic endings, nor has GABA-energic synapsis in the CNS of mammals been demonstrated. There is no doubt, however, that GABA is a specific substrate of CNS and that it supports oxidative phosphorylation.[258-260] Jovanovic *et al.*[261,262] established that in X-irradiated CNS the glutamate and GABA content decreased at 2–3 days after irradiation. When ^{14}C-U-glutamate is administered to 950 r whole-body gamma-irradiated mice, the ^{14}C appears in glutamate with high specific activity; in GABA and glutamine, it is low. During the subsequent days glutamate specific activity decreases and that of GABA and glutamine increases. The authors believe that the glutamic-GABA and the glutamic–glutamine systems are different as regards their radiosensitivity.[263]

It may be added that the *in vitro* respiration of CNS of whole-body, 5, 50, and 500 r gamma-irradiated rats with glutamate and/or GABA as substrates varies according to the CNS area and the dose; further, that there is an increase in respiration in general (5 and 50 r) at 1mM and more at 10 mM GABA concentration; with 500 r a decrease appears.[264]

It is impossible to judge the radiation effect on GABA as neurotransmitter per se. It is, undoubtedly, a postulate which needs better neurobiochemical antecedents.

D. 5-Hydroxytryptamine

The effect of radiation on this central neurotransmitter candidate has been studied along one basic line: the CNS 5-HT content. Ershoff *et al.*[265] do not find significant CNS 5-HT variations after 900 r of X- and gamma-radiation as compared to untreated pair-fed control rats; however, the values of these groups are lower than those of unexposed and ad libitum-fed controls. Renson *et al.*[266] reported that the 5-HT hypothalamus content of adult rat after 1000 r (whole-body) X-irradiation, decreases. Speck[267] verifies a decrease of 5-HT rat CNS following 500, 9000, and 20,000 r X-ray exposure with subsequent recovery at 48 hr. Randic, Supek, and their group[268-272] have reported that with a dose of 900 r whole-body X-irradiation, there is no significant change in 5-HT rat brain content. At greater doses, 6000–12,000 r, there appears a significant 5-HT increase in mice brain. Lott *et al.*[273] conclude that a high dose of irradiation (10,000 r) is necessary to produce in rat brain slices *in vitro* alteration in the 5-HT content, except for the decrease in thalamus and pons/medulla. In turn, Komesu *et al.*[274] find no effects at exposures between 600 and 1200 r whole-body X-irradiated animals. Chernov *et al.*[275] worked on guinea pigs (500 r whole-body X-ray) and rats (700 r); in the former an immediate decrease of the content appears; in rats, however, there is a marked increase. In both, levels become normal in the days following the exposure. Finally, Egaña *et al.*[276-277] with β-internal radiation with the dose already indicated in this chapter, determined the 5-HT content in rat brain cortex, "diencephalon", hypothalamus, olfactory bulbs, midbrain, and spinal cord. Hypothalamus and midbrain are the most radiosensitive structures, and respond with a 5-HT increase at a very low dose; at the high dose, 500–2400 r, the content generally decreases (Table X).

TABLE X

5-HT Content of CNS Levels, μg/g CNS Level Fresh Weight[a]

	Brain cortex	Diencephalon	Hypothalamus	Olfactory bulbs	Midbrain	Spinal cord
Normal	0.130 ± 0.01 (32)	0.240 ± 0.03 (32)	0.440 ± 0.03 (35)	0.290 ± 0.06 (32)	0.210 ± 0.02 (32)	0.110 ± 0.001 (30)
Irradiated (rad) total-dose whole encephalon						
Single dose 0.4	0.120 ± 0.03 (18)	0.250 ± 0.03 (20)	0.490 ± 0.12 (20)	0.280 ± 0.06 (20)	0.230 ± 0.02 (20)	0.140 ± 0.02 (20)
3.3	0.160 ± 0.02 (16)	0.340[b] ± 0.06 (16)	0.550[b] ± 0.09 (16)	0·310 ± 0.05 (16)	0.300 ± 0.04 (16)	0.140 ± 0.01 (16)
8	0.140 ± 0.001 (16)	0.280 ± 0.03 (16)	0.520 ± 0.04 (16)	0.280 ± 0.03 (16)	0.220 ± 0.01 (16)	0.120 ± 0.01 (16)
14	0.140[b] ± 0.01 (16)	0.230 ± 0.02 (16)	0.490 ± 0.01 (16)	0.310 ± 0.03 (16)	0.270 ± 0.03 (16)	0.130 ± 0.01 (16)
31	0.140[b] ± 0.01 (16)	0.270 ± 0.006 (16)	0.530[b] ± 0.08 (16)	0.300 ± 0.05 (15)	0,260 ± 0.04 (16)	0.150 ± 0.01 (16)
56	0.130 ± 0.02 (16)	0.310 ± 0.04 (16)	0.400 ± 0.04 (15)	0.270 ± 0.03 (15)	0.230 ± 0.02 (16)	0.140 ± 0.01 (16)
410	0.120 ± 0.01 (16)	0.200 ± 0.03 (15)	0.350[b] ± 0.03 (15)	0.210[b] ± 0.02 (16)	0.210 ± 0.02 (16)	0.110 ± 0.001 (16)
One month 683	0.100 ± 0.009 (18)	0.108[b] ± 0.004 (16)	0.380[b] ± 0.03 (16)	0.220[b] ± 0.09 (15)	0.200 ± 0.02 (15)	0.110 ± 0.001 (16)
Four doses 1470	0.100[b] ± 0.02 (18)	0.200[b] ± 0.02 (18)	0.380[b] ± 0.05 (18)	0.220[b] ± 0.05 (16)	0.190[b] ± 0.03 (15)	0.090[b] ± 0.001 (16)

[a] From Egaña *et al.*[226,227] 5-HT determination, Vane's technique.[282] All the values have been corrected by a factor of 20% (80% recovery). Mean, S.D., and number of determinations in triplicate.

[b] $P < 0.01$ between normal and irradiated.

It is evident that from the foregoing experimental data no conclusions can be obtained, because, apart from the fact that the conditions used by the different authors are varied, the results commented upon give no information on the synthesis, transport, and subcellular binding of 5-HT and on its release in relation to the state of permeability of the neuronal and glial cell membranes.

V. CONCLUDING REMARKS

The present state of knowledge of the effects of radiation on the biochemistry of CNS is rudimentary and is based mostly on analytic facts, and lacks explanations of the mechanisms of these events. This is why merely enunciations of very elementary significance are formulated.

1. It seems that the CNS is a highly radiosensitive system, at least from a radiobiochemical point of view. Responses quantitatively different from normal are known at doses as low as 0.6 rad (β-internal radiation). As they are biochemical "lesions," they are, as a rule, reversible. At high doses, the alterations are, generally, of an inhibitory character, and thus the lesions become permanent, as the morphopathological sequel follows. It is conceivable that a great majority of the biochemical processes of the CNS are radiosensitive. Perhaps those most affected are the ones tied to production, liberation, and utilization of energy. In this sense, the noteworthy decrease of CNS ATP could be mentioned. On another level, it could be assumed that the enzymes and the enzymatic systems which catalyze the intermediary reactions of the metabolism and the synthesis of macromolecules would be per se very radioresistant. An observation may be added to certain findings quoted in the text; a certain form of uncoupling, e.g., between the increased respiratory process on one hand, and the lowering of ATP on the other (Table XI), which coincides with a diminished aero- and anaerobic glycolysis.

2. The growing CNS is more radiosensitive than the adult one, analyzed from a neurobiochemical angle. Apart from the aqueous structure of the developing CNS, this difference could be based on an eventual maturation of DNA and RNA and of certain enzymatic systems and subcellular components.

3. One of the noticeable events which probably take place during the irradiation process would be the alteration of the membrane permeability, not only of cellular structures, but also of the structures of intracellular compartmentalization and of the subcellular organization, especially mitochondrial, as well as of the vascular cerebral endothelium. The consequences of these alterations are self-explanatory.

4. There is no clear-cut criterion to estimate the ratio of dose rate and total dose and neurobiochemical effects. There are cases in which an area of CNS, e.g., hypothalamus, appears least-irradiated (β-internally),

TABLE XI

Ratio $\frac{\text{ATP content, } \mu\text{mole}}{\text{oxygen consumption, } \mu\text{mole}}$ Whole Encephalon[a]

Condition		ATP content	O_2 Consumption	$\frac{ATP}{O_2}$ Ratio (× 100)
Normal		3.39	150	2.3
Irradiated rad total dose whole encephalon				
Single dose	0.4	1.95	205.3	0.95
	3.3	2.64	172.3	1.53
	8	2.36	221.6	1.07
	14	2.48	186.5	1.3
	31	3	157.2	1.9
	56	3.31	226.2	1.5
	410	2.97	129.4	2.3
One month	683	2.40	130.4	1.87
Four doses	1470	3.23	118.2	2.73

[a] From Egaña.(105) β-irradiated (^{32}P); ATP determination.(280) Encephalon weight 1970 (±120) g for all groups, except in the 1470 rad irradiated ("four doses") 1850 (±150) g. The oxygen consumption of encephalon represents the sum of the O_2 uptake of all levels (from brain cortex to pons/medulla), measured *in vitro*.

but nevertheless shows radiobiochemical effects more notable than other CNS areas more intensely irradiated. These findings open the way to study the intrinsic radioresponse of each area of the neuroaxis. It is conceivable that the archiencephalic and the myelencephalic structures are more radioresistant than the neoencephalic ones. This statement is also subject to experimental proof.

5. It is assumed that chemical reactions are usually at the basis of all changes in biological material and consequently in irradiated living matter. In our case, it would be strange if this initial assertion were at fault. But, and here lies the problem, we are dealing with energy deposition of a very special kind, which is not localized in chemical bonds to be broken, but affects molecules as a whole and determines their fate. We have to deal with "molecules" or entities the existence of which is transient; they only live perhaps 10^{-9} sec. Do we have chemistry or biochemistry in 10^{-9} sec? Surely, all indications should lead to the former. Biochemical processes are slow, perhaps the fastest occurring at 10^{-4} sec. Thus, facing the cumulus of unanswered questions posed by radioneurobiochemistry, with statements based on techniques and concepts in use at present in "traditional" biochemistry, dedicated to studies at a molecular level, one is

tempted, following the ideas of Szent-Györgyi,[278,279] to explore the effects of radiations at a submolecular level, i.e., through the eventual movements of electrons within the molecules, movements which, it is suspected, take place in the course of the habitual intermediary metabolism of CNS on which the radiant energy could well have bearing, even at low doses.

ACKNOWLEDGMENTS

The author wishes to express his deep thanks to Dr. Laszlo Goth for the preparation of the manuscript in English; to Dr. Wanda Solodkovska for the translation of the Russian literature, and to Dr. Harold Behrens and Jorge E. Allende for helpful criticism of the paper.

VI. REFERENCES

1. *Effects of Ionizing Radiation on the Nervous System*, International Atomic Energy Agency, Proceedings Series; Vienna Symposium, Vienna (1962).
2. *International J. Neurology* **3**: 533–572 (1962). (Issue devoted to ionizing radiation and central nervous system). Montevideo, Uruguay.
3. T. J. Haley and K. S. Snider, eds., *Response of the Nervous System to Ionizing Radiation; First International Symposium*, Academic Press, New York (1962).
4. T. J. Haley and R. S. Snider, eds., *Response of the Nervous System to Ionizing Radiation; Second International Symposium*, Little Brown, Boston (1964).
5. N. N. Livshits, Physiological effects of nuclear radiations on the central nervous system, *Ad. Biol. Med. Phys.* **7**:173–248 (1960).
6. H. Gangloff and O. Hugh, The effects of ionizing radiation on the nervous system, *Ad. Biol. Med. Phys.* **10**:1–90 (1965).
7. A. V. Levedinsky and Z. N. Nakhil'nitskaya, *Effects of ionizing radiation on the central nervous system*, Elsevier Publ. Co. Amsterdam (1963).
8. D. J. Kumeldorf and E. L. Hunt, *Ionizing Radiation: Neural Function and Behavior*, Academic Press, New York (1965).
9. W. R. Stahl, A review of Soviet research on the central nervous system effects of ionizing radiation, *J. Ment. Dis.* **124**:511–529 (1959).
10. W. R. Stahl, Recent Soviet work on reactions of the central nervous system to ionizing radiation, *J. Ment. Dis.* **131**:213–233 (1960).
11. C. D. van Cleave, *Irradiation and the Nervous System*, Rownan, Littlefield, New York (1963).
12. P. T. Lascelles, *in Biochemical Aspects of Neurological Disorder* (Third Series), (J. N. Cumings and M. Kremer eds.), pp. 200–213. Oxford Blackwell Scient. Publ., Oxford (1968).
13. A. M. Jeliffe, *in Biochemical Aspects of Neurological Disorders* (Third Series), (J. N. Cumings and M. Kremer eds.), pp. 174–199, Oxford Blackwell Scient. Publ., Oxford (1968).
14. K. J. Altman, G. B. Gerber, and Sh. Okada, eds., *Radiation Biochemistry*, Vol. 1. *Cells*, Academic Press, New York (1969).
15. G. B. Gerber and K. J. Altman, eds., *Radiation Biochemistry*, Vol. 2, *Tissue and Body Fluids*, Academic Press, New York (1969).

16. W. Haymaker, *in The Structure and Function of Nervous Tissue* (G. H. Bourne, ed.). Vol. 3, pp. 441–518. Academic Press, New York (1969).
17. J. N. Yamazaki, A review of the literature on the radiation dosage required to cause manifest central nervous system disturbances from in utero and post-natal exposure, *Pediatrics* **37**:877–903 (1966).
18. L. V. Cherkasova, K. B. Fomidenko, T. M. Myronova, F. D. Koldovska, V. A. Kukuchkina, and V. G. RemCerCerg, *The ionizing radiation and the substrate metabolism*, Akad. Nauk, URSS, Minsk (1962) (Russian).
19. W. Geests, Influence des radiations ionizantes sur le developpement du systeme nerveux, *Acta Genet. Med. (Roma)* **16**:275–309 (1967).
20. L. C. Cherkasova, Biochemistry of the low dose of ionizing radiation, Akad. Nauk SSSR, Minsk. (1964) (Russian).
21. M. R. Sikow and D. D. Mablum, eds., *Radiation Biology of the Fetal and Juvenile Mammal*, AEC Symposium Series (Batelle Symposium, USAEC Division of Technical Information Extension, Oak Ridge (1970)).
22. S. P. Hicks and C. J. D' Amato, *in Advances in Teratology* (O. H. M. Woollam, ed.), pp. 196–250, Logos Press, London (1966).
23. R. Latarjet and L. H. Gray, Definition of the terms protection and restoration, *Acta Radiol.* **41**:61–62 (1954).
24. G. P. Swanson and B. Kihlman, in Ciba Found. Symposium *Ionizing Radiations and Cell Metabolism* (G. E. W. Wolstenholme and C. M. O'Connor, eds.), pp. 239–251, J. & A. Churchill, London (1956).
25. R. L. Platzman, An isotopic effect in the probability of ionizing a molecule by energy transfer from a metastable noble-gas atom, *Nature* **195**:790–791 (1962).
26. E. J. Hart and R. L. Platzman, *in Mechanisms in Radiobiology* (M. Errera and A. Forssberg, eds.) Vol. 1, Chapter 2, pp. 94–257, Academic Press, New York (1961).
27. R. L. Platzman, *in Symposium on Radiobiology* (J. J. Nickson, ed.), Chapter 7, John Wiley and Sons, New York (1952).
28. R. L. Platzman, *in Radiation Biology* (W. Claus, ed.), chapter 2, pp. 15–72, Addison-Wesley, Reading, Mass. (1958).
29. R. L. Platzman, *in Radiation Research* (Proc. Third Int. Cong. Radiat. Res., Cortina d'Ampezzo, Italy) (G. Silini, ed.), pp. 20–42, North Holland Publ., Amsterdam (1967).
30. M. G. Ord and L. A. Stoken, *in Mechanisms in Radiobiology*, Vol. 1, Chapter 3, pp. 259–331 (M. Errera and A. Forssberg, eds.), Academic Press, New York (1961).
31. S. Wolff, *in Mechanisms in Radiobiology* (M. Errera and A. Forssberg, eds.), Vol. 1, Chapter 6, pp. 419–475, Academic Press, New York (1961).
32. D. E. Lea, *Actions of Radiations on Living Cells*, Cambridge Univ. Press, London (1946).
33. N. W. Timofeeff-Ressovsky and K. G. Zimmer, *Das Trefferprinzip in der Biologie*, Hirzel, Leipzig (1947).
34. F. Hutchinson and E. Pollard, *in Mechanisms in Radiobiology* (M. Errera and A. Forrsberg, eds.), Vol. 1, Chapter 1, pp. 71–92, Academic Press, New York (1961).
35. W. Gordy, *in Proc. Int. Congr. Radiat. Res., Burlington, Vt.*, Suppl. 1, pp. 491–510 (1959).
36. H. Shield and W. Gordy, Electron spin resonance studies of radiation damage to the nucleic acids and their constituents, *Proc. Nat. Acad. Sci.* **45**:269–281 (1959).

37. W. Gordy, Free radicals from biological purines and pyrimidines, *Ann. N.Y., Acad. Sci.* **158**, Art. 1, pp. 100–123 (1969).
38. A. O. Allen, Radiation yields and reactions in dilute inorganic solutions, *Rad. Res. Suppl.* **4**:54–73 (1964).
39. M. Daniel and E. Wigg, Oxygen as a primary species in radiolysis of water, *Science* **153**:1533–1534 (1966).
40. E. Egaña, R. Valderas, and J. Suarez, *in Fisiopatología General* (E. Egaña, F. Ugalde, A. Valenzuela, and S. Bozzo, eds.), Chapter VIII, pp. 311–404, Editorial A. Bello, Santiago (1963) (Spanish).
41. S. P. Hicks, Radiation as an experimental tool in mammalian developmental neurology, *Physiol. Rev.* **38**:337–356 (1958).
42. S. P. Hicks and C. J. D'Amato, *in Disorders of the Developing Nervous System* (W. S. Fields and M. M. Desmond, eds.) Chapter 4, pp. 60–97, C. C. Thomas, Springfield (1961).
43. S. P. Hicks and C. J. D'Amato, Low-level radiation and changes in glia and neurons populations in developing brain, *Fed. Proc.* **23**, Part I, p. 128 (1964).
44. S. P. Hicks, C. J. D'Amato, and J. L. Falk, Some effects of radiation on structural behavioral development, *Int. J. Neurol.* **3**:335–348 (1962).
45. W. Haymaker, G. L. Laqueur, W. H. Nauta, J. E. Pickering, J. C. Sloper, and F. S. Vogel, The effects of ^{140}barium and ^{140}lanthanum gamma radiation, of the central nervous system and pituitary gland of macaque monkeys, *J. Neuropathol. Exp. Neurol.* **17**:12–57 (1958).
46. J. Klatzo, J. Miquel, W. Haymaker, C. Tobias, and L. S. Wolfe, *in Effects of Ionizing Radiation on the Nervous System*, International Atomic Energy Agency, Proceedings Series, Vienna Symposium, Vienna (1962).
47. W. Haymaker, M. Z. H. Ibrahin, J. Miquel, N. Call, and A. J. Riopello, Delayed radiation effects in the brain of monkeys exposed to X-rays and Y-rays, *J. Neuropathol. Exp. Neurol*, **27**:50–79 (1968).
48. W. Haymaker, *in Neurological Problems* (J. Chorobski, ed.), pp. 173–176, Pergamon Press, Oxford (1967).
49. A. Arnold and P. Bailey, Alterations in the glia cells following irradiation of the brain in primates, *AMA. Arch. Pathol.* **57**:383–391 (1954).
50. R. H. Brownson, D. B. Suter, and D. D. Diller, Acute brain damage induced by low dosage X-irradiation, *Neurology* **13**:181–191 (1963).
51. S. P. Hicks, K. A. Wright, and K. E. Leigh, Time intensity factor in radiation response, acute effects of megavolt electrons (cathode rays) and high and low energy X-rays with especial reference to brain, *AMA. Arch. Pathol*, **61**:226 (1956).
52. A. Goldfeder, Ionizing radiation as a tool in biological research, *Trans. N. Y. Acad. Sci.*, Series II, **20**:809–837 (1958).
53. B. Andersson, B. Larsson, L. Leksell, W. Mair, B. Rexed, and P. Sourander, *in Response of the Nervous System to Ionizing Radiation, First International Symposium* (T. J. Haley and R. S. Snider, eds.), pp. 345–358, Academic Press, New York (1962).
54. B. Larssen, L. Leksell, B. Rexed, and P. Sourandeı, Effects of high energy protons on the spinal cord, *Acta Radiol.* **51**:57–64 (1957).
55. M. M. Aleksandrovskaya, Comparative morphologic studies of changes of the rat brain, antenatal irradiation with 50, 150, and 200 r doses at the 12th day of embryological development, *Voprossi Nevkopath.* **10**:125–127 (1962) (Russian).
56. M. M. Aleksandrovskaya, Histologic changes in the central nervous system of animals exposed to ionizing radiation during intoxication and infections. *Trans.*

Inst. Higher Nervous Activity, Pathophysiology Series, *Izd. Akad. Nauk SSSR*, **4**:226–234 (1958) Moscow (Russian).

57. N. Schummelfeder, *in Response of the Nervous System to Ionizing Radiation*, First International Symposium (T. J. Haley and R. S. Snider, eds.) pp. 191–220, Academic Press, New York (1962).
58. W. Lierse and H. D. Franke, Ultrastrukturelle Frühreaktionen am Kleinhirn des Meerschweinchens nach ^{60}Co-Bestrahlung des Kopfes, *Strahlentherapie* **131**:595–602 (1966).
59. J. M. Ordy, T. Samorajski, L. A. Horrocks, W. Zeman, and H. J. Curtis, Changes in memory electrophysiology, neurochemistry and neuronal ultrastructure after deuteron irradiation of the brain in C57Bl/10 mice, *J. Neurochem.* **15**:1245–1256 (1968).
60. B. Kaack, Augmentation of bioelectric activity by gamma irradiation, *Am. J. Physiol.* **213**:625–628 (1967).
61. G. La Grutta, S. Abbadessa, S. Avellone, E. Natale, I. Di Gregorio, M. E. Montalbano, E. A. Ortolani, and M. T. Zagami, Azioni delle radiazioni ionizzanti su alcuna attivita biochemiche e bioelettriche del sistema nervoso centrale, *Arch. Fisiol.* **65**:271–312 (1967).
62. A. L. Carsten, Irradiation and functional pathogenesis of X-irradiation effects in the monkey cerebral cortex, *Brain Res.* **7**:5–27 (1968).
63 V. M. Anan'ev, A. M. Kogan, and V. A. Nazarov, Electro-physiological study of cortical activity in the rabbit under the influence of total ionizing radiation, *Biull. Eksp. Biol. Med.* **65**:32–36 (1968) (Russian).
64. T. Minamisava, T. Tsuchiya, and H. Eto, The effects of ionizing radiation on the spontaneous and evoked brain electrical activity in rabbits. 2. The effects of X-rays on the hippocampal spontaneous electrical activity, *Nippon Acta Radiol.* **27**:1243–1249 (1967) (Japanese)
65. A. J. Stepanov, Certain indices of the bioelectric cerebral activity in persons subjected to the action of ionizing radiation during work, *Grg. Sanit.* **4**:47–49 (1967) (Russian).
66. G. A. Antropov, V. P. Godin, and A. V. Kolesnikova, Effect of small doses of penetrating radiation on the cortico-subcortical interaction, *Grg. Sanit.* **7**:49–52 (1968) (Russian).
67. J. García, D. J. Kimeldorf, and E. L. Hunt, The use of ionizing radiation as a motivating stimulus. *Psychol. Rev.* **68**:383–395 (1961).
68. J. García, B. H. Buchwald, B. H. Feder, and C. Wakefield, *in Effect of Ionizing Radiation on the Nervous System*, International Atomic Energy Agency, Proceedings Series; Vienna Symposium, Vienna (1962).
69. V. Novakova and J. Sterc, Higher nervous activity in rats irradiated at different periods of postnatal development, *Physiol. Behav.* **2**:421–423 (1967).
70. W. R. Adey and R. L. Schoenbrun, Irradiation effects of brain wave correlates of conditioned behavior. Final Report (1960–1965) (UCLA# 34P-60) (N67-13272#) (1965).
71 L. Court, P. Magnien, M. Avargues, and P. Laget, Modifications de la vigilance chez le lapin adulte soumis a une irradiation gamma-globate non lethale *C. R. Acad. Sci.* (*D*) (*Paris*) **266**:1052–1055 (1968).
72. E. L. Hunt, H. W. Carroll, and D. J. Kimeldorf, Effects of dose and of partial exposure on conditioning through a radiation induced factor, *Physiol. Rev.* **3**:809–813 (1968).

73. J. M. Johnson, The effects of repeated low dose X-irradiation on discrimination and radiosensitivity, Ph.D. thesis, Florida State University, Tallahasee, Fla. (1968).
74. V. A. Khasabova, Changes in higher nervous activity in rhesus monkeys induced by a total gamma-irradiation, *Zh. Vysshci Nervnoi Deyatel'nost: im I. P. Pavlova* **17**:730–738 (1967) (Russian).
75. O. N. Voevodina, Distant results of X-rays influence on higher nervous activity of dogs, Leningrad, *Meditsina*, 146 pp. (1967) (Russian).
76. R. J. Young, P. H. Chapman, D. J. Barnes, G. Carroll-Brown, and C. Hurst, Behavioral and physiological responses of *Maceca mulata* monkeys to supralethal doses of radiation, N69-20576; AD68-746; SAM-TR68-73.
77. J. C. Martin, Spatial avoidance in a paradigm in which ionizing irradiation precedes spatial confinement, *Radiat, Res.* **27**:284-289 (1966).
78. W. S. Moos, H. Le Van, B. T. Mason, H. C. Mason, and D. L. Hebion, Radiation-induced avoidance behavior transfer by brain extracts of mice, *Experientia* **25**:1215–1219 (1969).
79. H. Brust-Carmona, H. Kasprzak, and E. L. Gasteiger, Role of the olfactory bulbs in X-ray detection, *Radiation Res.* **29**:354–361 (1966).
80. E. L. Gasteiger and S. A. Helling, X-ray detection by the olfactory system: ozone as a masking odorant, *Science* **154**:1038–1041 (1966).
81. G. P. Cooper and D. J. Kimeldorf, Responses of single neurons in the olfactory bulbs of rabbits, dogs, and cats to X-rays, *Experientia* **23**:137–138 (1967).
82. G. P. Cooper, Receptor origin of the olfactory bulb response to ionizing radiation, *Am. J. Physiol*, **215**:803–806 (1968).
83. G. P. Cooper, Response of olfactory bulbs neurons to X-rays, as a function of nasal oxygen concentration, *Science* **167**:1726–1727 (1970).
84. W. Zeman, *in Response of the Nervous System to Ionizing Radiation*, Second International Symposium (T. J. Haley and R. S. Snider, eds.) General Discussion of Part V., pp. 722–724, Little, Brown, Boston (1964).
85. Yu. K. Kudritskiy, Change in the excitability of a motor reflex under a summation of the effects of small doses of X-rays, *Vestn. Rentgenol. Radiol.* **6**:15–21 (1955) (Russian).
86. M. N. Livanov and D. A. Biryukov, Changes in the Nervous System following Exposure to Ionizing Irradiation, Reports at the Second V. N. Conference, Peaceful Uses of Atomic Energy, Vol. 22, U.N., Geneva (1958).
87. R. Pape and J. Zakovsky, Die Röntgen Strahlensensibilität der Retina, *Forsch. Gebiete Röntgenstrahlen* **80**:65–71 (1954).
88. F. N. Serkov, Ye. D. Dubovyy, and M. A. Yamnovsky, Electroencephalographic changes in polycythemia patients treated with radioactive phosphorus, *Vrach. Delo.* **10**:1010 (1956) (Russian).
89. Yu. G. Grigor'yev, Problems pertaining to the action of small doses of ionizing radiation on physiological functions, *Radiobiolog.* **1**:966–968 (USAEC Rept. AEC-tr-5427), pp. 236–241 (1961).
90. Yu. G. Grigor'yev and L. N. Krasnikova, Reception of ionizing radiation by organism, *Med. Radiol.* **8**:85–91 (1963) (abstract from *Nucl. Sci. Abstr.*, **17**:31908 (1963)).
91. S. P. Hicks and C. J. D'Amato, Low dose radiation of developing brain, *Science* **141**:903–905 (1963).
92. H. E. Himwich, P. Sykowski, and J. F. Fazekas, A comparative study of excised cerebral tissue of adult and infant rats, *Am. J. Physiol*, **132**:293–296 (1941).

93. H. E. Himwich, *Brain Metabolism and Cerebral Disorders*, Williams and Wilkins, Baltimore (1951).
94. J. R. Lott and J. F. Hines, Effects of X-irradiation on respiration in rat brain tissue slices, *Texas. J. Sci.* **19**:391–394 (1967).
95. E. Egaña and F. Ugalde, Energetic metabolism of the CNS, *Rev. Med. Chile* **80**:359–362 (1952) (Spanish).
96. E. Egaña and S. Candiani, Correlation between the symptomatology, blood sugar and metabolic indexes of the CNS during the experimental insulin coma; First International Congress of Neurological Sciences, Brussels, July 1957; Vol. V, pp. 152–167, Pergamon Press, London (1959).
97. G. M. Frank, Early reactions of the organism to irradiation in relation to the localization of the effect, *in Proc. Int. Conf. Peaceful Uses Atomic Energy*, 1st Geneva, 1955, Vol. 2, p. 93, I.D.S., Columbia Univ. Press, New York (1956).
98. G. M. Frank, N. A. Aladjova, and A. D. Snezhko, Some aspects of the biophysical analysis of radiobiological effects, *Proc. Int. Conf. Peaceful Uses Atomic Energy* 2nd Geneva, 1958, Vol. 22, p. 383, I.D.S., Columbia Univ. Press, New York (1959).
99. A. D. Snezhko, Change in the oxygen consumption of brain tissue, following X-irradiation. *Biofizika* **2**:67–78 (1957) (Russian).
100. A. D. Snezhko, Measurement of the free oxygen concentration in the normal cerebral tissue in chronic experimental conditions, *Biofizika* **1**:585–592 (1956) (Russian).
101. W. H. Florsheim, C. Doernbach, and M. E. Morton. Effect of X-ray on radioactive phosphorus turnover and oxygen consumption of brain, *Proc. Soc. Exp. Biol. Med.* **81**:121–122 (1952).
102. P. S. Timiras, J. A. Moguilevsky, and S. Geel, *in Response of the Nervous System to Ionizing Radiation*; Second International Symposuim (T. J. Haley and R. S. Snider, eds.) pp. 365–376, Little, Brown, Boston (1964).
103. J. A. Moquilevsky and S. Geel, Long-term effects of whole body X-irradiation on cortex and hypothalamus, Q_{O_2} in rats, *Fed. Proc.* **22**:272 (1963).
104. E. Egaña, *in Effects of Ionizing Radiation on the Nervous System*, International Atomic Energy Agency, Proceedings Series, pp. 267–282, Vienna Symposium, Vienna (1962).
105. E. Egaña, Some effects of ionizing radiations on the metabolism of the central nervous system, *Int. J. Neurol.* **3**:631–647 (1962).
106. B. Chance and G. R. Williams. The respiratory chain and oxidative phosphorylation, *Advan. Enzymol.* **17**:65–134 (1956).
107. J. K. Grant, ed., *Methods of Separation of Subcellular Structural Components*, Cambridge University Press, Cambridge (1963).
108. B. Chance, ed., *Energy-linked Functions of Mitochondria*, First Colloquium of the Johnson Res. Found., Academic Press, New York (1963).
109. A. L. Lehninger, *The Mitochondrion*, W. A. Benjamin, Inc., New York (1965).
110. J. M. Tager, S. Papa, E. Quagliariello, and E. C. Slater, eds., *Regulation of Metabolic Process in Mitochondria*, B. B. A. Library, Vol. 7, Elsevier, Amsterdam (1966).
111. B. Chance, ed., Control of Energy Metabolism, Colloquium of the Johnson Res. Found., Academic Press, New York (1965).
112. E. C. Slater, Z. Kaniuga, and L. Wojtczak, eds., *Biochemistry of Mitochondria*, Academic Press, New York (1967).
113. T. P. Singer, ed., *Biological Oxidations*, Interscience Publ., New York (1968).
114. D. E. Green and H. Baum, *Energy and the Mitochondria*, Academic Press, New York (1969).

115. L. G. Abood and L. Romanchek, Inhibition on oxidative phosphorylation in brain mitochondria by electrical current and the effect of chelating agents and other substances, *Biochem. J.* **60**:233–238 (1955).
116. R. Tanaka and L. G. Abood, Isolation from rat brain of mitochondria devoid of glycolytic activity, *J. Neurochem.* **10**:571–576 (1963).
117. A. A. Abdel-Latif and L. G. Abood, Biochemical studies on mitochondria and other cytoplasmic fractions of rat brain, *J. Neurochem.* **11**:9–15 (1964).
118. L. G. Abood, *in Handbook of Neurochemistry* (A. Lajtha, ed.), Vol. 2, pp. 303–326, Plenum Press, New York (1969).
119. E. Egaña and C. Oberti, CNS mitochondria morphology and irradiation (to be published).
120. E. Egaña and R. Valderas, The effect of β-internal irradiation on CNS mitochondria, Progress Report to the International Energy Agency, Vienna, Contract 95, RB/2 January (1963).
121. E. Egaña and B. San Martín, The effect of whole body-gamma (^{60}Co) irradiation on the CNS mitochondria respiration, First International Meeting of the International Society for Neurochemistry, Strassbourg, July/August (1967) (abstract).
122. K. F. Swaiman and J. M. Milstein, Oxidative decarboxylation of aspartate, alanine and glycine in developing rabbit brain, *Biochim. Biophys. Acta.* **93**:64–70 (1964).
123. K. F. Swaiman and J. M. Milstein, Oxidation of leucine, isoleucine and related ketoacids in developing rabbit brain, *J. Neurochem*, **12**:981–986 (1965).
124. J. M. Milstein, J. G. White, and K. F. Swaiman, Oxidative phosphorylation in mitochondria of developing rat brain, *J. Neurochem.* **15**:411–415 (1968).
125. E. Egaña, B. San Martín, and B. Oporto, Gamma-(^{60}Co)-whole body-irradiation and biochemical indexes of developing CNS mitochondria, Second International Meeting of the International Society for Neurochemistry, Milan, Sept. 1969 (abstract).
126. D. R. Dahl and F. E. Samson, Jr., Metabolism of rat brain mitochondria during postnatal developments. *Am. J. Physiol.* **196**:470–472 (1959).
127. B. Chance and M. Baltscheffsky, Spectroscopic effects of adenosinediphosphate upon the respiratory pigment of rat-breast-muscle sarcoma, *Biochem. J.* **68**:283–295 (1958).
128. E. S. Higgins, Factors influencing respiratory control in brain mitochondria, *J. Neurochem.* **15**:589–596 (1968).
129. M. Bacila, A. P. Campello, C. H. Vianna, and D. O. Voss, The respiration chain of the rat cerebrum and cerebellum mitochondria: respiration and oxidative phosphorylation. *J. Neurochem.* **11**:231–242 (1964).
130. D. E. Smith and J. F. Thomson, Physiological and biochemical studies on various species exposed to massive X-irradiation, *Radiat. Res.* **11**:198–205 (1959).
131. P. Mitchell and J. Moile, Stoichiometry of proton translocation through the respiratory chain and adenosine triphosphatase system of the rat liver mitochondria. *Nature.* **208**:147–151 (1965).
132. P. Mitchell and J. Moile, *in Biochemistry of Mitochondria* (E. C. Slater, Z. Kaniuga, and L. Wojtczak, eds.) pp. 53–74, Academic Press, New York (1967).
133. H. R. Mahler and E. H. Cordes, *Biological Chemistry*, Harper International, New York (1966).
134. W. H. Florsheim and M. E. Morton, Brain and liver phosphorus metabolism in the acute irradiation syndrome, *Am. J. Physiol.* **176**:15–19 (1954).
135. E. Egaña and R. Valderas, ^{32}P chromatographic studies in β-internal irradiated CNS, 2nd International Congress of Radiation Research, Harrogate, Great Britain, August (1962) (abstract).

136. K. Ozawa, K. Seta, H. Takada, and C. Araki, On the isolation of mitochondria with high respiratory control from rat brain, *J. Biochem.* (*Tokyo*) **59**:501–510 (1966).
137. K. Ozawa, K. Seta, and H. Handa, The effect of magnesium on brain mitochondrial metabolism, *J. Biochem.* (*Tokyo*) **60**:268–273 (1966).
138. G. Y. Gorodiskaya and O. N. Barmine *in Voprosy Biokhim. Nervnoi Sistemy* (A. V. Pallatin, ed.) p. 295, Acad. Nauk Ukr. SSR, Kiev. (1957) (Russian).
139. B. M. Grayevskaya and R. Y. Keilina, Changes of carbohydrate metabolism induced by whole body X-ray irradiation in the organism of animals, *Vest. Rentgenol. Radiol.* **4**:21–26 (1955) (Russian).
140. E. Egaña, N. Francotte, R. Valderas, and O. Gonzalez, Effects of gamma whole-body irradiation (^{60}Co) on the synthesis of some CNS "free amino acids" from glucose, *Nuclear Hematology*, November/December (1968).
141. E. Egaña and R. Rodrigo, unpublished data.
142. R. E. Kay and H. Chan, Effect of X-irradiation on glucose metabolism in rat cerebral cortex slices, *J. Neurochem.* **14**:401–403 (1967).
143. S. Nakazava, T. Hara, and K. Veki, The effects of X-irradiation on protein metabolism on the brain, *Can. J. Biochem.* **43**:1091 (1965).
144. J. de Vellis, Glycolysis in rat brain tissue slices following neonatal head X-irradiation: relation of regional differences to the LDH:GPDH ratio, *J. Neurochem.* **15**:1057–1060 (1969).
145. J. de Vellis and O. A. Schjeide, Time-dependence of the effect of X-irradiation on the formation of glycerol phosphate dehydrogenase and other dehydrogenases in the developing rat brain, *Biochem. J.* **107**:259–264 (1969).
146. J. de Vellis, O. A. Schjeide, and C. D. Clemente, Protein synthesis and enzymic patterns in the developing brain following head X-irradiation of newborn rats, *J. Neurochem* **14**:499–511 (1967).
147. V. Bonavita, G. Amore, S. Avellone, and R. Guarnieri, Lactate dehydrogenase isoenzymes in the nervous tissue. V. The effects of X-rays on the enzyme of the developing and adult rat brain, *J. Neurochem.* **12**:37–43 (1965).
148. D. Richter, *in Molecular Basis of Some Aspects of Mental Activity*, Vol. 1 (O. Walaas, ed.), Academic Press, New York (1966).
149. H. Schields and W. Gordy. Electron-spin-resonance of X-irradiated nucleic acids, *Bull. Am. Phys. Soc.* **1**:267 (1956) (abstract).
150. J. N. Herak and W. Gordy, ESR study of nucleosides bombarded with hydrogen atoms, *Proc. Nat. Acad. Sci.* **57**:7–11 (1966).
151. W. Gordy and H. Schields, Electron spin resonance investigation of the proteins, *Bull. Acad. Roy. Belge* (Classe Sciences) **33**:191–207 (1961).
152. J. Duchesne. Origine de la radiorésistence des proteins et des acids nucléiques, *Rapports de la Reun. Int. Physique*, pp. 1–4, Paris (1964).
153. A. van de Vorst, Effect des rayonnements ionizants sur la matière en phase solide, Doctoral thesis, Département de Physique Atomique et Moléculaire et Institute de Physique, Faculté de Sciences, Université de Liège, 131 pp., Liège (1963).
154. F. Shimazu and A. L. Tappel, Comparative radiolability of amino acids of proteins and free amino acids, *Radiat. Res.* **23**:203–209 (1964).
155. W. M. Dale, *in Actions Chimiques et Biologiques des Radiations* (M. Haissinsky, ed.), Masson et Cie., Paris (1955).
156. W. M. Dale, in Ciba Found. Symposium *Ionizing Radiation and Cell Metabolism* (G. E. W. Wolstenholme and C. M. O'Connor, eds.) pp. 25–34, Churchill, London (1956).

157. T. Hasseltrom, M. C. Henry, and B. Murr, Synthesis of amino acids by beta-irradiation, *Science* **125**:350–351 (1967).
158. B. B. Singh and M. G. Ormerod, Primary radical formation in irradiated proteins, *Nature* **206**:1314–1315 (1965).
159. A. Beloff-Chain, R. Cantarazo, E. B. Chain, J. Massi, and F. Pocchiari, Fate of uniformly labelled ^{14}C-glucose in brain slices, *Proc. Roy. Soc. B.* **144**:22–28 (1955).
160. R. Balazs, K. Magyar, and D. Richter, *in Comparative Neurochemistry* (D. Richter, ed.) pp. 225–248, Pergamon Press, London (1964).
161. R. Balàzs and R. J. Haslam, Exchange transamination and the metabolism of glutamate in brain, *Biochem. J.* **94**:131–141 (1965).
162. M. K. Gaitonde, D. R. Dahl, and K. A. C. Elliott, Entry of glucose carbon into amino-acids of rat brain and liver "in vivo" after injection of uniformly ^{14}C-labelled glucose, *Biochem. J.* **94**:345–352 (1965).
163. R. Vrba, M. K. Gaitonde, and D. Richter. The conversion of glucose carbon into protein in the brain and other organs of the rat, *J. Neurochem.* **9**:465–475 (1962).
164. A. Geiger, Y. Kawakita, and S. S. Barkulis, Major pathways of glucose utilization in the brain perfusion experiments *in vivo* and *in situ*, *J. Neurochem.* **5**:323–338 (1960).
165. J. E. Cremer, Amino acid metabolism in rat brain; studied with ^{14}C-labelled glucose, *J. Neurochem.* **11**:165–185 (1964).
166. E. Egaña and E. Szirmai, in Symposium der Arbeitsgemeinschaft für Strahlenbiologie in der Deutschen Röntgengesellschaft, Stuttgart, pp. 78–82, 11 Mai, 1969. Thieme Verlag, Stuttgart (1970).
167. E. Egaña, unpublished data.
168. S. N. Alexandrov. *Doklady Akad. Nauk SSSR* **106**:153 (1956), quoted in reference 5.
169. J. de Vellis and O. A. Schjeide, Effects of ionizing radiation on the biochemical differentiation of the rat brain, *Proc. Ninth Ann. Conf. Hanford Biol. Symp. Radiation Fetal Juvenile Mammal* (Batelle Symposium) (M. R. Sikov and D. D. Mablum, eds.), VSAEC Division of Technical Information Extension, Oak Ridge (1970).
170. M. K. Gaitonde and D. Richter, The metabolic activity of the protein of the brain, *Proc. Roy. Soc. B.* **145**:83–99 (1956).
171. F. N. Minard and D. Richter, Electroshock induced seizures and the turnover of brain protein in the rat. *J. Neurochem.* **15**:1463–1468 (1968).
172. F. Orrego and F. Lipman, Protein synthesis in brain slices; effects of electrical stimulation and acidic amino acids, *J. Biol. Chem.* **242**:665–671 (1967).
173. G. P. Talwar, S. P. Chopra, B. K. Goel, and B. d'Monte, Correlation of the functional activity of the brain with metabolic parameters. III. Protein metabolism of the occipital cortex in relation to light stimulus. *J. Neurochem.* **13**:109–116 (1966).
174. W. Florsheim, C. Doernbach, and M. E. Morion, Effect of X-ray on radioactive phosphorus turnover and oxygen consumption of brain, *Proc. Soc. Exp. Biol. Med.* **81**:121–122 (1952).
175. S. V. Gastevz and D. A. Chetverikov, The intensity of phospholipid exchange in the central nervous system of the rat during severe radiation sickness, *Doklady Akad. Nauk SSSR* **142**:1180–1183 (1962) (Russian).
176. M. P. Vinogradov, Effect of whole body X-irradiation upon turnover rate of lipids in the rat brain. *Radiologia* **2**:695–699 (1962) (Russian).
177. O. A. Smirnova. Influence of ionizing radiation on metabolism of brain phospholipids, *Ukr. Biokhim. Zh.* **33**:208–213 (1961) (Russian).

178. E. Grossi, M. Poggi, and R. Paoletti, *in Response of the Nervous System to Ionizing Radiation*, Second International Symposium (T. J. Haley and R. S. Snider, eds.) pp. 389–402, Little, Brown, Boston (1964).
179. A. Vernadakis, R. Casper, and P. S. Timiras, Influence of prenatal X-radiation on brain lipid and cerebroside content in developing rats. *Experientia* **24**:237–238 (1968).
180. O. A. Schjeide, R. I-San Lin, and J. de Vellis, Molecular composition of myelin synthesized subsequent to irradiation, *Radiat. Res.* **33**:107–128 (1968).
181. Ch. O. Lee, Radiation effects on the metabolism of phospholipids in the central nervous system of albino rats, Progress Report V., Nasa-CR 96228-PR-5., August (1968).
182. W. O. Carter, E. S. Redgate, and W. D. Amstrong, Changes in the central nervous system after 700 r total-body X-irradiation, *Radiat. Res.* **8**:92–97 (1958).
183. W. O. Carter and W. D. Amstrong, *in Effects of Ionizing Radiation of the Nervous System*, International Atomic Energy Agency, Proceedings Series, Vienna Symposium, Vienna (1962).
184. M. M. Aleksandrovskaya, Some data of ionizing radiation on the central nervous system of animals, *Trans. Inst. Higher Nervous Activity*, Pathophysiology Series, Izd. Akad, Nauk SSSR Moscow, **4**:211–217 (1958) (Russian).
185. M. M. Aleksandrovskaya, Histopathologic changes in the central nervous system exposed to ionizing radiation during intoxication and infections, *Trans. Inst. Higher Nervous Activity*, Pathophysiology Series, Izd. Akad. Nauk SSSR, Moscow (1957), USAEC Rept. AEC-tr-3661 (Bk 1) pp. 108–112.
186. M. I. Artynkhina, On the effects of ionizing radiation on the central nervous system of rabbits, *Transactions of the Institute of Higher Nervous Activity*, Pathophysiology Series, Izd. Akad. Nauk SSSR, Moscow, **4**:238–247 (1958) (Russian).
187. M. I. Artyukhina, Histopathological changes in the central nervous system and internal organs of white rats following various doses of ionizing radiation on the brain area, *Trans. Inst. Higher Nervous Activity*, Pathophysiology Series, Izd. Akad. Nauk SSSR Moscow, 1958, **4**:252–260 (1958) (Russian).
188. Y. L. Yamamoto, L. E. Feinendeger, and V. P. Bond, Effect of radiation on the RNA metabolism of the central nervous system, *Radiat. Res.* **21**:36–65 (1964).
189. E. C. Pollard, Degradation of ribonucleic acid in dilute solution by ionizing radiation, *Nature* **210**:1393–1395 (1966).
190. D. M. Climent and W. G. Overend, Examination of the effect of γ-rays on deoxyribonucleic acids, *Arch. Bioch. Biophy.* **122**: 563–568 (1967).
191. R. A. Cox, W. G. Overend, A. R. Peacoke, and S. Wilson, The action of γ-rays on sodium deoxyribonucleate in solution, *Proc. Royal Soc. (London) B.* **149**:511–513 (1958).
192. M. W. Warshaw and J. Tinoco, Jr., Absorption and optical rotatory dispersion of six dinucleoside phosphates, *J. Mol. Biol.* **13**:54–64 (1965).
193. H. de Voe and J. Tinoco Jr., The stability of helical polynucleotides: base contributions, *J. Mol. Biol.* **4**:500 (1962).
194. G. Scholes and M. Simic, Radiolysis of aqueous solutions of DNA and related substances; reaction of hydrogen atoms, *Biochem. Biophys. Acta* **166**:255–258 (1968).
195. B. B. Singh and M. G. Ormerod, Primary radical formation in irradiated protein, *Nature* **206**:1314–1315 (1965).
196. A. D. Lenherr and A. Charlesbay, Energy transfer in nucleoproteins, *Int. J. Rad. Biol.* **12**:51–60 (1967).

197. C. J. Dean, M. G. Ormerod, R. W. Serianini, and P. Alexander, DNA strand breakage in cells irradiated with X-rays, *Nature* **222**:1042–1045 (1969).
198. E. R. Walwick and R. K. Main, Stability of a deoxyribonucleic acid-synthesizing system to X-radiation, *Biochim. Biophys. Acta* **55**:225–227 (1962).
199. J. Merits and J. Cain, Rapid loss of labelled DNA from rat brain due to radiation damage, *Biochim. Biophys. Acta* **174**:315–321 (1969).
200. B. Pullman and A. Pullman, The electronic structure of the purine-pyrimidine pairs of DNA, *Biochim. Biophys. Acta* **36**:343–350 (1959).
201. B. Pullman and A. Pullman, *Quantum Biochemistry*, Intersc. Publ., New York (1963).
202. B. Pullman and A. Pullman, *in Comparative Effects of Radiation* (M. Burton, J. S. Kirby Smith, and J. L. Magee, eds.) pp. 105–123, John Wiley & Sons, New York (1960).
203. B. Pullman, *in Molecular Biophysics* (B. Pullman and H. Weisbluth, eds.), Academic Press, New York (1965).
204. B. Pullman, Quantum-mechanical calculations of biological structures and mechanisms, *Ann. N.Y. Acad. Sci.* **158**:Art. 1:1–19 (1969).
205. A. Szent-Györgyi, *Introduction to a Submolecular Biology*, Academic Press, New York (1960).
206. A. Szent-Györgyi, On the possible role of quantum phenomena in normal and abnormal mental function in ultrastructure and metabolism of the nervous system, *Res. Publ. Ass. Nerv. Ment. Dis.* **40**:325–336 (1962).
207. J. Duchesne, Biological Effects of Ionizing Radiation at the Molecular Level, International Atomic Energy Agency, Proc. Series, Brno Symposium, Vienna (1962).
208. J. Duchesne and A. van de Vorst, Origine de la radioresistance des acids nucléiques en phase solide, *Bull. Acad. Roy. Belg.* (Classe Sciences) **51**:778–809 (1965).
209. C. Alexander, Jr. and W. Gordy, Electron spin resonance of an irradiated single crystal of guanine-hydrochloride-dihydrate, *Proc. Nat. Acad. Sci.* **58**:1279–1285 (1967).
210. W. Gordy, Free radicals from biological purines and pyrimidines, *Ann. N.Y. Acad. Sci.* **158** Art. 1:100–123 (1969).
211. A. Müller, The formation of radicals in nucleic acids, nucleoproteins and their constituents by ionizing radiations, *Prog. Bioph. Mol. Biol.* **17**:99–147 (1967).
212. J. K. Setlow, *in Current Topics in Radiation Research* (M. Ebert and A. Howard, eds.) pp. 195–248, North-Holland, Amsterdam (1966).
213. M. I. Gregersen, Ch. Pallavicini, and S. Chien, Studies on the chemical composition of the central nervous system in relation to the effects of X-irradiation and of disturbances in water and salt balance. I. Chemical composition of various specific areas and structures of the brain in the dog and the monkey. *Radiat. Res.* **17**:209–225 (1962).
214. M. I. Gregersin, Ch. Pallavicine, and S. Chien, II. Effects of X-irradiation on the chemical composition of brain tissue in dogs. *Radiat. Res.* **17**:226–233 (1962).
215. S. Chien, Ch. Pallavicini, L. J. Cizek, and M. I. Gregersen, III. Effects of disturbances in water and electrolyte balance on the chemical composition of brain tissue in dogs. *Radiat. Res.* **17**:234–243 (1962).
216. G. B. Rowley and R. L. Dellenback, IV. Effects of X-irradiation on the chemical composition of rabbit brain, *Radiat. Res.* **17**:244–252 (1962).
217. T. Volcana, A. Vernadakis, and P. S. Timiras, Electrolyte content of the cerebral cortex in developing rats after prenatal X-irradiation, *Experientia* **22**:608–609 (1966).

218. E. D. Willis, The effect or irradiation on sub-cellular components metal-ion transport in mitochondria, *Int. J. Rad. Biol.* **11**:517–529 (1966).
219. R. Valderas and E. Egaña, Experimental criterion to study the relation of internal β-irradiation and biochemical effects in the CNS, *Nuclear Hematology*, June (1966).
220. M. Lefort, *in Actions chimiques et biologiques des Radiations* (M. Haissinky, ed.), Masson et Cie, Paris (1955).
221. H. E. Aebi, Detection and fixation of radiation-produced peroxide by enzymes, *Radiat. Res. Suppl.* **3**:130–162 (1963).
222. E. Egaña and M. I. Velarde, *in Comparative Neurochemistry* (D. Richter, ed.) pp. 275–278, Pergamon Press, Oxford (1964).
223. E. Egaña. Catalase effect on the respiration of normal and beta-internal irradiated rat CNS, Book of abstracts, *Third International Congress of Radiat. Res.*, Cortina d'Ampezzo, Italy, June/July (1966).
224. R. N. Feinstein, J. T. Braun, and J. Howard, Acatalasemic and hypocatalasemic mouse mutant, II. Mutational variations in blood and solid tissue catalases, *Arch. Biochem. Bioph.* **120**:165–169 (1967).
225. G. M. Gallagner and S. H. Buttery, Biochemistry of sheep tissues, enzyme systems of liver, brain, and kidney, *Biochem J.* **72**:575–582 (1959).
226. H. J. Adler, Catalase, hydrogen peroxide and ionizing radiation, *Radiat. Res. Suppl.* **3**:110–129 (1963).
227. R. N. Feinstein, Attempts at protection of mice against ionizing radiation by exogenous catalase, *Atompraxis* **6**:205–207 (1960).
228. H. Le Van and W. S. Moos, Possible effects of radiation-produced hydrogen peroxide on post-irradiation aversion in mice, *Experientia* **23**:749–751 (1967).
229. H. Le Van, W. S. Moos and D. L. Hebron, Direct and indirect effect on X-irradiation on conditioned avoidance behavior, *Med. Exp.* **18**:161–168 (1968).
230. D. D. Morris and R. N. Feinstein, Mechanism of mouse awareness of X-radiation, catalase and hydrogen peroxide required for detection of X-rays by mice, *Nature* **222**:688–689 (1969).
231. E. Egaña, ACh content of the brain and the C.S.F. during the states of rest and of optical nerve excitation, *Ann. Acad. Biol.* **1**:183–191 (1937) (Spanish).
232. P. B. Bradley, Synaptic transmission in the central nervous system and its relevance for drug action, *Int. Rev. Neurobiol.* **11**:1–56 (1968).
233. V. P. Whittaker, The isolation and characterization of acetylcholine-containing particles from brain, *Biochem. J.* **72**:694–706 (1959).
234. V. P. Whittaker and M. N. Sheridan, The morphology and acetylcholine content of cerebral synaptic vesicles, *J. Neurochem.* **12**:363–372 (1965).
235. V. P. Whittaker, *in Mechanisms of Release of Biogenic Amines* (U. S. von Euler, S. Rosell, and B. Uvnäs, eds.) Vol. 5, pp. 147–164, Wenner-Gren International Symposium Series, Pergamon Press, Oxford (1966).
236. L. W. Chakrin and V. P. Whittaker, The subcellular distribution of N-Me-H-acetylcholine synthesized by brain *in vivo*. *Biochem. J.* **113**:97–107 (1969).
237. V. P. Whittaker, *in Handbook of Neurochemistry* (Abel Lajtha, ed.) Vol. 2, Chapter 14, pp. 327–364, Plenum Press, New York (1969).
238. E. de Robertis, L. Salganicoff, L. M. Zieher, and G. Rodrigues de Lores Arnaiz, Acetylcholine and cholinacetylase content of synaptic vesicles, *Science* **140**:300–301 (1963).
239. E. de Robertis, G. Rodriguez de Lores Arnaiz, L. Salganicoff, A. Pellegrino de Iraldi, and L. M. Zieher, Isolation of synaptic vesicles and structural organization of the acetylcholine system within brain nerve endings, *J. Neurochem.* **10**:225–235 (1963).

240. G. Rodriguez de Lores Arnaiz, M. Alberici, and E. de Robertis, Gangliosides and acetylcholinesterase in isolated membranes of the rat brain cortex, *Biochim. Biophys. Acta* **135**:33–43 (1967).
241. E. de Robertis and G. Rodriguez de Lores Arnaiz (A. Lajtha, ed.), Vol. 2, Chapter 15, pp. 365–392, Plenum Press, New York (1969).
242. A. Silver, Cholinesterases of the central nervous system with especial reference to the cerebellum, *Int. Rev. Neurobiol.* **10**:58–109 (1967).
243. Ye. N. Petrovnina, Impairment of acetylcholine metabolism in the brain of white rats following polonium injury, *Byull. Radiats. Med.* **1**:52 (1958) (Russian).
244. E. Egaña, Beta-internal irradiation and cholineacetylase activity of encephalon, *Radiat. Res.* **12**:432 (1966) (abstract).
245. G. J. Maletta and P. S. Timiras, Acetyl- and Butyrylcholinesterase activity of selected brain areas in developing rats after neonatal X-irradiation, *J. Neurochem.* **13**:75–84 (1966).
246. G. J. Maletta, A. Vernadakis, and P. S. Timiras, Acetylcholinesterase activity and protein content of brain and spinal cord in developing rats after prenatal X-irradiation, *J. Neurochem.* **14**:647–652 (1967).
247. E. T. Browning and M. P. Schulman, ^{14}C-Acetylcholine synthesis by cortex slices of rat brain, *J. Neurochem.* **15**:1391–1405 (1968).
248. D. J. Palaic and Z. Supek, Liberation of brain 5-hydroxytryptamine and noradrenaline by X-ray treatment in the newborn and adult rats, *J. Neurochem.* **13**:705–709 (1966).
249. K. N. Prasad and M. H. van Woert, Dopamine protects mice against whole body irradiation. *Science* **155**:470–472 (1967).
250. J. L. Gray, E. J. Moulder, J. T. Tew, and H. Lensen, Protective effects of pitressin and of epinephrine against total body X-irradiation, *Proc. Soc. Exp. Biol. Med.* **79**:384–387 (1952).
251. E. Roberts and S. Frankel, γ-aminobutyric acid in brain, *Fed. Proc.* **9**: 219 (1950) (abstract).
252. E. Roberts and E. Eidelberg, Metabolic and neurophysiological roles of γ-aminobutyric acid, *Int. Rev. Neurobiol.* **2**:279–332 (1960).
253. E. Roberts, *in Molecular Basis of Some Aspects of Mental Activity* (O. Walaas, ed.) Vol. 1, Academic Press, New York (1966).
254. E. Roberts and K. Kuriyama, Biochemical-physiological correlations in studies of the γ-aminobutyric acid, *Brain Res.* **8**:1–35 (1968).
255. K. Kuriyama, E. Roberts, and Johan Vos, Some characteristics of binding of γ-aminobutyric acid and acetylcholine to a synaptic vesicle fraction from mouse brain, *Brain Res.* **9**:231–252 (1969).
256. N. Seiler and M. Wiechman, Zum Vorkommen der Gamma-amino-buttersäure und der Gamma-amino beta-hydroxy-buttersäure in tierischen Gewebe, *Hoppe-Seyler's Z. Physiol. Chem.* **350**:1493–1500 (1969).
257. E. Egaña, unpublished data.
258. G. M. McKahnn and D. B. Tower, Gamma-amino butyric acid: a substrate for oxidative metabolism of cerebral cortex, *Am. J. Physiol.* **196**:36–38 (1959).
259. Y. Tsukada, Y. Nagata, and G. Takagaki, Glucose metabolism and amino-acid in brain slices, *J. Biochem.* (*Tokyo*) **45**:979–984 (1957).
260. E. Egaña, Two types of metabolic influences of the central nervous system, *Second Latin American Congress of Neurology, Symposium on Metabolic Aspects and Function of the Nervous System*, pp. 373–386, University of Chile Press, Santiago, Nov. (1960) (Spanish).

261. M. Jovanovič and N. Svecenski, Effect of high dosage of ^{60}Co-gamma rays on gamma-amino butyric and glutamic acid content in brain tissue of mice, *Strahlentherapie* **125**:588–590 (1964).
262. H. Jovanovič and N. Svecenski, Effect of high doses of ^{60}Co gamma rays on the glutamic/gamma-aminobutyric acid ratio in brain tissue of mice, *Strahlentherapie* **129**:446–447 (1966).
263. M. Jovanovič and A. Cordic, Transformation of labelled glutamic acid in the brain tissue of irradiated mouse, *Strahlentherapie* **134**:533–535 (1967).
264. E. Egaña, unpublished data.
265. B. H. Ershoff and E. M. Gal, Effects of radiation on tissue serotonin levels in the rat, *Proc. Soc. Exp. Biol. Med.* **108**:160–162 (1961).
266. J. Renson and P. Fischer, Libération de 5-hydroxytryptamine par le rayonnement-X, *Arch. Int. Physiol* **67**:142–144 (1958).
267. L. Speck, Effects of massive X-irradiation on rat electroencephalogram and brain serotonin, *J. Neurochem.* **9**:573–574 (1962).
268. M. Randic, Z. Supek, and Z. Lovašen, *in Effects of Ionizing Radiation on the Nervous System*, International Atomic Energy Agency, Proceedings Series, Vienna Symposium, pp. 263–264, Vienna (1962).
269. M. Randic and Z. Supek, Influence of high doses of X-irradiation on 5-hydroxytryptamine in the brain of rats, *Int. J. Rad. Biol.* **4**:637–638 (1962).
270. Dj. Palaic, M. Randic, and Z. Supek, X-irradiation and 5-hydroxytryptamine in the brain of rats and mice, *Int. J. Rad. Biol* **6**:241–246 (1963).
271. Dj. Palaic and Z. Supek, Liberation of 5-hydroxytryptamine in the rat brain after X-irradiation, *Int. J. Rad. Biol.* **9**:601–603 (1965).
272. Dj. Palaic and Z. Supek, Drug-induced changes of the metabolism of 5-hydroxytryptamine in the brain of X-ray treated rat, *J. Neurochem.* **12**:329–333 (1965).
273. J. R. Lott and J. F. Hines, 5-OH- tryptamine content in rat brain tissue X-irradiated *in vitro*, *Texas J. Sci.* **20**:91–94 (1968).
274. N. Komesu and T. J. Haley, Lack of effect of X-irradiation on brain 5-hydroxytryptamine concentration, *Proc. West Pharmacol. Society* **11**:77–80 (1968).
275. G. A. Chernov and M. O. Raushenbakhn, The study of the role of serotonin (5-hydroxytryptamine) in the pathogenesis of acute radiation sickness. II. Changes in the serotonin content in the intestine and brains of guinea-pigs and rats in acute radiation sickness, *Prob. Gemat. Pereliv Krovi* **5**:3–7 (1960) (Russian).
276. E. Egaña and M. I. Velarde, Effects of β-internal (^{32}P)-irradiation on the 5-HT content of CNS levels, *Experientia* **23**:526–527 (1967).
277. E. Egaña and M. I. Velarde, *in Progress in Nuclear Hematology and Allied Fields* (E. Szirmai, ed.) pp. 149–157, Medical Section, The Institution of Nucl. Engin. London (1967).
278. A. Szent-Györgyi, Charge transfer and electronic mobility, *Proc. Nat. Acad. Sci.* **58**:2012–2014 (1967).
279. A. Szent-Györgyi, Intermolecular electron transfer may play a major role in biological regulation defense and cancer, *Science* **161**:988–990 (1968).
280. W. E. Cohn and C. E. Carter, The separation of adenosine polyphosphate by ion exchange and paper chromatography, *J. Am. Chem. Soc.* **72**:4273–4275 (1950).
281. E. F. McFarren, Buffered filter paper chromatography of *d*-amino-acids. *Anal. Chem.* **23**:168–174 (1951).
282. J. R. Vane, A sensitive method for the assay of 5-hydroxytryptamine, *Biol. J. Pharm. Chemother.* **12**:344–349 (1957).

SUBJECT INDEX

GPSR Compliance

The European Union's (EU) General Product Safety Regulation (GPSR) is a set of rules that requires consumer products to be safe and our obligations to ensure this.

If you have any concerns about our products, you can contact us on ProductSafety@springernature.com

In case Publisher is established outside the EU, the EU authorized representative is:

Springer Nature Customer Service Center GmbH
Europaplatz 3
69115 Heidelberg, Germany

Batch number: 10837688

Printed by Printforce, the Netherlands